Dictionary of Literary Biography

1 *The American Renaissance in New England*, edited by Joel Myerson (1978)

2 *American Novelists Since World War II*, edited by Jeffrey Helterman and Richard Layman (1978)

3 *Antebellum Writers in New York and the South*, edited by Joel Myerson (1979)

4 *American Writers in Paris, 1920-1939*, edited by Karen Lane Rood (1980)

5 *American Poets Since World War II*, 2 parts, edited by Donald J. Greiner (1980)

6 *American Novelists Since World War II, Second Series*, edited by James E. Kibler Jr. (1980)

7 *Twentieth-Century American Dramatists*, 2 parts, edited by John MacNicholas (1981)

8 *Twentieth-Century American Science-Fiction Writers*, 2 parts, edited by David Cowart and Thomas L. Wymer (1981)

9 *American Novelists, 1910-1945*, 3 parts, edited by James J. Martine (1981)

10 *Modern British Dramatists, 1900-1945*, 2 parts, edited by Stanley Weintraub (1982)

11 *American Humorists, 1800-1950*, 2 parts, edited by Stanley Trachtenberg (1982)

12 *American Realists and Naturalists*, edited by Donald Pizer and Earl N. Harbert (1982)

13 *British Dramatists Since World War II*, 2 parts, edited by Stanley Weintraub (1982)

14 *British Novelists Since 1960*, 2 parts, edited by Jay L. Halio (1983)

15 *British Novelists, 1930-1959*, 2 parts, edited by Bernard Oldsey (1983)

16 *The Beats: Literary Bohemians in Postwar America*, 2 parts, edited by Ann Charters (1983)

17 *Twentieth-Century American Historians*, edited by Clyde N. Wilson (1983)

18 *Victorian Novelists After 1885*, edited by Ira B. Nadel and William E. Fredeman (1983)

19 *British Poets, 1880-1914*, edited by Donald E. Stanford (1983)

20 *British Poets, 1914-1945*, edited by Donald E. Stanford (1983)

21 *Victorian Novelists Before 1885*, edited by Ira B. Nadel and William E. Fredeman (1983)

22 *American Writers for Children, 1900-1960*, edited by John Cech (1983)

23 *American Newspaper Journalists, 1873-1900*, edited by Perry J. Ashley (1983)

24 *American Colonial Writers, 1606-1734*, edited by Emory Elliott (1984)

25 *American Newspaper Journalists, 1901-1925*, edited by Perry J. Ashley (1984)

26 *American Screenwriters*, edited by Robert E. Morsberger, Stephen O. Lesser, and Randall Clark (1984)

27 *Poets of Great Britain and Ireland, 1945-1960*, edited by Vincent B. Sherry Jr. (1984)

28 *Twentieth-Century American-Jewish Fiction Writers*, edited by Daniel Walden (1984)

29 *American Newspaper Journalists, 1926-1950*, edited by Perry J. Ashley (1984)

30 *American Historians, 1607-1865*, edited by Clyde N. Wilson (1984)

31 *American Colonial Writers, 1735-1781*, edited by Emory Elliott (1984)

32 *Victorian Poets Before 1850*, edited by William E. Fredeman and Ira B. Nadel (1984)

33 *Afro-American Fiction Writers After 1955*, edited by Thadious M. Davis and Trudier Harris (1984)

34 *British Novelists, 1890-1929: Traditionalists*, edited by Thomas F. Staley (1985)

35 *Victorian Poets After 1850*, edited by William E. Fredeman and Ira B. Nadel (1985)

36 *British Novelists, 1890-1929: Modernists*, edited by Thomas F. Staley (1985)

37 *American Writers of the Early Republic*, edited by Emory Elliott (1985)

38 *Afro-American Writers After 1955: Dramatists and Prose Writers*, edited by Thadious M. Davis and Trudier Harris (1985)

39 *British Novelists, 1660-1800*, 2 parts, edited by Martin C. Battestin (1985)

40 *Poets of Great Britain and Ireland Since 1960*, 2 parts, edited by Vincent B. Sherry Jr. (1985)

41 *Afro-American Poets Since 1955*, edited by Trudier Harris and Thadious M. Davis (1985)

42 *American Writers for Children Before 1900*, edited by Glenn E. Estes (1985)

43 *American Newspaper Journalists, 1690-1872*, edited by Perry J. Ashley (1986)

44 *American Screenwriters, Second Series*, edited by Randall Clark, Robert E. Morsberger, and Stephen O. Lesser (1986)

45 *American Poets, 1880-1945, First Series*, edited by Peter Quartermain (1986)

46 *American Literary Publishing Houses, 1900-1980: Trade and Paperback*, edited by Peter Dzwonkoski (1986)

47 *American Historians, 1866-1912*, edited by Clyde N. Wilson (1986)

48 *American Poets, 1880-1945, Second Series*, edited by Peter Quartermain (1986)

49 *American Literary Publishing Houses, 1638-1899*, 2 parts, edited by Peter Dzwonkoski (1986)

50 *Afro-American Writers Before the Harlem Renaissance*, edited by Trudier Harris (1986)

51 *Afro-American Writers from the Harlem Renaissance to 1940*, edited by Trudier Harris (1987)

52 *American Writers for Children Since 1960: Fiction*, edited by Glenn E. Estes (1986)

53 *Canadian Writers Since 1960, First Series*, edited by W. H. New (1986)

54 *American Poets, 1880-1945, Third Series*, 2 parts, edited by Peter Quartermain (1987)

55 *Victorian Prose Writers Before 1867*, edited by William B. Thesing (1987)

56 *German Fiction Writers, 1914-1945*, edited by James Hardin (1987)

57 *Victorian Prose Writers After 1867*, edited by William B. Thesing (1987)

58 *Jacobean and Caroline Dramatists*, edited by Fredson Bowers (1987)

59 *American Literary Critics and Scholars, 1800-1850*, edited by John W. Rathbun and Monica M. Grecu (1987)

60 *Canadian Writers Since 1960, Second Series*, edited by W. H. New (1987)

61 *American Writers for Children Since 1960: Poets, Illustrators, and Nonfiction Authors*, edited by Glenn E. Estes (1987)

62 *Elizabethan Dramatists*, edited by Fredson Bowers (1987)

63 *Modern American Critics, 1920-1955*, edited by Gregory S. Jay (1988)

64 *American Literary Critics and Scholars, 1850-1880*, edited by John W. Rathbun and Monica M. Grecu (1988)

65 *French Novelists, 1900-1930*, edited by Catharine Savage Brosman (1988)

66 *German Fiction Writers, 1885-1913*, 2 parts, edited by James Hardin (1988)

67 *Modern American Critics Since 1955*, edited by Gregory S. Jay (1988)

68 *Canadian Writers, 1920-1959, First Series*, edited by W. H. New (1988)

69 *Contemporary German Fiction Writers, First Series*, edited by Wolfgang D. Elfe and James Hardin (1988)

70 *British Mystery Writers, 1860-1919*, edited by Bernard Benstock and Thomas F. Staley (1988)

71 *American Literary Critics and Scholars, 1880-1900*, edited by John W. Rathbun and Monica M. Grecu (1988)

72 *French Novelists, 1930-1960*, edited by Catharine Savage Brosman (1988)

73 *American Magazine Journalists, 1741-1850*, edited by Sam G. Riley (1988)

74 *American Short-Story Writers Before 1880*, edited by Bobby Ellen Kimbel, with the assistance of William E. Grant (1988)

75 *Contemporary German Fiction Writers, Second Series*, edited by Wolfgang D. Elfe and James Hardin (1988)

76 *Afro-American Writers, 1940-1955*, edited by Trudier Harris (1988)

77 *British Mystery Writers, 1920-1939*, edited by Bernard Benstock and Thomas F. Staley (1988)

78 *American Short-Story Writers, 1880-1910*, edited by Bobby Ellen Kimbel, with the assistance of William E. Grant (1988)

79 *American Magazine Journalists, 1850-1900*, edited by Sam G. Riley (1988)

80 *Restoration and Eighteenth-Century Dramatists, First Series*, edited by Paula R. Backscheider (1989)

81 *Austrian Fiction Writers, 1875-1913*, edited by James Hardin and Donald G. Daviau (1989)

82 *Chicano Writers, First Series*, edited by Francisco A. Lomelí and Carl R. Shirley (1989)

83 *French Novelists Since 1960*, edited by Catharine Savage Brosman (1989)

84 *Restoration and Eighteenth-Century Dramatists, Second Series*, edited by Paula R. Backscheider (1989)

85 *Austrian Fiction Writers After 1914*, edited by James Hardin and Donald G. Daviau (1989)

86 *American Short-Story Writers, 1910-1945, First Series*, edited by Bobby Ellen Kimbel (1989)

87 *British Mystery and Thriller Writers Since 1940, First Series*, edited by Bernard Benstock and Thomas F. Staley (1989)

88 *Canadian Writers, 1920-1959, Second Series*, edited by W. H. New (1989)

89 *Restoration and Eighteenth-Century Dramatists, Third Series*, edited by Paula R. Backscheider (1989)

90 *German Writers in the Age of Goethe, 1789-1832*, edited by James Hardin and Christoph E. Schweitzer (1989)

91 *American Magazine Journalists, 1900-1960, First Series*, edited by Sam G. Riley (1990)

92 *Canadian Writers, 1890-1920*, edited by W. H. New (1990)

93 *British Romantic Poets, 1789-1832, First Series*, edited by John R. Greenfield (1990)

94 *German Writers in the Age of Goethe: Sturm und Drang to Classicism*, edited by James Hardin and Christoph E. Schweitzer (1990)

95 *Eighteenth-Century British Poets, First Series*, edited by John Sitter (1990)

96 *British Romantic Poets, 1789-1832, Second Series*, edited by John R. Greenfield (1990)

97 *German Writers from the Enlightenment to Sturm und Drang, 1720-1764*, edited by James Hardin and Christoph E. Schweitzer (1990)

98 *Modern British Essayists, First Series*, edited by Robert Beum (1990)

99 *Canadian Writers Before 1890*, edited by W. H. New (1990)

100 *Modern British Essayists, Second Series*, edited by Robert Beum (1990)

101 *British Prose Writers, 1660-1800, First Series*, edited by Donald T. Siebert (1991)

102 *American Short-Story Writers, 1910-1945, Second Series*, edited by Bobby Ellen Kimbel (1991)

103 *American Literary Biographers, First Series*, edited by Steven Serafin (1991)

104 *British Prose Writers, 1660-1800, Second Series*, edited by Donald T. Siebert (1991)

105 *American Poets Since World War II, Second Series*, edited by R. S. Gwynn (1991)

106 *British Literary Publishing Houses, 1820-1880*, edited by Patricia J. Anderson and Jonathan Rose (1991)

107 *British Romantic Prose Writers, 1789-1832, First Series*, edited by John R. Greenfield (1991)

108 *Twentieth-Century Spanish Poets, First Series*, edited by Michael L. Perna (1991)

109 *Eighteenth-Century British Poets, Second Series*, edited by John Sitter (1991)

110 *British Romantic Prose Writers, 1789-1832, Second Series*, edited by John R. Greenfield (1991)

111 *American Literary Biographers, Second Series*, edited by Steven Serafin (1991)

112 *British Literary Publishing Houses, 1881-1965*, edited by Jonathan Rose and Patricia J. Anderson (1991)

113 *Modern Latin-American Fiction Writers, First Series*, edited by William Luis (1992)

114 *Twentieth-Century Italian Poets, First Series*, edited by Giovanna Wedel De Stasio, Glauco Cambon, and Antonio Illiano (1992)

115 *Medieval Philosophers*, edited by Jeremiah Hackett (1992)

116 *British Romantic Novelists, 1789-1832*, edited by Bradford K. Mudge (1992)

117 *Twentieth-Century Caribbean and Black African Writers, First Series*, edited by Bernth Lindfors and Reinhard Sander (1992)

118 *Twentieth-Century German Dramatists, 1889-1918*, edited by Wolfgang D. Elfe and James Hardin (1992)

119 *Nineteenth-Century French Fiction Writers: Romanticism and Realism, 1800-1860*, edited by Catharine Savage Brosman (1992)

120 *American Poets Since World War II, Third Series*, edited by R. S. Gwynn (1992)

121 *Seventeenth-Century British Nondramatic Poets, First Series*, edited by M. Thomas Hester (1992)

122 *Chicano Writers, Second Series*, edited by Francisco A. Lomelí and Carl R. Shirley (1992)

123 *Nineteenth-Century French Fiction Writers: Naturalism and Beyond, 1860-1900*, edited by Catharine Savage Brosman (1992)

124 *Twentieth-Century German Dramatists, 1919-1992*, edited by Wolfgang D. Elfe and James Hardin (1992)

125 *Twentieth-Century Caribbean and Black African Writers, Second Series*, edited by Bernth Lindfors and Reinhard Sander (1993)

126 *Seventeenth-Century British Nondramatic Poets, Second Series*, edited by M. Thomas Hester (1993)

127 *American Newspaper Publishers, 1950-1990*, edited by Perry J. Ashley (1993)

128 *Twentieth-Century Italian Poets, Second Series*, edited by Giovanna Wedel De Stasio, Glauco Cambon, and Antonio Illiano (1993)

129 *Nineteenth-Century German Writers, 1841-1900*, edited by James Hardin and Siegfried Mews (1993)

130 *American Short-Story Writers Since World War II*, edited by Patrick Meanor (1993)

131 *Seventeenth-Century British Nondramatic Poets, Third Series*, edited by M. Thomas Hester (1993)

132 *Sixteenth-Century British Nondramatic Writers, First Series*, edited by David A. Richardson (1993)

133 *Nineteenth-Century German Writers to 1840*, edited by James Hardin and Siegfried Mews (1993)

134 *Twentieth-Century Spanish Poets, Second Series*, edited by Jerry Phillips Winfield (1994)

135 *British Short-Fiction Writers, 1880-1914: The Realist Tradition*, edited by William B. Thesing (1994)

136 *Sixteenth-Century British Nondramatic Writers, Second Series*, edited by David A. Richardson (1994)

137 *American Magazine Journalists, 1900-1960, Second Series*, edited by Sam G. Riley (1994)

138 *German Writers and Works of the High Middle Ages: 1170-1280*, edited by James Hardin and Will Hasty (1994)

139 *British Short-Fiction Writers, 1945-1980*, edited by Dean Baldwin (1994)

140 *American Book-Collectors and Bibliographers, First Series*, edited by Joseph Rosenblum (1994)

141 *British Children's Writers, 1880-1914*, edited by Laura M. Zaidman (1994)

142 *Eighteenth-Century British Literary Biographers*, edited by Steven Serafin (1994)

143 *American Novelists Since World War II, Third Series*, edited by James R. Giles and Wanda H. Giles (1994)

144 *Nineteenth-Century British Literary Biographers*, edited by Steven Serafin (1994)

145 *Modern Latin-American Fiction Writers, Second Series*, edited by William Luis and Ann González (1994)

146 *Old and Middle English Literature*, edited by Jeffrey Helterman and Jerome Mitchell (1994)

147 *South Slavic Writers Before World War II*, edited by Vasa D. Mihailovich (1994)

148 *German Writers and Works of the Early Middle Ages: 800-1170*, edited by Will Hasty and James Hardin (1994)

149 *Late Nineteenth- and Early Twentieth-Century British Literary Biographers*, edited by Steven Serafin (1995)

150 *Early Modern Russian Writers, Late Seventeenth and Eighteenth Centuries*, edited by Marcus C. Levitt (1995)

151 *British Prose Writers of the Early Seventeenth Century*, edited by Clayton D. Lein (1995)

152 *American Novelists Since World War II, Fourth Series*, edited by James and Wanda Giles (1995)

153 *Late-Victorian and Edwardian British Novelists, First Series*, edited by George M. Johnson (1995)

154 *The British Literary Book Trade, 1700-1820*, edited by James K. Bracken and Joel Silver (1995)

155 *Twentieth-Century British Literary Biographers*, edited by Steven Serafin (1995)

156 *British Short-Fiction Writers, 1880-1914: The Romantic Tradition*, edited by William F. Naufftus (1995)

157 *Twentieth-Century Caribbean and Black African Writers, Third Series*, edited by Bernth Lindfors and Reinhard Sander (1995)

158 *British Reform Writers, 1789-1832*, edited by Gary Kelly and Edd Applegate (1995)

159 *British Short Fiction Writers, 1800-1880*, edited by John R. Greenfield (1996)

160 *British Children's Writers, 1914-1960*, edited by Donald R. Hettinga and Gary D. Schmidt (1996)

161 *British Children's Writers Since 1960, First Series*, edited by Caroline Hunt (1996)

162 *British Short-Fiction Writers, 1915-1945*, edited by John H. Rogers (1996)

163 *British Children's Writers, 1800-1880*, edited by Meena Khorana (1996)

164 *German Baroque Writers, 1580-1660*, edited by James Hardin (1996)

165 *American Poets Since World War II, Fourth Series*, edited by Joseph Conte (1996)

166 *British Travel Writers, 1837-1875*, edited by Barbara Brothers and Julia Gergits (1996)

167 *Sixteenth-Century British Nondramatic Writers, Third Series*, edited by David A. Richardson (1996)

168 *German Baroque Writers, 1661-1730*, edited by James Hardin (1996)

169 *American Poets Since World War II, Fifth Series*, edited by Joseph Conte (1996)

170 *The British Literary Book Trade, 1475-1700*, edited by James K. Bracken and Joel Silver (1996)

171 *Twentieth-Century American Sportswriters*, edited by Richard Orodenker (1996)

172 *Sixteenth-Century British Nondramatic Writers, Fourth Series*, edited by David A. Richardson (1996)

173 *American Novelists Since World War II, Fifth Series*, edited by James R. Giles and Wanda H. Giles (1996)

174 *British Travel Writers, 1876-1909*, edited by Barbara Brothers and Julia Gergits (1997)

175 *Native American Writers of the United States*, edited by Kenneth M. Roemer (1997)

176 *Ancient Greek Authors*, edited by Ward W. Briggs (1997)

177 *Italian Novelists Since World War II*, edited by Augustus Pallotta (1997)

Documentary Series

1 *Sherwood Anderson, Willa Cather, John Dos Passos, Theodore Dreiser, F. Scott Fitzgerald, Ernest Hemingway, Sinclair Lewis*, edited by Margaret A. Van Antwerp (1982)

2 *James Gould Cozzens, James T. Farrell, William Faulkner, John O'Hara, John Steinbeck, Thomas Wolfe, Richard Wright*, edited by Margaret A. Van Antwerp (1982)

3 *Saul Bellow, Jack Kerouac, Norman Mailer, Vladimir Nabokov, John Updike, Kurt Vonnegut*, edited by Mary Bruccoli (1983)

4 *Tennessee Williams*, edited by Margaret A. Van Antwerp and Sally Johns (1984)

5 *American Transcendentalists*, edited by Joel Myerson (1988)

6 *Hardboiled Mystery Writers: Raymond Chandler, Dashiell Hammett, Ross Macdonald*, edited by Matthew J. Bruccoli and Richard Layman (1989)

7 *Modern American Poets: James Dickey, Robert Frost, Marianne Moore*, edited by Karen L. Rood (1989)

8 *The Black Aesthetic Movement*, edited by Jeffrey Louis Decker (1991)

9 *American Writers of the Vietnam War: W. D. Ehrhart, Larry Heinemann, Tim O'Brien, Walter McDonald, John M. Del Vecchio*, edited by Ronald Baughman (1991)

10 *The Bloomsbury Group*, edited by Edward L. Bishop (1992)

11 *American Proletarian Culture: The Twenties and The Thirties*, edited by Jon Christian Suggs (1993)

12 *Southern Women Writers: Flannery O'Connor, Katherine Anne Porter, Eudora Welty*, edited by Mary Ann Wimsatt and Karen L. Rood (1994)

13 *The House of Scribner, 1846–1904*, edited by John Delaney (1996)

14 *Four Women Writers for Children, 1868–1918*, edited by Caroline C. Hunt (1996)

Yearbooks

1980 edited by Karen L. Rood, Jean W. Ross, and Richard Ziegfeld (1981)

1981 edited by Karen L. Rood, Jean W. Ross, and Richard Ziegfeld (1982)

1982 edited by Richard Ziegfeld; associate editors: Jean W. Ross and Lynne C. Zeigler (1983)

1983 edited by Mary Bruccoli and Jean W. Ross; associate editor: Richard Ziegfeld (1984)

1984 edited by Jean W. Ross (1985)

1985 edited by Jean W. Ross (1986)

1986 edited by J. M. Brook (1987)

1987 edited by J. M. Brook (1988)

1988 edited by J. M. Brook (1989)

1989 edited by J. M. Brook (1990)

1990 edited by James W. Hipp (1991)

1991 edited by James W. Hipp (1992)

1992 edited by James W. Hipp (1993)

1993 edited by James W. Hipp, contributing editor George Garrett (1994)

1994 edited by James W. Hipp, contributing editor George Garrett (1995)

1995 edited by James W. Hipp, contributing editor George Garrett (1996)

1996 edited by Samuel W. Bruce and L. Kay Webster, contributing editor George Garrett (1997)

Concise Series

Concise Dictionary of American Literary Biography, 6 volumes (1988-1989): *The New Consciousness, 1941-1968; Colonization to the American Renaissance, 1640-1865; Realism, Naturalism, and Local Color, 1865-1917; The Twenties, 1917-1929; The Age of Maturity, 1929-1941; Broadening Views, 1968-1988.*

Concise Dictionary of British Literary Biography, 8 volumes (1991-1992): *Writers of the Middle Ages and Renaissance Before 1660; Writers of the Restoration and Eighteenth Century, 1660-1789; Writers of the Romantic Period, 1789-1832; Victorian Writers, 1832-1890; Late Victorian and Edwardian Writers, 1890-1914; Modern Writers, 1914-1945; Writers After World War II, 1945-1960; Contemporary Writers, 1960 to Present.*

Dictionary of Literary Biography
Yearbook: 1996

Dictionary of Literary Biography Yearbook: 1996

Edited by
Samuel W. Bruce

and

L. Kay Webster

George Garrett, Contributing Editor

A Bruccoli Clark Layman Book
Gale Research
Detroit, Washington, D.C., London

Advisory Board for
DICTIONARY OF LITERARY BIOGRAPHY

John Baker
William Cagle
Patrick O'Connor
George Garrett
Trudier Harris

Matthew J. Bruccoli and Richard Layman, Editorial Directors
C. E. Frazer Clark Jr., Managing Editor
Karen Rood, Senior Editor

Printed in the United States of America

Published simultaneously in the United Kingdom
by Gale Research International Limited
(An affiliated company of Gale Research)

The paper used in this publication meets the minimum requirements
of American National Standard for Information Sciences—Permanence
Paper for Printed Library Materials, ANSI Z39.48-1984.∞ ™

This publication is a creative work fully protected by all applicable copyright laws, as well as by misappropriation, trade secret, unfair competition, and other applicable laws. The authors and editors of this work have added value to the underlying factual material herein through one or more of the following: unique and original selection, coordination, expression, arrangement, and classification of the information.

All rights to this publication will be vigorously defended.

Copyright © 1997 by Gale Research
835 Penobscot Building
Detroit, MI 48226

All rights reserved including the right of reproduction in
whole or in part in any form.

ISBN 0-8103-9972-5
ISSN 0731-7867

"Nobel Lecture 1996" by Wisława Szymborska
Copyright © 1996 by the Nobel Foudation

10 9 8 7 6 5 4 3 2 1

Contents

Plan of the Series vii
Foreword ix
Acknowledgements x

The 1996 Nobel Prize in Literature: Wisława Szymborska 3
   Stanisław Barańczak

The Year in Poetry 11
   David R. Slavitt

James Dickey, American Poet 31
   Ernest Suarez

The Year in Fiction 33
   George Garrett

The Year in Literary Biography 55
   William Foltz

The Year in Drama 78
   Howard Kissel

The Year in Children's Literature 90
   Caroline C. Hunt

The F. Scott Fitzgerald Centenary 105
   Judith S. Baughman
   Reports and discussion by Richard Anderson, Richard Bausch, Robert Bausch, Sydney Blair, Frederick Busch, George Garrett, Lloyd C. Hackl, Joseph Heller, Alan Margolies, Doc Rossi, and Budd Schulberg

The Ira Gershwin Centenary 144
   Philip Furia
   Tributes from Sheldon Harnick, Michael Lasser, Deena Rosenberg, and Lawrence D. Stewart

The John Dos Passos Centenary 172
   Richard Layman
   Tributes from Daniel Aaron, Ashley Brown, Lucy Dos Passos Coggin, Ellen Bromfield Geld, Townsend Ludington, Norman Mailer, Barry Maine, Lisa Nanney, Donald Pizer, David Sanders, and Roslyn Targ

Book Reviewing and the Literary Scene 188
   George Garrett

Conversations with Publishers IV: An Interview with James Laughlin 207

Reading Series in New York City 216
   Kelli Rae Patton

Conversations with Bookmen (Publishers) III: An Interview with Otto Penzler 219

Falsifying Hemingway 228
    Matthew J. Bruccoli

Die Fürstliche Bibliothek Corvey 231
    Nancy Emery and Charles Egleston

The Mercantile Library of New York 235
    Harold Augenbraum

The American Trust for the British Library 239
    Andrew Digby and Roy Sully

The McKenzie Trust 241
    Ian Gadd and Martin Moonie

The St. John's College Robert Graves Trust 243
    Patrick Quinn

The Canadian Publishers' Records Database 247
    Carole Gerson

The Book Arts Press at the University of Virginia 252
    Terry Belanger

The Glass Key and Other Dashiell Hammett Mysteries 254
    Mark Sutcliffe

Kingsley Amis 258
    Merritt Moseley

Joseph Mitchell 265
    Raymond J. Rundus

Orville Prescott 272
    Peter S. Prescott

Letter from London 276
    Julian Evans

The Booker Prize 279
    Merritt Moseley

Editorial 285
    Matthew J. Bruccoli

Statement of Correction and Response 287

Literary Awards and Honors Announced in 1996 289
Necrology 298
Checklist: Contributions to Literary History and Biography 300
Contributors 302
Cumulative Index 305

Plan of the Series

...Almost the most prodigious asset of a country, and perhaps its most precious possession, is its native literary product — when that product is fine and noble and enduring.

Mark Twain*

The advisory board, the editors, and the publisher of the *Dictionary of Literary Biography* are joined in endorsing Mark Twain's declaration. The literature of a nation provides an inexhaustible resource of permanent worth. We intend to make literature and its creators better understood and more accessible to students and the reading public, while satisfying the standards of teachers and scholars.

To meet these requirements, *literary biography* has been construed in terms of the author's achievement. The most important thing about a writer is his writing. Accordingly, the entries in *DLB* are career biographies, tracing the development of the author's canon and the evolution of his reputation.

The purpose of *DLB* is not only to provide reliable information in a convenient format but also to place the figures in the larger perspective of literary history and to offer appraisals of their accomplishments by qualified scholars.

The publication plan for *DLB* resulted from two years of preparation. The project was proposed to Bruccoli Clark by Frederick C. Ruffner, president of the Gale Research Company, in November 1975. After specimen entries were prepared and typeset, an advisory board was formed to refine the entry format and develop the series rationale. In meetings held during 1976, the publisher, series editors, and advisory board approved the scheme for a comprehensive biographical dictionary of persons who contributed to North American literature. Editorial work on the first volume began in January 1977, and it was published in 1978. In order to make *DLB* more than a reference tool and to compile volumes that individually have claim to status as literary history, it was decided to organize volumes by topic, period, or genre. Each of these freestanding volumes provides a biographical-bibliographical guide and overview for a particular area of literature. We are convinced that this organization — as opposed to a single alphabet method — constitutes a valuable innovation in the presentation of reference material. The volume plan necessarily requires many decisions for the placement and treatment of authors who might properly be included in two or three volumes. In some instances a major figure will be included in separate volumes, but with different entries emphasizing the aspect of his career appropriate to each volume. Ernest Hemingway, for example, is represented in *American Writers in Paris, 1920-1939* by an entry focusing on his expatriate apprenticeship; he is also in *American Novelists, 1910-1945* with an entry surveying his entire career. Each volume includes a cumulative index of the subject authors and articles. Comprehensive indexes to the entire series are planned.

The series has been further augmented by the *DLB Yearbooks* (since 1981) which update published entries and add new entries to keep the *DLB* current with contemporary activity. There have also been *DLB Documentary Series* volumes which provide biographical and critical source materials for figures whose work is judged to have particular interest for students. One of these companion volumes is entirely devoted to Tennessee Williams.

We define literature as the *intellectual commerce of a nation:* not merely as belles lettres but as that ample and complex process by which ideas are generated, shaped, and transmitted. *DLB* entries are not limited to "creative writers" but extend to other figures who in their time and in their way influenced the mind of a people. Thus the series encompasses historians, journalists, publishers, book collectors, and screenwriters. By this means readers of *DLB* may be aided to perceive literature not as cult scripture in the keeping of intellectual high priests but firmly positioned at the center of a nation's life.

DLB includes the major writers appropriate to each volume and those standing in the ranks behind them. Scholarly and critical counsel has been sought in deciding which minor figures to include and how full their entries should be. Wherever possible, useful references are made to figures who do not warrant separate entries.

Each *DLB* volume has an expert volume editor responsible for planning the volume, selecting the figures for inclusion, and assigning the entries. Volume editors are also responsible for preparing, where appropriate, appendices surveying the major periodicals and literary and intellectual movements for their volumes, as well as lists of further readings. Work on the series as a whole is coordinated at the Bruccoli Clark Layman editorial center in Columbia, South Carolina, where the editorial staff is responsible for accuracy and utility of the published volumes.

One feature that distinguishes *DLB* is the illustration policy — its concern with the iconography of literature. Just as an author is influenced by his surroundings, so is the reader's understanding of the author enhanced by a knowledge of his environment. Therefore *DLB* volumes include not only drawings, paintings, and photographs of authors, often depicting them at various stages in their careers, but also illustrations of their families and places where they lived. Title pages are regularly reproduced in facsimile along with dust jackets for modern authors. The dust jackets are a special feature of *DLB* because they often document better than anything else the way in which an author's work was perceived in its own time. Specimens of the writers' manuscripts and letters are included when feasible.

Samuel Johnson rightly decreed that "The chief glory of every people arises from its authors." The purpose of the *Dictionary of Literary Biography* is to compile literary history in the surest way available to us — by accurate and comprehensive treatment of the lives and work of those who contributed to it.

The *DLB* Advisory Board

Foreword

The *Dictionary of Literary Biography Yearbook* is guided by the same principles that have provided the basic rationale for the entire *DLB* series: 1) the literature of a nation represents an inexhaustible resource of permanent worth; 2) the surest way to trace the outlines of literary history is by a comprehensive treatment of our lives and works of those who contributed to it; and 3) the greatest service the series can provide is to make literary achievements better understood and more accessible to students and the literate public, while serving the needs of scholars. In keeping with those principles, the *Yearbook* has been planned to augment *DLB* by reflecting the vitality of contemporary literature and summarizing current literary activity. The librarian, scholar, or student attempting to stay informed of literary developments is faced with an endless task. The purpose of the *DLB Yearbook* is to serve those readers while at the same time enlarging the scope of the *DLB*.

The *Yearbook* includes articles about the past year's literary events or topics, as well as obituaries and tributes. Each *Yearbook* also includes a list of literary prizes and awards, a necrology, and a checklist of literary histories and biographies published during the year. This *Yearbook* continues the *Dictionary of Literary Biography Yearbook* Awards for the novel, first novel, poetry, children's book, and literary biography.

From the outset, the *DLB* series has undertaken to compile literary history as it is revealed in the lives and works of authors. The *Yearbook* supports that commitment, providing a useful and necessary current record.

Acknowledgments

This book was produced by Bruccoli Clark Layman, Inc. Karen L. Rood is senior editor for the *Dictionary of Literary Biography* series. L. Kay Webster and Samuel W. Bruce were the in-house editors.

Administrative support was provided by Ann M. Cheschi and Brenda A. Gillie.

Bookkeeper is Joyce Fowler.

Copyediting supervisor is Laurel M. Gladden Gillespie. The copyediting staff includes Phyllis A. Avant, Patricia Coate, Jeff Miller, William L. Thomas Jr., and Allison Trussell.

L. Kay Webster and Jane M. J. Williamson are editorial associates.

Layout and graphics supervisor is Pamela D. Norton.

Office manager is Kathy Lawler Merlette.

Photography editors are Julie E. Frick and Margaret Meriwether. Photographic copy work was performed by Joseph M. Bruccoli.

Production manager is Samuel W. Bruce.

Software specialist is Marie L. Parker.

Systems manager is Chris Elmore.

Typesetting supervisor is Kathleen M. Flanagan. The typesetting staff includes Melody W. Clegg and Patricia Flanagan Salisbury.

Walter W. Ross, Steven Gross, and Mark McEwan did library research. They were assisted by the following librarians at the Thomas Cooper Library of the University of South Carolina: Linda Holderfield and the interlibrary-loan staff; reference-department head Virginia Weathers; reference librarians Marilee Birchfield, Stefanie Buck, Stefanie DuBose, Rebecca Feind, Karen Joseph, Donna Lehman, Charlene Loope, Anthony McKissick, Jean Rhyne, Kwamine Simpson, and Virginia Weathers; circulation-department head Caroline Taylor; and acquisitions-searching supervisor David Haggard.

Dictionary of Literary Biography Yearbook: 1996

Dictionary of Literary Biography

The 1996 Nobel Prize in Literature
Wisława Szymborska
(2 July 1923 -)

Stanisław Barańczak
Harvard University

BOOKS: *Dlatego żyjemy* (Warsaw: Czytelnik, 1952);
Pytania zadawane sobie (Kraków: Wydawnictwo Literackie, 1954);
Wołanie do Yeti (Kraków: Wydawnictwo Literackie, 1957);
Sól (Warsaw: Państwowy Instytut Wydawniczy, 1962);
Sto pociech (Warsaw: Państwowy Instytut Wydawniczy, 1967);
Wszelki wypadek (Warsaw: Czytelnik, 1972);
Lektury nadobowiązkowe (Kraków: Wydawnictwo Literackie, 1973);
Wielka liczba (Warsaw: Czytelnik, 1976);
Lektury nadobowiązkowe, cz. 2 (Kraków: Wydawnictwo Literackie, 1981);
Ludzie na moście (Warsaw: Czytelnik, 1986);
Koniec i początek (Poznań: Wydawnictwo a5, 1993).

Collections: *Wiersze wybrane* (Warsaw: Państwowy Instytut Wydawniczy, 1964);
Poezje wybrane (Warsaw: Ludowa Spółdzielnia Wydawnicza, 1967);
Wybór poezji (Warsaw: Czytelnik, 1970);
Poezje (Warsaw: Państwowy Instytut Wydawniczy, 1970);
Wybór wierszy (Warsaw: Czytelnik, 1973);
Sounds, Feelings, Thoughts: Seventy Poems, translated by Magnus J. Krynski and Robert A. Maguire (Princeton, N.J.: Princeton University Press, 1981);
People on a Bridge: Poems, translated by Adam Czerniawski (London: Forest Books, 1990);
Lektury nadobowiązkowe (Kraków: Wydawnictwo Literackie, 1992);
Wieczór autorski (Warsaw: Anagram, 1992);

Wisława Szymborska (Reuters)

View with a Grain of Sand: Selected Poems, translated by Stanisław Barańczak and Clare Cavanagh (New York: Harcourt Brace, 1995); published in Poland as *Widok z ziarnkiem piasku* (Poznań: Wydawnictwo a5, 1996).

The decision to award the Nobel Prize in Literature to Wisława Szymborska could have come as something of a surprise to those critics, journalists, and other avid Nobel watchers who presume that the Stockholm jury picks its laureates each year in accordance with some principle of proportional distribution of the prize among continents, nationalities, genders, and literary genres. For such observers the statistical odds in the fall of 1996 certainly worked against Szymborska. She is, after all, not only the fourth Polish writer honored with the highest of literary distinctions (after Henryk Sienkiewicz in 1905, Władysław Stanisław Reymont in 1924, and Czesław Miłosz in 1980), but also the second Polish poet awarded the prize in a mere sixteen years. And yet, in the eyes of those familiar with contemporary Polish poetry in general, and Szymborska's work in particular, her statistically improbable victory only proves that members of the Nobel Committee are able to recognize true greatness when they see it. Along with Miłosz (1911-), Zbigniew Herbert (1924-), Tadeusz Różewicz (1921-), and the late Miron Białoszewski (1922–1983), Szymborska has been one of the dominating figures in the incomparably rich and lively arena of Polish poetry in the second half of this century. Her individual position, however, is due not merely to her being both a successor and a leading representative of a splendid literary tradition but, even more so, to her own hard-earned achievement.

Wisława Szymborska was born on 2 July 1923 in the small town of Kórnik (actually Bnin, which later merged with Kórnik to form a joint township bearing the latter's name) near the city of Poznań, the industrial and cultural center of the western part of Poland. She was one of two daughters of Wincenty Szymborski and Anna (née Rottermund); her father served as the steward of Count Władysław Zamoyski's family estate. When Szymborska was eight her family moved to the historic city of Kraków—as much an informal capital of the southern part of Poland as Poznań is of the western one—to settle down for good. Since that time Szymborska's entire life, if one does not count her infrequent and usually short travels, has been spent in Kraków. It was in that city that she attended high school before 1939. During the war and the Nazi occupation Szymborska continued her education in a clandestine study group and made her living as a clerk at the railway office, among other odd jobs. It was also in Kraków, at its ancient Jagellonian University, that she studied in 1945–1948, majoring in sociology and Polish literature. After graduation she worked for a while as an assistant editor in publishing houses until in 1953 she became the editor of the poetry section in the Kraków-based weekly *Życie Literackie* (Literary Life), a position she held until 1968. It was in that weekly that she published regularly over the years her *Lektury nadobowiązkowe* (Noncompulsory Reading), a series of brief, often humorous, essay-reviews of mostly nonliterary books dealing with subjects as diverse as ancient history, ornithology, the life of Casanova, gardening, jazz, home repairs, and hatha yoga. Collected in three consecutive book editions, these reviews (which after she left *Życie Literackie* continued to appear on the pages of other periodicals) constitute her entire literary work other than poetry or poetic translation (she is an excellent translator of French poetry, from Agrippa d'Aubigné to Baudelaire). In 1948 she married Adam Włodek, a poet and editor, whom she divorced in 1954. She has no children.

There is not much more to say about Szymborska's life, since up to her winning of the Nobel Prize it was, except perhaps for literary awards and other tokens of public appreciation, almost totally uneventful. As much as such things can be planned, that was mostly the result of her deliberate choice. The poet is known for her quiet way of life, unassuming manner, and unwillingness to embrace the status of a celebrity. She shuns public gatherings, rarely travels abroad, hates being photographed or interviewed, and refuses to be involved in politics (although she did lend her public support to numerous initiatives of the human rights and democratic reform movements during the 1970s and 1980s). As a private person, however, she is far from being a recluse; on the contrary, she is very much part of the cultural landscape of Kraków and maintains lively contacts with her small circle of close friends, who adore her for her human warmth and personal charm and for her conversational brilliance and quick sense of humor (there has been no better author of extemporized limericks in the history of Polish light verse). A large part of her personal legend has been built by her famous homemade postcards to friends, which consist of pasted newspaper cuttings of words and images combined in the manner of the quasi-surrealistic collage to a hilarious effect. Her dislike of being in a public spotlight is by no means a sign of antisocial inclinations; rather, it stems from her sober recognition that the larger part of writers' public functioning is an empty ritual and an unnecessary waste of their inner resources.

This line of argument finds its close parallel in the extraordinary sparseness of Szymborska's literary production. Just as she refuses to take part in activities in which she has no genuine interest, her chief creative commandment appears to be, quite simply, "Thou shalt not write irrelevant poems." The net result of this principle is that, as the seventy-three-year-old author of about two hundred individual poems, she is among the least prolific major poets of our time; it is likely that there has been no Nobel Prize–winning poet who has written less verse. Over the three pre-Nobel decades she published a few poems a year in the Polish literary press, and her slim collections came out at seven- to ten-year intervals. Just as her love of privacy has nothing to do with autistic self-centeredness, her small output is not a result of either laziness or writer's block. Rather, she publishes deliberately little because she holds the highest standards for herself. Asked in a recent interview what secret reason lurks behind her reluctance to publish more poems, she gave a typically brief and concrete answer: "I have a wastebasket in my study."

This wastebasket, by the way, swallows more than just drafts and unpublished manuscripts. Szymborska's self-criticism also works, so to speak, in reverse. She renounced a long time ago the first two of the nine collections of original verse that have been published over her fifty-two year career, and over the last three decades she has refused to let their contents be reprinted. It is not particularly hard to understand why. The year of her debut in the literary press, 1945, was probably as bad a moment as any in recorded history for young poets in Poland to learn their trade. The pull of the socialist-realist temptation was almost irresistible, and not just because of the prosaic fact that under Communist rule the writing of propagandistic verse naturally facilitated the poet's career. Another, and no less important, reason was that the traumatic experience of the unspeakable atrocities of the just-ended war pushed the typical young idealist to embrace impulsively the ideology that seemed—particularly when set against the backdrop of the "decayed liberalism" of the Western democracies—to provide the only efficient way of preventing the moral failure of humankind from happening again.

All these notions find their rather naive but also quite sincere expression in Szymborska's early verse. She made her debut on 14 March 1945—that is, in the final weeks of World War II—with a poem titled "Szukam słowa" (Searching for a Word), published in a supplement to the daily *Dziennik Polski*. In a sense the poem's title is symbolic of what her early writing was all about: a desperate search for a language and style strong enough to carry the burden of twentieth-century experience. The reason the search was, at that stage, largely unsuccessful comes down to the fact that the poet was not yet able to grasp the basic paradox out of which her art was soon to evolve: in the poetry of our age, the source of such strength is the author's ability to doubt rather than believe, to be skeptically ironic rather than blindly positive, to ask questions rather than make assertions.

From this point of view, it is interesting to compare the title of Szymborska's first collection, *Dlatego żyjemy* (That's What We Live For, 1952), with that of her second, *Pytania zadawane sobie* (Questions Put to Myself, 1954). It is in the semantic gap between these two titles that we can catch the first glimpse of the genuine Szymborska, the one who was to find her fully original voice with her third collection, *Wołanie do Yeti* (Calling Out to Yeti, 1957). The youthful self-confidence of the first book's title gives way to mature self-questioning and doubt; perhaps most significant, the plural *we* is replaced with the singular *myself*.

In one of the few interviews Szymborska has given in the course of her career, she said that in her early writing her chief mistake was to try to love humankind instead of human beings. One might add that the aesthetics of socialist realism demanded love for nothing less than humankind while at the same time, ironically, narrowing the multidimensionality of human life down to just one, social, dimension; however, in contrast, it is Szymborska's focus on the individual that allows her to view human reality in all its troublesome complication. The breadth and depth of her vision of human existence, with its inescapable entanglement in numerous different orders of being at once, can hardly be explained by interpreting it in the light of existentialist philosophy, as some Polish critics in the late 1950s and 1960s were fond of doing. Szymborska reveals more affinity with the ironic vision of a Montaigne than with any of the twentieth-century existentialists, if only because of the latter's notorious lack of any sense of humor. For her the result of the realization of the human being's simultaneous involvement in so many different dimensions of existence—from the biological to the civilizational, from spatial to temporal, from social to historical, from unique to universal—cannot be restricted to the revelation of the tragic; it also, perhaps first of all, includes the sense of awe, amazement, and amusement with the dissonant and yet somehow harmonious complexity of it all. Irony and humor are the poet's means to gain a distance from the human predicament, a distance that has nothing to do with detached indifference; rather, it is a position necessarily taken in order to see things more clearly. Even though the focus of Szymborska's poetry is decidedly anthropocentric, she often employs a lyrical voice that belongs to a nonhuman speaker, regardless of whether the place he occupies on the ladder of evolution is inferior, as in the case of an animal, or superior to ours, as in the case of some imagined extraplanetary observer:

> I, a tarsier,
> sit living on a human fingertip.
> ....
> My good lord is gracious,
> my good lord is kind.
> Who else could bear such witness if there were
> no creatures unworthy of death?
> You yourselves, perhaps?
> But what you've come to know about yourselves
> will serve for a sleepless night from star to star.
>
> ("Tarsier")

> Carry on, then, if only for the moment
> that it takes a tiny galaxy to blink!
> One wonders what will become of him,
> since he does in fact seem to be.
> And as far as being goes, he really tries quite hard.
> Quite hard indeed—one must admit.
> With that ring in his nose, with that toga, that sweater.
> He's no end of fun, for all you say.
> Poor little beggar.
> A human, if ever we saw one.
>
> ("No End of Fun")

There has been no shortage of profound poets and multidimensional worlds created by their imaginations in our century, yet what makes Szymborska's position in world literature quite unique is that, in some mysterious way, the uncompromising complexity of her ironic vision never prevents her work from being accessible. Since 1957 her popularity in Poland has been growing steadily, reaching staggering proportions with her most recent volume, *Koniec i początek* (The End and the Beginning, 1993). Some of her recent poems, such as the amazing and moving "Cat in an Empty Apartment" (in which the absence of someone who is dead is presented from the perspective of the house pet he left behind), have already acquired the status of cult objects among Polish readers. Moreover, even though her stylistic inventiveness makes her poems difficult to translate in spite of their clear logic and transparent construction, she has nevertheless successfully crossed the language barrier in numerous translations, and her work has for many years been enjoyed and admired by non-Polish audiences. All this is, to reiterate, highly unusual. We are used to the fact that a poet's popular appeal is a commodity purchased in exchange for some concessions, for the poet's renunciation of at least part of what constitutes the natural ambiguity of his or her individual self. In contrast, Szymborska seems to be endowed with an almost superhuman ability to be complex yet comprehensible, ambitious yet approachable, individualistic yet involved.

If this secret can be explained, the explanation will have to reach back to the moment early in her career when Szymborska realized that what attracts people to poetry today is not its potential for making statements, but rather its art of asking questions. The situation of the uncertainty of the individual mind is clearly the single most powerful mechanism of generating lyrical action in her work; correspondingly, the model of inquiry or self-inquiry, of asking "questions put to myself" as well as to others, makes its presence felt with striking frequency and insistence throughout her entire oeuvre. To say that, however, will not suffice to characterize her individual method; its crucial point is the apparent naiveté of the questions asked:

> "How should we live?" someone asked me in a letter.
> I had meant to ask him
> the same question.
> ....
> [T]he most pressing questions
> are naïve ones.
>
> ("The Century's Decline")

Significantly, the uncertainty as to the most basic dilemmas of existence is shared by the poet and her imagined reader. One fundamental reason for the accessibility of Szymborska's poetry is that the "pressing questions" she keeps asking are, at least at first sight, as "naïve" as those of the man in the street. At the same time, the brilliance of her poetry lies in pushing the inquiry much farther than the man in the street ever could or would. Many of her poems start provocatively, with a question, observation, or statement that seems downright trite, only to surprise us with its unanticipated yet logical continuation. What can sound more banal than proclaiming at the outset of a poem that "nothing can ever happen twice," that one cannot reenter the Heraclitean river? And yet the next three lines, by pursuing this thought to its end, offer a view of human existence that, unexpectedly, turns out to be pregnant with complex ironies:

> Nothing can ever happen twice.
> In consequence, the sorry fact is
> that we arrive here improvised

and leave without the chance to practice.

("Nothing Twice")

Similarly, the poeí that lent its title to *Koniec i początek* opens with a statement that sounds so disarmingly trivial that it seems not to contain any potential for revelation at all:

After every war
someone has to tidy up.
Things won't pick
themselves up, after all.

Yet it soon turns out that the "naïve question" implied in this poem (namely, the question of how our species can carry on at all if it is doomed to move within the vicious circle of destruction and reconstruction) concerns no less pressing an issue than the meaning of human history—or perhaps the senselessness of it. What makes this poem typically Szymborskian is that its initial naiveté almost imperceptibly moves to another plane. The action of "cleaning up the mess" turns, by metaphoric equation, into the process of forgetting. Just as one must remove the rubble after the war, one must remove the remembrance of human evil; otherwise, the burden of living would be unbearable. But this means that we never learn from history. Our ability to forget makes us, at the same time, repeatedly commit the same tragic blunders.

Szymborska's "naïve questions" would not perhaps be posed with such force, if not for the fact that they are actually reactions to statements. The assertive statement that provokes her inquiry may not be included in the poem, at least not in full and not as a direct quotation; nevertheless, it is always there, if only concealed in the question itself and waiting to be guessed by the reader. Implied thus or put forward more explicitly, such a statement is, as a rule, a widely shared opinion on an issue, and the "naïve question" is supposed to raise doubt about its validity. The opinion not only reflects some common belief or is representative of some widespread mind-set, but also, as a rule, has a certain doctrinaire ring to it. The outlook behind it is usually speculative, antiempirical, prone to hasty generalizations, collectivist, dogmatic, and intolerant. The poet's function as seen in Szymborska's poems is to prick the balloons of such self-confidence and blatant oversimplification by proposing an exception to the supposedly all-binding rule, an individual case that contradicts the supposedly universal law. The natural role for poets, then, is to be eternal spoilsports. No happy utopia needs their sadness; no serious undertaking of the movers and shakers of the world needs their mockery. And that is precisely why, to quote the concluding sentence of Szymborska's Nobel Lecture, "It looks like poets will always have their work cut out for them."

References:

Stanisław Balbus, *Świat ze wszystkich stron świata: O Wisławie Szymborskiej* (Kraków: Wydawnictwo Literackie, 1996);

Balbus and Dorota Wojda, eds., *Radość czytania Szymborskiej: Wybór tekstów krytycznych* (Kraków: Znak, 1996);

Edward Balcerzan and others, *Szymborska: Szkice* (Warsaw: OPEN, 1996);

Małgorzata Baranowska, *Tak lekko było nic o tym nie wiedzieć . . .: Szymborska i świat* (Wrocław: Wydawnictwo Dolnośląskie, 1996);

Urszula Biełous, *Szymborska* (Warsaw: Interpress, 1974);

Anna Legeżyńska, *Wisława Szymborska* (Poznań: Rebis, 1996);

Barbara Sienkiewicz, ed., *Poznańskie Studia Polonistyczne: Wokół Szymborskiej* (Poznań: Wydawnictwo WiS, 1995);

Anna Węgrzyniakowa, *Nie ma rozpusty większej niż myślenie* (Warsaw: Towarzystwo Zachęty Kultury, 1997);

Dorota Wojda, *Milczenie słowa: O poezji Wisławy Szymborskiej* (Kraków: Universitas, 1997).

NOBEL LECTURE 1996
Wisława Szymborska

The Poet and the World

They say the first sentence in any speech is always the hardest. Well, that one's behind me, anyway. But I have a feeling that the sentences to come—the third, the sixth, the tenth, and so on, up to the final line—will be just as hard, since I'm supposed to talk about poetry. I've said very little on the subject, next to nothing, in fact. And whenever I have said anything, I've always had the sneaking suspicion that I'm not very good at it. This is why my lecture will be rather short. All imperfection is easier to tolerate if served up in small doses.

Contemporary poets are skeptical and suspicious even, or perhaps especially, about themselves. They publicly confess to being poets only reluctantly, as if they were a little ashamed of it. But in our clamorous times it's much easier to acknowledge your faults, at least if they're attractively pack-

Front cover of the 1995 volume of Wisława Szymborska's poems in translation. The circular label was added in 1996.

aged, than to recognize your own merits, since these are hidden deeper and you never quite believe in them yourself . . . When filling in questionnaires or chatting with strangers, that is, when they can't avoid revealing their profession, poets prefer to use the general term "writer" or replace "poet" with the name of whatever job they do in addition to writing. Bureaucrats and bus passengers respond with a touch of incredulity and alarm when they find out that they're dealing with a poet. I suppose philosophers may meet with a similar reaction. Still, they're in a better position, since as often as not they can embellish their calling with some kind of scholarly title. Professor of philosophy—now that sounds much more respectable.

But there are no professors of poetry. This would mean, after all, that poetry is an occupation requiring specialized study, regular examinations, theoretical articles with bibliographies and footnotes attached, and finally, ceremoniously conferred diplomas. And this would mean, in turn, that it's not enough to cover pages with even the most exquisite poems in order to become a poet. The crucial element is some slip of paper bearing an official stamp. Let us recall that the pride of Russian poetry, the future Nobel Laureate Joseph Brodsky, was once sentenced to internal exile precisely on such grounds. They called him "a parasite," because he lacked official certification granting him the right to be a poet . . .

Several years ago, I had the honor and pleasure of meeting Brodsky in person. And I noticed that, of all the poets I've known, he was the only one who enjoyed calling himself a poet. He pronounced the word without inhibitions. Just the opposite—he spoke it with defiant freedom. It seems to me that this must have been because he recalled the brutal humiliations he had experienced in his youth.

In more fortunate countries, where human dignity isn't assaulted so readily, poets yearn, of course, to be published, read, and understood, but they do little, if anything, to set themselves above the common herd and the daily grind. And yet it wasn't so long ago, in this century's first decades, that poets strove to shock us with their extravagant dress and eccentric behavior. But all this was merely for the sake of public display. The moment always came when poets had to close the doors behind them, strip off their mantles, fripperies, and other poetic paraphernalia, and confront—silently, patiently awaiting their own selves—the still-white sheet of paper. For this is finally what really counts.

It's not accidental that film biographies of great scientists and artists are produced in droves. The more ambitious directors seek to reproduce convincingly the creative process that led to important discoveries or the emergence of a masterpiece. And one can depict certain kinds of scientific labor with some success. Laboratories, sundry instruments, elaborate machinery brought to life: such scenes may hold the audience's interest for a while. And those moments of uncertainty—will the experiment, conducted for the thousandth time with some tiny modification, finally yield the desired result?—can be quite dramatic. Films about painters can be spectacular, as they go about recreating every stage of a famous painting's evolution, from the first penciled line to the final brushstroke. Music swells in films about composers: the first bars of the melody that rings in the musician's ears finally emerge as a mature work in symphonic form. Of course this is all quite naïve and doesn't explain the strange mental state popularly known as inspiration, but at least there's something to look at and listen to.

But poets are the worst. Their work is hopelessly unphotogenic. Someone sits at a table or lies on a sofa while staring motionless at a wall or ceiling. Once in a while this person writes down seven lines only to cross out one of them fifteen minutes later, and then another hour passes, during which nothing happens . . . Who could stand to watch this kind of thing?

I've mentioned inspiration. Contemporary poets answer evasively when asked what it is, and if it actually exists. It's not that they've never known the blessing of this inner impulse. It's just not easy to explain something to someone else that you don't understand yourself.

When I'm asked about this on occasion, I hedge the question too. But my answer is this: inspiration is not the exclusive privilege of poets or artists generally. There is, has been, and will always be a certain group of people whom inspiration visits. It's made up of all those who've consciously chosen their calling and do their job with love and imagination. It may include doctors, teachers, gardeners—and I could list a hundred more professions. Their work becomes one continuous adventure as long as they manage to keep discovering new challenges in it. Difficulties and setbacks never quell their curiosity. A swarm of new questions emerges from every problem they solve. Whatever inspiration is, it's born from a continuous "I don't know."

There aren't many such people. Most of the earth's inhabitants work to get by. They work because they have to. They didn't pick this or that kind of job out of passion; the circumstances of their lives did the choosing for them. Loveless work, boring work, work valued only because others haven't got even that much, however loveless and boring—this is one of the harshest human miseries. And there's no sign that coming centuries will produce any changes for the better as far as this goes.

And so, though I may deny poets their monopoly on inspiration, I still place them in a select group of Fortune's darlings.

At this point, though, certain doubts may arise in my audience. All sorts of torturers, dictators, fanatics, and demagogues struggling for power by way of a few loudly shouted slogans also enjoy their jobs, and they too perform their duties with inventive fervor. Well, yes, but they "know." They know, and whatever they know is enough for them once and for all. They don't want to find out about anything else, since that might diminish their arguments' force. And any knowledge that doesn't lead to new questions quickly dies out: it fails to maintain the temperature required for sustaining life. In the most extreme cases, cases well known from ancient and modern history, it even poses a lethal threat to society.

This is why I value that little phrase "I don't know" so highly. It's small, but it flies on mighty wings. It expands our lives to include the spaces within us as well as those outer expanses in which our tiny Earth hangs suspended. If Isaac Newton had never said to himself "I don't know," the apples in his little orchard might have dropped to the ground like hailstones and at best he would have stooped to pick them up and gobble them with gusto. Had my compatriot Marie Sklodowska-Curie never said to herself "I don't know," she probably would have wound up teaching chemistry at some private high school for young ladies from good families, and would have ended her days performing this otherwise perfectly respectable job. But she kept on saying "I don't know," and these words led her, not just once but twice, to Stockholm, where restless, questing spirits are occasionally rewarded with the Nobel Prize.

Poets, if they're genuine, must also keep repeating "I don't know." Each poem marks an effort to answer this statement, but as soon as the final period hits the page, the poet begins to hesitate, starts to realize that this particular answer was pure makeshift that's absolutely inadequate to boot. So the poets keep on trying, and sooner or later the consecutive results of their self-dissatisfaction are clipped together with a giant paperclip by literary historians and called their "oeuvre" . . .

I sometimes dream of situations that can't possibly come true. I audaciously imagine, for example, that I get a chance to chat with the Ecclesiastes, the author of that moving lament on the vanity of all human endeavors. I would bow very deeply before him, because he is, after all, one of the greatest poets, for me at least. That done, I would grab his hand. " 'There's nothing new under the sun': that's what you wrote, Ecclesiastes. But you yourself were born new under the sun. And the poem you created is also new under the sun, since no one wrote it down before you. And all your readers are also new under the sun, since those who lived before you couldn't read your poem. And that cypress that you're sitting under hasn't been growing since the dawn of time. It came into being by way of another cypress similar to yours, but not exactly the same. And Ecclesiastes, I'd also like to ask you what new thing under the sun you're planning to work on now? A further supplement to the thoughts you've already expressed? Or maybe you're tempted to contradict some of them now? In your earlier work you mentioned joy—so what if it's fleeting? So maybe your new-under-the-sun poem will be about joy? Have you taken notes yet, do you have drafts? I doubt you'll say, 'I've written everything down, I've got nothing left to add.' There's no poet in the world who can say this, least of all a great poet like yourself."

The world—whatever we might think when terrified by its vastness and our own impotence, or embittered by its indifference to individual suffering, of people, animals, and perhaps even plants, for why are we so sure that plants feel no pain; whatever we might think of its expanses pierced by the ray of stars surrounded by planets we've just begun to discover, planets already dead? still dead? we just don't know; whatever we might think of this measureless theater to which we've got reserved tickets, but tickets whose lifespan is laughingly short, bounded as it is by two arbitrary dates; whatever else we might think of this world—it is astonishing.

But "astonishing" is an epithet concealing a logical trap. We're astonished, after all, by things that deviate from some well-known and universally acknowledged norm, from an obviousness we've grown accustomed to. Now the point is, there is no such obvious world. Our astonishment exists per se and isn't based on comparison with something else.

Granted, in daily speech, where we don't stop to consider every word, we all use phrases like "the ordinary world," "ordinary life," "the ordinary course of events" . . . But in the language of poetry, where every word is weighed, nothing is usual or normal. Not a single stone and not a single cloud above it. Not a single day and not a single night after it. And above all, not a single existence, not anyone's existence in this world.

It looks like poets will always have their work cut out for them.

—Translated from the Polish by Stanisław Barańczak and Clare Cavanagh

The Year in Poetry

David R. Slavitt
University of Pennsylvania

This broad conspectus of the poetry of 1996 involves certain strategies that are, to put it mildly, counterintuitive. We all learn early on what not to read. (Most Americans learn to avoid all poetry because their first encounters with it have been unfortunate, mostly in schools where, for the wrong reasons, teachers who don't really like it themselves use it, or more accurately, abuse it, for perverse pedagogical reasons. What poet, imagining all the reluctant and mostly wretched essays his elegant quatrain may beget, wouldn't think hard about letting it out of the house?)

What we learn, then, is not to read but to avoid, to skip this dreary, strident feminist, that political kvetch, the other boring, academic Surrealist. We learn to be suspicious of the inept but much honored literary pooh-bah or the newest model of Helen Vendler's arbitrary modish fabrication. We confine ourselves to reading our friends and those whom our friends have recommended. Sometimes, in magazines, there can be a performance graceful and accomplished enough to make it worthwhile at least to try to remember the name. And there are reviews at which we sometimes glance—not for what the reviewers actually have to say but for the snippets they quote. We try not to be too promiscuous, having learned how long the odds are against any sort of pleasurable outcome. But then, perhaps for some sins of an earlier life, one winds up doing something like this—taking at least a glance at as much as possible.

After such a wallow, it is all but impossible to suggest in any civil way the dismal level to which a great art has been reduced by people who have no apparent talent for it—or even interest in it—but have taken it up because it is indoor work that doesn't involve heavy lifting. Most of these poets hang around universities where you either have to know something (a Greek verb maybe, or the table of the elements) or else be willing to teach Bone-head Comp. or (the same course with a different listing) Advanced Creative Writing/Poetry. Poetry therefore becomes careerist, which is laughable. But the poems these people produce aren't funny. Or even interesting. What we get, over and over, are sordid, swinish, clumsy, self-indulgent reshufflings of the same tired tropes of last year of tenured movers and shakers. Or the endorsers are the editors of the series. Or they are the outside readers for the press. Or the judges of the Feinshmecker Award (which includes a reading at and publication by the press of the University of Very Far Away).

It is not a pretty sight. But it is worth doing, I keep telling myself, because it forces me to keep up with what's going on, and, from the good work, to get a sense not only of the poetry, but also of the country. Chaim Gouri, an Israeli poet who gave a reading in Philadelphia this year, told a story about the remark of an Egyptian poet (whose name escapes me) to the effect that the Yom Kippur War wouldn't have been such a surprise to the Israelis if their secret service had had a unit assigned to reading Arab poetry.

Here and there, I've found interesting work, some of it by people whose names I'd never heard before. And I don't think I'm fooling myself about seeing a continuation of the trend I've been watching in recent years. Despite all odds, the long poem persists, perhaps because such an undertaking excuses the practitioner from the depressing business of magazine submissions, or, for those who are "names," it offers a way of resisting the process that turns even the prettiest baubles into grubby classroom exercises in explication and deconstruction.

Many of these are at least in part funny, which is a further defiance of those Gradgrind workshops where nobody has any sense of humor (because if they did, they couldn't be doing what they do). The comedy can cut awfully close to the bone, as gallows humor does, and still be witty. It requires, moreover, a kind of dexterity, a knack for timing that not even all real poets can call upon. (Try, for example, to imagine John Milton's dinner-table repartee.) It also requires some faith in the art itself, which is almost as rare these days as is the ability to enjoy poetry. Lest they be forced to confront the possibility that they have devoted their lives to something frivolous, a good many critics and academics—and some writers, too, I am afraid—try to turn literature into something else and, presumably, better: philosophy, history, politics, religion even. Those who have the courage of their lack of convictions can read Robert Herrick, X. J. Kennedy, or Anthony Hecht and not worry that some of this is frothy and trivial. Such poetry is like meringues, which are no good if they're chewy.

What I elect to start with, then, is Richard Moore's *The Mouse Whole* (Negative Capability Press), which is an epic (mock, obviously) about a mouse (obviously). The title comes from the fact that this volume comprises "The Education of a Mouse," "Marriage of a Mouse," and "Apotheosis of a Mouse." There is a foreword by Howard Nemerov, from back in 1978, in which he discusses how the mouse epic has been kicking around for ten or a dozen years and joking about how little clout he has had in helping Moore find a publisher. There is a blurb from Robert Lowell. And the poem? It is, as these shades attest, poignant, witty, and quite wonderful. Here's a piece in which the mouse climbs out of the sewer in which he was born:

> One day I climbed to the shelf
> where the passage was bright and dry,
> and saw in the glittering distance
> dark bars against the sky—
> the sky whose very existence
> was to me completely unknown.
> How brilliantly blue it shone.
> Drawn on by the light, by a fate
> beyond mice, I crept to that grate,
> which burned with a glow so intense
> it seemed to shatter my sense.
> A booming shook in my ear
> and my body crawled with fear
> when a shadow above me passed
> with a rumbling deep and vast.
> Was it Heaven out there? Was it Hell?
> I was ignorant. How could I tell?
> And I fled back into my hole.
> Yet something began in my soul
> on that tumultuous day
> that was never to fade away.
> I'd often creep to the edge
> of the lowest part of the ledge
> and gaze at that grated sky.
> The passage was usually dry;
> but once when the sky was gray
> down the darkened passageway
> (which was, I learned later, a drain)
> came torrents of turbulent rain.
> Then suddenly I was aware
> as it soaked in my greasy hair
> and dampened my shivering flesh
> that the water was almost fresh.

It is a small epiphany from a masterful piece that goes on for 223 pages of witty, angry, funny, forlorn playing. But the mind's play is the soul's business, and it is being transacted here. Moore's tribulations getting this before the world are as epic as the adventures of his *riduculus mus*. Not surprisingly, the volume is dedicated to Sue Walker, its publisher, who is lionhearted compared to most other mousy purveyors of poetry.

Mark Rudman's success with his long autobiographical poem that began with *Rider* and continues now with *Millennium Hotel* (Wesleyan University Press) has been much greater. Livelier and more interesting than *Paterson* and almost as large in its concerns as the *Cantos*, this series of reports from Rudman's mental and psychic frontier has the associative structure and, more important, the scale of those modernist works. He can also work the same magic connecting the most intimate and personal observations—about himself, his childhood, his city, or his child—with larger and more generally accessible cultural artifacts, like the Rembrandt painting of "The Polish Rider" in the first volume, or, now, in this volume, chunks of Horace or Ovid or Heinrich Heine. I find what he does with a couple of Horace's odes to be most dazzling, but that may be a private matter, for Horace is the Latin poet upon whom I crack my head every now and then when I feel adventurous or masochistic enough. I can never get that suave, clubbable assurance, but Rudman's mad method of imagining Horace addressing the Elizabeth Taylor/Richard Burton romance and using the lines of the odes as touchstones is just breathtaking. It is impossible in a paragraph or two to suggest the wonderful intricacy of these poems, or, more accurately, this poem. But one can offer a snippet to show what the music can be like. At that level Rudman's virtues proclaim themselves clearly enough, as in the poem "Jury Duty":

> Walking here last May on my lunch break from the law
> the light hit the time-darkened, turbulent stones
> with such tender, fierce, erotic force
> like a match struck in a cave
> I ground to a halt
> and wanted to rub my skin
> in, on, against, within . . . , but rested, restlessly,
> content with the palm of my left
> hand on the rough granite and,
> lightly, both cheeks . . . ,
> and would have lingered
> had not my citizen-reminder-beeper begun
> its *this is not the time, this is not the place*
> routine, as if this confluence could happen any day,
> and our species were not endangered by the daily
> catastrophe of delay[.]

There is, I understand, a third volume that he's working on now. I can't wait.

Campbell McGrath's *Spring Comes to Chicago* (Ecco) is, technically, a collection, but with a long poem as its centerpiece. "The Bob Hope Poem" is, at sixty-eight pages, book length by my reckoning, and all but epic. I was attracted to the volume by its witty cover with a photograph of a bedspring in a snow-covered piece of urban wasteland. The impish energy of that is quite right for the rowdy and Whitmanian (even Whitmaniac) voraciousness of this remarkable performer. Astonishingly, "The Bob Hope Poem" is actually about Bob Hope, at least in part, because he is, inevitably, a piece of indigestible gristle in the American stew:

> young and beautiful Bob, smiling his smile of purest mastery,
>   leering the leer of an era untroubled by doubt or uncertainty.
>
> Bob agog with Bing on the golf course, Bob playing kissy with
>   string-starlets.
> Bob pushing Pepsodent, Bob shopping Oscars, Bob selling war
>   bonds with Eleanor Roosevelt.
> Bob in the jungles of Guadalcanal, Bob raising the flag on Iwo
>   Jima, Bob as Coyote, Bob as Loki, Bob in the ashes of Na-
>   gasaki.
> Bob as Cook, Bob as Darwin, Bob as Columbus, Bob as Buzz
>   Aldrin.
> Bob in Palm Springs with the ghosts of dead Presidents, Bob in
>   bronze at the Museum of American History, Bob teeing off in
>   the Sea of Serenity.
> Bob the body and Bob the shadow, Bob and the echo Bob the call.
> Bob the imperial envoy of the American system, Bob the corporate
>   janissary, Bob the mad jester of cultural hegemony.
>
> THANKS FOR THE MEMORY!

One of the incidental pleasures about this was the thought I had that my tepid feelings about Whitman may not be from any shortcoming in myself, but could be old Walt's fault. He hasn't got much sense of humor. It isn't just that, like John Milton or William Wordsworth, Whitman is so damned solemn, but that the solemnity misses a part of experience to which poets ought to be able to respond and to use. McGrath's range of response is astonishing and grand. His earlier books are *Capitalism* and *American Noise,* and they show where he was coming from. This one demonstrates where he was going, and it is his best work so far.

William Heyen's *Crazy Horse in Stillness* (BOA Editions) tells us about Crazy Horse and George Custer and the American West, and is Virgilian in its epic sweep. Grandeur is not what we're comfortable with these days, but Heyen's rueful admission of what terrible things we have done to one another and to the land gives his poem the same balance that Virgil achieved by admitting how badly Aeneas behaves most of the time. Heyen is a quiet poet whose name doesn't come up often among the speculators on the poetry exchange, but his is an enormous talent. His vision is steady and his craftsmanship is impeccable. This is

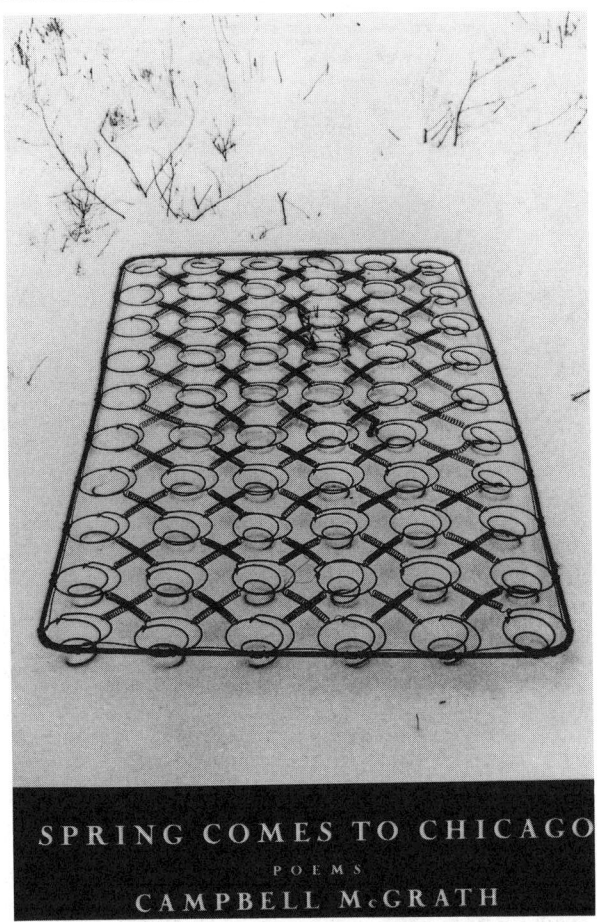

Front cover for Campbell McGrath's third volume of poetry, which includes "The Bob Hope Poem"

an exciting and profound book. Here's a sample piece, taken from "Forces":

> Crazy Horse & Custer rode through one another
> emerging on the other side.
> This happening in a warp of starlight
> too long ago in the future to predict
>
> or remember. For each of them,
> It was as though a wind made of locusts
> the size of particles of pollen, or smaller,
> atoms, or smaller, had swept them
>
> together, the exchange being ... inevitable,
> necessary, good. Crazy Horse wore
> a breastplate in the shape of a butterfly,
> Custer a red cravat, but these were insubstantial,
>
> pure. The starlight wrinkles. Their horses
> blink, separate, & reassemble. Then, here,
> the two warriors pause to estimate their final
> destinations, mount, & ride on.

Out of nowhere (or Fayetteville, which is one of its suburbs) a book by Joe Survant appears with the

Cover for William Heyen's volume described as "Virgilian in its epic sweep"

news that it is the winner of the 1995 Arkansas Poetry Award. *Anne & Alpheus 1842–1882* (University of Arkansas Press) is a series of intersecting monologues by a husband and wife out on the Kentucky frontier during the specified years, and it reminded me a little of Hugh Nissenson's odd and wonderful book, *The Tree of Life* (a novel, but similar in subject and even in tone). This isn't a huge undertaking like the Heyen book, but it has a miniaturist vividness and an underplayed eloquence that builds persuasively. The cumulative effect is difficult to demonstrate, but the appealing restraint of the diction in any given section comes through quite clearly. This is the start of a meditation by Anne on the death of their daughter Catherine:

> Too much killing.
> We live by death
> and now the messy
> deaths of animals
> have entered the
> house, and I cannot
> hide her. Four nights
> of fever and now
> a fifth. A dry cough
> wastes her little strength
> and nothing helps.
> Helpless we watch her
> blur and fade.
> I would give back
> all the blood and
> flesh and become
> like cattle grazing
> in the passive grain
> her mind flaring
> while the body dies.

David Mason's *The Country I Remember* (Story Line Press) is a collection of poems, but the fifty-six-page-long title piece is, like Survant's poem, a pair of voices, those of a father who fought in the Civil War at Chickamauga and survived a series of Confederate prison camps, and his daughter, who remembers the family's trek west to Washington State and her own adventures in Oregon and California. Mason is a cautious, conservative craftsman. His blank-verse version of these events is almost self-effacing in its modesty, but he has the knack some soft-spoken people do of commanding attention. Maggie remembers how it was just before she set out:

> When you leave a place it is more beautiful
> those last few days, the earth will open up
> secrets you never guessed at: the hushed grass,
> the bluish cottonwood that seems to wait
> and breathe with you in solitary union.
> Alone on a hill at night, you can feel
> the world was made for us to listen to.

That world of nineteenth-century America seems to have found several bards who, at the end of the twentieth, are determined to perform one of the basic duties of the poet: to remember and to help others remember along with them those things that are important and have made us what we are.

Ellen Bryant Voigt's *Kyrie* (Norton) is a sonnet sequence about the influenza pandemic of 1918–1919, a great plague that killed something like twenty-five million people. Historical, religious, and lyrical, it grapples with the most difficult questions we ever have to face: what is life, and what does it mean to be a survivor (as we all are, one way or another)? How can we confront the pitilessness of nature in a world where such things happen?

> Soon it was a farmer in the field—
> someone's brother, someone's father—
> left the mule in its traces and went home.
> Then the mason, the miller at his wheel,

and deep in the forest the hunter, the logger,
and the sun still up everywhere in the kingdom.

Susan Ludvigson's *Trinity* (Louisiana State University Press) is a triptych of pieces on art and God, shot through with a strange Gnostic passion that is commanding and mysterious. The first of these is a brooding poem about Saint Mary Magdalene (in whose honor there is a church in Rennes-le-Château in the French Pyrenees). Then there's a kind of dialogue between Emily Dickinson and God. And finally, there is a woman artist-pilgrim who visits Ste. Magdalen's church where it all comes together, more or less. It is a rich and complicated work, disturbing in the fervor of its yearnings:

> What does it mean, what conjecture
>
> might connect the pieces? I read of the Earth,
> the speculations of philosophers and physicists
> about the stars, our place, the search
>
> revealing order. Can a leaf, and eye, be chance? We insist
> that it can, it is. And yet the odds deny
> this cooled and rhymed universe. What if I stop resisting
>
> the vast intelligence? How should I ally
> myself? And what of chaos, the fiery fiend
> of artists? Does one truth make a lie
>
> of another? Who's at the heart of us? What bends
> like steel to belief? How do we make
> ourselves artists, and to what end?

And on a lighter, madder note, there is Wiliam Van Werlt's *Don Quickshot* (Livingston Press/University of West Alabama), a burlesque extravagance that is willing to take pratfalls and descend to the silliness of painful puns because that's the way to describe large swaths of our experience. And because it is fun. This is a loopy version of Don Quixote, obviously, but the "Don" part has turned him into a mafioso, and Van Wert's clownishness has taken him beyond the edge of the cliff where, like the Roadrunner, his legs whir and he hangs there through simple and simple-minded belief. Here is a small snippet, picked almost at random:

> They camped with an expert on critical
> theory, who gave them books
>
> for light reading around the bonfire:
> Todorov on the big toe, Bataille
>
> on the solar anus, Freud on the uncanny;
> Plato, Saussure, and someone on cups;
>
> positivist hysteria in the psychophants;
> Fish on swimming; Said on silence;
>
> Adorno on porno. The expert's name
> was Quine, and he was full of lax
>
> taxonomies.

This is an impressive number of book-length efforts, I submit, and that is without any reference to my own *A Gift* (Louisiana State University Press), a long poem about the life of Lorenzo da Ponte, Mozart's librettist, who is the least-well-known great poet (or the greatest unknown poet) in Western Civ. I shan't quote myself or try to characterize the book, but it would be altogether incredible and out of character if I neglected to note that it exists.

Among the year's largest collections, those new-and-selected volumes that are longer-term statements of poetic careers, there were some familiar names and some that were altogether obscure—to me, at any rate. And, of course, there was hardly any correlation between onomastic prominence and poetic and literary proficiency.

In Philip Appleman's *New and Selected Poems, 1956–1996* (University of Arkansas Press), there is an appealing earnestness and, within very narrow limits, a range of vers libre effects that can carry off an acute observation—often about Darwin and biological curiosities—but from which there isn't a lot of extra lift. Appleman never sounds better or smarter or funnier than his donnée. There can be a flat-footed charm, though, that one gets accustomed to:

> Prudence is a killing frost;
> Providence poisons the soil;
> Wisdom, always moving on, leaves
> a brown track in the grass.
>
> I will close my eyes and hope for
> some lucky drift of jasmine,
> low voices, and a dumb
> trembling in the groin.

Appleman's book deserves a place on the shelf. It may be like turmeric or juniper berries, those gastronomic embellishments for which one reaches only once or twice a year, but when one needs a dash of just that flavor, it's nice to have it at hand.

George Mackay Brown was born in Stromness in the Orkney Islands in 1921 and lived there for most of his life. His *Selected Poems 1954–1992* appeared last year from the University of Iowa Press. It is a strange, austere volume—almost too good to be true. For a time, I half hoped it might be the work of some literary prankster who had invented the entire thing, the poet (a face with a protruding chin that makes Jay Leno's look modest) who converted to Catholicism and hated all modern things—and the poetry, too. A Bennie Hill send-up of British regionalism? But no, it is all quite real and, from time to time, even impressive, if you can keep from giggling:

> The road to the burn
> Is pails, gossip, gray linen.
>
> The road to the shore
> Is salt and tar.
>
> We call the track to the peats
> The kestrel road.
>
> The road to the kirk
> Is a road of silences.
>
> Ploughmen's feet
> Have beaten a road to the lamp and barrel.
>
> And the road from the shop
> Is loaves, sugar, paraffin, newspapers, gossip.
>
> Tinkers and shepherds
> Have the whole round hill for the road.

The Collected Poems of Sterling A. Brown (Triquarterly Books/Northwestern University Press) originally appeared in 1979 from Harper and Row but was reissued last year, partly because it deserved to be in print, but more likely because with all these courses in African American literature there is a market now for Brown, who died in 1989 and was a contemporary of Langston Hughes, Claude McKay, and Jean Toomer. Brown's is an interesting body of work, a sly Audenesque reworking of blues and ballad forms, but about blacks and often in dialect. It is yet another version of the pastoral, and at his best, he was very good indeed. His "Long Gone," one of the important poems of our literary history, ends:

> I don't know which way I'm travelin'—
>   Far or near,
> All I know fo' certain is
>   I cain't stay here.
>
> Ain't no call at all, sweet woman,
>   Fo' to carry on—
> Jes' my name and jes' my habit
>   To be Long Gone.

Jorie Graham's *The Dream of the Unified Field: Selected Poems 1971–1994* (Ecco) came out late last year so that I missed it, and then it won the Pulitzer Prize. Graham, the recipient of a MacArthur Fellowship, is one of those discoveries (or inventions?) of Helen Vendler, and I confess that for all these and many other reasons, I find it difficult to read her work with the attention it probably deserves. My allergy was exacerbated by an academic performance last year by Christopher Ricks in which he found the drama of a Graham poem in the way in which the parenthetical phrases seemed to overflow their punctuation marks (he called them "lunulae"). Graham is not responsible for such extravagant silliness, but my mind is a jumble in which the compartmentalization is not so thorough as it ought to be. And her poems are dense enough and complicated enough so that they may be fairly said to invite strenuous interpretation. At the very least, I found myself asking, often and with some insistence, just what the hell was going on. Here's a swath of the title poem:

> Starting home I heard—bothering, lifting, then
>   bothering again—
> the huge flock of starlings massed over our
>   neighborhood
> these days; heard them lift and
> swim overhead through the falling snow
> as though the austerity of a true cold thing, a verity,
>   the black bits of their thousands of bodies swarming
>   then settling
> overhead. I stopped. All up and down the empty oak
> they stilled. Every limb sprouting. Every leafy backlit
>   body
> filling its part of the empty crown. I tried to count—
> then tried to estimate—
> but the leaves of this wet black tree at the heart of
>   the storm—shiny—
> river through limbs, back onto limbs,
> scatter, blow away, scatter, recollect—
> undoing again and again the tree without it ever ceasing to
>   be full.
> Foliage of the tree of the world's waiting.
> Of having waited a long time and
>   still having
> to wait. Of trailing and screaming.
> Of engulfed readjustments. Of blackness redisappearing
>   into
> downdrafts of snow. Of indifference. Of indifferent
>   reappearings.

In a gross way, this is clear enough. There are these birds in the tree, in the wintertime, and it's very affecting. But why that is so and what the precise terms of the fussy and energetic transactions are ("engulfed readjustments"?) is quite opaque. The nervous breaks in the lines and the parentheses (or "lunulae") and dashes (or "linulae"?) seem, moreover, arbitrary, willful, and even hostile. (If we don't get it, it must be our fault, who are so much less sensitive than she and so much less able to respond to such delicacy of observation!)

I spent some time riffling through the pages of this volume, trying to find something I liked better, something that was at least a little different, but it is all much of a piece and only interesting when it verges on self-parody: "A moonless enterprise consisting of towers not there to the naked eye. / Consisting of fountains, yes, but invisible, no?"

John Haines's *The Owl in the Mask of the Dreamer: Collected Poems* (Graywolf Press) is the work of a man who has spent a lot of time out in the Alaskan wilderness (or is that a pleonasm?). He is at his best in quick takes about nature, where he has the right toughness of emotion and miserliness with words. His "Prayer to the Snowy Owl" is this:

> Descend, silent spirit;
>
> you whose golden eyes
> pierce the gray
> shroud of the world—
>
> Marvelous ghost!
>
> Drifter of the arctic night,
> destroyer of those
> who gnaw in the dark—
>
> preserver of whiteness.

I'm not sure the "Marvelous" is either earned or even necessary, but the rest of it has a clean cold bite of authenticity. There are a good number of such pieces of modest but indubitable accomplishment.

Robert Hayden died in 1980, but his *Collected Poems* (Liveright) only came out last year, edited by Frederick Glaysher and with a graceful and informative introduction by Arnold Rampersad. I was glad to see this book appear, having met Hayden once under almost ideal circumstances—he was consultant in poetry at the Library of Congress and had invited Al Poulin Jr. and me to read there. That was almost twenty years ago and he and Poulin are both dead, so I am the repository of the memory of a fine evening for which I am still grateful. He was a black man to whom Auden had been friendly, and he had learned a great deal from W. H. Auden—elegance, deliberation, and a tough-minded authority I see again and again on the pages of this characteristically modest volume of not quite two hundred pages. Paradise Valley was the part of Detroit in which he grew up, and this is one of his elegies for it:

> Our parents warned us: Gypsies
> kidnap you. And we must never play
> with Gypsy children: Gypsies
> all got lice in their hair.
>
> Their queen was dark as Cleopatra
> in the Negro History Book. Their king's
> sinister arrogance flashed fire
> like the diamonds on his dirty hands.

> Quite suddenly he was dead,
> his tribe clamoring in grief.
> They take on bad as Colored Folks,
> Uncle Crip allowed. Die like us too.
>
> Zingaros: Tzigeune: Gitanos: Gypsies:
> pornographers of gaudy otherness:
> aliens among the alien: thieves,
> carriers of sickness: like us like us.

Samuel Hazo's unfortunately titled *The Holy Surprise of Right Now: Selected and New Poems* (University of Arkansas Press) is a book that is difficult to dislike, but I rouse myself. The trouble is that Hazo is obviously a nice guy, but poetry demands more than decency and affability. There has to be a willingness to risk unpleasantness if only for the sake of honesty and accuracy. The cheerfulness of the title is like a set of rose-colored glasses, distorting whatever Hazo looks at and blocking out the harmful rays of whatever he doesn't approve of. As he says in "King Nothing":

> The isms
> in the words explode like spittle
> in my face . . .
>   If all we do
> begins in our imagining, I should
> expect the worst more often than
> the best.
>   It sells.
>   It meets the needs
> of hate.
>   It happens on command.
> It lacks the lightning of the best,
> which comes when it comes to write
> the man the way a dance
> will dance the dancers when the dance
> is right.
>   Like love it lets
> the eyes become the face's
> sky again and takes us where
> it takes us—never soon
> enough but not too late.
> As long as that's worth
>   waiting for, I'll wait.

I imagine Philip Larkin reading that and his snort of contempt.

Shirley Kaufman's *Roots in the Air: New and Selected Poems* (Copper Canyon Press) is much influenced by the fact that Kaufman "started her life over again in Israel in 1973," as the jacket copy informs us. This makes for some appealing work, but it also sets her in a context where the competition is almost daunting—she writes on subjects that Yehuda Amichai has more or less made his own. That she holds up reasonably well in such glare is evidence of a considerable talent and, just as important, a quiet earnestness that is impossible not to like. When she is driven to it, she can even make fairly impressive leaps, as in "Stones":

> When you live in Jerusalem you begin
> to feel the weight of stones.
> You begin to know the word was made of stone, not flesh.
>
> They dwell among us. They crawl up the hillsides and lie down
> on each other to build a wall.
> They don't care about prayers,
> the small slips of paper we feed them between the cracks.
>
> They stamp at the earth
> until the air runs out
> and nothing can grow.
>
> They stare at the sun without blinking
> and when they've had enough,
> make holes in the sky
> so the rain will run down their faces.
>
> They sprawl all over the town
> with their pitted bodies. They want
> to be water, but nobody
> strikes them anymore.
>
> Sometimes at night I hear them
> licking the wind to drive it crazy.
> There's a huge rock lying on my chest
> and I can't get up.

She gets talky sometimes, and now and then postures a bit, but at best she is quite good. And the best is all that counts.

The late Jane Kenyon's *New and Selected Poems* (Graywolf Press) presents an even greater problem for anyone who might even think of saying something adverse. Let me work by indirection, then, and start out with Bill Moyers, whose towering vulgarity is the epitome of everything that is wrong with PBS. His special on Kenyon and Donald Hall, "A Life Together," and his segments on them in his folksy barbecue of American poetry, "The Language of Life," got everything wrong. He was interested not in the poetry but in the idealism that ought to prompt it, or in the suffering from which he serenely supposed it had to come. Hall had been flirting with death (it rejected him), and then Kenyon actually did die (of leukemia), and this is her posthumous volume, assembled, Hall tells us, on her deathbed, while he read aloud to her and they argued about what should go in and what shouldn't.

Music swell. Freeze frame. Wow!

You perhaps begin to intuit what I find myself rebelling against, chafing at, suspicious about. The poems aren't terrible. They are less bad, probably, than Hazo's. But they're so brave: she was depressed and lets us know it:

> Elavil, Ludiomil, Doxepin
> Norpramin, Prozac, Lithium, Xanax,
> Wellbutrin, Parnate, Nardil, Zoloft.
> The coated ones smell sweet or have
> no smell; the powdery ones smell
> like the chemistry lab at school
> that made me hold my breath.

Well, suck it up. Get over it! I know a dozen poets whose medicine chests offer some or all of the above, and they soldier on, somehow, without whining and holding out their quivering hands to show us their beating hearts. I like Jane Kenyon better when she's being brave, or trying to be:

> An oriole sings from the hedge
> and in the hotel kitchen
> the chef sweetens cream for pastries.
> Far off, lightning and thunder agree
> to join us for a few days
> here in the valley. How lucky we are
> to be holding hands on a porch
> in the country. But even this
> is not the joy that trembles
> under every leaf and tongue.

That it is Donald Hall she's probably holding hands with, and that the two of them are dreaming up earnest things to say to Bill Moyers, doesn't exactly help me, but I'm quite willing to admit that most of that is my fault.

Sydney Lea's *To the Bone: New and Selected Poems* (University of Illinois Press) is an agreeably burly kind of book. The title poem is about an accident Lea had in which he cut his leg with a chainsaw, "To the Bone" actually, and is a rambling, half-drugged performance into which he tried to pack virtually everything with all kinds of accidental clutter, but with a determination—or at least the fervent wish—to make some sense out of it all. He posits an array of random oddnesses but then admits:

> none of which
> you have the least awareness of   and why on earth
> reader   why on earth should you   no   I'm speaking rather
> of the sorts of signs that any body gathers
> in the lifelong effort to make a life cohere
> and give it worth
>
> and I have small doubt
> your signs are other far
> and yet do you not swear   then doubt yourself   then again
>   swear
> as the night air bears upon you
> the sounds outside   such as they are wherever you are
> oppressive   it all comes together somewhere[.]

It is a poetry large in ambition and generous in spirit, and if Lea's technical facility doesn't call attention to itself, that doesn't mean it isn't there. He is equal to the demanding tasks he sets before himself, and this volume is a most impressive achievement.

Jane Miller's *Memory at These Speeds: Selected Poems* (Copper Canyon Press) is glitzy and hyper and too strenuous for me. I can't imagine a mood in which my eye falls with pleasure on such a passage as:

> a diorama o'erpowering everything else in common lime-
> light—
> dykes on bikes, fag hags, drag queens, steroidal buffs, mid-
> night
> blue black semi-nudes, boytoys, unzipped sado-masochistic
> six-foot tricks, the semi-erect, the innocent, in gym shorts
> and in slips, tuxedos, T-shirts and cut-offs, jeans impaled
> at the crotch—godly, larger-than-life meanings assigned
> to them by messages spelled out on their chests, "Silence
> Equals Death," etc., until, so engorged, their numbers blur
> into a mass of energy that finally disperses into the missions
> and the tenderloins from whence they came[.]

Maybe I'm just too old for this. But I find myself wondering about the oddly poetic "o'erpowering" and the peculiar "from whence," which seems as awkward as "to whither" would be and bothers the grammarian in me.

Lisel Mueller's *Alive Together: New and Selected Poems* (Louisiana State University Press) represents thirty-five years' work of a poet I have been reading steadily and always with admiration for the economy of her diction and the precision with which she negotiates between passion and decorum. It is a fastidious sensibility to which one can repair when oppressed by raucousness and jangle, as one so often is in these times. Consider the elegant simplicity of this:

> When I am asked
> how I began writing poems,
> I talk about the indifference of nature.
>
> It was soon after my mother died,
> a brilliant June day,
> everything blooming.
>
> I sat on a gray stone bench
> in a lovingly planted garden
> but the day lilies were as deaf
> as the ears of drunken sleepers
> and the roses curved inward.
> Nothing was black or broken
> and not a leaf fell
> and the sun blared endless commercials
> for summer holidays.
>
> I sat on a gray stone bench
> ringed with the ingenue faces
> of pink and white impatiens
> and placed my grief
> in the mouth of language,
> the only thing that would grieve with me.

As we approach the millennium and people consider where we are as a culture, one of the painful questions that has to be asked is why Lisel Mueller isn't famous.

Leslie Norris's *Collected Poems* (Dufour Editions) is an extraordinary collection. Norris is a Welshman who taught for many years at Brigham Young, and his work has a dazzling formal facility. A representative snippet from "Siencyn ap Nicolas upon his Death-Bed" shows him playing with enjambment:

> I have learned tolerance, that soft word,
>
> however unwillingly. It is one of my late,
> one of my few, virtues. I hate
> the hypocrite who takes a pleasure
> in his honesty. Many a desolate
>
> truth I've undone with a kind
> lie. And hope to be forgiven. We can't stand
> all kinds of truth. If God calls for my
> account I must ask him to turn a blind
>
> eye to truth's ledgers, for there
> I have no credit[.]

I had never heard of Norris or seen any of his poetry. If I had, I'm sure I would have noticed and remembered, for he is memorable.

Robert Pinsky's *The Figured Wheel: New and Collected Poems 1966-1996* (Farrar, Straus & Giroux) is a quite different kettle of fish. Pinsky is famous—as poets reckon these things, anyway. His translation of *The Inferno* had a front-page review in *The New York Times Book Review*. (I thought it was OK, but not as good as the old Laurence Binyon version.) He is a poet of large gestures, as one might gather from the titles of his earlier volumes (*The History of My Heart, An Explanation of America*), and works by accretion and association, throwing in whatever comes to mind with an all but vaudevillian zest. One of his new poems, "Ginza Samba" starts out:

> A monosyllabic European called Sax
> Invents a horn, walla whirledy wah, a kind of twisted
> Brazen clarinet, but with its column of vibrating
> Air shaped not in a cylinder but in a cone
> Widening ever outward and bawaah spouting
> Infinitely upward through an upturned
> Swollen golden bell rimmed

> Like a gloxinia flowering
> In Sax's Belgian imagination
>
> And in the unfathomable matrix
> Of mothers and fathers as a genius graven
> Humming into the cells of the body
> Or cupped in the resonating grail
> Of memory changed and exchanged
> As in the trading of brasses,
> Pearls and ivory, calicos and slaves,
> Laborers and girls, two
>
> Cousins in a royal family
> Of Niger known as the Birds or Hawks.

I can see what it is doing and where it is going, and I can even admire it in a way, but it does seem overheated, a bit overbearing. Do we need to be told that *Sax* is a monosyllable? What would a Belgian imagination be—something that includes Sax and, say, Ysaye? But this is nit-picking. Pinsky has an interested and interesting mind, and one can't not read him.

Ruth Pitter's *Collected Poems* (Enitharmon; Dufour Editions) is a splendid book. Pitter, who died in 1992 at the age of ninety-five, was a prominent poet in her time, and this volume will remind those who remember her how good she was—and, I expect, astonish those who never heard of her. She was compared with Elizabeth Bishop and Marianne Moore but was like them only in her sharp, clear focus of vision. Her poetic practice was rather more conservative than theirs, which, oddly enough, makes her seem less dated than she might otherwise now appear. This new collection of her poems, with an introduction by Elizabeth Jennings, was published in London by the Enitharmon Press and in America by Dufour Editions in Chester Springs, Pennsylvania—a small but splendid outfit that also did the Leslie Norris volume and has been bringing British poetry over here for almost forty years now (I have on my shelf a slender volume of theirs by Geoffrey Hill from 1958). Ruth Pitter's poetry is likelier to endure than most of what passes for important in this short-attention-span culture. The way she makes simple words in unpretentious syntactical settings bear enormous weight is amazing. "Morning Glory," a relatively late poem, demonstrates in twenty lines this mysterious knack:

> With a pure colour there is little one can do:
> Of a pure thing there is little one can say.
> We are dumb in the face of that cold blush of blue,
> Called glory, and enigmatic as the face of day.
>
> A couple of optical tricks are there for the mind:
> See how the azure darkens as we recede:
> Like the delectable mountains left behind,
> Region and colour too absolute for our need.

> Or putting an eye too close, until it blurs,
> You see a firmament, a ring of sky,
> With a white radiance in it, a universe,
> And something there might seem to sing and fly.
>
> Only the double sex, the usual thing;
> But it calls to mind spirit, it seems like one
> Who hovers in brightness suspended and shimmering,
> Crying Holy and hanging in the eye of the sun.
>
> And there is one thing more; as in despair
> The eye dwells on that ribbed pentagonal round,
> A cold sidereal whisper brushes the ear,
> A prescient tingling, a prophecy of sound.

That move in the last stanza is as grand as any in Andrew Marvell's "Garden," but it doesn't call attention to itself or betray any sign of effort. It's altogether a marvel.

Leon Stokesbury's *Autumn Rhythm: New and Selected Poems* (University of Arkansas Press) has some interesting pieces in it, and Stokesbury has a lively sensibility. But he horses around more than he needs to and makes jokes and other such disarming gestures, as if he were embarrassed by his own talent. When he is caught up with a subject, particularly with a narrative on which he can rely comfortably, he rattles along with authority and charm and lets the words speak for themselves and be funny on their own. It would be hard to resist the donnée of "Evening's End" in which the speaker hears a Janis Joplin song on the radio:

> And it took me back to this girl I knew,
> a woman really, my first year
> writing undergraduate poetry
> at the Mirabeau B. Lamar
> State College of Technology
> in Beaumont, Texas,
> back in 1966.

When he's at the top of his form, as he is in that piece and several others, he has a distinctive sparkle.

Nancy Willard's *Swimming Lessons: New and Selected Poems* (Knopf) is a vexing book, often too cute for its own good. Willard can write perfectly accomplished pieces—the title poem of this volume is one of these—but she tends to gush about children and pets and righteous causes, and I find myself resisting. She writes in "Carpenter of the Sun," for instance:

> My child goes forth to fix the sun,
> a hammer in his hand and a pocketful of nails.
> Nobody else has noticed the crack.
>
> Twilight breaks on the kitchen floor.
> His hands clip and hammer the air.
> He pulls something out,

> something small, like a bad tooth,
> and he puts something back,
> and the kitchen is full of peace.
>
> All this is done very quietly,
> without payment or promises.

This is perhaps the dopiest version of Virgil's fourth Eclogue that has ever been attempted. If Willard didn't rise above this treacle now and then, it would be pointless to upbraid her. (Or, maybe, if this didn't have Knopf's borzoi on it and a series of acknowledgments to *The New Yorker,* it would be ungallant and excessive.) But she is good enough to take seriously, which means that she should be held to higher standards than those of the old Art Linkletter "Kids Say the Darndest Things" show.

This brings us, relatively late in the day I'm afraid, to the individual collections, of which there were, as in most years, some good, some terrible, and many middling. I begin with the work of the *nomenclatura*.

A. R. Ammons's *Brink Road* (Norton) is a generous collection—more than 150 poems that date from 1973 to the present, and if you know Ammons, you know what to expect. If you don't it is hard to describe, because the real business of the poems is not so much in the details, however carefully observed and meticulously presented they may be, as in the transitions, the leaps from this to that. There is also a ruminative quality to Ammons's work in which the language and the processes of thought become the subject with which he can dare, and usually get away with, a peculiar flat-footedness, as in "The Way of One's Desire":

> One not lost finds no way
> terror brightens what it sees:
> home's a destination one
>
> departs with to part with:
> okay never looks to be okay,
> and not-okay, looking, sees
>
> the only not-okay: you who
> know, even as if not knowing,
> tell me how does one err
>
> to find one's erring: where
> in the wild are the wiles
> that school the way back home?

Considering Joseph Brodsky's *So Forth* (Farrar, Straus & Giroux), what can one say about the work of a Gulag alumnus and Nobel laureate who died this year? Anyone with a shred of decency would praise him or stand mute, right? But I find this work excessively mannered and self-congratulatory, not to mention opaque:

> A burnt matchstick's residue, a myopic
> naked statue, a pergola looming wanly
> are excessively real, excessively stereoscopic
> since there's nothing they can turn themselves into. Only
> horizontal properties, in their fusion, can spawn a monster
> with a substantial fallout or follow-
> up. For an explosion-sponsored
> profile, there is no tomorrow.

The last four words are clear enough, but did Brodsky know that their primary association is that that's what sportscasters always say at some point during the Super Bowl?

Hayden Carruth's *Scrambled Eggs & Whiskey* (Copper Canyon Press) won the National Book Award for poetry last year, mostly, I think, because it is accessible and unintimidating. Carruth uses broad brush strokes for large effects and assumes a gruff persona to make a kind of calliope music—engaging, but hardly subtle. And yet, when there is a narrative to rely on, it allows for a surprisingly supple and interesting performance. In "Auburn Poem" he talks about meeting an ex-wife at the sickbed of their thirty-nine-year-old daughter who is suffering from incurable cancer. There is a note of self-reproach here that John Updike strikes in some of his best short stories. The connection and its limitation is finely balanced:

> You have been married
> for thirty years to another man, and I
> have been married to three other women
>
> and have lived with six whom I did not
> marry—a disgrace but there it is, done
> and irrevocable. We are old. You are
>
> sixty-nine and I am seventy. It would be
> sentimental folly to say I can see in you,
> or you in me, the lineaments of our
>
> loving youth. Yet it is true. Your voice
> especially takes me back. We are here
>
> because our daughter, whom we conceived
>
> one fine April night in Chicago long ago,
> is crucially vulnerable. We meet in agony,
> in wordless despair. We meet after years
>
> of separation and mildly affectionate
> unconcern.

For him to get away with such flat-footed declarations as "sentimental folly" or "crucially vulnerable" is remarkable, but, with the odd strategy he has worked out, I think it is fair to say that he carries it off. A kind of magic, then.

Louise Glück's *Meadowlands* (Ecco) appeals to a fairly exquisite taste. One must believe in wiftiness as evidence of psychic stress, and if that dramatic structure holds, then the poems aren't what they initially seem: that is, profoundly silly. There are dialogue poems out of a madhouse marriage, one of which, "Ceremony," I confess to liking for reasons I can't quite understand. It starts out:

> I stopped liking artichokes when I stopped eating
> butter. Fennel
> I never liked.
>
> One thing I've always hated
> about you: I hate that you refuse
> to have people at the house. Flaubert
> had more friends and Flaubert
> was a recluse.
>
> Flaubert was crazy: he lived
> with his mother.
>
> Living with you is like living
> at boarding school:
> chicken Monday, fish Tuesday.
>
> I have deep friendships.
> I have friendships
> with other recluses.

How much of this is a pose and how much is authentic, I can't begin to guess. At a reading Glück gave not long ago in Philadelphia, she announced earnestly that it had taken her years of therapy to get to the point where she could use contractions in her work. Or question marks.

Donald Hall's *The Old Life* (Houghton Mifflin) has the runner-up entry in the competition for the year's most embarrassing poem. There are thoughts gentlemen try not to admit to consciousness, or at least conceal from others. Hall has no compunctions, however, about letting us see him at his most squalid:

> In nineteen ninety-three
> I was up for the NBA in
> poetry. From the first
> day, when I reckoned up the judges
> and nominees, I claimed
> A. R. Ammons would win the award.
> Nevertheless, he won it.
> When Archie walked past our table
> toward the stage, I reached
> for his hand and shook it like a good sport
> At the reception, the judges
> touched my shoulder, dropped their eyes,
> and said my stuff

> was terrific. I went to sleep easily,
> mildly let down, and woke
> at three-thirty in murderous rage.

It's creepy enough to make one long for Hass's "despised disjecta."

Robert Hass's *Sun under Wood* (Ecco) is an extraordinary book. One way to account for it is that Hass's duties as the poet laureate of the United States have driven him quite mad. Disgusted by the pieties everyone around him is continually mouthing, he wants to spew forth something different. Or it could be that he sits there in his poet's office laughing as he writes such lines as:

> "Nosepicking," you imagine explaining
> to the upturned reverential faces, "is in a way the ground floor
> of being. The body's fluids and solids, its various despised disjecta,
> toenail parings left absently on the bedside table that your love
> the next night notices there, shit streaks in underwear or little faint
> odorous pee-blossoms of the palest polleny color, the stiffened
> small droplets in the sheets of the body's shuddering late-night loneliness
> and self-love, russets of menstrual blood, toejam, earwax,
> phlegm, the little dead milias of white corpuscles
> we call pus[.]"

Is he perhaps trying to get the office of poet laureate abolished? Another poem of his, "My Mother's Nipples," is as embarrassing and pointless as anything I've read this year.

Seamus Heaney's *The Spirit Level* (Farrar, Straus & Giroux) does what he has generally done, combining a set of rural references with a most urbane and sophisticated sensibility. As with the work of Robert Frost, there is a risk of self-imitation, but, as with Frost, Heaney manages to escape this most of the time. But he skates sometimes on awfully thin ice:

> Now it's St Brigid's Day and the first snowdrop
> In County Wicklow, and this is a Brigid's Girdle
> I'm plaiting for you, an airy fairy hoop
> (Like one of those old crinolines they'd trindle),
>
> Twisted straw that's lifted in a circle
> to handsel and to heal, a rite of spring
> As strange and lightsome and traditional
> As the motions you go through going through the thing.

I don't even mind having *trindle* (*Orbs.* or *dial.* to make round) and *handsel* (to make an inaugural gift or one of good omen) shoved down my throat, because the poem works and that last line is entirely successful.

Anthony Hecht's *Flight Among the Tombs* (Knopf) is similarly reliable, if slight. Indeed, the slightness is so unmistakable as to amount to a kind of statement. It is as if Hecht finds most of what he encounters in the culture and the age to be contemptible and he turns away. ("Meanwhile the mind from pleasure less / Withdraws into its happiness," Marvell tells us, and it's a reasonable program.) The volume is handsome, with a good number of reproductions of wood engravings by Leonard Baskin. The poems are deft, punctilious, and grandly civilized, and Hecht achieves his effects mostly by the suave manipulation of syntax and a dazzling facility with prosody. The first piece in the book, "Death Sauntering About," in which Death strolls around a racetrack, is first-rate. The last two stanzas are:

> The ladies' gowns in corals and mauves and reds,
> Like fluently-changing variegated beds
>   Of a wild informal garden,
> Float hither and yon where gentlemen advance
> Questions of form, the inscrutable ways of chance,
>   As edges of shadow harden.
>
> Among these holiday throngs, a passer-by,
> Mute, unremarked, insouciant, saunter I,
>   One who has placed—
> Despite the tumult, the pounding of hooves, the sweat,
> And the urgent importance of everybody's bet—
>   No premium on haste.

David Ignatow's *I Have a Name* (Wesleyan/New England) is the same old same old. Ignatow's reputation continues to be one of the great mysteries to me, what with his trite ideas, awkwardly expressed, in poems that wouldn't get a passing grade in most creative-writing workshops. "One Can" reads, in its entirety:

> One can fall in love as often as a tree grows leaves.
> It is perfectly natural but not free of guilt
> and complications, unless one takes oneself to be a tree.

Is it a woodie joke? I wish.

Carolyn Kizer's *Harping On* (Copper Canyon Press) is a jewel of a book. Kizer has been for some years one of my favorite poets—although for reasons she might object to. Twenty years ago, when liberalism was very much in control, there were a few pet conservatives who could get on the air, mostly because of their charm and wit. Kizer has been my pet liberal. She can sound the note of righteous indignation without being strident or boring. Charming and witty, she can get away with almost anything. And then, from time to time, I find myself agreeing with her, because she is by no means programmatic. Here's a small swath of "Fin de Siècle Blues" in which she addresses a clutch of dead poets:

> O you serious men and women
> who wrote your poems, met your classes,
> counseled your students, kept your friends
> and sent magic letters home,
> your lives are pillaged and rearranged
> by avid biographers who boast that they tell all,
> so it seems you always reeled in a mad whirl
> of alcohol, abandonment and sexual betrayal.
> (I sorrow for this stain on your memory,
> Anne, Randall, Ted, Elizabeth,
> Delmore, John, and Cal.)
>
> As writers, what are we to do?
> Our roles as witnesses ignored,
> our fine antennae blunted
> by horror piled on horror,
> our private matters open
> to the scrutiny of voyeurs.
> If we have wit and learning
> it's met with apathy
> of the ever-more-ignorant young.
> How do we hope to carry on
> in the last gasp of the millennium?

Maxine Kumin's *Connecting the Dots* (Norton) crossruffs between literary sophistication and descriptions of country life that all but demand parody. But then how can anybody parody a passage like this, from "Early Thoughts of Winter," where she reveals her praxis at its clearest?

> Spraddle-legged in the humbling steam
> of the manure pile I stand shoveling
> pickup loads to tuck the garden in
> dreaming beyond backache and tedium
> of February with each dip, lift and fling
>
> remembering Heidegger, his broad-jawed noun
> *Geworfenheit,* for the castaway's condition
> the shipwrecked seeker after news across
> the water, the burrower, gleaner I need to be.

My associations of Martin Heidegger and manure are that the two terms are roughly equivalent, and my hope, therefore, is that this is a joke, but I'm not sure. The rest of the poem is perfectly straight, ending up with how

> the wild turkeys, the bachelor moose
> and endearing cluster of juncos braving
> the barn floor, comrade castaways

Henry Taylor, whose Understanding Fiction *received the* Dictionary of Literary Biography Yearbook *Award for a Distinguished Book of Poetry Published in 1996 (photograph by Melissa Laitsch)*

  demand from February good news
across the water.

  I am quick to admit, though, that mine is a minority view, and some poets for whom I have high regard have high regard for Kumin. So. . . .

  W. S. Merwin's *The Vixen* (Knopf) is annoying but admirable. The annoyance arises from his odd notion that punctuation marks are an endangered species and that, by not using them, he is somehow protecting the ecosystem. He started out with ordinary pointing, and then, I think it was with *The Carrier of Ladders* in 1970, abandoned the periods and commas in a way that was old hat in the late 1930s. It's tiresome, and what is particularly vexing is that the poetry is good enough to make it worth one's while to endure this quirk of his. *The Vixen* is mostly about the time he spent in southwest France. As a sample, I offer the beginning of "Night Singing":

Long after Ovid's story of Philomela
 has gone out of fashion and after the testimonials
of Hafiz and Keats have been smothered in comment
 and droned dead in schools and after Eliot has gone home
from the Sacred Heart and Ransom has spat and consigned
 to human youth what he reduced to fairy numbers
after the name has become slightly embarrassing
 and dried skins have yielded their details and tapes have been
slowed and analyzed and there is nothing at all
 for me to say one nightingale is singing
nearby in the oaks where I can see nothing but darkness
 and can only listen and ride out on the long note's
invisible beam that wells up and bursts from its
 unknown star on on on never returning[.]

 I guess I should be grateful for the majuscules and the apostrophes and shut up.

 Henry Taylor's *Understanding Fiction* (Louisiana State University Press) is the book I choose for the *Dictionary of Literary Biography* Award for the year's best volume of poetry. This is a somewhat arbitrary selection. There are three or four books of outstanding accomplishment and dazzling mystery, and to pick among them is, finally, a matter of personal taste. Or elective affinity or neurological compatibility. One picks a poet as one picks a blend of coffee, a wine, a cheese, a cigar. The reasons for the choice aren't necessarily explicable to anyone else, or even to oneself. But that doesn't make it any the less a commitment. I delight in *Understanding Fiction* and am grateful that it exists.

 This leaves us with the work of ordinary mortals, poets who aren't especially eminent or prominent (not that most people in the produce section of your local supermarket would be able to identify more than one or two names in the foregoing list of bigshots and heavyweights). One reasonable way to begin is to mention various series that have either started up this year or have done particularly interesting work. The Peterloo Poets, an operation of Story Line Press, brings British poets on the list of that Cornwall publisher into the American market, pretty much the way Dufour Editions has been doing for some years. *Safe as Houses* by U. A. Fanthorpe is extraordinarily fine. Fanthorpe's name was not previously known to me, but I shall be hunting down her *Selected Poems* of 1986 and anything else I can get my hands on. A piece of "Sirensong," a poem about the blitz and its aftermath, demonstrates her remarkable acuity:

We were precocious experts on shrapnel and blast.
Things broken weren't replaced. What was the point?
Friends were lost, too. You didn't talk of it.

We knew how bombs sliced off a house's flank,

Uncovering private parts; how bedroom grates
Still stuck to wall through wallpaper flailed outside;

How baths slewed rudely, rakishly into view;
How people noted, and talked of what they saw;
How ours might be the next; and what they'd say.
Peace made no difference. Still too young to matter,
Someone still fighting somewhere, some children
Are invaded for ever, will never learn to be young.

We missed the jazz and swing of our extrovert parents,
The pyrotechnic raves of our groovy kids.
Our ground was never steady underfoot.

We had no wax to cancel the siren's song.
Lie low, lie low.

The University of South Carolina Press has launched its James Dickey Contemporary Poetry Series, the selections for which were not, as one might suppose, made by Dickey, who was still alive and picking books for the Yale Younger Poets Series, but by Richard Howard. This is in the way of a tribute to Dickey. There are four books a year, and my favorite of the first crop was *All Clear* by Robert Hahn, the president of Johnson State College in Johnson, Vermont. The idea of any university president relaxing in the evening by writing elegant meditative poems is quite bizarre, but charming. The other three volumes in the series published in 1996 were *The Land of Milk and Honey,* by Sarah Getty; *Traveling in Notions: The Stories of Gordon Penn,* by Michael J. Rosen; and *United Artists,* by S. X. Rosenstock. The next two volumes in the series, *Error and Angels: Poems by Maureen Bloomfield* and *Portrait in a Spoon* by James Cummins, are scheduled for publication in spring 1977.

The Carnegie Mellon Classic Contemporary Series, which I mentioned last year, continues to do a remarkable job, bringing back out-of-print early volumes by poets who are better established now. This year, for instance, they have done Ellen Bryant Voigt's 1983 volume, *The Forces of Plenty;* Carol Muske's 1981 *Skylight;* and Deborah Digges's 1986 *Vesper Sparrows.* For readers or for acquisition librarians who missed those books when they appeared—as it was quite easy to do, because Doubleday, for instance, which published the Muske, kept poetry in print for twenty-five minutes or so. Or, with collections such as Maria Flook's *Reckless Wedding,* first brought out by Houghton Mifflin in 1982, there is another shot for the considerable number of people who, like me, have never heard of her. She is an interesting, rather spooky writer and worth looking at.

Carnegie Mellon publishes new books, too, of course, and this year did Michael McFee's *Colander,* which verges dangerously on cuteness, but which, if

Dust jacket for one of the first volumes in the poetry series launched in 1996 by the University of South Carolina Press

you're in the right mood, is quite wonderful. The title poem:

Upside down, a holy helmet
crowning my son's impressionable head,
its fleet fierce horns.

A planetarium dome
sieving light in ideal patterns.

On its feet, an altar vessel,
six stars surrounding a larger seventh
through which water falls

and billows like the breath of God
from a battered censer.

A mask of sweet ether
my blinded mother lifted to her face
over the still-steaming sink.

Other books I liked enough to mention in the little space that remains:

Jack Agüeros's *Sonnets from the Puerto Rican* (Hanging Loose Press) is fairly rough-hewn, but the tension between the pain and anger on the one hand and the weird sense of humor on the other is enough to sustain this engaging book. The piece about the one-legged bicycle messenger is not a Monty Python routine but a legitimate and effective poem.

Talvikki Ansel's *The Shining Archipelago* (Yale University Press) was James Dickey's last selection for the Yale Younger Poets Series, which is reason to mention her. The poetry is spooky, a little breathless, often pleasing, but sometimes verging on silliness:

> When I can't sleep,
> I'll hold my hand as if I held
> a pear, my fingers mimicking
> the curve. The same curve as the newel post
> I've used for years, swinging
> myself up to the landing, always
> throwing my weight back. And always
> nails loosening, mid-bound.

These are the newel post's nails? Or hers?

Catherine Savage Brosman's *Passages* (Louisiana State University Press) is her fourth collection. She started out well and keeps getting better. The economy of her effortless transitions is a wonder. The suite of poems about food at the end of the book, "Artichokes," "Oysters," "Mushrooms," "Lemons," and "Asparagus," is a set of profound broodings about what one's sense impressions can prompt. "Mushroom" ends:

>   . . . I want
> to eat the moments that I cannot save,
> deny the ruddy loves of autumn bruised
> and decomposing in a bitter alchemy—
> all lymph and humus, like the membrane
>
> of the mushrooms, breaking underfoot,
> crumbling in my fingers—neither green
> nor bloody red, but blanched like skin
> along the thigh—organs of dumb desire,
> or fodder for the cold, devouring gods.

Why she hasn't been laden with honors and awards is an unfathomable mystery to me.

Juanita Brunk's *Brief Landing on the Earth's Surface* (University of Wisconsin Press) is a delicate book, but it has a secure strength. She is able to convey the whole range of emotions that one might experience in an Eric Rohmer movie in a few quick moves in a piece like "I Used to Be Unbelievably Sexy." This is her first book and is the winner of the Brittingham Prize, the judge this year being Philip Levine. She is someone to watch.

William Dickey's *The Education of Desire* (Wesleyan/New England) is a posthumous volume by a poet whose work I have admired ever since his debut with *Of the Festivities* (1959). As W. D. Snodgrass points out in a brief introduction, those early poems were the decorous and reserved productions one might expect from a mature poet, while these, boisterous as they are and taking risks as they do, seem to be the work of a young man. They are, indeed, a series of mystical, ecstatic, sometimes angry, always daring responses to his impending death of AIDS-related illnesses. It is an assertion of life and life's energy, and even with subjects such as the death of John Berryman, about which I am nervous (because what is left to say?), he is sharp and illuminating. It begins:

> Henry went over the edge of the bridge first; he always did.
> Then Mr. Interlocutor and Mr. Bones, then the blackface minstrels
> with their tambourines. You have to empty out
> all the contents before the person himself dies.

Stephen Dunn's *Loosestrife* (Norton) is domestic and confessional, although he often confesses to things that are more virtuous than vicious. But can one blame him for being a decent fellow, a good husband and father? His attempts at wickedness sometimes work well enough to yield interesting figures. In "After Making Love," for instance, he starts out by declaring that "No one should ask the other / 'What were you thinking?'" because the answers, if truthful, would mostly be unwelcome. But then he takes it further and suggests:

> Some people actually desire honesty
> They must never have broken
>
> into their own solitary houses
> after having misplaced the key,
>
> never seen with an intruder's eyes
> what is theirs.

At moments like that, he is not only very good, but good in a way that is characteristically and recognizably his own. What more can one ask of a poet?

Laura Fargas's *An Animal of the Sixth Day* (Texas Tech University Press) is one of those volumes that makes a chore like this seem worthwhile. A first book from Texas Tech Press is not where my hand might hesitate if it were at a bookseller's shelf—if, indeed, I could find a bookstore that stocked Texas Tech titles. But as I flip and riffle through one book after another, linguistic adroitness, physical and spiritual vision, and sheer talent occasionally announce themselves, as they do eloquently here. It is a delightful and delighting voice that she has, and she combines exuberance and learning to make one achieved poem after another of extraordinary quality. A short one that suggests some of these virtues is "Valhalla":

> The dead remember the smell of bread,
> what is was like when the orange orchards
> west of town bloomed into the breeze.
> They remember conversation, they try to talk.
> Gambling, sex, riddles. Grasping at
> the sound of one cricket under the steps.
> They don't know why exactly.

She isn't a teacher, but, like Wallace Stevens, a lawyer.

Carol Frost's *Venus & Don Juan* (Triquarterly Books/Northwestern University Press) is her sixth book, and she is a poet whose work I read with both pleasure and profit. I can see that her poems work—that's hard to miss—but I have trouble figuring out just how she makes them do what they do. In terms of cuisine, it's not just a set of recipes but a whole approach to food and cooking. Consider, for instance, the conclusion of "Homo Sapiens":

> In this lonely, varying light of dawn with the residue of desire
> like mist departing, I am walking. Was it in your eyes,
> where my elongated face shone, I saw for the first time—
> as if all the transparent fire in these trees had become palpable—
> a hunger that was not wholly animal?
> The need to tremble like dogwood, feeling the rain touch down.
> My strange blood rises, and I may see you, fair leaves slipping over you, half-hidden
> in the morning. With the beasts beside a pond,
> I conjure the inward sun to leap into my brain. What remains?
> Wild, beautiful petals all around.
> A beast's face. And something, something else.

Patricia Goedicke's *Invisible Horses* (Milkweed Editions) is a book I could have included in that first section of book-length works, because its subject, perception and cognition, is unified and unifying. It rests at the border between psychology and philosophy. Think of the surprisingly delicate remarks of Ludwig Wittgenstein or William James, and imagine them in the voice of an accomplished poet. The refinement is startling and sometimes just this side of nonsense, but it isn't nonsense, after all, to say "What the skin knows / the underside of one wrist / says to the other." She is clearer, later, in another poem, in "Door/Ways" where she returns to the subject (or, actually, she has never left it):

> What if we broke through the skin
> that sack of boiling juice
>
> the magnificent sphincters/those sliding
> oiled muscles inside
>
> how could we stand the raw
> dripping meat, the
>
> unmentionable thoughts oozing
> over love's cups[.]

Of the books in the pile, this is one I am sure I shall go back to many times.

Eamon Grennan's *So It Goes* (Graywolf Press) is not exactly adventurous, but he addresses the usual domestic subjects with the precise, tiny brush strokes of a Dutch genre painter. His book, too, I shall look at again and probably feel guilty about, because a proper appreciation of this kind of thing requires a fresher palate than I can claim at the moment. But I can see in "House" or in "Bubbles" that he is graceful and accomplished.

Jeffrey Harrison's *Signs of Arrival* (Copper Beech) is a book with a lot of travel poems in it, not just the usual high-culture spots in France and Italy but Alahan, Varanasi, Kathmandu, and even Anapurna. Harrison, whose first book was *The Singing Underneath,* had an Amy Lowell Travelling Poetry Scholarship from the Academy of American Poets, and he squeezed all the good out of it that he could. This experience gives his work an amplitude that is interesting and refreshing.

Brooke Horvath's *Consolation at Ground Zero* (Eastern Washington University Press) is a frequently rueful book. Along with the batterings most of us learn to expect, he has a child with Down's syndrome—and the child, naturally, has a Down's syndrome doll, Dolly Down. At that pitch, grief or even mere tetchiness takes on a new dimension. A good many of the things most poets are too shy to say directly, Horvath will just blurt out with all the bluntness to which the world has initiated him. As, for instance, in "Three Dead Birds":

> Birds, my daughter says, pointing.
> Eyes, mouth, feet—
> touching after each word
> herself.
>
> Baby birds, I say,
> as though baby birds
> always lay
> in a broken sprawl on the sidewalk,
> as though Watts was always burning,
> as though death were a lesson in diction.

Laurence Lieberman's *Dark Songs: Slave House and Synagogue* (University of Arkansas Press) is a collection of poems about life and history in the

Caribbean, where politics are both dreadful and comic. These horrors Lieberman describes are distant, on small islands where they're not at all threatening, but are still somehow interesting, if only because they deflate a great many humanist pretensions about the dignity of man and the sanctity of human life. His tone is finely calibrated, energetic, and just this side of wise-ass to convey his fondness for the islands that doesn't preclude any kind of philosophical disgust.

Paul Mariani's *The Great Wheel* (Norton) is also breezy, but there is a ballast of learning lightly worn and a gift for graceful and effortless prosody. In the title poem, a sestina, the circularity of the form becomes vertiginous as if the reader were actually on the Ferris wheel, and the experience is quite magical. He flirts with disaster—all sestinas do—but avoids it. Here is the last stanza and the envoi:

> For all our feigned bravado, we could feel the evening
> over us, even as we stared down upon the blur of leaves,
> our wives, our distant children, on all we would return
> to, the way shipwrecked sailors search for lights
> along a distant shore, as we began the last descent
> leaving the tents and Garden with its Great Wheel
>
> to return, my dear friend, to the winking lights
> along the boulevard, leaves lifting and descending
> as now the evening air took mastery, it & the Great Wheel.

Alicia Ostriker's *The Crack in Everything* (University of Pittsburgh Press) is a rich dialogue of doubt and faith in which the terms are sometimes theological, sometimes erotic, sometimes medical, and often all those at once. She takes stands (against the horrors of war, against suffering and death, for art and music) that are not exactly surprising, but she has a voice that takes off from earnest dither now and then to soar and swoop and make surprising connections. Or there are small but wonderful observations, as in "Neoplatonic Riff" that begins: "May, and after a rainy spring / We walk street gallant with rhododendrons." Gallant? Yes!

Elise Paschen's *Infidelities* (Story Line Press) is a first book, but Paschen is altogether accomplished and poised, with her strong, even violent emotion constrained by the whalebone of form, which she wears very well indeed. She manages an effortless sestina, even a ghazal, and is good at manipulating the mysteries of proper nouns. An elegant poem about skiing begins:

> This is Solitude where the altitude's
> over 10,000 feet, and I've been skiing
> Rhapsody, Concord, Vertigo, and Paradise.

She has—she is—the real thing.

Paul Ruffin's *Circling* (Browder Springs Press) is a collection of poems mostly about country life, farming, and fishing, and while the bucolic mode is hardly new, Ruffin gives it to us vividly and with an earned authenticity that makes it as fresh as it was when Theocritus and Virgil dreamed it up. This will not be an easy book to find, I'm afraid, so I give the address of the press: P.O. Box 823521, Dallas, TX 75382. It's worth the effort.

Maureen Seaton's *Furious Cooking* (University of Iowa Press) combines hysteria with a matter-of-fact diction that reminds me, I guess, of Tama Janowitz or any of her classmates in the school of young glitz. Her poetry threatens to fly apart but she keeps it, just barely, focused and coherent. A characteristic series of transitions from "A Constant Dissolution of Molecules" is:

> My lovers are so fragile! They love to fish and throw the fish
>   back in,
> to wash their hair in the trout stream—and look, every one of
>   them
> sits beside me as I write and says: "I love it when you write
>   about me."
> My lovers are systematic about control. One hoots out the win-
>   dow at women
> beneath the El train. One hints at a liaison with Calvin, an old
>   buddy
> who wears panty hose. One is chasing her ex-girlfriend
> up the stairs at an AA dance. We are all fighting now, my lovers
>   and I,
> fighting on Broadway in New York, Broadway in Chicago,
> Wherever there is a Broadway my lovers and I will be
>   fighting.

It's not exactly the poetry of meditation, but it's not at all bad.

Alan Shapiro's *Mixed Company* (University of Chicago Press), on the other hand, is a cautious series of exploratory steps, a careful treading that lets us know that it's a minefield the poet is traversing. This is Shapiro's fifth book, and its virtues are exactly those we have learned to expect. In a poem about an early sexual experience (the extent of which we are allowed to imagine for ourselves), he is sixteen and he is speaking to his memory of the girl who was the same age, but either a little more precocious or a little less backward than he. It was, as he recalls, a lesson in which:

> I was the kind of student
> eager to answer
> correctly
> everything he's asked
> in order not to please
> the teacher so much as
> to avoid all the attention
> being wrong can bring.

In other words, trying hard
to follow
everywhere you led
was really just for me
a way of being left
alone, however happily,
with all the pleasure
pleasing you
(or thinking anyway I had)
enabled me to take.

Tom Sleigh's *The Chain* (University of Chicago Press) is a fine book. Sleigh's name was new to me, although he's published *After One* and *Waking,* about both of which good things were written. What I did for my first take was glance at the title poem, written evidently sometime after his mother's death (it used to be called "The Chain," but he has changed it to "Eclipse"). I haven't read a better sonnet in a very long time:

When from among the dead the faces
Of her mother and father turn to her,
At first bright and blank as ice, then melting
to flow to see their only daughter

Miraculously restored, then the living
She loved and who loved her
Fade like fog and are forgotten;
Blank as a wall, her mind registers

The mutual play of light off those long-dead faces
Which, as her parents move to embrace her,
Glow blindingly bright—my final glimpse
Of her face eclipsed by the other dead coming near:

Stepping from every shadow to surround her
They lock hands in an unbreakable chain.

David Wagoner's *Walt Whitman Bathing* (University of Illinois Press) is a series of small, precise, efficient gestures that resonate and enlarge as they are intended to do. In books by other poets, one looks for the three or four outstanding pieces. With Wagoner, I look for lapses from his high standard, and can't find any. Perhaps it is this reliability of his that has caused him to be almost taken for granted. But scrutiny of any particular poem produces surprises and epiphanies—"Blindman," for instance:

He waits by the quiet street
His cane, his fingers,
And his listening face
Trembling, not out of fear
But alert with wonder
At what lies just beyond
The end of what he is

And all he remembers
As now he steps forward
Into his near future
As deliberately as a spider
Scuttling from stone to curb
To stairway where he climbs
Into the spell of his room,
Where his light hands
Lead him to lie down,
Where slowly his mind's eye
Out of a different night
Lifts open like a moon.

C. K. Williams's *The Vigil* (Farrar, Straus & Giroux) reminds me of Igor Stravinsky's remark about Antonio Vivaldi: "He had only one trick, but it was a good one." Williams's long lines, of ten and eleven stresses, suggest an intensity or fervor with which he invests ordinary observations with significance. "Another burst of the interminable, intermittently torrential, dark afternoon downpour" is a line I would object to as excessively adjectival, but Williams's voice—or prosody—legitimates it somehow.

Linda Zisquit's *Unopened Letters* (Sheep Meadow Press) seems slight at first blush, a series of breathy love poems and quick takes, but they stick in the mind. Zisquit lives in Jerusalem where the light does surprising things. That landscape suffuses her poems and may be, after all, what they are "about." She writes in "Alive:"

In this country
who can pretend innocence?
Even the no man's land,
or the poppy-covered desert
where I sat aching to brush
another man's flesh with a leaf
knows landmarks, signs.

Have I left out anyone? Well, of course, but some of those omissions were deliberate. There were, nevertheless, books of some worth that I never got my hands on. The Internet makes this job a little easier than it used to be, but that can be misleading. There are small presses that are impossible to find even for the most assiduous surfer, tiny outfits doing good books. And there is no hierarchy whatever. "Knopf" or "FSG" on the spine is no assurance of quality. They're not even looking for quality; what they want is the names with recognition that will command sales. Those august imprints are therefore slightly suspicious, even a little vulgar.

Still, as the foregoing florilegium suggests, the situation is not altogether hopeless. What we have here can't be characterized as a dearth.

Dictionary of Literary Biography Yearbook Award for Distinguished Volume of Poetry Published in 1996

Henry Taylor's *Understanding Fiction* (Louisiana State University Press) is a book I like for the modesty of its claims. Taylor never hectors or whines or blusters or tries to push poetry into some "higher" realm of theology or philosophy. Indeed, he finds all that–and there is so much of it around–distressing and distasteful. What he admires is mastery, as he explains in "Master of None," a poem that comes as close to a manifesto as he is likely to allow:

Cover for Pulitzer Prize winner Henry Taylor's fourth collection of poems

> An old story comes back
> from wartime, when steel was scarce, and broken parts
> of farm machines had to be put back together;
> on such an errand, once, my father stood around
> for half a day, waiting his turn with a welder.
> To kill time, he wondered aloud
> how hard welding was,
>
> and the shop foreman said,
> "That dumb son of a bitch out there learned to do it."
> So, out of necessity, he mastered it,
> and I watched him for years; now greed gnaws at me
> whenever I watch some close work well done
> and remember the shop foreman's line,
> or mysteries I absorbed
>
> from my life among horses,
> who taught me, in their way, most of what I know now.
> With less faith, I have acted in plays, taken up
> trim carpentry, writing computer programs,
> tuning Volkswagens ... these sidelines and others
> have drawn me toward mastery
> approached, or barely imagined.
>
> For the sake of odd skills
> I have shirked the labors I live by, and pursued
> one more curiosity for my collection, or puttered
> awhile with an old one, thereby evading
> the higher, more difficult arts, such as knowing
> at sight the life-changing moment
> when the right touch or word
>
> might turn aside the rage
> that careens through the house like electrical wind ... [.]

What pleases me about this is the elegance of its modulation. That was Dryden's great skill, the ability to elevate or deflate without any discernible shift in the gears of the poetry. The knowledge of life-changing moments is "close work," surely enough, but the gulf between the two is all but impossible to bridge, and the yearning from the one side of the metaphor to the other is all the more eloquent for having been so quietly implied.

This collection, his fourth, is not only a defiance of all the excess and vulgarity around us in the lit biz, but also at least a beginning toward a cure. Like many herbs with pharmacological capabilities, these flowers are the kind of thing tenderfeet are not likely to notice out there in the meadow. But for those who have the eyes to see, this is a patch of moly.

–David R. Slavitt

James Dickey, American Poet
(2 February 1923 – 19 January 1997)

Ernest Suarez
Catholic University of America

James Dickey, one of the major figures in American literature during the second half of the twentieth century, died in Columbia, South Carolina, on 19 January 1997. He had been suffering from pulmonary disease and other illnesses for almost two years.

Dickey enjoyed a controversial, varied, and brilliant career. Born on 2 February 1923 in the Atlanta suburb of Buckhead, Dickey distinguished himself as a high-school athlete at Atlanta's North Fulton High School. In 1942 he entered Clemson College, where he played in the backfield of the freshman football team for a season before enlisting in the U.S. Army Air Corps. During World War II Dickey logged almost five hundred combat hours, serving as a navigator with the 418th Night Fighters in the South Pacific. He was awarded the Air Medal, the Asiatic Pacific Ribbon, the Philippine Liberation Ribbon, and seven battle stars. The war both fascinated and horrified him; it changed his outlook on life, as he came to view existence from the perspective of a survivor, an attitude that later informed his creative practice. In 1945 he wrote his parents declaring that all he did was "lay around and reflect on how lucky I am to be alive." But Dickey also discovered another preoccupation during his military service: literature. While nights often threatened death, days included many monotonous hours, which Dickey soon began to fill with books. His interest in literature, particularly poetry, deepened as the war progressed. At one point Dickey wrote his mother, asking her to send him fifty-two books, which his letters show he devoured at an obsessional pace.

When the war ended Dickey entered Vanderbilt University with the goal of eventually becoming "a book reviewer for some small-town newspaper." At Vanderbilt he was encouraged to write by various teachers, including Monroe Spears, who became a lifelong friend. By the time Dickey arrived at the university, John Crowe Ransom, Allen Tate, Robert Penn Warren, Randall Jarrell, and Peter Taylor had departed, but the New Criticism was still very much in the air. Spears and others in-

James Dickey on the University of South Carolina Horseshoe (photo by Terry Parke)

structed Dickey on the technical and formal dimensions of poetry. Dickey published four poems in the *Gadfly,* Vanderbilt's literary magazine, and in 1949 graduated summa cum laude with a major in English and minors in philosophy and astronomy. Four days after being awarded the B.A., he applied to enter the M.A. program in English at Vanderbilt. He completed the degree in 1950, writing a thesis titled "Symbol and Image in the Shorter Poems of Herman Melville."

In 1948 Dickey married Maxine Syerson, with whom he moved to Houston, Texas, in 1950, when Dickey took a teaching position at Rice Institute. After one semester Dickey was recalled to the air force for service in Korea. Dickey called this the "most depressing period" of his life, as he was forced to leave Maxine, who was pregnant, and literature for another stint in the military. Dickey later remarked that the only thing that sustained him during this time was the acceptance of two of his poems by the *Sewanee Review*. In 1952 he returned to Rice, until 1954 when he received a *Sewanee Review* fellowship and departed to write in Europe. He accepted Andrew Lytle's invitation to teach at the University of Florida in 1955 but resigned in the spring of 1956 over a dispute concerning the sexual nature of a poem, "The Father's Body," that he had read to a group of faculty wives.

In April 1956 Dickey began a successful career as an advertising copywriter and executive in New York, and, subsequently, in Atlanta. Dickey also continued to publish poems in literary magazines. By this time he had developed a friendship with Ezra Pound, whom Dickey had visited in St. Elizabeths Hospital in Washington, D.C., in 1955. Dickey read Pound's essays "over and over again," using Pound's conception of the image to develop his own image-centered approach, which eventually resulted in his first book, *Into the Stone* (1960).

During the 1960s Dickey experienced one of the most remarkable streaks of literary accomplishment in American literature. On the strength of *Into the Stone* Dickey won a Guggenheim Fellowship, which allowed him to quit the advertising business and spend a year writing in Positano, Italy. A series of original books of poetry–*Drowning With Others* (1962), *Helmets* (1964), *Buckdancer's Choice* (1965), and *Poems 1957–1967* (1967)–won him wide acclaim and several awards, including the National Book Award, the Mellville Cane Award, and a prize from the National Institute of Arts and Letters. In the 1960s he taught at Reed College, San Fernando Valley State College, and the University of Wisconsin before being appointed Consultant in Poetry to the Library of Congress, a position he held from 1966 to 1968. In 1969 he joined the University of South Carolina and was subsequently appointed the First Carolina Professor of English.

With success also came a series of attacks, initiated by Robert Bly's declaration that Dickey was a "toady to the government" who supported the Vietnam War. Though Dickey was active in the presidential campaign of antiwar candidate Eugene McCarthy, Bly's assault was followed by virulent attacks from other critics, whose denunciations of Dickey's use of violence were propelled by their dissatisfaction with the Vietnam War. Dickey's poetry about World War II became interpreted through the lens of Vietnam War politics, and he was upbraided for what critics saw as his failure to condemn violence, a situation complicated by Dickey's exuberant lifestyle and his image as a "macho" poet.

Though upset by the attacks, Dickey refused to soften his themes or alter his way of living. In *Deliverance* (1970), which became a best-selling novel and a major motion picture, and in many poems Dickey continued his investigation of the psychological dimensions of primitivism. His method of creating differing situations and characters in order to probe the positive and negative aspects of the elemental allowed him to explore what various mixtures generate, but continued to confuse many critics attempting to assess Dickey's "ideology." However, Dickey still received widespread recognition–he was inducted into the National Institute of Arts and Letters in 1972 and was asked to compose and read a poem for President Jimmy Carter's inauguration.

In the 1980s and the 1990s Dickey, a relentless experimenter, continued to push his creative practice in new directions. In *Puella* (1982) and other books of verse he moved away from narrative, attempting to create still pieces with complex sonic structures. In *Alnilam* (1987), a massive lyrical novel, Dickey used the perspective of a blind man to explore the poetic process, as the protagonist is forced to construct reality through imaginative images. Subsequent publications include *The Eagle's Mile* (1990), *The Whole Motion: Collected Poems, 1945–1992* (1992), and his third novel, *To the White Sea* (1993). In 1988 Dickey was inducted into the fifty-member American Academy and Institute of Arts and Letters, and in 1996 was awarded the Harriet Monroe Prize by *Poetry* magazine for lifetime achievement. Despite illness he taught until five days before his death.

Dickey's first wife, Maxine, died in 1976. He is survived by his second wife, Deborah Dodson, with whom he had a daughter, Bronwen. He is also survived by two sons, Christopher and Kevin, from his first marriage.

Tributes to James Dickey will be published in the 1997 *Yearbook,* and he is the subject of a forthcoming volume in the *Dictionary of Literary Biography Documentary Series*.

The Year in Fiction

George Garrett
University of Virginia

"From the cascade of trivial works published this month, it seems that the novel has gone into a very deep sleep."

—Joan Mellen,
Baltimore Sun (12 May)

"Fiction does have an *appropriately* disturbing function, for its fundamental value is to make us see, to tease us out of our own minor fictions—those novels we thought we knew up close, our lives."

—Robert Gingher,
The World and I (December)

It was widely reputed and reported to be a bad year for literary fiction. Though the standard blockbusters, same crowd this time as any other recent year, fared well—and had to in order to earn back their enormous advances—literary fiction was said to be in deep trouble. This was reflected noticeably in the fact that the principal reviewers, newspapers, magazines of all kinds, radio, and television heavily favored nonfiction titles, all the more so since it was an election year, full of sound and fury and the loud noise of public events and issues. It was a busy year for biographies, memoirs, confessions of all kinds, historical and contemporary reports, and wonders and extraordinary events as we moved toward the end of the century and the millennium.

Fiction had a hard time keeping up with the fact. It might (in a literary sense, at least) be called the year of Harold Brodkey, who died in January as the year began, and whose posthumous account of his losing battle with AIDS, *This Wild Darkness* (Holt), was favorably reviewed by Michiko Kakutani in the daily *New York Times* ("Going to Die, but First There Is So Much to Say") on Christmas Eve. Steve Wasserman, book editor of the *Los Angeles Times,* argues that "Trade book publishing faces the most serious structural crisis in American history" and that good books of all kinds are in trouble. He blames the situation on conglomeratization, "short-term bottom line mentality," junk culture, and "the cult of celebrity, sensationalism, and gossip." Concluding: "Taken together, these factors threaten to leave us in a condition not unlike that which greeted the Chinese people at the end of the Cultural Revolution: ignorant of tradition, contemptuous of habits of quality and excellence, unable to distinguish the good from the bad." (See the *National Book Critics Circle Journal,* December 1996.)

Still, novels and collections of stories kept appearing, fighting each other for the limited review space and the hoped-for possibility of a few bright minutes of public attention. In spite of everything, there were enough new fiction titles, not to mention plenty of reprints, for anyone to be aware that many good books, more than we care to imagine, were lost in the scuffle and shuffle. There are plenty of good books, certainly, not mentioned here or anywhere else except maybe *Books in Print*. And, by the same token, there must be any number of novels and collections of stories which, out of our inevitable ignorance and the inability to keep up, we overvalue and overpraise. There is no justice in any of it; hard to imagine how there could be, this side of heaven's gate; but we must begin somewhere and do what we can do with what can be done.

As usual, the literary year stumbled and staggered to its closure with the making of many lists, personal and institutional, the best-of and worst-of this and that published during the year. Like last year, this year's lists did not have as much overlapping (even among the short lists for the major prizes) as one might have expected and imagined. This may well be because, in the case of fiction, for example, everybody's lists treat and concern literary rather than commercial books. And in the matter of literary fiction, seldom if ever these days on best-seller lists, personal tastes and highly subjective judgment play a major part. Add ignorance to the equation. None of us has a full grasp of the real situation, not when thousands of volumes jostle for limited space and attention. All of us miss a large part of what appears and disappears. So why should I here and now come forward with my own listing of the best and the brightest works of fiction, almost exclusively literary, for this year? I do so with this important reservation: that, unlike the makers of

The Year in Fiction   DLB Yearbook 1996

Dust jacket for Richard Bausch's seventh novel

most lists (and I insist on including the judges for the major and minor prizes and awards), I not only know that mine is strictly limited to the books I have been able to read or have skimmed, based usually on some preliminary reviews somewhere or other, but also I am willing to admit my limitation publicly, even cheerfully. Meaning this much: there are plenty of books of fiction that I did not read or, indeed, know anything about during 1996. But, at the same time, there are a great many books (too many probably; one becomes a little jaded) that I have actually encountered and experienced this year.

What follows first, then, is my own private list of the best works of fiction (in my best judgment) published in America in 1996.

DLB HONOR ROLL OF NOTABLE FICTION

Laurie Albert, *The Price of Land in Shelby* (University Press of New England). Born in New Hampshire, mostly raised in Boston, and now living in Vermont, Laurie Alberts is the author of a story collection and two novels. This new one follows the fate of the five Chartrain siblings of Vermont, telling their story from 1965 to 1994, in shifting third-person point of view. It is a story, as writer Bret Lott asserts, "big in scope and big in heart"—"The landscape was familiar, but familiar in the way of things you didn't know you had forgotten, a grove of beech with the smooth grey skin of elephants, the same crumbling roof of an old hunting camp that had been built without a foundation and rotted inch by inch into the ground; the rusting gas can from a long-ago logger who'd given up, stumps that were mossy and broken down that she recalled as topped by flat pale circles, the chainsaw's mark."

Margaret Atwood, *Alias Grace* (Doubleday). Her twenty-ninth book, her ninth novel, *Alias Grace* is based on the true story of Grace Marks, who was directly involved in a double murder in Canada on 23 July 1843. The story alternates between a first-person recollection of Grace and third-person sections of Dr. Simon Jordan, a student of mental illness: "We went up to Belfast in a cart hired by my uncle, which was a long journey and very jolting, but it did not rain much. Belfast was a large and stony city, the biggest place I had ever been in, and clattering with wagons and carriages. It had some grand buildings, but also many poor people, who worked in the linen mills day and night. The gas lamps were lit as we arrived in the evening, which were the first I ever saw; and they were just like moonlight, only greener in color."

Andrea Barrett, *Ship Fever: Stories* (Norton). Winner of the National Book Award, *Ship Fever* is a brilliant and highly original collection of eight stories which have in common a concern with the history of science. It weds the truths of science and storytelling: "The mouth of the Amazon was like a sea, and could be distinguished from the ocean only by its extraordinary deep-yellow color. The Rio Negro was as black as the river Styx. Jet-black jaguars and massive turtle nests, agoutis and giant serpents; below Baiao, a crowd of Indians gathered, laughing and curious to watch Alec skinning parrots."

John Barth, *On With the Story: Stories* (Little, Brown). A dozen stories about husbands and wives, told by the author/narrator to his own wife while the couple is at a resort ("the last resort"), the whole self-conscious, self-reflexive, and uniformly brilliant, this is Barth's first book of stories since *Lost in the Funhouse* (1968): "But in redefining the story, I find I've lost its point. Better to lose than to be lost? That depends. Better the key lost than the car? For sure, in this instance anyhow. But although a *roman a clef*, for example, ought to be at least reasonably intelligible without its clef, a code without its key is as meaningless as a key without its code. And it, moreover, as has elsewhere been proposed, the key to the treasure may *be* the treasure, then."

Richard Bausch, *Good Evening Mr. & Mrs. American and All the Ships at Sea* (HarperCollins). An outstanding novel by one of our finest working novelists, Bausch's eleventh book (seven novels and four collections of stories), set in the 1960s, tells the funny/sad story of the fate and fortunes of an idealistic innocent, Walter Marshall, who dreams, among other things, of one day being the president and taking up the torch of the fallen JFK: "When he dressed, he stood for a minute in front of the mirror, assuming the poses of Kennedy at a news conference, pointing, choosing another questioner. He had the hair just about right now. He stood close and tried the smile. His teeth weren't quite straight enough. He stepped back and pointed again, looking serious, presidential. It would be a picture the news services would circulate, perhaps after he was assassinated."

Madison Smartt Bell, *Ten Indians* (Pantheon). Turning away (briefly) from the historical world of Haiti, a trilogy of novels begun with last year's *All Soul's Rising,* Bell, in a quickly paced and unflinching story set in contemporary, inner-city Baltimore, follows Mike Devlin as he opens a Tae Kwan Do school in a black neighborhood. What happens is reported by a variety of well-realized voices and points of view: "So we sitting out front of the house, just me'n my homegirl Tamara and Gramma Reenie. Me'n Tamara had pulled out some chairs so we can sit and hang our feet out over that green metal rail that run our whole block of the Poe Homes. Gramma Reen, she just sat down there on the door sill. Her little knees poking up through that green dress, no bigger'n a bird legs."

Thomas Berger, *Suspects* (Morrow). *Suspects* is Berger's twentieth book since 1958, when his first, *Crazy in Berlin,* appeared. Master of all the known genres, Berger here exploits and explores the murder mystery, mostly through the viewpoint of Nick Moody, detective first-grade: "Unless he was working, Moody went to Walsh's most nights rather than go home early enough so that he could be bothered by his fellow tenants, whom the super, disregarding his request, had told he was a cop, which meant neighbors knocked on his door every evening, bringing problems from the trivial—noisy kids in the hallway—to grim: a young woman had been raped and beaten half to death on the tenth floor a month before he moved in. . . . But that was a Sex Crimes case and not one for Homicide. When he explained as much, people were annoyed to hear that policemen pretended the arbitrary distinctions among themselves were meaningful and not just TV jargon."

Wendell Berry, *A World Lost* (Counterpoint). This latest of more than thirty books by Berry is a slender, evocative story of young Andy Cortlett of Port William, Kentucky, whose Uncle Andrew was

Fred Chappell (photograph by Katherine Morse-Chappell)

murdered in July of 1944. Andy gradually discovers the true story of his uncle even as he recollects and comes to understand his own and his heritage: "He was on my mind forever too, as I now see. But I was a child; for me every day was new. I lived beyond my loss even as I suffered it, and without any particular sympathy for myself. And what I have grown into is not sympathy for myself as I was but sympathy for Grandma and Grandpa as they were."

Larry Brown, *Father and Son* (Algonquin). Widely reviewed and highly praised, Brown's third novel pits Glen, an ex-convict, against his crippled father, Virgil, and his illegitimate half brother Bobby Blanchard, who happens to be the county sheriff. Reviewing it for *Washington Post Book World* ("Southern Cross," 26 September), novelist James Hynes pinpoints the special qualities of *Father and Son:* "It doesn't take a diehard fan of Southern Noir to tell that this situation is tailor-made for violence; in synopsis, it reads like Jim Thompson. But a forced originality would miss the point here; Brown's consummate skill at storytelling rings the changes as satisfyingly as a Noh play, making it new."

Fred Chappell, *Farewell, I'm Bound to Leave You* (St. Martin's). Chappell is the author of more than twenty books of poetry, fiction, and criticism. This novel is the third in a series devoted to the Kirkman family of Wind Mountain in North Carolina and narrated by young Jess Kirkman, whose dream is to become a writer. As Jess and his father wait for the death of Jess's grandmother, he recollects eleven tales that came to them from both mother and grandmother, lyrical tales with deep roots in folk song and folk art and in the storytelling of all ages: "This is a tale with four tellers. My grandmother told it to me a number of times and at other times my mother gave me the account in very different terms. Sometimes they would talk about it simultaneously, shooting each other puzzled or affirming glances as they ranged from point to point."

Tracy Daugherty, *The Woman in the Oil Field* (S.M.U. Press). This book is a collection of ten diverse short stories by the author of two novels. A student of the late Donald Barthelme, he is already known for the tension and dramatic contradiction "between longing for and subverting the literary and socio-historical past." Critic Lance Olsen (in a jacket blurb amid a throng of blurbs) continues: "The outcome is splendid fiction, often funny, always moving, that pushes beyond the by-now orthodox landscape of postmodernism into a fresh and psychologically rich new country."

Guy Davenport, *The Cardiff Team: Ten Stories* (New Directions). Intellectual, witty, erudite, adventurous, innovative, and busily productive (author of seventeen books), Davenport is the creator of (in the words of his publisher) "a wondrous collage of persons, events, and ideas from cultural history." He is a master of many voices and tones of voice, as in "Dinner at the Bank of England": "Their friendship was a sweet mystery. The British explain nothing, and do not like to have things explained. The captain had doubtless told his friends that he'd met this American who was dashedly friendly when he was in Boston, had even given him a book about Harvard College, where he was a professor wallah. Followed sports, the kind of rugger they call football in America. Keen on wrestling and track. Speaks real French and German to waiters, and once remarked, as a curiosity, that he always dreams in Spanish."

Donald Davidson, *The Big Ballad Jamboree* (Mississippi). Recently discovered, this only novel by the poet and critic Donald Davidson is set in summer 1949 in the imaginary Appalachian town of Carolina City and is, among other things, a love story centered on country singer Danny MacGregor and music scholar Cissie Timberlake: "If I could figure out, for sure, what Cissie wanted, then I could figure out the choice. But every time I thought I had the answer to what Cissie wanted, I bumped into a big question mark somewhere. Maybe it came down to the plain fact that I didn't count one bit in what Cissie wanted. Maybe I just wasn't good enough—or she didn't trust me—or anybody."

Joan Didion, *The Last Thing He Wanted* (Knopf). Didion's tenth book and her first novel in twelve years received mixed notices. "Despite the Iran-Contra context, the novel has no politics," wrote Adam Begley in the *New York Observer* ("She's in the Pantheon, So Why Sweat for Fiction?," 9 September). "Despite the thriller plot, the novel has no action. Or almost none, and that little bit is filtered through halting reconstruction and analysis." Maybe so. But, in its own terms, this novel about gunrunning and Western Hemisphere hanky-panky, witnessed by the sensitive and intelligent Elena McMahon, is elegantly economical, almost quintessential Didion: "She had some cash, there were places she could have gone. Just look at a map: the unnumbered other islands there in the palest blue shallows of the Caribbean, careless islands with careless immigration controls, islands with no designated role in what was going on down there."

Carl Djerassi, *Marx Deceased* (Georgia). In this, his fourth novel and tenth book, the distinguished scientist Carl Djerassi tells the quasi-metafictional story of novelist Stephen Marx, who goes to considerable, imaginative trouble to pretend to be dead and thereby to know what, if anything, people really think about him. In his "Preface," which works as a part of the experience of the novel, Djerassi compares the search for peer approval that is a "compulsive drive" for both scientists and writers: "Most writers also display a need for approbation by their peers. Novelist Stephen Marx's preoccupation with his own image is not very different from describing a scientist's hunger for peer validation. In each instance, that urge is both the nourishment and the poison of a creative mind."

Andre Dubus, *Dancing After Hours* (Knopf). This collection, Dubus's first book in ten years, earned him America's leading short-story award, the REA Award, which, in addition to the honor and attention, brings $30,000. Judges Barry Hannah, George Garrett, and Jayne Anne Phillips selected him and his book from a long list of impressive competition, their feelings aptly if accidentally contained and confirmed in Jack Sullivan's review, "Picking Up the Pieces," in *Washington Post Book World*: "The light does come in these often dark stories, glimmers of grace and dignity in the shadows of hardships. The flat irony, mechanical cleverness, and self-conscious cynicism of so much contemporary fiction have no place here. *Dancing After Hours* solidifies Dubus's position as one of America's wisest and most eloquent storytellers."

Charles East, *Distant Friends and Intimate Strangers* (University of Illinois Press). This new addition to the extensive Illinois Short Fiction series is East's second collection of stories in more than thirty years. Fourteen elegantly crafted stories set in the new South ("in the age of TV soap operas and radio call-in shows"), all marked by the presence of strong, distinctive, and wholly credible characters: "The town is exactly as he remembered it, the streets deserted in the summer sun, the drugstore closed on a Wednesday afternoon—he can't rouse the pharmacist or the doctor. The doctor has gone to Memphis to play in a golf tournament."

Ralph Ellison, *Flying Home and Other Stories* (Random House). This collection consists of thirteen stories, six published posthumously, all of them written between 1937 and 1954, put together by Ellison's literary executor, John F. Callahan, who includes a full-scale introduction to the collection. Of his choices of these thirteen stories from among others, Callahan writes: "The present collection chronicles Ellison's discovery of his American theme. In technique and style, subject matter and milieu, the thirteen stories show the young writer's promise and possibility in the late thirties and his gradual ascent to maturity in the mid-forties when, unbeknownst to him, he was about to conceive *Invisible Man*."

Percival Everett, *Watershed: A Novel* (Graywolf) and *Big Picture: Stories* (Graywolf). One of the most gifted and productive (these are his ninth and tenth books) of the younger generation of American writers, Everett offers, in the words of his contemporary, Madison Smartt Bell, a "fine combination of humor, satire, and well-founded outrage." To which should be added the rare quality of powerful integrity. *Watershed* is a mystery, and it begins with a violent confrontation between the government (FBI) and natives of the (imaginary) PLATA Indian Reservation and goes backward and forward from there. The nine stories of *Big Picture* are various in setting and situation: "The moon was unrelenting as Joseph stared out the bedroom window. The cornbread globe, just shy of full, sang a glow of restful light, but Joseph was cursing it. He went to the window and looked down at the bay mare in the pasture. He couldn't climb back into bed. He couldn't lie between the sheets with that woman; he wouldn't have her foot brush his leg or her hair tickle his shoulders."

Sebastian Faulks, *Birdsong* (Random House). British author Faulks offers a powerful and moving World War I story, following the fate of Stephen Wraysford, a young Englishman who finds himself in the trenches (and *under* the battlefield in dark tunnels beneath the German lines): "Of their original platoon only he, Brennan, and Petrossian were still at the front. The names and faces of the others were already indistinct in his memory. He had an impression of a weary troupe of greatcoats and grimed puttees, of cigarette smoke rising beneath helmets. He remembered a voice, a smile, an habitual trick of speech. He recalled individual limbs, severed from their bodies, and the shape of particular wounds; he could picture the sudden intimacy of revealed internal organs, but he could not always say to whom the flesh belonged. Two or three had returned permanently to England; the rest were missing, buried in mass graves or, like Reeves's brother, reduced to particles so small that only the wind carried them."

Lucy Ferriss, *Against Gravity* (Simon and Schuster). This third novel by Lucy Ferriss is centered on Gwyn "Stick" Stickley, growing up in upstate New York, trying out a dancing career in Manhattan, returning to care for her injured father, finding, finally, a life: "Afterwards I stood out on the corner by Carnegie Hall and just breathed in the city—the rush of people, the heat from the buildings. My solution was working, I thought sometimes. I was forgetting. Once a week, because Frankie insisted, I even took voice lessons, way down in Tribeca with this alto who looked like Morticia Addams."

Tim Gautreaux, *Same Place, Same Things* (St. Martin's). Twelve short stories in this debut collection by a south Louisiana writer who has already earned advance praise from James Lee Burke, Shirley Ann Grau, Robert Olen Butler, Susan Dodd, and Andre Dubus. He deserves it fully, here displaying all the solid virtues of the finest fiction: "I'm going to tell you about the last time I went to confession. I met this priest at the nursing home where I work spoon-feeding the parish's old folks. He noticed I had a finger off, and so he knew *I* was oil field and wanted to know why I was working indoors."

James W. Hall, *Buzz Cut* (Delacorte). The poet and professor Hall's seventh novel, his fourth in the successful South Florida series, starring the hero known as Thorn, is a fast-paced thriller in which he and his friend, Sugarman, try to unravel a complex and dangerous scheme involving extortion and violence and a cruise ship called *Eclipse*. Hall's work has, justly, begun to receive serious attention from the reviewers: "The lawyer's twelve-year-old son warned Thorn if he touched either him or his sister, both of them would damn well litigate his ass off. The girl said she was one quarter away from the all-time record score on her machine and her brother maybe three quarters away. Thorn watched them work their joysticks for half a minute then he

Dust jacket for the twenty-seventh book by the leading practitioner of the novel in dialogue

stooped down and yanked the plugs on both machines. They all watched the numbers fade from the scoreboard like stars against a dawn sky."

Barry Hannah, *High Lonesome* (Atlantic Monthly). Thirteen stories in this volume, the eleventh book by the hip and highly praised southern writer, are various in style and content, subject, and treatment, but all are linked by Hannah's inimitable voice, his singular slant of vision: "I'll remember him there before the next moment, loved and honored and looking ahead to a breakout, on that little beach. He could be taken for a real man of the world, interested even in puppets, even in fine fabrics. You could see him—couldn't you?—reaching out to pet the world. Too long had he denied his force to the cosmos at large. Have me, have me, kindred, he might be calling. May my story be of use. I am meeting the ocean on its own terms. I am ready."

Donald Harrington, *Butterfly Weed* (Harcourt Brace). Here is the latest in the remarkable series of novels about the mythical and imaginary place, Stay More, Arkansas, and its beings (not only people) told by Harrington, an art historian by trade. Here the story chiefly concerns Doc Swain, the self-taught physician of Stay More, and his life history and his loves: "Alonzo Swain, learning of his son's continu-

ing virginity, told Colvin that he ought to grow some upper lip hair in order to make himself more desirable to women. At twenty, Colvin still looked like a teenager. He was handsome. He was tall and dark. But he looked like a kid. All of the other men in Stay More, without exception had full mustaches. It was the vogue of the day, perhaps inspired by the nation's current president, Teddy Roosevelt. Alonzo's own mustache covered his entire mouth and often got chewed up at mealtimes. And women could not resist him."

George V. Higgins, *Sandra Nichols Found Dead* (Holt). This is the twenty-seventh book, beginning with *The Friends of Eddie Coyle* (1972), and the fourth "Jerry Kennedy novel," featuring his lawyer protagonist. ("I do criminal. My name is Jerry Kennedy and I don't take civil cases.") This time, however, he has to, and to solve a complex mystery as well, in *Estate of Sandra Nichols v. Peter Wade*. Prolific, inventive, and productive, Higgins has developed into the consummate professional, one of the handful of popular writers of crime fiction who have earned the critical respect usually reserved for literary novelists: "Ordinarily I'd say that the age of sixteen would probably bar any further reference to that person as a child. But that was what Maggie Cameron still clearly was, midway through the afternoon that I first met her. She was a girl-child dozing in the shade of the enormous overspreading old black maple, her soft, frizzy blond hair disheveled against her pale skin, her torso femininely formless under a white mesh short-sleeved polo shirt, her legs still thin in her Levi's."

John Irving, *Trying to Save Piggy Sneed* (Arcade). Irving's ninth book, "a dozen short works"—three memoirs, six short stories, and three critical pieces ("essays of appreciation")—plus sixteen photographs, is odd enough to explain how it appeared and swiftly disappeared without more than minimal and very mixed public attention. But Irving deserves some credit for the risks he took here, risks inherent in the bravado of simple self-exposure and the assertion of the obvious or of strongly held, cantankerous opinions: "The concept of a celebrated critic is an oxymoron to me; nevertheless, I feel I must explain to my fellow Americans that German literary culture is quite different from our own. Our literary culture is small and contained; our writers are of no political influence in our society. One happy result of the relative unimportance of writers in the U.S. is that literary critics are of even *less* importance to us. (Try to imagine *any* critic on the cover of *Time* or *Newsweek*!)"

Josephine Jacobsen, *What Goes Without Saying* (Johns Hopkins). This handsomely made volume brings together thirty stories from poet and critic Jacobsen's four earlier collections of short fiction, the first of which was published in her seventieth year. Eight of these stories were included, over the years, in the O. Henry Prize Stories collections, and one collection, *On the Island,* was nominated for the PEN/Faulkner Award. Varied in setting, place, and subject, the stories are unified by the precision of the author's vision, a quietly authoritative verbal felicity, a strong sense of character, and a concern with the shape and architecture of a tale, line by line, paragraph by paragraph: "He was a passionate gardener. In a space in which nature, left unattended while one's back was turned, shot up great leaves, sprang vines a foot ahead, pushed bushes up to block the sea, he happily battled. Wearing, against the sudden violent bouts of rain, knee-high boots caked with earth, he slashed and clipped, pruned and repressed—he had even formed a sort of ledge of truncated awning from the offshoots of the great vine with a trunk thick as his arm, which grappled the roof of the concrete porch. The huge swatches of color—bougainvillea, hibiscus, oleander, ixora—responded with curbed glory against the dazzling backdrop of sea" ("Mr. Meadow's Cup").

Greg Johnson, *I Am Dangerous* (Johns Hopkins). This third collection of short stories by Greg Johnson consists of thirteen stories, varied in setting and diverse in voices and points of view, most of them originally published in outstanding literary magazines. Johnson is also a poet and the author of three volumes of literary criticism: "She had a sly, teasing style with Matt that had always excited him. Tonight, as during those raw, bewildering nights of their first passion, she could sit back, flick her long hair behind her bare shoulders and gently laugh at him. Her laughter was like murmuring, he thought; it seemed to come from underwater."

William Kotzwinkle, *The Bear Went Over the Mountain* (Doubleday). It takes clear vision and a skilled hand to create classic satire in these grungy and surreal times. But Kotzwinkle has done it before and does it again, wonderfully, with this story of (among other things) a bear from Maine who finds a new and different life as the author of a best-selling novel, *Destiny and Desire.* Everything is subjected to the catharsis of laughter. Nothing sacred or profane about the way we live now is spared: "The publishing party for the bear's book was held in a downtown warehouse that'd been gutted to house the newest, hottest disco. The walls were brick with exposed electrical conduits. Brick pillars supported the cavernous interior; catwalks on four sides formed balconies; posters of James Dean and Marilyn Monroe hung behind the bar, while living legends moved about on the crowded floor. The bear's frowning face with its suspicious, beady eyes was on the latest *GQ*. There'd been interviews in *Publishers Weekly* and *The Village Voice*."

Jeanne Larsen, *Manchu Palaces* (Holt). The poet and translator (of several Chinese works) Larsen presents her third novel in a series dealing with Chinese history (the preceding volumes are titled *Silk Road* and *Bronze Mirror*). Though following the life story of Lotus, a courtesan in eighteenth-century Manchu, this lyrical book touches on the conventions of metafiction, allowing ghosts outside of ordinary time, including a woman aviator from the 1930s, to play parts in the complex story. There are poems and documents and a variety of narrative styles and voices, all of which work to justify the metaphor of the intricate, convoluted architecture of Manchu palaces: "The power of the exotic is a strong thing, is it not? Just look at that immodest neckline, those beaky noses and blatant legs, the freakishly discolored hair? Marvel at that little dollhouse of a temple, the funny pigtailed chap groveling facedown—as surely no civilized man would lower himself to do!—before an impossibly ornate throne. How bizarre! How charming! What fun to gawp and wonder if true human feeling can exist within!"

Starling Lawrence, *Legacies: Stories* (Farrar Straus). The senior editor at Norton, Lawrence also writes fiction of the highest quality. *Legacies* is his first collection of stories and includes award-winning fiction (the Lytle Prize and the Balch Prize), eight stories, varied in subject and setting but linked by attitude and tone of voice. A significant contribution to the American story: "But he hurried on, drawn by curiosity and by the reverberation of that other phrase that had come to him unbidden: the northern lights. He had been small enough for his father to hold him effortlessly against his hip in the savage cold of another night with the unearthly light pulsing above them, veils of strange colors that shimmered among the bright stars from a point up near the top of the heavens down almost to the far horizon of hills."

John le Carré, *The Tailor of Panama* (Knopf). The story of a British immigrant, Harry Pendel, a tailor and would-be spy in contemporary Panama, described (on the book jacket) as "a Casablanca without heroes, a hotbed of drugs, laundered money and corruption." The book joins together the author's own ways and means with the entertainment tradition and conventions of Graham Greene: "Other white-gloved hands took Pendel's suitcase and passed it through an electronic scanner. He was beckoned to a scaffold. Standing on it, he wondered whether spies in Panama were shot or hanged. The gloved hands returned the suitcase. The scaffold declared him harmless. The great secret agent had been admitted to the citadel."

Elmore Leonard, *Out of Sight* (Delacorte). Leonard's thirty-third book has all his old virtues—the snap, crackle, and pop of pace, scene, dialogue, and memorable characters, together with a new twist: its protagonist is a woman—Deputy U.S. Marshal Karen Sisco. "The smell of sauerkraut in the manager's first-floor apartment. His watery wide-open eyes as she assured him the residents wouldn't be disturbed. Telling him this as she imagined their reaction to the *SWAT* team invading the place. The senior citizens in the lobby, mostly women, sweaters over their shoulders, bifocals shining, real fear in their eyes at the sight of black uniforms and jackboots, the helmets, the ballistic vests with FBI in yellow, big, on the backs of the vests, the automatic weapons at port arms, the *SWAT* team coming through the lobby like a troop of Darth Vaders."

Mark Lindensmith, *Short-Term Losses: Stories* (S.M.U. Press). The ten stories in this collection, though set in heartland America, are various and marked by a virtuoso diversity of form and a clear, compelling voice. This collection introduces an important new talent: "Her saying his full name, Lloyd Christensen, was like an incantation. It was as if by just saying the name out loud she was able to capture some unintelligible essence of him and clasp it to her breast so that he couldn't—didn't want to—escape. She was a shaman in an ancient rite, and he would atone."

Talking Horse: Bernard Malamud on Life and Work (Columbia), edited by Alan Cheuse and Nicholas Delbanco. This collection is a mixed bag of previously unpublished essays, speeches, notes, and stories, all directly involved with the body of work Malamud created: "If the writer is not enlightened, his work has taught him nothing. It must teach him to understand his work, as it will if he is listening; teach him—ultimately—to understand himself."

Thomas Mallon, *Dewey Defeats Truman* (Pantheon). Set in the campaign year of 1948 in Owosso, Michigan (birthplace of Thomas E. Dewey), the novel, Mallon's fourth, follows the love-story/triangle of Anne MacMurray; Peter Cook, "the town's number-one up-and-coming Republican"; and Jack Riley, war hero and labor leader. Reviewing the book in *The New Criterion* (December), Peter Schwendener ("Precincts of Elsewhere") identifies the special qualities of Mallon's fiction: "The symmetrical pairing of the love triangle and the national election gives the novel something of a fairy-tale quality. We are not in Updike territory where a hallucinatory vividness seizes every bicycle, cocktail napkin, or disordered bedsheet, but in a more generalized setting where the emphasis falls on moral, emotional, and political choice."

David Markson, *Reader's Block* (Dalkey Archive). A highly original and altogether satisfying novel, Markson's fifth (he also lists three "Entertainments" in his list of works), *Reader's Block* takes the form of a notebook or commonplace book in which Reader jots down brief aphorisms, quotations, facts of the lives of artists and writers, and notes for a novel in progress:

> "Nonlinear. Discontinuous. Collage-like. An assemblage. Or of no discernible genre?
> A seminonfictional semifiction? Cubist?"

At any point, adroit as he is, Markson could have failed at his self-assignment by cuteness or self-pity or bathos. But he falters nowhere. The book is an astonishing artistic success and, even more astonishing, deeply moving in the ways that only the finest fiction can be. This brave and gifted writer has been highly praised by writers as distinctly different as William Kennedy, Ann Beattie, David Foster Wallace, and Gilbert Sorrentino. But his work does not need endorsements. It speaks for itself in its own inimitable language:

> "Philip Larkin:
>
> I wouldn't mind seeing China if I could come back the same day.
> *Kristallnacht.*
> Edna St. Vincent Millay died at the first light of morning after having sat up all night reading a new translation of the *Aeneid.*
> I have a narrative. But you will be put to it to find it.
> None of Andrew Marvell's best poems was published in his lifetime."

Hilary Masters, *Home Is the Exile* (Permanent Press). Masters's eighth novel is a rich and complex story of our times, told in a fascinating double narrative—part of it told by pilot Roy Armstrong, back from the Spanish Civil War in 1939, and Walt Hardy, former campaigner for Robert Kennedy in 1968 and in the 1980s on the cover of *Time* as one of those involved with the Iran-Contra scandal and Lt. Col. Oliver North; the former is told in the first-person of a journal, the latter in a third-person narration, brilliantly executed. The work becomes the stories of lapsed idealists. How the stories intersect and, finally, become one whole is the magic trick of this moving novel. Masters grows better and goes deeper and darker with each new work: "Then, if you're Walt Hardy, you understand something about power; how even if you want to give it up, even if it might be taken from you, some of it, the glow of it will stick to you; the way the smashed splendor of lightning bugs had struck to your fingers on summer evenings in Pittsburgh; cold and bright and mysteriously long after the insect was dead."

Jill McCorkle, *Carolina Moon* (Algonquin). Author of five novels and a collection of short stories, McCorkle here tells a richly detailed story ("six parallel love stories *and* an unsolved murder mystery") set in the eastern Carolina town of Fulton. *Carolina Moon* received mixed reviews but earned the praise of other southern novelists like Madison Smartt Bell, Lee Smith, and Beverly Lowry. Its mixture of comedy and pathos, of three-dimensional individual characters, and an equally realized sense of community is uniquely her own: "Now the memory makes him shudder as he sits on the damp and cold sand; the tide is coming in, claiming his past. Someday, when he is forty-one, he will breathe a sigh of relief because he will know there isn't something genetic to make him go out and buy a gun. Who's to say it couldn't happen? Who's to say there isn't something perched and waiting in his brain, something that would send him out of this world in a goddamned burst of flames."

Don McNeill, *Submariner's Moon* (Oberon). Seven stories, all set in the world of Newfoundland, from World War I to the present, *Submariner's Moon* is a first collection by a native of Newfoundland who went far and wide as a journalist, serving as bureau chief in Moscow and later as Middle East correspondent for CBS News. In a statement about the stories of this book, Ann Beattie praised them for "the careful, deliberate way they slowly unfold. They're about intensely personal moments of revelation—understated stories about souls in crisis, framed nicely by events in the outside world."

Stephen Millhauser, *Martin Dressler: The Tale of an American Dreamer* (Crown). With this volume, his seventh and most straightforward book, Millhauser has found himself on the short list for almost every fiction prize in the country. It tells the story of a young entrepreneur in late-nineteenth-century New York City who "walks a haunted line between fantasy and reality, madness and ambition, art and industry": "The newspaper reports were on the whole favorable, though Martin detected a frequent note of puzzlement or bewilderment: the critics, while admiring particular effects, seemed uncertain when the question arose of what exactly the Grand Cosmo was. Some called it a hotel; a Jew, taking a

Dust jacket for Stephen Millhauser's well-received novel set in late-nineteenth-century New York

hint from the ad campaign, called it an experiment in communal living."

Alice Munro, *Selected Stories* (Knopf). Drawn from the seven collections of short stories Munro has published since 1965, these twenty-eight stories, arranged in roughly chronological sequence, make a large and abundant volume (550 pages). The author is widely honored in her native Canada as well as in the United States and abroad. Mostly set in Ontario, the stories are not thereby limited. *The New Yorker* (16 December) said that Munro's "work is boundaryless," and that its themes are the classic ones: the sorrows of love; the difficulty, especially for women, of forging a self; the upwelling urge to violence in man and nature alike; and the narrative tricks played by history."

Lewis Nordan, *Sugar Among the Freaks: Selected Stories* (Algonquin). With the aid and comfort of bookseller and Nordan fan Richard Howorth, who made the selections and the arrangement, Algonquin has republished fifteen of Nordan's stories described as, like his novels, "emotionally complex and literally bizarre." Nordan's world is his own: "Daddy had propped a square mirror in front of the kitchen window. On the counter were four half-pints of Early Times and four shot glasses. (He

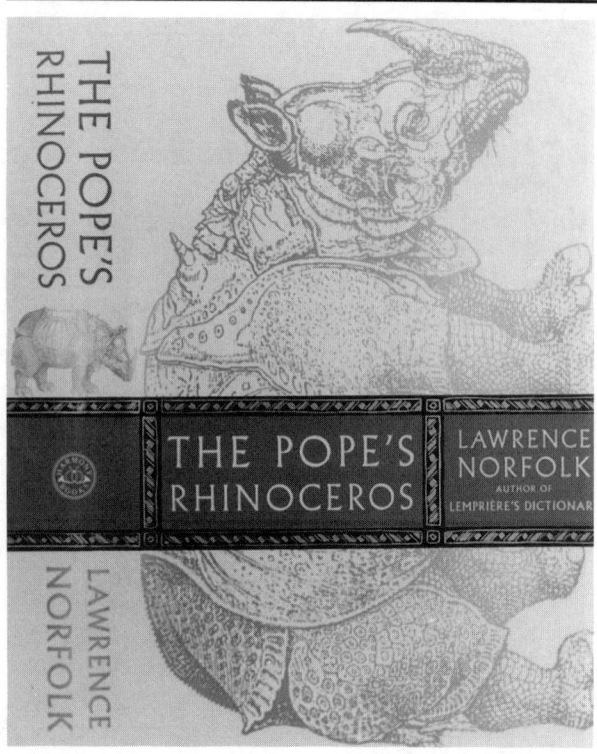

Dust jacket for British novelist Lawrence Norfolk's account of a sixteenth-century Pope's private zoo

called this 'shooting fours.') Mama was saying something to him. I knew what. She was saying to him that he was an artist, that he was special and perfect and magic, that his pain was special. She was telling him she wanted to carry his pain for him. I had heard it a hundred times, she meant every word. In front of the bare fig trees I grieved and celebrated my invisibility."

Lawrence Norfolk, *The Pope's Rhinoceros* (Harmony). His second novel, set in the early sixteenth century, mostly in Rome, Norfolk's tale involves Pope Leo X and the competitive search by Spain and Portugal to bring a rhinoceros to his menagerie. Reviewer Steven Moore ("The Beast in the Vatican," *Washington Post Book World,* 15 September) writes: "Norfolk has expressed admiration for the work of Thomas Pynchon and has a similar gift for displaying a casual mastery of the most outlandish historical materials." Richly convoluted and with an easy, leisurely elegance of style, the novel creates a complete world where details jostle for the reader's attention: "Let it rain powder, a fine volcanic ash to a depth of three inches, in which the tailored shoes and boots of the players make fine, crisp prints. Here—these tiptoed heelless ones—these are Venturo's as he creeps in the back door of the palazzo. Hoofprints record the Orator's cortege as it plods from stable to stable, and these more martial impressions signal a militia of some kind. There are indecipherable scuffs and skids, meaning violence, perhaps, and then there are the looping and overlapping tracks that lead from the palazzo, swerving south of Navona, north of the Campo de Fiori, taking alleys and backstreets westward to where the Tiber bumps up against the city."

Joyce Carol Oates, *We Were the Mulvaneys* (Dutton). One of her best novels (there are twenty-six), this one, told by Judd Mulvaney, concerns the happy family of which he was a part at High Point Farm in Mount Ephraim, New York; how they fell apart in terrible crisis; and what became of them: "The terrifying possibility came to Patrick: our lives are not our own but in the possession of others, our parents. Our lives are defined by the whims, caprices, cruelties of others. That genetic web, the ties of blood. It was the oldest curse, older than God. *Am I loved? Am I wanted? Who will want me, if my parents don't?*"

Janet Peery, *The River Beyond the World* (Picador). Following her highly regarded first book, *Alligator Dance* (S.M.U. Press), a collection of stories, Janet Peery's first novel, nominated for the National Book Award, is set in the Texas border country and deals mainly with the complex, intersecting lives of two women—Luisa Cantu and Eddie Hatch—and covers roughly half a century in its time scheme: "Twenty years should count for something. She had walked the steps between houses as many times as there were stars at night. Antonia said the stars were there in the daylight too, but hiding in the light. How many more were there she couldn't see? That was how many steps she'd made."

Alexandra Ripley, *A Love Divine* (Warner). Not often do we find a reason to honor a work of unabashed popular fiction. Ripley, the author of a series of romances and of *Scarlet,* the daring and hugely successful (and widely trashed by reviewers) sequel to Margaret Mitchell's *Gone With the Wind,* has here produced a large-scale, well-researched, powerfully imagined, deeply moving narrative of the life and times of Joseph of Arimathea. Palestine, Rome, and Britain are summoned up for the setting. The ways and means are those of popular genre fiction—commercial fiction—but the story is solidly plotted, and the telling is perfectly consistent: "Rufus was reluctant to leave the city on the Loire. Joseph saw it, and pretended not to. He knew that Rufus had found female companionship in every place that they had stayed, but he had never offered any information about his carnal adventures and Joseph hadn't asked for any. He was an old man now, over seventy, but he could remember the needs of the young, and Rufus was still in his twenties."

Susan Fromberg Schaeffer, *The Golden Rope* (Knopf). Schaeffer's tenth novel tells the complex and mysterious story of identical twins, Doris and Florence Meek, told in first-person narrative by Doris, who describes herself as the "shadow of Florence, the famous one, whom everyone thinks he loves": "So many people spend their lives trying to become what we were from the moment of our birth, the soul mates, the doubles, the ones who are never alone, never lonely. It didn't last, not for us, but at least we had it."

The Collected Stories of Arno Schmidt (Dalkey Archive). This book is volume three in the Dalkey Archive Edition of the distinguished and influential German writer who died in 1979. There are thirty-five stories from three collections—*Tales from Island Sheet, Sturenburg Stories,* and *County Matters.* Translator John E. Woods warns the reader that he or she "will be entering a most unconventional world, where Joycean ploys do somersaults across the page": "He was already leaning out over his domain, in a cloud of smoke: shoulders the color of baking chocolate, a belly of old copperplate, feet like those of dear departed Queen Louise—famous for the size of hers. (A little shorter than I? Maybe by about the thickness of a tram ticket.)"

Sam Shepard, *Cruising Paradise* (Knopf). It isn't easy for a ruggedly handsome movie actor and Pulitzer Prize–winning playwright to earn serious and favorable attention with a more literary venture, this diverse collection of forty "tales," for example. Reviewing *Cruising* for the *Washington Post* ("Smooth 'Cruising'," 29 July), C. J. Rawlins found many correlations with the tone and the ways and means of contemporary literary fiction but justly concluded that Shepard's delivers far more than low-rent grit and looming despair. True to his gifts, he presents these tales in straightforward and uncondescending ways, with transcendent wit, and the sum is compassion.

Christopher Tilghman, *Mason's Retreat* (Random House). This distinguished first novel by the author of the highly praised story collection *In a Father's Place* tells the story of a family on the Eastern Shore of Maryland beginning just before World War II. Greeted with extraordinary advance praise ("Tilghman's first novel places him securely in the ranks of our most accomplished writers"—*Publishers Weekly*), *Mason's Retreat* was widely reviewed and justly praised. It is my selection for the best first novel of the year: "Patience over the days and weeks and months. Edward was dancing for time through this winter, consumed with work, proud of what he was trying to do, proud of his efforts at home, dancing as spring began with its feathery

Christopher Tilghman (photograph by Marion Ettinger)

buds on the tulip trees, in its warming of the Bay, its sputtering and bubbling of the wet soil."

Barry Unsworth, *Morality Play* (Doubleday). Unsworth, whose *Sacred Hunger* (1991) shared the prestigious Booker Prize with Michael Ondaatje's *The English Patient,* here tells a fourteenth-century murder story, as witnessed and reported by Nicholas Barber, a lapsed priest who has joined a company of traveling players: "It was now that it came to me—a lesson that was to be learned over again in the days that followed—that the player is always trapped in his own play but he must never allow the spectators to suspect this, they must always think that he is free. Thus the great art of the player is not showing but concealing."

John Updike, *In the Beauty of the Lilies* (Knopf). This novel, the prolific author's seventeenth, is a kind of family saga, covering four twentieth-century generations, beginning in 1910 with the Presbyterian clergyman Clarence Wilmot and ending in 1990 with his great grandson, Clark. Large (almost five hundred pages), ambitious, and marked by Updike's special stylistic gifts and his inimitable eye for pertinent details, this book takes as its theme the testing and trials of the American Dream in this century. Alma, the granddaughter (Clark's mother), was, of all perfect things, a movie star: "As ardent on the seventh take as on the first, she felt in [Gary]

Cooper's arms the full edge of her much greater youth and energy and desire. Yet, seeing the first rough cut of the film (they would never let an ingenue see the rushes) . . . she was astonished at how Cooper dominated the screen—at how his leathery face, with its baleful Nordic eyes and slightly frozen mouth, so inert-seeming in the cluttered glare of the sound stage, possessed a steady inner life beside which her own apparition was flickering, nervous, discontinuous."

Bruce Wagner, *I'm Losing You* (Villard). The screenwriter and novelist (*Force Majeure*) comes close to the ultimate Hollywood novel with a tightly written, tightly plotted story that includes (in the words of his publisher) "porn stars in love, scheming dermatologists, cell-phone conversations that never quite connect, and dying men who wear million-dollar watches." Elements of a possible roman à clef are heightened by the appearance of more than one "real" figure: "Although for the sake of verisimilitude, certain public figures do make incidental appearances or are briefly referred to in the novel, I have included them without their knowledge or cooperation. . . . " ("Author's Note"). A variety of styles, all of them relentlessly hip, give the story some crackling energy: "Zev and the boys out by the pool, talking cock. Alfred the Steward long since airborne, black box and portable flotation device intact. There's Yon Koster, the trainer who wrestled muscle from the liposuctioned abs and flabby arms of a classic endomorph: Zev, with that unwieldy, oddly over-developed, dressed-for-success praying mantis Thorax, like Jeremy Irons's in the third *Die Hard*."

Tom Whalen, *Roithamer's Universe* (Portals Press). Poet, critic, translator, and author of short fiction and novels, Tom Whalen is one of the most gifted and prolific of contemporary American avant-garde writers. This novel (R. H. W. Dillard announces in a blurb, "It's the *Moby Dick* of mice!") features the Encyclopedia Mouse who appeared in *The Camel's Back* (1993) and, in the hopeful words of its small publisher, "offers the reader parodies, puns, allusions, monologues, meditations, poems, and a wealth of myth and madness." Whalen is a master of this genre: "*Mouse memory:* When I reached the remains of the Eastern Washington University Library my brothers and sisters had already wantonly consumed the open stacks, leaving this curious gray ball of timidity and tomfoolery without a decent meal. I licked spider webs off me, but little nourishment found I now in this spectral stuff."

Mitch Wieland, *Willy Slater's Lane* (S.M.U. Press) Centered on two eccentric, reclusive brothers in rural Ohio, this powerful and elegantly written first novel has earned high praise. The novelist and nonfiction writer John Keeble writes: "Mitch Wieland is a natural storyteller. His sense of the commonplace macabre is reminiscent of Sherwood Andersen, as is his wise, quiet, remarkably graceful style." A native of Sugarcreek, Ohio, and author of short stories, Wieland now lives in Idaho: "His hands were bone-white on the wheel. He opened each finger one at a time. The Buick sat sideways on the road—if someone rounded the corner he would be killed. He backed from the guard rail and pulled quickly away. The pond beyond the trees shone like a pane of dark smoky glass."

OTHER SIGNIFICANT TITLES

Relatively speaking, the year in fiction may well prove to have been a slow one, a quiet one, especially in terms of literary fiction. Nevertheless, in summing up it is clear that there was a wide variety of novels and short-story collections brought out (by a wide variety of publishers, large and small), some, as usual, by established masters, others by newcomers and apprentices. It was a year for collections of short stories. *New Yorker* writers William Trevor and Mavis Gallant received considerable attention. Trevor's *After Rain* (Viking) has eleven stories, six of which focus on female protagonists. Critic Brooke Allen, writing in *The New Criterion* (November), praised his special skills: "Trevor's extremely unobtrusive writing is, in fact, the finished product of a lifetime's toil. He is one of the rare serious writers who is not afraid to let his characters' voices obscure his own; it is a mark both of his skill and his humanity." Mavis Gallant's *Collected Stories* (Random House) has fifty-two stories in nine hundred pages, coming from a forty-five-year writing career, many of them dealing with the lives of expatriates. Her stories were celebrated by *The New Yorker* as "marked by a Flaubertian elegance and didacticism."

New Directions brought out *The Collected Stories of William Carlos Williams,* and Hill and Wang published *Short Stories of Langston Hughes*. British feminist Sara Maitland was represented by a reimagining of a variety of tales, fables, and myths with *Angel Maker: The Short Stories of Sara Maitland* (Holt). Poet Ted Hughes brought out his first collection of short stories, set in the 1950s and 1960s, *Difficulties of a Bridegroom* (Picador). Joyce Carol Oates received strongly favorable reviews for her latest collection, *Will You Always Love Me?,* as did George Saunders for his first collection, *Civilwarland in Bad Decline,* a novella and stories mostly set in theme parks of the future. Ana Castillo wrote of Mexican American lives and loves of every kind in *Loverboy* (Norton). David Algahori's stories, in *Words are Something Else* (Northwestern) are set in his native Serbia. In *The Sexual Life of Savages* (St. Martin's), Stokes Howell offers a variety of story forms for stories set in rural Missouri and New York City. Mary Morissy's *A Lazy Eye* (Scribners) deals with contemporary Ireland. Fred Pfeil's *What They Tell You to Forget* (Pushcart) was the winner of the fourteenth annual Editor's Book Award. One of the most widely reviewed and highly praised books of stories was Tobias Wolff's third collection, *The Night in Question* (Knopf), fifteen stories written over the past decade. Will Self produced a collection of diverse and satirical stories, *Grey Areas* (Atlantic Monthly), described in the *Review of Contemporary Fiction* (Fall) as "*Saturday Night Live* skits with a Master's Degree."

M. Evelina Galang's *Her Wild American Self* (Coffee House) is a first collection concerned with the lives of Filipina American women. Matt Corry's *Monolith* (Xurban Press) consists of *very* short stories—thirty-one of them in ninety-six pages. Occasionally the big commercial houses bring out books that are openly experimental, like the fifty-two "fictions" that compose David Shields's *Remote* (Knopf). Pulitzer Prize winner Robert Olen Butler created a cycle of twelve stories all based, one way or another, on tabloid headlines in *Tabloid Dreams* (Holt). Holt also published the twelve-story collection of Irish writer Colum McCann, *Fishing the Sloe-Black River*. *Unlocking the Air* (HarperCollins) by Ursula K. LeGuin received mixed notices but was short-listed for the Pulitzer Prize. Gina Berriault's *Women in Their Beds* (Counterpoint) consists of thirty-five short-short stories. The late Shirley Jackson (1919–1965) was represented by fifty-four previously uncollected stories in *Just an Ordinary Day* (Bantam). In "Distress Signals," *New York Times Book Review* (29 December), Joyce Carol Oates maintains "that only six or seven of the stories in the collection merit publication."

One of the best-received story collections was British novelist Julian Barnes's first, *Cross Channel* (Knopf), ten stories free-ranging in time and space. Another first collection, this one by naturalist and nonfiction writer Jonathan Maslow, *Torrid Zone* (Random House), has seven stories all set on the Gulf Coast and some involving historical figures such as Osceola and Jean Laffite. Another important posthumous collection is *Burning Your Boats: Collected Short Stories* (Holt) by Angela Carter, collecting stories from her four published volumes together with six previously uncollected stories, some realistic and some distinctly innovative, as in her retelling of old fairy tales with a contemporary spin. Sun and Moon

Dust jacket for novelist Julian Barnes's first collection of stories

published *Djuna Barnes Collected Stories,* comprising forty-three tales written between 1914 and 1942. Jim Shepard's *Batting Against Castro* (Knopf) and Bret Lott's *How to Get Home* (Blair) earned respectful attention from reviewers. Chinese American writer Ha Jin's collection *Ocean of Words: Army Stories* (Aoland) comes out of the author's own military experience in China.

University presses more and more frequently favor the publication of collections of stories. Among those published during this year are: Lester Goran's *Tales From the Irish Club* (Kent State), stories about the Order of Hibernians in Pittsburgh during the 1950s; Edward Falco's *Acid* (Notre Dame), described by Bruce Allen in *The New York Times Book Review* (21 April) as containing a little of Raymond Carver and a lot of Andre Dubus; Lori Baker's highly experimental *The Lost Scrapbook* (Illinois State); Wendy Brunner's *Large Animals in Everyday Life* (Georgia); Robert Franklin Gish's *Bad Boys and Black Sheep* (Nevada), stories of the West; Ian Stavans's *The One-Handed Pianist* (New Mexico); Layle Silbert's third collection, *New York, New York* (St. Andrew's College Press); Vasily Shukshin's *Stories From a Siberian Village* (Northern Illinois); Botho Strauss's *Couples, Passersby* (Northwestern); and Jerry Bumpus's *The Civilized Tribes* (Akron).

With literary fiction perceived as being in serious trouble and with the lists of literary fiction published by the big commercial houses shrinking, if not yet completely disappearing, the fate of literary fiction is increasingly in the hands of university and small presses. The former have gradually become serious players in three ways: first by publishing original fiction, novels as well as stories; by publishing translations; and by reprinting good works of literary fiction that have gone out of print. Exemplary of the effort to find and to publish outstanding literary fiction is the S.M.U. Press, which published both novels and collections of short stories during the year: *The Woman in the Oil Field: Stories* by Tracy Dougherty; *Circle View: Stories* by Brad Barkley; *Short-Term Losses: Stories* by Mark Lindensmith; *Border Dance* (novel) by T. L. Toma; *Bitter Lake* (novel) by Ann Harleman; *Acceptable Losses* (novel) by Edra Ziesk; and *Willy Slater's Lane* (novel) by Mitch Wieland. The State University of New York Press published two new novels in its Contemporary Continental Philosophy Series, both by David Farrell Kress and both basically biographical novels: *Nietzsche* and *Son of Spirit,* a novel about G. W. F. Hegel's son, Louis, also known as Ludwig Fischer. SUNY Press also published a first novel, *Memoirs of a Terrorist,* by Sally Patterson Tubach. Northwestern brought out *Don Quixote in Exile,* an experimental and autobiographical novel by journalist Peter Furst. Chicago published Franco Ferrucci's *The Life of God (As Told By Himself)*. University Press of New England brought out W. D. Wetherell's *Wherever That Great Heart May Be* and veteran novelist Kit Reed's *J. Eden.* The University of Tennessee Press published the latest novel by veteran story writer and novelist David Madden—the lean and evocative *Sharpshooter.* Georgia published, to the highest sort of critical praise ("One of the best books I have ever read . . . a masterpiece."—Fred Chappell), *The Sweet Everlasting* by Judson Mitcham. Meanwhile, Illinois, Pittsburgh, Iowa, and Johns Hopkins Press, among others, continued to publish original fiction.

Perhaps more important, however, are the reprints, books of quality and merit being brought back into print in trade paperback editions by university presses. For example, the University of South Carolina Press has brought out all but one of the novels of Mary Lee Settle (*Choices* was already in print in a Harcourt Brace trade paperback), as well as her memoir, *All the Brave Promises,* all in uniformly handsome paperback editions, each with a newly written introduction by the author. Johns Hopkins has reprinted the two sequential Civil War novels by Mary Johnston—*The Long Roll* (1911) and *Cease Firing* (1912). L.S.U. Press's major reprint series, Voices of the South, has, at this writing, published twenty-six titles, the latest of which, this year, give a sense of the variety of the series: Fred Chappell, *It Is Time, Lord* (1963); Shirley Ann Grau, *The House on Coliseum Street* (1961); Allen Tate, *The Fathers* (1938); George Garrett, *An Evening Performance* (1985); Sheila Bosworth, *Almost Innocent* (1984); and Evelyn Scott, *The Wave* (1929). North Carolina reprinted *The Bitterweed Path* by Thomas Hal Phillips, described by its publisher as "a subtle engagement of homosexuality and cross-class love." The University Press of Mississippi reprinted a major earlier novel of Elizabeth Spencer, together with (for the first time) six of her short stories set in Italy, *The Light in the Piazza and Other Italian Tales,* with a critical introduction by Robert Phillips.

Returning to the short story, there is the matter of anthologies. *Prize Stories 1996: The O. Henry Awards* (Anchor) is the last of the series to be edited by William Abrahams, who has been the editor/judge since 1967. This year's volume, except for the novelty of awarding the first prize to Stephen King for his story "The Man in the Black Suit," has few surprises in its twenty stories; some good solid tales by established story writers such as Ellen Douglas, Jane Smiley, William Hoffman, Joyce Carol Oates, Alice Adams, and Ralph Lombreglia; and some "discoveries," new names on the block such as Becky Hagenston, Lucy Honig, and Frederick G. Dillen. The competition, *Best American Short Stories 1996* (Houghton Mifflin), had as its rotating editor John Edgar Wideman, whose choices have little overlap with *Prize Stories.* Only Alice Adams and Joyce Carol Oates and Akhil Shamara appear in both, and only Akhil Shamara is represented by the same story—"If You Sing Like That For Me." Combing much the same list of literary magazines, Wideman selected more stories from the little magazines, offering a different galaxy of stars—Rick Bass, Robert Olen Butler, Stephen Dixon, Stuart Dybek, Mary Gordon, David Huddle, Melanie Rae Thon—and the new names and faces are different, too: William Henry Lewis, Jason Brown, William Lychack, Angela Patrinos. The third of "the big three" annual anthologies, *New Stories from the South* (Algonquin), edited by Shannon Ravenel, does offer at least one surprise—a story by William Faulkner which, until it appeared in *The Oxford American* (May / June 1995) as "Rose of Lebanon," had never been published. Ravenel defends "Rose of Lebanon" as being as "new" as any of the fifteen stories in her gathering.

Among the busy host of other anthologies, anthologies of all kinds for all kinds of groups and audiences, I select three for special mention. First, *CrossConnect: Writers of the Information Age* (CrossConnect) is a gathering of recent stories from four electronic issues published online since 1995. There are forty-one contributors of poetry and prose, most of them relatively unknown in print, though there are exceptions. Clayton Eshleman, Aaron Even, Michael Knight, Doug Lawson, David McNair, Helen Norris, Barry Spacks, and Robert Sward have literary reputations in print as well as on the internet. Another interesting gathering of poetry and prose, both fiction and nonfiction, is *The Sacred Place: Witnessing the Holy in the Physical World* (University of Utah Press), edited by W. Scott Olsen and Scott Cairns. This volume includes poems by Richard Wilbur, Charles Wright, Kelly Cherry, Annie Dillard, Cathryn Hankla, and Jorie Graham; and prose by Scott Russell Sanders, John McPhee, Barry Lopez, Rick Bass, Bret Lott, Carol Bly, and Katherine McNamara.

Fascinating in matters of form is *Micro Fiction: An Anthology of Really Short Stories* (Norton), edited by Jerome Stern, a collection of fifty-four stories by fifty-three writers, all of them required to be 250 words or less, the ultimate short-short. Aside from the fun of the form, it is of interest who among living and working American story writers elected to play the game. Some well-known talents are included: Joy Williams, Rick DeMarinis, Pamela Painter, Fred Chappell, Francois Camoin, James Kelman, Stuart Dybek, Russell Edson, Ron Carlson, Ursula Hegi, Amy Hempel, David Bottoms, Padgett Powell, Janet Burroway, Sam Shepard. Stern, the writer and director of the writing program at Florida State University, died shortly before the anthology was published. Rest in peace. Finally, one should make mention of a great writer, one not ordinarily thought of as the author of short fiction, the late Samuel Beckett, whose shorter prose pieces have been collected in *Samuel Beckett, The Complete Short Prose 1929–1989* (Grove).

This was a year which saw (once again) public announcements that the novel is dead or, best-case scenario, has managed to outlive its time and usefulness. An outline and summing up of various pieces on this subject is in David Streitfeld's "But I Saw the Movie...," *Washington Post Book World* (8 September). Streitfeld quotes Arthur Krystal, who wrote in *Harper's* that "there comes a time when one outgrows novels." He cites a special issue of *The Review of Contemporary Fiction* (Spring), "The Future of Fiction," which, Streitfeld writes, "was full of queasy writers worried that they were talking only to themselves." He notes another special issue, this one of the *Hungry Mind Review,* which dealt with the same concerns. Finally he turns his attention to another article in *Harper's,* "Perchance to Dream," by novelist Jonathan Franzen, who "explored at length his angst over a career that no one seemed to care about." Also of interest in this context is "The Novel Is Decaying into News and Narcissism," by Joan Mellen, *Baltimore Sun* (12 May): "Seldom are there large issues in these books (the future of a society, an alternative social order or politics, the way history shapes an individual). Instead they wallow in a dead end of questions of ethnic identity and sexual betrayal and a timid reliance on the suffocatingly personal, as if the experience of the individual, in or out of a family, were the entire canvas available to the novelist. Their version of the news is the news of narcissism."

Nevertheless, living or dying, sick or well, the novel continued to appear and to hold forth on every imaginable subject and in all kinds and forms. There were, for example, two first novels about Amelia Earhart–*Hidden Latitudes* (Scribners) by Alison Anderson and *I Was Amelia Earhart* (Knopf) by Jane Mendelsohn. The latter became a best-seller and jumped from an in-print total of 30,000 to 225,000 as a result of being recommended by radio personality Don Imus, estimated to have ten million regular listeners ("Get Don Imus to Read Your Next Book," *Publishers Weekly,* 21 May). There were several books which dealt with the sinking of the *Titanic. Down With the Old Canoe* (Norton) by Steven Biel is a nonfiction cultural history. British author Beryl Bainbridge told a first-person story through an imaginary nephew of J. P. Morgan in *Every Man for Himself* (Carroll & Graf). Norwegian writer Erik Fosnes Hansen weighed in (to excellent reviews) with *Psalm at Journey's End* (Farrar Straus), which tells the tale from the points of view of seven members of the ship's orchestra. There was also *Maiden Voyage* (Villard) by Cynthia Bass; but readers were spared *Titanic* by James Cameron, which was cancelled in October by HarperCollins. For the record, Jess Lee Kercheval told the same story in 250 words in "Carpathia" (*Micro Fiction*). All of the above (except "Carpathia") occasioned a literary essay by John Updike in *The New Yorker* (14 October) titled "It Was Sad."

Major and minor stars of the contemporary American literary scene had new novels during the year. *The Last of the Savages* (Knopf) was supposed to bring back Jay McInerney to center stage; but, except for the Book-of-the-Month Club's not entirely neutral description ("a mature, reflective and masterful tale of two friends whose lives intersect even as they travel in radically different directions"), it was savaged by the majority of reviewers. Mary Morris's *House Arrest* (Doubleday) centers on Maggie Conover, a writer for Easy Rider travel guides, who finds herself in a lot of trouble involving the disappearance of a dictator's daughter. Jessica Hagedorn, in a jazzy, quick-cutting style, follows the life of her Filipina protagonist, Raquel "Rocky" Rivera, in the Philippines and San Francisco and New York in *The Gangster of Love* (Houghton Mifflin). Louis Begley's *About Schmidt* (Knopf) is the story of the aging corporate lawyer Albert Schmidt and a piece of valuable property at Bridgehampton in the Hamptons. Marge Piercy, in *City of Darkness, City of Light* (Fawcett) takes on the French Revolution and a cast of "real" historical characters. In Ron Hansen's *Atticus* (HarperCollins), short-listed for the National Book Award, Atticus Cody travels to Mexico to try to find out about the life and death of his prodigal son, Scott. A. M. Homes's *The End of Alice* (Scribners), a story of pedophilia and other perversions, succeeded in shocking a lot of book reviewers. Siri Hustvedt's *The Enchantment of Lily Dahl* (Holt), her second novel, concerns a nineteen-year-old protagonist who works in the Ideal Cafe in Webster, Minnesota. Mainly she has an intense love affair with the mysterious New York artist Edward Shapiro.

Ex-con Don Judson's first novel, about an ex-con named Holgate, *Bird-Self Accumulated* (N.Y.U.), is a nonlinear first-person story that won the Mamdouha S. Bobst Literary Award. Jeff Vandermeer's short novel *Dradin, In Love: A Tale of Elsewhen and Otherwhere* (Buzz City Press) received a laudatory notice in *The Review of Contemporary Fiction* (Fall) from veteran reviewer Lance Olsen, who celebrates the cast of characters: "a dwarf tatooed head-to-foot with river maps, creatures called Mushroom Dwellers you just don't want to know too much about, a laughable and sad alcoholic father, a geophagist mother, and a masturbating saint." Jonna Scott's *The Manakin*, nominated for several awards, is a vaguely gothic novel set in and centered around a decaying old mansion near Rochester, New York. Cult hero William T. Vollman's *The Atlas* (Viking) is, in fact, a gathering of fictions, memories, and odds and ends, aptly described by Michael Hemmingson (*Review of Contemporary Fiction Summer*): "These works are not fictions, but essays, or news broadcasts from hell."

Speaking of such broadcasts, much ado was made over a huge (1,079 pages) novel by another cult figure, David Foster Wallace, *Infinite Jest* (Little, Brown), and its failure to be short-listed for the National Book Award or the Pulitzer Prize. Paul West, a leading creator and critic of adventurous, sometimes avant-garde fiction, was less than enthusiastic about Wallace's enterprise: "It is heartening to associate one's self with these preposterous singularities, but this time around the jest is on the author. Some readers will find the concluding 'Notes and Errata' pages, 983–1079, more daring than the novel proper." ("Reams of Acute Modernity," *Washington Post Book World*, 24 March). Another veteran of the avant-garde, Robert Coover, put a new spin on an old-fashioned theme—that small towns of the American heartland may not be quite as wholesome as they outwardly and visibly seem—in *John's Wife* (Simon and Schuster). Award-winning E. Annie Proulx received mixed reviews for her sequence of related tales concerning the life and hard times of an accordion in America—*Accordion Crimes* (Scribners).

Brian Moore's *The Statement* (Dutton) is, in part, a thriller devoted to Pierre Brossard, a former member of the *Milice* of Vichy France, who has been in hiding for forty years and now is hunted by assassins. Matthew Jones's *A Single Shot* (Farrar Straus) was described by its publisher as "a literary thriller." It follows John Moon, who accidentally kills a young girl while hunting. Susan Straight, the author of two novels and a collection of stories, is white but writes successfully from the points of view of black characters and often in their dialects. Her latest, *The Gettin' Place* (Hyperion), earned respectful attention.

John L'Heureux's seventh book, *The Handmaid of Desire* (Soho), is a satirical, academic novel. Another satire, though wide-ranging, all-purpose, is Tama Janowitz's *By the Shores of Gitchee Gumee* (Crown). Elizabeth McCracken's *The Giant's House: A Romance* (Dial), nominated for the National Book Award, is set in Cape Cod and tells of the love affair of Peggy Cort, a librarian, with a twelve-year-old boy suffering from gigantism. McCracken was included in the *Granta* magazine list of the best American writers under forty. Mona Simpson, widely advertised and earning mixed reviews this time, offered her novel of domestic drama, *A Regular Guy* (Knopf), as her first work written in third-person. Paul Theroux published a metafictional autobiography, a blending of facts and fictions in the tradition of recent work by Philip Roth, about a writer who is, as it happens, named Paul Theroux, in *My Other Life* (Houghton Mifflin). This Paul Theroux flirts with despair—"A bank of phones now looks to me like a wailing wall." More traditionally, Louis Auchincloss, in *The Education of Oscar Fairfax* (Houghton Mifflin), presents a biographical novel of upper-class WASP life, told by an eighty-year-old narrator.

At the other end of the artistic rainbow is punk novelist Kathy Acker, who plays riffs on themes of Robert Louis Stevenson in *Pussy, King of* the Pirates (Grove). Judith Merkle Riley's *The Serpent Garden* (Viking) is the story of an ambitious female painter in the courts of sixteenth-century France and Enland. Kit Reed's *J. Eden* (New England) brings together three families from the Upper West Side to spend a summer together in upper New England. Ivan Doig's latest, *Bucking the Sun* (Random House), is set during the Depression in his familiar landscape of Montana and is chiefly concerned with the building of the Fort Peck Dam. Louise Erdrich's *Tales of Burning Love* (HarperCollins), her sixth novel, is set in North Dakota and posits what could happen if the four widows of Jack Mauser were trapped in his red Explorer during a blizzard and pass the time telling stories to each other. Finally, the late Henry Roth's *From Bondage*, the third volume in the *Mercy of a Rude Stream* series, continues the life story of Ira Stigman.

African American writers continued to make an increasing impact on the literary scene. Walter Mosley's *A Little Yellow Dog* (Norton), the latest of his Easy Rawlins series, is set in Los Angeles in 1964 and jumps the gap that often separates genre from mainstream fiction. Terry McMillan's *How Stella Got Her Groove Back* (Viking), dealing with an affluent forty-two-year-old single parent in San Francisco, was a considerable commercial success. Clarence Major's *Dirty Bird Blues* (Mercury House) is the story of Manfred Banks, a blues musician in Omaha in 1950. John Edgar Wideman's *The Cattle Killing* (Houghton Mifflin), his most complex and ambitious novel so far, richly full of sound and fury, is centered on an itinerant eighteenth-century black preacher in Philadelphia. Performance artist Sapphire talks to her readers in Ebonics through the medium of Claireece Precious Jones in *Push* (Knopf). Ernest Hill created the life of Jamie Ray Griffin, a student involved in the early efforts to integrate the high school of Pinesboro, Louisiana, in *Satisfied With Nothin'* (Simon and Schuster). He earned a bit of a tomahawk chop from Patricia Elam Ruff in *The Washington Post* ("An Unsatisfactory Journey," 20 August): "The question is, why did he write the book? If he is truly satisfied with it, then his title is perfect." Jamaica Kincaid's *Autobiography of My Mother* (Farrar Straus) is set in Dominica and is the story of Xuela Claudette Richardson, a woman seething with hatred. It was judged by Cathleen Schine ("A World As Cruel As Job's," *New York Times Book Review*) to be "a brilliant fable of willed nihilism."

Southern novelists continued to be an active part of the literary scene. Padgett Powell returned to the story of Simon Manigault in the short (145 pages) novel *Edisto Revisited* (Holt). Susan Shreve

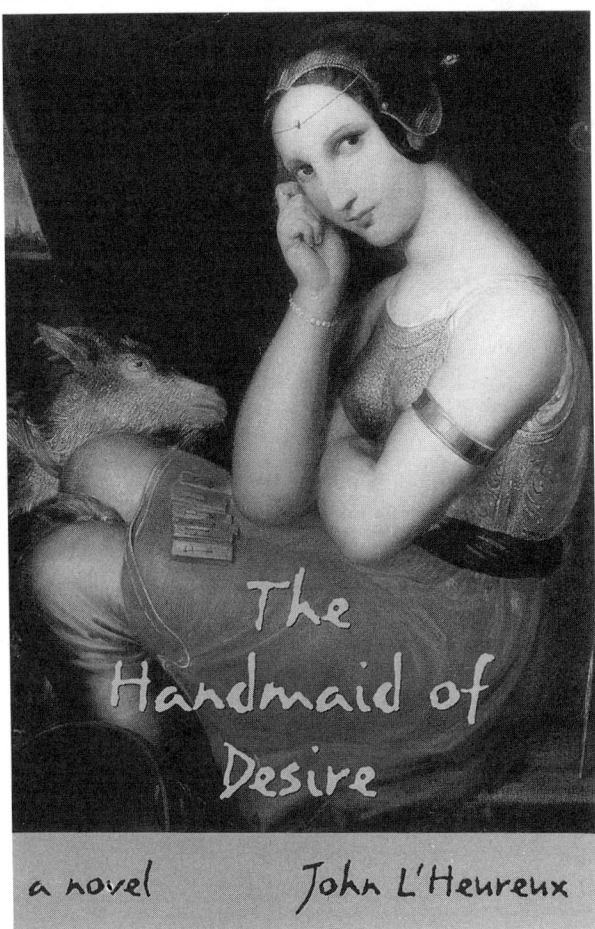

Dust jacket for John L'Heureux's seventh book, an academic satire

set her latest, *The Visiting Physician* (Doubleday), in the Midwest in the fictional village of Meridian, Ohio, telling a kind of thriller involving the abduction of a child. Vicki Covington's latest, *The Last Hotel for Women* (Simon and Schuster), concerns the early days of the civil rights movement and a southern family, the Fraleys. John Gregory Brown tells the story of the life of a black painter in New Orleans in *The Wrecked, Blessed Body of Shelton Lafleur* (Houghton Mifflin). Shelby Hearon's *Footprints* (Knopf) crams a lot of problems and issues into a slender novel, as noted by Robb Forman Dew ("Modern Maturity," *New York Times Book Review*, 31 March): " 'Footprints' takes on an extraordinary number of issues: the ethical and emotional implications of organ transplants, adultery and its effect on marriage, the compromise women make who sacrifice their own career to that of their husbands, the relationship of parents to children over all and in almost every permutation, the importance and implications of women's friendships, the dichotomy of mind versus heart, and more." And all in a frame of 191 pages! Nancy Peacock's first novel, *Life Without Water* (Longstreet), lets Cedar, the child of hippies, tell the story of a North Carolina commune.

Not surprisingly, especially since British entrepreneurs have heavily invested in American publishing, there were an ample number of British titles on the market in 1996. If you expand the limits to include not just British, but English-language, books, the impact is greater, even critically. Of the four works of fiction chosen by the editors of *The New York Times Book Review* as the best of the year ("Editors' Choice," 8 December) only one is a novel and none is by an American writer. *The Moor's Last Sigh* (Pantheon), by Salman Rushdie, is the novel; the others, collections of stories, are: *After Rain* (Viking) by William Trevor, *The Collected Stories of Mavis Gallant* (Random House), and *Selected Stories* (Knopf) of Alice Munro. Not a banner year for Americans.

Doris Lessing's *Love, Again* (HarperCollins) is the story of Sarah Durham, a sixty-five-year-old grandmother and manager of a theater company in London who falls in love again with not one, but two younger men. In *A White Merc with Fins* (Pantheon), James Hawes tells a caper story set in London, described by *New York Times* reviewer Michiko Kakutani (1 March) as "an antic satire of greed, class warfare and sexual high jinks." Kakutani, in her early rave review, calls Salman Rushdie's *The Moor's Last Sigh* "a huge, sprawling, exuberant novel" and "many books at the same time" (28 December 1995). It tells the story (among many other things) of the da Gama-Zogoiby clan and particularly the last of this family, Moraes ("Moor") Zogoiby, who is cursed with a genetic disorder that causes him to age twice as fast as is normally the case. In their citation for the novel in "Editors' Choice," the editors of *The New York Times Book Review* pulled out all the stops: "A biting parody of the family saga novel; a celebration of Bombay in its cosmopolitan years; a masterly rendition of jokey, punning Indian English; a dark assessment of an Indian religious nationalism; and a mordant reflection on the future of serious art."

John Derbyshire's *Seeing Calvin Coolidge in a Dream* (St. Martin's) offers the multinational spectacle of a British author writing the story of a Chinese immigrant in America. Irishman Roddy Doyle tells the story of thirty-nine-year-old Paula Spencer, the victim of an abusive husband, in *The Woman Who Walked Into Doors* (Viking). In Graham Swift's *Last Orders* (Knopf) four men meet at the seaside town of Margate to scatter the ashes of their friend Jack Arthur Dodds, a butcher. Australian Rod Jones set his story in America in 1892 and 1893, using historical characters such as Frederick Jackson Turner and the eponymous *Billy Sunday* (Holt). A. S. Byatt's *Babel Tower* (Random House) is, at the least, highly self-reflexive. Reviewing it for *Washington Post Book World* ("Tongues of Fire," 12 May), Shashi Tharoor allowed: "*Babel Tower* is, in part, a book about a book that is itself about the nature of all books."

Irvine Welsh's *Trainspotting* (Norton) is the story of Mark Renton and his mates, dope addicts one and all, which served as the source for the successful and critically acclaimed movie. Patrick O'Brian's *The Yellow Admiral* (Norton) continued the long, double-digit series telling the story of Jack Aubrey, R.N. Canadian Matt Cohen in *The Bookseller* (St. Martin's) gives an account of Paul Stevens, a Toronto bookseller, who finds himself involved in an obsessive love affair. Fay Weldon's *Worst Fears* (Atlantic Monthly), her twenty-first novel, offers one more close study of the woe that is marriage and the delusions that are love.

D. M. Thomas's new novel, *Lady With a Laptop* (Carroll & Graf), concerns the sometimes hilarious hijinks (a murder also) at a holistic holiday center in the Greek isles. Ruth Rendell appears in person to solve the murder. Quieter and more straightforward is Anita Brookner's fifteenth novel, *Incidents in the Rue Laugier* (Random House), in which a young woman tries to reconstruct and fully imagine the life of her French mother. Emma Donague, the daughter of man-of-letters Denis Donague, tells a story of lesbian love in *Hood* (HarperCollins). *High Latitudes* (Farrar Straus), the fifth novel by James Buchanan, the financial correspondent for the *Times* (London), is a tale of corporate intrigue, told in an eccentric, offbeat way by Jane Haddon, who is trying to save the life of a Glasgow textile mill.

Spanish writers, both Latin American and European, continue to be published and influential in the United States. Argentine writer Tomas Eloy Martinez, with excellent timing, brought out *Santa Evita* (Knopf) a compendium of stories of the life and the strange "life-after-death" of Eva Peron. Peruvian novelist and public figure Mario Vargas Llosa received prompt and serious attention with his latest, *Death in the Andes* (Farrar Straus), the story of two police officers assigned to the mountain village of Naccos and threatened by the infamous *Sendero Luminoso*. The reception for *Death in the Andes* was mixed. Describing it in *The New Yorker* (15 April) as "some cloudy mountain mysticism and a lot of tough talk," John Updike concluded: "*Death in the Andes* is rich fare, hastily and confusingly served. Half as many strands, more fully developed, would have made a better novel." Mexican writer Paco Ignacio Taibo II published a new thriller, *Return to the Same City* (Mysterious). Although his detective hero, Hector Belascorra-Shayne, died violently in the earlier book, *No Happy Ending,* he is here revived and restored by popular demand. *Tinisima* (Farrar Straus), by Elena Poniatowska, is a biographical novel about the photographer Tina Modotti (1896-1942).

Modern and contemporary French writers continue to be translated and published by the University of Nebraska Press. They published Marcel Benabou's *Why I Have Not Written Any of My Book* and Maurice Blanchot's *Most High*. Also from Nebraska came *Slander* by Linda Le, a Vietnamese novelist who writes in French. Danish writer Peter Hoeg, whose *Smilla's Sense of Snow* was a surprise success, followed with *The Woman and the Ape* (Farrar Straus), which did not fare as well. In London a woman named Madelene Burden falls in with a gifted ape named Erasmus, who Robert Bernstein describes in *The New York Times* (4 December) as "supersimian, a cross between Arnold Schwarzenegger and Albert Schweitzer." Dutch writer Harry Mulisch's *The Discovery of Heaven,* set in World War II and after, is told in first-person by an angel. John Updike called it "huge in length, erudition, and wealth of incident." (*The New Yorker,* "Angels in Holland," 25 November).

From Germany came Herta Muller's third novel, *The Land of Green Plums* (Holt) about the German minority in Romania during the oppressive days of Nicolae Ceauşescu. Muller grew up among the German minority in Romania. Possibly trying to duplicate the success of *Smilla's Sense of Snow,* Doubleday published *Blackwater,* a mystery by Swedish writer Kerstin Ekman. At once lyrical, discursive, philosophical, and funny, Czech master Milan Kundera's *Slowness* (HarperCollins) asks, "Why has the pleasure of slowness disappeared? Ah, where have they gone, the amblers of yesteryear?" Another Czech novelist, Josef Skvorecky, produced, of all things, a novel of the American Civil War—*The Bride of Texas* (Knopf).

Jan Potocki, a Polish writer who wrote in French and killed himself in 1815, was Englished for the first time with a novel set in Spain in 1739—*The Manuscript Found in Saragossa* (Viking). Albanian grand master Ismail Kadare wrote of Egypt in the twenty-sixth century B.C. and of the construction of the Great Pyramid of Cheops in a political allegory of his own oppressed Albania in *The Pyramid* (Arcade). Kadare is a potential winner of the Nobel Prize. Egypt's Naguib Mahfouz has already won the award (1988); Doubleday brought out his 1959 novel, *Children of the Alley,* a religious parable which was taken with deadly seriousness by Islamic militants, who called it blasphemy. From the Persian came *My Uncle Napolean* (Mage), a 1970 novel by Iraj Pezesll Kzao about an upper-class Iranian family in the 1940s and afterward.

With the great wall down and the Cold War (at least for the time being) over, more Russian writers are beginning to be known in America. The winner of the first Russian Booker Prize, Mark Kharitonov's novel, *Lines of Fate* (New Press), tells the story of the lives of two men—Anton Lizavin, a doctoral student, and Simeon Milashevick, a writer in the early years of this century who is the subject of Lizavin's dissertation. Vladimir Makanin's *Baize-Covered Table With Decanter* (Reader's International) takes a long hard look at the old regime. Two new translations of Nikolai Gogol's classic, *Dead Souls,* appeared; one, by Bernard Guilbert Guerney for Yale University Press, the other for Pantheon by the distinguished team—Richard Pevear and Larissa Volokhonsky. Israeli authors, as usual, were very much part of the American literary scene. A. B. Yehoshua's *Open Heart* (Doubleday) is set in Israel and India and focuses on a young Israeli doctor—Benjy Rubin. In his eleventh novel, *Don't Call It Night* (Harcourt Brace), Amos Oz offers a pitiless, unblinking view of life in Israel. Yeshayahu Koren's first novel, *Funeral at Noon* (Steerforth), is about life in a small Israeli village in the 1950s.

Asia and the Pacific Rim interest American publishers and intrigue American readers. From Indonesia came *House of Glass* (Morrow) by Pramoedya Anata, which won the $50,000 Magsaysay Award given by the Philippine government. This is the fourth novel by Anata in a series about the days when Indonesia was a colony of the Dutch. From Japan there was Akira Yoshimura's *Shipwreckers* (Harcourt Brace), set in a remote seaside village in medieval Japan and told by Isaka, a nine-year-old village boy. Kobo Abe's *Kangaroo Notebook* (Knopf) was described in the *New York Times Book Review* as "just this side of utter madness, somewhere between Kafka's 'Metamorphosis' and William Burroughs's 'Naked Lunch.' " (David Slavitt, "Weird and Weirder," 28 April). Nobel Prize–winner Kenzaburo Oe, in *A Quiet Life* (Grove), presents a complex personal story which is ostensibly written by his daughter Ma-chan in diary form while she looks after her retarded brother, Hikari. Finally, as if in full circle, we are more and more aware of Chinese American writers and their distinctive points of view. Gus Lee's *Tiger Tail* (Knopf) tells of Capt.

Jackson Kan, U.S. Army, who is sent to Korea in 1974 to search for an American officer who has disappeared. *Mona in the Promised Land* (Knopf) by the popular Asian American Gish Jen is about growing up in suburban New York.

Sequels made news in 1996. Julia Barrett published a continuation of Jane Austen's *Sense and Sensibility* in *The Third Sister* (Fine); and Austen's great-great grandniece, Joan Austen-Leigh, produced *Later Days at Highbury* (St. Martin's), an epistolary novel which follows *Emma*. The exploitation of Margaret Mitchell continued. In May Scribners brought out a sixtieth-anniversary edition of *Gone With the Wind* and also published a South Pacific novel written when Mitchell was sixteen—*Lost Laysen*. Meanwhile, St. Martin's, which had advanced $4.5 million for *Tara*, intended to be the sequel to *Scarlett: The Sequel*, developed "irreconcilable differences" with the British author, Emma Tennant, who was the hired gun for the job. Three previously unpublished short works by Zora Neale Hurston were discovered in a 1925 sorority yearbook and will be published in due course. Coffee House Press has announced its intention to publish a comprehensive three-volume set of Paul Metcalf's *Collected Works*. Metcalf is Herman Melville's great-grandson.

Dictionary of Literary Biography Yearbook Awards for Distinguished Novels Published in 1996

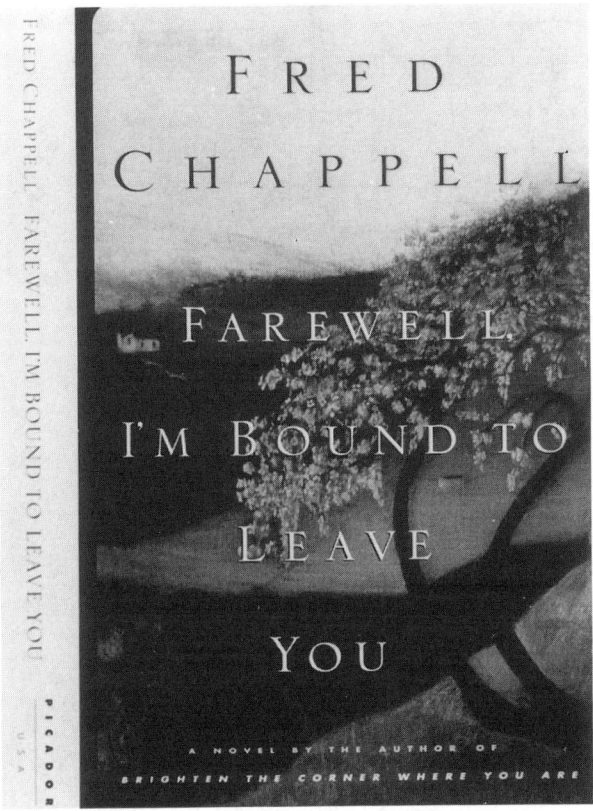

Dust jacket for the third novel in a series concerning the fictional Kirkman family of Wind Mountain, North Carolina

Dust jacket for the first novel by an admired short-story writer, set on the eastern shore of Maryland

NOVEL

Fred Chappell's *Farewell, I'm Bound to Leave You* (Picador), the third of his series about the Kirkmans of Appalachia, is a tightly linked gathering of thirteen tales under one roof and one sky, joining also Chappell's special talent in the evocation of place, the power and energy of the folk art tradition and idiom, and, above all, the delicacy and lyrical sophistication of a true poet's touch. Together with his mastery of many voices and the compassionate credibility of his characters, these qualities render the novel wonderfully and uniquely his own, a large achievement and a major contribution to American letters.

FIRST NOVEL

Highly and widely praised by a variety of critics and reviewers, *Mason's Retreat* (Random House) by Christopher Tilghman is, if anything, better than the sum of all its kudos. Its brilliance and its evocative, elegant style, its appropriately graceful pace, do not disguise nor contradict its power and the sense that these people, a cast of characters worth knowing, matter and that their joys and inevitable sorrows have meaning and resonance. In the end, the experience of *Mason's Retreat* is one of celebration, sadness, and mercy—an altogether extraordinary achievement.

Dictionary of Literary Biography Yearbook Award for a Distinguished Volume of Short Stories Published in 1996

Dust jacket for the volume of stories by the senior editor at W. W. Norton

The eight stories of *Legacies* (Farrar, Straus), by Starling Lawrence, various as they are in subject and situation, subtle and evocative as they may be in the compassionate discovery and development of memorable characters, rich as they prove in verbal texture and feeling, are joined together by an admirable voice, so deeply traditional as to seem brand-new, so careful and controlled as to seem at once stately and spontaneous. In the apt words of the distinguished novelist Stephen Becker: "Lawrence induces a growing and almost breathless tension by the simplest and most refined narration, and only after finishing does the reader become aware of the subtleties that have moved him."

The Year in Literary Biography

William Foltz
University of Hawaii

Bicentenaries are celebrations for most; reviewers, however, dread them. Fortunately, Ian McIntyre's *Dirt & Deity: A Life of Robert Burns* (HarperCollins) explains and does not just celebrate his subject on the two-hundredth anniversary of the poet's death. True, James MacKay's *Robert Burns: A Biography,* which remains the standard work, appeared in 1992, but it was in two volumes. One advantage of McIntyre's work—and it is an advantage—is its size: a compact four-hundred-odd pages that present the life, sane literary analysis, and a funny treatment of Burns. Documentation is light, but the endpapers have clear maps of north and south Scotland.

The Oxford edition of Burns's poetry by James Kinsey comprises three volumes (one of commentary): fortunately McIntyre resists the temptation to go poem by poem and attach the life to the works (in part this was MacKay's approach). Though McIntyre is forced at times to rely on some frankly bad poems, he makes sure the reader knows their faults. Burns's letters are cited at length.

The poet's childhood remains incompletely documented: Burns's later recollections of his sexual rite of passage, his extensive reading, and his democratic social conscience are inflated. What is sure is his impoverished father's insistence that his son be as well educated as possible. The early chapters of McIntyre's book, though brief, provide essential information on two important matters: land tenure and theology; the former allows for interpretation of Burns as exhausted plowman, the latter as religious satirist.

McIntyre gives fine attention to the satiric poems, such as "Holy Willie's Prayer" and "Holy Fair." McIntyre avoids "owlish exegesis," providing instead the biographical and social background. He even provides the text of those poems he calls "ingenious essays in obscenity." Burns, usually, and with good sense, decided not to write sham Augustan verse (his letters are another matter). When he did add the merely poetic to his songs, McIntyre is quick to point out their faults.

Burns's life becomes more thoroughly documented after he reaches the age of twenty-seven (he would live another ten years); records are so sparse on the preceding years that by page 78 of McIntyre's biography Burns has sired two bastards and published *Poems, Chiefly in the Scottish Dialect* (1786; second edition 1791). In the space of a year and a half Burns had two women pregnant at the same time and had fallen in love with the famous "Highland Mary." To the first bastard he assigned his copyright, with his brother Gilbert as trustee. He finally married the second woman, Jean Armour, in 1788, after she had borne him another four children. After marriage, Burns failed as a farmer but succeeded as an excise man, or customs inspector.

McIntyre also quickly reviews the legends about Highland Mary: was she "something of a lightskirts" (W. E. Henley's phrase) or ideal maiden? Did she die bearing Burns's love child? Her name was Margaret, not Mary. Burns may have intended to immigrate to Jamaica with her, but she died of typhus. Further, the two poems written about her in 1786 are in stilted eighteenth-century diction. True, three years after her death Burns wrote a poem to her, but it is maudlin; a week later he is writing bawdry.

McIntyre's last chapter, "Apotheosis," traces the poet and his followers. An amusing argument about Burns's children is documented: HarperCollins and the biographer, following a suggestion of James MacKay, wished to compare a lock of Highland Mary's hair with the infant remains buried in Greenock's Cemetery (the reputed tomb of Highland Mary and her bastard). Since these remains were exhumed in 1920—an extra skull was also found—it seemed reasonable to the biographer to try to match genetic samples. The authorities steadfastly refused permission, however. The Greenock Burns Club proves that all politics is local: it has triumphed over the Convenor of Monuments and a department of health and has the Inverclyde District Council firmly in its sporran.

McIntyre's presentation of Burns's affectional side is balanced and sane, and the biographer has

Dust jacket for the biography published in the bicentennial of Robert Burns's death

made good use of the recent 1984 edition of the poet's letters. These letters, like much of Burns's poetry, can be fulsome, witty, ingenuous, conniving, and indecent; the best are the few in broad Scots. But it was in clear Saxon that Burns expressed his affection for the grass widow Agnes McLehose; he is "Sylvander," who will desert her, the hopelessly amorous "Clarinda." McIntyre takes more than sixty pages to recount this affair. A fat volume of letters to a Mrs. Dunlop also survives—she was maternal; Burns fairly honest. While corresponding with both, he revisited Jean Armour in February 1788. He found her, compared with McLehose, tasteless, vulgar, and mercenary. She was also pregnant with twins and had been turned out of her father's house. Two weeks later he consoles her, however: "Oh, what a peacemaker is the guid weel-willy p[int]le!" Burns finally married her in April. His two lady epistolary friends would learn of his marriage later. Burns's *Common-Place Book* shows him actuated by justice, humanity, and a fear of the afterlife to marry Jean. Whatever his reasons for marriage, they were varied, and McIntyre presents the contradictions of a complex man.

It was not with justice and humanity that Burns treated Jenny Crow, a twenty-year-old servant who bore his child while he exchanged sentiments with McLehose. When she was dying three years later, he asked McLehose to send her five shillings; he had spent six shillings on window repair a month earlier. Jenny died two months later. McIntyre lets the reader draw the conclusion. Burns's wife, who survived him by a whole generation, must have put up with a great deal: she even adopted one of his bastards in 1791. By that time, however, Burns was on the road collecting revenues and inspecting tanneries.

The biography is good on Edinburgh and the "hard-drinking wits and jinks" of its clubs and associations (the phrase is once again from Henley). Burns rose in society as he met his country's literary and scientific notables: the rhetorician Hugh Blair, the author John Home, the moral philosopher Dugal Stewart, the geologist James Hutton, and even the limping six-year-old Walter Scott—"You'll be a man yet," Burns consoled him. Burns's remarks on his tour of Scotland are balanced with the earlier observations of James Boswell and Samuel Johnson;

the two latter, writing about a more traditional Scotland, are more interesting socially. Burns's journal is more lively, as maiden aunts defend their youthful nieces. One virgin survivor, married a month later, had a son who later became Director of the Russian Imperial Mint. This and other similar facts are undocumented and seem irrelevant.

In the last few years of his life, Burns becomes more fuddled than Tam O'Shanter. These years make for sad reading. Whether in his cups or not, Burns deliberately and imprudently courted political disfavor during the time of the French Revolution: like Samuel Taylor Coleridge he was denounced for his beliefs. However, he was buried in a fancy though unpaid-for uniform.

Burns's genius was twofold: as author and redactor. After the second edition of his poems, by which he made between $20,000 and $50,000, Burns's career as a poet had ended, but not his genius. Besides improving "Green Grow the Rashes," "John Anderson My Jo," and "Auld lang syne" he wrote on his own "Flow Gently, Sweet Afton," the patriotic "Scots Wha Hae Wi' Wallace Bled," and the revolutionary "Is There for Honest Poverty." Burns has a lyric for everyone. Readers will agree with the biographer's judgment: Burns "lacks the intellectual ferocity of a Carlyle and the philosophical cloudiness of a Wordsworth and we like him the better for it."

The subject of John Williams's *William Wordsworth: A Literary Life* (St. Martin's Press) is not so much cloudy as nebulous. This biography supplements but will not replace Stephen Gill's *William Wordsworth: A Life* (1989). This is due not to its length, though Williams's work is less than half of Gill's, but its narrower focus. Williams aims not to remove the personal myth, but to disentangle the protean persona of the poems from the actual circumstances of the writer. To succeed he must first examine the often intrusive split between Wordsworth's private and public voice; the latter rhetorical mode first emerged in *Descriptive Sketches* (1793).

He then needs to determine the poet's audience, which can be the single self-conscious muser, the select few, or the great British public. Williams does this mainly with the early shorter poems; he is less successful in his treatments of *The Prelude* (1805) and *The Excursion* (1804), in which he is clearly indebted to Kenneth Johnston's *Wordsworth and "The Recluse"* (1984). The Wordsworth of Williams's biography is a personal, autobiographical poet; Wordsworth may repudiate the social theories of William Goodwin in the preface to the second edition of *Lyrical Ballads* (1800), but Williams does not mention the poet's rejection of an equally important doctrine, the psychological associationism of David Hartley.

Williams's first chapter sets up this theoretical framework. The biography proper, interspersed with literary criticism of future poems, begins with the second chapter. The trauma of the death of Wordsworth's parents was accompanied by the boy's removal to an uncongenial setting; his uncle, a man of strong political opinions, became his guardian. Slowly, the child realized that his father died a Tory and that he must free himself from his uncle's emotional and political control. This picture of a politicized child, prepped for radicalism, rather than the six-year-old darling is, Williams admits, speculative.

Williams reviews the academic and political milieu of Hawkshead Grammar School more convincingly. Wordsworth's education was classical, but the classical doesn't place, as Williams thinks, a fixed obligation on the poet to present his society as the mirror of a sanitized classical world. The Roman Juvenal attacked his society in his satires, as did Samuel Johnson in *London* (1738). Yet Hawkshead did have a reputation for liberal thinking: it had more science and mathematics than Samuel Taylor Coleridge's school, encouraged modern literature and poetry writing, and was surrounded by Dissenters and even Quakers. In such a setting, Wordsworth came to equate writing with liberty and liberalism.

The Wordsworth who left Cambridge in 1791 is either the solitary wanderer or the conspicuous public commentator. Williams wishes to fuse the two as his subject searches for a voice from 1792 to 1797. At times too many Wordsworths are offered, and the reader wonders which one impregnated the twenty-five-year-old Annette Vallon after a few weeks' acquaintance. Was it—and this is Williams's list—the twenty year-old apprentice poet, the sullen political rebel, the devoted nephew touring France, the burned-out wanderer, or the "unhatched" Wordsworth of *The Prelude*? When he leaves her is it because of political events (as a Girondist rather than a Jacobin), the economic situation (he needed a curacy to support Annette's pregnancy), or fear for his life? Williams concludes that all of the above were motivating factors. This is not satisfactory.

Upon his return he met Coleridge, and wrote and rewrote. Poems such as "The Ruined Cottage" freed him from the guilt he felt in deserting both the Revolution and Annette and helped establish a new narrative voice (one separate from the subject's feelings, unlike the unredeemed bleakness of "Salisbury Plain"). After a trip to Germany, Wordsworth would display a new confidence and a new subject.

Dust jacket for the unfinished biography, published a decade after Baker's death

Williams argues the sojourn in Goslar was central for Wordsworth's development as a poet, that it was the crucible of Wordsworth's modernity: there, secure and isolated, a slumber stole his spirit so much that it "could not feel / The touch of earthly years." In this reading of what is usually seen as a "Lucy" poem, both Lucy and Wordsworth's late mother evaporate in the frigid air of Lower Saxony. Two Wordsworths (and his sister, Dorothy) returned to England from Goslar in 1799, one depressed and uncertain about his future, the other confident in his finances and sure of literary success. Wordsworth rejoined Coleridge and formed what Williams calls the coterie, for which *The Prelude* was about to be written. With this new, small audience Wordsworth could rework earlier poems. "The Discharged Soldier" of 1798, becomes part 4 of *The Prelude,* a poem that substitutes the general human condition for the earlier particular political event.

Wordsworth's darkest years, 1807–1815, receive good treatment. They were dark in his finances (Francis Jeffrey trashed his 1807 *Poems*), in his family (two of his children died), in his friends (Coleridge became a moral degenerate), and even in his homes (chimneys didn't draw for months); these misfortunes make Williams sympathetic to what most critics have seen as his subject's growing political conservatism. But that the 1798 Pedlar of "The Ruined Cottage" became the pious and redeemed 1807 Wanderer of *The Excursion* does not really suggest any radical change in Wordsworth's theological and political views, even though the suffering female vagrant of the earlier poem had a radical political fervor one hundred pages earlier. Nor is Wordsworth's hustling for a government job any indication that his social radicalism has changed. Williams's concluding three chapters on Wordsworth's supposed Toryism will prove contentious.

This short biography has many advantages. The author considers poems rarely examined. The unfamiliar "To the Daisy" is used to explicate the selflessness of the subject of "The Happy Warrior," who is surprisingly a poet and also a patriot. This warrior's propensity for "storm and turbulence" slips into the honest, rural toil of the notorious "To the Spade of a Friend." "Peter Bell" is self-parody; its companion piece, which gets equal treatment, is "Tintern Abbey."

The bibliography for the student's guide is too current, however: of the fourteen books listed in "Further Reading," only two were published before 1986. The index is a mystery: a Jebb and a Paley are mentioned along with a William Frend on page 32, none is identified or in the index. If William Wilberforce can make it to the index, why not Richard Heber? Why aren't all the poems that are examined listed in the index? Williams's biography does have its merits, but stick with Gill.

Carlos Baker's *Emerson Among the Eccentrics* (Viking) was substantially finished in 1986, a year before his death. Baker's daughter has helpfully updated her father's notes, supplemented by recent editions of Ralph Waldo Emerson and Margaret Fuller's letters; James R. Mellow, the 1992 biographer of Ernest Hemingway, provides an introduction and epilogue that praise Baker as biographer as much as the biographer praises his subject.

Baker's work makes seamless and skillful work of letters and journals, but since this biography is organized by decades and by characters, it moves by fits and starts. It often starts with an essay on one character, goes forward a few years, and then winds up in reverse as a new character and new chapter start all over. It's as though the reader has to learn of Julius Caesar only from Plutarch's lives of Mark Anthony, Cicero, and Pompey. If Plutarch was Emerson's model after 1820, why not Baker's? Because six hundred pages are not twenty—*Emerson Among the Eccentrics* has the chronological overlap, a fault Baker ascribes to Emerson's work.

This Emerson has no childhood: by the seventh page of Baker's biography he is in his late twenties with a deceased first wife. He quickly becomes an adult marked by extraordinary generosity and friendship to the gifted and the mad, the incompetent and the savant—but not toward his second wife—here the unsatisfactory, shadowy figure of earlier biographies.

These series of lives increasingly and amusingly present evidence of Emerson's charity. If to him shadow and sunlight were the same, as he announced in "Brahma" (1857), the shadow sheltered those in mental darkness. Both the odd poet Jones Very and the glare of Margaret Fuller receive disinterested sympathy from Emerson and from Baker. Margaret Fuller, like Henry David Thoreau, receives four separate and separated chapters; Theodore Parker and Nathaniel Hawthorne, three; Very and Walt Whitman, two. Sixteen other subjects or topics, including the Civil War, get a single chapter.

The chapters on Fuller comprise fifty-five pages and are a splendid example of character study: her letters, her friends, her effusions are as confused as the objects of her unconsummated affection toward both sexes. The reader awaits her sexual awakening. Though the facts have been clear earlier, Baker dramatically and skillfully proposes several potential admirers: the Italian patriot Giuseppe Mazzini and the Polish poet Adam Mickiewicz. The latter's line was that Fuller should lose her virginity (to him) and thereby liberate herself and her sex. Finally, she achieved "liberation" in her thirty-seventh year with Marchese Giovanni D'Ossoli. She maintained silence on the subject, however—she might share the passions of her soul with Carlyle and Emerson, but not the details of her sexual liaison. In this case Baker's narrative delay works well. The child, Angelo, is born slightly comically under the "generative powers of Ceres." Baker has a fine sense of suspense: the reader waits for more than ten pages for Emerson's response, and then Ossoli, Margaret, and the infant immediately perish in a shipwreck off Fire Island.

Only after the reader learns that Emerson sent Thoreau to find out details and any manuscripts does Baker tell of the Ossolis' last days in the siege of Rome in May 1849 a year before his death, and his and Fuller's final month in Florence—with Fuller's premonitions of a babe lost at sea. This is narrative skill at its finest. But how many of the forty-seven entries in the index present new information, especially in the light of Joan Von Mehren's 1994 *Minerva and the Muse,* which treats Fuller's friends and her editorial work at *The Dial* and the *Tribune?*

With Jones Very, Emerson was in a bind: how can one exercise editorial control over poems dictated by the Holy Ghost? Emerson seems to have succeeded with Very, but he would not be as successful with Margaret Fuller and Thoreau. Very's reputation has risen in the last ten years, and readers will note the similarities between his "The Wild-Flower" and Emerson's "Rhodora," although Very's poem is not representative of his Old Testament adaptations and meditations.

Praise of Very leads to praise of William Ellery Channing and then to praise of Walt Whitman. Emerson's discovery and indiscriminate praise of the latter is not that remarkable: he was equally fulsome with Very and Channing. This time he got it right. Nathaniel Hawthorne appears, as he must, and remains aloof from Emerson, though not from Herman Melville. What Baker sees as Hawthorne's "sneers" at Emerson could be plain common sense.

Literary comment is fairly light; there are scattered references to the books and poems (seven pages at a stretch). The essays receive more examination. Better are the excerpts of Thoreau's, Fuller's, and Theodore Parker's criticism of Emerson's writings, and Charles Eliot Norton on his ossified optimism. Readers wishing to know what Emerson thought are better served by Robert Richardson's *Emerson: The Mind on Fire.* (1995). As with any good biography, however, Baker's work changes the reader's opinions and pictures: the vivid Rockwell Kent woodcut of the maddened John Brown was not the freedom lover who first visited Emerson six months earlier: he grew the beard later for the strike at Harpers Ferry.

The later portions of the biography after the Civil War move away from a focus on individuals. These chapters, with the exception of Emerson's visit to California, read sadly: year by slow year, death by death. Baker's telling of the Emersons' trip to Europe and Egypt reads like a nonsatiric version of *The Bostonians:* who they met, where they stayed, what they saw. Lacking Baker's never-written final chapter, "Exeunt omnes," this biography fades as does its subject's voice and failing memory.

Six years after Emerson's death, Nicholas Murray's great-uncle was at Matthew Arnold's funeral in 1888; the great-nephew has written *A Life of Matthew Arnold* (Hodder and Stoughton). But after Park Honan's *Matthew Arnold: A Life* (1981), for whom is Murray's book written? Perhaps for the general reader and the general undergraduate, those who have forgotten their Arnold or don't know him. The biography falls into two nearly equal parts: Arnold as poet and then as critic. This is the usual split and the usual biographical weakness. The poetry section is better: Murray shows skill in weaving the life with some of the poems—particularly in his thirty-page treatment of the Marguerite poems and her lover.

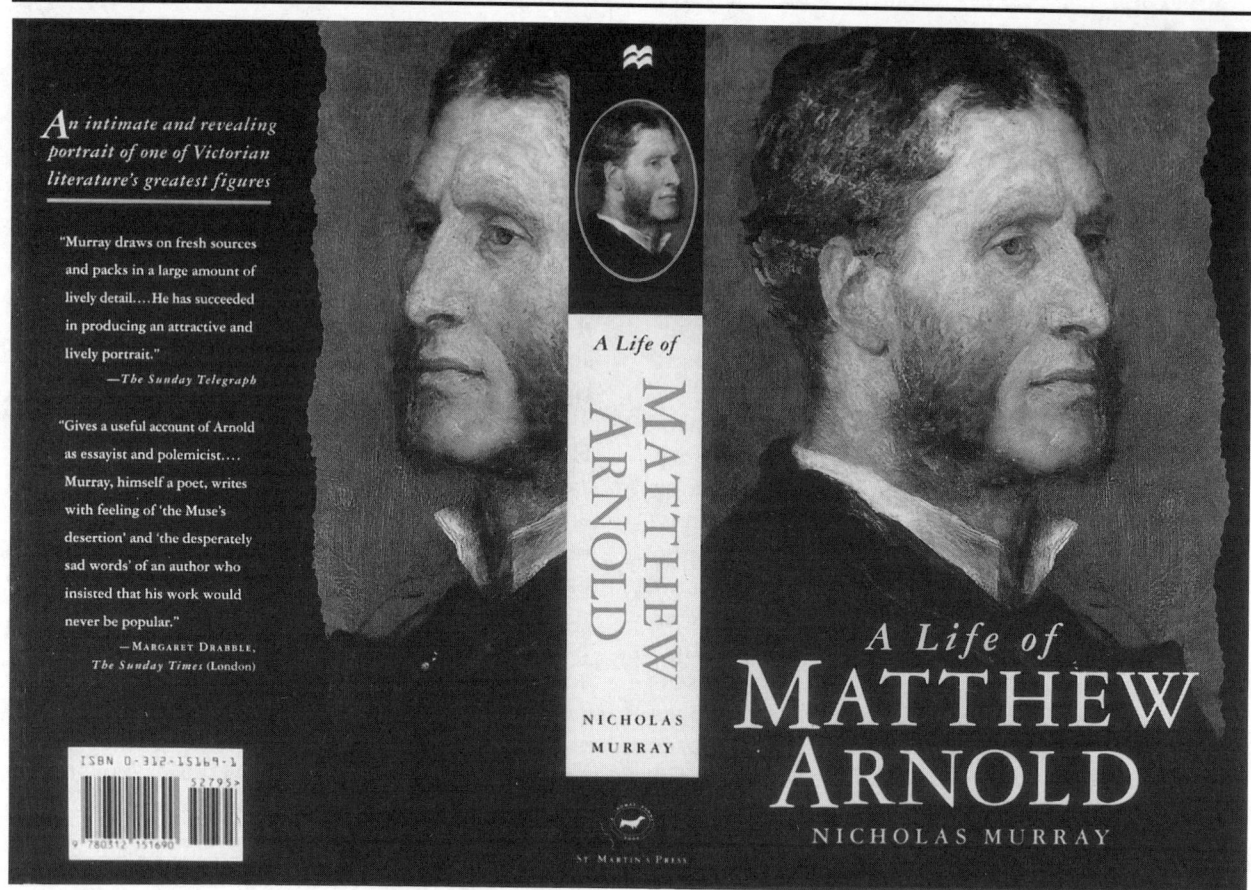

Dust jacket for the biography of the Victorian poet-critic-educator

The last thirty-five years of Arnold's life were fairly uneventful: he wrote; he published; his cook died. This may account for Murray's rehashing the poems and especially the critical works. The chief internal events are family mortality: Arnold's father, Thomas, died in middle age, and some of his children died young. Thomas watched over their beds, correcting examinations and writing a preface to *Culture and Anarchy* (1869) as they died. Arnold had good reason both in youth and middle age to become the most elegiac lyricist of his times. Murray fleshes out these later years with summaries of the essays.

Concerning the Marguerite poems, Murray argues that the general sincerity of Arnold's poetry and his letter of September 1848 to Arthur Hugh Clough establish the truth of an affair that ended unhappily. It was Arnold's fault—he was too sensitive to her spotted past. Arnold also admitted that he lacked two strengths: a "trenchant force" and a "will like a dividing spear." Murray decides that Arnold's vacillations in political emotions were reflected in the unhappy conclusion of this affair. That there may have been an earlier alliance in Paris in 1847 is merely hinted at.

So Arnold married over the initial objections of his wife's family. The biographer does not like Arnold's wife, Flu: she is modified by the loaded adjective "Belgravian," which places her firmly in London's fashionable districts. One of the thirty-eight photographs in the biography is of the Arnolds' joint tombstone. Only her name and her father's can be read: "Third Daughter of the Late Honble Mr Justice Wrightman." After his marriage and for the following thirty-five years Arnold inspected schools and wrote a little poetry, much good criticism, and too much fuzzy theology. Murray is good with facts: in one year Arnold investigated 173 schools, 117 educational institutions, and 368 teachers.

Despite the burial of Arnold's emotional life, his unhappiness in love, and his disgust with English politics, the biographer sees links between his subject's farewell to poetry and his acceptance of new social vocation: both John Keats and education lacked order. Arnold's insistence that Keats lacked order leads to a psychological insistence for national standards—including the suppression of Welsh as a language of instruction. This is a telling point. Murray insists, against his modern detractors, that Arnold was no elitist: he showed compassion for the ragged poor in the face of Parliament's indifference and opposed intellectual pabulum for the oppressed, but he did so with an urbanity that infuriated many readers. Depending how one thinks about educational standards, the detractors may have a point. The biographer admits this: Arnold used Marcus Aurelius to justify force-feeding the masses politically, morally, and philosophically. That Edward Gibbon and John Stuart Mill made a different use of this Roman emperor has escaped the biographer's attention.

Arnold's theological status is like his father's: amateur. John Henry Newman wondered if Arnold was a Christian in the 1840s; years later he was equally puzzled by Matthew's praise (Arnold liked his style). Murray has read the inchoate proto-Bultmann religious essays of Arnold's fifties and the earlier essays on school reform. His summaries will save the student much time.

It is useful to be reminded that Arnold was the first layman to hold the post of professor of poetry at Oxford since 1708, and the first to lecture on English literature, from 1857 to 1867. But Murray is wrong when he states Arnold succeeded his father's saintly contemporary John Keble in that appointment: Keble, who Thomas included among the "Oxford Malignants," lectured from 1831 to 1842.

Murray has made impressive use of family letters at the Bodleian Library, especially those to Arnold's sister "K." This biography is very lightly annotated. Usually this wouldn't be a problem with a poet of few poems, but when an author's prose works run to more than ten volumes, readers expect something. With the often-assigned and perhaps even read *Culture and Anarchy,* readers shouldn't expect full scholarly treatment of his ideas in a four-hundred-page work, but a reference to David Delaura's *Hebrew and Hellene* would have helped, as would more than six pages of excerpts from this central essay. And certainly it should have been made plain that the essay's overcited phrase "sweetness and light" was, as Arnold acknowledged, that of Bishop Thomas Wilson. The reader could also expect some information about those Arnold wrote about: Bishop John Colenso, whose calculations on the amount of plover eggs Israel ate in Sinai would cement all arteries, is merely a name—and is on the wrong page in the index. The religious writings of Arnold are his least-read essays today; but even so the reader never learns, though Murray promises, what Arnold thought of his niece Mrs. Humphrey Ward's *Robert Ellsmere* (1888), that tediously important religious novel. That Arnold wrote only one essay on a novelist (Leo Tolstoy) is not surprising; that he read Charles Dickens's *David Copperfield* thirty years after its publication is. Murray has not established Arnold as "a keen reader of novels."

Literature and Dogma sold ten thousand copies, but Arnold didn't make much money from it. One great advantage of Murray's work is that he keeps Arnold's income firmly in sight: money worries plagued Arnold (or perhaps his wife). He married on £800 ($32,000, in terms of today's economy) a year in 1851; in 1858 he hit £2,200 ($88,000) and bought a house in Belgravia. In 1870 he made £1,200 (almost $50,000); thirteen years later he began his American lectures to pay off his son's gambling debts. His delivery never matched Dickens's 1868 success; he made £1,000 but had to deliver the same lecture twenty-nine times. Still, $1,400 per lecture isn't too bad. His writings never matched his salary as a school inspector (he often rose at 5:00 A.M. to correct papers). Fortunately—and unlike Dickens—the lecture tour didn't kill him, and Arnold moved toward his end not entirely happily—for he recalled his father's early death—but busy and social. Like many a Victorian, Arnold had an intellectually turbulent youth, an introspective affair, wrote some poems, and then settled into domesticity and prose.

Thomas Hardy is almost the reverse of Arnold: unhappy domesticity and unhappy poetry. Some fifteen years ago James Gibson edited a selection of his poems; this year his *Thomas Hardy: A Literary Life* (St. Martin's Press) has appeared. Though he edited the poems, Gibson wisely devotes twice as much space to the novelist as to the poet. Like *Tess of the D'Ubervilles* (1891), Hardy's life is presented in phases: preparation, the novelist, and the poet.

Since Gibson is also the president of the Hardy Society his knowledge of topography is unexcelled. This is, with important exceptions, a two-hundred-page civilized and nonpolemical work: readers expecting a guide to life, or Thomas the Doubter, will, one hopes, be disappointed. Gibson gives no footnotes, merely a reference to a source—and sometimes not even that ("it has been pointed out"). Years of devoted scholarship make his judgments confident; his familiarity with texts is cleverly displayed. The literate insider realizes that Gibson's reference to the "cliff-hanger" of *A Pair of Blue Eyes* means Elfride's petticoats. His bibliography is a simple list. Not surprisingly, Gibson prefers the later editions of the novels; his discussions of some revisions of 1895 and the Wessex Edition of 1911 are clear and telling.

All Hardy biographers have it rough: the writer burned many of his papers and was reticent, especially about his social origins. No letters to his father or mother survive, and servants' recollections are unreliable. Among recent biographies Gibson prefers Michael Millgate's 1982 book, whose chronology he adopts. There is no mention of Martin Seymour-Smith's massive 1994 biography. Robert Gittings's *Young Thomas Hardy* (1975), Gibson asserts, is selective in its evidence, puritanical, and ill-tempered. Gibson may be crotchety, but forgives his subject.

Hardy wanted to be a poet first, but it is from the novels that Gibson constructs his subject's life, which begins with the writer's childhood in Dorset and his love of music, and moves him at the age of twenty-two to London, where he worked as an architectural assistant planning a course of self-improvement: French, music, and painting. He reminds the reader—and such is Gibson's intention—of Jude of Christminster. As with Jude, the flesh called. Perhaps with prostitutes, but more likely—and more like Jude—Hardy learned about sex and class structure from a maternal cousin who was a servant. From 1863 to 1867 he was engaged to Eliza Nicholls, the daughter of a coast guard; Gibson believes that their passion remained unconsummated, as London offered little privacy. It is Hardy's relationship with another cousin that has caused an uproar during the last thirty years. The incest in 1867 with Tryphena Sparks—his cousin at best or niece at worst—produced a baby with the appropriate name of Randy. Relying on Millgate's biography, Gibson rejects the niece theory but accepts the possibility of Sparks being Arnold's cousin: these things happen.

Gibson's portrait of Hardy just before he wrote his first novel is fourfold: architect, autodidact, lower-class man-about-town, and nascent poet. From John Ruskin he learned that architecture was a moral reflection of an organic community. Hardy, who returned to his roots in Dorset in 1867 (the same year he apparently broke with Eliza Nicholls—which Gibson doesn't emphasize), realized that poetry wouldn't pay and started to write the series of novels that assure his fame.

After he started his career as a novelist, the outward life of Hardy was a given: in London for the season, occasional trips to the Continent, the rest of the year in Dorset. In London, Hardy met the literary and fashionable world, relationships that are not developed in this succinct biography. Hardy's out-of-wedlock birth occurs halfway through the book, rather than in the first chapter. His closeness to his dour, protective mother is not immediately apparent. She died at ninety-one; his first wife didn't even go to her funeral. This action, a significant violation of Victorian custom, is not explained. There are other odd events, including the death of Henry Moule, Arnold's childhood friend and mentor. Moule failed mathematics, ruined his chance for a fellowship, and drank; his repulsed advance to Hardy may have driven him to suicide.

Hardy's first wife, Emma, seems to have been a gelid snob. Their initial love and her encouragement of his writing—here Gibson relies on the later poems, insisting that they're not posthumously nostalgic—began in 1870. On one page Gibson suggests that Emma's family conspired to bring the couple together, on the next that her father saw him as a "low-born churl," but that the churl's 1874 success with *Far From the Madding Crowd* allowed them to marry the same year. If Hardy's wife was so cold, one wonders how she reacted to Hardy's first novel, *Desperate Remedies*. Could she have made a fair manuscript of a novel that a contemporary review found full of "the dark ways of human crime and folly" and not be interested? Gibson will not speculate—but he should have.

Gibson does pursue the cooling marriage as it appears in Hardy's writings and through the words of visitors. He tracks its failure in the novels, finding the view of marriage most tragic in *The Mayor of Casterbridge* (1885). A decade after its first publication Hardy changed the novel's "atmosphere of domesticity," to one of "stale familiarity." After twenty years of marriage the Hardys shared the past and, Gibson adds, "Their love of animals." The biographer has some sympathy for Emma, but not much. Hardy was unreasonable about her failure to walk long distances. She seems to have lost her good looks quickly and become querulous, dowdy, and jealous of her husband's attraction to younger, married, and more-civilized women: "Altogether Eastern ideas of matrimony secretly pervade his thoughts, and he wearies of the most perfect, and suitable, wife," she wrote in 1899.

These imagined seraglios included Mrs. Arthur Henniker. Hardy was fifty-three (the reader never learns, irritatingly enough, Henniker's age); and the married daughter of General Pitt-Rivers. Not that anything seems to have come of these attractions: Hardy was not Dickens nor even a Thomas Trollope. There are no illustrations in the book—no pictures of these odalisques, of Emma, of his second wife, or even of Hardy himself.

In 1912 Emma died, and two years later Hardy married an equally querulous woman, Florence Dugdale, whom he had hired as an assistant in 1905. The biographer will not say whether or even when they consummated Arnold's passion: he was her senior by forty years, but still led a "full and active life." Then in 1910 "by design or accident" the querulous Emma met the querulous Florence and hired her as her own secretary. The biographer is too polite toward his subject, but not toward his subject's second wife, whom he clearly dislikes. The last fourteen years of Hardy's life are marked by visitors and volumes of new and recycled old verse. He died in January 1928. His wife had his heart cut out in his bedroom, his body cremated and buried in Westminster Abbey. His heart was buried in the parish cemetery.

Gibson is good with figures. Hardy was initially imprudent with copyrights; he sold *Under the Greenwood Tree* (1872) for £30 in 1871, but by 1875 he got £700 for *Far From the Madding Crowd*. In 1901 he received £400 just from a paperback edition of *Tess*. Multiplying by forty for today's dollar yields nearly $16,000 for paperback rights. By the end of his life Arnold's annual royalties were nearly $250,000.

Sometimes Gibson's rescue of the minor novels seems too generous, but he pulls in other critics to aid him. In one case his judgment seems odd: *The Return of the Native* (1879) is a "noble failure." In another he is portentous: surely Arnold realized love was mankind's predicament in the universe before Hardy did. Gibson's advice, that one should read the youthful *Under the Greenwood Tree* while still young will come too late for most readers. Yet generally one is inclined to agree with the biographer. Hardy does show off when paintings of Holman Hunt and Édouard Manet become tableaus in his novels too obviously.

Gibson realizes he must make the case for the poems that he sees—as did Hardy—as more autobiographical than the novels. He even notes that more than 150 poems begin with the first person—a useful fact. The poetry is serviceably integrated with Hardy's early life. To do this Gibson posits that many poems, though written much later, are nostalgic, and nostalgia creates the most meaningful mode of Hardy's verse.

The novelist became a poet, not because *Tess* or *Jude* was unpopular, but because he could afford to publish his poetry. His early interest in the Napoleonic Wars, the subject of *The Dynasts,* goes back to childhood families, continued in the 1870s as he interviewed veterans. The *Dynasts,* despite the biographer's appreciation, will probably remain unread and not because "modern culture is made up of sound-bites." Gibson, as the editor of the *Collected Poems,* has mastered the bibliographical confusion of his subject. But of the 124 poems in the index, most group themselves too densely in the last fifty pages of this biography. Gibson argues correctly that Dorset was best for Hardy. His best novels come alive there, and the reader hears the dialect of William Barnes; the urban setting and speech of *A Laodicean* is a failure.

Bernard Shaw was a contemporary of Hardy's but knew, or made himself known to, many more literary and political figures. The reader is told twice in Stanley Weintraub's *Shaw's People: Victoria to Churchill* (Pennsylvania State Press) that these twelve essays, though they have appeared in print before (except the one on Winston Churchill), have been revised and are the result of thirty years' scholarship. This is true: 150 pages of periodical publications have become a book some 100 pages longer. Weintraub's scholarship is assured; his publications on Shaw include a guide to research and a study of his subject's art criticism; he also edited Shaw's autobiography (1969–1970) and diaries (1986). The illustrations in *Shaw's People* are admirable.

The biographical sketches collected in this book, for the most part, are either expanded annotations, the detritus of research that couldn't appear as extended appendices, or succinct distillations of earlier work. The great advantage of these studies is their meticulous documentation; their disadvantage is their occasionally peripheral nature. This is not true, however, of the two fine essays on T. E. Lawrence and Frank Harris. The former receives a new emphasis in sixteen pages: the trial press run of *The Seven Pillars* (1922) served Shaw as an emotional source for *Saint Joan* (1924) and even provided some dialogue. Even more importantly Lawrence is not only the model for Private Meek of *Too True to Be Good* (1932), but also for its Aubrey Bagot. Shaw's treatment of Frank Harris, his coeval in fame during the late 1880s and early 1890s, is a bit uneven. Harris, as editor of *The Fortnightly Review* and a year later *The Saturday Review,* was the first to assign Shaw to cover music and then drama. Shaw happily did so until 1898, just before Harris went bust and became a literary confidence man and perhaps a blackmailer. Toward the end of his life, Shaw attempted to help Harris with miscellaneous literary projects that were often demoralizing and even trivial.

Shaw appears at his best in his relationship with Sean O'Casey: lending him money, cosigning a lease, and finding a producer to film *Juno and the Paycock*. He also tried, unsuccessfully, to quiet a man almost as cantankerous as Frank Harris. The biographer notes O'Casey's indebtedness to *Heartbreak House* (1921) and *Major Barbara* (1905), and presents an eloquent defense of Shaw's poetic elements.

Shaw's championship of the disgraced Oscar Wilde results from his fellow countryman's encouragement of the young and unknown Shaw. After Wilde's exile Shaw suggested him for membership in a proposed Academy of Letters (the other playwright he pushed was Henry James). Weintraub's argument that Wilde's earnestness moved into *You Never Can Tell* (1895) is convincing. The friendship between another Celt, William Butler Yeats, and Shaw was uneasy and ambivalent, as was their literary relationship. Yeats did help Shaw get a banned play performed in Dublin, but later exiled him to some peculiar phase of the moon in the astrological speculations of *A Vision* (1926).

Weintraub's speculation that the Molly Bloom of James Joyce's *Ulysses* (1921) was initially Shaw's Sally Lunn in *Overruled* (1916), an obscure play, reveals the biographer's unmatched familiarity with Shaw and Joyce. Weintraub also excavates the substrata of Joyce's *Finnegan's Wake* (1939) to discover Joyce's hidden, if not obscure, admiration. They never met. Nor did Shaw meet H. L. Mencken, though seventeen years after Mencken's adulatory *George Bernard Shaw: His Plays* (1905), the Baltimore journalist did walk near Shaw's flat, but by that time he was no longer a Shavophile. Mencken had become responsible for what Weintraub sees as the typical American idea of Shaw: a "satirist with a slapstick."

Three minor essays flesh out this book. One concerns the actress Edith Adams, who assisted in her husband's suicide in 1893–(he lives on as Lord Dubedat in *The Doctor's Dilemma* [1906] as does she as Jennifer. Then there is Gen. William Booth, with obvious links to *Major Barbara* [1907], and less obvious: Weintraub has used the Salvationist archive at Texas. There is also the insufferable Seigfried Trebitsch, Shaw's nonidiomatic German translator, whom Shaw put up with for half a century. Even the biographer is at a loss to explain this situation.

More than thirty pages, the longest and previously unpublished essay, is devoted to Churchill. Despite an almost sixty-year political antipathy, Churchill and Shaw remained admirers of each other's style. They knew each other in Edwardian days and agreed on the misconduct in World War I (though not on the war itself). Politically, Churchill appears the more sane of the two men: he recognized Shaw as thinker and clown; Shaw himself acknowledged that his comic talents injured his economic proposals. Despite Weintraub's sympathy he cannot ignore the extent to which Shaw let himself be duped by Joseph Stalin. By the end of this essay Shaw recounts summaries of the more-clowning suggestions of this octogenarian.

As Churchill ends this volume, Queen Victoria begins it. The chapter on Victoria, though indebted to Weintraub's *Victoria: An Intimate Portrait* (1987), suggests that the domestic turmoil of the young British queen is seen as a Cleopatran Victoria's idolization of a Caesarian Melbourne. More amusing is Victoria's representation in Shaw's play *Man and Superman* (1903), some two years after her death. The play exhibits what Shaw called "an awful and majestic spectacle," an English lady in mourning. Victoria last surfaces in *The Apple Cart* (1928), in the reign of her grandson, George V. His son's reign saw Shaw's death.

John Davidson is known mainly as a poet. John Sloan's *John Davidson, First of the Moderns: A Literary Biography* (Oxford) confirms that reputation despite the biographer's knowledge of Davidson's plays and novels. This is not the poet of J. Benjamin Townsend's *John Davidson: Poet of Armageddon* (1961), a poet whose reputation still suffered from William Butler Yeats's personal animus: Yeats excluded him from *Oxford Book of Modern Verse* in 1936. Nor is this Townsend's irritating displaced Scot and Nietzschean: instead Sloan presents a hardworking, underpaid, hypersensitive, touchy, tenderhearted but rude professional writer in three hundred densely factual pages.

Davidson, despite the assistance of Shaw; the encouragement of George Gissing, his exact contemporary; and a Civil List pension (£100, and then only in 1906 when he was almost fifty), committed suicide. Despite Sloan's sympathy, he has written a biography of a man who was ill-mannered for little reason, self-pitying, and invariably imprudent in self-defense. Davidson's letters to his sister-in-law, unpublished material of his friend Rudolf von Leibich about his years in Scotland and entry into London's literary scene of the 1880s and 1890s, a thorough knowledge of late-Victorian periodicals and presses, Davidson's Reader's Reports, and even obscure church pamphlets have all made Sloan's book a well-documented biography, though it is an expensive one—priced at seventy dollars.

Whether Davidson was the first of the moderns depends on a definition of modernism, a term that, like Romanticism, requires discrimination. Sloan's definition includes a democratic subject matter, especially the rootlessness of the dingy urban scene, and a new poetic idiom that describes it. To this he adds the use of fragmentary allusion and literary montage. If this sounds like T. S. Eliot's *The Waste Land* (1922) it is: Eliot becomes Davidson's heir. Eliot's preference for *Thirty Bob a Day* fixed Davidson in anthologies. Sloan wants to expand the canon and adjust the life.

Davidson's father was a dissident Congregational and Sabbatarian minister, his mother severe.

Much of his youth was spent in the urban squalor of Greenock. Well-read in Scott, Thomas Carlyle, William Shakespeare, and John Bunyan, Davidson spent his childhood in bleak poverty; these conditions, united to the rigors of late-nineteenth-century religious debate, led to his breakdown at eighteen: like John Stuart Mill he had to leave his father's creed. In this event, Sloan argues, are revealed the stirrings of Davidson's later belief of the exceptional man.

At Edinburgh University, well trained in straight philology but bored, he dropped out of college and began teaching. For ten years he held various posts in Scotland, met not only the crazed Algernon Swinburne, whose assistance he, too, optimistically expected, but Oscar Wilde's friend, John Barlas. Davidson was finally fired, for no good reason except his temperament, from a post in Perth. By that time he had fallen in love, slept with, and married Margaret McArthur in 1885.

They went into debt; she became pregnant. Davidson wrote his first novel, *The North Wall* (1885), an apparent attack on the sentimental novel rather than on aestheticism; finished an improbably titled play *Smith: A Tragedy* (1888); began a music-hall farce, *Scaramouch in Naxos,* whose use of literary montages, rather than Victorian pantomime, anticipates modernism and Hugo von Hofmannsthal's *Ariadne* (1912).

In 1889 Davidson arrived alone in London with manuscripts of poems and three hundred copies of his plays he had had privately published in Greenock (his wife, temporarily left in Scotland, was pregnant again). Initially, only George Meredith seems to have liked the volume; Meredith got Davidson a job reviewing for the *Academy* and the *Speaker*. The *Academy* reviewed his *Plays,* and Davidson realized he had to support himself with journalism. His professional life was chronicled by his friend George Gissing in *New Grub Street* (1891); his private life was that of *The Whirlpool* (1897). Unwin published an edition of the *Plays:* it sold only fifty copies.

His earlier meeting with John Barlas and newspaper contacts made him a ghostwriter for the fortunately obscure Charles James Wills's novels. Hoping to survive by ghosting, Davidson sent for his family to join him in London. The idea was not unreasonable, despite London's expense: his *Perfervid* (1890) got better reviews than Wilde's *The Picture of Dorian Gray* (1890). Davidson wrote one-third of Wills's *His Sister's Hand* (1892), for which he received £15; Wills received £200. He met Oscar Wilde; joined the Rhymer's Club in July of 1890, when the club was still Celtic; and translated Montesquieu's *Persian Letters* in 1891.

At the Rhymer's Club meetings Davidson kept his independence but not his temper: to him Yeats was a pemmican-watercress eater who couldn't hold his liquor; Ernest Dowson, a rosebud poet; and Lionel Johnson, an absinthe anorectic. The Scottish Davidson was not welcomed but got back at the club in his *A Full and True Account of the Wonderful Mission of Earl Lavender* (1895)—and the club members got back at him. Yeats could not understand Davidson's search for a subject matter and called him, in print, a young poet whose heart was "full of an unsatisfied longing for the commonplace." Davidson called Yeats's early poetry "effeminate pedantry," but Yeats blamed and never forgave Davidson for an insult he never wrote—an anonymous reviewer in the *Daily Chronicle* proclaimed of Yeats's *The Countess Kathleen* (1892) that the poet's verse would never "linger in the memory." Outliving Davidson by thirty years, Yeats effectively buried his posthumous reputation in his *Autobiographies* (1938): "And now no verse of his clings to my memory." Not until T. S. Eliot's "Tribute" of 1957 was Davidson reread seriously. Rather than join the Celts in an 1891 anthology, Davidson self-published his first remarkable book of poems, *In a Music-Hall*. It was largely ignored.

Davidson's next and best-known work, *Fleet Street Eclogues* (1893, with a second series in 1896), was begun shortly after his father's death in 1891 (Davidson inherited little and was embittered). Sloan's speculations on the connection between Davidson's father's death and his new poetic voice are cogently argued. His argument that Davidson's reading of Gibbon produced in his novel a London that is the Rome of the late Roman Principate is convincing. The combination of Gibbon's pessimism, the death of Davidson's father and sister, his constant underemployment, and his sensitivity to slights all combined to his belief in hereditary decline. Davidson would often rent a room to write in peace; other times he would simply disappear. He found it difficult to be a husband, a father, and a writer in the same house.

Davidson's relations with John Lane of Bodley Head, who published the *Eclogues,* marked an improvement in his fortune: three hundred copies were published and well reviewed; the reviews were reprinted in a second edition a few months later. Davidson, ever the reviewer, soon and anonymously praised other Bodley Head titles and his friends in the *Speaker*. Was this puffing? Sloan suggests mildly that it was not, yet later admits Davidson's *St. George's Day* was "not unlikely" aimed at procuring its author the vacant laureateship. (Alfred Austin got it instead.)

The first half of Sloan's biography covers the next few years until 1897, which were Davidson's high point of fame, compensation, and quarrels over compensation with Lane. His acquaintance came to include Edmund Gosse, Max Beerbohm, and the artist William Rothenstein. Then arrived the famous *Yellow Book*. Sloan rehearses the controversies over this periodical's inception. To the first issue of five thousand copies Davidson had contributed two poems; Aubrey Beardsley, a drawing; Beerbohm, an ironic essay. The reviewers contributed their scorn (fueling Lane's success). The second issue included Davidson's often and only anthologized poem, "Thirty Bob a Week," whose diction reflects that of the music hall. Lane then republished Davidson's plays with Beardsley's illustrations.

Ballads and Songs followed in 1894 (Davidson received £20 for the manuscript); it was well reviewed and sold well. Davidson tried his hand at drama again, as a translator, as an adaptor, and as a playwright. His first performed play, a Balkan melodrama, met with some success but with Shaw's disapproval: it was written in blank verse. His later plays, since they ignored popular taste, failed. In 1906 Shaw, for unclear reasons, offered to finance a play, but this project also died. Shaw insisted that Davidson write it in prose and condemned the attempts Davidson sent him. In Davidson's view, Shaw was simply jealous. Unfortunately the manuscript of the play was destroyed after Davidson's death. Shaw was probably correct in his assessment, however; Davidson's revisions and adaptions were also losers: Alphonse Daudet's novel *Sapho* (1884) became *Fanny Legrand* and was never performed.

The second half of this biography records Davidson's fall into poverty again, his sensitivity to the point of paranoia, and his manic-depression. Davidson believed he suffered from late-nineteenth-century neurasthenia, a morbid nervousness brought on by social rebellion in a degenerate age. This rebellion figures in his later poems that call for a firm leader; what was Carlylean moved toward the Nietzschean. Sloan, convincingly, finds this new strain prefigured by his patriotic poems and his temperament and not simply the result of reading Nietzsche secondhand.

Though his early journalistic style is overblown Carlyle and Johnson, and too much of his 1890s verse looks back to the practice of Scott and the Pre-Raphaelites, it was his subject matter, the anomic city, which was new and modern. His pastoral treatment of the urban is ironic, inverted, and therefore modern. Sloan provides close readings of many poems; an especially fine reading of *Ballad of a Nun* makes it more than a slap at Adelaide Anne Proctor's *A Legend of Provence* (1858). By the new century a newer theme appears in Davidson's *Testaments:* the overreacher who alters history.

The last three years of his life were spent at Penzance, reviewing books, revising his *Testaments* and impossible plays that attacked Christianity while praising Genghis Khan, Tamerlane, and Attila. Critics attacked him: the *Academy* wrote of his "ghastly and bumptious enormity" and of his "dangerous and stupid twaddle." Then he fell ill with the flu. Sloan's account of Davidson's death and its aftermath is masterly: the poet shot himself while at sea, but his body wasn't found until five months later, in September 1909. Headlines announced a mystery; Arthur Conan Doyle, an old friend, told the family that Davidson might have committed suicide, but he was not sure. The poet William Watson claimed in *The Times* that Davidson was another done-in Keats. Sloan suggests his suicide was the final assertion of an imperious will.

Sloan is good with expenses, better than Davidson, who willfully failed to understand what a contract is. Thirty bob a week is what Davidson claims to have lived on. Yet—and Sloan doesn't see this as extravagance—he would spend half a guinea for a supper with a good claret. That he brought his own cigars with him and chose the "most appropriate" cognac probably did not console his family. One of his last things he wrote was a menu in French.

In the late 1890s Davidson was making six pounds a week, a schoolteacher's salary, but he had to support his mother and sister, pay school tuition for two boys, and assist his brother. Hence he wanted a pension: if William Watson ("A garden is a lovely thing, God wot!") could get £100, why not himself? The Royal Literary Fund came up with a £250 grant in 1898 (thanks to Gosse). The prime minister thought that £50 a year would be enough, but he finally received £100 which, Sloan argues, conferred neither esteem nor independence.

Sloan's book is valuable in presenting the 1890s beyond Wilde and the Decadents. This, after all, was the decade in which George Gissing suffered and wrote; the publisher John Lane's aspiring lower-class Englishman bought books; Tess's child, Father Time, died in *Tess of the D'Ubervilles;* and the nineteenth century ended.

If the *Duino Elegies* of Rainer Maria Rilke represents this century's major poetic achievement, a major biography is a necessity. Ralph Freedman has provided one—of more than six hundred pages—in *Life of a Poet* (Farrar, Straus and Giroux). The biographer succeeds in linking Rilke's early poems to his later. The reader observes the poet's gradual development, radically altered by changes in tone, diction, and subject, an alteration that springs from Rilke's changing eye and ear, his change of residence, and his changing emotional alliances.

Place is the most confusing element of Rilke's life. In this area, Wolfgang Lippemann's 1981 biography has a slight edge, with a chronology that tracks the poet from his birth in Prague, to various Bohemian castles, Venetian palaces, Parisian apartments, and Spanish resort hotels, to his final Swiss tower—Rilke embraced places like a faithless lover. Freedman's scholarship is comprehensive but never intrusive, and the documentation is thorough. Rilke probably wrote too many letters, but this biographer has read them all and cleared out the archives. Freedman is aware that his subject's letters, like his poetry, are both honest and fictive confessions.

Rilke's problems begin early. Born between two classes, his mother, Phia, used to dress him as a girl. The family decided to send him off to a military prep school, effectively cutting him off from the literary events of his adolescence. Freedman has little sympathy for Rilke's mother, an inchoate feminist shipwreck on the shoals of pious sentimentality. Rilke's childhood experiences adumbrate the themes of his later poetry, the masculine military (death) and the feminine poetic (eros). Both served him well.

His *Cornet* (1900), which transmuted his schooldays into an early prose poem of chivalrous glory, kept him from the front after the failure of German offensives in 1915: Rilke sought deferment as a cultural monument. He was called up, edited dispatches in Vienna six hours a day, dined well at dinner parties, and served six months.

The feminine was sought and often found in a series of women whom he later exiled from his bed, but not from his correspondence. Even his first love (but not his first experience; there was a servant earlier), Valerie von David-Rhônfeld, remained devoted to his memory. In 1893 she was his muse, briefly his fiancée, and then the financier of his first book (she sold the family lace). The book was dedicated to a noblewoman, but Valerie got an inscribed copy consecrated with a signed poem—a poem Rilke recycled for another woman eight months later. The noblewoman never replied.

For years the chief among these feminine adorers was a Petersburg resident, Lou Andreas-Salomé, whom Rilke met in 1897, when she was thirty-six and married to a prominent Iranist. Their marriage was never consummated. Freedman says she "made of eroticism a spiritual iconography." Though the meaning of this statement is unclear, it is evident that this happy Oedipus (now calling himself Rainer rather than Maria) found a lover, an audience, and a confidante. Andreas-Salomé also provided entrée to Russian artistic circles for the first of Rilke's two trips there; under her tutelage he saw Russian art. His poetic skills would become pictorial, particularly in *The Book of Monkish Life* (1899), one of a series of *Books (of Pictures, of Pilgrimage, of Poverty and Death)*. He also met twice with Leo Tolstoy, who was more interested in Andreas-Salomé's Persian-speaking husband than in an obscure German-speaking Czech.

Rilke also formed other odd and enduring links in Moscow and Petersburg. The young Boris Pasternak was relieved to see Andreas-Salomé and Rilke depart after they called on his father, Leonid. In 1926 Rilke would praise the son's poems to the father, then in exile in Berlin. Among this biography's forty photographs, Leonid's portrait sketch seems best to capture the brooding writer. Boris, older and impressed by this praise, asked Rilke to send the *Elegies* to Maria Tsvetaeva, a Russian expatriate poet who reminded Rilke of his earlier trips to Russia. He wrote her a poem that she shared only with Pasternak. She also offered herself to Rilke in hyperthyroid euphemisms. He demurred.

After their second trip to Russia, Rilke broke with Andreas-Salomé; she later claimed that her emotions for Rilke were too much for her. During a trip to Florence, fleeing Andreas-Salomé, Rilke met and ingratiated himself with Heinrich Vogeler, who provided entrée to the art colony Worpswede, outside Bremen. There in 1910 Rilke married Clara Westhoff, a sculptress and student of Auguste Rodin, and the event led to Rilke's division from their mutual friend Paula Becker and further from Lou Andreas-Salomé. Freedman argues persuasively that as Andreas-Salomé led him to the pictorial, Clara via Rodin led him to form.

Rilke's family finally decided that such events were not the acts of a college student and cut his allowance. As a married couple Rilke and his new wife were doomed: husband and wife, he wrote, "should guard each other's solitude." But their marriage survived as something else. Freedman sees it as a continuing epistolary event of distance closeness—a paradox, but so was Rilke. The couple, centered in themselves and their work, neglected and nearly deserted their daughter, Ruth. Seeing Paula Becker's death in 1908 as the passing of another sacrificing and self-centered artist, Rilke wrote *Requiem for a Friend*. If he learned form from Clara, it was Becker's love of Paul Cézanne that gave color to his verse.

In Paris, after finishing a monograph on Rodin in 1912, Rilke ran away from Clara back to Andreas-Salomé, who kept him at bay two years with letters. Rodin had fired Rilke as his secretary, but as the poet's reputation grew, Rodin took up with him again, but the now-prominent Rilke broke with the sculptor in 1913. The ostensible cause was Rodin's broken promise that Clara do a bust of Rilke; the real reason was—and again Freedman's honesty is impressive—that the sculptor's refusal injured Rilke's self-esteem. He was also trying to divorce Clara at this time.

There was an affair with Marthe Hennebert when Rilke was thirty-six and she sixteen. Was he Pygmalion or Humbert Humbert? Was she a sex object who became a surrogate daughter and, like his daughter Ruth, an emotional and physical exile? He certainly saw both infrequently: he was away from Hennebert for an eight-year gap, from 1911 to 1919; and he missed his daughter's wedding in 1922 (but lit four candles for her in his favorite hillside chapel). He even seems to have drawn on Ruth's education funds. Ruth, however, like so many of his women, would dedicate her life to his memory.

There were other women. Most were lovers at first, then confidantes. In 1912 Helene von Nostitz first recognized the poet's eroticized landscapes. For a few weeks he was with the actress Hedwig Berhard, then as 1914 began with the pianist Magda von Hattingberg. He began to write poems to her nightly; this affair lasted three months. Freedman lists them all, not as items on a scorecard but to explain how the fervor of the lover's letters became the intensity of his subsequent poetry.

Later there was Loulou Albert-Lasard, an artist whom Rilke met in a Bavarian spa in August 1914. Her angry memoirs (1952) fill in the war years. Both Albert-Lasard and Rilke were able to persuade her husband to rent a separate apartment for her: this way the two could be alone but not on their own. This becomes and always was Rilke's pattern: a muse, but only on call.

Also on call and occasionally living with him (until his cleaning lady objected) was Baladine Klossowa, who found him a Swiss tower to dwell in. After 1920 she replaced Andreas-Salomé. Like the other women in Rilke's life, she was both the inspiration and obstacle to composition. She was the emotional midwife of his *Elegies,* and even of *The Sonnets to Orpheus.*

Rilke's social climb began with the aristocratic patronage he came to expect as he took refuge in the castles of the great. In 1909 Princess Marie von Thurm and Taxis was in her fifties, Rilke in his thirties. In a few months, she became his mother confessor, succeeding in part Andreas-Salomé. The princess offered him shelter and good food at her own castle on the Adriatic. She was not the only one to extend these invitations: Rilke lived like an academic superstar cosseted by patrons rather than foundations.

The reader can get an almost clear idea of Rilke's response to these women by comparing the letters announcing the completion of his *Duino Elegies* in 1922. To Klossowa he's romantic, caressing the tower of Muzot he rented on a sweetheart lease; to Andreas-Salomé he is distant, factual, and self-quoting; to Princess Marie he recalls the hurricanes of his spirit first felt at Duino, her castle. This is one of the few times the biographer's consistently detached sympathy swerves toward the great poet rather than toward his confidantes.

It was in Duino in 1912 that Rilke began the first of the *Elegies,* upon which he labored for a decade in Spain, Paris, and Switzerland. Freedman cuts through the afflatus of origins and proceeds to offer incisive readings of all ten *Elegies*. He does this successfully, for he never neglects those earlier works that had first explored the connection between salvation and art, and sexuality and death. Even if the earlier poems are "amazingly immature" they led to lyric greatness. The 1899 images of carnal maidenhood and an eroticized Christ prepare the reader for the achievement of Rilke's later poems. It is a mark of Freedman's scholarship that he can place Rilke not only in the baroque tradition of sexualized divinities but the Renaissance one investigated by Leo Steinberg in his *The Sexuality of Christ* (1983). By the time of the *Fourth Elegy* the phallus can coexist with death as a "stiff corpse." The verdict is still out, Freedman suggests, on Rilke's French poems, the writing of which infuriated many Germans in the 1920s.

Despite many dry spells in comfortable rooms, it is amazing how much Rilke could write when he wished. Freedman's chronology has to be worked out, but most of the *Fourth Elegy* was apparently written in one day; half a week later he began and finished the second part of *The Sonnets to Orpheus*. This beats even John Keats, the other lyricist most obsessed with Eros and Thanatos. Even earlier in Rilke's career, verse poured from him. Returning from Russia in 1901 he wrote thirty poems in one week; they would become *The Book of Pilgrimage*. A week later in Viareggio he composed the thirty-five poems of *The Book of Poverty and Death*. Since many readers will come to this biography with some knowledge of German, it is unfortunate that the volume presents only Freedman's clear translations and not the German text. Line numbers would also have been helpful.

Freedman has avoided the temptation to write a hatchet job, surely a strong enticement with a subject like Rilke. The reader can observe pretense, snobbery, and disloyalty on his own. This doesn't prevent Freeman from accuracy. Rilke's claim that even at sixteen he spoke with the prince of Thurm and Taxis is just that; he and Andreas-Salomé "chased celebrities wherever they found them" in Russia; his notes to Rodin are "hyperbolically flattering"; the letters are "sycophantic." Poems are "manufactured" to serve his career; their dedications are "strategic."

Freedman wants to believe that Rilke's politics grows out of his poetic sensibilities; as they change, so do his social beliefs. At first he supports the war, then becomes pacifistic; he is attracted to the revolution in postwar Munich, then flees the dangers of the counterrevolution. He only fled to write better. His friends tried in vain to cool his ardor for Benito Mussolini. The touching anecdote that Rilke died of sepsis from picking a rose is told—and rejected. Rilke died in a Swiss clinic of leukemia.

The correspondence of Ezra Pound and William Carlos Williams has been ably edited by Hugh Witemeyer in *Pound / Williams: Selected Letters* (New Directions). The volume is another in the series of letters from authors connected with James Laughlin's New Directions books, a firm that kept and still keeps America literate. The editor has already edited Laughlin's and Williams's letters and written a study of Pound's poetry.

From the surviving five-hundred-plus letters covering half a century, from 1907 to Pound's death in 1963, Witemeyer has selected 169 examples, mostly from the collections at Buffalo, Bloomington, and Yale. Perhaps a fourth of this volume is made up of annotation to the individual letters and some seventy-seven one-paragraph biographical notes. Though the footnotes are extensive, they allow the reader to pursue a topic in other works. Some of the biographical notes seem superfluous—for example, Martin Van Buren and Benito Mussolini—but all figure prominently if oddly in the correspondence.

There are some gaps and omissions: from 1907 to 1920 no letters from Pound survive; from 1921 to 1932 fifty letters are evenly split, but there is a two-year gap during which Pound worked on his *Cantos* and Williams visited Europe; from 1933 to 1951 (except during the war) they exchanged about ten letters a year, Williams writing a bit more than Pound, but much more during Pound's incarceration. The exchange ends with Pound's death in 1963; Pound was recovering his strength as Williams was losing his during the last decade. A short introduction precedes each section. Witemeyer's notes to Pound's economic theories illuminate what still appears to be blather, but even blather needs clear explication.

Though one might expect extensive literary comment in these letters, there is not much. Williams's *Paterson* (1946) figures briefly, his novels fitfully; Pound's *Cantos* (1925–1960) appear mostly as titles. Periodicals such as *Blast* (1933–1934) and movements such as vorticism flourish and are canceled. Publishing problems, especially those of Williams, predominate over the works. What the reader finds instead of literary matters—and just as well, since Pound has been overannotated—is a suffering friend and an insufferable correspondent. The former tries to understand; the latter often can't be understood. Williams's final letter forgives the shortcomings Pound never acknowledged he had—even in prison.

There is also a debate on the place, literally, of the American Writer. Should it be the Garden State or the Roman Borghese? Williams preferred native flora. Pound, relying on a German anthropologist, suggested that remaining in America would limit Williams's poetic vision; to cure Williams's colonial attitude Pound provided reading lists. They never agreed on the subject. Williams resisted the expatriate urge and Pound lamented that they were "sequestered and divided by the fuckin buttocks of the arse . . . wide atlantic ocean" as early as 1920.

Pound's invective is always telling, for example: "Amy Lowell's perfumed cat-piss wd. be putrid if it had been done by a Pueblo indian." Williams is not without a similar gift: the French surrealist crowd spent the war years living off "the snot that runs from Peggy Gugenheim's nostrils which they lick up and thrive on." The metaphor continues, with T. S. Eliot as a "pot of addled ewe's snot." Pound was more tolerant of Eliot. Ironically enough, what Pound felt about Eliot is what Williams came to think of Pound. When Eliot returned to Harvard in 1931 as Norton Professor of Poetry, Pound suggested a cordial and amiable opposition be formed to Eliot's "private predilections" for kings and episcopacies but that he be welcomed for his literary discernment. Williams decided to take his advice, in part.

Williams takes a strong and consistent stand over Pound's treason (and it was that), rejecting the defense of mental incompetence. He forgives Pound for what he is but not for what he did. "And if you're shot as a traitor what the hell difference should that make to you?" he asks in one letter,

Dust jacket for the biography of the leader of the surrealist movement

since Pound's reputation will remain undamaged. Though at first Williams couldn't bring himself to visit the incarcerated Pound, he did write a series of long, familiar letters to cheer him up. Pound's replies are painfully succinct for this period (1945-1951).

If Pound is the better poet, Williams was the better person—a physician appalled by the broken pieces of men he treated, a man worried about his sons, his wife, and his mother, a friend to a nearly intolerable genius.

As early as 1919, in the first of many surrealistic periodicals, *Littérature,* André Breton insisted that the artist justifies the work, not the work the artist. For a biographer this means a life of Breton is a history of Surrealism. Such is Mark Polizzotti's *Revolution of the Mind: The Life of André Breton* (Farrar, Straus & Giroux); the two cannot be separated, nor can one nicely distinguish Breton's private life from his public positions. This makes for a thorough work of more than 750 pages: some 100 of which are clear notes. The biographer has interviewed more than fifty surrealists and their genetic or ideological heirs.

Polizzotti has surveyed 125 written works—the tracts, magazines, manifestos, denunciations, periodicals, and pronouncements by surrealists and their allies—and examined nearly one hundred works by Breton himself, correcting misprints. He has redated letters, mastered the complex bibliography of this movement, found scraps of material at Austin, and read exhibit catalogues. Because of this copious research, the official nativity of the movement in 1924 is not reported until one-third of the way into the book. It was officially disbanded in February of 1969, three years after the death of its founder, organizer, and leader.

Polizzotti likes and sympathizes with Breton and despises his enemies: the painter Francis Picabia, who broke with Breton, foreshadows the "model jet-setter." Two of the earliest founders of Surrealism, Louis Aragon (fifty-seven items in the index excluding specific works) and Paul Eluard, the first husband of Salvador Dalí's Gala (thirty-nine items excluding specific works), both receive the biographer's scorn.

One comes to admire Breton's uncompromising fanaticism—a fanaticism that prevented him from writing, as Aragon and Eluard did, Stalinist epinicians and from hustling the outré like Dalí—a fanaticism that even makes more likable a man who can change his erotic alliances quickly and cruelly. In the 1930s Breton, like Andrei Vishinsky, held his show trials, but exile from a café is not a sentence to the Gulag. After the war Eluard deserted Zavis Kalandra, their Czech host from the tour of Prague in 1935. Breton defended Kalandra, who was shot.

The provincial Breton saw service as a physician in the Great War, where he met Jacques-Pierre Vaché, from whom Breton acquired his detached, cool irony. Breton saw Vaché's suicide (or drug overdose) as an artistic statement, but denied it took place in the company of "winged androgynes"—all his life he detested homosexuals, and Jean Cocteau received his acerbic scorn. As Vaché exited, Tristan Tzara entered, the Sol Hurock of Dadaism who also later traded his iconoclasm for Stalinist orthodoxy.

Dada and communism interrupt the narrative flow of this biography, but they must. Breton's first *Manifesto of Surrealism* (1924) predates Dada, and Polizzotti rightly reemphasizes his subject's point: the precursors of Surrealism are writers, not painters. Among these writers, for Breton, was the Comte de Lautréamont (actually Isodore Ducasse) who replaced Paul Valéry in Breton's esteem. One of Lautréamont's works, *Poésies,* was so obscure that Breton had to read the one surviving copy in the Bibliothèque Nationale. The origins of Breton's metaphors begin with the earlier poet's well-known simile from 1860: "Beautiful as the chance meeting on a dissecting table of a sewing-machine and an umbrella." Metaphors move to deeds: "the Head of Vercingetorix" is not a mixture of Charles Dickens and Julius Caesar, but a sexual position in a Celtic version of the Karma Sutra.

The October Revolution of 1917 did not affect the nascent surrealists; yet in the 1920s and 1930s they had to confront the political. Breton sidestepped it: he and Aragon were interested in the party in 1920 (as late as 1925 Aragon dismissed the Russian Revolution as "a vague ministerial crisis"; by 1930 he was a fervent believer). Breton was primarily attracted to the political creed by Leon Trotsky as a stylist in 1925, but if the party concentrated only on the material instead of freeing the irrational, Breton would and did criticize it and its organs, especially *L'Humanité*. Polizzotti does a good job with what might be seen as the changes in Breton's politics. Breton removed the hard-line political surrealists from the movement and himself from Aragon in 1934. Four years later he traveled to meet the exiled Trotsky in Mexico. One of the forty-nine photographs in this biography shows Breton with the exiled stylist and Diego Rivera.

The final break occurred at the 1935 International Congress of Writers organized by the Party: the Soviets, already cool to Surrealism, became more so when Breton slapped Ilya Ehrenburg; Breton went on to attack the Franco-Soviet Pact and Socialist Realism. Surrealism was no longer in the service of the revolution (the title of an earlier pamphlet), for the revolution no longer served Surrealism. Breton kept his ideological purity, but his power in the surrealist movement waned when he was kicked out of the International Writers' Conference.

Breton's purity is evidenced again in his break with Pablo Picasso in 1946. In Polizzotti's view, Picasso, like Dalí, was an opportunist during the war. Even earlier Picasso kept his distance carefully and never formally joined the surrealist movement—this, too, accounts for Breton's break: the personal invades the philosophical.

In March 1941 the forty-five-year-old Breton fled France with his wife and child to America. Polizzotti defends this controversial action: Breton needed a free platform, feared prisons, and wasn't a warrior by temperament. Among his fellow refugees was Claude Lévi-Strauss, who introduced him to New World art and mythology. Arriving finally in New York, Breton was socially, linguistically, and philosophically isolated. He accused Dalí of being a fascist; Dalí demurred but admitted to being an "opportunist"; Polizzotti had earlier suggested that Dalí was attracted to Eluard's wife, Gala, because bedding the wife of a famous surrealist would give him entrée to Parisian art circles.

Dalí and his estate went on to prosper, but Breton was rarely comfortable. Hired by Gaston Gallimard in 1920 as a secretary, Breton made 400 francs a month; he also made 50 francs per session for helping Marcel Proust proofread *The Guermantes Way;* he missed two hundred errors. In 1950 he turned down an award of 300,000 old francs, yet he solicited pictures from well-known artists only to sell them.

If the political infighting was complex, so was Breton's emotional life. Polizzotti's explanation, and a convincing one, for Breton's changing enthusiasms, whether for works of art, writers, or women, is that he would begin in an ardor that would later cool to indifference and possibly dislike. Of course, he might stop and reverse the pattern anytime. He was surreal.

When he fell in love, he fell into an abyss that exalted him. This happened with three wives: the first, Simone Kahn, admired the poet semiplatonically; his attraction to the next, Jacqueline Lamda, a painter and exotic underwater dancer, lies in a pun (he called her his "night of the sunflower"); the third, adumbrated, Breton claimed, in his 1924 *Prolegomena to a Third Surrealist Manifesto,* was Elisa Claro, whom he married in 1944 in Reno. They all figure in his works, as do others, especially Nadja, who gave her name (her last name is not known) to his most famous book. This demimondaine was most attracted, Breton wrote, to that characteristic no reader of his works would ever suspect he had: his simplicity. Others can be pursued under the index's "romantic entanglements." Given the importance of women in Breton's real and imaginative life, Polizzotti argues that his marriages ended, not because of infidelities, but because Breton, inspired by a child-woman who provided him with attention, sex, and support, would awake to find that his muse was merely human.

It is difficult to draw conclusions about the place of women in Surrealism. Polizzotti admits this. One woman, Leonora Carrington, who was the lover of Max Ernst in the late 1930s, concluded that Surrealism was "another bullshit role for women"—but she said this fifty years later. From the 1920s almost to the 1950s, women were integral to the movement, not as subordinates, but as colleagues and lovers. Dante's Beatrice died; Breton's beloved women insisted on their independence.

Where there are questions or controversies, Polizzotti provides possible answers and explanations. Who was it who actually swung from a chandelier? Was Freud's influence direct or filtered through his French followers, as Anna Balakian argued twenty-five years ago? Polizzotti admits it is difficult to decide: Breton himself is inconsistent. Even the sincerity of the automatic writers—and to Breton Surrealism began with this event in 1919—is

The Year in Literary Biography

Dust jacket for the first book-length biography of the Polish writer who variously participated in Surrealist, Dadaist, and Futurist movements

suspect. To Breton this seems to have made no difference.

More surreal than automatic writing are the surreal events. Surrealism, despite Breton's dictionary definitions, is a slippery term; it is a revolution of the mind, a perception that issues in events, events that were to instruct, not just to delight. The "happenings" of a generation ago were not surreal. This biography is richly and correctly anecdotal: if Surrealism is the outward sign of an inward perception, then events predominate. Breton had not an instinct (he was too intellectually consistent) but a talent for the superlatively inappropriate. In a Jane Austen novel such would be a serious character flaw; in Breton it's serious policy. Dalí would deliver a lecture wearing a diving bell (and nearly suffocate when it stuck) more to amuse himself than make a point. Not so Breton. Readers may not be amused at the interest he took watching someone call Cocteau's mother to announce her son's fictitious death, nor at his followers' later muggings to avenge both Breton and Gloria Swanson. Not all of his followers were violent; they retained an eye for the press event and an ability to mock it: journalists invited to celebrate Breton's sixtieth birthday at a luxurious hotel arrived instead to a coal miners' convention.

One of Polizzotti's later chapters, "I Am Surrealism," corrects the usual picture of Breton—the vehement radical of the 1920s and 1930s—becoming a devotee of the esoteric. That he wasn't is Polizzotti's point: the forms change, the focus remains: alchemy replaces automatic writing, as Lautréamont replaced Valéry, as the social theorist Charles Fourier replaced Trotsky, as the night of the sunflower replaced his first wife.

In the late 1950s and 1960s Breton became that which he never wanted to be: a given. There was a break in the tradition (if something so fluid has a tradition). The postwar young inherited Surrealism and felt they had to live up to a heritage that Breton had created. But it is too much to see the 1968 student demonstrations or riots in Paris as a result of manifestos from the 1930s calling on the young. In 1946 Breton's lectures in Haiti may have catalyzed a change of government. It was not, as Polizzotti believes, a "revolution." The United States retained fiscal control of Haiti.

Polizzotti is often witty, sometimes coy, and, upon occasion, precious: the original Nadja is "obnubilated" over time; Aragon changes sex with advancing age like the swordtail fish but these are minor flaws. Readers have needed a book that makes sense of a confusing movement. Polizzotti has written it.

Aleksander Wat began as surrealist, Dadaist, or futurist—in Poland the definitions shift. Tomas Venclova's *Aleksander Wat: Life and Art of an Iconoclast* (Yale) is the first full-length study in English of Wat, a Polish writer first introduced to English speakers by the Nobel laureate Czeslaw Milosz. The author's "Select Bibliography" lists nineteen articles in English, most a page long; there are two previous books that mention him, for a total of fifteen pages. Consequently Venclova's work is perhaps 20 percent biographical and the rest critical. This is a sane approach for a person most non-Polish readers have never read—or, alas, even heard of—until recently.

Venclova may have had little choice over the amount of biographical data in his book. There is little, he says, on Wat's communist activities in the 1930s and 1940s, even less about his deportation to Kazakhstan after his arrest in 1940. The biographer is forced to rely on *My Century*, a transcript of thirty-nine recording sessions Wat made with the help of Milosz at Berkeley in 1965, two years before his death. Not all of this oral memoir has appeared in English; Venclova provides new translations of excerpts.

The publications in Polish from 1985 to 1988 of Wat's fiction and essays (see Milosz's English version in *Lucifer Unemployed,* 1990), a 1992 edition of his poems and two volumes of translated poetry (*Mediterranean Poems,* 1977, and *With the Skin,* 1989; both translated by Milosz), have secured Wat's place as one of the outstanding "cultural" figures of the twentieth century; it is Venclova's aim to remedy the Western world's ignorance of him. The biographer further claims that Wat's output is more provocative than that of his Polish contemporaries, which "appear dated to the discerning modern reader and are doomed to oblivion"; as a critic of Stalinism, Venclova claims, Wat is equal to Aleksandr Solzhenitsyn. The former claim is large and partisan—only a specialist could refute or support it, since even Slavicists have only a small acquaintance with Wat's works. The latter remains to be seen.

This biography proceeds in an orderly fashion, presenting the life of an aesthetic radical who, attracted to communism, joined the party (though Wat denied it). This puts him in the ranks of Bertolt Brecht, Louis Aragon, Paul Eluard, and especially Vladimir Mayakovski—with an important exception. Unlike Mayakovski, Wat did not "opt for an exit into death," but recovered to criticize his initial conversion to communism. That religion played some part in his recovering seems probable, but what sort of religion remains is either too obscure or too personal to understand. Venclova, especially in the later chapters, presents more the results of Wat's political and theological conversions than the process that produced them. This is unfortunate.

The reader learns of Wat's scholarly Jewish background and of a nanny who introduced him to the liturgy of the Roman church. Most of his immediate family would perish in the Holocaust. Wat had his family secretly baptized in 1954 and saw himself as a suffering Christ.

As a child Wat was precocious, reading everything from Miguel de Cervantes to Coventry Patmore. His studies at the University of Warsaw were erratic; he seems to have drunk a good deal; read Arthur Rimbaud and Charles Baudelaire; avoided the Scamandrite School of poetry (happy and tepid); and embraced Futurism and the activities of Futurism and Dada. On one occasion, drunk, Wat stripped, proclaimed himself an ostrich, and announced he would lay an egg. After having served in the Russian-Polish War of 1920, he, along with his school friend Anatole Stern, published a manifesto, *gga.* Wat's contribution was the autoerotic poem *Flying Petticoats.* The public was outraged; *gga* was censored. To survive, Wat wrote advertising copy—parodically, Venclova assures the reader, citing a letter written forty years later. In 1922 Wat became a member of the Union of Polish Writers; in 1926 he joined PEN. His poetry from this time is cryptically vivid: "In the cabins cuddled up into violins / the widows were overgrown by moss. Red and sultry as July, the cows prayed in the clouds" (*The Countryside*).

So far Venclova's narrative seems to be moving in an orderly and chronological fashion, but next the premier Polish futurist work, *Pug Iron Stove,* takes up some forty pages of criticism, yet this work predated *gga* by two years. The reader who wishes to discover Wat's life must fast-forward until months and date appear again. This happens repeatedly in the biography.

The reader now learns that perhaps as early as 1920 Wat was attracted to the purifying cruelty of Russian Communism. A marriage with Paulina Lew, "Ola," in 1927 cured him of suicide's attraction, which figures in *Pug Iron Stove.* His major source of income was translation, of Fyodor Dostoyevsky, Georges Bernanos, and O. Henry; his short stories, collected in *Lucifer Unemployed* (1927), were also well reviewed. The stories seem as surreal as his early poetry: the plots are peculiar. One of them "A Story," mixes pedophiles, Old High German, and a brass larynx together in an ovoid room. Wat met Mayakovski in 1927, and there may have been a secret undocumented meeting in April 1929, shortly before the gloomy Russian's suicide. This is new and important information.

Marshal Joseph Pilsudski, the Polish chief of state, censored Wat's "facto-montage" of skits on Poland's proletarian troubles. He traveled to the West to find writers for his communist journal, *Literary Monthly.* Was this Wat's moral nadir or a happy time? Venclova is unsure. Wat's literary criticism at the time was rigidly Stalinist: Erich Maria Remarque's *All Quiet on the Western Front* (1929), for example, is described as sentimental bourgeois writing.

In 1931 Ola gave birth to a son, and the Polish government arrested Wat for subversive writings (an attack on prison conditions was the formal cause). He spent three months in a Lvov prison. Released, he worked for a publishing house, wrote a few poems, and followed the party line in the Great Purge of 1937–1938, when five thousand Polish communists were executed in the U.S.S.R. As the war began, Wat fled to Soviet-occupied Lvov with other leftists, organized an official writers' group, and wrote a celebration of Lvov's annexation to the Ukraine. Here even Venclova questions Wat's later explanation that he wrote out of fear.

The Soviets arrested him as a Trotskyite, Zionist, and Vatican spy in January 1940; in April,

Dust jacket for the sketches by Virginia Woolf's nephew of eighteen members of the Bloomsbury group

Ola was arrested and deported to Kazakhstan. There Wat was reunited with his family ten months later, after being interrogated in the Lyubyanka and then freed when Hitler invaded the U.S.S.R. The former Dadaist began to question the system. Among those he met in Alma Ata, such as the poet Constantine Paustovsky, the satirist Mikhail Zoshchenko, the literary theorist Victor Shklovsky, and the film director Sergei Eisenstein, only the last thought Stalinist dictatorship would survive the war. Wat refused to take a Russian passport.

Wat's publicly expressed doubts first appear in a 1942 obituary of a Polish journalist. He was arrested again after the U.S.S.R broke with the Polish government-in-exile. Freed in 1943, Wat spent the next three years in Kazakhstan. Venclova has not located much material about these years, but he demonstrates that the language of Wat's poetry began to change during this period. Wat returned in 1946 to a destroyed Warsaw, found a good job in a publishing house, rejoined PEN, and at its various international meetings praised the tolerance of the new regime. Venclova finds these accolades excessive and notes that not once did Wat mention the suffering of the Jews under Hitler: his position was straight party dogma.

In 1947 Wat suggested that socialist realism was not perfect, that art need not just mirror life—ideas he repeated in various speeches over the following few years. Ironically, it was Wallenberg's syndrome (vertigo, disequilibrium) that saved Wat: he was allowed to travel freely for his health in the West. Wat seems to have believed his affliction was divine vengeance. Wat was secretly baptized in 1953 (out of respect for the sacrament, Venclova suggests, though readers may suspect prudence), and his poetry began to reflect the equation of the suffering Christ with the poet's experiences.

Wat's book of poetry, *Verses,* sold out quickly in 1957; the *Partisan Review* mentioned it favorably. Two years later he decided on permanent exile. In Paris the Wats shared an apartment with Roman Polanski, and Milosz helped with money and employment as Wat began to attack communism. His *Mediterranean Poems* were well received; he was invited to UCLA and arrived ill in 1963. Suffering from writer's block, Wat was assisted by Milosz in recording his memoirs (translated, in part, as *My*

Century: The Odyssey of a Polish Intellectual, 1988). He returned to Paris depressed, took forty sleeping pills, and died.

Wat's life takes up perhaps a fifth of Venclova's study. The biographer admits that there are gaps in documentation. The index is too light. Some events will always be unclear, but there remains room for speculation. Wat's sister, Seda, was once engaged to the future Gen. Wladyslaw Anders, who led the army of the Polish government in exile. To what extent did this affect Wat's life? Venclova does not say. The book would also have benefited from photographs, of which there are none.

As a sensitive literary critic, Venclova has introduced readers to an extraordinary poetic voice. That is the great value of this work. Wat himself admitted that his poetry is difficult and obscure. Fortunately Venclova's background in Middle European cultural history illuminates many of the obscurities. His explanations are thorough: the Sicilian landscape, Empedocles, Orestes and the Furies, and lexical rarities are all employed to explain Wat's work. There is jargon.

Upon rare occasion Venclova misses: despite his assertion to the contrary, Jewish tradition does assign a name to Cain's wife (Awan, in the *Book of Jubilees*). His claim that Wat anticipated the surrealists' discovery of automatic writing is shaky: it looks to be a tie between Wat and Breton in early 1919. But any scholar who brings a talent to light deserves the gratitude of the curious. Aleksander Wat is obscure and difficult; Venclova's book makes him less obscure.

Quentin Bell, the nephew and biographer of his aunt Virginia Woolf and the author of earlier books on Bloomsbury, continues and bids farewell to eighteen of his subjects in the two hundred pages of *Bloomsbury Recalled* (Columbia University Press). What started as an autobiography has become willowy sketches of eighteen people, Bell's "elders and betters" associated with Bloomsbury by aesthetic and sexual alliances; the latter, as one might expect, are thoroughly confusing. Some years ago, to prove the potential dangers of exuberant sex, *People* magazine drew an erotic genealogical chart of Hollywood stars. Bell's volume could use the same.

This does not mean one cannot recover the life of Quentin Bell from the chapters, but the admirable index is the better guide. The best guide to this volume is the author's *Bloomsbury* (1968), a work that provides masses of information that, unstated in this work, color his opinions. To read all the chapters at one sitting is disquieting: the reader keeps going back in time to 1914 from one chapter to the next.

Certain expectations are, thankfully, not met: by now Woolf's treatment by her half brothers has entered the *acta feminarum*. Bell treated this sensational more-than-brotherly love twenty years earlier. Here, on one page, he encapsulates and tones things down: half brother Gerald's inspection of the five-year-old Virginia's genitals is a schoolboy's misdemeanor; half brother George did not copulate with his sister, though his actions could have been a "prelude" to rape.

Lytton Strachey figures briefly; there is more on his sisters; much more on all of them in Michael Holroyd's biography. "Carrington" of movie fame is merely one of the "Bloomsbury Bunnies," Woolf one of "Cropheads." Yet Strachey did arrange for Bell to meet Henri Matisse, whose ego was as innocent as it was vast.

Generally, Bell's sketches are best when he recalls artists: the chapter on Roger Fry, who was Bell's mother's lover when he was born, demonstrates that Fry's imagination could even persuade T. S. Eliot of a dark star's imminent collision with the earth. Later and less famous critics and artists, such as Claude Rogers and Lawrence Gowing, are rarely as amusing. Lady Ottoline Morrell has a few pages that are worthwhile, however. Only a papal sedillia could match her plumes, her jewelry, and her train as she walked in the Trastevere in 1931.

Bell's narration is often curiously disinterested. Recalling that Wilde's *De Profundis* (1905) was published five years before Bell's birth, the reader can expect this volume, some three generations later, to acknowledge rather than defend Wilde's orientation. Bell is surprising, even with the recent change in sensibility: when Leonard Woolf hired John Lehmann at the Hogarth Press in 1934 (Virginia approving both of his adorable curls and intelligence), the reader learns, in a subordinate clause, that Lehmann was, for a short time, Bell's lover. The quarrels between Leonard Woolf and Lehmann then become "a case of Greek meeting Greek," ostensibly referring to financial disputes, but too cleverly to something else. Bell also states that spouse beating, if in private, is a matter of mutual consent, but public humiliation "is, it seems to me, a different matter." Understatement has its uses, but it is overused here: a soldier "found himself marching, as young soldiers do." Such phrases are what remains of Bloomsbury cleverness.

Then there are the "catamites" of John Maynard Keynes, one of whom, "the toast of the British Sodom," reverts etymologically to Ganymede, endures the rapture of an elderly American lady, and finally is rescued and returned to his rightful owner: a Mary Butts who claimed she was Bell's half sister (by his father, Clive). Since they looked so alike, she suggested they exchange clothes and saunter out into Parisian society (they didn't).

The illegitimacy of Bell's half sister, Angelica, by Duncan Grant and Vanessa, is announced sotto voce after he attacks her doll. There are eleven Bells in the index: Angelica appears only under her husband's name: David Garnett. In the light of Angelica Garnett's *Deceived with Kindness: A Bloomsbury Childhood* (1985) Bell's book is paying off some old scores.

Bell consistently calls his parents by their Christian names, Clive and Vanessa. Vanessa comes off looking better. After the death of Vanessa's brother, Thoby, Clive is finally accepted by his wife, and from this marriage comes a nascent Bloomsbury: Roger Fry, Duncan Grant, John Maynard Keynes, Desmond and Molly MacCarthy, the Woolfs, and the Strachey family.

All of the above get their own chapter, generally of less than twenty pages. Clive suffered his grandfather's tyranny, Quentin the "desperate womanizing" of his father's affairs. Vanessa, after Clive had removed himself, took up with Roger Fry and then Duncan Grant; her marriage, like her affairs, was involved with art. The second affair was complicated by Grant's escaping, at times, to the brawny arms of some proletarian lover. Bell's admiration for his mother's anticipation of modern sensibilities is more honest than filial: "faced by the need to call a cunt a cunt, she discovered she could not quite abandon the proprieties."

This author has his anecdotes, and some are amusing: for example, Leonard Woolf, before going to bed, walks to the garden to urinate on a bust of his wife's grandfather, a colonial secretary. Readers interested in the personalities of the 1920s and 1930s should read earlier studies of Bloomsbury or any ten pages of Sir Osbert Sitwell's five autobiographies.

This year's biographies travel in the English-speaking world from the bawdry and lyrics of Scotland to the attenuated sensibilities of Bloomsbury, with occasional quick stops for Wordsworth, Hardy, and Arnold. More valuable are the studies of André Breton, Ranier Maria Rilke, and even of Aleksander Wat. These expand their subjects' lives to include those winds of doctrine that blew through the first half of this century: Futurism, Dadaism, and Surrealism.

Dictionary of Literary Biography Yearbook Award for a Distinguished Literary Biography Published in 1996

Dust jacket for the major biography of the great twentieth-century German poet

Insofar as most novels reflect the writers' exterior worlds and poems the interior, the life of a poet is often more difficult to write than that of a novelist. Ralph Freedman of Princeton University has written studies of those novels that partake of both the interior and the exterior, the works of lyrical novelists such as Virginia Woolf, André Gide, and the subject of his 1978 biography, *Hermann Hesse: Pilgrim of Crisis.* This year he succeeds again with his *Life of a Poet: Rainer Maria Rilke,* the first thorough biography of the poet in English.

It cannot have been an easy task: Rilke's interior seems to have been unpleasant, his exterior boorish. He betrayed his lovers, vacillated in his friendships, ignored his family, and toadied up to his patrons. It seems anyone injured by Rilke wrote about him in memoirs and letters; most appear as blind to his faults as they are aware of his genius. This is half true of his biographer. Freedman coolly presents the worst side of Rilke as a man while remaining keenly aware of his poetic brilliance. Rilke's egoism is embedded in his letters, and Freedman shows the poet both at his best and in his least-flattering moments.

Life of a Poet is a well-written biography, meticulous in documentation, clear in narration, and somber in judgment. As a literary critic Freedman demonstrates how the romantic excess of early, if not juvenile, verse evolved into the tight lyric control of *The Sonnets to Orpheus* and the *Duino Elegies.* As a biographer Freedman uncovers those patterns of behavior that sustained Rilke's literary talents, even as the poet excluded those closest to him in emotional alliance.

The Year in Drama

Howard Kissel
New York Daily News

When Eugene O'Neill's *Strange Interlude* was published in 1928, shortly after its New York premiere, it sold one hundred thousand copies. The number is significant. *Strange Interlude,* after all, was considered an avant-garde play in its time. You arrived at the theater at 5:15, spent nearly three hours, had an hour for dinner, returned at nine and finished shortly after eleven. (Robert Benchley called it "just an ordinary nine-act play.")

Despite the rigors it involved for the audience, it was O'Neill's most successful play. It ran 441 performances on Broadway, and two companies performed it on the road for three seasons. This may account for why its central innovation, having the actors speak directly to the audience, was well enough known that it could be satirized a few years later in the Marx Brothers' *Duck Soup* (1933), which was hardly intended for avant-garde audiences.

Strange Interlude has its sensational aspects, but they were of a subtler sort than those of, say, *Desire Under the Elms* (1924) or *Anna Christie* (1921); it became known for its aesthetic daring rather than its prurience, which makes the sale of one hundred thousand copies all the more remarkable.

In the ensuing seventy years, and especially during the last few decades, there have been almost no "best-selling" playwrights. Many highly regarded plays of recent years have not even been published in hardcover, but rather in quality paperback editions—as Tony Kushner's *Angels in America* was. The assumption is that even the best or most talked-about new work in the theater is only likely to be purchased by a limited market.

This year, however, there is a different way of measuring the impact of the most publicized theatrical event in many years, the musical *Rent:* by the sales of its original cast recording.

An updated version of Giacomo Puccini's *La Bohème* set in New York's East Village, *Rent* received an unusual amount of attention and publicity because of the unhappy circumstances surrounding its opening. Only hours after the final dress rehearsal, Jonathan Larson, who conceived the show and wrote its book, music, and lyrics, died of an aortic aneurysm at the age of thirty-five.

Larson was already regarded as a promising writer. A year earlier a theater piece for which he wrote music and lyrics, *J. P. Morgan Saves the Nation,* had been performed in the financial district by the site-specific theater group En Garde Arts. Larson was considered a protégé of the most influential figure in contemporary musical theater, Stephen Sondheim, who was said to admire the fact that, unlike most young "serious" theater composers, Larson had an affinity with and a talent for rock—most young theater composers, Sondheim felt, were out of touch with the true "pop" music of our time.

When *Rent* opened on 13 February, a few weeks after its author's death, at the New York Theater Workshop on East Fourth Street in the heart of the East Village whose artists it depicts, it received—for the most part—reviews that could only be described as ecstatic. In some ways the reviews were eulogies for the young composer whose talents would never be fully developed.

In no time plans were being made to move *Rent* to Broadway, an ironic development in view of its vehemently antibourgeois stance. Record companies were vying for the rights to make the cast recording—the winner was one of the most powerful figures in the recording industry, David Geffen, and there was considerable speculation as to which of the hot artists in his stable would make "singles" of numbers from the score.

Shortly before it opened on 29 April at the Nederlander Theater on Broadway, *Rent* won the Pulitzer Prize for drama, making it only the sixth musical to win the prestigious award in the eighty years it has been given. Just after the opening, Bloomingdale's announced it was opening a *Rent* boutique to market at uptown prices the funky garments one might buy in the actual East Village for a pittance.

Rent was featured on the cover of *Newsweek*. The *Wall Street Journal* devoted a front-page story to *Rent's* financial odyssey from an early workshop presentation to its Broadway apotheosis. It told how

 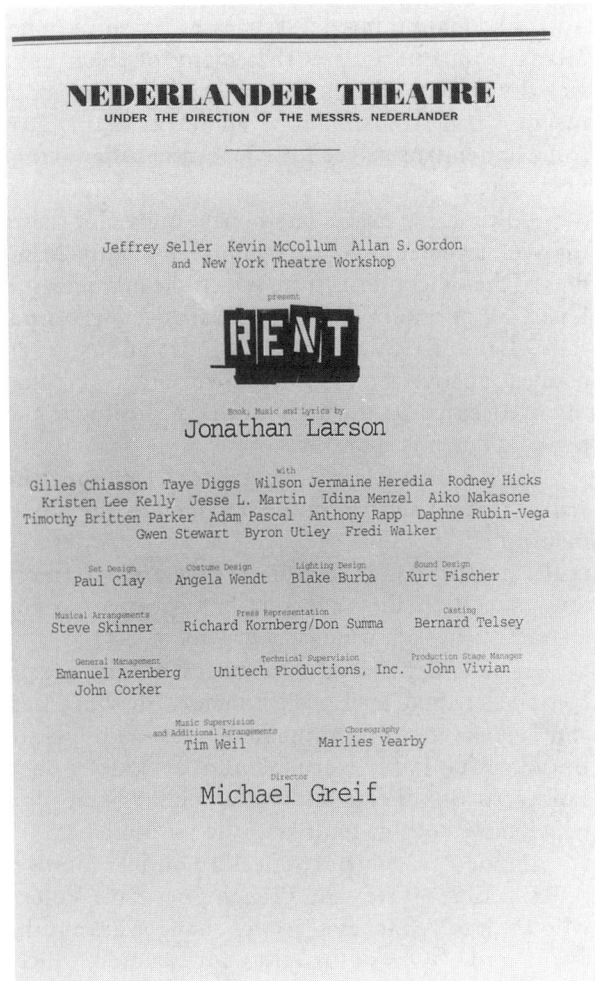

Program for the East Village version of La Bohème

an initial investment of $4,000 on the part of two young producers blossomed into a project worth millions.

All this brought *Rent* to the attention of a much larger audience than normally becomes aware of new pieces of theater. Nevertheless, the original cast album—which was released with much fanfare and started on the *Billboard* album charts at number 19—sank within ten weeks to number 193 despite the final cut, which featured Stevie Wonder joining the cast to sing the most eloquent number, "Seasons of Love." (Ultimately none of the "hot" artists who were going to turn *Rent* into a crossover hit recorded anything from the score; Stevie Wonder, I have been assured, has not been "hot" for some time.) By the end of the year 227,000 copies of the record had been sold—impressive for a musical, of course, but a drop in the bucket for a pop album.

Ordinarily a musical—even a musical that won the Pulitzer Prize—would not warrant this much space in a serious discussion of the year's theater, but it is rare that any theatrical offering becomes as much a "phenomenon" as *Rent* has. It does seem worth noting, though, that to judge by the album sales—however popular and lucrative the play was, however much publicity it garnered—*Rent* generated less interest in our culture than *Strange Interlude* did in the America of Calvin Coolidge.

Ultimately this is not so surprising, since *Rent* is not quite as prepossessing a work as one might suppose from all the hoopla. In many ways it is standard *epatisme,* though even its most extreme attempts to startle the bourgeois audience come across, in our shockproof times, as comical rather than unsettling. In Larson's bohemia, for example, Mimi is not the shy, demure creature of Puccini. She performs in a bar catering to sadomasochists, and, for all we know, the match she requests from her neighbor, a video artist, is to help her freebase cocaine rather than light a candle in her apartment.

The main thing that separates Larson's bohemians from Puccini's is that many of them have wealthy parents in suburbia who are all too eager to rescue them from financial embarrassment. Larson's bohemians starve for effect rather than out of necessity.

Ultimately *Rent* is less honest in its plot twists than its nineteenth-century model. Puccini's Mimi dies of the tuberculosis from which she has been suffering all opera long, giving the work its great pathos. Larson's Mimi makes a miraculous, even comical, recovery on her deathbed, suggesting that the preoccupation throughout the musical with impending death is spurious.

If the text seems inconsistent, so does the score, though several numbers, such as the aforementioned "Seasons of Love," show considerable musical skill. As for the lyrics, they are extremely uneven. Often they are simply recitative, as the characters explain what they are doing.

The most pretentious lyrics are in a self-conscious tribute to "La Vie Bohème," in which Larson drops a lot of fashionable names—"To Uta, to Buddha / To Pablo Neruda" plus obligatory obeisances to Sondheim and Susan Sontag—most of which have nothing to do with "la vie bohème."

Some of the lyrics reflect a political mindset fixed in time—it was the French poet Paul Valéry who declared that "everything changes except the avant-garde." One of the lyrics, for example, which, except for the fact that answering machines were less prevalent, might just as easily have been written thirty years ago, goes, "We're living in America / At the end of the millennium. / Leave your conscience at the tone / You are what you own."

There is a lyric that suggests how Larson's work might have matured. Roger, the video artist who narrates the show, sings about "the 3-D I-Max of the mind." First he sings, "That's poetic." Then he reconsiders: "That's pathetic." A more seasoned Larson might have done this thinking in private. Sadly, we will never know how his budding talent might have developed.

Rent was frequently compared with the 1968 *Hair* as a groundbreaking event. Rock itself was still novel—at least in the theater—back then. So was the formless amateurism of *Hair* at a time when musical theater was still structured and professional. *Hair* had a few great songs that conveyed the mood of the time and, most important, it had nudity, which made it commercially attractive. The songs and the title still evoke nostalgia, although the show itself is no longer viable, as was demonstrated a few years ago when a London revival closed quickly and a Broadway-bound revival closed on the road.

Rent has more substance than *Hair,* and it is possible that "Seasons of Love" will establish itself as a perennial, but whether the show will be regarded as more than a fashionable curiosity remains a question.

Given the flashiness and excess energy of *Rent,* it is hard to imagine a more dramatic contrast than the most distinctive play to open in 1996, Brian Friel's *Molly Sweeney,* which opened at the Off-Broadway Laura Pels Theater on 7 January. For the new play by the author of *Philadelphia, Here I Come* (1964) and *Dancing at Lughnasa* (1991), the basic set consists of three chairs, one for each character. The actors never address each other directly. They speak only to the audience—in monologues.

Nor do they move around much. Rarely do they stray far from their appointed chairs. Only in the last few minutes of the play does the title character—a blind woman who regains and then loses her eyesight—move any great distance. Her walk around the back of the stage past a mural evoking flowers in the muted but provocative palette of Odilon Redon is the most movement we see all evening. And yet this brief, literal walk is the least momentous of the journeys Molly Sweeney takes in the course of this absorbing and entrancing play.

Born in rural Ireland, Molly Sweeney became blind before her first birthday. But her father taught her to "see" with her fingers and her ears, and her first monologue, in which she describes the garden of her childhood, is quite rapturous, a world "appareled in celestial light."

As a young woman Sweeney finds employment as a masseuse in a health club. There she also finds a husband, a sweet, wide-eyed man who pursues various hopeless schemes with the zeal of a utopian. At the age of forty she comes to the attention of an eminent albeit alcoholic surgeon who restores her sight.

He "performs" the surgery (he has a lovely speech noting the theatricality of surgical terminology), knowing it will probably help his reputation more than his patient's well-being. His fears are well founded. The disparity between the actual world and the world she has built in her imagination over forty years is so great that it throws Sweeney into a depression. She then loses her fragile sight, which turns out to be a blessing. Blind again, she can recapture "the glory and the freshness of a dream."

Sweeney's is an unsettling odyssey. At several points her husband notes that the scientific terms for her condition parallel certain religious terms. The inability of a newly sighted person, for example, to reconcile reality with her previous perception of it makes her an "agnostic." Her condition of impaired

Program cover for Brian Friel's play about an Irish woman who has been blind from infancy

vision is, technically, "gnosis," a term pertaining to mystical religion. Presumably her return to blindness constitutes a regaining of faith. This religious "teasing" recalls Friel's *Wonderful Tennessee* (1993), in which four friends sail to an island whose history involves a brutal ritual murder—the journeys Friel's characters take are toward light, but it is often the darkness that illuminates them. It is not only the blind who have trouble reconciling reality with their inner visions, and this fact makes Sweeney's journey universal and moving, not just a medical curiosity.

The play was performed with unusual sensitivity and radiance by Catherine Byrne as the title character, Jason Robards Jr. as the alcoholic doctor (one of the subtlest and richest characterizations in his distinguished career), and Alfred Molina as the hapless husband. Although they spoke in monologues the effect was surprisingly dramatic. There was a powerful counterpoint in Friel's artful juxtaposition of the poetic visions of Sweeney, the despairing, sober realizations of the doctor, and the loopy optimism of the husband. The overall effect is luminous.

Program for Baitz's play about the family of an American diplomat

Of the homegrown plays the most powerful was Jon Robin Baitz's *A Fair Country,* which opened on 20 February in the Mitzi Newhouse, an Off-Broadway venue that is part of Lincoln Center Theater. The title of the play was drawn from "Refugee Blues," by W. H. Auden: "Once we had a country and we thought it fair, / Look in the atlas and you'll find it there: / We cannot go there now, my dear, we cannot go there now."

Baitz's "refugees" are the family of an American diplomat whose career stalled on the not very promising continent of Africa in the 1970s. There Harry Burgess's duties consisted largely of arranging tours for American arts troupes—like a San Francisco theater company, all ex-convicts, performing *Idiot's Delight*—through the Third World.

Harry is married to the bitterly unhappy Patrice, who has given up a promising career as an art historian at the Phillips Collection in Washington, D.C., to be a diplomat's wife, a position for which she has no talent whatsoever. They have two sons—Alec, an angry young journalist who has made contacts with radical blacks in South Africa, and Gil, a confused boy trying to help his severely disturbed mother. Except for Patrice and Gil, the family is seldom together.

The two most exciting scenes involve all the Burgesses—where all the spoken and unspoken tensions and resentments come to the fore. This is particularly true in the second confrontation, when the family entertains an elderly Dutch diplomat in a smart apartment in Amsterdam on New Year's Eve. His dry wit makes an elegant counterpoint to the savage thrusts around him. The scene has the ferocious vitality and the painful, sad humor of a Shostakovich string quartet.

The weakness of the play is the political freight with which Baitz weighs down the family. The dysfunctional Burgesses are meant to symbolize America and its failures in diplomacy, especially in Africa. When, near the end of the play, Harry laments, "To have imagined that we could do anything decent with the world . . .," we are to understand it as deep hypocrisy.

Baitz has not portrayed Harry as harshly as he did the comparable father in his 1993 *Three Hotels.* The plot, however, turns on Harry's betrayal of Alec and his young radical friends, which he justifies as an attempt to help Patrice, but which we know is intended mainly to aid his own flagging career. This sellout reflects a naive understanding of both human nature and politics. It reduces an otherwise sophisticated play to the level of Group Theater agitprop. Nevertheless, *A Fair Country* was gripping theater because of the great cast—especially Judith Ivey as Patrice, who turned what could have been merely a bitchy, destructive sniper into a moving, tortured soul. The play, despite its mechanical plot, has an acerbic and pungent music that marks Baitz's as a distinctive and important voice.

The winner of the New York Drama Critics' Circle Best Play Award for 1996 was August Wilson's *Seven Guitars,* which opened on 28 March at the Walter Kerr on Broadway. If many of Wilson's plays have the feeling of jazz, this one is like a muted trumpet in the wee hours of the morning. Unlike some of his other plays, where the riffs are dazzling and the energy explosive, this one is full of quiet truth.

One simple reason that the play is so full of musicality is because several of its characters are musicians. Set in 1948 in a poor black neighborhood in Pittsburgh, *Seven Guitars* takes as its theme the great postwar migration of country black folks from the hostile South into the seemingly more receptive but ultimately treacherous terrain of the urban North.

The play shows us the last few days of Floyd (Schoolboy) Barton, who has just had his first hit record. After recording in Chicago, Barton has come back home to Pittsburgh just in time for his mother's funeral. He has spent all of his newfound money—has even pawned his guitar—to buy flowers. Arrested on trumped-up vagrancy charges, he has spent three months in the workhouse. When, however, we first see Barton, played with tremendous flair by Keith David, he leaps onto the stage with so much swagger, so much disarming charm, that we cannot doubt he will make good.

The first act is pure talk as Barton and his friends sit in the backyard and chat. The tone is tangy blues except when one of the characters, a poignant scripture-quoting prophet lost in his own fantasies, injects the sound of spirituals. The conversation meanders but the effect is mesmerizing.

In the second act, alas, Wilson decides things must happen, and the play becomes forced. The violence that brings Barton down seems entirely artificial. The densely plotted second act is far less compelling than the seemingly unstructured, talky first.

Wilson and Baitz are two of the strongest voices to emerge in the theater in recent years. There was a time when young writers made their first marks by writing plays, then graduated to writing screenplays. In recent years the theater has no longer been a stepping stone—young writers move directly into writing screenplays for Hollywood or television. It is far more lucrative (and intellectually less demanding) than writing for the theater.

When a young writer of talent does take the theater seriously it seems worth noting. Such was the case with Kenneth Lonergan, whose *This Is Our Youth* was a droll and telling assessment of the generation (presumably his own) that came of age in the 1980s. The play was presented by what is consistently the most impressive of the Off-Off-Broadway subscription theaters, the New Group, at the Intar Theater on 30 October.

The two young men Lonergan depicts are the sons of men who subsumed their 1960s idealism in the vigorous pursuit of wealth. Their sons, as if in some unconscious attempt to recapture their fathers' lost spirit, let money slip through their fingers.

Dennis is an entrepreneur, dealing drugs with great panache. His friend Warren is simply hopeless. He has stolen wads of money from his father and run away from home with a huge suitcase filled not with clothes but with "collectible" toys and a rare 1950s toaster. Dennis tries to realize a profit by

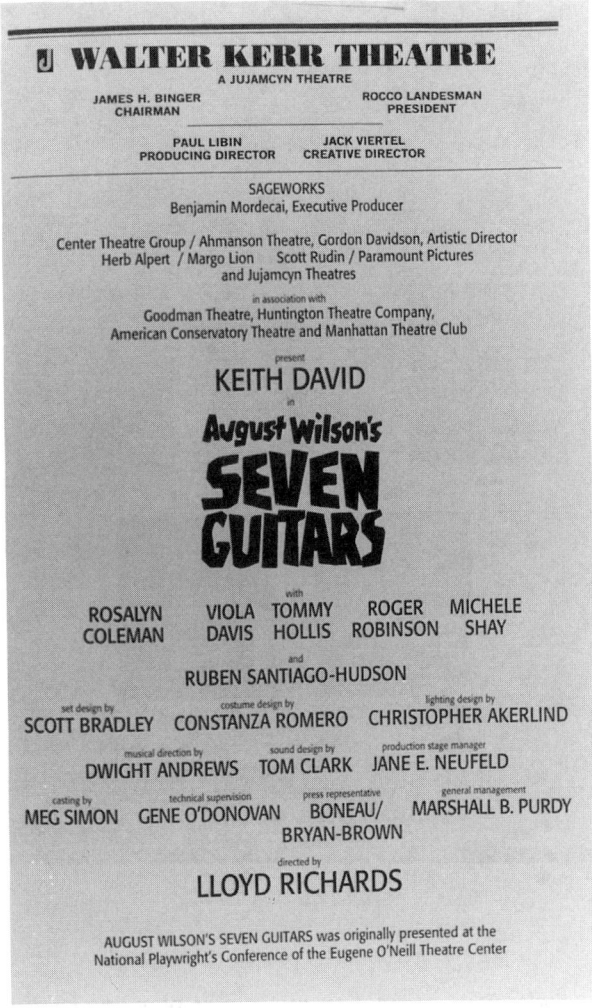

Program for the play that won the New York Drama Critics' Circle Best Play Award in 1996

selling Warren's toys and the presumably valuable toaster, but well into the play it becomes clear that Warren will never be able to repay the money he has stolen from his father. The uproarious comedy of the two young men scrambling to make money takes on an oddly moving poignancy.

Lonergan's play is funny and touching but ultimately unsettling—an impressive combination. He has written screenplays, but *This Is Our Youth* is unmistakably a work for the theater, and Lonergan seems potentially a major talent.

As usual, much of the nonmusical drama New York sees is imported from elsewhere, generally London. Sometimes the plays that arrive here seem a little parochial, like David Hare's *Skylight,* a skillfully written encounter between Tom, a London restaurant tycoon, and Kyra, the young woman who was his assistant—and his mistress—until his dy-

August Wilson in 1986

ing wife discovered their affair. The play opened on 19 September at the Royale on Broadway.

Tom pays Kyra a surprise midnight visit after having been out of touch with her for several years. She has abandoned her lucrative career as his employee to live in the slums of London and teach underprivileged children. What ensues is a fairly predictable argument between Kyra, with her admirable ambitions, and Tom, whose cynicism is boundless. She is the daughter of a prosperous lawyer—he describes her childhood sneeringly as "being pushed by nannies beside stormy English seas." He accuses her of suffering from the middle-class tendency to sentimentalize and idealize the poor. She has transformed her life, he charges, into "an act of denial." She in turn taunts him about "those carpaccio and ricotta-stuffed restaurants of yours."

What makes the play seem "foreign" to American audiences is the underlying sense that Hare is still railing against Thatcherism and its powerful impact not only on the British economy but also on its culture. To be fair, Hare does give some of his wittiest lines to Tom, but the play rarely rises above settling scores and trading insults.

That it made for an unusually entertaining evening owed a great deal to the two English actors who performed it, Michael Gambon and Lia Williams. Gambon is best known to American audiences from the PBS miniseries *The Singing Detective*. He is considered one of England's most important actors, and it seems extremely odd that it has taken him so many years to make his American debut.

Another auspicious debut was that of playwright Stephen Bill, who has had a steady if unspectacular career in England. He made his American debut with *Curtains,* also presented by the aforementioned New Group at the Intar on 7 April. *Curtains* is a dark, acerbic comedy about our indifference to the elderly and death. At one time these subjects were shrouded in piety, even reverence; now our attitudes toward both are as unsentimental and ruthless as our attitudes toward sex.

Curtains begins with a lower-middle-class English family celebrating the birthday of its elderly, ailing matriarch, Ida. Her rigid stare suggests she is only vaguely conscious, and her daughters shout directly into her face to communicate even the most trivial things. The celebration is desultory, to say the least. When the birthday song has been sung and attempts have been made to shove some cake down her throat, she is left alone with her daughter Katharine. Katharine, who has vowed to save her mother from unnecessary suffering, first gives her an overdose of pills, which Ida spits out. Next Katharine tries to suffocate her with a plastic bag, then finally succeeds in smothering her with a pillow.

The death scene is remarkable not only because Laura Esterman, who played Katharine, was so painfully ambivalent about her task but because Kathleen Claypool, who played Ida, was able to project the feelings of her mute character merely with her eyes and her arms. The smothering takes some time, playing with our own ambivalence about whether she should survive or not.

The second act is almost anticlimactic as the members of the family discover first the death, then how it occurred. Their responses are minimal in either case. The play is both comic and brutal, its implications as funny as they are disturbing. It seems odd that Bill's work is so unknown here.

A writer whose work invariably finds productions in New York but is often considerably less interesting than Bill is Caryl Churchill, the author of such plays as *Cloud Nine* (1979), *Top Girls* (1982), *and Serious Money* (1987). Her most recent play, *The Skriker,* which opened on 15 May at the Off-Off-Broadway Joseph Papp Public Theater, was an attempt to find latter-day equivalents for the medieval

English fairies, who, unlike the amiable lot in *A Midsummer Night's Dream,* were a rather malevolent bunch.

We are familiar with fairy tales in which ugly, brutal creatures demand a kiss, in exchange for which either they or we become princes. The Churchill characters who demand professions of love are social outcasts—homeless, mad, or both. The plot, such as it is, concerns two young women. One of them has murdered an infant; the other is pregnant, then the mother of an infant. Both are haunted by the Skriker, who, when not involved in evil deeds, is given to rambling monologues that are supposed to be the ravings of a mentally ill woman but are full of the free-associative puns of James Joyce.

The play was well-directed and mounted, full of haunting stage pictures and sound effects, but it came across as the worst sort of artsy-craftsy twaddle.

A category that seems to grow from year to year is that of plays concerning the Holocaust. This year there were several plays on this tricky subject, including a theatrical debut by the acclaimed novelist Cynthia Ozick. The most substantial of these efforts was Tim Blake Nelson's minimalistic *The Grey Zone,* which was presented by the Off-Off-Broadway Manhattan Class Company and opened at their theater on 10 January.

As a genre Holocaust plays are irritating, because they so often rely on predictable responses and simpleminded evocations of an unimaginable, incomprehensible horror. (The worst example of this was a play called *Anne Frank and Me,* in which an American high-school student, some of whose friends are Holocaust-deniers, makes an imaginary journey back to Nazi-occupied Paris and ultimately, in a cattle car bound for Poland, meets Anne Frank—the plot was insulting and the dialogue on the level of a sitcom.)

By contrast *The Grey Zone* makes this well-worn subject fresh by limiting its focus. Set in Auschwitz, the play is about a squad of men who transfer bodies from the gas chambers to the crematoria. Nelson's play captures the horrifying reductionism of the death camps.

In those factories for the production of corpses, the death camps, life was stripped of everything until it seemed barely more desirable than death. Nelson conveys this powerfully in the image of a teenage girl who miraculously survives a gassing, probably because the corpses above her were so densely packed they prevented her from breathing the gas. However much a miracle her survival seems to the normal observer, her life is a severe liability to everyone around her. The decisions that

David Hare, author of Skylight *(Hulton Getty Picture Collection, Ltd.; photograph by Loveridge)*

must be made about what to do with her—the most sensible thing, of course, would be to kill her—dramatize the tenor of daily life in Auschwitz in an enormously pungent way.

The most interesting character in *The Grey Zone* is a Jewish doctor, Miklos Nyiszli, on whose memoir it is based. His main job is to perform "prep" work on twins for Dr. Mengele's experiments. It is a telling commentary on this bleak, barren landscape that he serves as a moral beacon within it.

The writing throughout is simple and direct. Sometimes visual images speak volumes, as when two men don gas masks to begin their "work." The play also had stunningly powerful sound effects—the two men in masks wait for their cue to begin working, which is the cessation of a frightening grinding noise. In its sparseness and rawness *The Grey Zone* illuminates a chilling subject hauntingly.

Jon Marans's *Old, Wicked Songs* deals with the subject less originally and somewhat artificially, but ultimately it, too, is quite moving. It was first presented Off-Off-Broadway by the Jewish Repertory Theater in 1995 and was moved to the Off-Broadway Promenade Theater, where it opened on 5 September. The title is drawn from a song by Robert Schumann, part of a cycle called "The Poet's Love," which a headstrong young American is studying with a crotchety old Viennese teacher.

An American piano prodigy named Stephen Hoffman is emotionally blocked and unable to con-

Michael Gambon and Lia Williams (photograph by John Haynes)

tinue his career. He has come to Vienna to study with a master, Professor Josef Mashkan, who forces him instead to do preliminary work with an eccentric teacher who, rather than focusing on the piano, forces Hoffman to learn the Schumann songs. The young American has to relearn how to handle his emotions. So does Mashkan, who, we discover (to no great surprise) is Jewish despite his seemingly fascistic attitudes.

In some ways the most disturbing moment in *Old, Wicked Songs* was the young American's realization that Dachau, which seems to exist in another universe, was in fact only a twenty-minute bus ride from Munich, suggesting that the horrors the Germans claimed not to have known about were taking place right next door.

The most glaring artificiality in the play was that what ought to have been a climactic moment—when the old teacher finally agrees to describe what happened to him during the war—takes place in pantomime. This decision injects the odor of Art into what was otherwise a naturalistic play.

Another treatment of the Holocaust theme that has a patina of classical music is Ronald Harwood's *Taking Sides,* which opened on 17 October at the Brooks Atkinson on Broadway. Harwood dramatizes the painful historic moment when the great conductor Wilhelm Furtwängler had to clear himself of charges of collaboration with the Third Reich.

There is abundant evidence that Furtwängler did what he could to help Jewish musicians escape. Harwood, in his introduction to the play, notes that the English were quite willing to clear him, but for reasons still not explained, part of the American team was resolutely opposed to his "de-Nazification." In *Taking Sides* Furtwängler, beautifully played by Daniel Massey, is opposed by a pit bull of an American who was an insurance claims adjustor before the war and brings the suspiciousness and tenacity of his profession to bear on a situation that requires greater sensitivity and nuance.

All that relieves the repetitive confrontation between the two chief characters is the occasional arguments provided by ancillary figures, particularly a young American soldier, born a German Jew, who heard Furtwängler as a child in Berlin and understands and forgives the compromises he made. As the relentless American interrogator, Ed Harris brought surprising sympathy to a largely unsympathetic, monochromatic role, but the character's inability to sound more than one note diminished the potential of the drama.

In some ways the most disappointing of the Holocaust dramas to appear this year (with the obvious exception of the inane *Anne Frank and Me*) was a play by Cynthia Ozick, which opened at the Off-Off-Broadway Jewish Repertory Theater on 24 June. *The Shawl,* based on her novella, concerns two women who survived Auschwitz—Rosa, who owns an antiques store as a way of clinging to remnants of the past, and her niece Stella, who studies psychology at the New School, hoping to find salvation through Sigmund Freud.

Garner Globalis, a Dapper Dan from the Midwest, arrives midway through the first act. He dresses and acts like a traveling salesman from an earlier time (when snake oil was still commercially available) but claims to represent a think tank. In the last half hour of a play that lasts nearly three we learn that Globalis is a Holocaust denier. He hunts down survivors, pretending to help them assuage their grief. His real aim is sinister—he wants to erase whatever traces he can find that provide evidence that the Holocaust existed. Rosa moves to Florida, where one of her neighbors has an extensive library of Holocaust research—Globalis buys it with the intention of making its factual material unavailable.

He mesmerizes Stella into allowing him to kiss the number tattooed on her arm, assuring her that this is the beginning of the erasure of her pain. We do not see or hear how he exorcises the traumas that have haunted Rosa, though this, too, seems to involve a physical embrace. Once she has cleared her mind of admittedly horrific wounds involving a child she bore in Auschwitz, Globalis gets her to sign a paper stating that her memories were a fantasy.

Program cover for Ronald Harwood's play about the "de-Nazification" of conductor Wilhelm Furtwängler

Globalis's success with these haunted Jews is all too facile, and the notion that they would part with their obsessions so readily is ultimately offensive. There are any number of reasons why non-Jews might be ready to deny the Holocaust, but the notion that Jews would surrender their past without a struggle does not resonate.

The one telling point Ozick makes is the easy, comical alliance Globalis forms with a young woman who manages the Miami building in which Rosa and other elderly Jews live. Uneducated, self-absorbed, a total airhead—an all too successful and unnerving caricature of many young people today—she makes a natural, unwitting accomplice for Globalis.

The Shawl, directed by the eminent film director Sidney Lumet, was distinguished by powerful performances, especially by Dianne Wiest as Rosa, but the play was deeply unsatisfying, especially given the expectations raised by so eminent and imaginative an author.

The year was notable for several illuminating revivals, particularly of Edward Albee's 1966 *A Delicate Balance,* which was presented by the Lincoln Center Theater on Broadway at the Plymouth, opening on 21 April. When the play was first performed audiences were puzzled by what seemed an artificial plot device—a suburban couple so unnerved by some unnamed terror that they ask to move in with their closest friends. In the last thirty years nameless phantoms seem less contrived than they did in those more complacent times. With a cast that included Rosemary Harris, George Grizzard, Elaine Stritch, and Mary Beth Hurt, Albee's play seemed far less a piece of iconoclasm than an elegant classical work.

Two Sam Shepard revivals fared less well. *Buried Child,* for which he won the Pulitzer Prize in 1979, was mounted in an overblown Broadway production–his first ever to reach the Main Stem. It opened at the Longacre on 30 April. This parable about a family ever at war with itself and determined to destroy its young seemed terribly dated and adolescent. Even less successful was a revival of his 1972 *The Tooth of Crime.* The latter, one of the first of his plays to receive critical adulation, invoked the mythology of the Old West. The showdown it depicts does not involve two gunslingers, but rather an aging pop icon and a cocky challenger. Although the play overflows with verbal innovation, its human core seems quite empty. It opened on 22 December at the Lucille Lortel, presented by two Off-Off Broadway companies, Second Stage and Signature, whose mandate is to revive recent plays rather than classics. Signature devoted its entire 1996–1997 season to exhuming Shepard's early plays.

There was a slew of musical revivals, the most impressive of which was Christopher Renshaw's total rethinking of Rodgers and Hammerstein's *The King and I,* which opened on 11 April at the Neil Simon on Broadway. Renshaw, an Englishman who has spent considerable time in Asia, gave his production an intense Thai flavor. That the musical could support this overlay of authenticity suggests it is not the condescending treatment of the Third World it has often been criticized for being.

Donna Murphy was a stern, troubled Anna, not the bubbly Lady Bountiful we usually see. Astonishingly, a young actor named Lou Diamond Phillips found ways to play the King that let him claim a role that has hitherto seemed to belong eternally to Yul Brynner. Phillips's King, despite his undeniable sensuality, had a boyish innocence that made his "foreignness" and autocratic manner easier to forgive. One came away with a new appreciation of the brilliance of "Shall We Dance," the reticence of whose lyrics and gestures barely contain the potency of the two characters' feelings for one another.

In an election year it was inevitable that politics would make itself felt. Douglas McGrath, who used to write an extremely droll column satirizing President Clinton in the *New Republic,* wrote a one-person play, *Political Animal,* almost totally devoid of humor. Christopher Durang, many of whose plays (*A History of the American Film* [1978], *Sister Mary Ignatius Explains It All For You* [1982]) seem merely to be extended skits, wrote a numbingly unfunny work, *Sex and Longing,* which was a feeble response to the assaults of the religious Right on the arts. (In the play, Jesus himself attends congressional hearings and is dismissed by right-wing zealots.)

Similarly strained was Freyda Thomas's adaptation of Molière's *Tartuffe* to the contemporary Bible Belt. In *Tartuffe: Born Again,* which opened at the Broadway subscription theater Circle in the Square on 30 May, Tartuffe is a TV evangelist. The action takes place in his TV studio where Dorine, a wily servant in the original, is the floor manager. In a good production of *Tartuffe* you can believe the relationships between the tyrannized family, their myopic father, and the opportunistic Tartuffe. Here it was purely a cartoon, Elmer Gantry in Dogpatch.

Politics of another sort are evident in Suzan-Lori Parks's *Venus,* which opened Off-Off-Broadway at the Joseph Papp Public Theater on 2 May. This is the sort of play that wins plaudits in academia and a fair share of grants from corporate entities eager to demonstrate their political correctness. Parks's play is about a nineteenth-century phenomenon, an African woman exhibited in London in 1810 under the title "The Venus Hottentot." She was later bought by a French doctor, who published an anatomical analysis of her. Presumably he also used her for his pleasure, though the only way Parks depicts it is to show him masturbating at the side of their bed.

In some ways the subject recalls that of *The Elephant Man.* Parks, however, has nothing to say beyond the most obvious—she was victimized because she was a woman and because she was black. What gives the play marginal interest is Parks's depiction of her as a carnival attraction, the focus of a world that can regard her only as a freak. This concept gives the play a strong visual style, but it is an idea that is interesting for about ten minutes.

Venus was staged by one of the stalwarts of the avant-garde, Richard Foreman. The production had such Foreman trademarks as strings transversing the stage at various angles and a red lightbulb, somewhat out of place in the nineteenth-century primitivist decor. Happily he did not employ his customary frightful-sounding buzzer, usually a signal for the actors to repeat the previous scene. (Such repeats are here part of the actual text.)

In some ways the text itself is the most interesting thing about *Venus.* Much of it is spelled in a style that suggests poor black southern dialect. The word *any,* for example, no matter which character says it, is consistently spelled *inny*. *Is* is generally spelled *iz*. In an early speech one of the circus characters says,

> I regret to inform you that the Venus Hottentot iz dead. There wont b inny show tonite.

One of the Venus Hottentot's last speeches is written as follows:

> Tail end of the tale for there must be uh end.
> Is that Venus, Black Goddess was shameless, she sinned, or else
> Completely unknowing thuh Godfearin ways, she stood showing her ass off in her iron cage.
> When Death met Love Death deathd Love and left Love tuh rot
> *Au naturel* end for thuh Miss Hottentot.
> Loves soul, which was tidy, hides in heaven, yes, thats it
> Loves corpse stands on show in museum. Please visit.

A chorus then chants:

> Diggidy-diggidy-diggidy
> Diggidy-diggidy-diggidy-dawg!

If Ebonics becomes an academic discipline, *Venus* may qualify as an instant classic.

In the category of freak shows falls *Cakewalk,* a play by Peter Feibleman about his affair with the great poseur (poseuse?) Lillian Hellman. It opened Off-Broadway at the Variety Arts on 6 November. It is possible that Tennessee Williams in his prime might have made us believe that an old, alcoholic, chain-smoking, ill-tempered crone could ensnare a handsome stud at least twenty years her junior, but even *Sweet Bird of Youth* (1959) doesn't go that far.

The amazing thing, of course, is that this did happen. Feibleman, a young writer, did somehow fall under the spell of Hellman, who was a friend of his family in New Orleans. They had an affair that lasted decades, which the play is unable to make plausible.

Hellman was, in her own way, an imposing figure, but in this version she is largely a wisecracking Jewish grandmother who occasionally uses swear words. In the hands of the skillful Linda Lavin, Hellman came across as smart, peckish, and domineering, but ultimately amiable. Since none of her fictional creations was as interesting or arresting as the heroic identity Hellman fashioned for herself, this airbrushed version seems a disappointment.

The Year in Children's Books

Caroline C. Hunt
College of Charleston

It was a year of contradictions. Familiar juvenile imprints continued to disappear into the maws of huge conglomerates; meanwhile, new small houses sprang up with specialized lists. Though the total number of publishers shrank, the number of new titles grew. Politics and public taste alike seemed more conservative, but children's and young adult books on highly controversial topics abounded. The lines between children's book categories blurred: there were picture books for young adults, tales halfway between fiction and nonfiction, and distinctly postmodern stories in which the author appeared as a real or imagined character. The most talked-about high fantasy book of the year, *The Golden Compass,* by Philip Pullman (Knopf), had a female protagonist and an anti-authoritarian aura, but it was widely compared to J. R. R. Tolkien's *The Lord of the Rings* (1954–1955). The two most talked-about picture books of the year were *The Story of Little Babaji,* by Helen Bannerman, with illustrations by Fred Marcellino, and *Sam and the Tigers,* by Julius Lester, with illustrations by Jerry Pinkney. Both dealt with a resourceful boy from non-European culture, but they were, avowedly, adaptations of a story first published nearly a century ago.

A mere seventeen corporations produced 84 percent of the juvenile titles in 1995, according to an exhaustive survey in *Emergency Librarian* for September 1996. The U.S.-based corporations included National Amusements (9.9 percent), Houghton Mifflin (6.8 percent), Hearst Publications (6.3 percent), Advance Publications (5 percent), Scholastic (4.4 percent), Harcourt General (4.2 percent), Time Warner (3.7 percent), and Walt Disney (1.7 percent). Familiar imprints owned by non-U.S. companies include Cobblehill, Dial, Dutton, Lodestar, and Viking, which belong to the U.K. corporation Pearson PLC (12.2 percent); HarperCollins, which is owned by another U.K. corporation, News Corporation; Scarecrow Press and Children's Press, owned by the French Lagardere Groupe; and Putnam, which belongs to Japan's Matsushita (percentage figures refer to the 1995 supplement to the *Children's Catalog*).

Sales were steady—a trend begun in 1995 after a worrisome downward slide. Retailers hung onto their 1994–1995 gains in market share; teachers, critics, and analysts noted with mixed feelings the shift from libraries to bookstores as the primary market for children's books. However, a lengthy article in the *Christian Science Monitor* of 26 September observed that the number of children's titles published annually had more than doubled (to fifty-five hundred plus) in recent years, accompanied by a rise in specialty children's bookstores from seven to approximately seven hundred. The number of small houses putting out specialty lists also increased markedly (these do not appear in the *Emergency Librarian* survey, partly because most are too new and partly because their individual market shares are so small).

Censorship efforts also held steady: the annual People for the American Way report on school censorship mentioned numerous challenges to Lois Lowry's Newbery-winning *The Giver,* on grounds ranging from offensive language to unacceptable themes (such as euthanasia and "mind control") and use of words such as *clairvoyance* and *transcendent,* vocabulary attacked by one challenger as part of "occult New Age practices." Only one review committee—in Brecksville, Ohio—is known to have removed the book completely, but several others restricted its use, for example forbidding reading it aloud or assigning it in class; agreeing to give parents lists of books in advance was another compromise. One striking development was the increased use of psychologists and other expert witnesses in favor of censorship. A University of Kansas professor, for example, filed a complaint about M. D. Hahn's *Wait 'til Helen Comes* on the grounds that its supernatural phenomena would terrify children; several censorship attempts to ban books on homosexuality also used expert witnesses.

Meanwhile, a spate of anniversaries showed how much remained unchanged in the world of children's books. The American Library Association (ALA) marked seventy-five years of Newbery Medal presentations for best children's books by

Two new versions of Little Black Sambo *published in 1996*

American authors. A handsome poster depicts the front covers of all the winners to date. The 1996 award, announced at the ALA's midwinter meeting, went to Karen Cushman's second medieval title, *The Midwife's Apprentice* (Clarion), no surprise to the many professionals who applauded the previous year's Newbery Honor book, *Catherine, Called Birdy* (Clarion). Shorter and more tightly constructed than its predecessor, the winning book appeals to middle grades and young adults alike. Strong sales of *Catherine* in 1995 continued into 1996 unabated. The surprise award of the year was the Caldecott Medal for best picture book, which went to Peggy Rathmann for *Officer Buckle and Gloria* (Putnam), an unpretentious tale of a police officer and his humorous dog. The cartoonlike illustrations went over well with young readers, and soon there were not only tapes, Braille versions, and the usual accompaniments, but even rubber stamps from Kidstamps.

The Detroit Public Library celebrated the centennial of its children's services in May with a week of activities. Among publishers, a survivor also celebrated an anniversary. William Morrow's spring catalogue sported a balloon logo with the legend "Morrow Junior Books–Fifty Years"; the back cover included a tear-off anniversary bookmark and photographs of famous Morrow authors (such as Jerry Pinkney and Steven Kellogg) as children, with the authors' names beneath the pictures in scrambled form ("RYJER KNNEYIP," "EVENTS GELLKOG").

Some celebrations were more controversial. The twentieth anniversary of Bruno Bettelheim's *The Uses of Enchantment* provoked considerable comment among scholars, who still do not agree either about Bettelheim's ideas or about the value of fairy tales. The seventieth anniversary of the Winnie-the-Pooh books was observed in 1996. By ironic coincidence Christopher Milne, the original Christopher Robin, died in April–more than two decades after memorializing an unhappy real-life childhood in *The Enchanted Places* (1974). At least one Milne biographer has disputed his memories.

The passing of Christopher Robin, the aging of Pooh, and the increasing criticism of many children's classics–for example, in such works as Herbert R. Kohl's *Should We Burn Babar?* (1995; widely distributed in paperback, 1996)–heightened the end-of-an-era feeling that permeates the 1990s. The deaths of an unusual number of children's writers and illustrators during 1996 also contributed to this sense. P. L. Travers (Helen Lyndon Goff) died at age ninety-six, within a few days of Christopher Milne. Mimi Kramer's appreciation in *Time* on 6 May commented on the major innovation of Mary Poppins: "The invention of an authority figure who was imperfect though invariably wise and right was a novel idea." Another favorite British children's author, Leon Garfield, died on Sunday, 2 June, at the age of seventy-four. Best known for his historical novels, he was also praised for his retellings of classical and biblical stories; *The God Beneath the Sea*

Dust jacket for the winner of the Caldecott Medal for the Best Children's Picture Book of the Year in 1996

won the 1970 Carnegie Medal. On 25 March the world of children's book illustrators also lost one of its award winners, Leo Politi, who received the Caldecott Medal for *The Mission Bell* (1953). Politi was eighty-six years old. Another loss among illustrators was Garth Williams, known for *Stuart Little* and *Charlotte's Web* as well as scores of other books; he died on 8 May at age eighty-four.

Several veterans of the children's book scene left long and varied legacies. Myra Cohn Livingston, whose credits include more than eighty volumes of poetry for children as well as good observations about children and verse, died on 23 August at the age of seventy. Eleanor Cameron, known for both her children's books (including the Mushroom Planet series) and her criticism (*The Green and Burning Tree*), died on 11 October at age eighty-four. Paul Heinz, editor of *The Horn Book Magazine* from 1967 to 1974, died in April of cancer. Heinz championed the study of children's literature before it became popular; in addition to founding Children's Literature New England, he wrote many influential articles, taught children's literature courses, and produced a translated version of *Snow White* with illustrations by Trina Schart Hyman. Glenn Estes of the University of Tennessee died on 16 July at sixty-one; he had taught children's literature for many years, served on ALA award committees, and edited three volumes on children's literature for the *Dictionary of Literary Biography*. Estes also pioneered in the dissemination of children's book information on the World Wide Web; his page remains accessible. Ruth Morris Graham, a former missionary and teacher who wrote several children's books, died at age ninety-five on 24 August. An even longer-lived children's author died on 26 June: Delia Goetz, age one hundred, who had gathered material for fiction through her travels as a government education specialist.

A sadly premature death was that of Pam Conrad, who succumbed to cancer on 22 January; the author of *Prairie Songs* and *Stonewords* (both multiple award winners) as well as numerous books for younger children was only forty-eight. *Zoe Rising* (Laura Geringer/HarperCollins), a sequel to *Stonewords*, was released posthumously during the summer. Another posthumous sequel was *The Beduins' Gazelle* (Richard Jackson/Orchard), by Frances Temple, who died at the age of forty in 1995. In this book Etienne, the young pilgrim of *The Ramsay Scallop*, returns as a student in Fez; there he meets Atiyah, of the desert Beni Khalid, who has been sent to

Fez by his scheming uncle Saladeen. In spite of a somewhat sentimental ending, the story of the two teenagers from different cultures in the year 680 is a worthy memorial.

Not all posthumous books were "serious." A predictable best-seller was the posthumous Dr. Seuss title, *My Many Colored Days* (Knopf), with splashy illustrations by Steve Johnson and Lou Fancher. The dedication, by the author's widow, reads simply "to Ted, who colored my days . . . and my life." *Days*-related illustrations enlivened the Seussville web site, where at Christmas a Seuss holiday game was popular.

Many favorite authors are still with us after decades of popularity. Astrid Lindgren delighted old fans and new with the first new Pippi Longstocking in three decades, *Pippi's After-Christmas Party*, with pictures by Michael Chesworth (Viking); the original Pippi book turned fifty last year. The longest-running continuous staff member of *The New Yorker* also produced a new book. William Steig's *The Toy Brother* (Michael di Capua/HarperCollins) is unusual in featuring human characters rather than animals. The story concerns a medieval family of alchemist, wife, and two brothers, one of whom inadvertently shrinks himself in his father's laboratory. Steig's earlier masterpiece, *Abel's Island*, marked its twentieth anniversary. Bernard Waber's long and successful career as an author/illustrator continued with the engaging *A Lion Named Shirley Williamson* (Houghton Mifflin), in which an animal named by mistake becomes the most popular exhibit in a zoo. Waber's Lyle the crocodile will observe his thirty-fifth anniversary in 1997, having debuted not in *Lyle, Lyle Crocodile* (1965), as one might think, but in the earlier book *The House on 88th Street* (1962), later made into an animated musical.

Some children's authors are famous not for their books but for other reasons: the former Duchess of York, for example, is not best known for her Budgie books. Similarly, it is fortunate that Jimmy Carter's reputation does not rest on his 1996 children's title, *The Little Baby Snoogle-Fleejer* (Times Books), with pictures by Amy Carter. Eric Idle, of Monty Python fame, reworked Edward Lear's best-known poem in *The Quite Remarkable Adventures of the Owl and the Pussycat* (Dove Kids), with illustrations both from Lear and, additionally, by Wesla Weller. In pun-laden prose Idle narrates the quest to save the Bong Tree from the evil Fire Lord. Mary Chapin Carpenter's *Dreamland*, illustrated by Julia Noonan, offered in print form a song from the compact disk *'Til Their Eyes Shine: The Lullaby Album* (HarperCollins). Jamie Lee Curtis's picture book, *Tell Me Again About the Night I Was Born* (HarperCollins), began as a story for her own adopted child and was instantly popular with parents and critics alike; *U.S. News and World Report*, in its 16 December roundup of children's books, called the tale "funny and warm."

Two name-brand poets came out with long-awaited new volumes. Shel Silverstein's *Falling Up* (HarperCollins), his first new book of poetry in twenty years, maintains the high standard of his previous works and, inevitably, reached the top of the children's nonfiction best-seller list. Jack Prelutsky's *Pizza as Big as the Sun*, with pictures by James Stevenson (Greenwillow), reached number three on the same *Publishers Weekly* list. Reviewers praised the oddball perspectives and rhymes in both books. Prelutsky's collaboration with Peter Sís, *Monday's Troll* (Greenwillow), was also popular.

Unquestionably the most famous of famous poets to grace the juvenile scene in 1996 was Emily Dickinson. *Poems for Youth*, chosen and illustrated by Thomas B. Allen (Little, Brown), provides attractive black-and-white drawings for some of Dickinson's poems. A purist might wish that Allen had not used (presumably for copyright reasons) the 1918 Bianchi edition; young readers may be quite surprised to meet the same poems later with dramatically different punctuation.

At the opposite extreme from the promotion of famous authors, some publishers submerged writers completely in favor of a uniform marketing format. Yes, the vogue for series continued without relief. The ingenious Magic School Bus series, with text by Joanna Cole and illustrations by Bruce Degen, continued its decline with *The Magic Schoolbus Inside a Beehive* (Scholastic), a contrived tale in which both children and bus develop yellow stripes and wings. The American Girl books, with matching toys, were still to be found everywhere, and imitators were still arriving on the scene. Scholastic's Dear America series debuted in 1996: small volumes with dark, serious-looking covers. Each book also has a matching sewn-in ribbon bookmark. Among the first titles in the series was *When Will This Cruel War Be Over?*, subtitled *The Civil War Diary of Emma Simpson, Gordonsville, Virginia, 1864*. In keeping with the authentic-looking presentation, the book carries the name of the author, Barry Denenberg, on the title page and copyright page but not on the cover. *The Winter of Red Snow: The Revolutionary War Diary of Abigail Jane Stewart, Valley Forge, Pennsylvania, 1777* also downplays the author, Kristiana Gregory, an experienced historical writer for young audiences. A parallel series from Simon and Schuster, the Girlhood Journeys Collection, included protagonists from different countries as well as different eras: fifteenth-

Dust jacket for Shel Silverstein's first book of verse in twenty years

century Nigeria, medieval England, and prerevolutionary Paris, for example.

Continuing the spate of creepy series that sometimes seem to have taken over the 1990s, at least judging from mass-market bookstores, several paperback series focused on the strange or disgusting, rather than the simply scary. Bone Chillers, from Harper Paperbacks, debuted in 1994, scaled back slightly in 1995 (from four titles to three), and continued in 1996 with *Welcome to Alien Inn, Attack of the Killer Ants,* and *Slime Time.* The bug theme surfaced again in a series for slightly older readers, The Bug Files from Berkley, an "all-new series of insanely frightening adventures starring all kinds of creepies, crawlies, cooties, hairies, stinkies, and slimies that know only one thing: Human beings are delicious!" The first volume in the series, *Squirmsters!,* by David Jacobs, involves "genetically engineered super caterpillars" breaking loose from their container on a small airplane and pursuing seventeen-year-old Kelly Wade. She is pictured on the cover surrounded by the super caterpillars and accompanied by the caption "The *ultimate* airline food!" Finally and most spectacularly, the Barf-o-Rama series from Bantam Young Readers (*The Great Puke-Off, The Legend of Bigfart, Mucus Mansion,* and so on) topped all its 1996 debut competitors for carefully crafted salable grossness. Incredulous readers may check its web site at http://www.bdd.com/barforama for details. Ann Doss Helm's column in the Danbury, Connecticut, *News-Times* described this and other Goosebumps competitors on 31 July, covering Animorphs, Spine Chillers, and the TV tie-in X-files series in addition to Barf-o-Rama. (The Animorph series also has its own web site.) As the year ended, Goosebumps was losing ground to several of the rival series, especially Animorphs; librarians reported record numbers of Goosebumps titles on their shelves.

Though series books seemed to be everywhere, the market for traditional children's genres did not seem to suffer. The picture books of 1996 continued the trend of the last few years toward multiculturalism, with a particularly strong showing of distinguished books by African American writers and illustrators. Walter Dean Myers teamed up with Synthia Saint James in a trickster story for the earliest grades, *How Mr. Monkey Saw the Whole World* (Doubleday). Alan Schroeder, author of last year's well-regarded *Carolina Shout!,* returned to the theme of his first book, *Ragtime Trumpet,* a multiple award winner; *Satchmo's Blues* follows young Louis Armstrong up through the purchase of his first dented trumpet from a pawn shop and is handsomely illustrated by Floyd Cooper, who has won several Coretta Scott King honor designations for previous books. Cooper's other big picture book, *Mandela* (Philomel) tells the South African leader's life in spare prose and richly colored pictures. In a new partnership, Grace Hallworth selected traditional Caribbean rhymes, and Caroline Binch, known for *Amazing Grace* and *Hue Boy,* illustrated them with a lively group of Caribbean youngsters. The collection, *Down by the River,* is subtitled *Afro-Caribbean Rhymes, Games, and Songs for Children* and comes from Scholastic.

For her second picture book, *Big Wind Coming!* (Albert Whitman), Karen English teamed up with Cedric Lucas (who has four previous books to his credit). Text and illustrations together give a convincing portrayal of an African American family facing a hurricane.

Evelyn Coleman's third picture book, *White Socks Only,* illustrated by Tyrone Geter (Albert Whitman), concerns a little girl's experience with a "whites only" drinking fountain in Mississippi; not understanding the sign, she carefully removes her black patent shoes before drinking from the fountain. The peaceful demonstration and mysterious victory that follow are told convincingly as a story to the little girl's granddaughter many years later. This is an outstanding book for any age. Going further back in time, Tony Johnston's *The Wagon,* with paintings by James E. Ransome (Tambourine), follows a Carolina family from slavery to emancipation.

Daniel and Kimberly Adlerman, who collaborated on *It's Raining, It's Pouring* (1995) under the name Kin Eagle, used their real names for the attractive picture book *Africa Calling, Nighttime Falling* (Whispering Coyote Press). The story ends with the lines "Slumbering through the darkest night, / I sleep protected till morning light. // Africa calling / nighttime falling / warmly beaming / peaceful dreaming," as the evocative wild animals dwindle away into friendly looking stuffed toys on a child's bed.

Rosemary Breckler related the childhood memories of her friend Jessica Huong Dang in *Sweet Dried Apples: A Vietnamese Wartime Childhood* (Houghton Mifflin), illustrated by Deborah Kogan Ray. Minfong Ho, author of *The Clay Marble,* teamed up with a new illustrator for *Hush! A Thai Lullaby* (Orchard Books); Holly Meade experimented with media for this book by using cut-paper and ink instead of her usual torn-paper collages.

Liz Rosenberg, author of last year's highly acclaimed *Carousel,* returned to her Eastern European family roots in *Grandmother and the Runaway Shadow* (Harcourt, Brace), illustrated by Beth Peck; author and illustrator dedicated the book to two young Jewish immigrant women—their grandmothers. Patricial Polacco's 1996 offering also dealt with Eastern European immigrants: based on a memory of her childhood, *The Trees of the Dancing Goats* (Simon and Schuster) tells of a winter during which her family made Christmas trees for neighboring families under quarantine. Polacco's own grandmother died during 1996, and the book is dedicated to her. Polacco had another 1996 title, *I Can Hear the Sun* (Philomel), featuring a park caretaker, two homeless people, a blind goose, and a settlement-house child named Fondo who is "not much on learning" and is about to be sent away because of this; like *The Trees of the Dancing Goats,* it ends with a satisfying surprise. Polacco's *Aunt Chip and the Great Triple Creek Dam* (Putnam) was also well received.

Some picture books defied easy classification. Tim Ladwig provided stunning illustrations for the classic hymn by Eleanor Farjeon in *Morning Has Broken* (Eerdmans). Tones of green and gold unite the joyful pictures. Also from Eerdmans, *Come Sunday,* by Nikki Grimes, illustrated by Michael Bryant, combines verse and luminous watercolors. Lisa Westberg Peter's poem of the seasons, *October Smiled Back* (Henry Holt), personalizes each month; Ed Young's dynamic collage-and-colored-pencil illustrations match perfectly. Dav Pilkey had two 1996 books: *God Bless the Gargoyles* (Harcourt, Brace) tells in verse of the respected origins of gargoyles, their decline from favor, and the partnership with angels that enables them to be joyful again; *Paperboy,* in contrast, takes place on earth.

David Wisniewski's *Golem* (Clarion) tells the story of the creature from the Prague ghetto; after a slow start, the book was widely admired by critics. An unusual picture book was *Jack, Skinny Bones, and the Golden Pancakes,* by M. C. Helldorfer, with pictures by Elise Primavera (Viking). This traditional-sounding trickster story is actually an amalgam of separate plot devices used in earlier sources as diverse as Uncle Remus, the Grimms, and Chaucer. Another synthetic folktale was Tony Johnston's *The Magic Maguey,* illustrated by Elisa Kleven. When little Miguel finds a way to save the great maguey plant that has given the villages so many useful things, even the rich landowner who wanted to cut it down has to agree to leave it in place. From Orchard, the unpretentious but delightful *Hold the Anchovies: A Book about Pizza,* by Shelley Rotner and Julia Pemberton Hellums, with photographs by Rotner, clearly illustrates the ingredients and making of pizza. Equally unpretentious, Margaret Wild's *Old Pig,* with pictures by Ron Brooks (Dial), offers a gentle look at mortality. Another sleeper came from a relative beginner. Philemon Sturges, a practicing architect, came to the world of picture books in 1995 with *Ten Flashing Fireflies* (North-South). His 1996 title is *What's That Sound, Woolly Bear?*, illustrated by Joan Paley (Little, Brown), a wonderful account of a caterpillar going quietly on her way as many kinds of other bugs make a series of satisfyingly vivid noises. Continuing the I Spy series, Jean Marzollo's *I Spy Spooky Night*, with photographs by Walter Wick (Cartwheel), climbed quickly to number one on the *Publishers Weekly* picture book best-seller list, beating out the posthumous Dr. Seuss title *My Many Colored Days*.

Several award winners showed they had not lost their touch. One of the best-received picture books of 1996 was *Lilly's Purple Plastic Purse*, by Kevin Henkes (Greenwillow). Readers who saw Lilly last in *Julius, the Baby of the World* will welcome Lilly's starring role as she struggles with school and her admiration for her teacher, Mr. Slinger, who wore "artistic shirts," "glasses on a chain around his neck," and "a different colored tie for each day of the week." Barbara Cooney's *Eleanor* (Viking) casts in picture-book format the story of Eleanor Roosevelt's lonely childhood. In a fanciful mode, *Grandmother's Pigeon*, by Louise Erdrich (Hyperion), relates the story of a highly unusual grandmother who takes off for Greenland, leaving among her collections a nest of eggs that hatch into passenger pigeons. Jim La Marche's luminous painting sets off the air of this story perfectly. *The Leaf Men and the Brave Good Bugs* (Laura Geringer/HarperCollins) was a favorite with reviewers, librarians, and children. William Joyce's magical tale concerns an old lady's garden, neglected when she falls ill but rescued by a brigade of courageous doodlebugs, willing to climb a tall tree to fetch the Leaf Men to the rescue.

Imports were also of high quality. Graeme Base ("a.k.a. Rowland W. Greasebeam, B. SC.") added to his successful picture books with *The Discovery of Dragons* (Harry N. Abrams). Lars Klinting's *Bruno the Tailor* (Henry Holt) shows how Bruno the Beaver makes himself a new apron. The book, published in Sweden (also in 1996) by Alfabeta Bokförlag AB, includes step-by-step directions that children can follow to make their own aprons; the appealing illustrations are in watercolor and colored pencil. Another fine import was John Burningham's *Cloudland* (Crown), an attractive fantasy with the author's unmistakable illustrations.

Several picture books were clearly designed for readers a bit older than the preschool set. The all-star partnership of poet Donald Hall, versatile illustrator Barry Moser, and Harcourt Brace's Browndeer Press imprint produced the delightful *When Willard Met Babe Ruth,* an intergenerational story that begins with young Willard and his father pulling Ruth's car out of a ditch in 1917 and ends nearly two decades later when Will takes his daughter (named Ruth) to see one of the Babe's last great games. Continuing her long record of distinguished picture books, Emily Arnold McCully's *The Bobbin Girl* (Dial) follows ten-year-old Rebecca's experiences as a mill girl in Lowell, Massachusetts, in the 1830s. A strike (based on an actual occurrence) provides an exciting plot line around which the details of young women's and girls' employment can be plausibly described.

One book was also better suited to older readers, but for a different reason: David Pelletier's *The Graphic Alphabet* (Orchard), not really an alphabet book but a meditation on the relationship between form and meaning. *B* is illustrated by a blue ball bouncing in the form of a slanted *b: k* by a knot formed of two slender pipes in the shape of a *k,* and *t* by a capital *T* falling over, precipitated by a small round object at its foot. Some of the letters might scarcely be recognizable without their captions—"rip" and "web," for instance—while others, such as "step," are so much altered that they might confuse the smallest readers. As a thoughtful adaptation of a traditional genre, this book with its computer-generated images is utterly delightful.

The most talked-about picture books of the year were two new versions of Helen Bannerman's classic (and controversial) *Little Black Sambo.* Bannerman's own version, a tiny book with her own illustrations, is still available for purchase, though many libraries have either weeded it out or put it into historical collections because of its racist reputation. *The Story of Little Babaji,* using Bannerman's original text but replacing the illustrations with new ones by Fred Marcellino, attempts to sanitize the story by making Sambo (rechristened Babaji) clearly an Indian, not an African as he appears in the original pictures. Aside from the name changes from Sambo, Mambo, and Jumbo to Babaji, Mamaji, and Papaji, the text remains Bannerman's; the title page shows Bannerman's and Marcellino's names inscribed on a green circle, with the title in the middle and four tigers running around the outside. Julius Lester's version, *Sam and the Tigers,* illustrated by Jerry Pinkney, retells the story in the Southern oral tradition. Lester has been convincing in his explanations of why and how he reworded the story and has discussed the book extensively on at least one online discussion group. Every critic in the English-speaking world weighed in on the rival

Julius Lester and Jerry Pinkney, author and illustrator of Sam and the Tigers

Sambo books; Adam Gopnik of *The New Yorker* devoted several columns to "the bizarre proliferation of alternative Sambos" in a dyspeptic survey of children's books on 18 November titled "Grim Fairy Tales," bearing the subhead "It's the end of the century and the little ones are in for it."

Many picture books deal with myth and folklore, and most of these are for a wide range of ages. One of the most ambitious entries in the folk/myth category was Marilyn McFarlane's *Sacred Myth: Stories of World Religions,* from Sibyl Publications of Portland, Oregon. Inspired to compile an anthology of tales from different faiths by the lack of such a book for her grandchildren, McFarlane selected five narratives from each of seven major traditions: Buddhism, Christianity, Hinduism, Islam, Judaism, Native American, and Sacred Earth. The book is lavishly illustrated with computer-generated images and printed in Korea on glossy paper stock. An interesting contrast to this was the more academic *One World, Many Religions,* by Mary Pope Osborne (Knopf), a nonfiction approach lucidly presented and illustrated with photographs.

Uma Drishnaswami retold a series of Indian tales in *The Broken Tusk: Stories of the Hindu God Ganesha,* illustrated by Maniam Selven (Linnet Books). Demi's *The Dragon's Tale, and Other Animal Fables of the Chinese Zodiac* (Henry Holt) illustrates the twelve signs of the Chinese Zodiac in beautifully integrated double-page spreads, each containing a large picture on one side and a hollow picture ring on the left, with a short narrative and a moral inside. *Eleven Nature Tails: A Multicultural Journey,* by Pleasant DeSpain with illustrations by Joe Shlichta (August House), is precisely summarized by its title. The publisher's American Folklore and Storytelling series, to which this volume belongs, includes a greater variety of titles than one might expect–from the well-received *Race with Buffalo and Other Native American Stories for Young Readers,* by Richard and Judy Dockrey Young, to the same authors' *Favorite Scary Stories of American Children,* Jackie Torrence's *The Importance of Pot Liquor,* and Ed Stivender's wonderfully titled *Raised Catholic (Can You Tell?).* The *Dial Book of Animal Tales from around the World,* by Naomi Adler, with pictures by Amanda Hall, appeared in the United States (Dial) and Britain (Barefoot Books). Adler is involved in storytelling, mask making, drama, and many other activities, and this background shows in the oral quality of the writing. Verna Aardema retold a Masai story, *The Lonely Lioness and the Ostrich Chicks,* with splendid stylized pictures by Yumi Heo. Commendably precise, the copyright page explains that the story is retold from an earlier anthology by the same author, now out of print, and that the source of that telling was "The Story of the Ostrich Chicks," in *The Masai: Their Language and Folklore,* by Claude Hollis (1905)–both long out of print.

European folklore was not forgotten. Marie Charlotte Craft's *Cupid and Psyche* retelling, beautifully illustrated by Kinuko Y. Craft (Morrow), figured on many "best" lists. So did *Midsummer Night's Dream,* a simplified version of Shakespeare's play with text by Bruce Coville and pictures by Dennis Nolan. Omar Rayyan revised pictures from a 1504 sketchbook by Kimmelino dePerugia–"reconstruction" is the word used by Rayyan himself–to provide illustrations for Eric Kimmel's version of *Count Silvernose* (Holiday House), retold from the Italian of Italo Calvino. A more eclectic offering was Jane Yolen's *Here There Be Angels,* illustrated by David

Wilgus (Harcourt Brace), which follows her previous books *Here There Be Dragons, Here There Be Unicorns,* and *Here There Be Witches.* The clear winner in the folk category, though, came from one of the century's great champions of children and their reading: Iona Opie edited *My Very First Mother Goose,* with illustrations by Rosemary Wells. The artist departs from her usual style almost in the direction of Sendak, with very satisfactory effect.

The peculiar category known as early chapter books attracted some fine professionals. For the Viking Easy-to-Read series, Betsy Byars produced *My Brother, Ant,* with illustrations by Marc Simont. Ant is short for Anthony, whose long-suffering big brother (unnamed) narrates the three amusing stories in the book. The equally prolific Dick King-Smith contributed to Hyperion Chapters, a series of beginning chapter books for second and third grades, with *Jenius, the Amazing Guinea Pig,* illustrated by Brian Floca. Though slight, the story retains the author's trademark gentle humor. For a wider readership, Candlewick published *Dick King-Smith's Animal Friends: Thirty-one True Life Stories,* illustrated by Anita Jeram in watercolor and ink. From Dodo the dachshund to the pigeon that stowed away on a cruise ship, these James Herriot–like stories will delight animal lovers of any age.

Nonfiction remained high in quality, as it has been throughout the 1990s. Diane Stanley's *Leonardo da Vinci* (Morrow), copiously illustrated both by Stanley and by a selection of Leonardo's own sketches, emerged as an outstanding juvenile biography. In the modern period, a family's Holocaust experience is recounted in *Four Perfect Pebbles: A Holocaust Story,* by Lila Perl and Marion Blumenthal Lazan (Greenwillow). From the small town of Hoya in northwest Germany, the Blumenthal family's war years took them first to Westerbork and then to Bergen-Belsen before liberation and an eventual move to the United States.

Biographies of female role models in a variety of fields figured prominently in 1996. Among the more traditional, Marie Tennent Shephard's *Maria Montessori, Teacher of Teachers* (Lerner) gives unusual emphasis to Montessori's early career (she was Italy's first female doctor) and is illlustrated copiously with excellent photographs. Previous works from Lerner include biographies of Margaret Bourke-White, Marie and Irene Curie, and Lillian Hellman. Also from Lerner, Jeff Savage's *Julie Krone: Unstoppable Jockey* is part of a series called The Achievers. Though clearly told, the story suffers from weak exposition: "Racehorses are expensive to buy and to maintain. The owner . . . hires a trainer to take care of it. The trainer hires a jockey to ride the horse in a race. A jockey hires an agent to find riders," etc. Another small press, Enslow Publishers of Springfield, New Jersey, and Aldershot, United Kingdom, added Barbara Kramer's *Toni Morrison: Nobel Prize–Winning Author* to its African American Biographies series. Unlike many juvenile biographies, the volume on Morrison includes scrupulous documentation, a nicely done chronology, and a carefully selected list of books for further reading. *Black Artists in Photography, 1840–1940,* by George Sullivan (Cobblehill), profiles eight photographers: Jules Lion, Augustus Washington, James P. Ball, the three Goodridge brothers, Cornelius M. Battey, and Addison Scurlock. Sullivan, author of more than one hundred juvenile titles, saw the need for this book when doing research for a volume on Matthew Brady.

A Strange and Distant Shore, by Brent Ashabranner, describes the exile of seventy-two Great Plains Indians to Saint Augustine, Florida, in 1875. The documentary account includes, as is usual for Ashabranner, meticulously selected photographs. Much of the material about the development of modern Indian art is largely unfamiliar. Russell Freedman's *The Life and Death of Crazy Horse* (Holiday) offers an unsanitized account of the legendary warrior.

From Davis Publishers of Worcester, Massachusetts, came *Jacob Lawrence: American Scenes, American Struggles,* by Nancy Shroyer Howard. This innovative volume, a Closer Look Activity Book, integrates Lawrence's own paintings with photographs, biography, capsule art-appreciation lessons, and a series of activities clearly marked at three different levels. In the series Milestones in Black American History, Chelsea House Publishers brought out Andrew Frank's volume *The Birth of Black America: The Age of Discovery and the Slave Trade.* Unflinching realism, excellent pictures, and meticulous documentation (with a nicely selected bibliography) make this one of the best volumes in the series to date.

Flood: Wrestling with the Mississippi, by Patricia Lauber (National Geographic Society), meets the high standards set by the author's previous award-winning books on Mount Saint Helens, fossils, and the origin of dinosaurs. Though the book focuses in particular on the floods of 1927 and 1993, the nature of the mighty river in normal times is equally compelling. Dorothy Hinshaw Patent teamed up once again with the photographer William Munoz to produce *Prairies* (Holiday House). From Carolrhoda Books in Minneapolis came *Powerhouse: Inside a Nuclear Power Plant,* by Charlotte Wilcox, with color photographs by Jerry Boucher.

Two nonfiction books had somewhat unlikely subjects. *Accidents May Happen: Fifty Inventions Discov-*

ered by Mistake, by Charlotte Foltz Jones, with illustrations by John O'Brien (Delacorte), covers a wide variety of fortunate errors under such headings as "Fed Up" (Wheaties and ice cream sodas), "Things to Write Home About" (Liquid Paper), and "Explosive Discoveries." The explanation of "ether and nitrous oxide" begins like this:

> Do you need to have a tooth pulled, an appendix removed, or a cut stitched up? A couple of centuries ago surgery was a pretty grim prospect.
>
> If you couldn't stand the pain (and who could?), there were several options. You could be:
>
> frozen,
> beaten senseless,
> asphyxiated,
> pumped full of alcohol,
> or given a piece of wood to bite down on.

The article goes on to chronicle, accurately, the disagreement over who actually "discovered" the safe use of anesthesia in the United States. A second improbable but fine nonfiction book is *The Inside-Outside Book of Libraries,* by Roxie Munro and Julie Cummins (Dutton). Libraries from small (Ocracoke) to large (New York Public) are described, including some in schools, prisons, and the military. The section on the Andrew Heiskell Library for the Blind and Physically Handicapped even contains a short excerpt in Braille from *My Side of the Mountain,* transliterated in the margin for sighted readers.

Fiction for the middle grades was as varied as usual. Michael Dorris's *Sees Behind Trees* (Hyperion) combines a Native American coming-of-age story with a lesson about physical handicaps. The main character is shortsighted but able to see things others cannot; his journey takes him to the land of waters and home again, with an infant he has rescued. The book was favorably reviewed but has not sold particularly well. By contrast, Phyllis Reynolds Naylor's *Shiloh Season,* a sequel to her Newbery-winning *Shiloh,* which picks up the thread of the little beagle's warped former owner, Judd Travers, hit the best-seller lists promptly. Though the young hero has earned the right to keep his dog, Travers remains a threat to that ownership and also to the whole family through his illegal hunting and drunken driving. This melodramatic, sentimental, and predictable book will doubtless remain almost as successful as its overrated (and Newbery-winning) predecessor. Also attempting to capitalize on a previous success was E. L. Konigsberg with *The View from Saturday* (Simon and Schuster).

Katherine Paterson, author of Jip: His Story *(photograph by Samantha Loomis Paterson)*

Several first novels were noteworthy. By an apparent coincidence, most of them concerned the leading fictional theme of 1996: the search for missing or unknown parents. *Family Tree,* by Katherine Ayres (Delacorte), tells of eleven-year-old Tyler's attempts to unravel her ancestry for a school project; by book's end she has met an Amish uncle and her non-Amish grandmother for the first time and brought some degree of reconciliation to her family. For mid-grade readers at the upper end of the age bracket, Suzanne Freeman's debut juvenile, *The Cuckoo's Child* (Greenwillow), deftly re-creates the feelings of Mia Veery, transported suddenly from Beirut in 1962 to Ionia, Tennessee, when her parents disappear and are presumed dead in a plane crash.

An eagerly awaited book was Nancy Farmer's *A Girl Named Disaster* (Richard Jackson/Orchard), detailing the quest of a Mozambican girl on the brink of puberty to locate her father in neighboring Zimbabwe. After a journey downriver, a sojourn on an island inhabited by baboons, and many hardships, she is reunited with her father's family—though it is clear that this is not a permanent solution. The scenery is lovingly rendered and the peripheral characters finely drawn.

Katherine Paterson's latest protagonist is also searching for his parents. Set in Vermont in 1855–1856, *Jip: His Story* (Lodestar) picks up when a lunatic, Put, arrives at the "poor farm" to which Jip has also been consigned after falling from a wagon as a young child. When lucid, Put contributes much to Jip's gradual maturing; so does his adored Teacher. Only toward the end of the book does the reader learn that Jip is the son of a slave and that Teacher is Lyddie Worthen of Paterson's earlier book, *Lyddie*.

Other fine tales of young people coming to terms belatedly with families came from writers with more exposure than Ayres or Freeman but less than Paterson. One is *The Fire Pony*, by Rodman Philbrick (Blue Sky/Scholastic), in which Roy and his maverick older brother, Joe Dilly, are hired at the Bar None ranch. Another is Nancy Hope Wilson's third juvenile novel, *Becoming Felix* (Farrar Straus), which chronicles the story of a twelve-year-old in West Farley, Massachusetts. Torn between music and farming, JJ gradually comes to understand his similarity to his grandfather Felix, after whom he was named. Melrose Cooper, author of the popular *Life Riddles*, brought out a "companion novel" called *Life Magic* about a middle child, Crystal, living with two talented sisters and an uncle dying of AIDS.

Many successful authors of books for the middle grades continued series that have sold well. Phyllis Reynolds Naylor added to the Bessledorf mystery series with the competent but predictable mystery *The Bomb in the Bessledorf Bus Depot* (Jean Karl/Atheneum), in which Bernie Magruder's suspicions turn toward his own family's involvement in a mysterious bombing next door. The versatile Susan Fletcher completed her trilogy with *Sign of the Dove* (Jean Karl/Atheneum), the sequel to *Dragon's Milk* and *Flight of the Dragon Kyn*. Fletcher is also known for her earlier comic novel, *The Stuttgart Nanny Mafia*. The inimitable Margaret Mahy continued her travel/adventure stories with *Tingleberries, Tuckertubs and Telephones* (Viking; British publication 1995). Saracen Hobday, a "particularly shy orphan," is transformed by a series of improbable adventures from a "limp lettuce leaf in the great salad of life" to a "crisp cucumber" therein. Following the success of *Lost in Cyberspace,* Richard Peck continued the adventures of Josh Lewis and Aaron Zimmer in *The Great Interactive Dream Machine* (Dial). One such author ventured into juvenile territory after a long career elsewhere. *Never Cry "Arp!" and Other Great Adventures* (Henry Holt) continues the series of outdoor humor stories from Patrick McManus. Better known as a humorist for adults, McManus has long had a "secret audience" of young people, according to the publisher; here at last is a book especially for them, twelve short stories about the mishaps of young Pat in the great outdoors.

Some authors surprised by turning up in an unfamiliar market. Sid Fleischman's autobiography, *The Abracadabra Kid: A Writer's Life* (Greenwillow) proved just as exciting as his fiction. A switchover from the Young Adult (YA) ranks was Cynthia Voigt, whose *Bad Girls* (Scholastic) chronicles the adventures of fifth-graders Mikey and Margalo. Joan Bauer, known for her hilarious YA titles *Squashed* and *Thwonk,* also moved into middle-grades fiction with *Sticks* (Delacorte), set in a family pool hall. *The Kid Who Ran for President,* by Dan Gutman (Scholastic), and *Nerd No More,* by Kristine Franklin (Candlewick), chronicle nonconformist boys in the years immediately before adolescence. Straddling the line between fiction and nonfiction, Holly Littlefield's *Fire at the Triangle Factory,* with illustrations by Mary O'Keefe Young (Carolrhoda Books), relates the friendship and shared peril of two fourteen-year-old factory workers, one Jewish, the other Catholic. Avi offered the first of a two-volume historical adventure that begins during the potato famine: *Beyond the Western Sea: The Escape from Home* (Orchard).

A distinguished import was Kazumi Yumoto's *The Friends,* translated by Cathy Hirano (Farrar Straus Giroux). Yumoto, who has a music degree and has written opera scripts, won a Recommended Book Prize from the Japan School Library Book Club for this 1992 story of two school friends who, fascinated by death, become acquainted with an elderly man and gradually realize that his life is more interesting than his impending death. A more controversial import was the Australian title *Sleeping Dogs,* by Sonya Hartnett; though shortlisted for the older reader category of the Australian Children's Book Council Book of the Year Award, it was criticized for its violence. Mette Newth's *The Abduction* was translated from the Norwegian and, like *Sleeping Dogs,* generally marketed for a YA audience in the United States.

Imports, however fine, are hard to sell. In an innovative move, Dutton attempted to deal with this problem by putting the full text of an import from Denmark, Bjarne Reuter's *The End of the Rainbow,* on its web site (http://www.penguin.com/usa/buster).

As has happened several times recently, the finest new fiction for middle grades was at the upper edge of the age category. In her 1996 offering, *The Ballad of Lucy Whipple* (Clarion), Karen Cushman departs from the medieval settings that won her a Newbery Honor for *Catherine, Called Birdy* and the Newbery itself for *The Midwife's Apprentice* and turned to gold-rush-era California. If the struggling one-parent family in the raw West sometimes seems reminiscent of the television show *Dr. Quinn, Medicine Woman,* still the articulate and often unhappy young Lucy (whose real name is California) is an appealing character.

Another leading contender for 1996 awards was *The Music of Dolphins,* by Karen Hesse (Scholastic). Raised by dolphins, Mila becomes a ward of the U.S. government and gradually learns to communicate with humans as they study her and attempt to learn dolphin language from her. As the entire story is told by Mila herself, the first chapters are necessarily very limited in vocabulary and syntax; appropriately, they are printed in huge primerlike letters. Gradually Mila becomes very articulate, but when she realizes that she is in effect a prisoner, she relapses—or rather chooses to go back to her former self, as the language and typography show. Not merely a new adaptation of Daniel Keyes's *Flowers for Algernon,* as might at first appear, this is a serious look at consciousness, language, and identity.

Parrot in the Oven: Mi Vida (HarperCollins) is a stunning first novel by Victor Martinez. It depicts a Mexican American boy coming of age in the projects.

Young Adult books continued much as usual: predictably controversial. Rich Wallace's impressive first novel, *Wrestling Sturbridge* (Knopf), introduces a 135-pound wrestler named Ben, from a small town in Pennsylvania where high-school wrestling is almost a religion. Despite its superficial likeness to Terry Davis's *Vision Quest,* this well-rounded tale is among the year's best. Marilyn Levy's *Run for Your Life* (Houghton Mifflin), a more upbeat sports-oriented story, is based on the true story of Darrell Hampton and chronicles the rise to success of an inner-city track team. Walter Dean Myers's *Slam* (Scholastic) turned the predictable ingredients of a basketball story into an excellent study of a young man in conflict.

A gratifying number of books for this age group showed a real sense of place. One of them was a YA debut, Liz Rosenberg's *Heart and Soul* (Harcourt Brace), a finely written story of a young Jewish girl coming of age in Richmond, Virginia, with an alcoholic mother, a chronically absent father, a worse-than-homely friend named Malachi, and an unquenchable talent for music. Rosenberg is well known as a poet and a writer of books for younger

Philip Pullman, author of The Golden Compass

children, most recently *Carousel*. Using a setting very unlike Richmond, Ruth White continued her steady output of YA novels set in rural Virginia with *Belle Prater's Boy* (Farrar, Straus and Giroux), a tale of a coal town where two children on the verge of puberty come to terms with family tragedies of the past. Harper Trophy published an original paperback trilogy by Chris Lynch, the Blue-Eyed Son books: *Mick, Blood Relations,* and *Dog Eat Dog*. The first one, *Mick,* uncompromisingly portrays the narrow and brutal world of a fifteen-year-old Irish American boy. This is a side of Boston one does not often see in juvenile books.

The most talked about YA title of the year was *The Golden Compass,* by Philip Pullman (Knopf), an import published a year earlier in Britain. In a departure from his fine Victorian thrillers, Pullman initiated a fantasy trilogy in which the fate of worlds may rest upon Lyra Belacqua's ability to read an ancient alethiometer, the instrument to which the title refers. Readers and critics alike have applauded Pullman's depiction of the relationship between humans and dæmons, their indispensable companions.

Several YA books successfully exploited the vogue for the supernatural. Jeanette Ingold's *Window* (Harcourt, Brace) follows the progress of Mandy, blinded in the accident that killed her mother. Leaning out the window of her grandparents' isolated Texas house, Mandy can, however, "see" the events of long ago in a small, localized time warp; understanding what happened two generations ago helps her come to terms with her own situation. Another time warp novel is *Mr. Was,* by Pete Hautman (Simon and Schuster), in which a young boy with an abusive father and a passive mother steps through a closet door in his grandfather's house to see the village as it once was and to learn, too late, the truth about the tangled identities in his family. In an unusual and disturbing werewolf story, *The Blooding,* by Patricia Windsor (Scholastic), teenage American Maris finds her job as an au pair in England to be more than she expected when the wife dies and the husband leads her into lycanthropy. Headed home by plane at story's end, she retains her double nature even after her mentor has been killed.

Louise Lawrence added to her list of successful YA science fiction with *Dream-Weaver* (Clarion), about the unlikely partnership of Troy, a raised-in-space young colonist, and Eth, an equally young dream weaver from Arbroth, a planet about to be colonized. Eth's tangerine-colored eyes are a particularly nice touch. Jane Stemp's first YA novel, *Waterbound* (Dial), is a dystopian portrayal of a future society in which less-than-perfect individuals are banished to a river world under the city. The author, who has cerebral palsy, is an activist for disability rights. Stephanie Tolan's *Welcome to the Ark* takes place at the turn of the twenty-first century and concerns a group of four "misfits" in an experimental program. Tolan, who is an advocate for exceptionally bright children, uses her passion to good advantage here.

Several YA novels examined the power of adults over teens, though in very different ways. *Evil Encounter,* by Sonia Levitin (Simon and Schuster), portrays an evil guru who takes over the lives of those in the therapy group that Michelle attends after her parents separate. This novel marks a change in Levitin's focus; she has been best known for her fine and award-winning books on Jewish themes such as *The Return.* Gregory Maguire's *Oasis* (Clarion) follows the fortunes of thirteen-year-old Hand after his father's sudden death and his long-absent mother's return. The "oasis" of the title is a rundown motel where all the action occurs.

Francesca Lia Block's *Girl Goddess #9* (Joanna Cotier/HarperCollins) is a collection of short stories about young women. Several of the stories concern sexual identity: one appeared earlier in the highly successful collection *Am I Blue? Coming Out from the Silence* and another in *When I Was Your Age* (Candlewick). In a variant on the year's popular find-the-parent theme, "Dragons in Manhattan" concerns a girl's search for her missing father—who turns out to be one of her two mothers, a transsexual who fathered her before undergoing a sex change.

At year's end, publishers and critics were scrambling to predict awards for the year and compile lists of the best. With the exception of the Opie/Wells Mother Goose collection, there was far less overlap on the various lists than usual; some "experts" spoke of a good year in children's books, some of a disappointing one. In retrospect, it remained a year of contradictions.

Dictionary of Literary Biography Yearbook Award for a Distinguished Children's Book Published in 1996

Dust jacket for Philip Pullman's novel, which has been compared to J. R. R. Tolkien's Lord of the Rings

In its second year, this award from the *Dictionary of Literary Biography Yearbook* once again diverges from recent tradition governing children's book awards in the United States. The 1995 award went to a nonfiction work, Jim Murphy's *The Great Fire* (Scholastic); the 1996 award recognizes an import, Philip Pullman's *The Golden Compass* (Knopf). While nonfiction is for various reasons overlooked by award committees, imports are actually excluded from many prize lists or relegated to separate, minor award categories such as translation. Thus, even the finest imported children's books in English receive less attention than they deserve. *The Golden Compass* deserves better.

Originally published in Great Britain in 1995 as *Northern Lights,* Pullman's volume is the first in a trilogy called *His Dark Materials.* In the classic tradition of high fantasy, the story centers on a quest, a young hero, and an assistant. In line with Pullman's earlier books, the main figure is a resourceful young woman, Lyra Belacqua, and her quest—a complex one that includes finding out the true moral nature of a mysterious parent. In a wholly new departure for Pullman, the assistant is a "dæmon," a spirit in animal form that accompanies every human being in the world of *The Golden Compass*. Lyra's dæmon, Pantalaimon, is able to change his shape at will; the dæmons of adults have settled into one particular form. Human and dæmon are inseparable, sharing a relationship closer than any other—a relationship that Pullman not only describes convincingly but also uses to fuel the plot at several points in his complex novel.

Beginning in Oxford, the novel broadens out to London, then Bolvangar, and finally Svalbard. Lyra's allies include "gyptians," the balloonist Lee Scoresby, a clan of witches, and a great armored bear named Iorek Byrnison. Ranged against them are other witches, other armored bears (including the corrupt usurper, Iofur Raknison), and Lyra's morally suspect parents, Lord Asriel and Mrs. Coulter. Pullman's narrative moves steadily forward, interspersed with set pieces: some chilling interviews between Lyra and Mrs. Coulter; the pathetic description of a child "severed" from his dæmon; fine battle scenes; and the combat to the death between Iorek Byrnison and Iofur Raknison.

Pullman's books are all grounded in fully realized settings–notably the Victorian scene of the Sally Lockhart trilogy and the Welsh town in which the mixed-race protagonist of *The Broken Bridge* struggles to find her identity. The detail and plausibility of Lyra Belacqua's world, part of "a universe like ours, but different in many ways," combine with the riveting story to lift the volume well above the usual run of high-fantasy offerings for the young. Under its British title this first volume of the trilogy won the Carnegie Medal for best children's book published in Great Britain during 1995; Pullman's subsequent remarks were widely quoted both in British papers and elsewhere. In an approving online review of the book, the Canterbury Public Library of Christchurch, New Zealand, excerpted the most controversial of those remarks:

> In adult literary fiction stories are there on suffrance. Other things are felt to be more important: technique, style, literary knowingness. The present day, would-be George Eliots take up their stories as if with a pair of tongs. If they could write novels without stories in them, they would.

Whether in New Zealand, in Britain, or in the United States, young readers are evidently enjoying Pullman's story. By the end of 1996 *The Golden Compass* appeared on American best-seller lists and was stocked in quantity by retailers such as Books-a-Million despite the fact that it was only available in hardback. The release of the second volume in the trilogy is expected by the summer of 1997.

F. Scott Fitzgerald Centenary Celebrations

F. Scott Fitzgerald's permanent stature as a major American writer was predicted by Stephen Vincent Benét in his 1941 *Saturday Review of Literature* review of *The Last Tycoon:* "You can take off your hats now, gentlemen, and I think perhaps you had better. This is not a legend, this is a reputation—and seen in perspective, it may well be one of the most secure reputations of our time." That now preeminently secure reputation was confirmed by the 1996 celebrations in the United States and abroad of the one hundredth anniversary of F. Scott Fitzgerald's birth.

Lecture Series and Readings

The Merc: In early 1996 the Mercantile Library of New York and the F. Scott Fitzgerald Society sponsored a centennial lecture series, "The Romantic American: F. Scott Fitzgerald and His World." On 10 April, John Callahan, Odell Professor of Humanities at Lewis and Clark University and author of *The Illusions of a Nation: Myth and History in the Novels of F. Scott Fitzgerald,* spoke on Fitzgerald and the 1920s. The following week Ruth Prigozy, professor of English at Hofstra University and author of *The Stories and Essays of F. Scott Fitzgerald,* offered thoughts on the public and private lives of the Fitzgeralds. On 25 April, Matthew J. Bruccoli, Jefferies Professor of English at the University of South Carolina and author of the standard biography, *Some Sort of Epic Grandeur,* discussed Fitzgerald as a professional writer in the literary marketplace. During the first week in May, Alan Margolies, professor of English at John Jay College, CUNY, and editor of *F. Scott Fitzgerald's Saint Paul Plays,* examined Fitzgerald's career in Hollywood.

The Players: On the evening of 8 October, The Players sponsored a dinner and celebration of Fitzgerald's life and work in New York City. Among the participants in the post-dinner program were Matthew J. Bruccoli, who discussed the reasons for Fitzgerald's extraordinary reputation; actor Keir Dullea, who presented a scene from John Kane's *The Other Side of Paradise,* a one-man show about Fitzgerald; actress Beth Lincks, who read from Zelda Fitzgerald's writings; writer-producer Gwinn Owens, who spoke about making his 1963 Fitzgerald documentary, *Marked for Glory;* writer Budd Schulberg, who recalled his experiences with Fitzgerald during his last years in Hollywood; Ring Lardner Jr., who talked about his father's friendship with Fitzgerald in Great Neck; and newspaper columnist Sidney Zion, who commented on the impact Fitzgerald's work had on him in his youth.

Westhampton Beach, New York: On 6 December the Westhampton Beach Performing Arts Center and the Westhampton Cultural Consortium cosponsored a staged reading from Budd Schulberg's revised theatrical adaptation of *The Disenchanted.* A fund-raising event for the Performing Arts Center, which is renovating an Art Deco movie theater for its productions, the reading was held in the ballroom of a private Gatsby-like home in Westhampton Beach. Schulberg offered preperformance remarks, and actors presented scenes featuring Manly Halliday (based on Fitzgerald), Jere Halliday (based on Zelda Fitzgerald), Shep Stearns (based on Schulberg himself when he was a young screenwriter), and Victor Milgrim (based on Hollywood producer Walter Wanger, for whom Fitzgerald and Schulberg had worked—with disastrous consequences—on the movie *Winter Carnival*).

Celebrations and Conferences
FitzFest '96, Montgomery, Alabama

Sponsored by Huntingdon College, the first F. Scott and Zelda Sayre Fitzgerald Festival took place in Montgomery, Alabama, on 28 June–1 July 1996. FitzFest, as it quickly became known, was designed not as an academic conference, but as a broader celebration, particularly of the Fitzgeralds' ties to Montgomery and the South.

Using the Huntingdon campus as a base, FitzFest's organizers recruited support from the Montgomery community, including the F. Scott and Zelda Fitzgerald Museum, the Alabama Shakespeare Festival, and the Montgomery Museum of Fine Arts. Participants attended lectures;

Fitzgerald statue by Michael B. Price, Saint Paul, Minnesota

shared meals, picnics, receptions, and a "Roaring Twenties" costume ball; and toured Fitzgerald sites in Montgomery. The Montgomery Museum presented an exhibition of Zelda Fitzgerald paintings, and the Alabama Shakespeare Festival staged a private reading of *A Piece of Paradise,* a new play about the Fitzgeralds in Montgomery, by Montgomery author Wayne Greenhaw. A Festival on the Green provided 1920s-era music, a modern dance performance, and an exhibition of automobiles from the 1920s and 1930s.

Presenters for FitzFest '96 were Wayne Flynt of Auburn University, who spoke on Montgomery at the turn of the century; Judith Paterson and Jackson Bryer of the University of Maryland, James L. W. West III of Pennsylvania State University, and Alice Hall Petry of Southern Illinois University at Edwardsville, all of whom treated the fiction and careers of the Fitzgeralds; and Frances Kroll Ring, Fitzgerald's last secretary, who spoke about her relationship with him.

Encouraged by the turnout and by the spirit of celebration caught in the 1996 FitzFest, Huntingdon College and other sponsors have agreed to make the Fitzgerald Festival a recurrent event. The 1997 FitzFest will be held 19–21 June; thereafter it will become a biennial celebration.

Richard Anderson
Huntingdon College

F. Scott Fitzgerald Centennia Celebration, Saint Paul, Minnesota

Measured by the quality of events and the number and variety of participants, Saint Paul's F. Scott Fitzgerald Centennial Celebration surpassed all previous efforts to honor the city's native son. Under the leadership of Garrison Keillor, hundreds of volunteers, backed by generous grants from area foundations, made the weeklong celebration the success it was.

The week began with the broadcasting on Monday and Tuesday, 23 and 24 September, of a two-part, locally produced documentary, "Fitzgerald in Saint Paul," narrated by Keillor on the local PBS radio station. The two-hour program included selections from Fitzgerald's writing read by Keillor and comments by Fitzgerald scholars, people who knew Fitzgerald, and relatives of Fitzgerald's Saint Paul friends.

The lighting of a birthday cake at Landmark Center in the heart of downtown Saint Paul on 24 September, Fitzgerald's birthday, marked the official beginning of activities. For three days participants attended movies based on Fitzgerald writings, heard period music, learned dances from the 1920s, and listened to actors and writers read from Fitzgerald's work. These readings sparked controversy when it was learned that the festival's organizing committee had edited the readings to omit Fitzgerald words and expressions that might offend some listeners. Over the next several days published letters to the editor of the *Minneapolis Star Tribune* attacked the committee's action.

On Friday a "Literary Festival" attended by more than three thousand students and others offered concurrent activities, where festivalgoers attended sessions with writers Michael Dorris, Donald Hall, Patricia Hampl, Joseph Heller, Eleanor Lanahan (a granddaughter of the Fitzgeralds), Bobbie Ann Mason, Jane Smiley, Tobias Wolff, Bill Holm, Robert Bly, and Scott Donaldson. Other venues included sessions with Fitzgerald's secretary Frances Kroll Ring, two separate presentations on "Fitzgerald in Saint Paul"—one by Lloyd Hackl and the other by John Koblas and Dave Page—and a presentation on "The Man Who Invented the Jazz Age" by Robert Sayre.

At noon on Friday a ceremony marked the U.S. Postal Service's first-day issuance of the F. Scott Fitzgerald commemorative stamp and the dedication of the bronze statue of Fitzgerald that stands in Rice Park downtown. The statue is a slightly-larger-than-life-size rendering of Fitzgerald, a hatless figure with a coat casually thrown over his arm. In his remarks sculptor Michael B. Price observed that he had placed the statue on ground level so that Fitzgerald would be accessible to everyone. During the evening Keillor moderated a panel discussion among festival participants, followed by a dinner and a parade of vintage automobiles around Rice Park and the Fitzgerald statue.

Saturday featured a marathon reading of *The Great Gatsby* by visiting writers and others in the Fitzgerald Theatre, formerly the World Theatre, which had been renamed in Fitzgerald's honor in 1995. From 10:00 A.M. until 3:00 P.M. readers mounted the stairs to the stage and took turns reading from the novel. In the evening a special Fitzgerald edition of the PBS radio show *The Prairie Home Companion* was followed by a Great Gatsby Ball in the Roy Wilkins Auditorium featuring playwright-actor Sam Shepard, actress Jessica Lange, and jazz pianist Butch Thompson.

The festival was accompanied by the publication of two books: *F. Scott Fitzgerald and St. Paul: "Still Home to Me"* (Adventure Publications), an illustrated biography by Hackl, and *Toward the Summit: F. Scott Fitzgerald in Saint Paul* (North Star Press) by Koblas and Page.

Beginning in early September, twenty-five-cent trolley tours of Fitzgerald's Saint Paul neighborhood were conducted every Saturday, with special tours operating on Tuesday, Wednesday, and Thursday of the festival week. Throughout the month first-edition Fitzgerald books, Jazz Age photographs, and other memorabilia were on display. From 19 September 1996 to 12 January 1997 the University of Minnesota, Saint Paul campus, hosted an exhibition, "Coming Apart at the Seams: Style and the Social Fabric of the 1920s," featuring the "Fitzgerald look" and "telling the story of the 1920s as expressed in its dress and decorative arts."

Fitzgerald's "return" to his hometown in September 1996 was more than a conventional literary event. Writers, scholars, and general readers of Fitzgerald were in attendance, as were thousands of high-school and college students from throughout the state. This format seemed a fitting tribute to Fitzgerald, who, as his Saint Paul friend Alexandra Kalman once recalled, "Always had time for young writers."

Lloyd C. Hackl
Century College

Fitzgerald Society/Princeton University Centennial Conference, Princeton, New Jersey

The F. Scott Fitzgerald Society's 1996 conference, cosponsored by Princeton University, was held September 19–21 at Princeton University. Some 250 people attended. Among the guests were former senator Eugene J. McCarthy; Eleanor Lanahan, one of Fitzgerald's granddaughters; Frances Kroll Ring, Fitzgerald's secretary in Hollywood; Robert Westbrook, son of Sheilah Graham; Honoria Murphy Donnelly, daughter of Sara and Gerald Murphy; and author Budd Schulberg. Conference directors were Jackson R. Bryer (University of Maryland, College Park), president of the Fitzgerald Society; Alan Margolies (John Jay College, CUNY), vice president; and Ruth Prigozy (Hofstra University), executive director.

Fitzgerald's novels were the subjects of two sessions on 19 September. The first, moderated by Roger Lathbury (George Mason University), was devoted to *This Side of Paradise*. The session featured papers by Kirk Curnett (Troy State University), Robert A. Martin (Michigan State University), Walter Raubicheck (Pace University), and Stephen L. Tanner (Brigham Young University). The second session, "*The Great Gatsby*: New Approaches," was moderated by Heidi Kunz Bullock (Randolph-Macon Woman's College). Speakers included Matthew Elliot, who discussed national identity and race in the novel; Richard Kopley (Pennsylvania State University, Du Bois), who spoke about correspondences between *The Great Gatsby* and nineteenth-century American literature; and Rama Nair (Osmania University, Hyderabad), who offered an Indian perspective on the novel.

Fitzgerald's short stories were also the subject of two sessions, both moderated by Peter L. Hays (University of California, Davis). At the first session, Kegan Doyle (Simon Fraser University) talked about "May Day"; Frances Kerr (Durham Technical Community College) discussed "Gretchen's Forty Winks"; S. S. Moorty (Southern Utah University) analyzed "The Swimmers"; and Mark Shipman

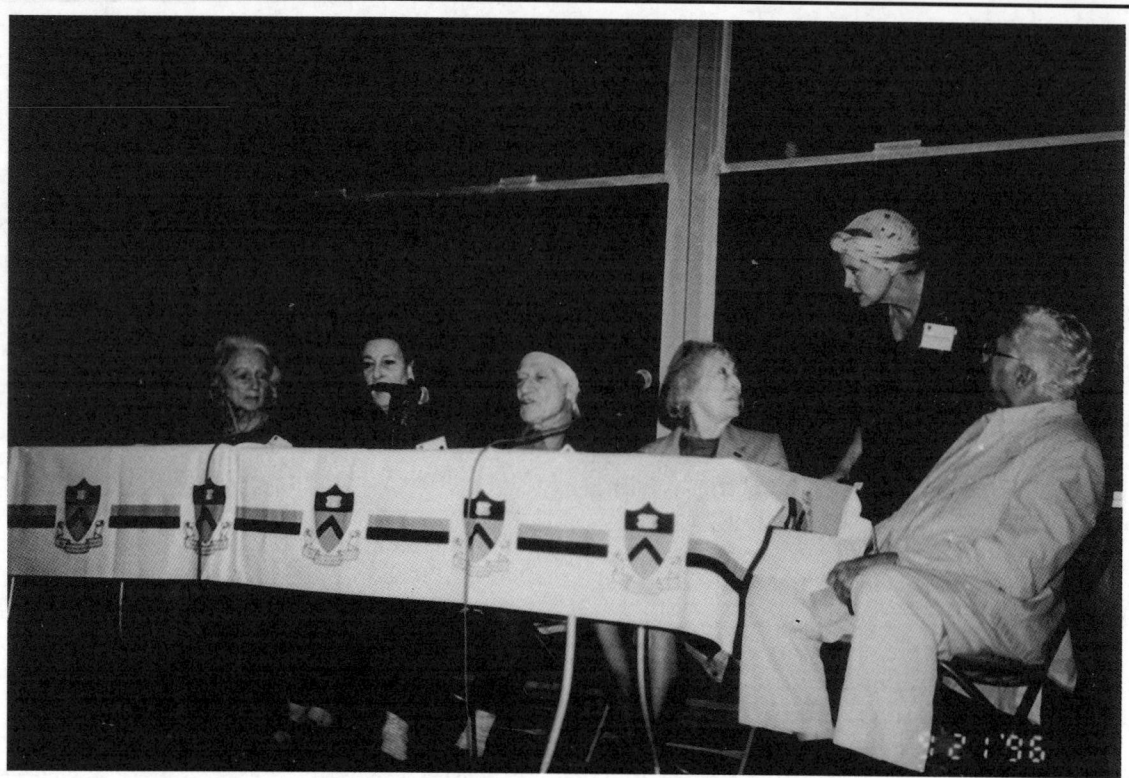

At the Princeton University Conference. Left to right: Honoria Murphy Donnelly, Fanny Myers Brennan, Tony Buttitta, Frances Kroll Ring, Linda P. Miller, and Budd Schulberg

(Tarleton State University) commented on "Crazy Sunday." During the second session Robin Gajdusek (San Francisco State University) offered an approach to *Tales of the Jazz Age;* B. McMullen (Oxford University) talked about "Absolution"; Mary McAleer Balkun (Seton Hall University) spoke about the "Josephine" stories; and Sawako Taniyama (Hyogo-ken, Japan) discussed the theme of covetousness in three Fitzgerald stories.

"Fitzgerald and Other Writers," chaired by George Wickes (University of Oregon, Eugene), featured Douglas E. LaPrade (University of Texas–Pan American) discussing Petrarchism and pragmatism in Fitzgerald's works, John M. Gill treating the relationship of Herman Melville's "Bartleby the Scrivener" to *The Great Gatsby,* Jane Vogel (Ithaca College) defining sources for Gatsby's innocence, and Robert D. Cowser (University of Nebraska, Lincoln) describing the shared world of Fitzgerald and Edith Wharton. "Theoretical and Psychological Approaches to Fitzgerald," a session chaired by Bryant Mangum (Virginia Commonwealth University), featured Andrew R. Grobman (Northeastern University), who offered a psychoanalytical analysis of Amory Blaine; Bruce Gilman (Salem State College), who discussed narcissism in Fitzgerald's youthful characters; Michael Nowlin (University of British Columbia, Vancouver), who talked about a castration motif in *The Beautiful and Damned;* and Jonathan Fegley (Macon College), who commented on Bakhtinian indeterminacy and open-endedness in Fitzgerald's novels.

The "Biography" session moderated by H. R. Stoneback (SUNY, New Paltz) also convened on 19 September. Seymour I. Toll spoke about the relationship between Fitzgerald and Judge John Biggs Jr.; P. Keith Gammons (University of North Carolina, Greensboro) discussed the influence of the South on Fitzgerald; and Jonathan Schiff traced the motif of the "replacement child" in Fitzgerald's fiction. A session devoted to "Theater and Film," moderated by Jeanne Fuchs (Hofstra University), featured Edward J. Rielly (Saint Joseph's College, Maine), who commented on Fitzgerald's dramatic purpose, and Richard Pioreck (Hofstra University), who spoke about *The Vegetable;* as well as a paper by Richard A. Davison (University of Delaware) that focused on Budd Schulberg's *The Disenchanted* and an essay by Wheeler Winston Dixon (University of Nebraska, Lincoln) that examined Fitzgerald's cinematic vision.

Other presentations on the first day of the conference included a talk by Frances Kroll Ring about her memories of Fitzgerald; a showing of the television film about Fitzgerald, *Marked for Glory,* with an introduction by the writer-producer Gwinn Owens; a showing of the film *An Author's Mother,* by Mark Axelrod (Chapman University), introduced by Ruth Prigozy; and a dramatic presentation of scenes from a stage adaptation of *Tender Is the Night* by writer and producer Simon Levy.

On 20 September, Scott Donaldson (College of William and Mary) moderated a session on *Tender Is the Night* with the following speakers and paper titles: Betty H. McFarland (Appalachian State University), "Nicole's Angle: From 'Dicole' to Warren Woman"; Stephen M. Brauer (New York University), " 'Diving into the Wreck': Intersections of Crime in *Tender Is the Night*"; Peter S. Taback (City College, CUNY), "Why Dick Diver Can't Keep Help"; and Ned Sparrow, "When Romantic Swords Cede to World War Gunshots." Papers in a session on *The Last Tycoon,* prepared by Robert Merrill (University of Nevada, Reno) and moderated by Alan Margolies, included "Fitzgerald's *Last Tycoon* and Final Style," by Milton R. Stern (University of Connecticut, Storrs); "*Tycoon,* Scott, Mummy, and Me: New Edition, Old Aesthetic," by Jeffrey Carroll (University of Hawaii, Manoa); "The Fall of a Stahr: The Function of the Myth of Icarus in F. Scott Fitzgerald's *The Last Tycoon,*" by David J. Partie; and "Hollywood: A New Dreamland for Fitzgerald's America," by Aiping Zhang (California State University, Chico).

"The Short Stories and 'The Crack-Up'" was the subject of a session moderated by Peter L. Hays and featuring papers by Nancy Van Arsdale (East Stroudsburg University), Joyce B. Anderson (Millersville University), D. Quentin Miller (University of Connecticut, Storrs), and J. Gerald Kennedy (Louisiana State University, Baton Rouge). During a concurrent session, "Textuality/Intertextuality and F. Scott Fitzgerald," moderated by Ronald Berman (University of California, San Diego), David Stouck and Janet Giltrow (both from Simon Fraser University) talked about echolalia in *Gatsby;* Veronica Makowsky (University of Connecticut, Storrs) discussed William Faulkner and Fitzgerald; and Gautam Kundu (Georgia Southern University) commented on the possible influence of Willa Cather on Fitzgerald.

Two sessions on "Fitzgerald and American Culture" were also offered on 20 September. The first, moderated by Benita Moore (Teikyo Marycrest University), included "Thalia Does the Hoochie-Coochie: Humor in the Fiction of F. Scott Fitzgerald," by D. G. Kehl (University of Arizona, Tempe); "Out of Minnesota," by James D. Bloom (Muhlenberg College); and "*The Great Gatsby*: A Musical Soundtrack," a presentation with music by Anthony J. Berret (Saint Joseph's University, Pennsylvania). The second session, moderated by Neila C. Seshachari (Weber State University), presented M. Thomas Inge (Randolph-Macon College) speaking on "Fitzgerald in the Funny Papers: Commentary of Mickey Mouse and Charlie Brown"; Dana Brand (Hofstra University) discussing "Fitzgerald, Elegance, and the Aesthetics of Modernity"; Al Elmore (Athens State College) talking about "Magic as Metaphor in Fitzgerald's Fiction"; and John B. Chambers (American University of Bulgaria) treating "Fitzgerald's Heroes and American Society."

"Fitzgerald and War" was the subject of a session moderated by Donald Noble (University of Alabama) with Bickford Sylvester (University of British Columbia) as respondent. The session offered papers by Todd H. Stebbins (William Penn College), Frederick Wegener, Kim Moreland (George Washington University), and Diane Isaacs (Fordham University). Another session, moderated by George Wickes, was devoted to "Fitzgerald and Other Writers." John F. Callahan (Lewis and Clark College) spoke about Fitzgerald and Ralph Ellison; Lawrence Broer and Gloria Holland (both of the University of South Florida) compared Fitzgerald's *Great Gatsby* and Norman Mailer's *An American Dream;* Dianne Timblin (Caldwell Community College) commented on Fitzgerald and Raymond Carver; and Toshifumi Miyawaki (Seikei University, Tokyo) discussed Fitzgerald and Haruki Murakami.

Other events on 20 September included "Fitzgerald at Princeton," an hour-long presentation with slides by Anne Margaret Daniel (Princeton University) of Princeton life during the years that Fitzgerald was in attendance; "Fitzgerald in Baltimore," a slide presentation by Joan Hellman (Catonsville Community College); and "A Conversation" among John Kuehl (New York University), Eleanor Lanahan, and Robert Westbrook. A reception at Cottage Club catered by Laura Donnelly featured foods that her grandparents, Sara and Gerald Murphy, served at similar parties. An evening session, "Contemporary American Writers Talk About Fitzgerald," moderated by Jackson R. Bryer, spotlighted authors Thomas Flanagan, George Garrett (University of Virginia), Edmund Keeley (Princeton University), and Hugh Nissenson.

On 21 September *The Beautiful and Damned* was the subject of a session moderated by Catherine Burroughs (Cornell College) and featuring papers by Barry Gross (Michigan State University) and Timothy Martin (U.S. Air Force Academy). A second session on Fitzgerald's novels, "*The Great Gatsby:* Language and Literary Style," was moderated by Richard Anderson (Huntington College) and included papers by Dan Coleman (Cornell University), Betty J. Cortright (University of South Florida), Joan Hellman, and Gail D. Sinclair. A session on the Pat Hobby stories moderated by Lauraleigh O'Meara (Arizona State University) offered four papers: "Pat Hobby–'A Good Man for Structure,'" by Christopher Ames (Agnes Scott College); "Tune in Next Month: Fitzgerald's Pat Hobby and the Popular Series," by Timothy Prchal (University of Wisconsin, Milwaukee); "Fitzgerald's Hemingway: The Debunking Last Word of 'Two Old-Timers,'" by Greg Metcalf (University of Maryland, Baltimore); and "Hollywood and Illusion in F. Scott Fitzgerald's *The Pat Hobby Stories* and *The Love of the Last Tycoon,*" by Douglas G. Baldwin (Yale University).

Fitzgerald's foreign reputation was the topic of two sessions moderated by Linda Stanley. Speakers in the first session included A. D. Hook (University of Glasgow), Sergio Perosa (Universita degli Studi di Venezia), Qing Qian (Beijing Foreign Studies University), and Claus Secher (Danmarks Biblioteksskole, Copenhagen). Those who participated in the second session included Udo Hebel (University of Potsdam), A. D. Hook, Somdatta Mandel (University of Calcutta), Miriam Mandel (Tel Aviv University), Sergio Perosa, Qing Qian, Claus Secher, Kiyohiko Tsuboi (Okayama University), and Svetlana Voitiuk (University of L'viv). The panel, "Editing Fitzgerald," moderated by John Bryant (Hofstra University), featured James L. W. West III (Pennsylvania State University) and Horst H. Kruse (Westfälische Wilhelms-Universität, Münster). A third panel on "Fitzgerald and Other Writers," moderated by George Wickes, included Edward Gillin (SUNY, Geneseo), who spoke on "Princeton, Pragmatism, and Fitzgerald's Sentimental Journey"; Ted Billy (Saint Mary's College, Indiana), who talked about "Lawrencean Subtext in *Tender Is the Night*"; and Steven Goldleaf (Pace University), whose topic was "A Twice-told Tale: Fitzgerald's 'Three Hours Between Planes' and O'Hara's 'Trouble in 1949.'" "Individual Responses to Fitzgerald," a session prepared by Donald Junkins (University of Massachusetts, Amherst) and moderated by Marie Ahearn (University of Massachusetts, Dartmouth), included papers by John T. Irwin (Johns Hopkins University), Howard R. Wolf (SUNY, Buffalo), and Eduardo Ribiero (University of Porto, Portugal).

Other presentations on the final day of the conference included a session titled "Memories of Fitzgerald," moderated by Linda Patterson Miller (Pennsylvania State University, Ogontz). Speakers were Fanny Myers Brennan, Tony Buttitta, Honoria Murphy Donnelly, Frances Kroll Ring, and Budd Schulberg. Morris Dickstein (Queens College and Graduate Center, New York) spoke about Fitzgerald during a lunch session; Alan Margolies moderated "On Teaching Fitzgerald," a discussion session prepared by James J. Martine (Saint Bonaventure University); and Honoria Murphy Donnelly and Fanny Myers Brennan participated in a slide presentation, "Fitzgerald on the Riviera."

The conference closed with a reception at Firestone Library, where an exhibition of Fitzgerald papers and associated materials had been on display since May 1996. The reception was followed by a banquet at which former senator Eugene J. McCarthy was speaker. On the morning of 22 September, those still in attendance participated in tours of the Princeton campus (with emphasis on Fitzgerald) led by Alfred L. Bush (Princeton University Library).

Alan Margolies
John Jay College, CUNY

First Annual F. Scott Fitzgerald Literary Conference, Rockville, Maryland

The First Annual F. Scott Fitzgerald Literary Conference, part of Rockville's yearlong F. Scott Fitzgerald Centennial Celebration, took place at Montgomery College, Rockville, Maryland, on 28 September 1996. Although Fitzgerald never lived in Rockville, as a boy he visited his father's relatives at their nearby farm, Locust Grove, in Montgomery County, and in 1903 he was a "ribbon holder" at his cousin Cecilia Delihant's wedding at her home in Randolph Station, south of Rockville. In 1931 he traveled from Paris to attend his father's funeral at Rockville in Saint Mary's Church. His father was buried beside other members of the family in Saint Mary's cemetery. F. Scott and Zelda Fitzgerald were first buried in Rockville Union Cemetery. In 1975 they were reinterred at Saint Mary's cemetery, and in 1986 their daughter, Scottie, was buried in the family plot.

The conference began with a showing of *Marked for Glory,* a 1963 documentary film about Fitzgerald, introduced by producer and writer Gwinn Owens. The documentary was followed by a panel discussion, "F. Scott Fitzgerald at 100," moderated by Jackson R. Bryer (University of Maryland, College Park), with panelists Earl Harbert (Northeastern University), James L. W. West III (Pennsylvania State University), Ruth Prigozy (Hofstra University), and Kim Moreland (George Washington University).

Events during the afternoon included fiction workshops with writers Alan Cheuse (George Mason University), Maxine Clair (George Washington University), Patricia Browning Griffith (George Washington University), William Loizeaux (The Writer's Center, Bethesda, Maryland), and Susan Richards Shreve (George Mason University), as well as a slide presentation, "F. Scott Fitzgerald in Baltimore," by Joan E. Hellman (Catonsville Community College). These sessions were followed by a panel discussion, "How to Get Published," moderated by Allan Lefcowitz (The Writer's Center, Bethesda, Maryland), with panelists Marie Arana-Ward (deputy editor, *Washington Post Book World*), literary agent Timothy Seldes (Russell and Volkening, New York City), and Jack Shoemaker (editor in chief, Counterpoint Publishers, Washington, D.C.). Scheduled for the same hour was a showing of *Three Comrades,* the only movie for which Fitzgerald received screenwriting credit; the movie was introduced by Alan Margolies (John Jay College, CUNY).

The final events of the day were the presentation of the First Annual F. Scott Fitzgerald Award to author William Styron and the announcement of the winner of the F. Scott Fitzgerald Short Story Contest, Jeff Minerd for "Stepping Off." Minerd read his story, and the conference concluded with Styron's reading of a brief tribute to Fitzgerald as well as a section from his 1979 novel, *Sophie's Choice.*

Alan Margolies
John Jay College, CUNY

"Fortunes and Misfortunes in the Life and Times of F. Scott Fitzgerald," Rome, Italy

On 11 and 12 October an international conference to celebrate the centennial of the birth of F. Scott Fitzgerald was held in Rome. Conceived by Doc Rossi of John Cabot University in Rome and by Mattia Carratello, a Fulbright scholar and doctoral candidate in American Studies at the University of Rome III, the conference aimed to delineate Fitzgerald's status among Italian readers and critics, as well as American and international commentators. The conference's title accommodated papers treating social history, cinema, music, and other arts

Front of schedule for the Centennial Conference, Rome

in relation to Fitzgerald's work, as well as biographical and literary topics. The conference was organized jointly by Rossi, Ugo Rubeo of the University of Rome "La Sapienza," and Daniele Fiorentino, director of the American Studies Center in Rome.

The conference was opened by Biancamaria Pisapia, chairman of the English Department at "La Sapienza," where the first session, chaired by Rossi, was held. Caterina Ricciardi of the University of Tuscia-Viterbo presented a paper on Fitzgerald's response to Rome. Recalling the writer's two unpleasant visits to the Eternal City, Ricciardi suggested that Fitzgerald's time there marked the end of his dream of Rome. In "The Decadent Legacy of *The Beautiful and Damned,*" Patrick Quinn of Nene College at Northampton examined Fitzgerald's early novel in the context of the nineteenth-century English decadent tradition reflected in Oscar Wilde's plays and Aubrey Beardsley's short stories.

"'The Money Swing': The Life and Times of F. Scott Fitzgerald in the Poetry of Anne Sexton" by Emma Marras Giannone examined the connection Sexton makes between her own life and times and those of Fitzgerald and his generation. In "Two Gatsbys: Translation Theory as an Aid to Understanding," Iain Halliday of the University of Catania and the University of Warwick considered two Italian translations of *The Great Gatsby,* Fernanda Pivano's of 1950 and Tommaso Pisanti's of 1989. Halliday declared that Fitzgerald's acute awareness of the contradictions of life and language renders his writing particularly appropriate for translation theory, which is concerned with the multifaceted potential of language. The first session was closed by Pietro De Logu of the University of Padova and the University of Venice, who treated "F. Scott Fitzgerald: Mythic Hero of the Jazz Age" and gave a synopsis of Fitzgerald criticism in Italy up to the early 1960s and the publication of Sergio Perosa's *The Art of F. Scott Fitzgerald.* De Logu argued that, following in the tradition of nineteenth-century novelists, Fitzgerald sought to provide a moral interpretation of his time, an emphasis that attracted many Italians—for political and literary reasons—to American literature during the post–World War II years.

After a reception at Villa Mirafiori, the afternoon session, chaired by Rubeo, was held at the National Library, where Italian translations of Fitzgerald's works were on display. This exhibition, curated by Maria Grazia Villani, also included photos, reviews, and bibliographical information showing the literary fortunes of Fitzgerald in Italy. Journalist and critic Piero Sanavio began the session with a paper on Fitzgerald's politics, which he depicted as mainly confused, growing out of naiveté or lack of concern on the writer's part. This view was countered by Winifred Farrant Bevilacqua of the University of Torino, whose "Chronotopes in *The Great Gatsby*" used Mikhail Bakhtin's theories to explore Fitzgerald's worldview and vision of humanity in *The Great Gatsby*. In "Fitzgerald's Novelistic Modes in *Tender Is the Night*," Charles Etheridge of McMurry University, maintained that the often-noted "divided vision," or dual voice, in Fitzgerald's fiction is directly attributable to competing novelistic modes present in his work. Closing the session, Karin Badt of the American University of Paris argued in "The Desecration of American Culture in F. Scott Fitzgerald and 1940s American Cinema" that the femme fatale in Fitzgerald's novels is similar to the portrayal of the amoral woman in the Hollywood film noir.

Three events on the first evening were sponsored by the Comune di Roma and held in the small theater of the *Palazzo delle Esposizioni*. A showing of the 1949 motion picture adaptation of *The Great Gatsby* was followed by the presentation of an ambitious new literature and arts magazine called *Praz!,* named after the noted Italian scholar Mario Praz. Managing Director Giorgio Minuti was joined in a discussion of the magazine's design and future plans by Pisapia, Paola Colaiacomo of "La Sapienza," and Viola Papetti of Rome III. The final event of the evening was an homage to Fitzgerald directed by Francesca Gatto that featured music of Scott Joplin and George Gershwin and that concluded in readings from "Afternoon of an Author," letters to and from Scottie Fitzgerald, and Fitzgerald's fiction.

The morning session on the second day, held at John Cabot University and chaired by Pisapia, began with Rossi's "Fantasies and Fools: Allegory in Fitzgerald's Short Stories." Rossi argued that for Fitzgerald the material of a work is not as important as its tone, air, or aspect, its imaginative handling of details and their nuances. Rossi pointed out that Fitzgerald tried to make readers aware of a "suggested world," not by writing allegories per se but by suggesting allegorical interpretation through fragmentation and parody of allegory. The next five speakers presented their papers in Italian, all of them concerned with different aspects of *The Love of the Last Tycoon.* Rubeo presented "*The Crack-Up*: autobiografia d'un moderno" (*The Crack Up*: Autobiography of a Modern), which focused on the central role these "narrative pieces" play in the final works of Fitzgerald's career, particularly *Tycoon.* Seeing the first-person narrators Nick Carraway and the voice of *The Crack-Up* as direct precursors of Cecilia Brady, Rubeo argued that Fitzgerald envisioned Brady as the most complex of his narrators because she has the capacity for a modernistic, disjunctive assemblage of the action (reflecting the frequent fragmentation and time shifts in Carraway's recollection) as well as the capacity to look at oneself with absolute coolness (reflecting the narrator of *The Crack-Up*).

The poet Anna Cascella, who published the 1995 critical work *I colori di gatsby: lettra di fitzgerald* (*The Colors of Gatsby: Readings on Fitzgerald*), read "Alcuni riflessioni su *The Last Tycoon*" (Some Reflections on *The Last Tycoon*), which concentrated on air—and what is carried in it—as a symbol of the "internal space" of travel and on water as a symbol of both destruction and regeneration. Using Fitzgerald's linguistic structure to examine the semiotics of light, sound, and heat waves in the text, Cascella drew attention to his use of the word *fuselage,* which links the internal space of Stahr's house with that of the airplane, or "flying seed," of the first chapter, foretelling Stahr's death in a voyage "by air."

Continuing the session, Alessandro Gebbia of "La Sapienza" delivered "*Shoot!* and *The Last Tycoon*," a paper comparing a novel by Luigi Pirandello and the unfinished novel by Fitzgerald. Drawing attention to structural and thematic resemblance between the two works, Gebbia maintained that Pirandello and Fitzgerald seized the cinema's secrets and techniques, which they appropriated into their own aesthetics for the novel. Michele Bottalico of the University of Bari presented "L'illusione del mito: Hollywood in *The Love of the Last Tycoon*" (The Illusion of the Myth: Hollywood in *The Love of the Last Tycoon*), which argued that Fitzgerald's portrayal of Hollywood in this work is almost completely negative. To close the morning session, Agostino Lombardo of "La Sapienza" presented "Il capolavoro incompiuto di F. Scott Fitzgerald" (The Unfinished Masterpiece of F. Scott Fitzgerald). Lombardo argued for seeing *Tycoon* as a masterwork even in its incomplete state.

After another reception the conference reconvened at the American Studies Center, where Fiorentino introduced the keynote speaker, Matthew J. Bruccoli, Jefferies Professor of English at the University of South Carolina, whose appearances at the Rome conference and at a 14 October symposium at the University of Milan were sponsored by the U. S. Information Service. Responding to the title of the conference, Bruccoli suggested that Fitzgerald's misfortunes were few: his alcoholism, which has been overemphasized, and Zelda's health problems, which colored his entire career. Fitzgerald's fortunes, on the other hand, were many. Calling him a supreme stylist of American literature, Bruccoli described him not as a modernist experimenting with form or time, but as a master of narrative shaped by nineteenth-century values, who—as Lionel Trilling observed—depicted life committed to a purpose or thrown away for the sake of an ideal. Taking a stand against the notion that Fitzgerald's short stories were nothing more than hack work intended to support his serious work of novel writing, Bruccoli portrayed Fitzgerald as a serious artist operating in a commercial world, juggling the claims of art with the claims of the marketplace.

After Bruccoli's talk Lombardo chaired a roundtable discussion that included Bruccoli, Quinn, Etheridge, De Logu, and Sanavio. The two primary issues that emerged were critical approaches to *The Love of the Last Tycoon* as a novel, since Fitzgerald left it unfinished, and Fitzgerald as modernist or continuer of the nineteenth-century narrative tradition. While Etheridge pointed out that Fitzgerald's technical experiments with point of view place him firmly in the modernist camp, Bruccoli maintained that Fitzgerald was a storyteller whose narrative technique is far removed from that of, say, James Joyce.

The conference produced papers ranging from traditional literary scholarship to interdisciplinary topics, giving an idea of the variety of interests engaging European scholars of American literature and culture. Speakers examined aspects of Fitzgerald's style and narrative technique, noting the critical voice that comes out of his juxtapositions of characters and scenes. Moreover, continuing the work of early Fitzgerald champions such as Edmund Wilson, Malcolm Cowley, Lionel Trilling, and Sergio Perosa, scholars are rediscovering the command Fitzgerald had over his art.

Doc Rossi
John Cabot University

F. Scott Fitzgerald Centenary Celebration, The University of South Carolina, Columbia

The F. Scott Fitzgerald Centenary Celebration, hosted by the University of South Carolina, was conceived not as a scholarly conference but as a celebration of the author's life and work. Funded by the Office of the Provost, the College of Liberal Arts, USC's Division of Libraries and Information Systems, the Department of English, the Alumni Association, and NationsBank, the Centenary Celebration scheduled its major events for 24–26 September but also supported outreach programs to the Columbia community and the state throughout the fall of 1996.

Introducing young readers to Fitzgerald's fiction was a primary concern of the Centenary organizing committee. Matthew J. Bruccoli donated Fitzgerald books to regional campuses of the USC system. Sandra Thomas of the S.C. State Department of Education and Harriett S. Williams of USC's College of Applied Professional Sciences spearheaded a campaign to encourage the teaching of Fitzgerald's novels and stories in all South Carolina high schools during the 1996–1997 school year. Williams authored *A Young Reader's Guide to F. Scott Fitzgerald*, which provided reading and writing assignments fulfilling State Department of Education objectives for senior English classes and which assembled documentary material on Fitzgerald's life and work. Produced by the Department of Education, the *Young Reader's Guide* was distributed to all South Carolina high schools. At least ten thousand South Carolina secondary students read *The Great Gatsby* during the fall of 1996 (one book dealer reported an order for nineteen hundred copies of the novel from a single Columbia high school), and at least two area schools sponsored Gatsby or Roaring Twenties dress-up days to coincide with the 24 September interactive teleconference broadcast, the kickoff event of the celebration itself.

One of fifteen billboards in Richland and Lexington Counties, South Carolina

Similarly, USC's annual First-Year Reading Experience, under the direction of Interim Provost Donald J. Greiner, chose *The Great Gatsby* as the text to be read by 650 incoming freshman who met on 19 August in groups of ten to discuss the novel with one of sixty-five USC professors from across disciplines. The students also heard a brief talk by Bruccoli, who presented all participants with keepsake facsimiles of Fitzgerald's entry from the Class of 1917 Princeton University yearbook. During the evening students and professors saw Jack Clayton's 1974 motion-picture adaptation of the novel. Richard Rose of the USC Art Department sponsored a Gatsby poster contest for members of his graphic-design class. Dawn Pyron's poster was chosen as the winner, and all of the posters were exhibited at USC's Museum of Education at the request of its director, Craig Kridel.

During September both the Richland County Public Library and an independent book dealer sponsored Fitzgerald-related activities that drew enthusiastic responses from the Columbia community. At the invitation of Sarah Linder, director of Main Library Services, Bruccoli and USC graduate students Robert F. Moss and Park Bucker presented lectures and led discussions on Fitzgerald in the library's ongoing "Let's Talk About It" program. The downtown branch of the library also provided the venue for the showing of movies with Fitzgerald connections: *Three Comrades* (1938), for which he received his only screen credit; *The Last Tycoon* (1976); and the two surviving film versions of *The Great Gatsby* (1949 and 1974). At the Happy Bookseller, Rhett Jackson hosted a promotion of Fitzgerald on Saturday, 14 September, by giving free copies of *The Great Gatsby* to customers. The two hundred copies of the Scribner paperback were provided by Carolyn Reidy, president of Simon and Schuster's Trade Division.

USC's McKissick Museum, directed by Lynn Robertson, held two pre-Centenary events. On the evening of 19 September the museum opened its exhibition, "Double Vision: Fitzgerald's World of Realism and Imagination," which included clothing, movie memorabilia, and other material documenting social trends during the 1920s and 1930s. Participants, many of whom wore 1920s attire, were entertained by Dick Goodwin's trio. On Sunday, 22 September, McKissick staff used the grounds of the center-campus Horseshoe as the site for croquet, cocktails, and high tea.

A 30 September post-Centenary event was hosted by George Terry, dean and vice provost for Libraries and Information Systems, in the Rare Book Room of Cooper Library. Lawrence Jordan, Columbia postmaster and an avid reader of Fitzgerald, presented cancellation cards of the Fitzgerald stamp t guests; he also gave large stamp reproductions or framed displays of the card and stamps to officials and departments of the university.

An ongoing and permanent feature of USC's Centenary Celebration is the F. Scott Fitzgerald Centenary WebPage, which can be accessed at http://www.sc.edu/fitzgerald/index.html. The site was originally designed by Ed Breland of USC's Department of Distance Education and Instructional Support, under the direction of Susan E. Bridwell. The WebPage is now maintained by Miriam Mitchell of the university's Department of Computer Services, with content determined and provided by Bruccoli and his graduate students. Abundantly illustrated, the Fitzgerald Centenary WebPage presently includes a Fitzgerald biography and chronology; primary and secondary bibliographies; essays and facts about Fitzgerald and quotations from him; the texts of several Fitzgerald stories; descriptions of items in the Fitzgerald Collection at USC; a history of the House of Scribner by Charles Scribner III; and a voice recording of Fitzgerald reading poetry by John Masefield, John Keats, and William Shakespeare. The site provides links to other major literary websites throughout the world. Between June and December 1996 USC's Fitzgerald Centenary WebPage had 16,600 "hits" and never fewer than seven hundred per week.

The formal program for USC's Centenary Celebration began on 24 September with "Project Discovery Celebrates the 100th Birthday of F. Scott Fitzgerald," a live television program broadcast nationwide and produced by S.C. Educational Television (SCETV) for the S.C. Department of Education. Hosted by Bette Jamison of the Department of Education and Doug Keel of SCETV, the broadcast included a brief tour of the Fitzgerald Exhibition at Thomas Cooper Library, interviews with James Dickey and Joseph Heller, and a question-and-answer session during which high-school students from throughout the state phoned in questions to Dickey and Heller. Asked to define their visions of Fitzgerald, Heller responded that he regarded Fitzgerald as "a heroic figure in literature, an inspiration," and Dickey remarked that Fitzgerald "brought to fiction essentially a poet's insight." Replying to a question about whether Fitzgerald was disillusioned with America, Dickey said that problems with money and prestige clearly bothered and fascinated the author of *The Great Gatsby;* Heller stated that Fitzgerald, like all serious novelists, evaluated and often criticized the ethics of the society in which he lived. Tracing influences beyond Fitzgerald on their own writing, Heller cited Evelyn

Poster by Dawn Pyron that won the graphic-design competition at USC

Waugh, Vladimir Nabokov, and Samuel Beckett; and Dickey mentioned James Agee, Malcolm Lowry, and Ezra Pound.

The date of 24 September was the one hundredth anniversary of Fitzgerald's birth, and the University Bookstore, managed by Bruce Darner, held a champagne-and-birthday-cake party, with jazz by the Dick Goodwin trio, in Fitzgerald's honor. Guests included fiction writers George Garrett, Richard and Robert Bausch, Sydney Blair, Vance Bourjaily, and Heller, all of whom signed copies of their books. Because the birthday party was the initial public event of the celebration proper, it brought together for the first time university personnel and the many Centenary participants from across the United States and overseas. The presence of twelve Japanese scholars and two Germans testified to the enduring reputation of Fitzgerald in those countries. Perhaps the most arduous

Becky and Kevin Lewis receiving a free copy of The Great Gatsby *from Rhett Jackson at the Happy Bookseller, Columbia, S.C.*

pilgrimage in honor of Fitzgerald was made by a student from Whitman College in Washington State; he had endured a three-day bus trip to attend the Columbia events and was gratified when Kevin and Becky Lewis, masters of USC's Preston College, offered him free lodging at the student residential college.

The evening banquet was sponsored by the Thomas Cooper Society, the university library friends group. Held at the Capital City Club in downtown Columbia, the banquet featured a keynote speech, "The Literature of Despair," by Joseph Heller. The nearly three hundred guests, who included John Baker of *Publishers Weekly* and Michael Rogers of *Library Journal,* saw the unveiling of the limited-edition Fitzgerald Centenary poster, created and signed by Kimberley Hamner of USC's Department of Publications and Printing; the poster is available from USC Press. Each banquet guest received a copy of *F. Scott Fitzgerald: 24 September 1896 to 21 December 1940,* a keepsake that printed tributes to Fitzgerald from twenty-seven writers: Jeffrey Archer, Margaret Atwood, Paul Auster, Richard Bausch, Robert Bausch, Thomas Berger, Sydney Blair, Vance Bourjaily, Frederick Busch, Nicholas Delbanco, Don DeLillo, James Dickey, Annie Dillard, Irvin Faust, Leslie A. Fiedler, George Garrett, George V. Higgins, John Iggulden, John Jakes, John le Carré, Norman Mailer, William Maxwell, Budd Schulberg, Charles M. Schulz, Mary Lee Settle, Tony Tanner, and Arnold Wesker. The keepsake was produced by Nancy Washington for the Thomas Cooper Society and the Thomas Cooper Library and was designed by Mary Arnold Garvin and Kimberley Hamner of Publications and Printing. Following Heller's speech he was presented with the Thomas Cooper Medal for Achievement in the Arts by Dean George Terry. The evening concluded with an exhibition of dances of the 1920s, choreographed by Susan Anderson, Stanislav Issaev, and Cynthia Flach of USC's Department of Theatre, Speech, and Dance, and by music of the period performed by the Dick Goodwin Quintet. Events of the banquet were covered by a CNN television crew, and both Heller and Bruccoli were interviewed by Jim Lehrer for PBS's *The News Hour.*

Events of Wednesday, 25 September, began with a reading at USC's Russell House Theater by Vance Bourjaily. Bourjaily's story, "Fitzgerald Attends My Fitzgerald Seminar" (originally published in *Esquire* in 1964 and reprinted as chapter four of his 1976 novel, *Now Playing at Canterbury*), was a favorite of Scottie Fitzgerald Smith, Fitzgerald's daughter, and also proved a favorite with the Centenary audience.

Following Bourjaily's reading USC President and Mrs. John M. Palms hosted a luncheon in honor of Budd Schulberg. Held in the garden of the president's on-campus home, the luncheon drew its menu from a 1924 meal shared by Fitzgerald and his Scribners editor, Maxwell Perkins, at the Biltmore Hotel. After lunch Schulberg delivered brief remarks about his relationship with Fitzgerald.

Frederick Busch was the featured speaker at the official opening of "F. Scott Fitzgerald and the Profession of Authorship," an exhibition of material from the Matthew J. and Arlyn Bruccoli Collection of F. Scott Fitzgerald at the Thomas Cooper Library. Busch commented on the importance of the USC Fitzgerald Collection and observed that it chronicles "Fitzgerald's long, deep battle with talent."

The initial rationale for the collection was to assemble every printing of every English-language edition of every book by Fitzgerald—supplemented by his contributions to books edited by others and his periodical publications. Inexorably the scope of the collection enlarged to include material about Fitzgerald and his times, as well as books by and about his literary friends. While writing about

Fitzgerald and editing Fitzgerald's works, Bruccoli built the most comprehensive working library for Fitzgerald research.

The proper function of an author collection is to provide the evidence for the study of the development of his reputation and the transmission of his texts. The Bruccoli Collection has some three thousand volumes of writings by Fitzgerald, including all the editions, printings, states, and issues listed in Bruccoli's *F. Scott Fitzgerald: A Descriptive Bibliography, Revised Edition* (Pittsburgh: University of Pittsburgh Press, 1988). The printed material includes proof copies and review copies of books by Fitzgerald. There are also some three hundred volumes of translations, as well as an extensive collection of movie and television scripts, stills, lobby cards, and sound and video recordings.

Fitzgerald was a painstaking reviser and rewriter. The manuscripts, revised typescripts, and galley proofs of his work in progress reveal the process by which he fulfilled his intentions. The collection holds the only set of the original galleys of "Trimalchio" before it was rewritten as *The Great Gatsby* and the only galleys of the first serial installment of *Tender Is the Night*. Among the materials for short stories are the revised typescripts for "The Swimmers," "The Count of Darkness," and "The Kingdom of the Dark." The collection also includes Fitzgerald's pocket notebook for his unfinished last novel, *The Love of the Last Tycoon*.

Inscribed books are cherished because they provide both literary evidence and sentimental value. Fitzgerald's inscriptions often comment on his work, and they are characteristically witty. The Bruccoli Collection is notable for its more than sixty-five books inscribed by or to Fitzgerald. Among the recipients of Fitzgerald's inscriptions housed in the collection are Van Wyck Brooks, Ernest Truex, Harold Ober, and Lois Moran. The books inscribed to Fitzgerald by their authors include *Ulysses* and *A Portrait of the Artist as a Young Man* by James Joyce, *For Whom the Bell Tolls* by Ernest Hemingway, *Ash-Wednesday* by T. S. Eliot, *Prejudices Second Series* by H. L. Mencken, and *How to Write* by Gertrude Stein. The collection holds Fitzgerald letters, postcards, and wires, as well as letters sent to him.

The Bruccoli Collection has a complete run of *The St. Paul Academy Now and Then* for the period during which Fitzgerald attended the school, issues of *The Newman News* that appeared while he was at the Newman School, and extensive holdings of Princeton University's *Nassau Literary Magazine* and *Princeton Tiger*–both periodicals with contributions by Fitzgerald. A star item of Princetoniana is the

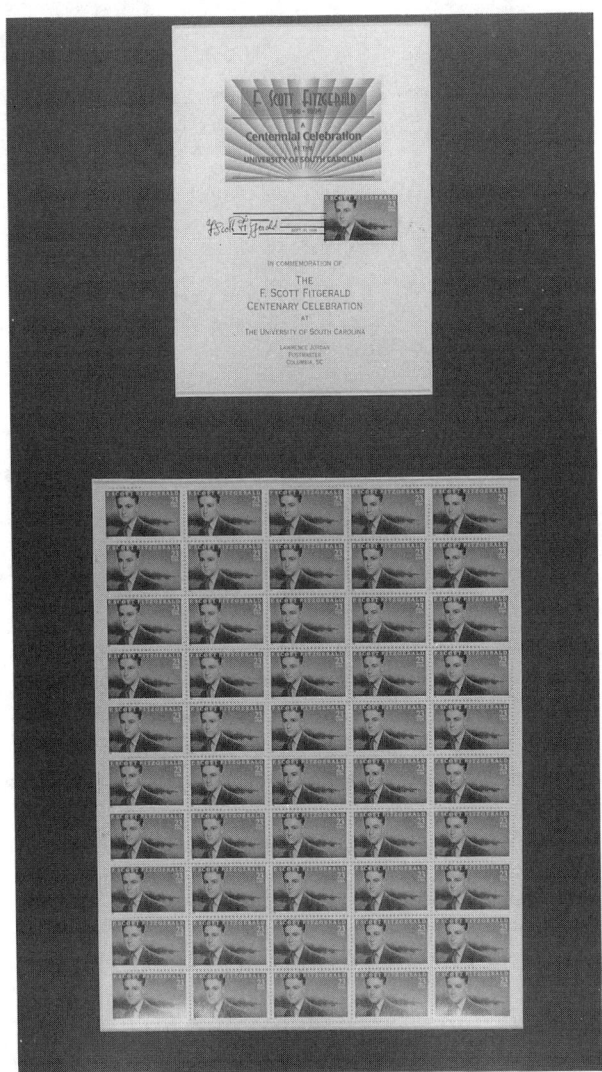

Keepsake and display presented by Columbia Postmaster Lawrence Jordan

printed acting script for Fitzgerald's first Triangle Club show, *Fie! Fie! Fi-Fi!*–with his additional lyrics. The collection includes Princeton class books, catalogues, and other publications relating to the university and the class of 1917.

Zelda Sayre Fitzgerald has become a subject of scholarly attention in her own right. The Bruccoli Collection includes specimens of her manuscripts and letters, with extensive holdings of publications by and about her. The Arlyn Bruccoli Collection of Zelda Fitzgerald's paintings was on loan to the Thomas Cooper Library for the Centenary exhibition.

Most of Edward Shenton's original pen-and-ink drawings for *Tender Is the Night* are in the Bruccoli Collection. Francis Cugat's preliminary sketches and paintings that developed into the celebrated dust jacket for *The Great Gatsby*, as well as Cugat's own

Matthew J. Bruccoli, Joseph Heller, and James Dickey in the Thomas Cooper Library Rare Book Room preparing for the interactive teleconference on 24 September. The poster is by Kimberley Hamner (photo by Gene Crediford).

duplicate painting of the final jacket art, were on loan to the Thomas Cooper Library from Arlyn Bruccoli for the exhibition. Mrs. Bruccoli also made available to the Thomas Cooper Library Gordon Bryant's original portraits of F. Scott and Zelda Fitzgerald.

The Bruccoli Collection has assembled a substantial amount of material relating to Sheilah Graham: her books, her correspondence with Bruccoli, and her account of Fitzgerald's death. The collection also features satellite collections of authors associated with Fitzgerald: Ernest Hemingway, Edmund Wilson, Budd Schulberg, Donald Ogden Stewart, and Ring Lardner.

The exhibition was mounted by Bruccoli, Judith S. Baughman, and Robert W. Trogdon, with the assistance of Special Collections staff members—Patrick Scott, Roger Mortimer, Jamie Hansen, Paul D. Schultz, and Mary Anyomi—and seven graduate students in Bruccoli's Fitzgerald seminar—Tracy Simmons Bitonti, Park Bucker, Michael Cody, Cy League, Catherine Lewis, Robert F. Moss, and Mary Sidney Watson—who also compiled the exhibition catalogue, *F. Scott Fitzgerald Centenary Exhibition*. A printed catalogue of the Bruccoli Collection is in preparation; material in the collection is listed in the USC Access Network (USCAN), which can be accessed through http://www.csd.scarolina.edu/research.html. Significant acquisitions are reported in the *F. Scott Fitzgerald Collection Notes*.

Small exhibitions for Budd Schulberg and Joseph Heller were mounted in conjunction with the main Fitzgerald exhibition. All the Heller material had connections with *Catch-22;* the Schulberg items emphasized his friendship with Fitzgerald.

The event immediately following the opening of the exhibition featured professional fiction writers discussing their indebtedness to Fitzgerald, who described himself as "in every sense a professional." The symposium, "The Influence and Example of Fitzgerald," was moderated by George Garrett, Henry Hoyns Professor of Creative Writing at the University of Virginia, who was introduced by USC English Department Chair Robert Newman. Garrett, in turn, introduced a panel of younger authors—identical twin fiction writers Richard and Robert Bausch and novelist Sydney Blair. The four writers then defined their responses to Fitzgerald's fiction, read and commented on fa-

Professors Emi Nagase and Matthew J. Bruccoli at the Fitzgerald birthday party hosted by the University Bookstore on 24 September

vorite passages from his work, and concluded with a discussion of Fitzgerald's impact on college-age readers today.

During the early evening of September 25 NationsBank and the Thomas Cooper Library hosted a reception on the patio of the Faculty House, after which guests walked to the nearby Drayton Hall Theater for the first performance in eighty-two years of Fitzgerald's 1914 Triangle Club show, *Fie! Fie! Fi-Fi!* Directed by USC graduate Richard K. Blair, the play was performed—as the 1914 production had been—by an all-male cast, in this case of USC students and community-theater actors. The script and lyrics by Fitzgerald and the energy of the actors-singers-dancers provided an entertaining evening for the audience, which, as Theatre, Speech, and Dance Department Chair Thorne Compton remarked, may have been the first "primarily sober" audience to see the show. *Fie! Fie! Fi-Fi!* sold out all performances of its four-night run. Videotapes of the opening-night performance are available from USC Press for fifty dollars.

The final day on the formal Centenary schedule began with one of its emotional highlights, a speech at the Russell House Theater by Budd Schulberg, who had worked with Fitzgerald and become his friend during the final years in Hollywood. Introduced by Lester Lefton, dean of Liberal Arts at USC, Schulberg detailed events surrounding his trip with Fitzgerald to Dartmouth to work on the movie *Winter Carnival*. Thoroughly bored by the movie's subject matter, the two writers avoided working on their script and instead endlessly talked: about writing and literature, about differences between their two generations, about their own lives and dreams. They also drank heavily, which led to their being abruptly fired on the Dartmouth campus by producer Walter Wanger. Schulberg ended his talk with a tribute to Fitzgerald's endurance as a writer, which he described as "the fruits of his good, hard labor and genius."

After Budd Schulberg's presentation, invited guests went to Harper College on the USC Horseshoe, where Dean George Terry and McKissick Museum director Lynn Robertson gave a luncheon. At its conclusion two Fitzgerald scholars from overseas offered remarks. Horst H. Kruse of Westfälische Wilhelms-Universität in Münster, Germany, and

Kiyohiko Tsuboi, formerly professor of English at Okayama University and now president of Kobe Women's University Seto Junior College in Japan, discussed the reasons for Fitzgerald's high reputation in their countries.

The final events within the formal Centenary Celebration schedule were held at the Richland County Public Library on the evening of 26 September. Producer Ed Breland introduced and then premiered his documentary film, *Getting It Right,* which defines the relationship of book collecting and scholarship and which focuses on the work of Bruccoli. Following the documentary, three English department graduate students—Park Bucker, Lee Anna Maynard, and Cy League—presented readings from Malcolm and Margerie Bonner Lowry's unproduced screenplay for *Tender Is the Night* and from works by Fitzgerald that spanned his career. The readings were selected and staged by Park Bucker.

USC's F. Scott Fitzgerald Centenary Celebration was an expression of the university's—and of the world's—admiration for one of America's greatest writers. It is therefore appropriate that during the Centenary year the USC Press, directed by Catherine Fry, published four new Fitzgerald books: *Reader's Companion to F. Scott Fitzgerald's Tender Is the Night,* by Matthew J. Bruccoli with Judith S. Baughman; *F. Scott Fitzgerald on Authorship,* edited by Bruccoli with Baughman; *F. Scott Fitzgerald Centenary Exhibition* (University of South Carolina Press for the Thomas Cooper Library), the illustrated exhibition catalogue; and *Fie! Fie! Fi-Fi!* (University of South Carolina Press for the Thomas Cooper Library), a facsimile of the acting script and musical score of Fitzgerald's first Princeton University Triangle Club show, copies of which are in USC's Fitzgerald Collection. For an author who once declared, ". . . writing has been my chief interest in life," the books are a particularly fitting tribute.

Judith S. Baughman
University of South Carolina

Excerpts from Joseph Heller's USC Address, "The Literature of Despair"

It's been a nice evening, and I've been having a wonderful time—till just now, when called upon to speak. Suddenly I feel like the malevolent witch at the party, because the title and topic I've chosen for my speech do not seem appropriate for an occasion on which so many people are in such good spirits. If I were an expert at improvisation, I would change it, but I'm not. So we're faced with a talk with a title I'm almost embarrassed to say: "The Literature of Despair."

The subject is not going to be certain novels we all know about or the characters in those novels. Instead, it is focused more on the literature about the authors of those novels. It is the literature of biography—the biographies, the life histories, of so many of the authors we think so much about.

The thought of such a subject came to me several years ago when new biographies of three well-known authors appeared in a short span of time. These were a biography of Henry James, a biography of Charles Dickens, and a biography of F. Scott Fitzgerald. To read even just the reviews of the biographies of these three distinguished authors was to read accounts of the very sad turns their lives took before they were over.

I'd known about Fitzgerald, of course, about his drinking, his decline in reputation, and the mental problems of his wife, Zelda. And I was familiar with Dickens from an early biography—with his compulsion to control, his passion for theatricals, his autocratic family life, and, toward the end, his driven need to perform publicly. But Henry James struck me with surprise, for I had not known much about him. And it was hard to believe that this figure who spoke with so much certitude on so many subjects over so many years could himself be vulnerable to spells of deep melancholy, to depression. If it happened to him, I thought, it could happen to anybody, to anybody who wrote creatively as a career, and it has happened to a very great many.

At about that same time, William Styron published his book about his own bout with depression, and Kurt Vonnegut, in a collection of nonfiction pieces, made reference to a similar experience. About then, John Cheever's diaries were published, posthumously, and I can't recall ever reading a sadder personal testament.

The more I thought, the more I was impressed by the fact that so many writers do have at least some very serious emotional trouble in their lives. Let me give you a roll call. So far as I know, it may be true that Shakespeare never suffered an unhappy moment in his life. But in the last century we had Edgar Allan Poe, a drinker and user of drugs. Hawthorne and Melville had their spells of melancholy. Coleridge also. Joseph Conrad had his nervous breakdown late in life. As a young man he had tried to kill himself once by shooting himself in the chest, and missed. Jack London was an alcoholic. He wrote a novel called *The Dipsomaniac* about this addiction. Malcolm Lowry in *Under the Volcano* writes

USC Dean of Libraries George Terry speaking at the Centenary banquet; with President John M. Palms and Joseph Heller (photo by Michael Rogers)

about a character who is an unregenerate alcoholic, and he, too, was writing autobiography.

Moving into our own time, we can assemble a very long list. Among the heavy drinkers often disabling themselves we can find: Eugene O'Neill, Edmund Wilson, Sinclair Lewis, John O'Hara, William Faulkner (who died after a drunken fall, I believe), Theodore Dreiser, and . . . Joseph Heller? A regular drinker, as you may already have observed tonight, perhaps a borderline, but not yet so classified. And sadly, for this occasion, F. Scott Fitzgerald.

I've left out many. Among the British, after Malcolm Lowry, there was Evelyn Waugh, as notorious as a drunk as for his bad nature, and Kingsley Amis, who died about a year ago from a fall down the stairs, probably while drunk. Truman Capote fits in for his drinking and for his "sedative medications." Tennessee Williams certainly fits in. Hemingway, we know, was a suicide, and a very heavy drinker, too.

Among my contemporaries were Jerzy Kosinski and Richard Brautigan, both recent suicides. Earlier was Ross Lockridge, author of the huge best-seller *Raintree County,* who then killed himself, and that same year, I think, Tom Heggen, author of the play *Mr. Roberts,* another huge success.

Among women, Emily Dickinson was a suspicious recluse; Virginia Woolf, we know, killed herself; Anne Sexton and Sylvia Plath, each a suicide. I fear if I wanted to delve into poets I would not know where I'd be able to stop.

Of course, the question to be answered is WHY? What is there about the literary occupation that causes, or is the concomitant of, so much wretchedness among so many people who are successful?

And the answer I give you tonight is: I don't know.

I'm aware, as I'm sure many of you are, that the implications of what I'm saying are not sound statistically. There is, for example, no control group, and it could be—it is theoretically possible—that all of us here tonight are suicidal drunkards and depressives. But, seriously, I cannot think readily of another occupational group that has so many of its major celebrities suffering the distresses I've mentioned.

. . . .

What are some of the factors that might cause or contribute to these predilections toward unhappiness? Well, in a general—a most general—way, I can try to guess:

There could be something in the nature of the work, the uncertainty of success, the greater uncertainty of maintaining a peak of success and income, over a working lifetime. After all, a writer can be discovered only once, and after that, the scrutiny of critics grows more exacting.

Ursula Kruse, Arlyn Bruccoli, and Professor Horst Kruse at USC's Centenary banquet, 24 September

Or there could be something in the nature of the individual and the early family setting that influenced the person toward fantasizing, fictionalizing—a tendency toward daydreaming extravagant scenarios of accomplishment. These imply a wish to excel and, with those who turn to writing fiction, plays, or poems, an ambition to excel as a writer.

Most likely, there is an indefinable mix of both, and maybe half a dozen other factors I haven't mentioned.

Then add to these the Freudian discovery that the conflicts, feelings of loneliness, and disappointments we possessed that at the beginning led us toward fictionalizing are not entirely satisfied by success but instead remain. There has been no miraculous transformation, and in many ways the sensitive parts of us remain exactly the same.

....

Let me throw in one more element, the factor that feelings of failure are almost certain to enter into the life of the published author, even with works that appear to be triumphs. Ernest Hemingway's most popular success was, I think, *For Whom the Bell Tolls,* and probably instead of rejoicing, he was enraged by those literary critics who found it deficient in quality. And what of Fitzgerald with *The Great Gatsby?* The praise was generally high, but sales were disappointing, so what joy he experienced was diluted by that—and, of course, by his chronic, desperate need for money.

....

With F. Scott Fitzgerald, it would seem to have been a combination of just about all I've touched on. In the portraits of the personality one finds in his biographies are elements with which I am personally familiar from childhood, adolescence, and early adulthood, something which I'd wager all of the names I have mentioned have experienced to a strong extent: principally, that desire, that need, to shine, to excel. And for a person to imagine himself doing that much is already evidence of a tendency to imagine—to fantasize—about one's own self involved with relationships, emotions, and situations that are the essence of fiction.

From childhood through adolescence through college, Fitzgerald did seek constantly to accomplish, to stand out. He tried football, in prep school and in college, but was too small. He tried theatricals, and if he could not excel in the theater, he wanted to succeed in the writing of fiction, and he persisted.

And at the age of twenty-three, he did succeed. As a young man of twenty-three he became a celebrated American novelist. He married the girl he wanted—one he had been in love with and had courted for several years. He had an income that for him and for others at that time—for most of us today—was a very extravagant income. And he reveled, gloried in his success, as he should, as I would, as any other newly successful author would. Perhaps he reveled in it too much. If his life had ended just there, we would have a romantic story with a very happy ending in the Horatio Alger mode. And it would not be of much interest.

But his life didn't end then. And in a little while he was also celebrated as a cut-up, as a drunkard. He was a subject of anecdotes, was often good company, and was often a severe trial to his friends, with rowdy antics that would make some of us wish to cringe and look away.

....

One of the saddest entries in the notes he made in the latter part of his life is the one about himself relative to Ernest Hemingway, whose career he had generously helped start. It went something like this—I may not be quoting exactly: "I talk with the authority of failure—Ernest with the authority of success." It was not a man flushed with self-esteem who could frame such a depressing comment as that one.

But the reality is that for more than half his adult life, Fitzgerald had good reason to think of himself as a failure. He had not published a novel since *Tender Is the Night,* on which he had labored a longer time than ever before and which was not received as well as he hoped it would be. His marriage had turned into a tragedy of a kind other people suffer in other people's novels; that is, his wife, Zelda, was back in an asylum. He was anything but a success as a film writer. And he was, as almost always, in desperate need of money for himself, his wife, his daughter. And that was the state he was in when he died suddenly of his heart attack, and the state of his life for at least the preceding ten years.

. . . .

I remember another statement of his: "There are no second acts in American lives."

If we had a question period now—and I'm thankful there won't be—one of the questions you might logically ask is: "What about you?" The answer I can give you is: "Not yet." And not likely, because of my age—I've already passed those dangerous decades of the others—and not likely soon, because after this occasion tonight, I think I'm going to be in a very good mood for a good long time.

Fitzgerald would be in a good mood, too, if he could see how his reputation has risen to so extraordinary an extent for someone who was virtually ignored and forgotten by 1940 or so. From three hundred copies of all his works sold then to three hundred thousand sold annually now.

And he'd be extremely pleased if he could be here tonight to watch and to hear: pleased, I think, but not surprised, because like all authors he had the tremendous vanity that is typical; and rather than feeling surprised, he'd feel justified, ratified, vindicated. Unfortunately, he isn't here tonight because, in his words, "There are no second acts in American lives."

Excerpts from Frederick Busch's USC Remarks

. . . Professor Bruccoli's . . . happy combination of scholarship and publishing acumen brings us closer to the great documents of modern American writing. Thinking of Fitzgerald, and of Professor Bruccoli, and of this afternoon, I reread Fitzgerald's story "Jacob's Ladder" . . . [a]nd I was struck by how Fitzgerald, a master of literary control, wrote again and again about being swept away or, as he often puts it, borne away. That tension—the control in art and craft exerted upon language about control being lost—seems to me as good an emblem for Fitzgerald as I, anyway, require.

I'll remind you that "Jacob's Ladder" is a story about an artist (here a singer) who has lost his gift (in this case, his voice). It concerns a man who, "like so many Americans . . . valued things rather than cared about them." Jacob falls in love with a beautiful younger woman whose career as an actress he makes possible. He isn't happy about her until she makes him miserable. She *becomes* a force he cannot resist. And we know this, as surely as we know such forces in, say, *Gatsby,* because Fitzgerald often immediately translates large forces—history, unguidable emotion, unconquerable weakness, the rushing-on of one's fate—into cars.

Just as those automobiles speeding between Manhattan and Long Island in *Gatsby* add up to destiny, which no one can adequately steer, here in the 1927 short story, Jacob and Jenny Delehanty jump not into bed but into a car they ride in to dinner in upstate New York, a car that rolls back into Manhattan with Jenny asleep on Jacob's shoulder, a car in which Jacob, later, "was borne along dark streets and light toward a future of his own which he could not foretell."

Jenny's future is discussed by Jacob in a taxi with a film director; her success is celebrated with a kiss in Jacob's car; his love is stated—"I won't marry you unless you love me," he says—in his car, driving from the movie studio; Jenny describes her anguish over an affair by telling Jacob how she got into a man's car and then, Jacob says for her, "It just—swept over you. . . ."

When he understands that he has lost Jenny, "the heavy tide of realization swept over him," and later, trying still to possess her while knowing that he never will, Jacob buys a movie ticket to her dazzling film and "found himself a place in the fast-throbbing darkness."

. . . That "fast-throbbing darkness" looks forward to *Tender Is the Night,* to Nathanael West's *Day of the Locust,* to Norman Mailer's *The Deer Park,* to Robert Stone's *Children of Light,* and to John Updike's *In the Beauty of the Lilies,* those explorations of the enfleshment on film of American dreaming. That throbbing darkness is about mental sex, at which Fitzgerald was brilliant. It is also Fitzgerald making love to the same mistress—her name here is Great Writing—as Hemingway. For we have to think of *Jake* Barnes in his cab in Madrid with Lady Brett, and *Jacob* in his cab with Jenny—each man confronting what to him is the harsh joke of love, each saying farewell for us to romantic ideas of love—as a salute from Fitzgerald to his cruel, disloyal brother in art whom Fitzgerald knows to have retained *his* voice and whose writing about Jake Barnes he recognized from the start as a level of art he might not again achieve: that "real thing" about which he wrote to Maxwell Perkins.

. . . young Matthew Bruccoli was struck by a speeding Fitzgerald in the Bruccoli family car. Hearing a radio treatment of "The Diamond as Big as the Ritz" in 1949, he was changed forever; and so, in truth, was American scholarship. In partial and happy consequence, Fitzgerald's long, deep battle with talent is here for us to see, in its particularities, in the splendid collection of Arlyn and Matthew J. Bruccoli available to us in the Thomas Cooper Library of the University of South Carolina. Amid the thousands of items in this literary cornucopia, we can read, in Fitzgerald's proofs of *Gatsby,* about the force on which we are borne back ceaselessly into the past, and the grinding gears and scary speed of the cars that are fate's messengers.

. . . The Bruccoli Collection is tangible memory of the failed love, the failed art, the failed writer who is also, Professor Bruccoli has reminded us again and again, the man who kept his love alive as well as he could, who toiled at his art some long time after it had given signs of forgetting *him,* and who labored in the face of huge, dark, countervailing forces to remain an artist and a decent man. . . .

Matthew J. Bruccoli's Response

I didn't know that this was going to happen; I have no prepared comments. The people who have been committing perjury for the past couple of days have said generous things about my Fitzgerald work. But it has been the result of monumental selfishness: it wasn't hard; it was fun. During those forty years there was always another trip to make, another lead to pursue, another dealer to find, another collector to meet. There were all the friends met in bookstores. Above all, there was my friendship with Frazer Clark, who became my partner in book collecting and other bookish endeavors. There were all those checks I wrote on Friday after the banks closed. You could buy anything on Friday afternoon, just so long as you covered the check on Monday morning. There were the times Fraze and I drove all night to cover our checks, laughing all the way.

It isn't over. I've been the target of commiseration about "Now you have nothing to do." The hell I don't! As long as they can wheel me into a bookstore I'll be buying books and improving the Fitzgerald Collection at the Thomas Cooper Library. That is what I do. I will still be looking for two Fitzgerald items that I stupidly failed to buy when I could have acquired them. I recently received a tip that one of them may be sold by the estate of the collector who got it. I had nothing to do with his demise. If the item is put up for sale, we'll get it. The lesson I've learned over and over again—but we don't always live up to our own best lessons—is that you never regret buying a book. There is no such thing as overpaying for a book. The only ones you regret are the ones that you failed to buy for some foolish, cowardly, irresponsible reason.

I repeat: building this collection was mostly fun and laughter. Much of the laughter was generated by my friendship—a weak word for my enduring feelings about her—with Scottie Fitzgerald. I didn't know F. Scott Fitzgerald, but working with Scottie was just as good. Her father would have found the words for her generosity; I can't.

Part of the fun that Scottie and I had during our "Daddy projects" resulted from the circumstance that we both enjoyed running gags. One of her running gags was about Gatsby's car. Scottie couldn't tell a Model A from a Rolls, but she kept spotting old cars and telling me that "This was the car Daddy gave Gatsby." It would always be impossible, but the gag went on. Shortly after Scottie died, a crate arrived at my house with a stained-glass representation of a 1930s Mercedes convertible. Scottie had seen it on her last outing before she died and arranged for it to be sent to me after her death. The gag ended with Scottie having the last laugh.

The name of the collection—the Matthew J. and Arlyn Bruccoli Collection—is not a convention. It truly is—was—half Arlyn's. She was informed of every purchase and spent much of her marriage in bookstores. With utter honesty I can state that my Arlyn never objected to the price of anything. She did complain about the boxes of books on the floor. I am deeply grateful to Dean Terry and President Palms for working out the arrangement that got the books off the floor and here where the collection will continue to grow for the use of students and researchers. We have acquired smashing items since the collection came to Thomas Cooper Library. We are going to continue acquiring material to make this F. Scott Fitzgerald Collection the best working archive for an American author.

Again: No Arlyn, no collection.

Excerpts from Budd Schulberg's USC Presentation

. . . When I saw Scott for the last time (I remember very well it was on the first day of December in 1940, and I was going east; I'd been working on my first novel); I went to say goodbye to Scott;

James Dickey and Budd Schulberg at the USC presidential luncheon, 25 September (photo by Michael Rogers)

and Scott was in bed. He lived in a sort of simple, fairly plain apartment right in the heart of old Hollywood off of Sunset Boulevard around the corner from Schwab's Drugstore, which was the hangout for everyone in the neighborhood. Scott had this desk built for him to fit around him in the bed, as he was pretty frail and feeling weak and at the same time found he could write in bed for two or three hours every day. And I brought with me a copy of *Tender Is the Night* that I really loved reading and still do. Scott inscribed it to my daughter, Vicky, saying, " . . . in memory of a three day mountain-climbing trip with her illustrious father–who pulled me out of crevices into which I sank and away from avalanches — ."

When Scott died, I was shocked, because that day, on the first of December–sitting there in bed, looking very pale but animated–his spirit was very, very alive. I had the feeling during that period that his mind, his intellectual energy, was high and that only his body was not up to what the rest of him was. First, I asked how his novel was coming. I didn't know exactly what the novel was about; he hadn't told me what the book was about. I told him that I was trying to write about Hollywood, and I had a sense that he was also, because over the last two years that I knew him he asked me so many questions about Hollywood–what it was like growing up there.

It was my hometown, and my father [B.P. Schulberg] had been running a big studio, Paramount Studio. It was the rival studio to M.G.M, which was run by Louis Mayer and Irving Thalberg. Mayer and my father in earlier days, the early pioneer days of the early 1920s, had been partners in a funky little studio called the Mayer-Schulberg Studio, where young, very young (he looked like a high-school boy) Irving Thalberg had first come to work for Mayer. And Scott was fascinated by the movies. He was almost obsessed by the movies and intensely curious about what the people were like, what my impressions were like. He asked me a lot about my father. My father and Irving Thalberg were considered the two intellectuals in the town, but I think my father was better read than Thalberg. He read all the classics and made a lot of movies based on the classics, *Crime and Punishment* and *Jennie Gerhardt* and *An American Tragedy.* . . . Irving was fastidious, except for smoking, and he was conscientious, hard-working, intense in his work. There was something almost princely about him. He was absolutely held in reverence by the whole Hollywood community, which was a very small world.

I always thought of Hollywood like a principality of its own. It was like a sort of a Luxembourg or a Liechtenstein. And the people who ran it really had that attitude. They weren't only running a studio; they were running a whole little world. Their power was absolutely enormous, and it wasn't only the power to make movies or to anoint someone or make someone a movie star or pick an unknown director and make him famous overnight. They could cover up a murder. The district attorney, Buron Fitts, was completely in the pocket of the producers. You could literally have somebody killed, and it wouldn't be in the papers. They ran this place, and, as I said, Scott was very, very interested in all of this.

. . . .

Some of it is in *The Last Tycoon*; the producer's blood is in *The Last Tycoon*. Writers just can't help taking things like that from what is around them. That last morning, when Scott showed me the opening chapter of *The Last Tycoon* and I read the opening paragraph and scanned the first page, I have to confess that I was a little bit hurt. For a moment my feelings were hurt. The book had opened with—well, there was an odd feeling of reading my own words, as I had said to Scott that, being raised in Hollywood, there didn't seem to be anything glamorous about it. It was a town that turned out a product. Instead of automobiles or tires or steel, in our town we turned out cans of film. And we had to turn out so many a week to keep the wheels turning. I said we live in a company town, just as much a company town as any coal town in West Virginia. And it's about that glamorous. That was all there, and I thought, "Gee, I thought Scott really liked me, but I guess I was just something that came along that was very handy for him."

Scott saw the look on my face. He was one of the gentlest, kindest, most sympathetic and generous writers I've ever met. At the same time, of course, he couldn't stop lifting something somebody else said, because that's the profession he was in. He looked at me and said, "Yes, Budd, I guess in Cecilia (the narrator of *The Last Tycoon,* a film magnate's daughter, who's gone off to Bennington, which is about as close to Dartmouth as one can get) I did combine you with Scottie." At the same time my emotions were mixed because I had the feeling, reading a page or two, that Scott was really onto something. And frankly he was onto something beyond what I had perceived with all my Hollywood experience.

When he died, well-known writers in Hollywood—Dottie Parker and John O'Hara and Donald Ogden Stewart and Robert Benchley—were aghast at the obits. If you look back, you'll see that the obit writers of *The Times* and *The Herald Tribune* and many other newspapers confused Scott with the Jazz Age. He had been the poet of the Jazz Age; he had been in many ways the voice, the spirit of the Jazz Age, and in a way almost the creator of the Jazz Age. And he himself had for a while seen himself that way. But now in the Depression years, those years of the 1920s—of the saxophones and the bathtub gin and the Charleston—were looked down on as a wasteful period, as a period when our country had really gone wrong, really gone off on a wild spree. And somehow they blamed him. Instead of seeing Scott as he was, as a remarkable interpreter of that time, they blamed him for it, as if the 1920s were all Scott's fault, and now that the 1920s were gone, it was a little like "good riddance" because that was Scott and he's gone and the times are gone and we're glad and goodbye. That was very much the tone of those obits.

So a bunch of us (I was sort of a tagalong; I was a kid along with them, although I knew them very well) approached *The New Republic* about doing a memorial issue, which came out in March 1941 and in which O'Hara and Edmund Wilson and many other of his supporters on both coasts wrote tributes to Scott's memory. Just a short piece of what I wrote was as follows: "When I was leaving for New York a few months ago I said goodbye to Scott and asked him how his novel was coming. It was the end of the day and he looked weary, for the writing didn't come so easy any more. It was a page a day now, but a *good* page, no matter what the fortunately anonymous *Times* and *Tribune* reporters and the unfortunately bylined Westbrook Pegler think. 'Oh slowly,' Scott said. 'But I'm having a good time with it. The first draft will be finished by the time you get back. You can read it then, if you like.'"

Anyway, a few weeks later I happened to be back in Hanover and I was in the Hanover Inn, and a professor . . . very casually said, "I'm so sorry to hear about Scott." And I hadn't. It was in the obits, I guess, that morning, and I hadn't seen it yet. I said, "What?" It was [a professor] who told me that Scott had died. And I felt a terrible sadness. I thought about the book. At that moment I didn't foresee what would happen to *The Last Tycoon,* and how it would find a published life of its own. . . . I just thought, "My god, what a loss that is. Why did that have to happen?"

. . . .

Scott had great dreams about Hollywood. It was not just the money. Most of the writers I

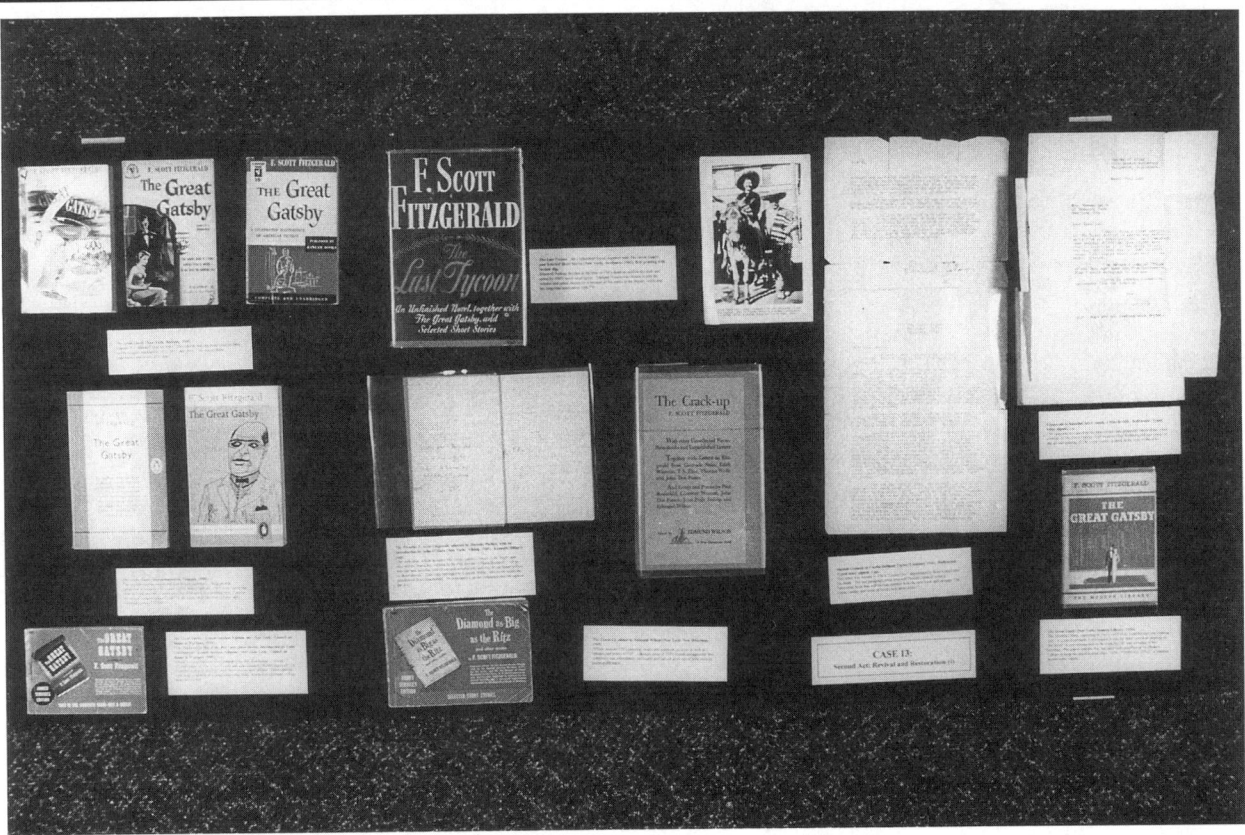

Exhibition case for "Second Act: Revival and Restoration," Thomas Cooper Library (photo by Keith McGraw)

knew—Faulkner and the others—just wanted to get the money and get out. Scott was different. He believed in the movies. He thought it was a great medium, and he thought it was one that any writer had to know, to learn. He thought it was the storytelling of the future, and he admired the moviemaker, as you can see from the remarkable vigor he had in portraying Monroe Stahr in *The Last Tycoon*.

....

I did know that Scott was going to write a novel; I didn't know what it was about. As it turned out, he was writing his and maybe *the* Hollywood novel. We stayed pretty close through that year. I would see him quite often. I would go out to Encino where he was living in this guest house of Edward Everett Horton, a very successful comedian of that time who'd made a lot of money in the movies. . . . I have memories of Scott finishing a Pat Hobby story and running down those outside steps and reading it out loud in that hopeful, eager way. There was something about Scott Fitzgerald that I am sure you would all feel: There was something about him that could make you cry, something about everything he had gone through, all the agony, all the pain. And yet he had this kind of youthful enthusiasm that was just so much a part of him that, no matter how life might batter him, it just couldn't beat it out of him. He was a wonderful friend that way.

When my daughter was born in May 1940 at Cedars of Lebanon, Scott saw me, asked me the name of the child, and I told him "Victoria." Scott said, "In Victoria's honor I'm going to change the name of the child in 'Babylon Revisited.'" He was working at that time for Lester Cowan, writing a screenplay based on that wonderful short story. In the next draft, indeed, there was Victoria in place of Honoria. Also when Vicky was born, Scott went into a monologue about a father and a daughter, about this little tot and how she grows and how he enjoys her beauty, and then there comes a time when she moves on to another man, and so he went on with this story. And he was charming about it, he really was.

Scott's reputation started with a trickle first in 1941 with the Edmund Wilson *Last Tycoon* and then on to 1945 with the O'Hara *Portable* and then in 1951 the Mizener biography, *The Far Side of Paradise*. And my *Disenchanted;* it helped, also. People would stop me on the street, honest to God, and say, "I hear that book is about Scott Fitzger-

A portion of the Fitzgerald Centenary Exhibition at the Thomas Cooper Library, USC (photo by Keith McGraw)

ald. I've never read him. I can't find any books by him. How do you find a book by Scott Fitzgerald?" That's the way it was, or seemed, in 1951. And then came all the contributions of the 1960s, the 1970s, and on . . . until now as we go into the twenty-first century, nobody has to champion Scott Fitzgerald. His reputation will only grow and grow and grow.

I have felt in this celebration an enormous joy for Scott. I can't help thinking—it may be a bromide—but I can't help thinking, Oh, God, if Scott could only see this. If Scott could only see the adulation. If Scott could only see that he accomplished so much of what he was after. And I do think that Scott with all of his sense of failure and also his sense of immortality ── . He did say to me, word-for-word, "I had a beautiful talent once, maybe, and I still think I have enough left for at least one more novel and maybe two."

I think the celebration that we've been experiencing in the last two days is a tremendous contribution. I think that this university has done so much to not only keep this flame alive, but to make it something that will remain lit into the twenty-first century. If I believed in an afterlife (I'm working on it), I know that Scott would be enjoying the fruits of all his good, hard labor and genius. And if it's only up to us, then we—we who read Scott, who love Scott and his work, and who are about to run over into a new generation—we here, in this room and in the time to come, are Scott's immortality.

George Garrett, Richard Bausch, Robert Bausch, and Sydney Blair at USC: "The Influence and Example of Fitzgerald"

George Garrett: What we are fixing to do, with a certain amount of flexibility, is that first I'm going to introduce these gifted younger writers who are our primary participants, and then I'm going to read a couple of paragraphs . . . that I wrote to represent my generation's attitude, which will be slightly different, I think, from theirs. Though I'm not as close in touch with direct memories of Fitzgerald as, say, Budd Schulberg, time has a way of overlapping. And anyway I feel at this point that I can speak for my generation in that matter.

On my left, Sydney Blair, novelist and short-story writer. Her novel *Buffalo* was published a couple of years ago to great and wonderful critical reviews and response. It's available in paper. She's just finished a new novel (which we're all looking forward to) with the promising working title of "Jan Gets a Dog." On my right ── . This is probably the only time in your lives when you'll sit in a room

Exhibition catalogues

face-to-face with identical twin American novelists, but here they are. On my immediate right, Richard Bausch (I think), the author of—prior to this fall—six novels and three collections of stories, all of them admired and well received. And on this day he has received the thing that lots of people wait for lifetimes without getting, a roaring, rave review in the daily *Times* for his new novel, *Good Evening, Mr. & Mrs. America and All the Ships at Sea.* . . . Anyway, we're delighted and feel it's a good luck omen for all of us. To his right is Robert Bausch, author of three novels—most recently, *Almighty Me,* which to the joy of his friends and others was purchased by the Disney people, and we keep hoping that they will go ahead and make the film. His most recent book is the one that was available yesterday in the bookstore, a very handsome and elegant collection of short stories called *The White Rooster and Other Stories.*

What else to say. I'll just say a few words myself, and then we'll come back and work across. Sydney has prepared a short piece, and the Bausch boys, acting as sort of a pair of stereophonic speakers, will wing it a little bit for you. They're good at that.

I wanted to say a couple of things that record my own experience of things. You would think, if you were familiar with my work at all, that I have little or nothing in common with the art of F. Scott Fitzgerald. Different world, different subjects, different ways and means of storytelling: It's all too

Fie! Fie! Fi-Fi!: *Finale, Act II. Front: Cholmondeley, Celeste, Fi-Fi, Mrs. Bovine, Del Monti, Clover Blossom, Dr. Blossom; Back: bandits, Colonel Pompine (photo by Lisa Martin-Stuart)*

Fie! Fie! Fi-Fi! *reception at the Faculty House; sponsored by NationsBank and the Thomas Cooper Library (photo by James Hardin)*

Program for the production

Luncheon honoring scholars from overseas, 26 September; left to right, Professor Kiyohiko Tsuboi, Vance Bourjaily, USC Press Director Catherine Fry, and Dean George Terry

true. But we do have some things roughly in common. Both of us went to Princeton, and it was basically the same kind of Princeton in those days, even though we were years apart. Both of us played freshman football. Fitzgerald was cut from the team, and I wasn't. I ended up hanging around the playing fields of Princeton long enough to be permanently injured, disabled really, in a modest way. Both of us spent our time in the U.S. Army—Fitzgerald in Alabama, and myself in Trieste, Austria, and Germany. Both of us served a time or two—he perhaps more seriously—in Hollywood as scriptwriters. He was better at it—pretty good, in fact. I had the fortune—or misfortune—of having most of my scripts produced. That's probably the worst thing that can happen to you. It's better if they're known to be wonderful scripts that no one can produce. One of my scripts is *Frankenstein Meets the Space Monster*. It's the only award-winning film I've been associated with. It won a Golden Turkey Award as one of the hundred worst films of all times. Anyway, the Fitzgerald script for *Gone With the Wind* was impossible to do because of the limits of time and budget at the time Selznick proposed it, but Fitzgerald had a wonderful script. Then there's a personal element. I had an uncle who was a scriptwriter out there who knew him and an aunt who was engaged to somebody who shared an apartment with him before he was married. My next-door neighbor was from the same entryway in the same class in college at Princeton and joined the service at the time. Jim Bettes, who joined the Marine Corps, not the army, . . . told me that if you made it through your third year at Princeton at the time of the First World War, you could get your degree—you could actually get it or were authorized to get a degree—if you enlisted in the service. It was an inducement for people to join the service in World War I. Everybody in that entryway enlisted, figuring to save a whole year of college and have fun in the army and Marine Corps. Most of them didn't get overseas, and I think that is the case with Fitzgerald.

The real connection from my point of view was discovering Fitzgerald, together with Faulkner (whose ninety-ninth birthday is today) and Hemingway, at Princeton. As a student in 1947 I read and reread his shining works, especially *The Great Gatsby* and *Tender Is the Night,* over and over again. I was in Princeton the summer of 1948. It was a green and shady summer. It was the last year that Princeton ever had a summer school, and it was the first year that they ever taught anything in their English Department that was written since 1900. And this was a great, great thing for them. I have known many writers for whom Fitzgerald is not by any means the primary model or influence of their own work, but I have yet to meet any writer who does not take Fitzgerald as an undeniable influence and a vital example. His direct influence is then wider and deeper than it might seem. It is an influence that has been beneficial to a whole generation of American writers. You can always learn from the finest works of Fitzgerald. It is "awesome," as the younger generation says; it is also encouraging.

I'd like to tell you one other thing that we've going to do because it's in the spirit of celebration. When we've done all our little things and before we open up to questions, we're each going to read a short passage of something we really like—short and to the point. As it turns out, everybody picked *Tender Is the Night* and *Gatsby.* Sydney?

Sydney Blair: Here are my ten minutes worth of prepared remarks on F. Scott Fitzgerald, quite anecdotal, be forewarned.

When I knew I was coming to this celebration, I conducted an informal survey among my friends and acquaintances and the occasional stranger of their impressions of F. Scott Fitzgerald. The survey, although random and wildly unscientific, was nonetheless telling. I approached some-

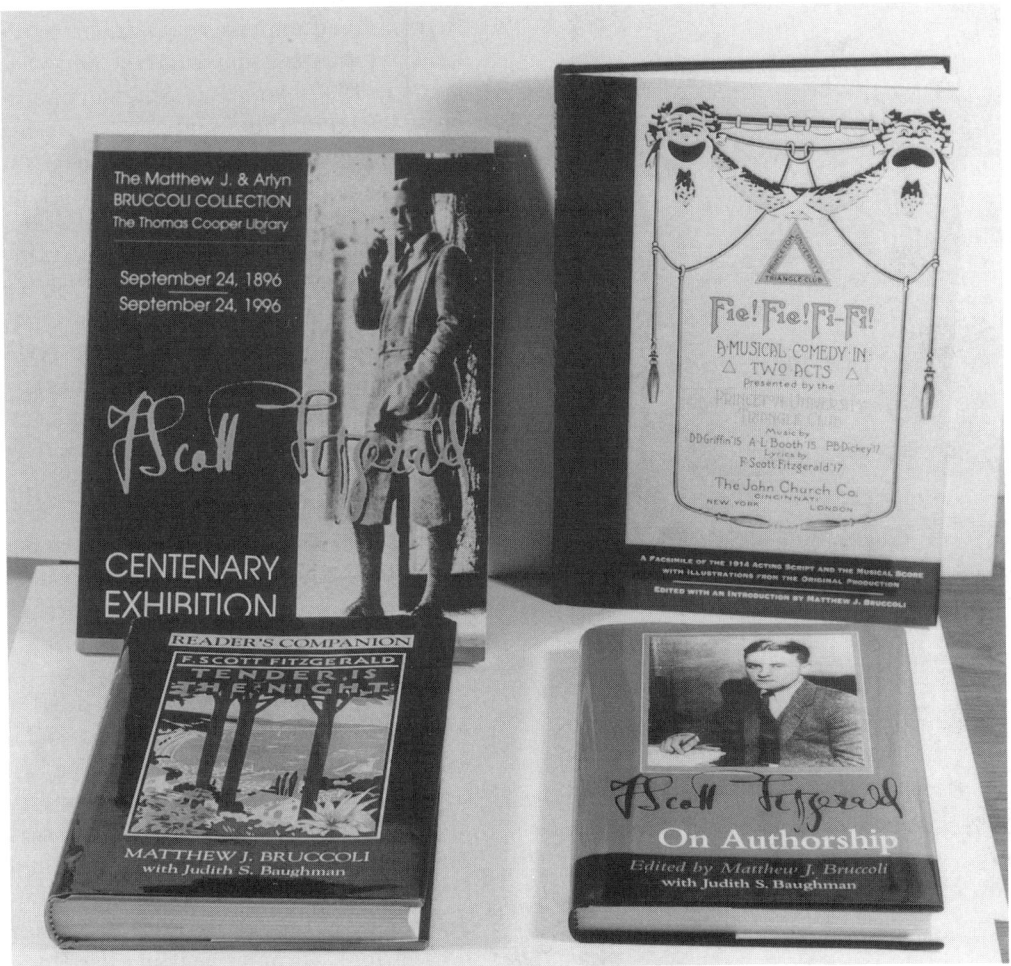

USC Press Centenary volumes

body explaining the purpose of this gathering and then asking what came to mind when they thought of Fitzgerald. Most of the people I talked to were readers, though not necessarily writers.

One, misunderstanding, said, "Wow, what are you going to wear?" He was under the impression that we were attending a weeklong costume party or something. This person said I should wear a beret and dark glasses and have a cigarette dangling from my fingers and a long, sleek cigarette holder. And I should look bored and every now and then should turn to someone and say, "You fool!" Which I thought was pretty funny. But these things do happen, and thus the man lives on. Recently I read about a Gatsbyesque lawn party being thrown (George and I live in Charlottesville) in a nearby town, designed to suggest 1920s champagne and hors d'oeuvres in a Victorian garden followed by a Southern clambake. Eaters, not readers, I figured. But I like to believe that Fitzgerald would have been pleased and amused that he lives on in this way at that clambake in Orange, Virginia.

But to get back to the survey. My teenaged son surprised and impressed me by rattling off the names of the novels most of us in this room have read, and when I asked him how he knew all this, he said from the "Authors" card game that we have had lying around the house for years. You may remember the original game—a sort of literary "Go Fish" that had these beautiful, romantic renderings of famous writers like Mark Twain and William Shakespeare, Henry Wadsworth Longfellow, et cetera. Louisa May Alcott, looking kind and homely, was the only woman, and Nathaniel Hawthorne was far and away the most handsome man, with flowing yellow hair and a light blue coat that matched his eyes. Playing with that deck of cards with everybody looking in the pink of health and good fortune was enough to make any child want to be a writer when he or she grew up. That's the deck that I knew.

Joseph Heller speaking at the USC Fitzgerald Centenary Banquet, 24 September (photo by Michael Rogers)

But it was the newer game that my son was referring to—the updated American version—which, as with most sequels, falls far short of the original. It depicts, among others, Ernest Hemingway; Theodore Dreiser (I don't mean falls short in terms of the writers, but in terms of the execution of the cards themselves); Thomas Wolfe; William Faulkner; Willa Cather, the only woman in this batch, looking perplexed; and the handsome F. Scott Fitzgerald. But in this deck the writers look sketchy and cartoonish and gaze out at you in dark-edged, bleak-eyed despair. There's little chance that a young cardplayer would yearn to follow in their footsteps. And if the pictures aren't enough to scare you off, the bio blurbs—not present in the older version—will do the trick. They have these little lines at the bottom of the card. Fitzgerald's concludes by noting that he lived expensively with large debts, drank heavily, was self-destructive, and died in Hollywood in 1940 bankrupt, sick, and alone.

Whether these statements are factually true doesn't much matter, I suppose, to youngsters playing the game. The pictures are what matter; winning the most books matters. We here know that the truth lies somewhere between these two games. And though nobody I know has yet become a writer as a result of playing "Authors," it's possible that once my son encounters those fluid, lyrical lines of Fitzgerald's, once he's exposed to his keen mind and expansive heart, he might, in a weak moment, want nothing more than to experience the magic of stories by taking a stab at writing one himself.

My mother's response to the survey: An avid reader, she was born the same year as Scottie, which would make her seventy-five. She said things like, "He went to Princeton, didn't he? Did he ever graduate? And wasn't he in the army for a while, but that didn't go very well?" By now I'm sensing a tone of mild disapproval. "He ended up in Hollywood, didn't he, but without Zelda." Then she said that the Fitzgeralds always reminded her of her own parents—pranksters, always laughing and drinking and smoking and carrying on, too clever by half. She said she'd read the books but couldn't remember much about them, except that they were sad, and she wasn't particularly partial to sad books.

A high schooler recalled her English teacher, because they were reading *The Great Gatsby,* surprising them all by coming to class dressed as a flapper and proceeding to play a lot of old music and demonstrating the Charleston. This girl loved the book and thought her teacher was a bit odd.

Another friend remembered, as a midshipman at the Naval Academy in the 1960s, that when his class was reading Fitzgerald his English professor herded the tiny literature class over to the library, called Mahan Hall, where a grand piano sat in one of the marble-floored alcoves outside the stacks. The professor sat down and played tunes from the 1920s to the young men who were strewn around on the floor in their uniforms listening, surrounded by flags of captured enemy frigates. The friend said they all thought it was great, a nice change from fluid dynamics. When he finished playing, the professor told them he couldn't look at the moon in the same way anymore, knowing it was littered with space junk.

Other reactions. From friends who drink too much: "Didn't he drink too much?" From poet friends: "Didn't he drink too much?" Actually from all writer friends. And, then with admiration, from a poet friend: "His novels have that elegiac tone to them, melancholic, mourning the loss of youth and romance." From friends suffering from writer's block (whatever that may be): "Don't you think he wasted his talent on those stories? He should have saved himself for the important work." From friends struggling to keep the faith while their best work sits languishing on some editor's desk: "He wrote those great novels, and then he wrote hundreds of commercial stories to support the novel-writing habit. He was a real writer, living and dying by the pen."

And a few final short responses: "Natty dresser" (I don't think that person had read anything), and "Rich people, they're not like us," and "I loved *The Crack-Up*. *The Crack-Up* was me." So, something for everyone, and everyone has something to say. He strikes a variety of chords in a variety of people. He's an inclusive writer. He invites us all in.

Nobody talked much about the books' gorgeous, silky prose, and the way Fitzgerald manages to combine a sharp attention to physical detail with a penetrating probe of the human heart and mind, to produce passages and descriptions and insights as luscious as a basket of peaches, to paraphrase another friend. He has a poet's ear for the rhythm and dance of language, a painter's eye for rich visual detail, and he can move from that dazzling detail to significant introspective musing in a heartbeat, rendering, in a way that seems effortless, some of the most wonderfully exquisite prose around. That's what I remember when I first encountered him in the haze of the turn of the decade from the 1960s to the 1970s—that and, what the people in my survey were also drawn to, the stories themselves, the complications and mysteries enveloping those romantic, racy, wandering, often unhappy people. The jams they got themselves into through willfulness or foolishness or dreams, and whether they'd be able to extricate themselves from them by book's end. Most, of course, at least in the novels, didn't escape unscathed. Some ended up dead, all the more deliciously tragic.

Fitzgerald described in his earlier novels the Jazz Age in a way that made you want be there. As Ken Kesey said, "You're either on the bus or off the bus." And when you read Fitzgerald—at least when you're young, which is probably the best time to read him for the first time anyhow—you're definitely on the bus. He embodied, as the card said, the Jazz Age; he was the Jazz Age. A spokesman for the so-called Lost Generation of the 1920s, whose youth spanned the generation between the two world wars, he wrote about what it was like to be carefree and young and irresponsible and without answers, and later he wrote about what it was to lose all that glit-

Frederick Busch at the 25 September opening of the Fitzgerald Centenary Exhibition, Thomas Cooper Library, USC

tery light and to struggle through the darkness that inevitably, in his case anyhow, follows. He celebrated youth in a way young writers up to then had not.

The young were to be seen and not heard: that was the prevailing notion at the turn of the century. Too much fun was bad for your health. We have an old children's book lying around the house, along with the cards, that was my grandmother's, not the one "too clever by half" (she probably would have burned this book or made paper airplanes of it), but the other one: the one born a year before Scott. The book is called *Slovenly Peter; or, Cheerful Stories and Funny Pictures*. But take my word for it, it's about as cheerful as a guillotine, as it relates in verse (which seems to make it even worse) what happens to children who get their clothes dirty or don't clean their plates at supper or can't stop eating or play hooky from school or have a sweet tooth or suffer from pride or fidget, et cetera, et cetera. Children, something tells me, like Scott and Zelda, children like all children. The stories are accompanied by the most terrifying illustrations that make the spook children in *Village of the Damned* look like Tinkerbell, not a bit cheerful and nowhere near funny.

But if such books (which were common fare in Victorian households) served as moral guides for children as independent as Scott and Zelda were, for example, then I can see how once out of the house, the only way to go would have been up, over, and out. Rebel and revel in it; you had little to lose. And

The Fitzgerald Room at the Thomas Cooper Library (photo by Keith McGraw)

I can imagine the world—golden and gleaming and present as never before at the close of World War I, promising everything and nothing to those eager young men and women.

This tradition of acknowledging and thereby honoring the self-conscious recklessness and glorious foolishness and irresponsibility and vitality of America's youth (upper- and middle-class youth, I should probably say) has been carried on by other writers in the decades since. I remember reading J. D. Salinger's famous novels and stories, as well as Kerouac's, whose restless young hitchhikers fled convention in the East in the 1950s and 1960s and headed for whatever spiritual experiences they might be lucky enough to stumble across. For sheer unbridled exuberance Kerouac was hard to beat. Then, when I focus on my own reading here, it was Ann Beattie's wry, laconic characters who, from the mid 1970s on, seemed to have the edge on the rest of the world. In the late 1980s Jay McInerney and company's literary brat-packers dominated the scene with their hip talk and cocaine habits and lively sex and with upscale jobs so necessary in support of those habits. Though Beattie's cool, precise prose owes more, technically speaking, to the spare lines of Hemingway than to Fitzgerald's rich prose and though, especially in her early books, her characters have more of a tendency to mourn the passage of the carefree 1960s from the comfort of their taupe couches than to actually get up and do something about it, she was as intimately connected to the vague yearnings of America's youth as Fitzgerald was in his time. And McInerney's characters aptly convey their particular brand of self-centeredness, where everybody suffers from ennui or is either too laid-back or too driven or too zonked to be bothered with trying to figure out why they feel so bad.

We can speculate on who will stand the test of time and be read in the year 2050, say. It's been over seventy years since *The Great Gatsby* was published;

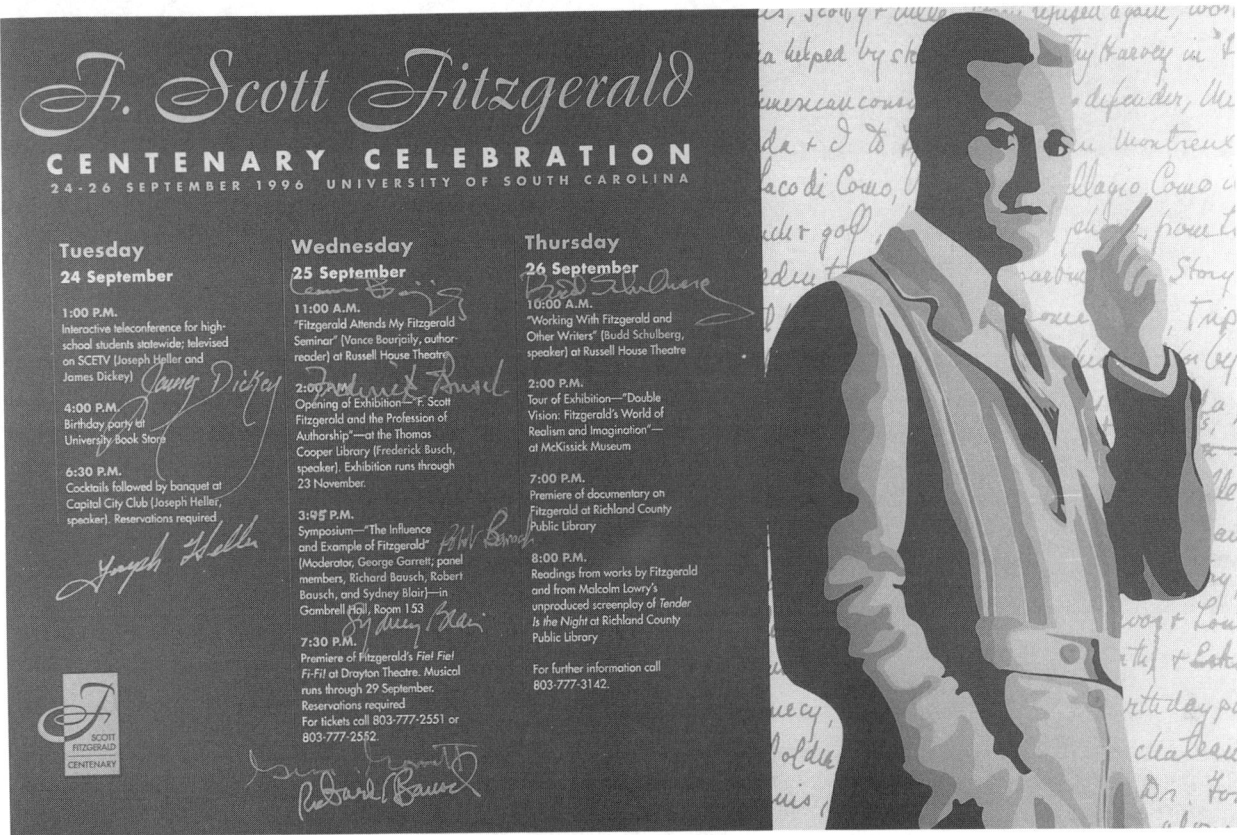

Schedule of events signed by the participants

over sixty since *Tender Is the Night,* if that's any help. So it's a fairly safe bet that Fitzgerald will still be around.

Compared to the later work and to his contemporary Hemingway, who was covering some of the same territory, Fitzgerald's work seems at times tenderly "old-fashioned," in the best sense of the word—not only in its elegant prose but also in the passion and drive of the characters themselves. These people acquire through the pages a certain breathless and giddy weightlessness, but they also carry heavily the strengths of their convictions, never quite escaping the snags and blurred complexities of duty and honor and integrity and love. These characters are not so much resigned to their respective fates as unable to escape them. And most of the time they seem to be trying, however shakily and however myopically, to do the right thing, as did Fitzgerald himself, if his *Notebooks* and essays and letters are any indication.

It is this restless, spirited quality of his characters and the fine writing that brings them to vibrant and complicated life, which enthralled me upon first reading him and which enthralls me still as I continue to read him. He been called a "born writer," the words flowing easily and effortlessly, but we also know he was a tireless re-writer, holding himself in his best work to standards of the highest order when it came to art and craft and truth.

He was also a wickedly funny writer whose wit arises here and there in the talk and deeds of his people. In *Tender Is the Night,* Dick Diver is killing time with the boring Collis Clay, and he's talking about his preference for living in—read, being "entertained by"—France rather than Italy, which is where they are when Collis insists, "I like Rome. Why won't you try the races?" And Dick says, "I don't like the races." "But all the women turn out." "I know I wouldn't like anything here," says Dick. "I like France where everybody thinks he's Napoleon. Down here everybody thinks he's Christ." Almost as irreverent as John Lennon's remark about the popularity of the Beatles, and probably too clever by half. But wonderful, and you wouldn't have missed it for anything.

Richard Bausch: I know you all have this [holds up a copy of *F. Scott Fitzgerald: 24 September 1896-21 December 1940*], but I want to say this aloud. I'm going to read what I have written here and then speak very briefly about my experiences with Fitzgerald.

Title page of the USC Centenary Celebration keepsake

I reread Fitzgerald perhaps more often than any other writer, and one of the saddest things I know of is the letter to Perkins, where he speaks of himself in the past tense: "In my own way I was an original." My God, that always makes me hurt. How much he would have reveled in the history of his work, though, and *Gatsby* stands right up there with the best of everything. This slender volume that manages to be inclusive, and to express the peculiarly cruel effects of our oddly materialistic brand of optimism, better than so many tomes, so many roundhouse attempts to be as big as the country. I hope there is a writers' heaven, and that he's sitting on a bar stool there, toasting, as he would be, everybody who sold him short.

The first time I ever heard Fitzgerald's name I was twenty years old, and there was a song—I was thinking I was going to be a songwriter and learning how to play guitar—and there was a song of Bob Dylan's called "The Ballad of the Thin Man," in which he has this verse: "You've been with the professors, and they've all liked your looks / With great lawyers you have discussed lepers and crooks / You've been through all of F. Scott Fitzgerald's books / You're very well read, it's well known." And for about two years, I resisted reading F. Scott Fitzgerald because there seemed something vaguely damning about having read F. Scott Fitzgerald. But I was reading Hemingway, and I was reading Faulkner, and I finally picked up a copy of "Babylon Revisited." That was the very first Fitzgerald I read. And what I found, of course, was what everyone finds—fresh life, every single time anyone picks it up. And I went from there, of course, to all the others.

My favorite remains *The Great Gatsby*. It's the book I admire most by about anybody who's ever written a book. But I reread both *Gatsby* and *Tender Is the Night,* and I reread the stories. I love teaching the stories. There are some stories that are neglected, you know, that don't get taught very often that are amazing—amazing portraits of being young. And the felt life in them—the sense of being alive in 1925, of being in a street in a city in 1925, the way it sounds and feels—is astonishing, no matter what the story's subject matter.

There's a story called "First Blood," for instance, that I've always loved, where a young woman, spoiled and romantic in the worst sense of the word at her age, convinces herself that she is in love with a man who's engaged or married (it's been about seven or eight years since I've looked at it). What she does is reduce him to the point of confessing his love to her while she's lying ill in a room, and he says, "I'm going to see what I can do. I've fallen in love. I love you." And he leaves the room. And Fitzgerald has her turn slightly and look at the door and say, "Gosh." You have this horrifying revelation of what this poor man is in for. And there are moments like that in all of the stories that are dazzling in their brilliance. So I read him for nourishment.

You read as a writer slightly differently from someone reading just for pleasure. You're always reading for pleasure—it's always pleasure because that's how you got started—but you're also trying to mine it, you're also plundering it, in a way. You read it, you look for ways of incorporating the best parts of it, to somehow infuse your own work with this kind of life. I think it's a very healthy thing for writers to do. But I never read Fitzgerald without learning something new every single time, which to me is the mark of a work that is literature and that will last. I have no worries about whether or not it's going to last, because I see students every day who are half my age and who are just as passionate about it as I was, and am, and who love it as much. It's in their eyes when they talk about it. So that's how I feel about it.

I did have a student come to me—a young woman—who said ——. I'd asked them to write what they read. "What are you reading? so that we'll have some common ground as we share some of the stuff." And this young woman raised her hand and said, "I couldn't begin to list all the books. I'm an avid reader. I mean I just read loads of books." And I said, "Well, just then talk about writers that you're reading." The first sentence of the piece she gave me

The Thomas Cooper Library, USC, Columbia, S.C.

said that her favorite writer was Epscott Fitzgerald—E-p-s-c-o-t-t Fitzgerald. Which was clear evidence that she'd never picked up a book in her life but that she'd heard the name and thought that she heard it one way and, of course, it was not what she should have heard.

There's a story—I think it's apocryphal, but I love to tell it—of Fitzgerald coming back from Paris. There are two stories about Fitzgerald I love to tell. And I use Fitzgerald as an example for writers dealing with the sense of writers celebrating other writers. Fitzgerald gave a story by this unknown neophyte named Hemingway to Morley Callaghan and wanted Callaghan to be as impressed with the story as Fitzgerald was. Here's a writer who is world famous at the time that this happens. This is a writer who everyone knows his name: he has the dream of every young writer, and he's still young, very young. He gets this story that he's very enthusiastic about from another writer, unknown, completely unknown. Morley Callaghan is unimpressed with the story. And Fitzgerald stands on his head in the living room of this place. "Does this impress you, Morley? Does this impress you?" I always have loved that story for the astonishing generosity that's in it—that this writer could be so passionate about the work of a contemporary that no one knew.

I love to tell that story, and I love to tell the story because this is the way we live our lives as writers. And Fitzgerald coming back just after publishing *Tender Is the Night,* and he's in a cab talking to the cabdriver. The reviews had been slow in coming and disappointing in general, because we had entered the 1930s and everybody had to be writing about communism, or else there was no sense in writing anything at all. And here was this book about these people who had a lot of money and were wasting away on the Riviera. Fitzgerald's talking about this to the cabdriver, and finally he looks at him and says, "You've never read anything I've written, have you?" And the cabdriver says, "Not unless you wrote *Black Beauty.*" I say I don't know if this happened; it's probably apocryphal. But I've always loved the story because it has to do with the writer's life in America.

I will end with a story that I know to be true. It happened to a writer named Jon Hassler. He was working in his cabin in Minnesota and developed trouble with the bathroom, the toilet. It stopped up—the septic tank. It really exploded, and there was shit all over the walls. He called the plumber who came to work, and the plumber was up to his ankles in this stuff, washing it off the walls. He looked over his shoulder at Hassler and said, "You're that writer guy from Minneapolis, right?" Hassler said, "Yeah."

And the plumber shook his head with real consternation and said, "I don't know how you can do that kind of work."

Robert Bausch: Whenever I talk about the influence of a writer, especially a writer I admire so much, I always come at it from two points of view. One is, as Dick was saying, as a writer. I'm a writer, and my first impression of Fitzgerald really was from his life, not his work. I had this image in my mind of what I wanted to be as a writer, and the two writers who most impressed me were Sherwood Anderson, for the way he dressed, and Fitzgerald, for the way he lived. Anderson with the flannel shirt: It seemed like he always had the Levis on and the long underwear sticking out under his sleeve. And Fitzgerald: As a young man I sort of incorporated the picture in my mind of the writer as the hard-living, hard-drinking fellow. So I took it first from the image of Fitzgerald, and later from the work.

But I also come at Fitzgerald as a teacher. I rarely teach books I love because I'm afraid my enthusiasm will ruin my skill as a teacher. But I am lately more and more terrified (I think "terrified" might be an accurate word) at what is happening in this culture in terms of the language and the art of writing. I'm afraid that our students in colleges and universities across this country are being lied to every day about what's going to be demanded of them in life. And I don't mean about their work. I mean in life. They are not learning how to be the best they can be when life rots in the beam and puts us in extremities of grief and pain. They don't learn that because they're not reading the right books. They're not being given the right books to read, and when they are given them to read, a lot of them don't do the reading as seriously or as consciously as they should.

I have students whom I admire. I don't like to be one of those teachers who sounds like he has contempt for his students. I don't. In fact, I have affection for them, real affection. And I want them to be exposed to the best works. So I find myself in the last few years teaching *Gatsby* to all my students. My creative-writing students have to read it, and I also want my composition students to read it, and I also have my American-literature students read it. These students come to me when I tell them they're going to read *Gatsby,* and of course the composition students want to know why. And I explain that Fitzgerald's work is so purely unquirky. I once did parodies of the various writers I teach for my students. I wrote parodies of Sherwood Anderson, the easiest thing to do; Hemingway was very easy; Faulkner was fun and easy. And I took months trying to write a parody of Fitzgerald. I couldn't do it. I ended up writing an imitation of some of those sentences in *Gatsby,* because I couldn't write a parody of him.

I realized that the reason I couldn't was because this was the real article. There's nothing quirky about this work. So I tell my students that. If you want the language used to its best uses, then this is what I want you to read. You want to learn to write better? This is what I want you to read. I don't talk about Latinate phrases or the King's English or anything like that; I just say, "If you want to learn how to write, I want you to read this book." And frequently I have students who have read the book in high school—have been asked to read the book in high school—and who then roll their eyes and say, "Ah, I'm not going to read that book again."

As a teacher you're always a researcher; you're always engaging in research. The best teachers are always stirring things up in their students and trying to find out what's in their minds and learning from that. So I'm always asking, "Why do you have that reaction to this book?" And my first impulse is—and this is one that I obviously can't act on—is that I want to slap them with the book. But I also want to know why they're rolling their eyes. And I find out that frequently they roll their eyes because they've been given the litany about this book in high school. People are teaching this book as if it were about the American Dream. It is. But it's not just about the American Dream. That's why it's a great book. The book is about the Human Dream. It's about the desire of all human beings to have in the future what they don't have now and to be somehow an engine in that, to be able to bring that about. And that's why it has survived and, I think, will survive for generations. It's not just about the American Dream; it's not just about the 1920s.

Something that I think Matthew Bruccoli said in one of his essays—I may be quoting somebody else, but I think I'm quoting Matthew—he talked about how the future is always brighter, that the past is used up. It's a golden past that we remember and cherish and the future is bright before us and the present is always —— . I think the phrase he used was "empty milk cartons and fruit rinds" and so on.

This is what we're living in now. If you could look closely at that future, it's really a kind of past that's the future we long for—and I remember how important it was in the 1960s to know this. I knew it from reading *Gatsby,* not from teaching it. In the 1960s when everybody was saying we had to get back to the garden, I was saying, "You ought to read *The Great Gatsby*. It's sort of about that."

But finally I force my students to read the book because I think it's great writing obviously, but I force them to read the book because I want to stir them, I want them to be thinking about literature and what it offers, I want them to back away a little bit from the screen and the keyboard and be again a human being holding a book that has the truth in it. I want to have them discover the truth there, not in a classroom and not with lecturers and not with anything I might say or anyone else might say. And so I find myself for the first time in my life forcing my students to read a book I love. And I'm enjoying it.

George Garrett: Now, since this is a celebration of the works of F. Scott Fitzgerald, we thought we'd read a paragraph or two of some favorite thing. Dick?

Richard Bausch: I guess another thing about Fitzgerald that is important—and I think it's true of every really great writer—is that you read a passage and it changes the way you see something the rest of your life. I never get on a train without thinking about this, and I rode many trains when I was in the Air Force. The way I like to travel is by train. I've traveled in the Midwest by train, and I never get on a train without thinking of this:

> One of my most vivid memories is of coming back west from prep school and later from college at Christmas time. Those who went farther than Chicago would gather in the old dim Union Station at six o'clock of a December evening with a few Chicago friends already caught up into their own holiday gayeties to bid them a hasty goodbye. I remember the fur coats of the girls returning from Miss This or That's and the chatter of frozen breath and the hands waving overhead as we caught sight of old acquaintances and the matchings of invitations: "Are you going to the Ordways'? the Herseys'? the Schultzes'?" and the long green tickets clasped tight in our gloved hands. And last the murky yellow cars of the Chicago, Milwaukee & Saint Paul Railroad looking cheerful as Christmas itself on the tracks beside the gate.
>
> When we pulled out into the winter night and the real snow, our snow, began to stretch out beside us and twinkle against the windows, and the dim lights of small Wisconsin stations moved by, a sharp wild brace came suddenly into the air. We drew in deep breaths of it as we walked back from dinner through the cold vestibules, unutterably aware of our identity with this country for one strange hour before we melted indistinguishably into it again.
>
> That's my middle-west—not the wheat or the prairies or the lost Swede towns but the thrilling, returning trains of my youth and the street lamps and sleigh bells in the frosty dark and the shadows of holly wreaths thrown by lighted windows on the snow. I am part of that, a little solemn with the feel of those long winters, a little complacent from growing up in the Carraway house in a city where dwellings are still called through decades by a family's name. I see now that this has been a story of the West, after all—Tom and Gatsby, Daisy and Jordan and I, were all Westerners, and perhaps we possessed some deficiency in common which made us subtly unadaptable to Eastern life.

Robert Bausch: It occurs to me that in some ways Fitzgerald is like Gatsby and we're all gathered at this huge, incoherent house, and after we're done, we're going to pull away from the driveway. But I love this section of the book. Frequently when I want to sell my students on this book, when I finally give in and tell them, "Goddamn it, it's good!" I read this section:

> Gatsby's house was still empty when I left—the grass on his lawn had grown as long as mine. One of the taxi drivers in the village never took a fare past the entrance gate without stopping for a minute and pointing inside; perhaps it was he who drove Daisy and Gatsby over to East Egg the night of the accident and perhaps he had made a story about it all his own. I didn't want to hear it and I avoided him when I got off the train.
>
> I spent my Saturday nights in New York because those gleaming, dazzling parties of his were with me so vividly that I could still hear the music and the laughter faint and incessant from his garden and the cars going up and down his drive. One night I did hear a material car there and saw its lights stop at his front steps. But I didn't investigate. Probably it was some final guest who had been away at the ends of the earth and didn't know that the party was over.
>
> On the last night, with my trunk packed and my car sold to the grocer, I went over and looked at the huge incoherent failure of a house once more. On the white steps an obscene word, scrawled by some boy with a piece of brick, stood out clearly in the moonlight and I erased it, drawing my shoe raspingly along the stone. Then I wandered down to the beach and sprawled out on the sand.
>
> Most of the big shore places were closed now and there were hardly any lights except the shadowy, moving glow of a ferryboat across the Sound. And as the moon rose higher the inessential houses began to melt away until gradually I became aware of the old island here that flowered once for Dutch sailors' eyes—a fresh, green breast of the new world. Its vanished trees, the trees that had made way for Gatsby's house, had once pandered in whispers to the last and greatest of all human dreams; for a transitory enchanted moment man must have held his breath in the presence of this continent, compelled into an aesthetic contemplation he neither understood nor desired, face to face for the last time in history with something commensurate to his capacity for wonder.

Sydney Blair: I could listen to them read Fitzgerald all day long.

And as I sat there, brooding on the old unknown world, I thought of Gatsby's wonder when he first picked out the green light at the end of Daisy's dock. He had come a long way to this blue lawn and his dream must have seemed so close that he could hardly fail to grasp it. He did not know that it was already behind him, somewhere back in that vast obscurity beyond the city, where the dark fields of the republic rolled on under the night.

Gatsby believed in the green light, the orgastic future that year by year recedes before us. It eluded us then, but that's no matter—tomorrow we will run faster, stretch out our arms farther.... And one fine morning —

So we beat on, boats against the current, borne back ceaselessly into the past.

George Garrett: I'm going to read two paragraphs—the first two paragraphs—of *Tender Is the Night,* which were the first two paragraphs of F. Scott Fitzgerald that I ever read. The purpose there, as he knew very well—and, thank goodness, we got away from that revised Malcolm Cowley edition, which ruined this—the purpose is to hypnotize the reader, and a whole series of things works hypnotically to do that: Some of the same things that you heard in Dick's reading about the trains in the Midwest, where a series of sensory-affective details bring this to life in a certain tone and exactness. There's one great moment in the second paragraph. Many of you have heard readings by Jim Dickey, and some of you remember how when a good line was coming up he'd warn you in advance: "It's comin'! It's comin'! Here it is! Wow, we'll be wreckage forever!" This passage has one, and I will print it out in neon when we get there, because it's just remarkable—it's that moment of genius in the middle of an otherwise spellbinding performance.

On the pleasant shore of the French Riviera, about half way between Marseilles and the Italian border, stands a large, proud, rose-colored hotel. Deferential palms cool its flushed façade, and before it stretches a short dazzling beach. Lately it has become a summer resort of notable and fashionable people; a decade ago it was almost deserted after its English clientele went North in April. Now, many bungalows cluster near it, but when this story begins only the cupolas of a dozen old villas rotted like water lilies among the massed pines between Gausse's Hôtel des Étrangers and Cannes, five miles away.

The hotel and its bright tan prayer rug of a beach were one. In the early morning the distant image of Cannes, the pink and cream of old fortifications, the purple alp that bounded Italy, were cast across the water and lay quavering in the ripples and rings sent up by sea-plants through the clear shallows. [*Here comes the moment* —.] Before eight a man came down to the beach in a blue bathrobe and with much preliminary application to his person of the chilly water, and much grunting and loud breathing, floundered a minute in the sea. When he had gone, beach and bay were quiet for an hour. Merchantmen crawled westward on the horizon; bus boys shouted in the hotel court; the dew dried upon the pines. In another hour the horns of motors began to blow down from the winding road along the low range of the Maures, which separates the littoral from true Provençal France.

I love that man coming out; I don't think anybody else would have thought of it—to suddenly humanize the scene. And then, if you were doing it in shots—which is one way he thought; he really was a very visual writer; no wonder he liked working with film—moving out to a big, panoramic view with boats in the distance. But in order to do that, he set that little man in the blue bathrobe; it's just wonderful.

Richard Bausch: Just the words themselves: "with much preliminary application to his person," "floundered a minute in the sea." That's such an amazing line.

George Garrett: So, now that we've been brilliant for an hour, you all are probably getting restless. Somebody might have some questions to ask one of the writers.

Questioner: Do any of you use Fitzgerald as a direct model for your writing?

Richard Bausch: I imitate his style every time I write. One of the proudest things in my life is that I was once introduced by somebody at Georgetown University at a writers' conference. He said, "How do I introduce Richard Bausch? How do I describe Richard Bausch? Imagine that F. Scott Fitzgerald had been raised by the Snopes clan." I'm very proud of that.

Robert Bausch: As I said, I tried to imitate Fitzgerald for my students, tried to do a parody of his work, and found it impossible. I ended up imitating that passage I read in *The Great Gatsby,* basically. Not very successfully. And I always took Fitzgerald's advice. Fitzgerald wrote about writing, too. I took his advice. He always said that you should never read only one writer while you're writing. You should always read five or six different authors at once. I'm not sure where he wrote that, but I remember ——

Richard Bausch: It's in a letter to Scottie. He said you should absorb six good authors a year. And we just found out, to our horror, while looking at the exhibition, that Scottie never seriously read the letters. She felt that they were a bombardment of parental intrusion on her life, these great letters.

Robert Bausch: Well, it's great advice. Because if you're reading six or seven different authors at once, you really end up sounding like yourself, which is what every writer wants to do. You don't end up sounding like the writer you're reading. I went through phases when I was a kid just starting. I wrote my first novel when I was in the eighth grade, and it sounded a whole lot like Edwin O'Connor because that's who I was reading then. Every writer I was reading I ended up sounding like. Some writers I was reading I didn't want to sound like, enough so that I stopped reading them. Henry James just completely ruined me; I just couldn't read him without sounding prissy when I wrote. But I don't think I could sound like Fitzgerald, because, as I said, I don't think his style is quirky. It's almost as if it were composed by the arbiter of the English language—whoever it was who said that this is how it should be spoken.

Sydney Blair: That's true. I mean, how hard it is to reproduce his style. But I do find that if I read writers that are as wonderful prose stylists as he is—just the writing alone—sometimes in an occasional wonderful moment it will seep through into what I'm doing, which I'm always grateful for.

George Garrett: There's one other aspect that has to do with the different kinds of influences we're talking about, influence or examples. Those of us—and there were several who came to Princeton right after World War II in the postwar years, an overlap of a decade for people who had started Princeton and their lives had been interrupted by World War II—those in the late 1940s. From that group came any number of writers, lots of poets, and some novelists. Probably the most famous one while we were there—who really was extraordinarily famous, perhaps in some way that even Fitzgerald had never been in his lifetime—was Frederick Buechner. We saw him wandering around carrying an attaché case and going places and mysterious ways. He's a novelist that I've come to admire greatly.

But I think that one of the things Fitzgerald did for those of us who were there at that place and shared that experience (with some differences) with him, one of those things was to help us to look to do other things. He did what he did—and I don't think it's applied more generally, but we're looking for an excuse to go in different directions anyway—he did what he did so well that it liberated writers coming along behind him. It was a model of excellence, and yet not a model to be imitated. He opened up the possibility of other doors, of other ways of telling stories. The only one of the writers I can think of was Edmund Keeley, who's better known as a translator of contemporary Greek poetry. But he's a very good novelist and wrote a very Fitzgeraldian novel called (the title comes from the Scott Fitzgerald epigraph) *The Gold-Hatted Lover,* a very good book. He is the only one I know who followed in Fitzgerald's footsteps.

It was Fitzgerald's excellence that helped define our difference. I don't know if that makes any sense to anybody, but it sometimes works that way, too, and, I think, in a different way for southern writers. Faulkner helped to do the same thing—to define what's different. There's no purpose in imitating his work; as in the case of Fitzgerald, it inspires. It's there, as something not to be imitated, but as a superior example for a writer and therefore very encouraging. It is not dismaying, no matter how despairing it might be. It's not dismaying at all; it's encouraging.

The Ira Gershwin Centenary
(6 December 1896 – 17 August 1983)

Philip Furia
University of North Carolina at Wilmington

MUSICAL PRODUCTIONS: *A Dangerous Maid,* Atlantic City, Nixon's Apollo Theatre, 21 March 1921; closed, on the road;

Two Little Girls in Blue, Boston, Colonial Theatre, 11 April 1921; New York, George M. Cohan Theatre, 3 May 1921;

Be Yourself, New York, Sam H. Harris Theatre, 3 September 1924;

Primrose, London, Winter Garden Theatre, 11 September 1924;

Lady, Be Good!, Philadelphia, Forrest Theatre, 17 November 1924; New York, Liberty Theatre, 1 December 1924;

Tell Me More, Atlantic City, Nixon's Apollo Theatre, 6 April 1925; New York, Gaiety Theatre, 13 April 1925;

Tip-Toes, Washington, National Theatre, 24 November 1925; New York, Liberty Theatre, 28 December 1925;

Oh, Kay!, Philadelphia, Shubert Theatre, 18 October 1926; New York, Imperial Theatre, 8 November 1926;

Strike Up the Band, Philadelphia, Shubert Theatre, 5 September 1927; closed there;

Funny Face, Philadelphia, Shubert Theatre, 11 October 1927; New York, Alvin Theatre, 22 November 1927;

Rosalie, Boston, Colonial Theatre, 5 December 1927; New York, New Amsterdam Theatre, 22 November 1927;

That's a Good Girl, London, Hippodrome, 5 June 1928;

Treasure Girl, Philadelphia, Shubert Theatre, 15 October 1928; New York, Alvin Theatre, 8 November 1928;

Show Girl, Boston, Colonial Theatre, 24 June 1929; New York, Ziegfeld Theatre, 2 July 1929;

Strike Up the Band, Boston, Shubert Theatre, 25 December 1929; New York, Times Square Theatre, 14 January 1930;

Girl Crazy, Philadelphia, Shubert Theatre, 29 September 1930; New York, Alvin Theatre, 14 October 1930;

Of Thee I Sing, Boston, Majestic Theatre, 8 December 1931; New York, Music Box Theatre, 26 December 1931;

Pardon My English, Philadelphia, Garrick Theatre, 2 December 1932; New York, Majestic Theatre, 20 January 1933;

Let 'Em Eat Cake, Boston, Shubert Theatre, 2 October 1933; New York, Imperial Theatre, 21 October 1933;

Life Begins at 8:40, Boston, Shubert Theatre, 6 August 1934; New York, Winter Garden, 27 August 1934;

Porgy and Bess, Boston, Colonial Theatre, 30 September 1935; New York, Alvin Theatre, 10 October 1935;

Ziegfeld Follies of 1936, Boston Opera House, 30 December 1935; New York, Winter Garden, 30 January 1936;

Lady in the Dark, Boston, Colonial Theatre, 30 December 1940; New York, Alvin Theatre, 23 January 1941;

The Firebrand of Florence, Boston, Colonial Theatre, 23 February 1945; New York, Alvin Theatre, 22 March 1945;

Park Avenue, Boston, Colonial Theatre, 23 September 1946; New York, Shubert Theatre, 4 November 1946.

MOTION PICTURES: *Delicious,* Fox Film Corporation, 1931;

Shall We Dance?, RKO, 1937;

A Damsel in Distress, RKO, 1937;

The Goldwyn Follies, United Artists, 1938;

North Star, RKO, 1943;

Cover Girl, Columbia, 1944;

Where Do We Go from Here?, 20th Century–Fox, 1945;

The Shocking Miss Pilgrim, 20th Century–Fox, 1947;

Ira Gershwin (Ira and Leonore Gershwin Trusts)

The Barkleys of Broadway, Metro-Goldwyn-Mayer, 1949;
Give a Girl a Break, Metro-Goldwyn-Mayer, 1953;
A Star Is Born, Warner Brothers, 1954;
The Country Girl, Paramount, 1954;
Kiss Me, Stupid, United Artists, 1964.

"Lyric writing," Ira Gershwin reflected at the height of his career in 1931, "has become a profession." With characteristic modesty, he added that it was "a precarious profession, no doubt—one that the east side marriage broker has as yet put no valuation on, one that is looked down on as a racket in some literary fields, but one which nevertheless requires a certain dexterity with words and a feeling for music on the one hand, and, on the other, the infinite patience of a gemsetter." Today, when performers usually write their own words and music, the profession of writing lyrics has all but disappeared, but in Ira Gershwin's day it was a highly specialized and collaborative art. To answer that perennial question, "Which came first?," it was usually the music, or at least the musical germ of a song. "I hit on a new tune," George Gershwin once explained with characteristic ease, "and play it for Ira and he hums it all over the place until he gets an idea for a lyric. Then we work the thing out together."

It was that simple—but, of course, it was also extraordinarily complex. Ira Gershwin compared writing lyrics to creating a mosaic—finding the precise syllables, words, and phrases that fit the notes of a given melody. Sometimes, it took him all night to find a single word. Once, he had to check into a hotel room for three days to find a setting for his brother's tricky string of staccato notes for the melody of "Embraceable You"; what he came up with—"Come to Papa, come to Papa, do"—shows why, among songwriters, Ira Gershwin was known as "The Jeweler." It was always the lyricist who had to adapt his verbal art to the musical idiom of his collaborator, and while we usually associate Ira Gershwin with his brother's music, he also collaborated over the course of his career with many composers, each with a distinctive musical style: Harold Arlen, Jerome Kern, Vincent Youmans, Harry Warren, Burton Lane, Arthur Schwartz, Kurt Weill, even Aaron Copland.

The lyricist's words were the glue that fused music and language into the hybrid art of song. It was the lyricist who made a song "memorable" by stitching together subtle rhymes that weave themselves into the listener's mind:

> *Some* day he'll *come* along,
> The man I love ...
> He'll look at me and *smile—I'll*
> underst*and and*
> in a little *while*
> He'll take my *hand*[.]

Another trick of the lyricist's trade was to make the lyric "singable" with long, open vowels a singer could lean on:

> He'll build a little *home*,
> Just meant for *two*,
> From which I'd never *roam*,
> *Who* would? Would *you*?

Above all, the lyricist had to give the song a distinctive "curve" that would set it apart from thousands of other love songs. In "The Man I Love," that curve came in the skewed verbs of the opening line. A lesser lyricist might have written "Some day *he'll* come along—the man *I'll* love," but Ira Gershwin makes the lover of the future seem already present: "Some day *he'll* come along—the man *I* love."

A lyricist like Ira Gershwin, who wrote for Broadway musicals and Hollywood films rather than the straight popular market, also had to tailor a song to specific characters, dramatic occasions, and performers. Ira came to think of his lyrics as "lodgments" that were designed to fit into a particular moment in a musical production, and that particularity is often what inspired the distinctive curve for a lyric. For Ginger Rogers's Broadway debut as a tough-talking cowgirl in *Girl Crazy* (1930), Ira started a romantic lament with a pugnacious, "Old Man Sunshine, listen you—don't you tell me dreams come true—just try it and I'll start a riot!" For another newcomer in the same show, Ethel Merman, he crafted a colloquial line whose last syllable could be held for a breathtaking eight bars: "Who could ask for anything more?"

When Ira Gershwin traced the genealogy of his profession, he looked first to the Elizabethan songwriters, such as John Dowland and Thomas Campion, but was careful to point out that for them the "words always came first, even though many of these highly talented men were also fine composers." Instead, he turned to the "satirists and parodists" of the period "who, discarding the words of folk song and ballad, penned—quilled, if you like—new lyrics to the traditional tunes." This "practice of putting new words to pre-existent song" culminated in the eighteenth century, and Ira found his true forebear in John Gay, who took "sixty-eight short airs" from "the great store of English, Irish, and Scottish melodies and set new words to them for *The Beggar's Opera*."

It may seem odd that Ira Gershwin and most of the other lyricists who continued this English lyrical tradition were the children of Jewish immigrants who were born within a year of each other in 1895 or 1896: Lorenz Hart, E. Y. "Yip" Harburg, Irving Caesar, and Leo Robin. Ira was the oldest child of Moishe and Rosa Gershovitz, who, after arriving in New York from Russia, changed their names to Morris and Rose Gershvin. Although his birth name was Israel, his parents always called him "Izzy" and by the time he learned what his real name was he had long since gone by "Ira." His brother George (whose real name was Jacob) was born in 1899, and when he quit school at age fourteen to work on Tin Pan Alley (the music-publishing industry then centered on Broadway and Twenty-eighth Street), he changed his name to Gershwin, and the entire family followed suit.

Like most children of immigrants growing up in New York City, Ira and George learned their English on the streets. Little impresses upon a child the power of language more than his ability to speak it better than his parents, particularly in a city where, by 1900, three-fourths of the people were immigrants. Ira, bespectacled and studious, was made the family representative who conferred with teachers and principals about his brother George's problems in school. Ira's facility with language was nurtured at Townsend Harris High, where he developed a love of poetry. Poetry was then a daily exercise in dramatic, oral performance. One by one, students would troop to the front of the room to recite "The Village Blacksmith" or "The Wreck of the Hesperus," replete with dramatic gestures. Not only did students have to memorize, recite, and perform poetry, they learned to write it. "We received a rigorous training," his classmate and fellow songwriter Yip Harburg recalled, "in the classical poetic forms. We were well-versed in the ballad, the triolet, the rondo, the villanelle." How many high-school and college students today—even those who write poetry—would know those poetic forms, much less be able to write them? What made those forms so demanding was that they came from the poetic tradition of a rhyme-rich language. In French, there are fifty-one rhymes for *amour;* in English, *love* rhymes only with *dove, above, glove, shove,* and, as a last resort, *of.* Trying to master the intricate rhyming patterns of the villanelle or rondeau in such a language as English was rigorous training indeed for a future lyricist who would have to say "I love you" in the fifty to seventy words allotted to him in the standard thirty-two bar chorus of a popular song.

In Ira Gershwin and Harburg's youth, poetry was not confined to the classroom. All the big New York papers carried poetry columns, such as F.P.A.'s (Franklin Pierce Adams) *Conning Tower* in the *New York World.* "We were living," Harburg exulted, "in a time of literate revelry in the New York daily press—F.P.A., Russel Crouse, Don Marquis, Alexander Woollcott, Dorothy Parker, Bob Benchley. We wanted to be part of it." Together, Yip and "Gersh" edited a poetry column for the high-school newspaper and then, when they went on to the City College of New York, continued it as "Gargoyle Gargles" in the campus newspaper. One of Ira's light-verse efforts from 1916 foreshadows his later work as a lyricist:

> A desperate deed to do I crave,
>   Beyond all reason or rhyme;
> Someday when I'm feeling especially brave,
>   I'm going to bide my time.

In 1930, for the musical *Girl Crazy,* Ira would recast this somewhat stilted and padded bit of verse into the charmingly elongated syllables of

> I'm bidin' my ti–me,
> 'Cause that's the kinda guy I'–m.

It was their apprenticeship in light verse, with its "metrical discipline and rhyming virtuosity," that later enabled Gershwin, Harburg, and other lyricists to match syllables to intricate musical patterns with ease.

An obvious influence on such nascent lyricists was the team of Sir William Schwenck Gilbert and Sir Arthur Seymour Sullivan, but it is significant that they first knew Gilbert as a poet. One day in high school, Harburg confided to Ira that his favorite book of light verse was Gilbert's *Bab Ballads.* Ira cooly inquired, "You know, of course, those 'poems' are song lyrics?" "There's music to them?" gasped Harburg. "Sure is," Ira told him and then invited his friend home to listen to *H.M.S. Pinafore* on the Gershwin family Victrola. "There were all the lines I knew by heart," Harburg exclaimed, "I was dumbfounded, staggered." That Ira and George Gershwin, Lorenz Hart and Richard Rodgers, Yip Harburg and Harold Arlen, and other American lyricists and composers would go on to write songs that rival those of Gilbert and Sullivan should have been impossible, given their completely different methods of collaboration. When Gilbert and Sullivan wrote songs, the words always came first, with Sullivan creating melodies to fit Gilbert's light verse. When he was asked, "Which came first?" Sullivan invariably snapped, "The words—of course." It seemed the only conceivable way to get quality in lyrics.

In American songwriting, however, the music came first and, as that music became more rhythmically intricate under the influence of jazz, finding words to fit it became more and more difficult. Yet it was those very difficulties that inspired lyricists like Ira Gershwin. When his brother started out with a ten-note phrase, Ira responded with a ten-syllable line:

> They all laughed at Christopher Columbus[;]

Then George's melody shortens to seven syllables, and Ira truncates his lyric accordingly:

> When he said the world was round.

Now George repeats the same initial ten-note phrase, and Ira follows with a parallel ten-syllable lyric:

> They all laughed when Edison recorded[;]

But where another composer might then have given his lyricist another parallel seven-note line, George abruptly stops short on one note—and Ira stops with him:

> sound!

While Ira sometimes complained that his brother's tricky rhythms gave a lyricist little room to "turn around," he could use those tiny confines as inventively as Edison.

Plying his wit within the constraints of musical rhythms, Ira Gershwin (and other lyricists of his generation) created an extraordinary hybrid between light verse and popular song. Just as these young versifiers were making the transformation to songwriting, they were inspired by an extraordinary series of musicals that were produced between 1915 and 1917 at the small Princess Theatre. Unlike bloated European operettas or Florenz Ziegfeld's lavish revues, the Princess shows had scaled-down casts, orchestras, and sets, placing their emphasis instead upon "smart" dialogue and sophisticated songs. Kern wrote the music for the Princess shows and demonstrated to aspiring composers, such as George Gershwin and Richard Rodgers, that the simple formulas of American popular song could be handled with subtlety and invention. Similarly, Kern's lyricist, the English writer P. G. Wodehouse, showed would-be lyricists such as Ira Gershwin that by letting the music come first, they could create artful fits—and sometimes even more artful misfits—that rivaled the best of Gilbert and Sullivan.

Far from restricting him, Wodehouse found that Kern's music enabled him to create effects he could not have achieved in light verse. Citing one instance of how Kern's sequence of "twiddly little notes" in "Till the Clouds Roll By" inspired him to come up with the subtle rhymes of "What bad *luck! It's* coming down in *buckets*," Wodehouse said, "I couldn't have thought of that in a million years." By letting the music take the lead, Wodehouse also found he could make a lyric sound more colloquial—less like poetry and more like conversation. When Kern threw a musical triplet into the melody for "Bill," Wodehouse used it to puncture romantic effusion with a vernacular shrug: "I love him because he's—*I-don't-know*—because he's just my Bill."

No one embraced the Princess shows more eagerly than Ira Gershwin. His diary recorded his enthusiastic responses to performances of *Leave It to Jane* and *Miss 1917,* and he also listened over and over to recordings of the songs of Wodehouse and Kern. It was in these years that his diary also records his first fledgling attempts at writing lyrics, though it took nearly six years of apprenticeship before Ira mastered the craft of wedding words to music. Working mostly with young composers other than his brother (who had already established himself as a major composer with his 1919 hit "Swanee"), Ira chose to write under the pseudonym Arthur Francis (the names of his other brother and his sister), so as not to seem to rely on George's success. By 1924, however, he had emerged as a full-fledged talent in his own right when he collaborated with George on *Lady, Be Good!,* the first of many successful Broadway musicals for the Gershwin brothers. Modeled on the Princess Theatre shows, *Lady, Be Good!* inaugurated a series of Jazz Age musicals that featured dynamic dances, snappy gags, and an overall rhythmic thrust that the Gershwin brothers captured perfectly in songs. When Ira heard his brother play the first number they needed for the show, he was stumped: "George, what kind of a lyric can I write for *that?*" Then, after a pause, he mused, "Still . . . it is a fascinating rhythm." Suddenly, he had his title phrase—always the hardest part of a song to come up with—but it still took days to work out the rest of the lyric. "There was many a hot argument between us," Ira recalled, but when "Fascinating Rhythm" was sung by the stars of the show, Fred and Adele Astaire, it sounded exactly like what Ira said a good lyric should: "rhymed conversation:"

> Fascinating rhythm—you've got me on the go!
> Fascinating rhythm—I'm all a-quiver!
> What a mess you're making!
> The neighbors want to know
> Why I'm always shaking
> Just like a flivver.

For the rest of the decade, in such frothy musicals as *Tip-Toes* (1925), *Oh, Kay!* (1926), and *Funny Face* (1927), Ira Gershwin took the American vernacular and made it sing. Noting the way Americans added suffixes to words, he concocted "Embrace me, my sweet embraceable you." Toying with the way we clip our syllables, he produced " 'S wonderful! 'S marvelous! / That you should care for me. / 'S awful nice! 'S paradise! / 'S what I long to see!" George's musical syncopation afforded him a chance to capture the way Americans collapse two syllables onto one:

> That certain feeling
> [beat] *The first* time I met you!—
> I hit the ceiling!
> [beat] *I could* not forget you!

Ira took his lyrics, he explained, "from thin air, literally and figuratively, by listening to the way Americans spoke to each other—their slang, their clichés, their catchphrases." An expression normally associated with finding one's mate in the arms of another—"How Long Has This Been Going On?"—became, in Ira Gershwin's hands, the euphoric exclamation of two jaded lovers experiencing their first truly passionate kiss. The argot of the 1920s, an era intoxicated with language, turns up in his most romantic lyrics, from "It's all bananas" in "But Not For Me" to "The world will pardon my mush" in "I've Got a Crush On You." Where other lyricists, such as Hart and Cole Por-

Catfish Row crap game in Porgy and Bess *(1935)*

ter, rhymed pyrotechnically, Ira did it quietly, within the most ordinary expressions:

> Although he may not *be*
> the *man some*
> Girls think of as *handsome*
> To my heart he'll carry the *key* . . .[.]

At times, he found he had to abandon rhyme altogether to let one of George's percussive melodies, as he put it, "throw its weight around":

> I got rhythm!
> I got music!
> I got my man—
> Who could ask for anything more?

Such a minimalist lyric took Ira Gershwin weeks to create, but, belted out by Ethel Merman in the 1930 production of *Girl Crazy,* it sounded like the perfectly natural exclamation point to end the Roaring Twenties.

During the Depression, what few Broadway productions there were took on a satiric edge, and here the Gershwins again led the way with *Of Thee I Sing* (1931), which made fun of the American presidency (and even more of the vice presidency). Modeled closely on Gilbert and Sullivan's spoofs of Victorian institutions and manners, this operetta gave Ira the chance to write other kinds of lyrics—recitatives, patter songs, and choral numbers—besides the standard romantic ballads he had come to feel were so confining. Even the few love songs he was required to supply added to the satire: "Love Is Sweeping the Country" captured the absurd hoopla of political campaigns, while "Who Cares?" managed to laugh at the Depression itself:

> Who cares what banks fail in Yonkers—
> Long as you've got a kiss that conquers!

The first Pulitzer Prize awarded for a musical went to Ira Gershwin and playwrights George S. Kaufman and Morrie Ryskind (George Gershwin was deemed ineligible for the purely "literary" award). A sequel to *Of Thee I Sing*, the 1933 *Let 'Em Eat Cake*, was regarded by both critics and the public as too darkly pessimistic, despite such brilliant nihilism as "Down with Everything That's Up":

> Down with music by Stravinsky!
> Down with shows except by Minsky!
> Down with books by Dostoyevsky!
> Down with Boris Tomashevsky!
> Down with Balzac! Down with Zola!
> Down with pianists who play "Nola"!

When George undertook an opera with DuBose Heyward, the South Carolina poet and novelist who had written *Porgy*, Ira turned to satiric revues. With his old friend Harburg and a young composer, Harold Arlen, he wrote songs and comic sketches for *Life Begins at 8:40* (1934), where "Gersh" and Yip indulged their long-standing love of wordplay:

> You're a builder-upper, a breaker-downer,
> A holder-outer, and I'm a giver-inner;
> Sad but true, I love it, I do,
> Being broken by a builder-upper like you.

With Russian composer Vernon Duke (formerly Vladimir Dukelsky) Ira also wrote songs for Fanny Brice and other luminaries for the *Ziegfeld Follies of 1936*, but the best song in the show went to newcomer Bob Hope. A "list" song in the style made popular by Cole Porter, "I Can't Get Started" weaves a lyric out of contemporary allusions, each "topping" the other in cleverness:

> I've flown around the world in a plane,
> I've settled revolutions in Spain . . .
> When J. P. Morgan bows, I just nod,
> *Green Pastures* wanted me to play God . . .[.]

In between these stylish revues, Ira was called in to help with *Porgy and Bess*, at first polishing Heyward's words for songs like "Summertime" so they fit with George Gershwin's music, then collaborating with Heyward on such lyrics as "I Got Plenty o' Nuttin'," as well as writing lyrics for several songs himself. Ira's lyrics for "It Ain't Necessarily So" and "There's a Boat Dat's Leavin' Soon for New York" helped make the ominous yet comical character of Sportin' Life much more central to the opera. His overall contribution was to fuse word and music together into a seamless whole that makes *Porgy and Bess* the most enduring of American operatic works.

Like most songwriters, the Gershwins found relief from the Broadway Depression by heading west to Hollywood. Ever since Al Jolson had sung Irving Berlin's "Blue Skies" in the 1927 "talkie," *The Jazz Singer*, it became vividly apparent that movies could be a new showcase for songs. Close-ups and special lighting made it possible to present a song more intimately than it could ever be performed on stage, and the realistic nature of film placed a premium on informal, colloquial lyrics. That same realism, however, presented studios with a problem: how would audiences accept ordinary characters suddenly bursting into—and out of—song, without even the applause that cushioned the transition from dialogue to song in stage productions? Their solution was to make musicals about singers—Broadway hoofers or neighborhood kids putting on a show—so that actors sang on screen because they were playing the roles of singers.

Gradually, movies experimented with integrating songs into the story of a film so they were presented not as performances but as the lyrical expression of a character's feelings. The fullest integration of lyrics into the dramatic fabric of a film came in the films Fred Astaire began making with Ginger Rogers in 1933. Astaire and Rogers moved as effortlessly from talking to singing as they did from walking to dancing, and lyricists could write for them in a more thoroughly natural style. When lyricists wrote for the stage, they had to supply long, open vowels so singers could sustain and project phrases to the back of the balcony. In Hollywood, however, where singers sang into a microphone (and frequently lip-synched on screen to their own prerecorded songs), lyricists could revel in the short vowels and clipped consonants that are more native to the English language. Writing for Fred Astaire in the 1937 films *Shall We Dance?* and *Damsel in Distress*, Ira Gershwin could use such thorny phrases as "Let's Call the Whole Thing Off," "Stiff Upper Lip," and "Nice Work If You Can Get It."

The Gershwins also used the introductory verse of a song to bridge the gap from spoken dialogue to the singing of the refrain. In the verse for "Things Are Looking Up," Ira had Astaire apologize for the transition from talking to singing:

> If I should suddenly start to sing
> Or stand on my head—or anything,
> Don't think that I've lost my senses;
> It's just that my happiness finally commences . . .[.]

Even in the refrain, Astaire still seems to be chatting rather than singing:

> Things are looking up,
> I've been looking the landscape over,
> And it's covered with four-leaf clover . . .[.]

In such movie lyrics Ira Gershwin captured the urbane nonchalance that was the hallmark of Astaire's style.

Writing together in Hollywood, Ira and George Gershwin's collaboration grew as intertwined as their lives. "Give me an Irish verse," Ira would ask his mercurial brother, then marvel as George instantly played a melody that had exactly "the wistful loneliness I meant." When Ira fitted that melody with the words, "A foggy day in London," George offered to add another note, which Ira set with "Town," giving the lyric the same quaint loveliness as the melody. When George played a four-note phrase for "They Can't Take That Away from Me," Ira asked for two more notes to accommodate the line "The way you wear your hat" and then added other, wryly tender images, such as "The way you hold your knife" and "The way you sing off-key." For "Our Love Is Here to Stay," Ira asked George for only one extra note, deftly placed here and there in the melody, so that he could add the simplest of words:

> The radio *and*
> the telephone *and*
> the movies that we know
> May just be passing fancies *and*
> in time may go . . .[.]

Ironically, this lovely song about loss was their last. Before they could complete it, George Gershwin died suddenly, from a brain tumor, at thirty-eight years of age.

Ira, who had always guided and protected a younger brother he regarded as the true genius in the family, was devastated by the loss. For years, he could not write at all, then gradually friends such as Kern and Arlen eased him out of his slump. Finally, on New Year's Day of 1940, he received a call from playwright Moss Hart, inviting him to collaborate on an experimental "play with music" with composer Kurt Weill. *Lady in the Dark* (1940) concerned a brilliant and successful businesswoman, played by Gertrude Lawrence, who finds herself coming apart emotionally and seeks out psychiatric help. As she lay on the analyst's couch, the lights would dim, the stage would revolve, and realism would give way to musical fantasy. For the surreal scenes set in her subconscious, Ira wrote such bizarrely brilliant lyrics as "Tchaikowsky," where newcomer Danny Kaye rattled off the names of forty-nine Russian composers. Throughout her therapy sessions, Lawrence struggles to recall the words to a melody remembered from childhood, and Ira Gershwin saw in her effort to bring the deepest layers of the unconscious to consciousness a parallel to the art of the lyricist, who similarly struggles to find words that will articulate the emotional significance buried in a musical pattern. When Lawrence finally recalls and sings the simple but haunting song, "My Ship," it concludes the most fully integrated score Ira Gershwin had ever written for a Broadway musical. Such integration anticipated Richard Rodgers and Oscar Hammerstein's 1943 production of *Oklahoma!,* which made "integration" between song and story the watchword on Broadway from then on. Ira Gershwin, however, played no further role in the development of the American musical. His only two other shows, *The Firebrand of Florence* (1945) and *Park Avenue* (1946), failed primarily because of weak books. Unlike Hammerstein and such younger lyricists as Frank Loesser and Alan Jay Lerner, Ira Gershwin never wrote both book and lyrics for a musical. The consummate specialist and collaborator, he said he "never tried his hand at a script" because he was "no expert at the other man's game."

Instead, he remained in California, writing songs for Hollywood movies. Some of these songs, such as "Long Ago and Far Away," which he wrote with Kern for Gene Kelly's 1944 film *Cover Girl,* were successful, and others marked genuine advances in the Hollywood musical. For the 1945 movie *Where Do We Go from Here?;* he and Weill wrote a miniature operetta that included such witty patter numbers as "Song of the Rhineland":

> Where the wine is winier
> And the Rhine is Rhinier
> And the Heine's Heinier
> And what's yours is minier.

While he worked with many great composers in Hollywood, however, Ira never again established the long-term collaborative relationship he had had with his brother. For some movies, such as *The Shocking Miss Pilgrim* (1947), he even tried setting lyrics to previously unused or fragmentary melodies George had left behind in manuscripts and notebooks. Such a ghostly collaboration produced some tenderly romantic ballads, such as "For You, For Me, For Evermore," as well as a witty satire on Boston manners in "The Back Bay Polka":

> No song except a hymn—
> And keep your language prim:
> You call a leg a "limb"
> Or they boot you out of Boston!

Ira Gershwin's great work in movies culminated with *A Star Is Born* in 1954, for which he worked with composer Harold Arlen and playwright Moss Hart to integrate songs fully into the story. Even though Judy Garland plays a singer and presents each song as a performance, lyrics such as "The Man That Got Away" are closely tied to her character and the dramatic context:

> The night is bitter,
> The stars have lost their glitter,
> The winds grow colder
> And suddenly you're older[.]

"The Man That Got Away," one of Ira Gershwin's greatest lyrics, was also his swan song. By the mid 1950s, a new musical style was taking over popular music, and the Hollywood studios dismantled their musical production units in the face of competition from the new medium of television. Until his death in 1983, Ira Gershwin spent his time being what his brother George had always called him—"Ira, the scholar." He edited manuscripts and other Gershwin papers, compiled a collection of his own lyrics, and helped establish archives at the Library of Congress and the Museum of the City of New York.

Like his fellow lyricists, Ira Gershwin started out as a poet, and now, from a modern perspective a century later, that is what he has become. When many people, at some emotional moment in their lives, seek for words, they are likely to echo song lyrics. Finding ourselves hopelessly at odds with another, we shrug, "You say *potato* and I say *potahto;* you say *tomato,* and I say *tomahto.*" Struck by another's success, we express our admiration (as well as our American belief that it can be emulated) with "Nice work if you can get it—and you can get it if you try." Our most familiar equivalent of Robert Herrick's and Andrew Marvell's formulations of mutability, finally, may well be Ira Gershwin's:

> In time, the Rockies may crumble,
> Gibraltar may tumble—
> they're only made of clay.

Such songs, like those of Cole Porter, Rodgers and Hart, and the other songwriters of their generation, are, like the most enduring poetry, "here to stay."

References:

Philip Furia, *Ira Gershwin: The Art of the Lyricist* (New York: Oxford University Press, 1996);

Ira Gershwin, *Lyrics on Several Occasions* (New York: Knopf, 1959);

Edward Jablonski, *Gershwin* (Garden City, N.Y.: Doubleday, 1987);

Jablonski and Lawrence Stewart, *The Gershwin Years* (Garden City, N.Y.: Doubleday, 1958; revised, 1973);

Robert Kimball, *The Complete Lyrics of Ira Gershwin* (New York: Knopf, 1993);

Kimball and Alfred Simon, *The Gershwins* (New York: Atheneum, 1973);

Deena Rosenberg, *Fascinating Rhythm: The Collaboration of George and Ira Gershwin* (New York: Dutton, 1991);

P. G. Wodehouse, *Author! Author!* (New York: Simon & Schuster, 1962).

A Tribute

from Sheldon Harnick

As a theater lyricist working in the waning years of the twentieth century, I can only look back with envy, incredulity, and melancholy at the flourishing state of the Broadway theater in the 1920s. In the "record-breaking Christmas week" of 1927, according to theater historian Robert Kimball, "20 offerings premiered during a seven-night period, 11 on the evening of December 26."

Ira Gershwin, only twenty-four years old at the beginning of the 1920s, was in the right place at the right time: lyricists who could write effectively for the musical theater were in demand. In the years 1920–1929 he had the opportunity, astounding by today's standards, to contribute more than 275 lyrics to some thirty musicals, with music by a dozen composers, including his younger brother George. It was an extraordinary opportunity for a gifted versifier to master his craft.

The first song that Ira Gershwin placed in a musical was "The Real American Folk Song (Is a Rag)," performed by Nora Bayes in *Ladies First* (1918). Written with George when Ira was twenty-two, this lyric already displayed many of the qualities that characterized his work throughout his creative life: fresh and inventive rhymes; supple phrasing; intelligence and wit in both the choice of theme and its subsequent development; and, finally, his transparent love for words, his contagious joy in rhyming, and his obvious pleasure in the manipulation of language. Yet to come were the sweetness and wry simplicity of his later love lyrics, of both the requited and unrequited varieties. Since I cannot wait to start quoting, here is a superb example from a verse of "But Not for Me" (1930): "Old Man Sunshine–listen, you! / Never tell me Dreams Come True! / Just try it – / And I'll start a riot . . . / I never want to hear / From any cheer – / Ful Pollyannas, / who tell you Fate Supplies a Mate – / It's all bananas!"

From the outset Ira was a master of unforced rhyme. (It is not that difficult to put together a string of rhyming words; the difficulty is to find rhymes that are intelligible and seem natural even when used in surprising combinations.) Here is Ira, age twenty-five, in a passage from "Dancing Shoes" (1921), tossing off a complicated rhyming sequence while preserving the airy, playful, jazzy quality of the lyric. Referring to those shoes, he writes: "They will make you frolic, / lose the diabolic, / weary melancholic blues."

Here he is again at his concise, inventive best in the song "Weaken a Bit" (1925): "Happy-faced, I'd feel much worse if I'd / Not embraced the life diversified." Or take these lines from the hilarious "Song of the Rhineland" (1945): "Where the wine is winier / And the Rhine is Rhinier / And the Heine's Heinier / And what's yours is minier!"

Ira's adventurous mind (and impeccable ear) constantly sought out new sources for fresh rhymes and wordplay. His lyrics abound in deft devices to strike the ear and divert the imagination. A few examples may help to indicate the range of his inventiveness. One device was the use of foreign words, sounds, and accents. Here is my favorite couplet from the pseudo-Italian "Bambino" (1920): I'll climb to your window like in da romance. / I no give a rap if I rip-a-da pants." He switched to pseudo-Russian for this droll device from "Katinkitschka" (1931). In refrain one: "Popitschka, Momitschka, / Will not sleep a winkitschka, / Thinking of Katink, Katink, Katink, Katink, Katinkitschka!" In refrain 2 Ira supplies a happy ending: "Popitschka, Momitschka, / Now can laugh and singitschka, / Since Katink, Katink, Katinka has a wedding ringitschka!"

Yet another source of wordplay, arising out of the slang of the period, is the truncating of words, as in "Sunny Disposish" (1926): "It's absolutely most ridic', / Positively sil'. The rain may pitter-patter – / It really doesn't matter – / For life can be delish / With a sunny disposish."

One of Ira's most extraordinary gifts was his ability to create words and syllables that, out of context, would be meaningless but that, in context, make perfect, amusing sense. Take this quatrain from "Sentimental Weather" (1936): "Blizzards may be blizzing, / Blustering winds may bluss – / Long as we're together, / It's sentimental weather for us."

Sometimes Ira's delight in the discovery of a source hitherto unexplored is almost palpable. In "Second Fiddle to a Harp," Ira is like a child with a new toy (a better analogy might be a painter with an entirely new set of colors) as he plays with terms from the world of music. "I'm in love fortissimo, / But she takes me pianissimo / though I'm all amoroso, / She gives me, 'No, no, no!' So / That's why I'm, head to toe so / Furioso!" Similarly, in "You're a Builder-Upper" (1934), there is an explosion of dazzling verbal dexterity as Ira happily constructs a string of hyphenated neologisms. "You're a builder-upper, / A breaker-downer, / A holder-outer, / And I'm a giver-inner; / Sad but true, / I'm a saperoo, too, / Taking it from a taker-over like you."

Ira Gershwin's professional life was an unremitting battle against clichés. In several lyrics he actually managed to make clichés work for him. In the remarkable, daring, and wholly original lyric "Blah, Blah, Blah" (1931), written as a "mild spoof" (Ira's words) of Hollywood love songs, he takes the clichés of all too many film ballads to an absurd extreme: "Blah, blah, blah, blah, moon, / Blah, blah, blah, above, / Blah, blah, blah, blah, croon, / Blah, blah, blah, blah, love," et cetera. In a song numbering only ninety syllables, actual words count for only twenty-nine of them; the rest are nonsense. What makes the song work, in addition to the breathtaking impudence of the concept, is that Ira was shrewd enough to utilize a graceful, appealing melody that George had written several years earlier as a romantic ballad.

Lyrics, as a rule, are written to be heard, not read. No one was more aware of this than Ira. Some of his love lyrics, which tend to look flat and unemotional on the page, assume an entirely new dimension when sung. However, a good many of Ira's lyrics read like light verse. Try "I Can't Get Started" (1936) with stanza after stanza of delightful verse, for example, "Good grief! I'm not exactly a clod! / *Green Pastures* wanted me to play God! / The Siamese twins I've parted – / But I can't get started with you."What becomes clear as one reads and listens to Ira Gershwin's lyrics is his growth, both intellectual and emotional, and his endless search for unusual rhymes and fresh ways to express familiar thoughts and feelings.

The True Collaborator
A Tribute from Michael Lasser

There is no such thing as a Gershwin song, only songs by the Gershwins. We first made the mistake in the 1920s and 1930s when George and Ira were writing together, and we continued to make it as late as the 1980s, when *Crazy for You* opened successfully on Broadway, subtitling itself *The New Gershwin Musical,* as if Ira were merely along for the ride.

I suppose George *is* more important than Ira. Certainly he was more ambitious, quicker, deeper, more sophisticated, more precocious—an extraordinarily talented dazzler and a serious composer of established reputation. Ira served George's genius while George lived, and continued to do so after he died. He saw himself as the less gifted of the two. Whether or not he was, we have joined him in enshrining George. One ironic result has been to reduce the stature of one of our greatest lyricists.

Much of the writing about the brothers' achievement describes George's contribution to popular music, and then mentions the fine supporting lyrics Ira wrote. There is more to it than that. It is important to ask if George would have made those same contributions without Ira—if the songs would have been as brilliant, as wonderfully realized, as well sustained throughout the fourteen years they had together. It even seems possible to ask if they were, in fact, George's contributions. Maybe they were the Gershwins' rather than George's. It seems to me they were.

Before they began to collaborate almost exclusively with one another in the mid 1920s, George worked with some very good lyricists and wrote some winningly youthful songs: "Swanee" and the regrettably forgotten "Nashville Nightingale" with Irving Caesar, as well as the deliciously naughty "Do It Again" with Buddy DeSylva. After George's death, Ira worked with some great composers. He wrote "My Ship" and "The Saga of Jenny" with Kurt Weill (and recorded them as well), the exquisite "Long Ago and Far Away" with Jerome Kern, and "The Man That Got Away" with Harold Arlen. But it was never again the way it had been—that joining in the blood that seemed so effortless even when it was hard work, that apparently perfect marriage of sensibilities.

Marriage is not too strong a word to use. It was a marriage of sorts, the joining of certain kinds of opposites, rooted in love, to create something permanent and intimate. The compatibility grew somehow from differences in temperament, character, and talent that were essential to the collaboration. You even sense it in their photographs. At first, you think they do not look anything alike. George is tall and intensely self-aware, with a long face and a large nose; Ira is stocky with a broad face and jowls, and he usually wears a benign expression. George often has on a coat and tie; there is something dashing about him. Ira, regardless of what he wears, is slightly unkempt. George always looked a little feral; Ira always looked as if he was about to smile. But then you notice the similar curl of the mouths, the heavy chins, and the steady looks; these are brothers, after all.

Once Ira established himself independent of George (by writing under the pseudonym of Arthur Francis), the brothers began their collaboration in earnest. They had written "The Real American Folk Song" together in 1918 and had a hit with "I'll Build a Stairway to Paradise" written with DeSylva four years later, but their first major work was the score for *Lady, Be Good!* in 1924. Songs like "Fascinating Rhythm," "Little Jazz Bird," "The Half of It Dearie Blues," "Hang on to Me," and the title number are not the sounds of George Gershwin alone. It should be clear from nothing more than the titles that these are the sounds of the Gershwins together: the jazzy, staccato rhythms of the music matched in kind by the slangy pizzazz of real talk; the mellow extended line of the ballads completed by romantic yet suggestively vulnerable lyrics. Ira was masterful at writing lyrics that sounded new yet somehow familiar, always hitting the ear as true:

> I've got the You-Don't-Know-the-Half-of-It-Dearie Blues.
> Will I walk up the aisle or only watch from the pews?
> With your permission
> My one ambition
> Is to go through life
> Saying, "Meet the wife."
> I've got the You-Don't-Know-the-Half-of-It-Dearie Blues.
> ("The Half of It Dearie Blues")

Sometimes, when they worked, Ira would give George a title or a line for a lyric. Sometimes, George would give Ira a bit of melody. But more often than not, they would sit together at the piano, George's untamed musical imagination turning out scraps of songs, one giddier than the next, from some bottomless fount, while Ira listened. Eventually, he would react to one. It sounded like something he could fit a lyric to. George would work on it. Ira would make suggestions. He would begin to fit a few words to a musical phrase. George would make suggestions back about a word here, a phrase there.

Ira wrote in *Lyrics on Several Occasions,* "One night I was in the living room, reading. About 1 a.m. George returned from a party ... took off his din-

Words and music: Ira and George at work (Ira and Leonore Gershwin Trusts)

ner jacket, sat down at the piano. . . . 'How about some work? Got any ideas?' 'Well, there's one spot we might do something about a fog . . . how about *a foggy day in London* or maybe *foggy day in London Town?*' 'Sounds good. . . . I like it better with *town*' and he was off immediately on the melody. We finished the refrain, words and music, in less than an hour."

So George would write the music and Ira the words, but the resulting song was often a collaboration in the truest sense. The division of responsibilities was clear, but the contributions were blurred.

Ira was less assertive than George about nearly everything. He was quite content to take the reflected glory; he did not have George's restless need for the spotlight. He was the loving older brother, the helpmate, who, like George and the rest of us, may have occasionally confused precocity with genius. George's genius was real enough; it did not need the distortions that came from youthful fame or Ira's devoted service.

As much as anything, George and Ira's songs caught the spirit of New York in the 1920s, when New York became the city America fell in love with. What they wrote, whether for Broadway or Hollywood, embodied a peculiarly New York sensibility that from the 1920s into the 1950s became an American sensibility. In a strange way, New York was America, because Tin Pan Alley, Broadway musicals, and even Hollywood musicals made it so.

From the time popular music became a major cultural industry early in the twentieth century, our songwriters looked out on America from the east side of the Hudson. They imagined themselves in Indiana and Alabama. They even imagined themselves returning to these far-off places they had never been. The spreading fields and small towns of the Midwest and the South were an essential part of American music, but they were almost always the places we had left behind, remembered, and longed to return to. The vantage point was New York. As early as 1914, Irving Berlin had a husband crow, "My wife's gone to the country! Hurrah! Hurrah!" while he remained behind in town. A decade later, Berlin spoke for the millions of New Yorkers who headed out to the country every summer seeking nothing but the chance to be lazy, carrying "a great big valise full / Of books to read where it's peaceful."

George and Ira were not the only ones to express this New York point of view, nor were they the first. Perhaps George M. Cohan began it in earnest. He was the slangy spokesman for a new America of burgeoning cities, extraordinary technology, expansive patriotism, and the churn of the melting pot. This cock-of-the-walk New Yorker strutted down Broadway, singing in a democratic language that sounded like real people talking. He once told the youthful Berlin, "The words must jingle, Irving. The words must jingle." When he was starting out, George Gershwin's hero was Kern; in some ways, it should have been Cohan.

Berlin learned the lesson Cohan taught. So did Ira. Though he probably learned wordplay in large part from W. S. Gilbert (whom Johnny Mercer calls "the father of us all"), Ira lifted his own lyrics to a level of playful wit that gave them their own uniquely jazzy jingle. It made them both fresh and familiar, conversational yet clever at the same time. His language was immediately accessible, yet satisfyingly witty:

> Don't mind telling you,
> In my humble fash,
> That you fill me through
> With a tender pash.
> When you said you care,
> 'Magine my emosh;
> I swore, then and there,
> Permanent devosh.
> You made all other boys seem blah;
> Just you alone filled me with AAH!
>
> (" 'S Wonderful")

Ira Gershwin almost never wrote about New York directly. He rarely wrote any of the formulaic songs about the buildings or the streets or even the sidewalks. He never wrote about Dixie or the banks of the Wabash. His songs were rarely nostalgic. Instead, he was writing about individuals and how they looked at the world. They were, like jazz, like anything improvisational, rooted in the present. The audiences he knew did not sit in front of radios in Dubuque dreaming about perfect love or even radios in Manhattan pining for the old homestead. The characters who sang his songs were the up-to-date young men and women, the innocent sophisticates, who populated musical comedy in the 1920s. New York is a matter of attitude, possibility, and optimism—the romantic readiness that F. Scott Fitzgerald embodied tragically in Jay Gatsby and that Fred Astaire portrayed so happily—and effortlessly—on stage and screen for half a century.

To get the point about the Gershwins and New York, you need to imagine the Manhattan skyline as the clarinet wails the opening notes to *Rhapsody in Blue,* perhaps in that classic movie shot across the Hudson from Jersey, with the skyscrapers seeming to jut up out of the river as ocean liners and tugboats make their way past, especially at dusk as night rises and light re-creates the city as a dream—a strange cacophonous dream never before imagined in all of human history, as noisy as it is romantic. A place where people came to make their fortune and find life's full zest. They found heartbreak as often as happiness, both essential to the spinning of legends. So they came to a city that chose to believe everything was possible, and then acted as if it believed the dream. New Yorkers were cocky and confident and a little smug, and New York was a place to fall in love with if you had the guts:

> Grab a cab and go down
> To where the band is playing;
> Where milk and honey flow down;
> Where ev'ry one is saying
>   "Blow that Sweet and Low-Down!"
>
> ("Sweet and Low-Down")

George's music caught the sound and the feel of New York in the 1920s, and Ira's lyrics were their perfect verbal counterpart. Together, they wrote the songs that became the world they lived in. Their songs were quirky and fragmented in rhythm, assertive and urbane in attitude, youthful and optimistic in sentiment. This was New York and anything was possible, and George and Ira Gershwin captured the point of view in popular songs. Their first show together, *Lady, Be Good!,* began with a brother and sister thrown out of their apartment for not paying the rent. So they set up housekeeping on the sidewalk and tell each other everything will be just fine. They are not crazy and they are not fools, but they possess the resilience New Yorkers still like to think is unique to them as well as the can-do optimism the city used to give off like an irresistible scent:

> If you'll hang on to me
> While I hang on to you
> We'll dance into the sunshine
> Out of the rain—
>   Forever and a day.
>
> ("Hang on to Me")

In the 1920s George and Ira Gershwin gave musical comedy a new American sound. Ann Douglas has written brilliantly in *Terrible Honesty: Mongrel Manhattan in the 1920s* about how white America absorbed

Brother geniuses (Ira and Leonore Gershwin Trusts)

black culture. George and Ira became part of the transformation when George introduced the rhythms of jazz to Broadway. Performers could no longer stand center stage, rolling their *r*s and biting off every letter in every word. You needed to ride the song, sliding along with its rhythms, finding your own buoyancy within the song–the way Fred Astaire did. But even Astaire could succeed only if the lyrics enabled him to. With George, the music caught up to the lyrics. With Ira, we completed the process begun with Cohan of learning to sing talk.

In the 1920s in New York, Ira Gershwin was, in effect, writing jazz lyrics. He had learned how to improvise from a riff, much as his brother seemed to when he wrote a melody. Ira had the discipline and the freedom to take the lyric on its own ride within the form dictated by what George wrote. The logic of the lyric was more complex and more purposeful than the random quality of most conversations, but it still retained the vitality, surprise, and breathless danger of spontaneity. Ira broke sentences for the sake of a telling rhyme. He used nonsense syllables. He repeated words and sometimes abbreviated them in the name of comic exuberance and to underscore his point:

> When a bounder starts to hiss,
>   You must give him blow for blow.
> Make the blighter cry, "What's this?
>   'Ullo, 'ullo, 'ullo, 'ullo, 'ullo!"
>   ("Stiff Upper Lip")

He repeated expressions and paused where you least expected it, always to match George's snappy rhythms. He managed to tell his own story and cast his own emotional spell to the music he was handed. Like few other lyricists, he learned to be true to the music and yet follow his own path in language. Only if you know George's melody and rhythm of "Beginner's Luck" can you hear how brilliantly Ira repeats words, pauses unpredictably between words and even pauses within words to suggest the delight the singer feels and the spontaneity of his expression:

> This thing that we've begun
> Is much more than a pastime,
> For this time is the one
> Where the first time is the last time!
> I've got beginner's luck,
> Lucky through and through,
> 'Cause the first time that I'm in love,
> I'm in love with you.
>
> ("Beginner's Luck")

The fragmented sound was characteristic of George and Ira's songs and was a perfect musical response to the frenetic quality of New York life, where so many things were always possible. The rhythms of New York were unique to themselves, compulsive and compelling. You hear them in the music and you hear them again in the words. Ira's lyrics were the sound of Harlem moving downtown, changing into something else. Suitably, it found its first early expression in a song about itself. There's not a single reference to love; this is a song about rhythm:

> Got a little rhythm, a rhythm, a rhythm
> That pit-a-pats through my brain;
> So darn persistent,
> The day isn't distant
> When it'll drive me insane.
> Comes in the morning
> Without any warning,
> And hangs around me all day.
> I'll have to sneak up to it
> Someday, and speak up to it.
> I hope it listens when I say: . . . [.]
>
> ("Fascinating Rhythm")

These irresistible, driven rhythms also appear more subtly in more-conventional songs where they lend an edge to the tune, just as Ira's clever lyrics, rooted in the rhythms of everyday speech, match George's music tangy bit for tangy bit. The jazzy rhythms match the ironic bite in the expression of sentiment in a song like "Nice Work If You Can Get It":

> Holding hands at midnight
> 'Neath a starry sky . . .
> Nice work if you can get it,
> And you can get it—if you try.

The infectious optimism of the Gershwins' songs was inseparable from the life of the city they knew so well. Taxis careened while showgirls paraded. Jimmy Walker strolled along Fifth Avenue and people danced the night away from Harlem to Greenwich Village. Everything was new and everything was possible. Yet even though Ira's words are optimistic and often tender, their sensibility is almost always adult rather than adolescent. They discern a kind of pleasure in love that does not depend only on hormones. There is a crowing confidence in some of the songs because the character goes far beyond reasonable expectations without ever losing believability, perhaps because their knowing way suggests that these are ultimately love songs for grown-ups. The romance is ardent but the lover also knows how fragile it is. He is aware of the world and of how it can intrude. This is no adolescent dream of idyllic perfection, but rather an assertion of hope. The fact that love may fail is no reason not to give yourself over to it, as long as your eyes are open. This is the last song George and Ira wrote together; in fact, Vernon Duke helped Ira finish the verse after George died. But in its affirmation of permanence expressed through the wittiest kind of hyperbole, this is a love song for adults:

> The more I read the papers,
> The less I comprehend
> The world and all its capers
> And how it all will end.
> Nothing seems to be lasting,
> But that isn't our affair;
> We've got something permanent—
> I mean, in the way we care. . . .
>
> In time the Rockies may crumble,
> Gibraltar may tumble
> (They're only made of clay)
> But—our love is here to stay.
>
> ("Love Is Here To Stay")

When the Gershwins wrote about New York, they aimed at the feel and pulse of the city. Consciously or not, they expressed in tight, pulsating little songs what it was like to be alive in a specific place at a specific time. We realize now how far George and Ira moved past the formulas of popular songwriting to create a collaborative voice uniquely their own, a voice that would speak with undimmed eloquence seventy-five years later. That it was also the voice of a jubilant New York City coming into the fullness of its power and magnetism only underscores the value and appeal of what they had written. It did not matter what the songs were about or where they were set, because in the inventive rhythms of their melodies and their words—in the guise of love songs—they were New York:

> We'll be like These Charming People
> Putting on the ritz,
> Acting as befits,
> These charming people:
> Very debonair,
> Full of savoir-faire.
> I hear that Mrs. Whoozis
> Created quite a stir:
> She built a little love-nest
> For her and her chauffeur.
> If these people can be charming,
> Then we can be charming, too!
> ("These Charming People")

"Ira the Words, George the Music"
A Tribute from Deena Rosenberg

> Their object all sublime
> They shall achieve in time
> To let the melody fit the rhyme
> The melody fit the rhyme

—Ira Gershwin, after Gilbert and Sullivan, 1941

"It was an age of miracles, it was an age of art, it was an age of excess, it was an age of satire," wrote F. Scott Fitzgerald of the 1920s. One highlight of this exuberant decade was an unprecedented number of musicals. There were more than four hundred Broadway shows between 1920 and 1930, not including revivals—a number never approached before or since. Perhaps the most important single event in American musical theater history occurred in 1924, when George and Ira Gershwin collaborated on their first Broadway show, *Lady, Be Good!*

By that time both brothers had written dozens of theater songs—but not with each other. George's reputation as a composer was growing, especially through the Al Jolson hit "Swanee" (1919) (lyrics by Irving Caesar). Ira, the lyricist, who initially went by the pen name Arthur Francis (the names of his younger brother and sister), was starting to be known for his work with Vincent Youmans. But as soon as George and Ira began their almost exclusive collaboration, something extraordinary clicked. From 1924 on they wrote classic after classic. It is safe to assume that they would not have written with the same range, versatility, and overall brilliance with other collaborators. As their composer colleague Arthur Schwartz wrote, "There was a chemistry between the two that touched off sparks, flames, explosions. Together they created musical [theater] history."

The brothers' personalities differed markedly. George was gregarious, a live wire who loved parties and traveling and wrote melodies rapidly, while Ira was reticent and easygoing, a homebody who slaved for days over a single lyric. Yet when they wedded words to music, they became like two halves of the same person. George heard music in Ira's lyrics; Ira heard lyrics in George's music; and their songs sound as if music and lyrics emerged from a single source. Their partnership produced over five hundred memorable theater and film songs, exquisite dramatic vignettes. They set high standards, moved their art form toward a new musical/lyrical vernacular, and paved the way for an American musical theater that insisted on closer integration of songs, plot, characters, and theme.

Undoubtedly the Gershwin brothers' collaborative genius owed much to what each brother brought to it. Words fascinated Ira. As a youngster, he read extensively and especially liked William Shakespeare, Samuel Johnson, James Thurber, Bernard Shaw, Theodore Dreiser, and light verse of every age—French rondeaus and triolets, W. S. Gilbert, P. G. Wodehouse, and Dorothy Parker. In high school and college Ira wrote and edited a light-verse and humor column with Yip Harburg, who also later became a celebrated lyricist.

Ira observed daily life closely. Diaries from his late teenage years have frequent notations like these:

> Sniffed in a day: Onions, whiskey, garbage, fur and camphor balls, fountain pen ink, fresh newspapers.

> Heard in a day: An elevator's purr, telephone's ring . . . a baby's moans, a shout of delight, a screech from a "flat wheel" . . . a hoarse voice . . . a match scratch on sandpaper, a deep resounding boom of dynamiting in the impending subway. . . (2 September 1916)

and

> The movies and their audience are a good means of studying. Yes. Psychology, ethics, fashions, manner. Manners. Would-be's. Have beens. Never weres. Can't be's. Impossibles. And here and there an occasional Is and Are" (15 May 1917).

Ira also wrote down the countless conversations he overheard that encapsulated how people commonly spoke.

While Ira read, wrote, and watched, George left the High School of Commerce to play the piano and listen. At the age of fifteen George became Tin

Pan Alley's youngest song plugger and made daily contact with popular composers such as Irving Berlin. At the same time he studied classical piano—Mozart and Debussy were lifelong favorites—and went often with Ira to hear music at places ranging from Carnegie Hall to the Cotton Club. George's acute ear matched Ira's vigilant eye: George said he listened to all kinds of music, "not only with my ears but with my nerves, my mind, my heart. I became saturated with the music. Then I went home and listened in memory. I sat at the piano and repeated the motifs over and over.... Most of my ideas arise from contact with people, from personalities and emotions of men and women I meet."

So, from the start, the Gershwin brothers combined an ear for the vernacular with a large measure of musical and verbal literacy. They shared a distinctively uncommon attribute: the ability to pick up and transmute the sound and spirit of their times in their work. George once spoke of "trapping tones," of hearing the rhythms and melodies in the continuous cacophony of everyday life. Ira talked about the importance of listening to casual conversation, to slang, to figures of speech he could appropriate and transform into lyrics. Both brothers were lifelong collectors of urban folklore, of colloquial content they then molded into lasting form.

Oddly enough, George and Ira connected through an instrumental work by George. *Rhapsody in Blue* premiered in February 1924 and brought the composer international renown. At first George thought to call his new work "American Fantasy" or "American Rhapsody." It was Ira who suggested *Rhapsody in Blue*. *Rhapsody* connects the work with the European classical music tradition; *in Blue* pulls it into contemporary America. The choice of the title reveals Ira's deep sensitivity to George's use of blue notes (made by flatting the third or sixth notes of the major scale to give a piece in a major, upbeat key a minor, poignant, or ambivalent flavor). Ira had discovered, in Isaac Goldberg's phrase, "the very psychology of the music."

Thus, *Rhapsody in Blue* was a crucial step in the brothers' work together. Once George had written it, Ira began to find words that brought George's distinctive musical idiom to life onstage. "Though George's . . . name became universal [quickly], Ira had a great influence on George," Harburg commented. "George's admiration for Ira . . . was lifelong, profound, and of the greatest significance in [George]'s growth, development and evolvement."

Immediately after *Rhapsody* was written, the brothers wrote two songs that became classics—"The Man I Love" and "Fascinating Rhythm" (1924).

"In the spring of 1924 when I finished the lyric to the body of a song—the words and tune of which I now cannot recall—a verse was in order," Ira later wrote.

> My brother composed a possibility we both liked, but I never got around to writing it up as a verse. It was a definite and insistent melody—so much so that we soon felt it wasn't light and introductory enough, as it tended to overshadow the refrain and to demand individual attention. So this over-weighty strain, not quite in tune as a verse, was, with slight modification, upped in importance to the status of a refrain.

Almost immediately Ira responded to the music with the line that contained the song's title and theme—"Someday he'll come along / The Man I love."

The melodic motif that begins the refrain of "The Man I Love" is almost identical to the first motif played by the solo piano toward the start of "Rhapsody in Blue"; it recurs throughout the work, and, in fact, concludes it. "The Man I Love" almost literally picks up where the "Rhapsody in Blue" leaves off. (See the musical example below.) The Greek root of *rhapsody* means stitching songs together. In George's words, his rhapsody is a kaleidoscope of modern Americans, clashing and blending. "The Man I Love" can be heard as the song of an individual detaching herself from those crowded into the *Rhapsody* to tell us about her specific predicament. As we will see, so can "Fascinating Rhythm," which begins with the same notes.

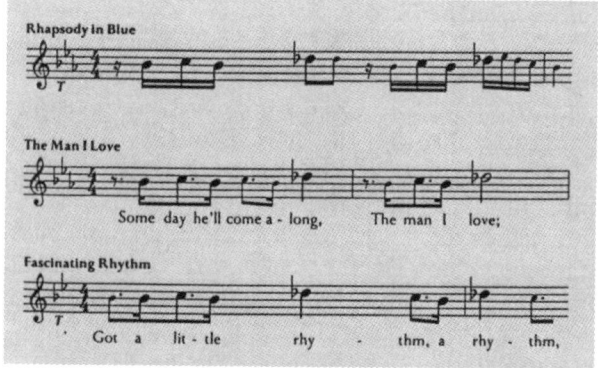

The opening musical fragment of "The Man I Love," which dominates the entire song, evoked from Ira a lyric that makes the phrase *the man I love* palpable and ever present throughout. The musical motif comes back again and again; so do the words *the man I love* and references in the lyric to the man as *he* and *him,* plus a preponderance of words that rhyme with *man,* such as *understand,* and *hand,* or that feature an *m* sound—*mellow moon, dream, theme, roam,* and *home.* Both in lyric and music, "the man I love" is never absent.

The character singing "The Man I Love" becomes three-dimensional because of the words and music together. The melody alone without the words would be sad, yearning; the words alone–simple, hopeful. Together they condition each other, coming together to make a person who is anxious and optimistic, the kind of human paradox the Gershwins saw in practically everyone.

The melodic fragment shared by "The Man I Love" and "Rhapsody in Blue" was not invented by George; in fact, it appeared in much of the popular music of the day, from W. C. Handy's "St. Louis Blues" to the famous sung gag line, "Good evening, friends." It took George's ingenious use of blue notes, his manipulation of richly suggestive melodic and rhythmic nuggets, and his provocative harmonies to transform an ordinary bit of musical vernacular into "Rhapsody in Blue." It took George *and* Ira to make it into "The Man I Love."

"The Man I Love" was the first great song written for a new genre–the American musical. It also represented an enormous leap in George and Ira's development as theater writers. With it, they discovered their symbiotic gifts and ability to make a song intrinsically dramatic through the wedding of music and lyric. With it, they discovered their voice.

"Fascinating Rhythm" is the second song by the Gershwins in which an individual seems to step out of the masses of *Rhapsody in Blue.* The start of the verse to "Fascinating Rhythm" strongly resembles the end of the *Rhapsody in Blue* and the beginning of the refrain to "The Man I Love." "I like to get the most effect out of the fewest notes," George told an interviewer. "This is getting back to the Mozart idea–simplicity in composition." What is striking is that songs like "The Man I Love" and "Fascinating Rhythm" could be so different–both in lyrics and music–and have so much in common.

With "The Man I Love," "Fascinating Rhythm," and the rest of the score to *Lady, Be Good!,* Ira began to be credited with the virtual invention of contemporary American musical theater lyrics. Lyric writing was a new profession that other contemporary giants such as Hart, Oscar Hammerstein, and Harburg would soon join. In most previous theater songs, except those by the Englishmen Gilbert and Wodehouse, the lyrics were considered secondary to the music. By the 1920s, Ira noted at the time, audiences had become "lyric conscious," listening with a critical ear.

"Fascinating Rhythm" is one of the songs that best epitomizes Ira's contribution. This song has striking musical/lyrical subtleties. In both verse and refrain the rhythm is driving, the key is unstable. The verse describes a persistent rhythm that "pit-a-pats through my brain," the hard vowel sounds sustaining the insistent image:

> Got a little rhythm, a rhythm, a rhythm
>   That pit-a-pats through my brain.

When the musical phrase is frenzied, its verbal counterpart is short, breathless, and equally frantic:

> Fascinating Rhythm,
> You've got me on the go!
> Fascinating Rhythm,
> I'm all a-quiver.
>
> What a mess you're making!
> The neighbors want to know
> Why I'm always shaking
> Just like a flivver.

When the music is broader and rhythmically more stable, the words form a longer, more descriptive sentence:

> Each morning I get up with the sun...
> To find at night no work has been done.

It is rare indeed for the words of a song to paint so vivid a picture of their musical complement.

Ira's favorite definition of song came from the *Encyclopaedia Britannica:*

> Song is the joint art of words and music, two arts under emotional pressure coalescing into a third. The relationship and balance of the two arts is a problem that has to be resolved anew in every song that is composed.

The Gershwins took up that challenge in every song and show they wrote. No matter what the character, situation or genre, they never duplicated themselves. Part of their creative makeup involved seeking new kinds of characters and shows–from musical comedies and satiric operettas to opera and films.

The Gershwins wrote songs for given moods, characters, and situations. Once they received a plot outline, they decided together on the general character of the melody and lyric. Often, the title came next. Ira found this part the most difficult. "The hardest part of lyric writing," he said, "is getting the basic idea, expressing it in a title of three words or

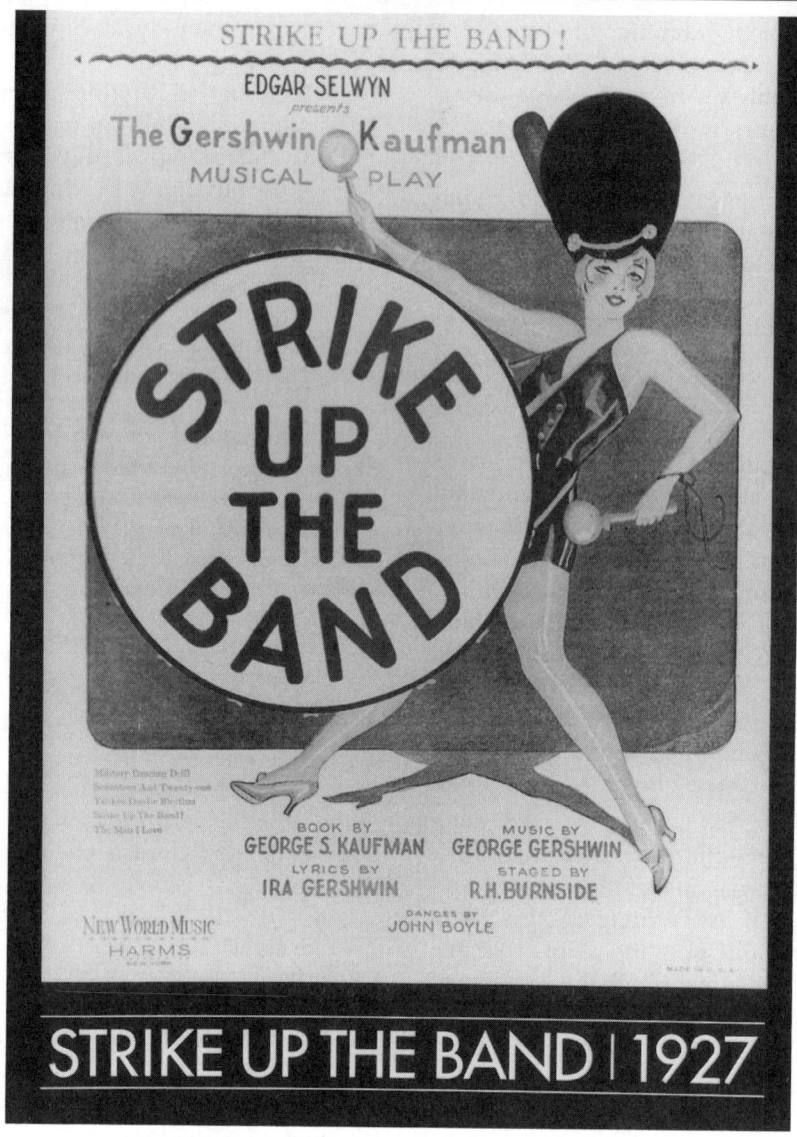

Front of sheet music for the Gershwins' 1927 show, which introduced "The Man I Love"

so and getting the first and last lines.... After getting the title... I skip nine times out of ten to the last line and try to work the title in again, with a twist, if possible. Every time I get a tune I sing it to myself to see how it sings." His procedure was "to come up with an idea that is consistent and complete, put it as a theorem in the title, and prove it QED to the listener's satisfaction." Thus a Gershwin title is really a short précis of an entire song.

Ira claims that specific titles "come from thin air, figuratively and literally," and that unlike good children, "titles should be heard, not seen. Listening to the argot in everyday conversation results in pay dirt for the lyric writers." Any number of titles prove his point: "They All Laughed," "I've Got a Crush on You," "I Can't Be Bothered Now," "Things Are Looking Up," "But Not for Me," "Isn't It a Pity?," and "Who Cares?"

But even the most colorful titles imaginable would mean little if not backed by a strong lyric conception to develop in the body of the song: "A seemly title is only half the battle," Ira observes. "What follows must follow through in the verse and refrain, whether the development is direct or oblique. In brief,"

A title
Is vital.
Once you've it–
Prove it.

Consider "But Not for Me," one of the Gershwins' few songs of unrequited love. As with most

Gershwin songs, the title sums up the theme of the lyric: The singer has missed out on something she thinks other people have. In the verse, she seems self-reliant, even defiant; her sparkle, courage, and sense of humor come to the fore:

> Old Man Sunshine–listen, you!
> Never tell me Dreams Come True!
>   Just try it–
>   And I'll start a riot.

She just might; *try it* and *riot* are set to bold rising intervals in contrast to the preceding descending scale motion.

The refrain melodies of "The Man I Love" and "But Not for Me" are similar in rhythm and shape; both start off the beat and pivot around a single note in a similar way. In both songs, an observation is made in the first line of the lyric and clarified in the second. Also, both second lines happen to be the title of the songs:

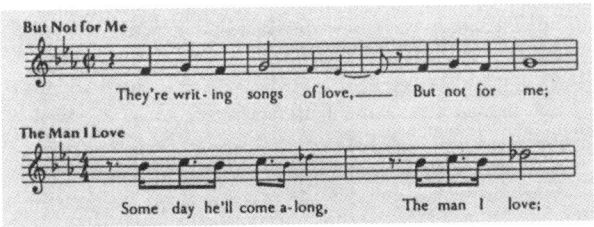

While the character singing "But Not for Me" speaks less optimistically than her counterpart in "The Man I Love," the music to the later song, more clearly in a major key than the earlier one, seems to imply that all is not lost. The tongue-in-cheek clichés throughout continually offset the sober thoughts.

> With Love to Lead the Way,
> I've found more Clouds of Gray
> Than any Russian play
>   Could guarantee.

After following through on this train of thought, the singer "laments" in conclusion:

> When ev'ry happy plot
> Ends with a marriage knot–
>   And there's no knot for me.

Ira commented on the last line, "This is a minor form of the trick or surprise ending, a structural device I have always liked. . . . Usually, when possible, a verbal twist or turn is striven for."

So in "But Not for Me" Ira achieved his stated aims: He summed up an idea in the title, traced it through the paradox of the character's sorrow and feistiness, and brought the title back with a twist at the end.

When the Gershwins first collaborated in the 1920s, the completed melody often preceded the completed lyric. Ira felt that a good lyricist must "first and foremost be in tune with the rhythm and accent of the music"; he should be able to make "the sound of the words parallel the musical line." This is harder than it sounds: "It can take years of experience to know that such a note cannot take such a syllable, that many a poetic line can be unsingable, that many an ordinary line fitted into the proper musical phrase can sound like a million." A lyricist must have "A fondness for music, a feeling for rhyme, a sense of whimsy and humor, an eye for the balanced sentence, an ear for the current phrase, and the ability to imagine oneself a performer trying to put over the number in progress."

In addition, Ira felt that a lyricist must be able to come up with words that "strike the ear right off with clean-cut snappy syllables. Good lyrics should be simple colloquial rhymed conversational lines." This notion has a lot to do with the kinds of melodies George was best at. Isaac Goldberg, George's first biographer, put it this way: "As Gershwin's hand becomes surer, as his feeling for the plasticity of the melodic lines grows more sensitive, his lines take on the contours of conversations. The born songwriter has a flair for colloquial lines. It is a way of making music talk, which is just what a song should do."

By the 1930s, according to Ira, his and George's process had gotten fluid and flexible. Ira said that George

> might have just the opening section of a tune and would wait for me to come up with some notion, and the words and music would then be developed almost simultaneously, or . . . I would have a title and/or a couple of lines he liked which he would tackle at the piano.

For instance, Ira wrote about the creation of "I Got Plenty o' Nuthin' " (1935):

> DuBose [Heyward, *Porgy and Bess* librettist] and I were in George's workroom. George felt there was a spot where Porgy might sing something lighter and gayer than the melodies and recitatives he had been given in Act I. He went to the piano and began to improvise. A few preliminary chords and in less than a minute a well-rounded, cheerful melody. "Something like that," he said. Both Dubose and I had the same reaction. "That's it! Don't look any further." "You really think so?" and, luckily, he recaptured it and played it again. A title popped into my mind. . . . "I got plenty o' nuthin'," I said tentatively. [And a moment later the obvious bal-

ance line, "An nuthin's plenty for me."] Both George and DuBose seemed delighted with it.

With the title and first line in mind, conditioning his muse, George finished the melody.

"I Got Plenty O' Nuthin' " is one of several Gershwin songs that deals with a salient aspect of the brothers' worldview. Another is the haunting "They Can't Take That Away from Me" (1937), whose seeming simplicity and conversational flavor make it so poignant. Many repeated notes and simple harmonies go with words like

> The way you hold your knife
> The way we danced till three,
> The way you changed my life—
> No, no! They can't take that away from me!
> No! They can't take that away from me!

Years later Ira modestly stated that occasionally he made usable suggestions to George about changing a note here and there. In "They Can't Take That Away from Me" it was Ira's idea to insert a pause, or rest, in the music before each line that begins: "[pause] The way you wear your hat." This rest is crucial to the song's impact, giving the music an urgency and a drama would otherwise be missing.

The Gershwins' sensitivity to life's more serious side also comes out in "Shall We Dance" (1937), another of the last songs they wrote together. This song points out an often unnoticed dimension of their worldview. The lyric goes, in part,

> Life is short; we're growing older.
>   Don't you be an also-ran.
> You'd better dance, little lady!
>   Dance little man!
>   Dance whenever you can!

Ira has observed that he wrote this lyric in direct response to the music: "The distinctive tune brings to the listener (me, anyway) an overtone of moody and urgent solicitude."

Ira recalls vividly how well he and George understood each other creatively when they worked on "A Foggy Day (in London Town)" (1937). The song literally poured out of the brothers simultaneously, based on the title and the situation. Here is one of the clearest cases of a classic Gershwin song whose music was inspired by the title/first line. That is, George wrote music that may never have emerged without Ira's initial input. Ira recalled,

> One night I was in the living room reading. About 1 a.m. George returned from a party... took off his dinner jacket, sat down at the piano.... "How about some work? Got any ideas?" "Well, there's one spot we might do something about a fog ... how about *a foggy day in London,* or maybe *foggy day in London Town?*" "Sounds good ... I like it better with *town,*" and he was off immediately on the melody. We finished the refrain, words and music, in less than an hour.

When the refrain was done, Ira recalled,

> All I had to say was: "George, how about an Irish verse?" and he sensed instantly the degree of wistful loneliness I meant. Generally, whatever mood I thought was required, he, through his instinct and inventiveness, could bring my hazy musical vision into focus. Needless to say, this sort of affinity between composer and lyricist comes only after long association between the two.

A Los Angeles writer put it another way in 1939, when he showed how Ira's gifts helped George:

> Ira contributed to George by recognizing which of his brother's musical phrases should be seized and written up. The composer threw the lyricist many little musical figures, and ideas which in passing George might have tossed aside were held onto because Ira caught them and pinned them down with words that could sing well and have meaning. Like all good lyric writers, Ira is musical, a musician whose instrument is words. It takes a musician to recognize what plays well on his instrument. Take "I Got Rhythm," three words on four notes. I wonder how well those notes would be remembered with other words?

Actually the initial notes of "I Got Rhythm"—which, like those of "The Man I Love," originated in *Rhapsody in Blue*—were the basis of many great Gershwin songs. They include "Of Thee I Sing," "A Foggy Day (in London Town)," "Nice Work If You Can Get It," and "They All Laughed," which features the Gershwins' last word on life. "Who's got the last laugh now?" they ask, meaning we all do, if we exercise our option.

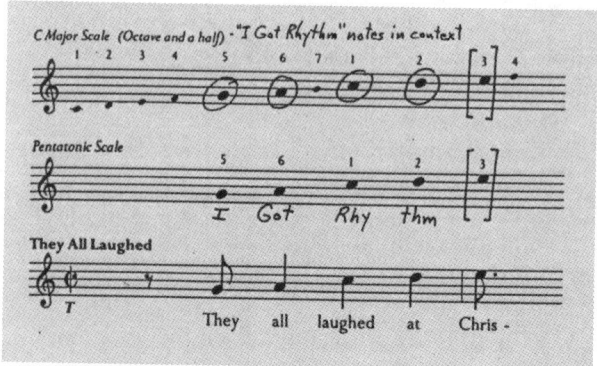

"I Got Rhythm" was a favorite of both Gershwins. George liked it so well that he wrote an orchestral piece based on it, *"I Got Rhythm" Variations,* which he dedicated to Ira.

A TRIBUTE

from Lawrence D. Stewart

This hitherto unpublished essay was written in the spring of 1954, while A Star Is Born *was still in production. Aficionados of that film will recognize small discrepancies between the movie's action as described here and what was ultimately seen on the screen. When I learned that Ira Gershwin was to write* A Star Is Born, *I asked him to save me the manuscript drafts of his lyrics. He was surprised that anyone would want such a thing, as he had never considered them as being manuscripts in a technical sense. He later gave me what he had salvaged, inscribing them with some embarrassment, "For anyone who's interested...." Outside of a few miscellaneous notes and drafts, these are the only rough drafts of his work he has ever saved. The rest of the material is based on my impressions, drawn from my visits to Roxbury Drive when the score to* A Star Is Born *was being composed. Ira Gershwin went over every word of this essay with me while it was "in work." Presumably he approved of this account, for in subsequent years he gave copies of it to scholars to cite in their own research. Ira Gershwin's working drafts for the lyrics of* A Star Is Born *are in the Harry Ransom Humanities Research Center of the University of Texas at Austin.*

IRA GERSHWIN AND "THE MAN THAT GOT AWAY"

Ten-twenty-one North Roxbury Drive in Beverly Hills is one of those reductions of the disorganized world into a highly crystallized state. It is the home of Ira and Leonore Gershwin, and into it come the creators of art, music, and literature who impose an order on the chaos of existence. Each mail brings letters from friends and fans around the world, together with review and presentation copies of books and recordings; throughout the night the phone rings steadily with calls from the cinematic community and from New York. Although the solipsistic, the egocentric, and the peripherally-perturbed drift in from time to time, most visitors are pleasingly engaging, while the man and woman who give the house its existence are themselves unpretentious, sane, and stable. So ordered is the image of the external world which has become this home, that Ira Gershwin himself seldom leaves the house to see the disorganized origin of this reflection. It is a rich house, rich in the comfortable furnishings of tradition, and it is generally evening there, where the heavy drapes conceal the world within from the world without, and everything shines in the glow of reflected light.

In the spacious living room, hung with Modigliani, Utrillo, Chagall, Soutine, and Rouault, Gershwin sits at a cardtable and works on his lyrics, setting, as F. Scott Fitzgerald might have said of him, "the rhythm of the year, summing up the sadness and suggestiveness of life in new tunes." The telephone may ring and divert his thoughts, but even then he sometimes jots notes for his rhymes on the phone pad when he is supposed to be considering more pedestrian matters, such as who will be coming to the Gershwin Plantation (as he terms it) that night for dinner or who will be arriving later for scrabble or billiards. In the same end of the room where the cardtable stands and Gershwin sits doodling as he sketches ideas for his lyrics is the Steinway at which George Gershwin wrote most of *Porgy and Bess.* When Harold Arlen arrives at about two-thirty in the afternoon with his old tan briefcase under his arm, the scene changes slightly: Gershwin moves to a cat-clawed green armchair nearer the piano, where he sits with Calliope, the Siamese, on his lap and pets her abstractedly as he smokes a cigar. He puts his glasses upon his forehead, leans back, and closes his eyes as he moves his hands in tempo with the music. Silence–discussion–the repetition of a melodic theme and the analysis of an idea. Arlen makes some phone calls while Gershwin debates with himself the use of a word. The return to the piano–the trial and revision of a line–the alteration of the melody–and then, when composition seems hopeless and confused, the sudden inverting of the tune or the lyric. The writing of another song has begun, and Gershwin is temporarily contented with the thought of making progress.

It is this Gershwin, Ira Gershwin, who is proving that his family name is a hallmark not only for music but for lyrics as well. During the years when George Gershwin composed, Ira was his chief collaborator, but the retiring lyricist moved aside to let the spotlight fall continually on the more glamorous composer brother. Both criticism and publicity urged an enchanted world to regard a Gershwin song as primarily a creation in melody. But in the years since 1937, Ira Gershwin's productions have confirmed what many earlier suspected: that the charm of a Gershwin song is not attributable to a wonderful tune and innocuous lyrics but to the amalgam of excellence both in music and in words.

Since George's death, Ira Gershwin has worked mostly on movies (*Cover Girl* and *The Barkleys of Broadway* were the most popular) and three stage

The poet at work (Ira and Leonore Gershwin Trusts)

shows, one of which (*Lady in the Dark*) raised to new significance the *genre* of the play with music. Although Gershwin dislikes all superlatives which refer to his own merit ("I'm good, but not the best," he will insist), his work on the new musical film, *A Star Is Born*, will cause many to regard his statement as exceedingly modest. His songs may not emerge as jukebox favorites, since they are strangely lacking in both maudlin and folksy types of philosophy; instead, they originate in the intelligence and clothe themselves in irony. They do not concern themselves with the statement of a moral but with the creation of an emotion; they are poetry and not doggerel verse.

Sometime early in December of 1952, Gershwin agreed to write the lyrics to the music of Harold Arlen and the script of Moss Hart for the re-make of *A Star Is Born*. The original version has long been a Hollywood standard, the apotheosis of the old myth that every girl not only yearns to be a movie star but that when the cinematic moguls smirk with beatific smile, all things are possible; the act of creation is not reserved to the deity alone. The re-written version was to be done by the playwright with whom Gershwin had worked on *Lady in the Dark*. As for the composer, it is difficult to criticize Arlen's ability to recreate the malaise of old-time, low-down jazz. Arlen too had collaborated with Gershwin when they had done *Life Begins at 8:40* (with Yip Harburg) in 1934. The three close friends were united for the first time, and they were working for one of their favorite performers, Judy Garland.

On Sunday, January 11, 1953, Hart drove up to Beverly Hills from Palm Springs, where he had been working on the script of the movie, and arrived at the Gershwins' for the first conference. This initial pilgrimage set the pattern for the musical work on the film, since Gershwin prefers to work at home instead of at the studio, and all concerned were willing to make 1021 North Roxbury the musical rendezvous point for work on the picture. At this session, Hart sketched the outline of the story and suggested seven places in the script for songs. The psychological motivation for, and the nature of, these compositions was discussed, and there was general talk about the particular devices which would be used in each situation. From this initial meeting, Gershwin emerged with a sheet of blue paper on which he had scrawled the briefest outline for the songs:

1 Shrine Auditorium Judy & band
2 *Dive Song*
3 Preparation in rehearsal (gay song) (then at preview complete #)
4 Song on Sound Stage (proposal with interruptions)
5 Motel Song (probably one to reprise later)

6 Prop # Tour de force (in Malibu sings all parts & imitations)
7 Reprise of 5 probably

These outlined descriptions ultimately became the following songs: 1. "Gotta Have Me Go with You"; 2. "The Man That Got Away"; 3. An untitled calypso commercial; 4. "Here's What I'm Here for"; 5. "It's a New World"; 6. "Someone at Last"; 7. Reprise of "It's a New World." One number, "Lose That Long Face" was not originally called for but was later added to the score, while the preview number of the third point caused the most difficulty in the picture. "Dancing Partner" was originally written for this position. Dismissed as unsuitable for the score, it was replaced with "Green Light Ahead," which was rejected in turn for "I'm Off the Downbeat." The latter was found unworkable for choreographic reasons, and as this article was written it was still uncertain what number would be used in this part of the movie. [The spot was filled with Roger Edens' "Born in a Trunk" and a medley of standards.]

Since Gershwin possesses an incredible memory, these notations were designed merely to recall certain fundamental requirements which his mind, in its action and reaction, would then remember completely. On Tuesday, January thirteenth, Arlen and Gershwin set to work. They attacked the Shrine Auditorium song first, and it was written with the usual difficulty, but it seemed to pattern itself more quickly than they had expected. (Since the songs were to be solos sung by Judy Garland, they were all to be kept quite direct without the obvious intrusion of profundity or sophistication.) By Thursday, January twenty-ninth, the first song was completed and the second (and favorite of both composer and lyricist), "The Man That Got Away," was well begun. By this time they had slowed down in their schedule and were spending whole days on the pursuit of *le mot juste*. One afternoon was required to evolve the phrase, "Good riddance, Good-bye," which is the first line of the bridge. It is this song which is of greatest interest in the manuscripts, and in its composition and structure it reveals much of the Gershwin-Arlen conception of the completely realized song.

"The Man That Got Away" is a song which has meaning on three different levels: within the picture itself, it establishes the acting ability of the star, Esther Blodgett, and convinces people within the picture that the star is not merely a singer, but is also a great actress. Within the character of Judy Garland herself, the song takes on personal and autobiographical implications which may be concealed from the general audience but not from those artists and theatrical people who know her. And within the song itself, it has a complicated and effective structure, revealing that traditional personal dilemma and psychological state of the person whose lamentation is called the blues. These three levels of meaning need separate exploration to demonstrate the complexity involved in what is regarded by many as only a trivial art: that of writing musicals.

The song occurs in the movie when Esther Blodgett (who in this version is a singer with a road band instead of the star-struck and aspiring actress) is with a small band in a night club. It is in the early morning hours after the customers have gone, and in the club there are only the tired band members rehearsing and the singer herself moving slowly about the bandstand, pouring coffee and singing, first under her breath and then to the room itself. Present at this "performance" is Norman Maine, who had become interested in Esther and had traced her to this hangout. The purpose of the song is to convince Maine, who hitherto had thought of the girl only as a generous person and a good band singer, that she possesses great dramatic talent as well. And it is on the basis of this "performance" that he urges her to enter the movies and become a star. Both Arlen and Gershwin agreed that Judy Garland's interpretation of the song came as close in execution to their conception as was legitimately possible to expect. What is impressive, too, is that the successful interpretation is due to Judy Garland herself and not to her vocal coach. She wanted to sing it loud and brassy, with the vibrating notes emphasizing the dramatic possibilities of her voice to carry out the story motivation. The vocal coach, however, wanted it "sweet" and in a lower key. After much difficulty, the song was recorded both ways, and Judy Garland's preference was adopted for the picture, because the lower and sweeter version lacked brilliance and all of those dramatic qualities which had to be developed in the song, not only in terms of itself, but more particularly in terms of its function in the story.

On Thursday night, March 25, 1954, Judy Garland saw for the first time a rough cut of *A Star Is Born*. Though the individual shots had been viewed throughout the shooting, they not been assembled before to reveal the general texture and structure of the picture. Early on the morning of the twenty-sixth, she called Gershwin much excited with the successful run-through, and said, "Oh, Ira, I know you and Harold wrote every one of those songs just for me!" And it is true that the individual songs were created not only with the theme in mind that would hold them in the picture, but with the possible application which would tie them to the life of Judy Garland herself.

In the same way that the Gershwin household and circle are mirrors of the external world but can live independently of it, so does the work produced in this microcosm have one meaning for the larger world and another which is restricted to it alone. It is a small world, somewhat resentful of intruders, and the creations of its citizens are thought of largely in terms of the response they find in each other. Whether Alan Lerner or Louis Calhern or Arthur Freed likes a song becomes of more immediate importance than whether it will appeal to the untheatrical citizen of the external universe; for these are the professionals whose reactions are the concern of the lyricist and the composer. In its own way, therefore, this concern is similar to some New York producers who think of their plays as being written for a small circle of newspaper critics. And since the songs have this special appeal, they are written with secret nuances and esoteric meanings known only to members of this limited world. Sometimes the meaning may be the origin of a song (as in the case of "Delishious," which originated in Leonore Gershwin and her father's pronunciation of "delicious.") Sometimes the meaning may be so secret that only the author and one friend will know (as in the case of "The Princess of Pure Delight," where "That will be twenty guilders, please!" is a secret joke between Gershwin and a close friend who was paying twenty dollars a session for analysis when *Lady in the Dark* was being written). But for *A Star Is Born* the songs acquire a different connotation. While the plot of the picture concerns the successful creation of a singer and movie star, the writers are thinking of it in terms of the successful *re*-creation of a singer and star: Judy Garland. The picture is to be her return to the screen after innumerable personal tragedies, and the songs and story are to have these secret overtones of her success and happiness. Admittedly, this will not be a literal translation of her experience from private life onto the cinemascopic screen. But in the emotional response which is created, every song is keyed to her experience: every song has a secret meaning for the citizens of the Gershwin world, and every song is designed so that only Judy Garland herself can sing it. While the words carry a logical meaning in themselves and will be sung by many other people, their full charge of meaning is restricted to those for whom the star is not just an actress and singer but a friend as well.

Since a song is, after all, a separate work of art, the meaning which ultimately emerges as its most significant is that which it possesses intrinsically. Long after the movie itself and the personal life of Judy Garland are forgotten, the song will continue to exist, and for that existence it must depend upon its internal order to give it permanence and aesthetic value. After days of work, the final version of the song was agreed upon by Arlen and by Gershwin:

> The night is bitter,
> The stars have lost their glitter,
> The winds grow colder
> And suddenly you're older
> And all because of
>   The Man That Got Away.
> No more his eager call;
> The writing's on the wall:
> The dreams you've dreamed have all
>   Gone astray.
>
> The man that won you
> Has run off and undone you.
> That great beginning
> Has seen the final inning
> Don't know what happened—
> It's all a crazy game.
> No more that all-time thrill,
> For you've been through the mill,
> And never a new love will
>   Be the same.
>
> Good riddance!–Good-bye!
> Ev'ry trick of his you're on to.
> But fools will be fools—
> And where's he gone to?
>
> The road gets rougher,
> It's lonelier and tougher.
> With hope you burn up,
> Tomorrow he may turn up.
> There's just no let-up,
>   The live-long night and day.
> Ever since this world began,
> There is nothing sadder than
> A one-man woman
> Looking for The Man That Got Away.
>   The Man That Got Away.

When Gershwin and Arlen began turning over ideas for a "dive song," Arlen suggested an eight-bar theme which he had had for some time: the music which later became "The night is bitter / the stars have lost their glitter." Gershwin listened to it and said, "Do you like 'The Man That Got Away'?" Arlen was struck by the appropriateness of this title, and they set to work on the song. Gershwin's conception was part good fortune and part inspiration, for finding a title is over half the work in writing, he feels. What is most interesting, of course, is that the title was suggested by the music, but that musical theme which accompanied the title in the score was not the theme which inspired the title itself. From this original suggestion, Gershwin worked out lines for a structural choral pattern:

> The song is played out.
> The moon is in a fade-out.
> The stars won't glimmer,
> The autumn wind is grimmer –
> And all because of
>   The Man That Got Away.

The manuscripts reveal increasing unhappiness with the glimmer-grimmer rhyme, while the "played-out," "fade-out" one belonged to a jargon which failed to emphasize the universal and professionally unrestricted emotion which the song was to convey. The aspect of using nature as a delineation of the singer's mood appealed to Arlen, if not to Gershwin himself (he particularly was uncertain about using the notion of "glitter" with "stars," as he felt it was too harsh a word). The lines were re-written to develop a rather conventional theme in terms of standard poetic imagery, the parallel between man's age and the time of the day and the year. The music for these lines was also provocative, in that it mounted steadily until the last syllable, which collapsed to the level of the first one–the rolling of the rock by Sisyphus translated into music. Only on the line, "The Man That Got Away," does the music continue to rise after a temporary fall, and there the music itself subtly reinforces the feelings of depression and confinement which limit the singer to her world. To be so confined is a portion of her tragedy. But there is a qualified happiness which would come to her if the one person she loved also shared these restrictions. He, however, is the one person who escapes them, and this makes her tragedy a double one. The fall of the music and the accent in these lines necessitated feminine endings; often the destruction of a poet, they are managed by Gershwin here to create a feeling of strength and resiliency. A similarly felicitous touch occurs in the use of "burn up," "turn up," and "let-up," in the concluding stanza. By breaking the rhyme itself, Gershwin has diverted attention from this repetition; and yet it is used to continue that note of despair which becomes an echo in the wishfully optimistic conclusion.

Arlen's music for the song can be described only in Gershwin's own term, the "sweet and low-down." The mournful theme, repeated and repeated in its rising climax, suggests the despair of the singer, while the last part of it each time it is repeated conveys all of the atavistic wailing of the latter-day daughter of Judah. This is another merger of the traditions of Jewish religious music and of New Orleans jazz into a modern formal construct. The elongated last line brings the song to its emotional height, while the repetition of the title at the conclusion (done at Gershwin's suggestion for a sort of coda) creates the effect of a resigned amen. Though there are religious influences and allusions in the song, they derive from the complaining aspect of all religion and not from adherence to any specific creed. The ecstasy of the mourner in his sackcloth and ashes is here addressed to personal and physical love.

Gershwin's manuscripts reveal how difficult it is to write a song such as this, where the language is the colloquial diction of the dive singer, and the rhymes, rhythms, and allusions must be consonant with her character. The title itself is the best example of this, where the "that" carries all of these connotations. Although Gershwin based it on the fishing expression, "the one that got away," his main concern was with delineating the character of the singer through her choice of diction. This deliberate use of informality was similar to the construction of "I Got Rhythm" and "'S Wonderful," which uninformed singers frequently nice-nelly to "I've Got Rhythm" and "It's Wonderful," much to the amusement and annoyance of the lyricist. In "The Man That Got Away," the allusions to the weather, the most common Biblical image (the writing on the wall), and to baseball–are precisely those things which would seem to be the accessible descriptive material of the dive singer. The fact that she might have forgotten the origin of "the writing on the wall" only subtly emphasizes her position as the reembodiment of a concept as old as time itself. Then there is an abundance of folk sayings and proverbs–the language of the ordinary person who, instead of speaking in self-governed arrangements of individual words, uses phrases which are intended to suggest the intensity of her feeling.

Gershwin's tendency is to write wittily, and the manuscripts show him drifting continually into smart expressions which he had to excise from the song itself; these worked their way into the margins of his manuscripts, as though he were compelled to be ingeniously inventive, even when the dramatic situation would not permit it. So there are humorous rhymes: "groovey, movie, and hotter than Vesuvi." There are cou-

Judy Garland singing "The Man That Got Away" in A Star is Born *(Ira and Leonore Gershwin Trusts)*

plet asides: "I rate a razzing / Perhaps he's Alcatrazing." (As Gershwin has pointed out, this could have been fine in "Boy, What Love Has Done to Me" in *Girl Crazy,* but is far too smart here.) Sometimes, lines were too obscure to be clearly understood by all, as in his allusion to golf: "There'll be no sun up, and misery is one-up." One version even headed towards nineteenth-century Victorianism: "There's just no sleeping / Your eyes are red with weeping / Though you know better / You're waiting for that letter." All of these were discarded as either intrinsically worthless or inadmissible in this song. One phrase, worked up in the various versions, appeared either as "What's black and bluer than," or "The blues grow blacker." These were rejected here, but the line "The blues black out" appeared in the next song that he and Arlen wrote, "Lose That Long Face." There is nothing to indicate that Gershwin rechecked the older drafts for new ideas (indeed, his practice is to keep notebooks on favorite unused lines, but he always forgets to use them when he's writing), and this suggests that his subconscious retains phrases which he feels have merit.

The conclusion of the song troubled both Arlen and Gershwin. Since they wrote the song together, sometimes a lyric line would be suggested and then music would be adjusted to it; sometimes the musical theme appeared, and then words had to be cut to fit it. Originally, Gershwin sketched the conclusion:

> There's just no let-up the livelong night and day,
> Till the one-man woman finds the man that got away.

This idea went through the following alterations:

> Oh tell me if you can
> What sight is sadder than
> The one-man woman looking for the man that got away.

***

> You always will be his.
> Oh what a sucker is

The one-man woman looking for the man that got away.

***

Oh since this world began
What is more dismal than
A one-man woman looking for the man that got away.

None of these was acceptable to them, and they considered a structurally different conclusion:

Oh since this world began
What's black and bluer than (or: What future's bluer than)
The gal that hopes to find the man that got away?
(or: The gal who's hoping to find the man that got away?)

Then there was the happy thought which brought them around to the original version, and the conclusion carried through its melancholy theme to a quiet end.

Arlen and Gershwin worked so closely together on this score that each made suggestions about the music and words. The resulting song is a product of reciprocal and mutual influences and responses. Each of the men said afterwards that this was one of the easiest and most pleasant collaborations he had ever undertaken; the successful songs revealed the perfect "wedding of the music and the lyrics"—lyrics and music being fused into a fine amalgam, so that one could never think of one without the other. As Gershwin said, "In composition, we both help on lyrics and music, so you can't really say that one did only one thing—but whenever you hear the song played (even if they play it without words), you don't say, 'Why don't they sing the lyrics?' but 'That's my song!'" Arlen agreed. It is the happy agreement of words and music that characterizes the excellence of this song and their other contributions to *A Star Is Born*.

John Dos Passos: A Centennial Commemoration

When John Dos Passos died in 1970 at the age of seventy-four, his critical reputation was at a low point. Readers who remembered the excitement his fiction evoked during the 1920s and 1930s had themselves either passed on or passed the time of their influence, and the new generation was too concerned with its own social upheaval to lavish attention upon radicals of the past—especially those who spoke to them in such confusing terms as Dos Passos did in his last years. He had seemed to align himself with conservative republicanism, and in the politically charged arena of literary criticism, that shift was construed as a baffling loss of the sociopolitical aesthetic that had distinguished his best work.

Dos Passos insisted that his course had never varied. His highest calling, he said, was as a teacher; his duty as a writer was to enlighten. Specifically, his goal was to help readers discover the enduring values that are the foundation of democracy, to measure contemporary society against them, and to offer readers a means of self-evaluation. Accordingly, his literary efforts were of two interrelated types: he wrote history to discover the character of the American tradition; and he wrote fiction to demonstrate by model how well contemporary Americans lived up to the principles that had molded their culture. He was never a revolutionary. Even when he was most critical of American institutions he did not advocate overthrowing them; he simply insisted that they measure up to the ideals upon which they were built. He defended the ideal of self-government and the preservation of individual freedom with the zeal of a fundamentalist, and he imposed it upon robber barons and labor leaders alike.

It has been nearly a third of a century since John Dos Passos died: time enough to consider the magnitude of his literary achievement. Clearly he was among the most important writers of his time, and he had particular significance as an American writer. He provided a broader picture of American society during his time than any other artist, and he conceived it with a full awareness of our shaping tradition. His work was characterized by informed perceptiveness, and it provides timeless lessons of fundamental value. He was a master of his craft.

The editors of the *DLB Yearbook* have asked people associated with John Dos Passos to offer tributes to him of their own design. Their contributions follow Dos Passos's 1948 letter to Frederick William Lowe Jr.

Lucy Dos Passos Coggin
Dos Passos's daughter

My father was fifty-four years old when I was born. He had always made up his own rules and suffered the consequences when they conflicted with prevailing trends. I suppose he wanted me to have the same opportunity. He chose to teach only indirectly by example and suggestion. The absolutes that readers hoped for in his books were nowhere laid down in rigid family decrees. Only after twenty-five years of making choices based on implied and invisible suggestions have I been able to identify a few themes.

There is no excuse for being rude.

You must maintain the self-discipline to be polite everywhere and always. This was often the first thing people noted when they met my father. One interviewer who came to the brick farmhouse in Virginia expecting to confront a fiery revolutionary moaned as he left, "He is even polite to the dog."

My father's politeness was essential in maintaining privacy and channeling energy into written opinions and conflicts. The freedom to be outspoken in support of opinions and ideas was never to be confused with the license to be needlessly offensive.

Listen carefully to what other people have to teach. Almost everyone has learned something from life. It is your job to find out what.

This rule had nothing to do with formal education or professional titles. He was as pleased to find out about eels from a crabber who used them as bait as to learn about Soviet medicine from an acknowledged expert.

Treasure your friends. They make life tolerable.

As the news of my father's death spread in the last days of September 1970, calls from all over the country came in to the rented apartment in Baltimore, Maryland. I was too dazed myself to remember all that was said, but I began to count as a constant theme developed. At least fourteen different individuals said they felt as though they had lost their best friend in the world.

Through the horrors of war and political turmoil my father had relied on friendship to humanize the world. It tempered the tragedies of the death of his first wife and his own illnesses and injuries. He built a personal web of affection across several oceans and continents.

Take a swim whenever you are near the water.

This held for pools, streams, lakes, rivers, and barely accessible ponds of doubtful purity. The rule functioned as a kind of metaphor for personal experience but was also meant to be taken literally. Exceptions were made only occasionally: once for water direct from a glacier in Alaska, and for several Brazilian lakes known to harbor schistosomiasis. The presence of piranhas and South American crocodiles posed no problem.

Never pass up a chance to travel.

Travel renews the edge on the powers of observation and helps ensure a sense of discovery. My father sometimes repeated that eight was the ideal age and that few people improved with additional years. Whatever served to revive the eight-year-old curiosity and eagerness was beneficial.

Never sit when you can stand. Walk whenever possible. Walk very fast.

Obliged to stay in one place to write, he encouraged vigorous activity whenever possible.

Quite a few unwritten rules involved food:

Avoid bad food. Seek out good food.

Food is an opportunity to magnify the enjoyment of every day. It is worth going to some trouble and considerable expense to keep up a fresh and varied menu. Early spring demands shad roe. Soft crabs glorify summer, and tender lettuce is a necessity all year long.

Never throw away edible food.

This rule was actually enforced potato by potato. Since my grandmother had seen starvation as a young girl in Petersburg, Virginia, during the Civil War, her son perpetuated a respect for usable nourishment as well as enjoyable food.

"Parsley on everything until the train leaves."

This phrase has been attributed to both sets of my godparents: Gerald and Sarah Murphy and Lloyd and "Diddy" Lowndes. The saying applied to all of them equally well. My father spent the month before he died in a small rented house in Head Tide, Maine, not far from Lowndes's house in Wiscasset. My mother dutifully ferried in live lobsters and fresh local produce despite the looming threat of my father's heart failure.

"I will retire when the undertaker comes through the door."

This was always my father's stated goal, and he did manage to write until his last days, even when an hour at the desk left him exhausted. He felt that men should be free to identify so completely with their chosen work that life without it would be an insult. Curiously, this gave women an advantage in being able to shape themselves as wives and mothers in long-term occupations with no recognizable end. At the same time, he somehow unwaveringly supported a woman's choosing a professional path and passing up their more traditional roles. Life for him did not seem to include a retirement clause. The duties of a free individual lasted until the last hour of the last day.

Norman Mailer
Author

John Dos Passos came closer to writing the Great American Novel than any other novelist. There is no doubt that *U.S.A.* is the largest, most ambitious, and most successful social portrait of America in the first half of the twentieth century that we have.

Ellen Bromfield Geld
Author, daughter of Louis Bromfield, family friend

The last time we saw John Dos Passos was in 1962, when, stopping by Pau D'Alho, he and his family collapsed for a few days in one place after a strenuous journey through an enormous country. He had been on a return trip to Brasília, where a few years earlier he had shared with President Juscelino Kubitschek de Oliveira the dream of a capital that would act as a heart to awaken the vast, scarcely touched interior of Brazil.

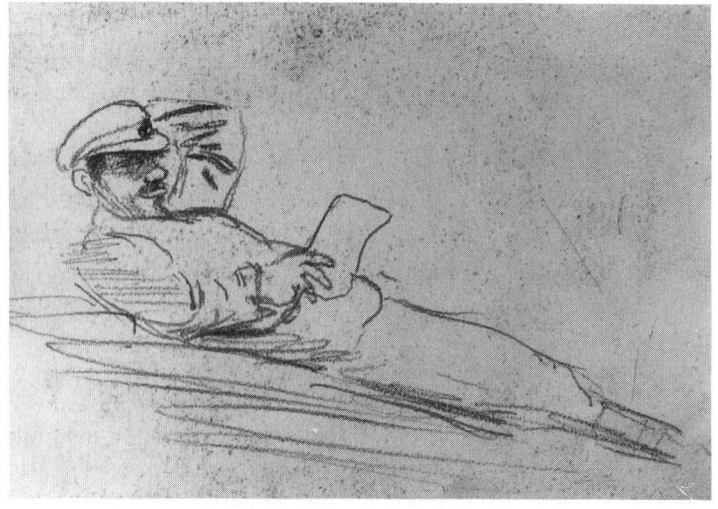

Sketches by John Dos Passos, circa 1918–1925

Though far from a failure, Brasília had certainly not solved the myriad troubles aroused by the sudden growth of industry thrust upon a weak and archaic agricultural base. Millions had "escaped" isolation in the country to pile themselves in the slums of cities that were not prepared for them. Brasília and other optimistic spending sprees had made inflation rampant. An object of the cold war—as it had never been of any serious foreign policy in other days—Brazil's precarious democracy was in dire straits.

With his ability to throw himself into a situation, ask, and listen, in *Brasil on the Move* (1963) Dos Passos captured much of this scene: the boundless enthusiasm of a Bernardo Sayaõ, who literally lived and died on the road he built from Brasília to the Amazon; the innocent zeal of students seeking answers in the works of Karl Marx, which they had read no more than they had the Bible, to the injustices they saw all around them.

Why were Brazilians as they were: at once cynical and optimistic, devoted, undisciplined, energetic, and enduring beyond credulity? How had Brazil itself survived as half a continent while Spanish America had quibbled itself into a multitude of bordering countries?

Going back to Brazil's beginnings in *The Portugal Story* (1969), Dos Passos attributed much to the Portuguese immigrant's ability to blend—marry themselves into the Indian population and later take slaves, both Indian and African, into their powerful, patriarchal families. He points out that the forces that twice defeated Netherlanders in Pernambuco were led by an Indian, a Portuguese, and a Negro. "The union of the three races in the war against the Dutch," he wrote, "foreshadowed the beginnings of Brazilian nationality."

These are works I became familiar with as a settler in Brazil, though decades before I had become a Dos Passos reader at Cornell. Now as then, it strikes me that this sensitive writer's contribution has always been that of giving history a perspective, providing a human picture of the makers of events and their victims.

Thus in the trilogy *U.S.A.* and in *Midcentury*, biographies rich in candor mingled with fictitious characters have given us the effect of such as Eugene Debs, John L. Lewis, Gen. Douglas MacArthur, Eleanor Roosevelt, and her husband on the lives of Wobblies, stenographers, assembly-line stewards, and farmers.

We follow these Americans from the beginning of the century, as part of a young and vital country in which space was not a problem. No matter how poor or how bad things were, there was a way to pursue ideals, to get ahead. Yet somehow in the scramble, their actions show us how ideals and ambition can become confused—as can patriotism and duty, when the innocent become cannon fodder in wars that are not of their making.

Through and beyond these wars Dos Passos gives us a view of ever-changing times, of swinging pendulums as space becomes a battleground of competition in which business brandishes Mothers' Day as a legitimate weapon. Unions—once struggling to give labor its leverage—come to use laborers themselves as tools in a battle for dominion.

The second war is won; the peace is lost. The victims are the millions whose lives are subjected for the next half century to a tyranny greater than any the world has yet known.

In Dos Passos's stories, masterfully and poetically told, history indeed repeats itself through the predictable acts of humanity. Thus we are reminded of how humans, capable of good, are eternally the perpetrators of evil wherever an excess of power exists. Our ability to be enlightened is our greatest means to progress, but constant vigilance and questioning are necessary.

Through the biographies Dos Passos provides the continuous thread that links the acts of past history-makers to those of the present, while his poetry and fictitious characters do the questioning. An enlightening job indeed, which perhaps only a combination of free speech and fine fiction can achieve.

Roslyn Targ
Literary agent for Dos Passos

I met Mr. Dos Passos at the end of the 1960s with his wife Elizabeth. He was extremely shy and very polite.

It is an honor to represent Mr. Dos Passos's foreign rights. His most important as well as his lesser works have continuously been published in France, where he is regarded as a classic author and respected as such, I believe, even more so than in the United States. Germany has the *U.S.A.* trilogy and *Manhattan Transfer* (1925) in print. Italy also has several works in print, and finally, I convinced a Spanish publisher to take *Manhattan Transfer*. Holland has published two volumes of the *U.S.A.* trilogy. At the moment, I am trying to interest Eastern European publishers as well as Chinese publishers in the major works.

Dos Passos made the time he wrote about come alive. He combined the events of the day together with imaginary characters as well as with famous people of the time. I believe as long as there are sensitive readers, his work will never die. I shall not give up finding new publishers for him around the world.

Daniel Aaron
Scholar, coeditor of the Library of America edition of the U.S.A. trilogy

The enduring Dos Passos is the driven explorer who took his readers, in Edmund Wilson's words, "behind the front pages of the newspapers," and whose inspired reportage animated institutional abstractions. The apogee of his work is *U.S.A.*

Recently republished by the Library of America (1996), it has inspired a spate of commentary about the "forgotten" or "neglected" author who sixty years ago was a coequal of Ernest Hemingway, F. Scott Fitzgerald, and William Faulkner. Much of this commentary is nostalgic ("how I felt when I first encountered *U.S.A.* in my callow youth") and spotted with questionable information about the writer and his political vagaries. The trilogy appears to have held its place as a quirky "masterpiece," notwithstanding those who consider it a petrified artifact of bygone days and dismiss the once admired "Newsreels" as a mere gimmick and the "Camera Eyes" as "swatches of avant garde *kitsch*." These so-called "modernist" experiments do creak a little, and so do Dos Passos's verbal affectations. Why then, has *U.S.A.* survived its blemishes? Why is this lengthy and complex novel so riveting? Why do readers, allegedly bored by anything that happened more than five years ago, yield themselves to Dos Passos's swirling narrative?

The lure, I suspect, isn't so much his mannered set pieces on notable or notorious American personages, emblems of something larger than themselves, nor his arty autobiographical monologues. Rather it's the pace and passion of *U.S.A.,* a collective chronicle of twelve unremarkable people as they scramble through the first three decades of the century, caught in the machinery of a society indifferent to their appetites and aspirations. They develop bad habits, behave generously and meanly, aren't very percipient about themselves or others, and plunge into destructive relationships. Dos Passos, the engaged reporter, conveys what they see and do and think with a fidelity made all the more convincing by his way of recounting their lives in their own locutions, almost as if they were prompting him. The America that emerges in Whitmanesque amplitude is sufficiently analogous to our own to make it accessible to 1990s readers: it is marked by economic and political struggle, an abrasive materialism, sexual exploitation, financial mismanagement and chicanery, and manipulative public relations.

U.S.A. is sui generis. None of Dos Passos's many imitators has managed to replicate the trilogy's mix of

John and Elizabeth Dos Passos at Spence's Point, Virginia, August 1964 (United Press International)

political passion, historical insight, stylistic exuberance, verbal energy, and corrosive comedy.

Ashley Brown
Critic

In 1950 John Dos Passos was living with his new wife, Elizabeth, in Westmoreland County, Virginia, where he had spent part of his boyhood. It was about this time that he visited Washington and Lee University. I was an instructor there and probably represented the modern point of view in a rather conservative department of English. Hence Dos Passos was put in my charge during part of his visit. In fact he was conservative in both literary and political matters by 1950, and I remember that he took some interest in the Vanderbilt Agrarians, Ransom and Davidson, who had been among my teachers. It seemed that he was trying to farm his land in Westmoreland County and clearly admired the Jeffersonian tradition; in 1954 he would publish a book called *The Head and Heart of Thomas Jefferson*. But his main business at Washington and Lee was to talk to those students who had literary ambitions. The boys naturally wanted to know what they should read for models; Dos Passos wasn't really very helpful; he suggested the Bible, William Shakespeare, and Miguel de Cervantes. One of the bolder students asked him about James Joyce; he replied that during the 1920s all the young novelists were reading *Ulysses* (1922), and they were often influenced by this book in various ways. Whether it should still be considered as a model he didn't say. He was a very polite, almost self-effacing man who looked upon writing as a full-time profession; probably this attitude was valuable for the students to know about. He generously gave us a new piece of his work for *Shenandoah,* a little magazine that several of us (faculty and students) had just founded. There, in the first number for Spring 1950, we published Dos Passos's Whitmanesque poem called "Century of Trials," alongside early work by Tom Wolfe and Cy Twombly, then students at Washington and Lee. I believe that "Century of Trials," though presented as a poem, would later be included in a new edition of Dos Passos's novel *Adventures of a Young Man.*

In 1950 Dos Passos was still a presence in fiction, and not just in the United States. A dozen years earlier Jean-Paul Sartre had called him the greatest living writer. Sartre's own unfinished tetralogy, *Les Chemins de la liberté* (1945–1949), so obviously imitative of Dos Passos, created almost as much stir in the United States as it did in France. But in another decade or so Dos Passos, still very productive, would lose much of the audience he had once had in his native country. Perhaps it was partly a matter of politics; his conservatism was then almost a liability, and during the 1960s literary youth turned elsewhere for their models.

Like many others, I had almost ceased to read Dos Passos. But in 1964 É went to live in Brazil for the first time; as I became acquainted with the literary scene there and elsewhere in Latin America, I began to hear about him again. As the grandson of a Portuguese immigrant to the United States, Dos Passos knew Portuguese and took considerable pride in this heritage. He had already visited Brazil and in 1963 published *Brazil on the Move,* an enthusiastic account of public life there. Rather more interesting, I thought, was that Latin American novelists such as Carlos Fuentes and Érico Veríssimo were finding Dos Passos's *U.S.A.* useful in writing novels about their large cities. His panoramic technique, adapted from certain chapters in *Ulysses,* was in turn adaptable to contemporary life in Mexico City and Pôrto Alegre (Veríssimo's city in southern Brazil). So any account of Dos Passos should emphasize something of his Latin associations; they are part of the larger interplay between North and South American literary culture during this period.

Donald Pizer

Scholar, author of Dos Passos' USA *and editor of* John Dos Passos: The Major Nonfictional Prose

I was in my teens, in the 1940s, when I first read *U.S.A.* I was living in a Brooklyn neighborhood, almost a village in its thin connection with the "outside" world. One sign of its insularity was that it lacked a public library, and thus one of the rituals of my boyhood was a weekly bicycle journey to a fairly distant (or so it seemed then) adjoining neighborhood, where I would draw out a supply of books from a tiny store-front library.

Like many of my generation, I was vaguely aware of *U.S.A.* as a book to read, and being young and a fast reader, I was not deterred by the immense size of the Modern Library edition with its thin, semitransparent paper. I read the work slowly at first, and then more and more rapidly, until I was fully immersed in it. I had somehow stepped out of my narrow segment of life into a universe of people and events that was America. I didn't know then why and how the book held me captive, but I was indeed its prisoner. Today, as a student of the complex achitectonics of *U.S.A.,* I know a great deal more about how it achieves its effects and what these effects are, but I have never duplicated, in reading it again and again, the excitement, the rushing compulsion to read and to continue reading, of that first encounter.

When, some forty years later, I told my teenage daughter the story of my weekly trip to the library on my bicycle, through rain and snow, she asked, "Did you realize the Lincoln parallel then or did it come to you later?" But when she herself then read *U.S.A.* out of curiosity because I was working on Dos Passos, she too disappeared from human intercourse for the few weeks it took her to encompass it. And so it will always be, I think, whatever the means by which we are led to it, for this great work of the imagination.

David Sanders

Dos Passos Scholar

John Dos Passos's lifelong subject was the U.S.A. As an expatriate child, he had no Yoknapatawpha County or so much as an American neighborhood as native soil. He wrote about his country's whole contemporary situation, sometimes by way of historical research, and in his trilogy—*The 42nd Parallel* (1930), *1919*

(1932), and *The Big Money* (1936)—he delivered a book unlike any other written by a twentieth-century American, something that resembles Walt Whitman's *Leaves of Grass* (1855). So far, in Dos Passos's centennial all toasts have begun with the celebrants' still green memories of their first readings of *U.S.A.*

U.S.A. is one of the triumphs of the modernist American writers who came out of World War I. It chronicles a national past from the turn of the century to the Great Depression, always questioning what the United States is all about. It is a design of forms—fictional narratives of ordinary Americans, poetic autobiography, montages of popular culture, and, starkly above these, the "portraits" from "Lover of Mankind" (Debs) through "Meester Veelson" to "Power Superpower" (financier Samuel Insull), all of it held together by Dos Passos's singularly brilliant use of the American language ("U.S.A. Is the slice of a continent.... U.S.A. is the letters at the end of an address when you are far away from home. But mostly U.S.A. Is the speech of the people"). The work does not rest in this unity, however, but moves in a tension between power and oppression that led the Camera Eye narrator—Dos Passos himself—to cry out: "all right ... we are two nations."

Three Soldiers (1921) and *Manhattan Transfer* (1925) are also about systems closing in on individuals; their anger and despair foreshadow the trilogy, but their techniques flicker in any comparison. Also in the 1920s appeared the travel collections *Rosinante to the Road Again* and *Orient Express,* jaunty records of Dos Passos's lifelong restlessness and curiosity that contain, incidentally, prescient observations of Spain and the early chaos of the Soviet Union. The way into *U.S.A.* now seems easier to trace than the abrupt changes in Dos Passos's work that came afterward.

After *U.S.A.* Dos Passos wrote incessantly for the rest of his life, publishing, aside from several collections, a second trilogy among eight novels, seven histories, four travel reports, and a memoir. None of these works increased his readership or critical reputation; they led, rather, to simplistic analysis of a sharp decline because of a change in politics. In fact, Dos Passos's passion for writing what he came to call "contemporary chronicles" collided with his persistent observation that everywhere man was opposed to ingenious systems he had created for enslaving himself. In *District of Columbia* and *Midcentury* he had nothing like the Camera Eye to shape a conflict much harder to grasp than the two nations of *U.S.A.* In writing history about Thomas Jefferson and his colleagues, Dos Passos sought, in his own phrase, ground to stand on, but, happily, a traveler like himself could not stay anywhere for long, and his last trip outside the United States took him very near the ends of the earth to appease a lifelong curiosity over the monolithic heads on Easter Island.

Townsend Ludington

Author of John Dos Passos: A Twentieth-Century Odyssey, *editor of* The Fourteenth Chronicle: Letters and Diaries of John Dos Passos, *and coeditor of the Library of America edition of the* U.S.A. *trilogy*

In 1974, the year after the publication of *The Fourteenth Chronicle: Letters and Diaries of John Dos Passos,* I was beginning to gather the additional materials I sensed I needed to write a full-fledged biography of the author. I knew of his friendship with Germaine Lucas-Championniere, a French woman whose brother, a doctor, he had met while driving ambulances in the Verdun sector in 1917. In 1919 he met her in Paris through Tom Cope, an ambulance driver whose friendship he had made in 1917 when he first arrived in France as part of the Norton-Harjes Ambulance Corps. She and Dos Passos quickly became close friends. He visited her at her family's apartment on the Rue de Clichy, and together they "did" the culture of Paris. "I was mad for music," she told me in 1974. Among the many musical events they attended were performances of Debussy's opera *Pelleas and Melisande;* she recalled that she had seen it more than sixty times herself.

While editing *The Fourteenth Chronicle* I had accepted Dos Passos's assumption that she and her family were somehow victims of World War II. He had tried to look her up while he was in Europe in 1946 but had not been able to locate her. I made the mistake of not investigating further. Then in 1974, knowing I would soon be in Paris, on impulse I took a look in the Paris telephone directory in the Wilson Library at the University of North Carolina at Chapel Hill and found her listed under the *L*s as living near Sacre Coeur, not particularly near the Rue de Clichy. If Dos Passos had inquired at her old address, he would have found no clues as to her whereabouts.

Without delay I wrote her and quickly received an answer. She invited my wife and me to dinner on the last evening we would be in Paris. When we arrived, we were greeted by a charming, elderly woman who sparkled with intelligence and wit—"une vielle demoiselle," or "a woman of a certain age," she joked about herself. The talk back and forth flowed fast as she tried to fill herself in about

Dos Passos in the years between her last correspondence with him in 1929 and his death in 1970.

Unceremoniously she brought out the letters and postcards he had written her in French until the end of the 1920s, 148 of them, if my count was correct, and certainly one of the major collections of his letters. She also showed us several chests of drawers filled with sketches and watercolors that he had made–my notes from that moment read, "Dos's paintings–particularly of NYC–cubist and expressionist"–and left with her during his many passes through Paris or through her family's vacation home in Brains, near Nantes.

She allowed me to take the letters and cards back to where we were staying that night. Some of them, of course, were brief, most notable for dates or some such matter, but others were full, telling letters in which Dos Passos revealed–amid his brand of humor, cynicism, romanticism, and frequent hyperbole–the themes and sometimes the techniques that were on his mind. After making notes and copying portions of as many letters as my wife and I could throughout the night, I took the metro early the next morning back to Sacre Coeur and returned the papers. I had not yet copied six of what seemed to me the fullest letters and said as much to Mlle. Lucas-Championniere. "Ah," she replied with a generosity I judged to be typical of her, "but what are six among 148? Take them with you." Which I did, and then I used their information in the biography I wrote. I present them here in my translation. [A sad note from a scholar's point of view: After Mlle. Lucas-Championniere's death I returned to France, visited her relatives who had inherited her possessions, and tried to convince them that the best use for the letters would be to keep the originals but send copies to add to the major collection of Dos Passos's papers at the University of Virginia at Charlottesville. (I pointed out that, while the relatives owned the physical letters, the contents were the property of the Dos Passos estate.) They sensed that there might be money to be made somehow, and my suggestions were for naught. As for the sketches and drawings, not knowing what they were, the relatives threw them out during a general house cleaning.]

The six letters presented below reflect well Dos Passos's ideas as he entered into the early and–many would assert–the most literarily successful portion of his long career. The first letter, about bullfighting, reflects his war weariness and scorn for the very thing that exhilarated his contemporary and sometime friend Ernest Hemingway. The next letters reveal his thoughts about cities, mass society, and American materialism, while the last shows his fascination with "the Russian experiment" in the wake of the political executions of Sacco and Vanzetti and reveals his interest in the new cinema, whose techniques played an important role in the crafting of his trilogy *U.S.A.*

The letters:

I

    Credit Lyonnais
    Alcala 8
    Madrid
    1–IX–19

Mademoiselle

The manner of a bullfight is the following.

There are moments in the life of a Spanish city when one notices that everyone follows a particular street, when everyone presses and jostles, that all the *landaus,* the *berlines,* the *coupets,* the *voitures de place,* the taxis, the limousines, the handcarts, the old women who sell the marvellously Greek jugs of water, the vendors of melons, of grapes, of fruit, the cats, the dogs and the pigeons, that the entire population goes in one direction. It is the hour for the bullfight. One enters, as in the last act of Carmen, sits down, and thinks about the struggles of the Roman gladiators, with disgust and scorn one looks at the reddish sand of the arenas, the wooden barricades painted red and yellow which protect the spectators, the barred gate through which enter the Christians, no, the bull.

One shrugs his shoulders. One is accustomed to all that. Then things begin.

Sound of a trumpet. Two men in black, holding themselves with great difficulty on handsome horses which prance in the crushing light of the amphitheater, enter, circle, and salute, with a magnificence spoiled a little by the difficulty they have holding themselves on their horses, before the royal box. Then the picadors and the matadors, the teams of mules that are going to pull off the corpses, enter in procession and salute the box empty of great personages. They are dressed in all colors, red, orange, purple, and the gold lace glimmers in the strong sun. Again the trumpets, and the bull enters, black, immense, leaping like the bulls that the caveman painted in the beginning of the world in the caves of Altamira. This is not a battle, it is a ritual, a sacrifice is carried out. One immolates a horse or two on the horns of the bull, and the disembowelled horses twist in grotesque attitudes on the sand reddened by the strong light. Then, with a superb gesture one puts the banderillas in the back of the already tired bull. Then there is a fierce cry of joy from the trumpets, the matador sets himself in the prescribed position and plunges the length of the sword into the thick and bloody neck of the bull. The bull hangs out his tongue from which frothy blood drips, and he turns his head from one side to the other in a bestial and fascinating manner like that of a little dog asking for some sugar. The matadors with their capes of red, green, and purple make a circle around him. He falls and rolls on the ground and becomes small and dirty on the sand of the arena. The teams of mules enter cheerfully and drag away the corpses to the sounds of bells and the snappings of whips. Another trumpet. The red, green and purple cloaks take their places, and it recommences.

It must have been a little like that when hundreds of bulls were sacrificed to the great gods at Knossos or Mycenae, or before the high walls of Ilion.

It is stupid, it is ugly, it is splendid—it is like a jumping contest or like the Russian ballet—But the nerves of the twentieth century, accustomed as they are to streams of blood spilling on the earth, find all that an interesting but disagreeable sensation.

I am enroute to Malaga.

The address at the head of this letter is permanent—

Au revoir
John R. Dos Passos

[All but several lines of this translation were published in *American Literature,* volume 60, May 1988, pp. 271-272.]

II

[Postmarked "London 17-6-20"]
Friday—London—day of rain and sky of lead.

It is today that you leave Paris. Bon Voyage. I would very much like to settle down somewhere in the country also. I live too much in cities. I forget the songs of the birds at dawn and the grand nights alone when one walks in the woods, in that murmuring silence of green things which are growing, of little animals which follow their paths running along without fear under the great protective shawl of the night—where life is ordered as are little people under the mantle of the dear Virgin.

And London is a little like cold beefsteak, solid but unattractive. I detest these long pale and washed out twilights in the suburbs of London, where each day ends in a small eternity of boredom, without color—deserted streets without life, tiny row houses of dull red which distort the gray sky as far as one can see with their incredibly similar roofs. [Shades of "The Love Song of J. Alfred Prufrock," which Dos Passos had certainly read] No, England is tedious, made by machines without having the fantastic and macabre atmosphere of my country.

I am writing in a small chilly room where a coal fire makes diminished noises like a sickly person. In front of the fire a little, fat and sickly dog is sleeping and snoring. A small clock whistles the half hours in a soft and frail voice. A petite old lady is reclining on a sofa— . . . And life is a very complicated ritual arranged by some little old ladies who have died among little old centuries. I am feeling, this afternoon, some of the desperate boredom of my childhood—I spent four years when I was very young in this corner of London—of these long pale afternoons where I had to remain in the house when my legs trembled with the desire to run, when I was overwhelmed with mad desires which tingled in my blood for reading or playing, and I thought of the life which I would one day live.

And now although I find myself in the promised land, is it worth yesterday's mad desires?

Sometimes I think that I have inside me one of those little gray rooms, furnished in the best taste, from which the true things of life, the sun and love and sweat and good work which overwhelms your arms—from where everything raw is excluded and where everything is able to enter only through literature, rules, forms,—Life must don a top hat and clean its feet before it can sit down in the little gray drawing room of the bourgeoisie.

And how many times have I destroyed it, this little gray room, but always—at the great moment, I find myself closed in there, and through the windows I watch great ponderous processions pass and disappear along the highway. God, if one could kill his grandfathers.

Write me—care of Morgan, Harjes—15 Place Vendome Paris

I hope that your mother is doing better—greet her for me

My miserable book [*One Man's Initiation—1917*], although all ready has still not seen the light.

Best to you
JRDP

[Portions translated in *John Dos Passos: A Twentieth Century Odyssey,* p. 196]

III

[Postmarked "Paris 27-7-20"]

In a cafe facing the Theatre de Cluny. Mild night, full of voices and of that despiriting premonition of autumn. A moon which hides and reappears constantly among the muslin clouds. A monotonous ragtime comes from a grinding orchestra. Do you know the story that Herodotus tells about King Mycerimus of Egypt? He consulted an oracle which told him he had only five years to live. With torches and banquets he mocked the gods by turning night into day. He no longer slept and when he died he had lived ten years. It is for me one of the most touching stories. The tales of St. Jean de Luz are full of marvels. At the moment I am lacking marvels. Yesterday when I went to bed a large brown cat with a twisted, Chinese appearance about it entered through my window. It looked at me with yellow, flaming eyes which had an incredible malice about them. Then it opened its mouth and let out a ferocious meow. Its topaz eyes became red as rubies, and with another cry of love or hate it fled into the night. Maybe it was a marvel. Let's hope so.

I have made a reservation on a ship which leaves from St. Nazaire for Mexico or Cuba on August 7. Maybe that will lead me toward some marvels. Like King Mycerimus of Egypt I want to double my life.

The fact is that your France which I love so much is too civilized for me at the moment. The unforeseen doesn't exist. Maybe it doesn't exist anywhere. French life is a beautiful ceremony where everything is accomplished according to a ritual established by ancient generations. Everything—for us other barbarians, men whose rituals are as yet incomplete—is indescribably gentle,—a person is like the lotus eaters. [In America] life plunges brutally, cruelly, toward new forms of organization. Our generation has put on the burning shirt of Nessus. It is a struggle to the death against vast mechanisms which are the slave systems of tomorrow. The battle will never end. It is lovely to pause under the fruit trees, to become intoxicated with the great slow rhythms of old cities, but the moment always comes when one can no longer resist the hot blood which always pushes us toward the battle, toward new paths.

Morning—all this seems very stupid. We should never reread our letters.

More marvels and the sound of the wind stirring the asphodels along the tops of the cliffs—All the best to you

J.R.D.P.

[Portions translated in *Odyssey,* pp. 196-197]

IV
23-9-20

I have never satisfactorily asked Mrs. Profit's pardon for having spilled the cream pitcher on her feet. Tell me if I am pardoned. I am a clumsy bear.

As to the ship sirens, I assure you that the more you listen to them, the more effect they will have on you. The sound of a far away siren, heard among the vast rocky aridity of cities, is the most moving sound I know. I ask myself why in New York one never hears the vibrant voice of sirens.

I am living in New York in a room that is furnished very grandly and heavily. It has ancient green wallpaper and a frightful red rug, a large puffed up sofa covered with a horribly bright velvet and a big round table deliciously covered with portfolios and notebooks. It is very pleasant to meet all my friends again after so many years. What is remarkable is that no one has changed. Everyone talks, eats, sings, writes, walks, criticizes absolutely the same as three years ago.—Only I myself seem to be more disoriented, more off the road, more wandering along ponderous paths in the valley of indecision than three years ago. I feel a little like I felt in my nightmares as a child, when I often dreamed that I was deep in the country and that everyone else was climbing into a carriage and that I was running from one side to the other without being able to find even a small place and that all the carriages were leaving at the sounds of ringing bells and that I was left alone. It is like that now. I have the illusion that all my old friends are climbing into one or another of the carriages which are following along in the great processions of life—and that I am left walking alone by chance along lost streets.

New York—after all—is magnificent—a city of cavedwellers with a frightful and brutal ugliness, full of thunderous noises of the groans of metal on metal, full of a never ending sound of wheels turning and turning on hard stones. People swarm meekly like ants along designated streets, crushed by the arrogant, pitiless things around them. I am reminded of Nineveh and Babylon, of Ur of the Chaldees, of the immense cities which hung like basilisks below the horizon in the ancient tales of the Jews, where temples rose as high as mountains, and people ran trembling through dirty little alleys to the constant sound of ships with hilts of gold. O for the sound of a brazen trumpet which, like the voice of the Baptist in the desert, will sing again about the immensity of man amid this nothingness of iron, steel, marble, and rock. Night time is especially marvelous and appalling, seen from the height of a Roof Garden, where women with raucous voices dance in an amber light, the blue-gray immensity of the city cut up by the enormous arabesques of electric billboards, when the streets where automobiles scurry about like cockroaches are lost in a golden powder, and when a pathetic little moon, pale and overwhelmed, looks at you across a leaden sky.

If I think back to the bell tower of Esconblac—I often regret that I did not sleep above the shrouded city. Maybe the dispossessed souls of ancient inhabitants would have whispered some rewarding dreams to me, or I would have had some bizarre, esoteric nightmares. The next time I will sleep at the foot of the belltower of Esconblac. But I am always so dazzled when I am offered some riches on the plate of life that I can't ever decide to take them until the last moment, when it is too late. I would have liked very much to have seen with you the Norman chiefs imprisoned by the brave Saint of the city of Guerande.

Another time I will tell you about the great melodrama of the deceits of editors and the machinations of literary agencies, adding melancholic-comic-historic-philosophic reflections about the G.A.B. (Great American Bluff) [a reference to the difficulties he was having to publish *Three Soldiers*].

All the best to you

D. P.

Good things to your very charming mother—hello to your brother and to the small nephews

[Translated in part in *Odyssey*, pp. 200–201]

V

[Postmarked "West Palm Beach Apr 3 (1924)"]

Isn't this paper chic? [a reference to the stationery with the letterhead "Sebring Development Co." and a border of oranges on it] I have just finished an extraordinary trip on foot through central Florida. After seven strenuous yet delicious days I found myself felled by fatigue, my face like an apple cooked by the sun, my nose like a ripe strawberry, in the train station of Venice in the Everglades. Happily I found a freight train there; then I took a motor canoe across huge Lake Okeechobee and now I am waiting for the train to Palm Beach, the winter station of senators and millionaires, which will take me to Key West where I am going, driven by my chronic islomania—

All of this country is fabulous and cinematic. There are cities built in three months. Ten years ago the interior was unknown except to the crocodiles. Now the poor crocodiles are all penned up in Alligator Farms and one has to pay twenty-five cents to see them. Everyone is sure to be rich in five years, except for those who make mistakes—But no one speaks about that. One arrives on foot, works for a year, from one's earnings buys an orange grove, in five years one travels in a limousine, in ten years is the founder of a city, a millionaire, a senator—it is the American Eden. An Eden without a serpent, where the fruit is never forbidden, except to those who cannot buy it, and it

John Dos Passos in June 1939, just after his first published attacks on literary communism

brings no dangerous wisdom to whoever eats it. But how bored these poor Eves and Adams are without the serpent to tease them.

When I am resting I read the Bible, a book that frankly I find very disagreeable, but devilishly well written, for the most part. But the Church is correct. This is not a book to put in the hands of Christians. The fierce tragedy of the Jews' mad egotism is not an edifying work. Its influence will always be against civilization and happiness. This book stands up with the ghastly and frightening sublimity of a mountain of graves across every highway. But it is time for the train—the best to you

Dos

[Discussed in *Odyssey*, p. 231]

VI

[No postmark but datable as mid-August 1928]

I have a very small cabin on a large ship which is descending the Volga. It is magnificent. My cabin smells of little apples which I bought in a village, outside there is the scent of the forest and a little of the passengers. I pass the time studying Russian until my head splits and drinking tea with pastries. Then I try

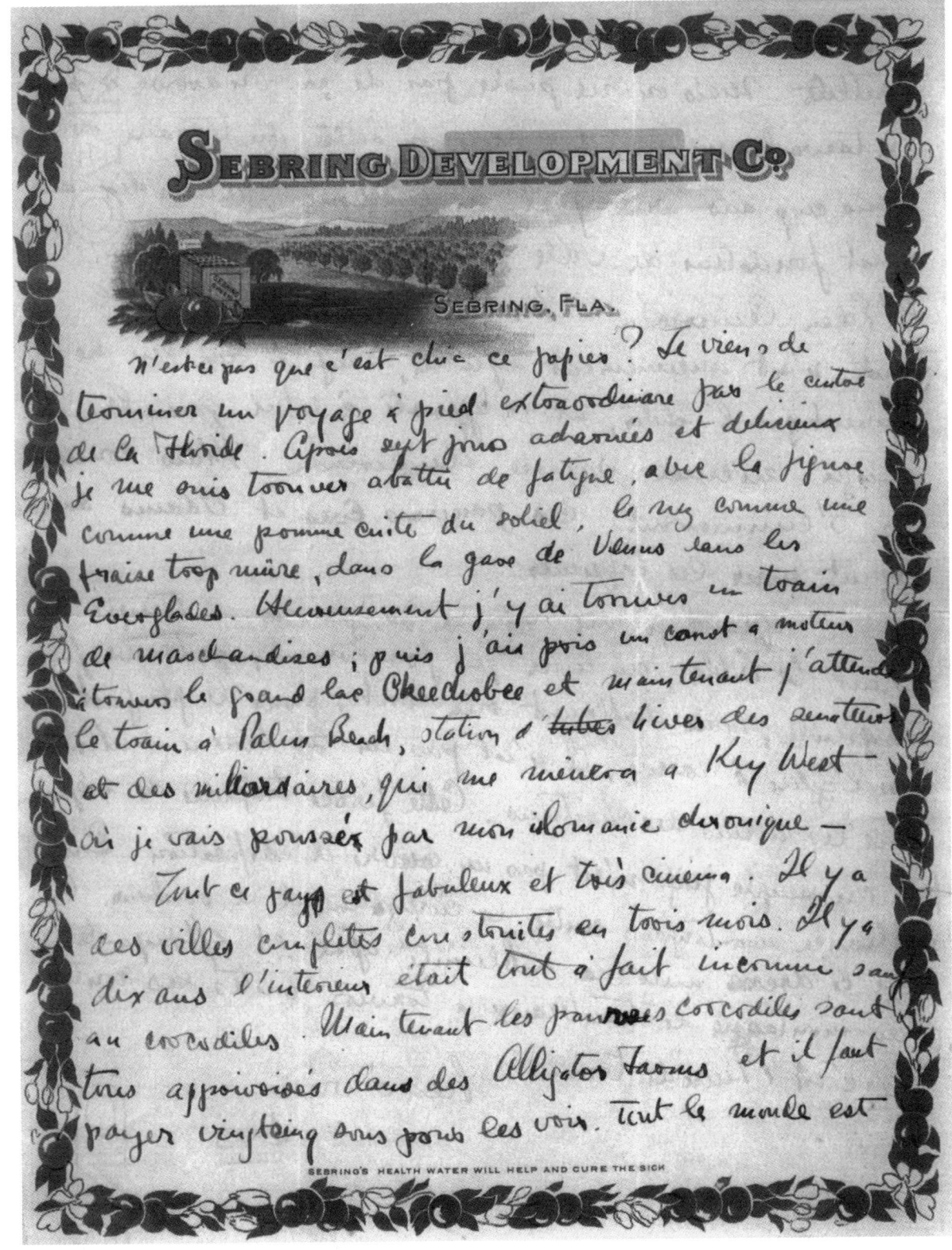

Dos Passos to Germaine Lucas-Championniere, 3 April 1924

to speak with the other passengers the Russian I have learned through the Hugo method, but they don't understand me at all. Then I take refuge in my cabin and again study the irregular verbs. I am in the midst of a trip to Astrakhan. Russia is truly magnificent. There is so much space everywhere, the rivers are immense, the buildings are immense, some facades measured in kilometers, everything has the crude and new sense of an American boomtown, even the antiquities. One eats and drinks a great deal and then has to fast a great deal. There is a lot of everything, people, conversation, fleas—and it is the country where foreigners are treated the best. Although no one has any money, you are treated with an extraordinary hospitality.

But life is so agreeable here, truly—you would not believe it—but everywhere there is a feeling of freedom and extraordinary energy.

I am writing seated at a small table on the bridge in the middle of gulping down a breakfast of tea and caviar 2 liters of tea, 2 liters of bread, 1 liter of butter, 1 liter of caviar. On the Volga there are large rafts made from big wood from the forest with small houses on top which slowly descend with the current as formerly on the Mississippi. Every day I bathe in the soft, brown water. It is the most pleasant trip I have taken in a long while. I feel happy, and my last foolish mistake is behind me. Lighting a cigarette I lit the box of matches, I don't know how, and I burned my nose and all my eyelashes. My nose looks like a cooked apple. Fortunately my eyes did not suffer—I see as badly as ever.

It is terrible that Mme. Bibi [Germaine's mother] broke her arm. And that Pamplona did not work out for you. I hope that now everything is fine and that you are swimming at St. Jean de Luz. But I fear that the healing will be slow. Poor Mme. Bibi has suffered a good deal because breaking a bone is not amusing. Please give her my best greetings.

I have not seen much theater although I have seen one piece by Meyerhold—magnificent setting—But I have seen a lot of film. The historical pieces about the revolution are especially good. The young assistant directors of the films here are the most interesting people I have met. Naturally they all say that theater is dead and that they adore photographic equipment as if it were a god—but they are full of energy and imagination. Also what is interesting are the popular theaters—in the workers' and union clubs where they make all sorts of half-improvised plays—under the names of Living Newspapers—They are very amusing, even the propaganda pieces for hygiene, etc. and all well acted by the people themselves.

Truly I am very happy to be here.

I am going to take a big walking trip in the Caucasus and then I will return to Moscow.

The best to you and all sorts of sympathies to Mme. Bibi. If she gets well fast I will bring her something pretty.

Dos
[Discussed in *Odyssey*, p. 270]

Lisa Nanney
Scholar

John Dos Passos wrote in his memoir *The Best Times* (1966) that his first awareness of modernist painting, when he visited the 1913 Armory Show in Boston, was a "jolt" that had profound effects on the work he was visualizing even as a student at Harvard. Likewise, my discovery that Dos Passos was a painter as well and my introduction to his sketches, pastels, watercolors, and gouaches brought a "jolt" of awareness to me as a graduate student just beginning to explore the possibility of writing a dissertation on Dos Passos as a modernist. When Dos Passos biographer Townsend Ludington showed his slides of some of the visual works in the course of a 1985 graduate seminar at the University of North Carolina, they revealed something of the writer's skill as a visual artist and hinted at the range of the techniques he tried. But the full scope and significance of his visual sense became clear when Elizabeth Dos Passos graciously allowed me to visit her at Spence's Point to research the original paintings. There, in the scores of compositions she showed me, were the visual equivalents of his evolution toward the innovative modernism of his greatest works: the pastel-washed recollections of his early travels; the stark, vivid impressions of wartime France swiftly sketched in cafes and in camp; the bright expressionist images of the motion and color of the city; the cubist evocations of New York harbor and the streets of Manhattan. His painting establishes that he understood as a practicing painter the dynamics of visual art that helped him create the formal innovations of his modernist novels.

Barry Maine
Scholar

John Dos Passos should be remembered (and taught) as a pioneer in American modernism, for his experiments in prose fiction in the 1920s and 1930s parted company with the emerging cult of personality (for example, Ernest Hemingway, F. Scott Fitzgerald, and William Faulkner) by employing expressionistic narrative devices in the service of a collective so-

Sketches by John Dos Passos, circa 1918–1925

cial history. He should also be remembered as the American writer most closely associated with the political left during that same period, walking picket lines with striking workers, helping to launch the *New Masses,* working and going to jail for the Sacco and Vanzetti Defense Committee, traveling to the Soviet Union to witness communist government in action (and returning more disillusioned than inspired), and writing novels that indicted monopoly capitalism for interfering with democratic ideals. But I hope he will be remembered (and read) as a writer passionately committed to ideals of social justice; for writing a trilogy of novels (*U.S.A.*) intended to evoke alarm and suspicion over assembly lines and bread lines, obscene profits resulting from speculation rather than disciplined labor, a mass media driven by yellow journalism, political rhetoric divorced from all connection to actual conditions, an entertainment industry feeding off the empty desire for glamour, growing economic inequality, special privilege, bigotry, political oppression, and intolerance; and for writing fiction with the power to move us from contemplation to action, the power to make readers want to work to make their country a better place to live. That is Dos Passos's legacy to American fiction as I see it.

Letter: Dos Passos to Frederick William Lowe Jr.
1948

The following letter from the Patee Library at Pennsylvania State University, provided courtesy of Charles W. Mann, chief of Rare Books and Special Collections, is a forthright statement of Dos Passos's political convictions a decade after his first denouncement of communism. The letter from Frederick William Lowe Jr. that prompted Dos Passos's response has been lost, but the gist of it can be inferred. In 1948 Lowe was a second-year instructor at Colgate University. He received his Ph.D. in American literature from Columbia University in 1957, having written a dissertation on Gertrude Stein.

Appendix A: John Dos Passos' letter to Frederick W. Lowe Jr. pages 1-2.

To Lloyd Lowndes
Palisades.

August 27 1948

Dear Mr Lowe,

As all the writing I have done since the particular period that seems to interest you most has been an effort to reorient my original boyish and possibly fanatical enthusiam for freedom and growth as against slavery and death, there is nothing much that I can add in a letter.

It is only because your letter seems to me to betray all the symptoms of the numb stupidity that has paralysed the thinking of so many people who hold the now fashionable "liberal viewpoint", and because I am very much afraid that in your position as an instructor you are infecting your pupils with that disease, that I am answering it at all. Any body of doctrines and prejudices that is not constantly checked against realities becomes numbing and debilitating. In the case of literate Americans this uncritical and incurious liberalism has grown to a point where it has become a real danger to our existence as a nation. If the United States goes under in the turmoil of this period of wars the whole cause of freedom is lost in the world. I just want you to sit down and ask yourself whether your reaction to the pieces you mention are the result of close thinking based on experience or whether they aren't the results of a purely emotional and easy acceptance of fashionable attitudes.

I havent found it so easy to stir the liberals of today who find it so easy to be stirred by the Sacco-Vanzetti case long after the fact to do anything to help the millions of Saccos and Vanzettis who were turned over by our liberal President and his liberal advisers to the slave gangs and torture chambers of the new despotism in Eastern Europe.

The cause of the new Saccos and Vanzettis is not fashionable; well neither was the cause of the two Italian anarchists when a few honest men stood up for them.

I write for Life because I find myself in sympathy with a good proportion of the editorial ideas behind it. It furnishes a large mass of thoroughly miscellaneous readers and I'm very grateful for the chance to set up an occasional pulpit in its clumns.

In my journalistic work as in my novels I try to state the truth as I see it. It seems to me that various sorts of socialism are setting up in the world a new form of exploitation of man by man, more ruthless and degrading to the human spirit than any of the old tyrannies. The only hope lies in finding some way of grafting into them some of our traditional methods of self-government and personal initiative. It seems to me that the most

Appendix A, cont.   6

[handwritten text, largely illegible: "...like it, you are to ...keep ... eyes open ... compose the rest of lives ..."]

2.

  You may think otherwise XX but you have no right to
apply to my writings or any other man's that peculiar smearing tone
for which the Communists have set the style , that impugns a man's
motives because you dont agree with his statements .

  Finally , I cant help saying something that I've wanted
to say to you college professors who teach contempory literature
every since I started writing . It seems to me that your business
is to teach the classics and to inculcate the standards of
civilized life . I can see how occasionally, in a composition
course, some good piece of current writing would be useful as an
example , but I am sure that the practice of teaching contemporary
novels is vicious and futile . Why cant you let the young people
read 'em and make up their own minds? And then the cream of the jest
appears when you get mad at a man because he wont stay in the coop
your professors constructed for him.

  My work hasn't any place on that shelf of yours, Mr Lowe,
until I'm dead and decently buried .

      Very sincerely yours ,

        John Dos Passos

Book Reviewing and the Literary Scene

George Garrett
University of Virginia

In keeping up with what is happening in the literary scene, one turns, naturally and regularly, to *Publishers Weekly* and to the major newspapers that follow the publishing world as a matter of business news—the *Wall Street Journal, The New York Times, The New York Observer* (excellent for literary gossip and "inside" stories), *New York Magazine,* and the *Washington Post,* especially the feature articles of David Streitfeld. Equally important to anybody trying to keep abreast of news and trends and covering all aspects of the literary scene is Campbell Geeslin's column, "Along Publishers Row," ongoing in the *Authors Guild Bulletin*. Little that is worth knowing about escapes Geeslin's scrutiny.

Here are some things, filtered through all these sources and others, that shaped the literary news and 1996. Start with money, dollars and cents. It may prove to have been a slow year for literary books of all kinds. The numbers are not in yet, and it has already been noted that the rate of returns from bookstores to publishers was unusually high: "For some titles, discarded books are spewing back to publishers at rates as high as forty percent of gross sales, a sobering trend that comes during a sluggish summer season where certain titles have piled up in Manhattan bookstore aisles in untouched stacks the size of large dogs" (Doreen Carvajal, "Many, Many Unhappy Returns: Publishers Are Awash in Unsold Books," *New York Times*, 1 August).

Nevertheless, the major publishers were busily offering large advances for potential blockbusters. Former publisher, now agent Joni Evans sold the rights to music man Quincy Jones's autobiography to Doubleday for $1.25 million. Dick Morris, former consultant to President Bill Clinton, received something in excess of $2 million for a book contract with Random House. According to *Newsweek* (22 July) Patricia Cornwell received $24 million–$27 million; Tom Clancey, $60 million; Danielle Steele, $60 million; Stephen King, $40 million; Dean Koontz, $25 million for multiple-book contracts, placing them (almost) in the category of successful slam-dunkers and rap-singing hip-hoppers.

Some literary authors earned a turn at the money trough as well—Salman Rushdie changed publishers and received $2 million for his next novel from Holt. Don DeLillo took home an advance from Scribners of $1 million for a novel of the Cold War called *Underworld*. George Stephanopoulos, former presidential aide-de-camp, earned a $3 million advance from Little, Brown. Ballantine coughed up $750,000 for a gay men's reference book—*Gaycare*. Warner Books paid $1 million to Nicholas Sparks for *The Notebook*. David Balducci got a $2 million advance for his new legal thriller, *Absolute Power* (Warner), and Random House paid Colin Powell $6 million.

Sometimes the dollars-and-cents news was all too typical. *Publishers Weekly* ("Ron Goldman's Family Signs With Morrow," 11 March) reported the advance to be $450,000. "Names & Faces" in the *Washington Post* (18 May) informed the public that a former reporter for that paper, Janet Cooke—who won a journalism Pulitzer Prize, then lost it when the article she wrote for the *Post* was discovered to be bogus—was paid a $700,000 advance against $1.5 million when and if a film about her, "Janet's World," goes into production. The rush of publicity concerning this event began with a full-scale piece in *GQ*.

Sometimes the news was of minimal interest to almost everyone, as in the same issue of *Publishers Weekly*—"Four Walls Lands Lish: Controversial Literary Figure Signs Unique 10-Book Deal with Small Publisher." No money was mentioned. On a somewhat larger scale than any of the above, American publishers claimed that during the year illegal copying of all kinds, worldwide, was costing them an estimated $352 million. China alone accounted for $125 million of this figure.

Sometimes literary writers earned money from a variety of prizes. David Malouf received $150,000 from the Dublin Literary Award, given for his book *Remembering Babylon* (Pantheon, 1995). Story writer Andre Dubus picked up $30,000 from the annual

Rea Award for his seventh and most recent collection of short stories, *Dancing After Hours* (Knopf). Meantime we learned how much Oprah Winfrey (who earns $171 million a year for her television talk show) means to American publishing. Mention and attention on her show put titles on the bestseller lists (David Streitfeld, "On Oprah: People Who Read," *Washington Post,* 26 September). There is a specific case in point: Oprah formed a book club for her television show, and one of the first results was instantly to send the first novel, Jacquelyn Mitchard's *The Deep End of the Ocean* (Viking), to the *Times* best-seller list (Paul D. Colford, "Praise from Oprah Propels Novels off the Shelves, onto Best-Seller Lists," *Newsday,* 10 November; see also Daisy Maryles, "Behind the Best-sellers," *Publishers Weekly,* 11 November, which reports: "All four books heading up *PW*'s bestsellers lists this week with a little help from Oprah").

More and more last year, television and even radio were being seen as an excellent opportunity to market books. Radio talk-show host Don Imus increased sales of Jane Mendelsohn's first novel, *I Was Amelia Earhart* (Knopf), tenfold, moving the book, briefly, to number eight on the *New York Times Book Review* best-seller list. Charlie Rose on PBS and Brian Lamb's "Book Notes" on C-Span are believed to be increasingly influential; and to a lesser extent, but important for literary books, both the program *Fresh Air* and Alan Cheuse's reviews on *All Things Considered* for National Public Radio are highly desirable to publishers—more than most printed book reviews.

It was a year of small changes and slight realignments. John Grisham's agent, Jay Garon, died and was soon replaced by David Gernet, Grisham's former editor at Doubleday. Saul Bellow became a client of agent Andrew "The Jackal" Wylie. In May Sonja Bolle stepped down as book editor of the *Los Angeles Times,* temporarily followed by acting book editor Claudia Luther, and finally replaced by Steve Wasserman, formerly of Random House. Larry Dark was named by Anchor Books to replace William Abrahams as editor for *Prize Stories: The O. Henry Awards.* Abrahams had edited the anthology for thirty years. Dark's announced goal was to be more competitive with Houghton Mifflin's series of prize stories and essays. Jumping shop from Penguin, literary star Paul Auster signed a three-book contract with Holt for two forthcoming novels (*Timbuktu* and *Dream Days at the Hotel Existence*) and a nonfiction memoir—*Hand to Mouth.* The Book-of-the-Month Club, whose use of literary judges was dropped in 1994, replaced its editor in chief, Tracy Brown, with the former marketing director, Richard F. Schnable (Doreen Carvajal, "Triumph of the Bottom Line," *The New York Times,* 1 April). Andrew Wylie ended his association with the agency Aitken, Stone, and Wylie. (For a profile of "The Jackal," see Frank Bruni, "The Literary Agent as Zelig," *New York Times Magazine,* 11 August.)

Eugene Walter, novelist, poet, and actor, received the Fine Arts Alabama Award for lifetime achievement from the University of Alabama. Short-story writer Tim Gautreaux was appointed Southern Writer-in-Residence at the University of Mississippi. The president of *Village Voice* announced that their literary supplement, *VLS,* will hereafter become a quarterly publication instead of appearing ten times a year. Critic Frank Lentricchia renounced and repented his former ways and means in "Last Will and Testament of an Ex-Literary Critic," announcing that he has given up a double and duplicitous way of life to become again a reader in love with literature: "In private I was tranquility personified; in public an actor in the endless strife and divisiveness of argument, the 'Dirty Harry' of literary theory, as one reviewer put it" (*ALSC Newsletter,* Fall).

Movie actor Ethan Hawke took eighteen months off from his acting career, with the help of a $300,000 advance, to write his first novel—*The Hottest State* (Little, Brown). He earned himself a lot of tomahawk chops, though Jeff Giles of *Newsweek* was kinder than most, calling it "endearing, if slight" ("Writing a New Chapter," 30 September). Holt editor Ray Roberts announced that the new Thomas Pynchon novel, *Mason & Dixon,* will be published in April 1997 in a first printing of two hundred thousand copies. It is set in colonial America and tells the story of Charles Mason and Jeremiah Dixon, British surveyors.

One little story that made a lot of news concerned the eighteen-year-old first novelist Jenn Crowell, whose forthcoming *Necessary Madness* (Putnam) is a huge success before publication: Book-of-the-Month Club, audio sales, film rights to Family Channel Pictures, foreign rights sold handsomely in England, Germany, Italy, Norway, Sweden, Finland, Denmark, and France. This book made waves in *The Hollywood Reporter* (17 June) and *Variety* (17 June), following a major piece in *The New York Times* (10 June)—Gayle Feldman, "Bet On a Young Author Pays Off in Foreign Sales." Crowell, still in high school and seventeen when she wrote the book, was a student of Madison Smartt Bell.

Like UFO sightings, there were any number of reported encounters with Salman Rushdie, especially in the Hamptons and in Key West. One of the rarest things in the world surely must be a *New York*

Contents page from the summer 1996 issue of Granta

Times front-page story dealing with poets and poetry. Robert Hass, poet laureate, found himself and his two-year battle against declining literacy on the front page of the 9 December *Times* (Francis X. Clines, "A Poet's Road Trip Along Main Street, U.S.A."). Of the experience of meeting with American businessmen in dozens of Rotary (and other) clubs across the nation, Hass allowed: "I had never been to a Rotary Club in my life, but now I've been to dozens. And you know, I had the prejudice that they're all Babbitts. But I discovered they're downtown business people who raise money for schools, most of them, and I made some friends." Another item, a long-standing question in literary history, seemed to have been settled once and for all when Dr. R. Michael Benitez persuasively argued in *The Maryland Medical Journal* (September) that Edgar Allan Poe died of rabies. ("Poe's Death is Rewritten as Case of Rabies, Not Telltale Alcohol," *New York Times,* 15 September).

Conflicts and controversies of various kinds made literary news. The case of Mumia Abu-Jamal, the convicted murderer appealing his conviction, and the efforts of a host of celebrities, including prominent writers such as Norman Mailer, Alice Walker, William Styron, and Maya Angelou, to help him gain his freedom continued. (Francis X. Clines, "The Case That Brought Back Radical Chic," *New York Times,* 13 August). When the anonymous author of the best-selling novel *Primary Colors* (Random House) was exposed (by a handwriting expert hired by the *Washington Post*) as Joe Klein, the *Newsweek* columnist and CBS commentator, much journalistic excitement followed, intensified by the fact that Klein had more than once flatly denied the truth. Media critic James Bowman in "Joe Klein's Tangled Web" (*The New Criterion,* September) summed up the situation: "Greed, envy and self-importance jostled for the upper hand as professional wordsmiths sought to find ever more grand-sounding rationales for their indignation at the fact that Klein had lied to them." Also worth seeing, though both quickly became irrelevant, are Bruce Weber, "Behind Every Great Anonymous Writer . . . ," *New York Times,* 21 March, a profile of the editor of *Primary Colors,* Daniel Menaker; and "No, Really, I *Am* Anonymous," by "Anonymous," *New York Times Book Review,* 19 May. See also Philip Weiss, "Joe Klein Doesn't Know He's Lying; Unmasked, Anonymous Comes Unglued," *New York Observer,* 29 July.

After acknowledging that they had somehow failed to print the epilogue in half a million copies of the paperback edition of Carl Hiaasen's *Stormy Weather,* Warner Books announced that they were ready to E-mail the four-page text to anyone who requested it. The nominations for the National Book Award in fiction, announced in October, displeased many critics and reviewers. David Streitfeld wrote, "The most newsworthy thing about the nominees for the National Book Award for fiction, announced yesterday, was what was missing, *Infinite Jest,* the gargantuan novel by *Wunderkind* David Foster Wallace that received some of the best reviews of this or any other year" ("Fiction Awards Overlook the Obvious: Nominations Exclude Critics' Darling," *Washington Post,* 3 October).

A battle that is not over yet began with the news, on page 1 of the *The New York Times* (18 May), that the distinguished physicist Alan Sokal had written a parodic, satiric article, "Transgressing the Boundaries: Toward a Transformative Hermeneutics of Quantum Gravity," richly laced with the tongue-twisting, lip-curling jargon of the die-hard deconstructionists and professors of cultural studies, and had submitted it to the unwitting (read: witless) editors of *Social Text,* who fell for the gag and proudly published it (Janny Scott, "Postmodern Gravity Deconstructed, Slyly").

The fifty-fourth issue of *Granta* (17 June) included a list of "The Best Young American Writers," vetted by fifteen regional judges and five final judges, and created something of a storm over who was in and who was out. The *New York Observer* ("Beefcake Granta!," 10 June) listed "Those Who Shouldn't Be There, But Are For Reasons We Will Try To Divine," a list that included Sherman Alexie Edwidge Danticat, Allen Kurzweil, Stewart O'Nan, Melanie Rae Thon, Madison Smartt Bell, Ethan Canin, and Kate Wheeler. Summing up the whole list the (anonymous) author opined: "What we've ended up with, we Americans, is gobs of family history and authentic feeling: a welcome-to-my-world (and that's all) collection that does justice to youth's immaturity and lack of wisdom, but not its energy."

Faced with the hard facts that black readers are now buying 160 million books a year and that only 3.4 percent of professionals in the publishing business are black, prominent black writers have begun to push for a change, demanding more executive and editorial positions for blacks ("An Emerging Prominence for Blacks in Publishing," by Doreen Carvajal, *New York Times*, 24 June). Literary battles over books about Whitewater and the morals and ethics of our First Family offered more public tomahawk chops than *The Last of the Mohicans*. Reviewing James B. Stewart's best-selling *Blood Sport* (Simon and Schuster) in *Harper's* ("To Make a Cow Laugh," July), Clinton loyalist Gene Lyons teed off on the author: "So profound and willful is Stewart's ignorance of all things that the only interesting question presented by his book is how so transparently a childish tale could be received with grave solemnity by the national press. What it shows, I think, is the enormous power of the trope 'Once upon a time'; also the care with which Steward has bestowed his flatteries on every reporter in Washington with a career stake in Whitewater." Turnabout is fair play, and Phil Gailey counterpunched in his review of *Fools for Scandal*, by Gene Lyons and the editors of *Harper's* ("A Conspiracy So Vast," *New York Times Book Review*, 4 August): "This is a nasty book, not because it challenges the reporting of the Whitewater story but because it assaults the integrity of the journalists who did the reporting. It is a smear job unworthy of any fair-minded critic of the press."

Another source of much print and journalistic hijinks was the civil suit between Random House and Joan Collins over whether or not they owed her the contractual advance on work they rejected. Under the specific circumstances the jury sided with Collins: "In a vote for verbal quantity over literary quality, a Manhattan jury yesterday said that Random House has to compensate the actress-author Joan Collins for a manuscript that its editors have already declared to be unreadable and unpublishable" (Jan Hoffman, "A Jury Has the Last Word, And It Favors Joan Collins," *New York Times*, 14 February). Gaining more attention than most novels and the source of much gossip and speculation was actress Claire Bloom's *Leaving a Doll's House: A Memoir* (Little, Brown), which painted a less than flattering portrait of her most-recent husband, novelist Philip Roth.

Near the year's end the newspapers and tabloids carried the mysterious story of the death by hanging (deemed most likely a suicide) of Chicago writer Eugene (Guy) Izzi, author of thrillers under his own name and various pseudonyms. The possibility of homicide is, at this writing, still being seriously investigated by police. Among the deaths of writers and the usual jostling battle for obituary place and space was one of special interest (*New York Times*, 21 August): "Geoffrey Dearmer, a poet of such abiding modesty that he had to live to a hundred to see himself lionized as one of England's poetic voices of World War I, died on Sunday at his home at Birchington-on-Sea, on the Kentish coast. He was 103." Speaking of living to a hundred, on 6 October the *New York Times Book Review* published a special issue, "100 Years," with book reviews from 1896 to 1996.

The book reviewing scene changed little from 1995. There seemed to be less attention paid to fiction, and highly favorable reviews seemed fewer than usual in the newspapers, magazines, and quarterlies. If there was no noticeable improvement in the space and the attention allocated to book reviewing, there was no discernible decline, either.

If reviews of literary books, especially fiction, were fewer and slower than usual, the predominant attention being lavished on a wide range of nonfiction, literary journalism, in the form of profiles, interviews, or topic- and issue-oriented pieces, was booming. In a sense many of these pieces replaced or, at any rate, confirmed book reviews. Probably the best all-around literary journalist at this time is David Streitfeld of *The Washington Post*. One need not always agree with his judgments or accept his sense of values, but he does an excellent, perhaps unique, job in keeping up with people, events, and trends for the *Post* and for the Sunday *Washington Post Book World*. What may have been his finest hour in literary journalism came with the discovery that Joe Klein was the author of *Primary Colors*. Streitfeld's story and byline ("Newsweek Writer Anonymous No More") moved up from its regular spot in the "Style" section to page 1 of the *Washington Post* (18 July): "Klein insisted that his motive in trying to remain anonymous had nothing to do with boosting sales of his book. Rather, he said, he feared reviewers would judge a book by Joe Klein on his reputation as a journalist—not on its literary merit. Also, he said that as a first-time novelist, he did not want to be embarrassed if he bombed as a fiction writer."

OUTSTANDING LITERARY JOURNALISM

*Madison Smartt Bell, "Unconscious Mind: The Art & Soul of Fiction," *AWP Chronicle* (May/Summer): "The lack of fixity, the flux of most creative writing classes, permits at least some kind of freedom—but freedom can be a spooky thing to handle. Precisely because the methods of instruction tend to be in a process of constant metamorphosis, it's important for the student to understand what the process is and where it may lead. Let the student beware, or at least, be aware."

*Michael Blowen, "Translating the Past into Present," *The Boston Globe* (4 August). Here is his profile of poet Richard Wilbur: "At the entrance to his study, there is an assortment of posters from the Wilbur-Molière productions. While the poet is not given to harsh verdicts on Molière interpretations, he does mention that he's not fond of directors who need the addition of codpieces to point out certain themes in 'The Misanthrope.' He also says that he loves Brian Bedford's work at the Stratford Theater in Canada. ('He's the best Molierest in North America.')"

*Frank Bruni, "The Literary Agent as Zelig," *New York Times Magazine* (11 August). Bruni presents this profile of literary agent Andrew Wylie: "But it is not enough. It is never enough. At the age of 48, Wylie is intent not on slackening but on maintaining his pace—on proving that the flamboyant sprinter who turned so many heads in publishing circles over the past decade can go the distance. He feels the eyes of his peers upon him once again, riveted by the recent unraveling of his trans-Atlantic partnership with two London agents, Gillon Aitken and Brian Stone. He is suddenly alone in the world."

*Arthur Crystal, "Closing the Books: A Once Devoted Reader Arrives at the End of the Story," *Harper's* (March): "Most readers, however, are afraid to question the experts—afraid of appearing too dense to appreciate a poem by James Merrill or Jorie Graham.... I don't know what's worse: the philistine who will not make the attempt to grasp the unfamiliar, or the half-smart, half-literary reader who, because *The New Yorker* or *The New York Review of Books* ... runs high-sounding critical essays, thinks that John Ashbery and Rita Dove are poets of the first rank."

*Jonathan Franzen, "Perchance to Dream," *Harper's* (April). Franzen's long article deals with the intense difficulty of being a novelist in this age and cultural climate. This one had a lot of people talking: "In his lifetime, Melville made about $10,500 from his books. Even today, he can't catch a break. On its first printing, the title page of the second Library of America volume of Melville's collected works bore the name, in 24-point display type, *Herman Meville*."

*Jennifer Howard, "Fred Chappell: From the Mountains to the Mainstream," *Publishers Weekly* (30 September). This interview states: "For all his publications and honors—including a Rockefeller Grant and the Best Foreign Novel Prize from the French Academy (for *Dagon*), Chappell is still unfairly considered by many to be a regional writer . . . 'I think I have always worked in relative obscurity,' he admits, 'and I've come to enjoy that. There's a lot of freedom in that. I always feel, when I sit down to write, the only person I really have to worry about failing is myself.' "

*Celia McGee, "The Incredible Dumbness of White Publishing," *New York Observer* (22 July). McGee presents a reaction and follow-up to a piece by Henry Louis Gates Jr. in *The New Yorker* (17 June) about the fact that book critic Anatole Broyard was an African American who "passed" for white: "The future of Broyard's literary reputation remains to be seen. Like many black literary observers, the critic and commentator Stanley Crouch finds that the posthumous attention paid Broyard by *The New Yorker* has mistakenly inflated his intellectual stature. 'It's not like discovering that Edmund Wilson was black,' he said, 'and black people further blaming whites for not knowing Broyard was black shows me that ethnic stupidity in this country is in a photo finish.'"

*Joan Mellen, "Academic Literary Theory is Killing Literature," *Baltimore Sun* (1 December): "You must use the word 'text' when you mean 'novel' because novel implies that a person, a writer with original thematic purposes, wrote that book. In place of discovering what the author had in mind, we are given—reader response theory' which says that a book means what a given reader at a particular time says it does. With the author declared dead or sentenced to death, who is to say otherwise? Anything goes."

*Mark Miller and Katrine Ames, "A League of Her Own," *Newsweek* (22 July). This piece profiles Patricia Cornwell: "'My personal life is not anybody else's business,' she said last week, sitting calmly in a Manhattan hotel suite, wearing jeans, a suede jacket, a 'Cause of Death' T-shirt and a cross. 'I don't believe people should be defined by their sexuality,' says Cornwell. 'People can think what they want. There's nothing I can do.'"

*Nick Paumgarten, "Young and Out for Literary Blood, The Baby Binkys Mean Business," *New York Observer* (22 April). This discussion of the new young agents comes at a time "when the agent has eclipsed the editor as the author's main crutch in the perilous, bottom-line-driven publishing world." Briefly profiled are Leigh Feldman, Sarah and David Chalfant, Sloan Harris, Nicole Aragi, Eric Simonoff, Kim Witherspoon, Heather Schroder, Lydia Wills, Laurie Liss, Theresa Park, Liz Ziemska, Laura Dail, and Jimmy Vines: "It's a competitive lot. There can only be so many Binkys and Esthers and only one Jackal. For now, on the record, these smooth-cheeked inheritors of big writers and big deals and big nicknames are being civil. It won't last."

*Salman Rushdie, "In Defense of the Novel, Yet Again," *The New Yorker: Fiction Issue* (24 June and 1 July): "The past fifty years have given us the oeuvres of–to name just a few–Albert Camus, Graham Greene, Doris Lessing, Samuel Beckett, Italo Calvino, Elsa Morante, Vladimir Nabokov, Günter Grass, Aleksandr Solzhenitsyn, Milan Kundera, Danilo Kis, Thomas Bernhard, and Marguerite Yourcenar. We can all make our own lists. If we include writers from beyond the frontiers of Europe, it becomes clear that the world has seldom seen so rich a crop of fine novelists living and working at the same time. . . . If V. S. Naipaul no longer wishes or is no longer able to write novels, it is our loss. But the art of the novel will undoubtedly survive without him."

*Peter Stevenson, "'Buffalo' Buford, Literary Gambler, Tries His Luck at Tina's *New Yorker*," *New York Observer* (15 January). Stevenson's piece is a profile-interview of editor Bill Buford: "Mr. Buford's work style has been somewhat slow to adapt to *The New Yorker*'s humming. 'Bill can only operate with a gun to his head,' said literary agent Amanda Urban. 'He's the most exasperating editor in the world to deal with, but also the most enlightened. Writers end up shaking their heads, saying, How can I love and hate this guy at the same time?'"

*David Streitfeld, "Authors Write of Passage: National Book Award Opens a New Chapter in Their Lives," *Washington Post* (23 November): "With the possible exception of the Pulitzer Prize, the NBA is the single most important award for American fiction. Given the right circumstances, it has genuine starmaking power–the literary equivalent of plucking the ingenue from the counter of Schwab's drugstore and casting her in a featured role in a big time movie."

*David Streitfeld, "The Latest Line on Pynchon," *Washington Post* (21 October). Streitfeld reacts to an announcement that a new novel by Pynchon, *Mason & Dixon* (Holt), will be published in April 1997: "Holt is planning for a first printing in the neighborhood of 200,000 copies, which means it believes the work will be a No. 1 best-seller. Even if the final figure is somewhat lower, the 59-year-old Pynchon is one of the very few members of the '60s avant-garde who continues to generate anything close to mass excitement. Kurt Vonnegut and William Burroughs are more or less in retirement, while John Barth and Joseph Heller no longer burn up the sales charts."

*David Streitfeld, "'Ship Fever' Wins Fiction Book Award," *Washington Post* (8 November). This is a report on National Book Awards for 1996: "Barrett, who has also published four novels, thanked the usual suspects in her acceptance speech: agent, editor, husband. More unusually, she acknowledged each of her fellow nominees and explained how much she liked their work. She also saluted the National Endowment for the Arts, which 'at a crucial time in my career made the writing of these stories possible at all, and is now in such danger.'"

*David Streitfeld, "Szymborska?: It Means Famous," *Washington Post* (4 October). Streitfeld profiles the Nobel Prize winner Wisława Szymborska: "At a luncheon yesterday at the Library of Congress for U.S. Poet Laureate Robert Hass, a number of poetry buffs confessed they either hadn't heard of her or hadn't read her. . . . Local bookstores had a single copy, at most, on hand yesterday; they were immediately gone. By noon Harcourt had orders for 12,000 more. 'For a Polish poet, that's not bad,' said Harcourt publicist Dori Weintraub."

*David Streitfeld, "A Writer Blocked," *Washington Post* (6 May). This is Streitfeld's profile of and interview with British author John Fowles: "Fowles, once so fertile he could draft a book in a month, has not published a novel since his illness. The stroke, he says on days he's feeling particularly grim, robbed him of his imagination."

*Egon Richard Tausch, "The Wonderful World of Porn," *Chronicles* (March): "Female career stereotypes live on in pornography; the publisher is not interested in the sexual exploits of over-sexed lady truck drivers or executives. If these books are a good indication of what men consider sexy, then feminism is doomed. If, on the other hand, women construction workers are as alluring to men as femi-

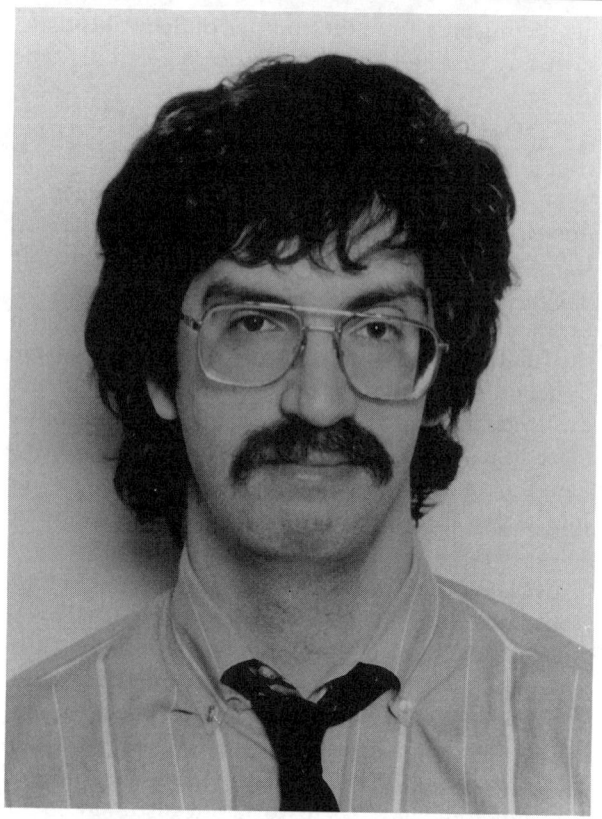

David Streitfeld, book reviewer for the Washington Post

nists say they should be, the word hasn't gotten to the porn publishers."

*Jay Tolson, "Art, War, and Shelby Foote," *DoubleTake* (Winter 1996): "Foote, like many writers, has always voiced discomfort with talk about the meaning of his novels. As he told his friend Walker Percy, with whom he engaged in a lifelong debate about the means and ends of art, his foremost goal was 'to teach people how to *see*. I want to impart a "quality of vision."' But if there is a recurrent theme of his work, it may be that life in the modern world is a gamble, a game of chance, a tournament, in which most people lose and are forced to adapt to defeat, sometimes in the most bizarre ways."

GOLD STAR REVIEWS AND TOMAHAWK CHOPS

One slight, though maybe significant change in 1996 is evidenced by the fact that "gold star" reviews (well-written and favorable reviews) were much harder to come by than during last year and earlier. "Tomahawk Chops" were plentiful and easy to find. Perhaps this situation is the result of the *New York Times Book Review*'s policy of being tougher on and more critical of new books, especially fiction.

Perhaps it follows from 1996 being a long, highly (at times wildly) rhetorical election year, a year of harsh words and clever put-downs. In any case there are far more tomahawk chops than usual, because there were many more to pick from. Note that among the gold star reviews the *Washington Times* magazine, *The World & I,* is represented several times, possibly because the reviews in *The World & I* are essay-reviews and there is usually no good reason or occasion in the context of that publication to mount sustained attacks. Their editorial policy seems to be highly selective, aiming for thorough reviews of good books. Of course, there are exceptions. See, for instance, Wayne C. Booth's review ("From a Roar to a Cough") of Leslie Fielder's *Tyranny of the Normal* (Godine) in the November issue of *The World & I:* "Most of what we find here does not probe very deeply into any one issue; rather we read reprinted talks on simple matters that, while engaging to him in 1973 or 1977 or 1980, now feel badly dated. Only one of the nine essays is dated after 1986, and even that one is full of examples that, though called recent, are discernible decades old."

GOLD STAR REVIEWS

*Adam Begley, "It's Gogol Meets Jim Carrey: A Tinseltown Tour in Hell," *New York Observer* (22 July). Review of *I'm Losing You* (Villard), by Bruce Wagner: "All Hollywood novels are about the vanity of human dreams. They get their juice from the juxtaposition of opulent wish fulfillment on the screen and ordinary despair in the dressing room. In this case substitute for ordinary despair abundantly creepy depravity. In the midst of fabled beauty, lots and lots of ugliness. Nothing ends well in *I'm Losing You*. Mr. Wagner won't allow even a milligram of saccharine."

*Doris Betts, "Growing Up on Wind Mountain," *The World & I* (January 1997). Review of *Farewell, I'm Bound to Leave You* (St. Martin's), by Fred Chappell: "If some of these mountain characters soar, if Chappell uplifts their down-the-road gossip to everything from tragedy to epic, he has come of age sufficiently to be unashamed of his passionate loyalties, his honest emotions, his lively guffaw. This cycle of folktales, musically written, though framed by the expectation of death on one end and by its bleak arrival on the other, swarms with the celebration of life."

*R. H. W. Dillard, "All for Love," *The World & I* (November). Review of *Belle du Seigneur* (Viking), by Albert Cohen: "To American readers it must surely be the most completely unknown major twentieth-century European novel, even to the most widely read among them. What those readers will have the opportunity to discover is certainly one of the largest, most perceptive, and most complex examinations of the dead (and deadly) seriousness of lust and, for that matter, of love ever written. But it is, at the same time, most assuredly a book that excites laughter."

*George Garrett, "The Fugitive," *New York Times Book Review* (12 May). Review of *The Big Ballad Jamboree* (Mississippi), by Donald Davidson: "In that sense, this first (and only) novel by Donald Davidson, written sometime in the 1950s but not published until now—indeed, not known to exist in a complete manuscript until it was found by the author's granddaughter—is an oddly apt parable speaking to this century's final decade. The problems just then becoming evident are still very much with us, only worse than all but the most pessimistic prophets imagined."

*Robert Gingher, "A Lesson in Violence," *The World & I* (December). Review of *Father and Son* (Algonquin), by Larry Brown: "Though his fiction shouts and draws startling figures, there is real craft and honesty here, less sensational than meaningful. 'Humble in the face of what is,' Brown exploits the full *power* of the mundane particular, his voice and vision finely attuned to the extent we resist or reinvent what is in order to survive."

*John Harper, "One Man's Dream Journey," Orlando *Sentinel* (29 September). Review of *Martin Dressler* (Crown), by Steven Millhauser: "And it is precisely this triangulation of Martin's dream-in-life with that of the author and reader that is most satisfying in this set piece of style and tone. *Martin Dressler* is a book for those who have grown weary of the old and are looking for something new."

*Stephanie Merritt, "Fear With a Southern Accent," *The* (London) *Times* (22 February). Review of *Omniphobia* (Louisiana State University Press), by R. H. W. Dillard: "Nineties nihilism and *fin-de-siecle* angst are commonplace themes for writers these days. Yet, standing out in sharp relief from this vast array of pessimistic fiction is *Omniphobia*. . . . *Omniphobia* is not a cheerful book, but it is immensely funny, provocative and ultimately optimistic."

*Carolyn See, "Easy Does It Again," *Washington Post* (19 July). Review of *A Little Yellow Dog* (Norton), by Walter Mosley: "God bless the day that Walter Mosley created Easy Rawlins! Easy, the cool dude with a stunning affinity for street life and a poignant yearning for respectability—a wonderful black guy with a brilliant mind and no opportunities, a guy who walked the mean streets of L.A. back in our parents' day, when a drive to the West Side could mean a trip to jail or worse."

*R. Z. Sheppard, "Living with the Ashes," *Time* (13 May). Review of *The Flaming Corsage* (Viking), by William Kennedy: "Kennedy, the Faulkner of upstate New York, again draws inspiration from Albany, the hometown he once described as an 'improbable city of political wizards, fearless ethnics, spectacular aristocrats, splendid nobodies, and underrated scoundrels.' The aforementioned now rub elbows and knock heads in a novel that once more demonstrates the author's passion for place and his skill as a literary magician."

TOMAHAWK CHOPS

*Brooke Allen, "Scheherazade's Exhaustion," *The New Criterion* (November). Review of *On With the Story* (Little, Brown), by John Barth: "This is the work of someone who has fallen so deeply in love with his own voice that fiction is at an end; all that remains is a narcissistic celebration of self. Barth is more interested in his own cogitations on art and narrative form than he is in any of his characters."

*Adam Begley, "When Brightness Falls and Mediocrity Calls," *New York Observer* (6 May). Review of *The Last of the Savages* (Knopf), by Jay McInerney: "If I were to hold my tongue and guard the secret, I might have to rehash instead the sad story of Mr. McInerney's fast-forward rise and slow-motion fall. Does anybody really need to hear again the cautionary tale about the young writer who traded literary talent for literary celebrity? It's not any edifying scenario, except, perhaps, for the poignant bit in the early 90s, a half-dozen years after *Bright Lights, Big City,* when the leader of the Brat Pack renounced the downtown scene and cranked out a big, earnest novel."

*Madison Smartt Bell, "Mountains of the Mind," *New York Times Book Review* (18 February). Review of *Death in the Andes* (Farrar Straus), by Mario Vargas Llosa: "The individual vignettes are often brilliant, but neither Lituma [the narrator] nor the reader nor perhaps the author himself can put them all coherently together. Perhaps a novel with so many characters and subplots and socioreligious-political complexities simply needs to be longer, a couple of thousand pages instead of a couple of hundred."

*Ginia Bellafante, "Sex, Lies and Psychopaths," *Time* (18 March). Review of *The End of Alice* (Scribners), by A. M. Home: "Why actually wade through the book when we know from the publicity what we're in for: a story that demands to disturb and repulse, a portrait of a sick mind filled with sexual imagery repellent enough to make Robert Mapplethorpe photos look like Tommy Hilfiger ads by comparison."

*James Bowman, "The Fantasy of Geography," *Wall Street Journal* (14 August). Review of *Dangerous Pilgrimages: Transatlantic Mythologies and the Novel* (Viking), by Malcolm Bradbury: "This is modernism from the point of view of the postmodernist. Mr. Bradbury sees nothing incongruous in putting Paloma beside Pablo Picasso, or Sylvester Stallone's Rambo on an equal footing with old Gunrunner Rimbaud. He is of that contemporary school that regards the taking of such pop-culture fluff seriously as a hallmark of profundity and the use of academic jargon as a sign of hipness."

*Phil Gailey, "A Conspiracy So Vast," *New York Times Book Review* (4 August). Review of *Fools for Scandal* (*Harper's Magazine*), by Gene Lyons and the editors of *Harper's Magazine*: "This is a nasty book, not because it challenges the reporting of the Whitewater story but because it assaults the integrity of the journalists who did the reporting. It is a smear job unworthy of any fair-minded critic of the press."

*Deborah Hornblow, "Trailer Is Ready to Sink in Mud," *Hartford Courant* (29 September). Review of *By the Shores of Gitchee Gumee*, by Tama Janowitz: "The novel plods and missteps from the start. At one point Janowitz sketches out the tedious details of an overblown mishap at the local library. A few pages later, one of her characters offers a wearying repetition of the same events.... The novel's few revelations appear in the form of clichés, something Janowitz first acknowledges and later seems to hope we'll ignore."

*James Hynes, "The Really Big Sleep," *Washington Post* (29 February). Review of *In the Presence of the Enemy* (Bantam), by Elizabeth George: "Elizabeth George's new novel, 'In the Presence of the Enemy,' is the longest, slowest, dullest book I have ever read. There's no point in being coy or oblique about it. There are about 200 pages of story here, played out over 519 pages of prose.... It makes Proust read like Elmore Leonard. I haven't slept so well in years."

*William Logan, "Gravel on the Tongue," *New Criterion* (June). One of the most able axmen in book reviewing circles, poet Logan peaks out with this verse chronicle, in which he clobbers any number of icons of the poetry scene: W. S. Merwin–"Though he describes the tangles of ivy, the flash of streams, these are dreams of experience, what experience would be like if it were prettier and bought retail, like wallpaper." Sharon Olds–"Olds is our poet laureate of oral sex, and as erotic as an old sock. She reports on her private acts with great elan, but it's like a scratchy old porn movie–there's something sad in the sleaze." Mark Doty–"Being gay and writing poems is an important occupation just now, but wasn't it an important occupation for Whitman, and Houseman, and Auden? Being gay didn't stop Auden from writing good poems; I don't know why it should stop Mark Doty." Louise Glück–"Louise Glück has taken sustenance from myth in a poetry starved of all else; her hollow-voiced language is as full of self-conscious angst as a Bergman movie." Seamus Heaney–"Heaney has become an institution now, and dangerous in the way institutions are: the Nobel Prize marks him wrongly as a spent force (Eliot called the Nobel 'a ticket to one's own funeral')."

*William Logan, "Old Guys," *The New Criterion* (December). Chronicle review of recent books by six poets–Charles Simic, A. R. Ammons, Robert Hass, Joseph Brodsky, C. K. Williams, and Anthony Hecht. Only Hecht emerges relatively unscarred. The others take a beating. For example: "Hass's dry self-examinations have a faintly puritanical edge, as if the reader's sublimation in pleasure were illicit. The midget Commissar at the center of his verse can never admit that even the breaking of illusion is an illusion–the writer was writing then, too. Many of the Poet Laureate's new poems, when they aren't about his own poems, are ragbag suites or odes, flitting from subject to subject (a poet tired of 'subjects' is often reduced to writing about writing), their concerns unified only by the writer's illusion of impulse–they are the progress of their own pathology."

*Bill Marx, "The Metafictionair Still at Play," *Boston Globe* (4 August). Review of *On With the Story* (Little, Brown), by John Barth: "Barth is like the glitzy magician who is so eager that you remember him that he gets in the way of his tricks; he celebrates the beauties of existence and the tenuousness of existence with a complacency that smacks of arrogance, of ego rather than agony."

*Patricia Elam Rugg, "An Unsatisfactory Journey," *The Washington Post* (20 August). Review of *Satisfied with Nothin'* (Simon and Schuster), by Ernest Hill: "The question is, why did he write this book? If he is truly satisfied with it, then his title is perfect."

*Carolyn See, "An Allegorical Tale Too Often Toad," *Washington Post* (23 August). Review of *The Frog* (Viking), by John Hawkes: "The jacket copy here assures us that John Hawkes is 'distinguished' (although whenever my publishers have called me 'distinguished' it means they are not going to give me much money). I'm sure this is a distinguished work. But I don't believe it was written to please an audience."

*Carolyn See, "Chick-Lit Lays an Egg," *Washington Post* (20 December). Review of *Chick-Lit 2* (Illinois State), edited by Cris Mazza, Jeffrey DeShell, and Elizabeth Sheffield: "Besides the rigid conventions of the avant-garde, the other thing to get depressed about here is the lack of material that the authors have to work with. It's an indictment of our own society, I guess, that young females are still locked in conflict with their moms and rage against their boyfriends and worry about cosmetics and food."

*Carolyn See, "Monkey Business," *The Washington Post* (6 December). Review of *The Woman and the Ape* (Farrar, Straus), by Peter Hoeg: "The righteous tone here is nearly unbearable. It's as though Hoeg himself had invented vegetarianism and the teetotalling way of life; as though he and he alone had discovered television might be bad for you or dead-end jobs a drag. What he enkindled in this reader was a passionate wish for a cheeseburger and a vodka martini. Better make both of them doubles."

*Charles Thompson, "Robert Penn McInerney," *Atlantic Monthly* (July). Review of *The Last of the Savages* (Knopf), by Jay McInerney: "Whatever McInerney's rank among modern novelists, he has yet to produce a work of real stature. The poet Peter Davison once remarked, unkindly, that if in McInerney's first novel, *Bright Lights, Big City,* one replaced every use of the word 'cocaine' with the word 'chocolate,' it would be a children's book."

*John Updike, "Tummy Trouble in Tinseltown," *The New Yorker* (5 August). Review of *I'm Losing You* (Villard), by Bruce Wagner: "Wagner's verbal animation rarely flags in his grisly tour of broken dreams and metastasizing careers, but this reader found himself counting the pages left to go in a wasteland so unrelievedly cratered. No character holds center stage for long, and relatively few engage our sympathies.... The entirety of 'I'm Losing You' is an outpouring of material indigestibly rich, presented in somewhat scattered fashion."

*Gore Vidal, "Rabbit's Own Burrow," *TLS* (26 April). Essay review of *In the Beauty of the Lilies* (Knopf), by John Updike. This piece is more a chain saw massacre than a tomahawk chop, leisurely in its unflinching savagery: "Although I've never taken Updike seriously as a writer, I now find him the unexpectedly relevant laureate of the way we would like to live now, if we have the money, the credentials and the sort of faith in our country and its big God that passes all understanding."

*Perhaps the ultimate Tomahawk Chop for 1996, even outdoing Gore Vidal on Updike, was writer Alexander Theroux's attack on the novel *My Other Life* (Houghton Mifflin), by his brother, Paul Theroux. In a review in *Boston* magazine (quoted in David Streitfeld's "Book Report," *Washington Post Book World,* 27 October) he calls his brother's novel "a failure" and adds this ad hominem critique: "We in the family don't mind his affected gentility, his smug and self-important airs, his urgent (expletive) insistence that he's a friend of lords and ladies, and only laugh at the fame he courts, the self-aggrandizement, inviting celebrities like Jane Pauley and Bryant Gumbel to his house, neither of whom, I believe, has ever quite managed to make it."

WHATEVER BECAME OF THE KING OF BABYLON?: THE SHORT SEASON OF ONE BOOK

"These are grim days to be a novelist, but it helps if no one's ever heard of you."

—Jeff Giles, *Newsweek* (3 June)

"I don't regret the years I put into my work. Perhaps I regret the fact that I was not two men, one who could live a full life apart from writing; and one who lived in art, exploring all he had to experience and know how to make his work right; yet not regretting that he had put his life into the art of perfecting the work."

— Bernard Malamud, *Talking Horse*

I have been warned not to write this piece. I have listened politely to the warnings and have to concede that those who have advised me not to do this have a point. It could all so easily be taken as (at best) a self-serving enterprise or maybe a good dose of sour grapes. That is certainly a real risk. On the other hand, I have now spent about ten years studying book reviewing and the American literary scene (among other things) for my annual article in the *DLB Yearbook*. During that same decade I have brought out a dozen books of my own and have written dozens of book reviews. That is to say that I have been much involved in the subject, one way and the other, like it or not. Why pretend otherwise? Why affect an innocence I do not quite deserve?

Besides, I am sixty-seven years old. By the time this piece appears in print, I will be sixty-eight. Isn't it just a little bit late in my life and in my writing career to be worrying too much about what other people might be thinking, how they might easily enough misunderstand, misconstrue, misinter-

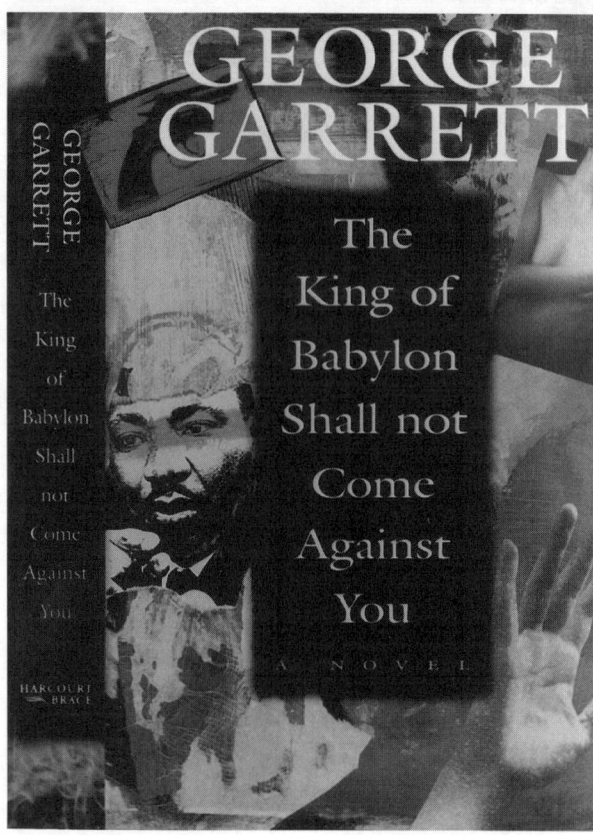

Dust jacket for Garrett's novel set in Florida on the day of the murder of Martin Luther King Jr.

pret what I am saying? What conceivable harm am I doing to anyone else?

Truth is, it does not seem all that risky. It is a challenge, sure, to try to achieve a balance and a relative neutrality, to be honest without either a private agenda or a hustle, when creating a kind of self-portrait: a portrait of the writer and his book during one brief season. It strikes me as an interesting idea. I cannot recall ever seeing such a thing before, though, of course, I could be wrong. There have been many critical and biographical pieces about the *reception* of a particular book or play or film. Who better to report the story than the writer, allowing that, beyond a certain point, the writer cannot claim to be a perfectly disinterested observer?

What I propose to do, then, is to follow the fate of my most recent novel, published in April 1996, *The King of Babylon Shall Not Come Against You* (Harcourt Brace), from its origins and beginning (as best as I can recall) through its publication, initial reception, and its place (if any) in the literary scene up to this writing–the end of 1996. The situation could change for better or for worse between now and publication of the *Yearbook* and even after that. The life of a book, though it were confined to the darkest inch the shelves of some libraries allow, goes on quietly beyond the life and times of both its author and its publisher.

All that I can do is to make it as honest (first, tell no lies) and worthwhile as possible, worth writing about and, I hope, worth reading. To tell the story, the truth, as I see it and understand it. To tell the truth *mainly* as Huck Finn once said about his author, allowing that, in spite of all candor and my best efforts, I might still be wrong about this and that.

I do not know where the "idea" for this story came from or precisely when, but I do know that it was first of all a short story called "To Whom Shall I Call Now in My Hour of Need?" Most likely written sometime in 1964, it was published in 1965 in *Red Clay Reader*. It bears a close kinship with a novel, *Do, Lord, Remember Me,* published the same year, first by Chapman and Hall in Great Britain, then by Doubleday in the United States. At this writing, the British version only is in print, as a volume in the Voices of the South Series of Louisiana State University Press. Both the British and the American versions were, in the view of everyone except my British agent, Gillon Aitken, too long by far. I was able, in both cases and versions, to cut the novel down to size–in half, actually–by removing a parallel story line and a large part of the comedy of the story. In a sense the story, "To Whom Shall I Call?" was a kind of outtake or spin-off from that novel. A little later the story, slightly revised and modified, appeared in a British collection of my short stories–*A Wreath for Garibaldi* (Rupert Hart-Davis, 1969).

And that seemed to be that. Except that the story still seemed to me, as more time passed, unfinished, or, anyway, incomplete. Something was missing. From time to time I thought about it and fiddled with it. In 1973 it reappeared as a novella called "The Satyr Shall Cry," in a collection of three novellas–*The Magic Striptease*. Soon it was clear to me that the story was still incomplete, that by opening up the short story and introducing some new complications, characters, and voices, I had really started something new. Old, too; for it included some elements, distorted to be sure, from the cut parts of *Do, Lord, Remember Me.* Anyway, it continued to puzzle and to trouble me. There were characters I wanted to get to know better. There were events, past and present, that kept asking for my renewed attention. There were old stories that seemed to ask for a new telling, a change of venue.

And the times, the fabulous 1960s, in fact and in fancy, kept changing, too, shape-shifting in memory. The original short story as well as the novel,

Do, Lord, were written in the early 1960s. But now all of the 1960s were behind us, history. The basic story seemed a part of that decade, but by now I was thinking more and more of the late 1960s, 1968 to be sure, that year of murders and losses and radical changes for better or for worse. How would these events of my story play off in and against the larger world of public events? Would there prove to be any connection? The only way to find out was to write and see. See from where? Well, from here and now (1994 as I was writing) looking back to 1968 as the then and there. Those became the two time lines of my single story—"here and now" / "then and there." Slowly, steadily it expanded itself (in several trial forms) into a novel. With an improbably working title of *The King of Babylon Shall Not Come Against You,* from Jer. 37:19: "Where are now your prophets which prophesied unto you, saying: The King of Babylon shall not come against you nor against this land?" It was a title that I seriously (wrongly) believed would not survive an editor's scrutiny.

Eventually the novel typescript went off to Harcourt Brace, which had changed hands and most personnel in the time since my last book, *Whistling in the Dark* (1992), had appeared. I doubted that they would take it, especially because of its form—told as it was in fragments and multiple voices and narration, all of this for the sake of narrative economy and because, as I began to realize, the theme of it all was (among other things) the impossibility of solving public or private mysteries from any one single point of view and on the basis of logic, cause and effect, alone.

I thought they might reject it outright, but they kept it, title and all. And they allowed me to continue with my editor, Cork Smith, who had retired but was freelancing. There was, as always, a little more work to be done, first of all some cutting, 150 pages or so; because the book was still too long, despite my economical methodology, and some restructuring, some rearrangement of parts. I had no problem with any of this. I think the cuts and revisions improved the pace of the story line considerably.

The only thing that I do regret is that, in the end, I did not include one little unit that I intended to. One of the devices of the story line, an effort to tell past history with strictest economy, is the use of "photographs," imaginary pictures out of the narrative past. Thus there are three such "photographs" of Martin Luther King Jr., based on well-known, published pictures and used to recount and meditate on his public life and times. And, as well, there are two imaginary pictures out of the local history of the Florida town that is the basic setting. One is a portrait of a Confederate officer. The other was a picture of a sawmill and its working crew. There was also to be a third picture taken from local history (the balance would have been three and three). But I cut it out, inadvertently and impulsively. A mistake. I could see that, too late, as soon as the book was published. It should have immediately followed page 296 in the book.

Well, then, here it is now, for the first and only time, once and for all:

Photograph: Flight Officer Royle
A boy on a bicycle, holding on for balance to one blade of the propeller of an early version (before the Merlin model) of the P51 Mustang. Only the nose is pictured, but, if nothing else, you cannot mistake the tilt and the look of it. The boy, slender and, for once in his life, evidently and perfectly serious, is identified as F/O Jojo Royle, Royal Canadian Air Force, and the plane is identified as his own—the "Dead Mule." The caption states that F/O Royle was awarded the DFC in 1943 shortly after this photograph was taken. He would have been, in spite of the fake birth certificate he used to enlist, in fact seventeen years old in 1943. He wears the heavy, wool, blue-gray British uniform with boots on, a visored garrison cap and an uninflated life jacket ("Mae West"). He and the "Dead Mule" are ready to fly anywhere in a matter of minutes. In this black and white photo, even the light is gray. Gray on gray....

He has credit at that time for shooting down three enemy aircraft with several others probably destroyed and a few damaged. He was likewise credited with the destruction of twenty locomotives during strafing raids. At the time of the photo, we are told, he was flying single aircraft night missions over France, and on a moonlit night near Rennes he shot down an ME110 preparing to land at a German airfield.

He has already killed, then, and surely by then knows that his own life is in the hands of gods and machines.

We are not told, presumably because we are expected to know such things, that the War is very far from won, indeed can easily enough be lost, in 1943. Between ourselves and that War there are only these young men, most as much children as F/O Royle regardless of their true age, ruthless killers whose lives and histories and futures are altogether expendable. They do not often allow themselves to think about that. Knowing Jojo you would say, and it's a safe speculation, that solemn or not (and so *thin* compared to the fat old man he now is), he is thinking of riding that bicycle to the village pub, wondering how much beer a man can drink and still bicycle back to the field down the narrow twisty hedgerowed lane. You would say he has his thoughts fixed on a young married woman in the village whose husband is half a world away in North Africa. Not bothering to think because he already knows it, that he may any time find himself in North Africa or Burma or anywhere else under the sun. He goes where he is sent, stays where he is told to.

What you might not guess, then or now, is how much he loves it, especially these night missions alone, alone in the widest deepest sky you can possibly imagine, hunched in his cockpit with the soft glow of the instruments matching the soft glow of fear and danger in his body. He will never feel anything like it, nothing so wonderful again. Half a century later he will still be dreaming himself alive in that airplane, in that desperate (though also often desperately boring) war. Will wake from his dream sad enough to weep, not for himself or anyone else near or far, but for the terrible gods and machines who blessed him once for a time before they forsook him forever and, with all their wealth of irony, left him to enjoy a long life and a slow death and the buried treasure of bitter memories.

While I was busy making the last revisions, there was yet another palace coup at Harcourt Brace—enough changes of partners to make even a veteran writer nervous. But except for moving headquarters back to New York and for scrapping the previous administrator's book jacket design (which, in fact, appears in their spring 1996 catalogue), nothing much changed (to my knowledge) as far as I and my book were concerned. It was still a "literary" novel on their spring list, more likely to be "review driven" than aggressively or expensively promoted. There would be a modest "tour," I was told, and depending on early sales and reviews and publicity, a limited amount of advertising. Nothing surprising there. Par for the course. See what happens. Sink or swim.

I had some hopes but precious few illusions. The making of a book, with all its abundant share of joy and tribulation, its deep doubts and sudden insights, is a wonderful experience. The process of publishing a book and, more and more in today's scene, *marketing* the product in person, is even ideally, at its very best, simply awful, almost obscene. Everybody knows this. But, strangely, everybody seems to have to learn it anew with each new book and its brief life and times. You have composed this work, as you do all your works, large and small, heavy or light, for no sake other than its own. You do not write to please a publisher. You already have a full-time job, so you do not have to write only for money. You do not write for any kind of real fame or celebrity. You are getting old but not yet crazy. Leave contemporary celebrity to fashion models and movie stars, to hip-hoppers and slam-dunkers. You write to and for an imaginary reader, one reader at a time. You and this reader come together and hold a conversation, a dialogue of give and take. You do not write anything aiming for "success" or fearing "failure" outside of the artistic successes and failures of the work itself. Nothing is ever as good as you hoped and wanted, as good as it ought to be. And when it is over and done with, your part, the making of the story, transformed by others from typescript into print, you will slowly begin, on your own, without needing book reviewers or literary critics to guide you, to understand better than anyone else can what the weaknesses and failures are and what did, in fact, work. All along, in the wild and wonderful process of making, of creation, you will have been making many choices, none of which is perfect, none of which is without some loss. Your hope is that the gains will outweigh the losses. In the end, later on, if you are lucky, you will come to respect the good choices and to learn something from the bad ones.

You are only one of many, a great many, literary, midlist American writers who have spent and are spending a lifetime learning and practicing your chosen craft and art. In these days, the 1990s, many of them are in trouble, have great trouble and difficulty finding publishers for their work. The marketplace is sorely vexed. In forty years you have moved from one publisher to another. It is only good luck and the ministrations of a good agent, Jane Gelfman, that have saved you and your novel this time. Next time who knows? In today's literary scene, with its limited market and interest in literary fiction, anything could happen. Next time depends a lot on this time.

Here I am, a professional with a body of work behind me and with plans and hopes for new and better work in the future. It has been a matter of ups and downs, give and take. But you are not prepared for retirement or resignation. For now, though, I must do what I can and follow the fortunes of *The King of Babylon Shall Not Come Against You,* hoping for some fair winds, at least a fair chance that this book may somehow meet up with my imaginary reader and prove worth that reader's time and attention.

Because Harcourt Brace was in the midst of radical changes, nobody there except Cork Smith had actually read the typescript when the time came for, among other things, flap copy. And, for reasons of timing and scheduling, they were in something of a hurry to get the book out. At that moment the San Diego office still ruled the Harcourt Brace roost. I was asked to furnish something I had never before been asked to do—to write a statement about the book that could be passed out to the salespeople at their meeting and that also might serve as a kind of guideline to the catalogue and jacket copy.

Here is the copy that I faxed to San Diego:

Attention: John Radiewicz

In the first week of April, 1968, a peak year in the turbulent sixties, the central Florida town of Paradise Springs was stunned by two murders, a suicide, a grotesque kidnapping, a deliberately set fire in a crowded public place, and other bizarre, interconnected occurrences. Subsequently, two members of an itinerant revival show were convicted of the murders. In the national scale these things were dwarfed by the murder of Martin Luther King, Jr. in Memphis (April 4) and the riots following that atrocity in more than one hundred American cities. In 1994 Bill Tone, an investigative reporter, goes to Paradise Springs to reconstruct the events and to place them in the context of the times. From documents and interviews with witnesses and participants, he seeks the truth. Moving back and forth from 1968 to the present, the story is told in the words of those people, among them a retired sheriff, a newspaper editor, a shady real estate developer, an African-American lawyer, a former nude model, a photographer, a waitress, a retired professor turned pornographer. Part detective story, partly an offbeat portrait of old and new South, *The King of Babylon Shall Not Come Against You* offers an original take on the immemorial theme of crime and punishment.

And here is how that copy was used (for better or worse) on the book's jacket flap. Close enough, in some ways improved; in others the nuances were . . . *different*. Of course the essential difference was that I was describing the book to the publisher. They, in turn, were trying to sell the book (through salespeople and bookstores) to that imaginary reader:

> In the first week of April 1968, the central Florida town of Paradise Springs is stunned by two murders, a suicide, a kidnapping, and arson. That same week the nation is shocked by the assassination of Martin Luther King, Jr. in Memphis and the riots that follow in more than a hundred American towns.
>
> Twenty-five years later, a slightly battered investigative reporter named Bill Tone goes to Paradise Springs to reconstruct these violent mysteries and to make some sense of them in the context of those mad times. In the course of his research Tone discovers that past and present reverberate, obscured and illuminated by unexpected sources. Indeed, one of the most unexpected—she was only four years old in 1968—is the town librarian, but Tone is also drawn into a maze of other stories from witnesses, participants, and even the departed. An ex-sheriff, a one-time nude model, a very successful African-American lawyer, a midget revivalist, his 300-pound common-law wife, and a retired professor turned pornographer are among the voices that sing Tone onto the rocks and into the whirlpools of the past.
>
> In the hands of one of our most accomplished novelists, the seeds of odd crimes in a small town a generation past bloom into a compelling and original version of the roots and ramifications of evil.

As for the matter of blurbs. What's to be said? Most people, with a few stellar exceptions, have to have them. I doubt if they really influence potential readers at all. But they do make the publisher feel a little better, maybe more secure about their judgment. I am not sure how many requests for blurbs (always an imposition on the valuable time of somebody else) went out. Four writers whose work and whose integrity I greatly admire came through—Madison Smartt Bell, David R. Slavitt, R. H. W. Dillard, and Richard Bausch. Their letters were excerpted, edited, and used on the book jacket. Some literary critics, each of whom I honor and respect, likewise sent in statements about the novel. Of these, to my disappointment and personal embarrassment, only a few sentences by one of them, Monroe K. Spears, were used at all, in this case as part of a press release accompanying review copies of the book: "*The King of Babylon Shall Not Come Against You* is a serious comedy, compulsively readable, continuously entertaining and highly meaningful . . . Highly irreverent, outrageously incorrect politically . . . it is truly a unique and exceptional work; I know of nothing like it."

What was most interesting to me was that each of these critics began not with a blurblike statement, but with a description of the story as he perceived it, building to a concluding affirmative judgment. In short, the statements of these critics were like highly condensed book reviews. Favorable, to be sure, but (I thought) useful for the publisher's purposes. Useful to me, as well, because I believe none of these men would, right or wrong, write anything they did not believe in. I am not sure why the publisher did not use any of these things except, of course, the brief excerpt from Spears. Perhaps ("The memory of American publishers is as brief as a May fly's."—Samuel Vaughan) nobody knew who they were. I was/am grateful to all of them. Here is what John W. Aldridge had to say:

> George Garrett is one of the most gifted, versatile, and productive writers of his generation. Over the years he has created an impressively original body of work—many novels, collections of his stories and poetry, plays, and literary commentary, culminating in his distinguished and widely acclaimed novels of the Elizabethan Age. In this new novel he has produced a brilliantly imaginative exploration of the complex fabric of emotions and events that led to a double murder and some other very bizarre occurrences in a small Southern town on the same day that Martin Luther King was assassinated. Garrett's narrative consists of interviews conducted by an investigative reporter with a variety of townspeople, most of whom finally form a cast of fully developed characters as the reporter's investigations proceed. Their collective voices become in turn a cho-

rus of commentary and reminiscence that effectively re-creates the state of mind of the town at the time the violent events occurred. The result is a superbly written, highly suspenseful Gothic mystery novel, one of the best books yet produced by a major writer working at the top of his form.

If you take out the laudatory stuff—this is, after all, intended to be the raw material for a traditional blurb—you have a general accounting of the ways and means of the book.

Soon enough blurb, jacket copy, all of it, became somewhat irrelevant. What matters for a contemporary, literary "review driven" book is the reviews. I am not at all certain how this system works or how it is intended to work. From the experience of reading book reviews in a great many newspapers and magazines over the past ten years in preparation for the annual *DLB Yearbook* article on book reviewing and the American literary scene, I observe that the ideal situation for a literary book is prompt and widespread attention with the added important variable of simultaneity. If reviews come soon following publication, in a wide variety of places, and if these reviews happen to appear at roughly the same time, give or take a week or fortnight, this becomes a triumph, earned or not, for the publisher. It does not seem to matter much what the substance of these prompt reviews may be, no matter (or not much, anyway) if they are positive or negative or mixed. The main thing is that they are there. Placement matters, too, at least in the Sunday papers. A well-placed mixed review, even a negative one, trumps a back-of-the-book rave.

Much of this makes a kind of sense when you pause to consider that most literary novels have a maximum scheduled shelf life of sixteen weeks at most, with six and eight weeks more and more common. Which means, among other things, that word of mouth has only a limited possibility of influence. By the time you have read a book and recommended it to someone else, chances are that the book is already off the shelf. Which, in turn, means that advance, prepublication reviews, such as the forecasts in *Publishers Weekly, Kirkus Reviews, Booklist,* and *Library Journal,* can be urgently important to the fate and brief lifespan of a literary novel. Blockbusters, at least as far as reviews are concerned, are relatively invulnerable. Best-sellers (usually blockbusters), once in place, are not easily harmed by bad reviews. Literary books can be wounded by bad or simply unsympathetic notices. Reviews can hurt, but can they help? Richard Bausch tells me that his excellent novel *Rebel Powers* (Houghton Mifflin, 1993) received outstanding advance reviews—starred in *Kirkus,* starred and boxed in *Publishers Weekly*. The book was promptly, widely, and almost simultaneously reviewed in all the major newspapers. And his publisher sent him out on a wide-ranging multicity reading and signing tour, resulting in many interviews and other kinds of publicity. When the hardcover season for *Rebel Powers* ended, he found that he had sold about four thousand copies, give or take. This was not, by any means, a disaster. There had been an early and profitable paperback sale. The visibility of the book and its author soon enabled him to find a new publisher with a better deal and better prospects.

Many other good midlist writers have not been so lucky. Lucky or unlucky, by and large our principal "business" effort is to survive long enough to get the lifework done and to find someone, a publisher, willing to publish it. Today that is not quite enough. If possible we must also go out on the road, reading and signing books and selling ourselves to strangers.

There are a lot of war stories about these road trips, readings, and signings. It's one of the things veterans share with each other. Hours of sitting at a table, pen in hand, while nobody buys a book. Sometimes they come directly to the signing table, face to face, almost nose to nose, to pick up cookies or cheese squares on toothpicks. Sometimes they pick up a copy of the book, glance at the jacket copy, flip a few pages, and put it back on the pile. . . . I, myself, have never failed to sell at least one copy, though sometimes that has been the only one. Usually, in the South, I have a cousin who will buy a book. We share stories, too, about the size of the audience for public readings. Many have read to a lone listener. One writer dined out on his reading to two people, who rose and departed after the first three minutes. Does that count as an audience of minus two?

These "author tours" have been routine for about ten years now. Writers uniformly agree that they are exhausting, time-consuming, and, most of the time, not cost-effective. The idea does not seem to be to sell books. There seem to be two practical publishers' goals of these tours. One is "visibility." A lot of the time, though not always by any means, a flying visit to a local bookstore can lead to publicity in the form of print interviews or even, more desirable, spots on radio or even television. This may or may not result in book sales, but, in any event, keeps the publisher's name alive and well and gives the publicity-and-promotion department tangible proof that they are actually doing something. Probably more important, sending out literary writers to bookstores here and there helps the relationship of the publisher to the bookstores. It is not this particular book that matters. It is the publisher's whole line, including the blockbusters. The publisher can deliver a real live writer. You, the writer, are working (for expenses) for the publisher. A kind of sales representative in disguise.

The King of Babylon Shall Not Come Against You and its writer did not (thank goodness) have a large-scale "tour." Which is just as well since I could not take time off from my teaching job for a real, sustained adventure. But there were a few events scattered across April and May: a book-and-author dinner, with other writers, at Richmond ("You will be driven to the Linden Row Hotel where you will mingle with the invited press"), readings-and-signings at a couple of Barnes and Noble superstores (Charlottesville, Virginia, and Orlando, Florida); at independent bookstores—Quail Ridge Books in Raleigh, North Carolina; The Regulator Bookstore in Durham; Mosswood Bookshop in Lakeland, Florida; Chapters in Washington, D.C.— a New York reading-and-signing at Rizzoli downtown. I was also sent, by another publisher (Louisiana State University Press) as it happens, to the Southern Festival of Books in Nashville. The Virginia Festival of the Book (Charlottesville, 28–31 March) was within walking distance. I do not mention here my regular readings and lectures at various schools and universities, of which there were a dozen or so during the same period of time and may, more or less, figure as book "promotion" also.

Even though the general subject, life in the late 1960s, seemed to me to be a natural for idle chatter, interviews turned out to be few and far between. *Publishers Weekly,* which had once been fairly friendly but had not interviewed this writer in more than a decade, decided to pass this time also. We ended up with one in the *Orlando Sentinel,* another in the *Roanoke Times,* and one in the *Knoxville News-Sentinel,* a radio interview in Tampa, Florida, and another with commentator Wayne Pond for the National Public Radio Program "Soundings," from the National Humanities Center. Not exactly what Harcourt Brace had aimed or hoped for. Indeed the novel generated a good deal less of this kind of attention than had *Whistling in the Dark,* a book of essays.

Since advertising was strictly limited—one ad, albeit a nice one, in *The New Yorker*—the book's chances to make its way in the world would have to depend almost exclusively on reviews. Of course, nobody knows how much advertising matters to a literary book. Maybe not at all. It matters to writers' egos, but may very well have no other value.

George Garrett

In terms of reviews there was one special problem for writer and publisher to worry over. Because the book's official publication date was set for 4 April 1996, for reasons of historical resonance and relevance, this did not allow for the usual lead time between the sending out of bound galleys and review copies to publications and reviewers. Literary books generally need a fairly long lead time. What this meant was that many, if not most, of the reviews of *King of Babylon* would probably appear well after publication date.

The early, advance reviews were mixed. First off, on 21 January, came *Publishers Weekly,* an unstarred notice mostly devoted to describing the general story line, though, at the conclusion, coming to judgment: "Clearly, Garrett's intent is to draw a picture of the moral breakdown of American society, using 1968 as the defining year that the culture changed for the worse. But because he seeks to assemble a wide spectrum of opinions, his capsule 'interviews' serve to fragment the narrative, draining his mordant social commentary of dramatic momentum." That reaction to the use of fragmentary and multiple narration, which I had deemed risky but necessary if I were to tell the story at all in a book shorter than, say, *Hurry Sundown* or *The Civil War: A Narrative,* would recur in many reviews. Some liked it and some didn't. One, the literary *Boston Review* (April/May) conceived of it as "old hat," but found a reason for it, writing that "though multiple narration is a somewhat worn-out device by now, it allows him to juxtapose a bittersweet recollection of the Old South with a view, tacitly accepting, of the far less regionally particular place it has become."

Kirkus weighed in a week later, generally more favorable in tone, though not unmixed either: "A provocative riff on the not-so-distant past, though more an entertaining colloquium on the state of the nation than a page-turning investigative yarn." The book made out a little better with *Library Journal* (15 February): "Though marred by some slow patches and a few too many lists, Garrett's latest is highly recommended for all libraries." Most favorable of all the prepublication notices was that of the *Booklist* (1 March): "As Garrett brilliantly plays out this theme within the novel—a grotesquely obese revivalist turns out to be the book's true spiritual center, an aging rock 'n' roller is the book's one true patriot—we are treated to a smart, cynical, exhilarating read."

In a sense, these earliest notices set the limits and the tone for the reviews that followed: a few (always welcome) raves, one or two strongly negative pieces, and the rest varieties of mixed responses, mostly in newspapers. Magazines were not much involved. I have, so far, seen only two notices in literary magazines, both by old friends, one by Mary Lee Settle in *The Texas Review* (Summer) and another in *The Sewanee Review* (Fall), by Walter Sullivan, both favorable and both of more than passing interest to the writer, because both are by first-rate novelists who know the drill and the score. Each read the novel a little bit differently, and what they found and responded to was, in a larger sense than the busy marketing of books, very helpful. Settle's conclusion is the kind of response that makes the whole thing, even the experience of the marketplace, seem worthwhile: "*The King of Babylon Shall Not Come Against You* is a book blessed with great humor, great anger, and great story-telling." There were, as ever, some rumors about the interest of other magazines, *Newsweek* for example, but these were nothing more than rumors. I heard, on pretty reliable authority, that a review was likely in *The World & I,* which had reviewed other books of mine in the past and for whom I had written book reviews. Since all their book reviews are essay-reviews, a notice there, with enough space allowed for some criticism and commentary, might prove interesting, even helpful. Nothing came forth from that corner.

On the other hand, Thomas Fleming, editor of the monthly *Chronicles,* produced a highly favorable review ("It's All Too Beautiful") for his August issue: "Part mystery novel, part social satire, *The King of Babylon* is more like *Tristam Shandy* than it is like postmodern fictions that are set in a nowhere populated by nobodies. It is also that rarest of rare book: an American novel that actually takes a close look at America." And there was, for the first time to my knowledge, my first review on the internet. I received a printout of a review by biographer Carl Rollyson, whose summation ought to make writer and publisher happy: "An ebullient work, full of eccentric characters and marvelous jokes, deftly probing contemporary culture and its connection to the 1960s."

Nevertheless, the long and the short of it, this time, would have to be the reviews in newspapers. The writer comes by these in several ways, none of them fully thorough or systematic. From time to time, mostly in the first heady days after publication date, the publisher sends along a package of clippings. One's friends across the country send along the things they see; and, more and more, reviews by the major papers are to be found on the internet. There is no escaping the matter. Even those writers I know who refuse to read their reviews, for various and sundry good reasons, inevitably stumble over them sooner or later.

Here is the sequence of newspaper reviews of *The King of Babylon* which have come to me so far. My guess is that it is fairly complete: *Dallas Morning News* (31 March); *Miami Herald* (31 March); *Baltimore Sun* (7 April); *Orlando Sentinel* (7 April); *Los Angeles Times* (8 April); *Kansas City Star* (14 April); *Washington Times* (21 April); *Roanoke Times* (21 April); *Cincinnati Enquirer* (23 April); *Atlanta Journal-Constitution* (28 April); *Richmond Times-Dispatch* (28 April); *Toronto Globe and Mail* (May); *Philadelphia Inquirer* (5 May); *Abilene Reporter* (12 May); *Staten Island Advance* (19 May); *Flint Journal* (19 May); *New York Times Book Review* (26 May); *Norfolk Virginian Pilot* (2 June); *Raleigh News & Observer* (2 June); *Washington Post Book World* (16 June); *Memphis Commercial Appeal* (30 June); *Chicago Tribune* (7 July); and the *Lincoln Journal Star* (4 August).

Of all these notices perhaps the most influential were those by Nancy Pate of the *Orlando Sentinel* and Margaria Fichtner of the *Miami Herald,* for these two were widely reprinted in syndication. Thus, for example, Fichtner's review reappeared in the *Milwaukee Journal Sentinel* (24 April), the *Omaha World-Herald* (21 April), the *Warren* [Ohio] *World-Herald* (21 April), and the *Columbus Ohio Tribune-Chronicle* (14 April). Pate's review reappeared in a wide range of newspapers, noticeably in the *Newport News Press* (21 April), Albany, New York *Times Union* (23 April); Fort-Worth *Morning Star-Telegram* (5 May); and the Columbia, South Carolina *State* (18 May).

Both of these reviews might be described as favorable with reservations. Couldn't ask for anything better than that. Or could you? Used to be, a while back, twenty or thirty years ago, the perfect book review was described by writers and publishers alike as favorable with some mild reservations. A fully formed rave was viewed at best as silly, at worst with serious suspicion. Not the case anymore. Just as grade inflation has bedeviled our institutions of higher learning, so review inflation seems to have changed the expectations (and needs) of writers and publishers. Fewer and fewer literary novels appear on the schedules and lists of the big commercial publishers.

Since there is, at best, only the most modest profit for successful literary fiction, the chief reason for publishing any of it is a kind of general promotion and advertising for the publishing house and its complete line. Prize winners (Pulitzer, National Book Award, National Book Critics Circle Award, and so on) have some job security–at least for a while. But the other literary books just deliver high visibility–that is, plenty of publicity and nothing less than rave reviews–to justify themselves. Which is one reason why a great many excellent and experienced American fiction writers are publishing their new books with small presses (if they are lucky enough to find one) these days. The pressure is on the writer to earn the best possible reviews. For some writers, at least, there is a serious temptation, every good reason not to take chances. Not to deal with risky subjects, not to tell the story in an unusual way or in any way that might be . . . misapprehended.

The overwhelming majority of reviews of *The King of Babylon* were in the favorable–mixed range. So far I have encountered only two that have to be called firmly negative. One–actually, all in all, it qualifies as "mixed," because the reviewer found a few things to say in favor of the book–was David Dawson's "Story of Changes Since 1968 Produces Complex, Muddled Puzzle for Readers," in the *Memphis Commercial-Appeal* (30 June). He argues that "the details of the murders are swamped by endless interjections about life in a small town during a cataclysmic period of history or about downright absurd opinions, sexual indiscretions and wild political and social theories of the residents of Paradise Springs. If this sounds confused, and confusing, it is."

The other one, more aggressively negative ("Erskine Caldwell Meets Rush Limbaugh. Cheap Fiction and Pulp Polemic Don't a Novel Make"), is more of a problem. It is the only review among all of them to take notice of and, briefly, to try to deal with the evolution of the novel. Reviewer Pierre Tristam, writing for the *Lakeland* [Florida] *Ledger* ("Garrett's 'New' Novel Misses Literary Mark," 19 May) is aware of the connections and kinship with the novel *Do, Lord, Remember Me* and sees the novella "The Satyr Shall Cry" as "an abbreviated first draft of 'Babylon.'" But he does not view these things as the gradual evolution of a story, arguing instead that the author's purpose was "only to recycle it, here with a central Florida peg, and a new publisher." There is no easy answer to that kind of charge, because you have to admit that somebody could honestly come to that conclusion. It would be (you hope) wrong, but it would not be illogical. In a classic sense, the conservation of literary material, the raw material of the story, has been deliberately recycled, though not for any purpose of deception. Here, then, is a review that did little harm, but set the writer thinking.

A couple of things are clear enough if you look at the list of papers and the dates. First, that we somehow failed to achieve the goal of simultaneity. Not counting the advance forecasts, the time of the newspaper reviews ran from the end of March to early July, scattered over the spring and early summer. From the writer's limited point of view, that would seem almost ideal, keeping the book alive beyond its usual shelf time. Not so. Publishers, as indicated, prefer timely and simultaneous reviews; and in any case the "major reviews," that is, the ones in what the publishers and most of the literary world take to be the most influential sources, should come as early as possible. Top of the line, for so many reasons, is the *New York Times Book Review*. Here publishers of fiction in 1996 were in a bit of a quandary. Lately the *Times Book Review* had been demonstrating, as if it were a matter of policy, a hard line on fiction of all kinds. Fiction reviews were significantly fewer than those of nonfiction books and slower to appear; and when they did appear they tended to be rougher on the writers and the books than had been the habit for many years. Perhaps the *Times* was trying to turn the tide of "review inflation." Perhaps, and more likely, the *Times* was hoping that lively and controversial reviews would interest more readers.

In any case, my publisher had mixed feelings about the prospect of the important review in the *Times Book Review*. Disappointed that they had not earned an early review, one that might have been useful during the book's prime-time shelf life, they were nevertheless nervous that there might not be any review at all or, paradoxically, that when it appeared, it would be devastating. Finally, seven weeks following publication date, came the *New York Times Book Review* with "False Prophets," by David Willis McCullough (26 May). It was, true to form, basically favorable, yet mixed. The only real complaint of the critic was an ineffable, if essential one—timing: "Mr. Garrett's can be shaky. In not shaping the inspired confusion he spreads so profligately, he causes his narrative to lose sharpness and bite, and his roaring dialogues and monologues too often go flat." The writer, at least, was happy to get such an easy landing in the *Times* during a season of tomahawk chops for fiction writers.

The other crucial review, from the publisher's viewpoint, came late, too. James Hynes's "Secrets of a Southern Town" (*Washington Post Book World*, 16 June) is favorable and mixed, questioning the validity of the "coincidence" of the Martin Luther King Jr. murder with the fictional events of the book, but giving the book more than merely its due: "The result is as much a magnificent piece of social history and cultural commentary as it is a novel, but more exciting, finally, and funnier, than any straight-arrow legal thriller." No writer I know of would have the arrogance and chutzpah to ask for anything more than the summation: "If there is a common thread through a body of work as diverse as Garrett's, it is that he is both uncompromising in his excavation of the truth and unconcerned that he might scare readers off. For the reader willing to engage with *The King of Babylon Shall Not Come Against You,* to meet it on its own terms, there are few recent novels as thought-provoking, as witty, as entertaining as this one."

The last of the majors, the review in *Chicago Tribune Books,* did not appear until 7 July. Bruce Cook's "George Garrett's New Novel of the South Is as Much History as Mystery" was predominantly favorable and carefully balanced in summation: "All of which is to say that if you are looking for a conventional murder mystery, with lots of local color, told in an unconventional way, you are bound to be disappointed. Local color is what it's about. And that, come to think of it, is the sort of novel one would expect from George Garrett." Some of this caution on the part of reviewers, pointing out that the book is not a conventional thriller, arose because in some places, some bookstores in particular, the book ended up shelved with the thrillers.

If there were many other reviews, they haven't come my way. As of now, by any standards, the original season of the book is over, though its story continues quietly. I have been invited to and will find myself reading from the book and talking about it at various places, including several book festivals, during 1997. I have no idea, at this writing, how well the sales went. There was a second printing, but all this happened I am told (and experience and evidence seem to confirm this) during a singularly slow season for fiction, with a higher rate of returns from all kinds of books, even the best-sellers and blockbusters, than usual. By the time this *DLB Yearbook* is in print, I ought to know some hard numbers. But I already know full well that *Babylon* was not a "breakthrough" book. It was not, I guess, exactly a disaster either. But that is not entirely relevant. In my terms, based on my forty years of experience as a published writer, the reception was positive and favorable. I have done better (more reviews more promptly and more uniformly favorable) and I have done worse. I have no legitimate complaints. Needless to say, I face considerable uncertainty in today's publishing scene. At my age and with my track record, I would expect most commercial publishers will have doubts and questions when my next book comes along. Nothing I can do about that except to write the next book and to try not to think about its future or mine.

As for the season of this book, well, it was . . . an *experience*. The publishing world and the literary scene keep changing (sometimes even for the better) often subtly, but enough so that it becomes a fresh experience each time. I have to look forward to the next book, the making of it if not the marketing. Meantime, no regrets. I wouldn't have missed it for the world.

Conversations with Publishers IV: An Interview with James Laughlin

DLB: The official purpose of the visit was to talk to you about publishing, but you indicated at lunch that you are no longer a publisher: you're now a full-time poet. Do you feel that the time you devoted to publishing and to the Alta ski resort deprived you of developing as a poet?

Laughlin: No. I kept on writing poems, but with little comprehension of what I was doing. The ski resort at Alta has been my life's chief work; I took a place that had nothing in it except beautiful mountains and snow and I converted it into one of the few ski resorts in the country that isn't a shi-shi hangout for drunks. On a good Sunday we get six thousand people riding our lifts—and these are serious skiers. The skiing is serious at all degrees of difficulty. I'm very proud of the development of Alta; it has done some good. That really did some good in the world.

DLB: It wasn't a hobby?

Laughlin: It began as a hobby. I'd been skiing a lot in my youth and then in 1939 my friend Dick Durrance told me, "You gotta come out here and see this place." My first day in Alta I saw it had marvelous dry snow. At first there was only one small lodge and one ski lift whose towers were old telephone poles, a bi-cable, if you can imagine that. One cable carried the chairs and the other pulled the rig. Today Alta has five lodges and seven modern lifts thriving. I've turned it all over now to my children; they enjoy it. They are blessed with wonderful managers.

DLB: You said a moment ago that at the time that you began writing poetry you didn't know what you were doing.

Laughlin: No, I didn't.

DLB: When did you figure out what you were doing, and how did you figure it out?

Laughlin: It took a long time to come to where it is now. You know the old trite story, of course, of Pound mauling my stuff and saying, "You'll never

James Laughlin

be a poet. You'd better be a publisher; it doesn't take any brains." It took a long time to figure out what I wanted to do. A long time. But gradually, the desire and the knowledge accreted. The collected poems volume runs to 560 pages long. There were many pitfalls on the way, but now I feel I'm closer to what interests me, and I'm enjoying the writing. That's the main thing.

DLB: This is a refutation of the rules that poets do their best work before they're thirty.

Laughlin: Certainly not in my case. My early work has a certain naive charm, but metrically it's too simple. The metric came from a suggestion made by Bill Williams—William Carlos Williams—whose books New Directions published. He said, "You can type; make your couplets as nearly the same length as possible." Bill was so practical. That's what I was doing for many years—making the couplet lines the same length. Then I loosened up and began to write free verse and quasi-learned verse laced with classical quotations. I had majored in Latin and Italian at Harvard. The title for my next book with New Directions is *The Secret Room,* but it might better have been "Looking Inward." In the new work there is a lot of looking inward, imagining what is inside me or imagining what is inside some girl. At the same time, I still write things that are straight remembrance, such as the ongoing long poem "Byways." Yes, there's some inwardness there; there is some fictional lying. But the main thing is to put down with as much clarity as I can stories about things that I remember. Then I've got on to a kick on the fivers, the pentastichs which are relaxation. If you get an idea, or read something, it's just very little work to shape it into five lines.

As I said, there is a lot of looking inward, inside myself and inside others. "Byways" assembles autobiographical, unrelated, non-chronological segments on people and places remembered. The thing is to put down as clearly as I can what happened in my life. Here the colloquial metric is one I borrowed from my friend the poet Kenneth Rexroth, whose *Dragon and the Unicorn* New Directions published in 1952. Rexroth called the meter his "busted trimeter." This past year I've been writing in pentastichs, a form that goes back to the Greeks. (*Penta*–five, *stich*–line: a five-line poem.) It's fun working with pentastichs because the subject matter may be anything: one's own thoughts, quotations from books read, bits from history. No rhymes necessary, just five lines of approximately the same length. With me, they rely often on the classics. Oh, here's one called "Good Philosophy." Plato, fourth century B.C. "When I give you an apple / if you love me from your heart / exchange it for your maidenhead / but if your feelings are what I hope they are not / please take the apple and reflect on how short lived is beauty."

Here's a nice one. This is called "An Exquisite Life." "Robert Montesquiou / the exquisite model for Proust's Charlus / kept pet bats in silver cages/ and for his famous receptions had each room of his dwelling / sprayed with a different suggestive perfume." I try to get some life into the pentastichs.

Here's one out of Lear: "His Hand: Act Five." "Now that he's old and foolish / his hand smells of mortality / wash it as he may / he can't regain the scent of the time when / lovely hands longed to touch and caress it."

This is called "The Crane." Lines from the Tamil, third century. "Go away crane, leave the garden / you have not told my love / the prince of the seashore / the torment that I suffer. / Go away crane, leave the garden." I wish I could read Sanskrit, but translation must suffice. Here's a nice one. This is out of Alain Danielou's Sanskrit. "The Invitation to Make Love." "Show her drawings of animals making love, then of humans / the sight of erotic creatures such as geese will make her curious / write amorous messages to her on palm leaves / tell her your dreams about her/ tickle her toes with your finger." That's from the *Kama Sutra,* the Indian classic of eroticism.

DLB: Have any of them been published?

Laughlin: Not really, no. I find that there's hesitancy. The editors think there's something odd about the form. Well, that's good. All my life I've tried to make New Directions odd. Bill Corbett, who took on the job of writing a book about New Directions, asked what was my intention in starting publishing. I'd say it was to be odd and different and "new directional." Nowadays, of course, it's very hard to find avant-garde writers. The avant-garde has been absorbed into the main marching body of literature. The most interesting poet who has turned up in years is Anne Carson. She is marvelous. A "new directional" writer. To do her kind of book was what New Directions was established for. When I started publishing, in 1936, many of the publishers were broke; they didn't want to do strange books; I did want to do strange books, so I had a wide-open field for many years. This continued up to a point—I couldn't give you the date of termination. It dawned on me suddenly when I saw Barthelme being published in *The New Yorker* and by Knopf; I saw that the avant-garde was no more a separate entity. It was absorbed into the language. Since then we've gone on; we have a wonderful staff in New York. Six people. A marvelous woman, Griselda Ohannessian, runs it for me. The editors seek to find novelty and nonconformity by looking in Europe. Some of the novels now come out of tips that we've had from the short reviews in the *TLS* and others by tips from publishers who have published some of our books; reciprocally, they point out to us good people abroad. Antonio Tabucchi, our Italian star, was pointed out to us by an Italian publisher. His novel *Requiem* is out of this world, it's stupendous. Just a small book. Not so massive as *Mann ohne Eigenschaften,* or anything like that. That's what the staff do now. They tell me what they're doing, and I holler. I complain about the jackets. I was always very classical in my approach to typography. But the production man found a person who could disobey every law of good typography, and it comes up looking like a smorgasbord. I fuss over such things as that, but otherwise I let them do what manuscripts they want. I read manuscripts as they send them up if my eyes are functioning—some days they aren't—and I'm very pleased with the way they're running the thing. It's making a tiny bit of money, which it seldom did in the past. It took twenty-three years for New Directions to be in the black. All that time I was living off my allowance and handouts from my family. With regret they would write another check to pay the printer, who was howling. That until, at Henry Miller's urging, I published Hermann Hesse's *Siddartha*—not a very good book—it's Buddhism sugared over, but the young people went wild over it and we sold a million copies. It took Henry a year to get me to have that book translated and published. Henry was full of philosophy as well as sex. His death wish was not fulfilled. He always dreamed that he would be buried in the Himalayas or in Tibet. Tibet, preferably. Instead of that he's in some box down there in Los Angeles, and the Japanese piano player ran off with all of his money. Dear Henry; he was a handful, but fun. He was a handful: dashing hither, dashing thither, nobody knew why or whither.

DLB: Mr. Laughlin, you started New Directions in 1936 as a Harvard undergraduate?

Laughlin: Right.

DLB: Would it be possible today?

Laughlin: No. Unless a person had a lot of money. You see, the cost of books when I started was nothing. I don't know whether you know that beautiful series that I got out, "The Poets of the Year." Those booklets sold for fifty cents. I got them printed for between two and three hundred dollars. If you went to that quality of printing today, you'd have to pay four or five thousand dollars, and it just won't work. I had great pride in that series, but I couldn't get people to subscribe. In the end we had only 660 subscriptions. What readers wanted to do was to pick and choose. You couldn't do that anymore. Impossible now.

DLB: So the basic reason that would prevent this adventure from being repeated is simply the cost of printing. Even though your family helped, and you lost money, you lost a controllable amount.

Laughlin: I was losing, I'd have to guess at it, but I was losing five to ten thousand dollars a year. The family met the deficit. I belonged to a wealthy family—I sent my mother a copy of every book because she was paying for many of them, and she had them proudly displayed in the living room at 104 Woodland Road in Pittsburgh. But the only person who ever touched those books was the parlor maid dusting them. My mother was afraid to read them for fear of what horror she would find in them. Her friends would call her up on the telephone and say, "I see James has published another Henry Miller book, Marjory. Why do you let him do that?" And she said, "Well, he wants to." And so that pristine set of books, not a fly speck on them, is now down in that beautiful glass tower at Yale in the Beinecke Library.

DLB: Why not Harvard?

Laughlin: I got kind of pissed off with Harvard. But the New Directions archive will go to Houghton Library; that is, the authors' letters and my carbons. It's appraised at two million.

Well, this is gossip, but the man is dead. I was put up for the Board of Overseers, but Joe Pulitzer blackballed me. Well, he's dead now. *De mortuis nihil nisi bonum.* But he was a supercilious, bright fellow. When he was living in Eliot House he had the largest room, with a grand piano, and he was keeping a girl at the Boston Ritz named Fleurie. And whether they will do anything with the letters, I don't know. They're awfully lazy about modern literature, you know. The Houghton Library is about old stuff. Every week there's some professor somewhere wants to ask a question about one of these writers, and if I can answer it, I answer it.

DLB: You were certainly very generous in answering every letter we ever sent you.

Laughlin: Some of the professors are ridiculous.

DLB: Speaking of jackpots, one of your jackpots along with the Hesse was Fitzgerald's *The Crack-Up*.

Laughlin: The Crack-Up went through about five printings, and we keep it in print. But it got me in Dutch with Edmund Wilson because the papers came from him and I didn't show him proofs until it was too late. He had systematically crossed out the name of every friend they'd had at Princeton, though the book said nothing bad about them. And I put the names back in, John Peale Bishop and this one and that one. Edmund wrote to me on one of his cards, "You are an impudent puppy."

DLB: As you know, *The Crack-Up* was originally offered to Maxwell Perkins, who not only turned it down but attempted to dissuade Wilson from publishing with anyone. Perkins was no doubt an editorial genius, but he was wrong in this case—it's a wonderful book. It's a key book.

Laughlin: It's good stuff. It's about a man going to pieces.

DLB: If any one book triggered that first phase of the Fitzgerald revival in 1945, it was *The Crack-Up*.

Laughlin: Wasn't it also *The Great Gatsby*?

DLB: You also had that in your modern classics series. With a wonderful introduction by Lionel Trilling. The question I was trying to sneak up on is, did Wilson have to persuade you to do *The Crack-Up* or were you enthusiastic from the start? Or receptive, if not enthusiastic?

Laughlin: It took me about two hours to see that this was a New Directions book. It was different. This was something odd and something serious. About two hours I'd say, and there were other books of that kind. We came by E. M. Forster because Forster had been late for lunch with Alfred Knopf at the Ritz in London. Dear old Alfred; I was very fond of him. There were so many accidents in publishing, you know, so many accidents.

DLB: For a long time, you were the sole judge of what would be published at New Directions. You didn't have a panel of readers or anything like that?

Laughlin: No, but here's what happened. I took a job for Bob Hutchins at the Ford Foundation. Bob was an old friend who had become a director of the foundation. Bob called me from Pasadena, which was then the foundation's base, asking if I "wanted to come out and spend a few million in cultural work." So I said, "Sure." And I was four years with the Ford Foundation doing things that I wanted to do and finally there was so much I wanted to do, but not what the board of directors wanted, that I was given a golden handshake of half a million bucks to depart. But it was a great experience. Under my own steam, I would never have gone to India; I would never have gone to Japan; I would never have gone to Khajarao or Rangoon, to Angkor Wat or Borabadur or Kyoto, or any of those great places, because I didn't know anything about the East. You see, my culture was strictly Presbyterian, Pittsburgh. I didn't know anything about Sanskrit. Thanks to Ford giving me these projects to do out in India, Japan, and other places, my whole life changed. I'm like what Rexroth was, a phoney Buddhist. I don't go through all the . . . Dear Gary Snyder, my friend, he does his breathing exercises every morning. I'm a phony Buddhist, but I like the idea of the Buddhist belief. I like it as an idea. It's colorful; it's appealing; to me metempsychosis makes much more sense than the idea that all of the Presbyterians or the Catholics are crowded up in heaven peeing on the angels. It just doesn't make sense to me whereas metempsychosis does. We were talking about who ran things at New Directions. Fortunately I was able to take this job for four years with Ford because Bob MacGregor turned up. He'd been working at Harper's. He was a friend, a very sensitive, intelligent man and a good publisher. He agreed to move his little business called Theatre Arts Books into one room in the New Directions office and then he would spend most of his time looking after New Directions. He was very faithful. And he began choosing books while I was away, and that established the tradition for others in the firm, choosing books.

DLB: During the four years when you were with the Ford Foundation, going to places that you said that you otherwise would never have visited, who are some of the authors you discovered?

Laughlin: First of all, Raja Rao wrote a very good book which is still in print, called *Kanthapura*, which is about South India. It tells what happened in a typical village when the agitators of Gandhi arrived to stir things up against the landowners. Another man I got to know was Alain Danielou, a Frenchman converted to Hinduism. He became one of my saints. He was gay but a perfect gentleman; he never laid a finger on me. He put me on to many marvelous works, such as early Tamil novels that are so poetic. (Tamil is the language of the southeastern part of India.) One of the novels he translated is the *Shillapadikaram* (*The Ankle Bracelet*), written in the tenth century by Prince Illango Adigal. A delicious romance full of native color. I got involved with a writer named P. Lal. He was an anglicized Indian who taught at a college in Calcutta. When I went there to see him he nearly killed me. He made me ride on the back of his motorcycle. Now if you know Calcutta, nobody in their right mind would do anything but crawl on all fours—the traffic is so bad. But anyway he was a wonderful guy. Lal translated the classic plays of Kalidasa for me. Then he attempted an abridged version of one of India's two great Sanskrit epics, the *Mahabharata,* which runs to twenty-two volumes. He tried, but there were too many characters. You can't condense it. A confusion of characters. There were other writers, and I got to know a number of sculptors and painters. One painting here in the house is a Telagu peasant glass painting of the god Krishna tooting his flute. The painting is done on the back side of the glass. I got interested in little Indian and Tibetan bronzes that have given me great pleasure. So it was a wonderful, wonderful experience, but what did me in at the Ford Foundation was paper work. Bob Hutchins was a real leader; he just said spend the money and then ask for some more. And you didn't have to do quarterly reports of what you were doing. You didn't have to do any groupthink. You could just do what you wanted. Henry Ford was worried about me and he said, "I need some insulation from this wild man." So he arranged with people such as Bill Casey to be directors of my fund, which was called Intercultural Publications. They protected Henry. Bob Hutchins got himself fired by being too honest. I mean he wouldn't kiss any babies. His replacement was a stooge named Rowan Gaither who just wanted me out. They were very careful. They knew they weren't dealing with a nobody, so they had a committee to go to Europe and Asia to study what I had done, and then write a report. That was the mentality, you see, but when it got to that stage, it wasn't fun anymore. It was dull.

DLB: And fortunately you had something all set to do.

Laughlin: So then I came back to New Directions, and Bob MacGregor, God bless him, stayed on. He was one of the best publishing people I ever knew. He had good taste and was conscientious and he could handle difficult authors. Henry Miller would get into a hemorrhage, but Bob could write him a soothing letter. And he knew how to handle Tennessee Williams, who was a dear friend. I couldn't stand the boys that Tennessee had around him—they were rancid. But Bob could; he would go to Tennessee's evening poker games where they did God knows what. Bob wouldn't even tell me what they did, and he kept Tennessee happy. Meanwhile, Tennessee and I had a very civilized relationship. Tennessee loved to have lunch at the Century Club which represented all the things that his mother, the ambitious mother, had aspired to, having her son with genteel people. So I was good friends with Tennessee, and he stuck with me. Lots of publishers were after him, but he always said, "I've got a publisher." He went along with me right until the end. Now we're bringing out—having them edited—a couple of his play drafts. These are things that he didn't quite finish.

DLB: I read the clips on you, and I wasn't surprised to find that you're the executor of several of the authors that you have published. We'd be very interested to have your ideas about what are the responsibilities of a literary executor.

Laughlin: You try to do the best you can for your author as if he were alive and do what he would do. Thomas Merton is the easiest one. We have three trustees: Bob Giroux of FSG, myself, and Tommy O'Callaghan, of Louisville, who used to sneak Tom out of the monastery to her house where she could feed him beer. Then, we have a superb woman trained by Alfred Knopf, Anne McCormick. Absolutely superb. She handles a big volume of correspondence with publishers all over the world who want to translate Merton books.

DLB: Is Pound one of yours?

Laughlin: I very carefully never got onto that board, because having lived closely with Pound in Europe, I knew all the troubles and complications of personalities. So what I am is an agent for the Pound Literary Property Trust. I have limited powers to grant permissions for things and to make deals. The number two lady at New Directions is a superb person, Peggy Fox. An absolute whiz on permissions and copyrights. She's done a great good for the Pound Trust by protecting copyrights and collecting money for the use of material.

DLB: William Carlos Williams?

Laughlin: That's been pretty informal. The two sons—Bill Jr. recently died, and that leaves Paul, who is really a business man. He ran Abraham and Strauss department store. Nice guy; both nice guys. That's been pretty informal. Peggy Fox handles the Williams brothers; they like her, and she likes them. When a point came up about a permission, she would have called Bill Jr., but now she calls Mimi, his wife. I don't know how that'll all come out; there are various Williams children and grandchildren.

DLB: Have you ever had to make decisions about publishing previously unpublished works or work in progress?

Laughlin: I can't quote you chapter and verse of cases, but well, yes: H. D. (Hilda Doolittle). That's a very interesting story. I can't go into her whole life, but she was supported in her later life by this very rich English woman, the heiress of steamship lines—Winnifred Ellerman. She left all of her manuscripts to the care of Norman Holmes Pearson of Yale. Nice guy. And he doled stuff out to us.

DLB: Pearson did the supplying, the deciding, as to which manuscript would be published?

Laughlin: Yes, Pearson decided what should be published in what order. About seven books came to New Directions, including the *Collected Poems,* edited by Louis Martz. It has been a best-seller year in and year out.

DLB: But you've never had to make the decision to publish or prevent publication of a work in progress, an unfinished work, that involved difficulty or unpleasantness with the family or with other interested parties?

Laughlin: Well, I don't remember such a case. That's not to say it didn't happen. I just don't happen to remember. I don't think we've had anything big. Pound occasionally turned up manuscripts which he had forgotten about. These I placed in magazines such as *Paris Review.* If something comes my way from a publisher, I just go ahead and pay the royalty to the estate. You can't be too finicky with a writer like Pound who spread himself in every direction.

DLB: In operating a publishing house such as New Directions you often had opportunities to publish authors who you knew would not be popular, publish works that you knew would not make money for your firm; conversely, did you also publish works you personally did not like because you felt they should be published?

Laughlin: There has been some of that over the years—duty publishing. A writer such as Robert Nichols is a good case in point. Poor sales, but he's so gifted, a remarkable imagination. I'd publish him. Then later when Bob MacGregor took charge of the business he liked to balance the books and have a plan and a budget.

DLB: You said that the purpose of the firm from the time you were a Harvard sophomore and you started this, was to emphasize the New Directions of literature.

Laughlin: Right. That's set forth in the preamble to the first number of our annual anthology, *New Directions in Prose and Poetry,* which ran for fifty-five numbers. Praising the "revolution of the word"—that was a quote from Eugene Jolas, the editor of the Paris magazine *transition,* who first published Joyce's "work in progress" and many pieces by Gertrude Stein. He was the godfather.

DLB: You felt that you wanted to do, had a compulsion to do, this kind of publishing, because if you didn't do it, the official custodians of literary taste wouldn't.

Laughlin: That's all different now. The barriers between commercial publishing and artistic writing have broken down.

DLB: Are there directions now that you would not take?

Laughlin: I stick to what I know best, what William Carlos Williams called "invention."

DLB: Derrida?

Laughlin: I can't understand most of the "theorists." Some of them seem to hate literature as we always knew it. But Sartre was understandable. We did several books by Sartre some time back.

DLB: The official canon makers, the official taste makers, the official custodians; who are they? Who are they now and do they really have any clout?

Laughlin: I don't read enough to answer that question—I have weak eyesight now. Sonny Mehta at Knopf has taste and terrific clout.

DLB: That's another publisher.

Laughlin: Oh. I thought you were asking for publishers.

DLB: Okay, let's do publishers that have clout.

Laughlin: Editors such as Joni Evans, and the husband of Tina Brown at Random House. These people have their banks behind them. They write a check for half a million dollars advance, and it corrupts literature.

DLB: They don't do what you did. They are looking for jackpots and blockbusters.

Laughlin: Well, it's their nature to do what they do. They are pigs feeding at the trough.

DLB: Same trough?

Laughlin: Same trough. I was crazy—everybody knows I was crazy—because I wanted to do things that interested me. People ask me, what is the New Directions line? How do you know what you want to print? I say, I want to print something that I enjoy, that excites me to read, that seems to me new and different. Do you read the poet-essayist Anne Carson? She'll take the top of your head off. She's an American teaching classics at McGill. Two books came out together, the Knopf book and our book, *Plainwater* (Knopf) and *Glass, Irony and God* (New Directions). She's really new and very exciting.

DLB: There are those who say that you were a hobbyist.

Laughlin: I used to work from eight in the morning to twelve at night, if that's a hobby. Whenever I was skiing I took with me correspondence and a book trunk. I remember the nights at Alta when people were upstairs looking at a corny ski movie, and I was down there reading a novel by H. D. in a manuscript sent to me by Pearson. That's not what I think a hobby is. I was obsessed. I had a need to do my thing. My thing was not to go into the family steel mill.

DLB: Was it ever a possibility?

Laughlin: I could have had a job there anytime. My family controlled the stock. My father used to take me down to the mill at Thanksgiving. I adored my father and he adored me. We'd go down at Thanksgiving to go through the mill, talking to the grimy workers over the sound of the fierce flames of the blast furnace. I wasn't going to go work there if I could avoid it.

DLB: Apart from the publishers that you mentioned, who are the literary czars now—the critics, the reviewers, the anthologists? Who, sir, are the reputation makers now? Outside of publishing?

Laughlin: It's awfully hard to say. In poetry there's only Herb Leibowitz, who publishes the best poetry magazine, *Parnassus*. He can make an impression. Then there's Joe Parisi at *Poetry* in Chicago. The only other thing is the *Times Book Review,* and they drag. I must say they get an awful lot of interesting people to review for them, and they don't hesitate to lay about them. That's the other maker. The anthologies, of course, are entirely in the hands of Norton, who put out the Norton books of this and Norton books of that. God knows they'd never put me in any of them, but I can't go in there and fuss. That is the most powerful anthology that the world has ever had.

DLB: The *Norton Anthology of American Lit* and *Norton Anthology of English Lit*.

Laughlin: There are a lot of them, enormous power. I don't know who runs them, who chooses the editors. They sometimes choose people of unknown name to compile these books.

DLB: But they're members of the academic network, which is what you would expect. Who is the best critic of poetry around?

Laughlin: Hayden Carruth and Richard Howard. They both make sense.

DLB: Updike?

Laughlin: There are some good academics who are influential; one in a thousand are good.

DLB: More and more university presses are launching their own poetry series; the University of South Carolina Press has just inaugurated its own poetry series. How do you feel about that?

Laughlin: The trouble with poetry is there are too many poets. One in a thousand. Kill the rest.

DLB: Who decides which ones get killed?

Laughlin: That's the problem. But I think anybody who looks at the situation—if Herb Leibowitz looked at the situation he'd say, "Too many mediocre poets getting published." The trouble is, poets want books to give to their friends; but it confuses—totally confuses—the bookstores. They don't know what to buy. After a name has been established, they now know that it's respectable to buy Pound, Williams, Dylan Thomas, Levertov, a few others that we've published. They're completely confused. I don't know what'll be the end of it. It's the reverse of what happened before. When I became a publisher, in 1936, the conventional firms had no will or money to publish anything new. Now all over the country there are these little mom-and-pop magazines and firms. Some of them are very good. I like best Copper Canyon Press and the Dalky Archive. They get really good stuff. And there are a couple of others that are really good. Otherwise there are just too many poets.

DLB: Shoot 'em.

Laughlin: Don't quote me on that. I have no right to tamper with a human being's dream. Here are all these ladies and gents wanting to be poets. I've done a poem about a fellow who is a poet, but he's never written a poem because he wants to be a poet. These people have a right to their dreams and their aspirations. So do I. I have a right. But I think I'm a little bit better than many of them. There are just too many poets.

DLB: How do you feel about the college creative-writing programs? Are they a good thing, apart from feeding the teachers?

Laughlin: Can you imagine Ezra, Williams, or Dylan Thomas enrolling in the Iowa School of Writing?

DLB: You enrolled in the Ezruversity.

Laughlin: But that was personal and private. There was no tuition. No, I think the reading and writing courses are often a racket. I don't know what degree they get: master of fox hunting. And they have to provide a living for them. And this helps to provide a living. They can get jobs as teachers of creative writing.

DLB: Regrets. What's the book that got away? What's the book you didn't publish that you wanted to publish, should have published, but something went wrong?

Laughlin: Murphy, by Samuel Beckett.

DLB: What went wrong?

Laughlin: I went off for the summer. John Slocum, my friend at Harvard, sent *Murphy* to me and said, "You ought to publish this." By the time I got back, Barney Rossett had gotten ahold of it for Grove. He did a fine job for Beckett. But I should have been the publisher. Beckett would have been an ideal New Directions author, because he was so quirky. Would have been fine. I don't know if there were others. . . .

DLB: Nabokov's *Lolita.* What went wrong there?

Laughlin: What you don't understand is that at the time he wrote it, this was a dirty book. I was very fond of him. He'd spent the summer with me in the Alta Lodge, out hunting butterflies. His nice wife was there and the child, the innocent child. I wrote saying: "Volya, you are so sophisticated, you may not realize the effect that this book is going to have on the college community of Cornell if you publish it. Your wife will be ostracized; stones will be thrown at your child." He said, "Well, maybe. What'll I do with it?" I suggested that he send it to Gerodias at the Obelisk Press in Paris. College communities are so awful, some of them, in that respect—no tolerance. Maybe they're better now; I don't know. I remember that I used to visit with L. R. Lind at the University of Kansas. Before we could have a drink he had to pull down the shades in the room. This was my slant on the college communities.

DLB: Is there a book you wish to hell you hadn't published? You were wrong about it?

Laughlin: That's hard to tell. You don't tend to think of the things you didn't like. A number of poets that the ladies at the office like and publish don't do much to me. One of the good ones we're publishing now is Michael Palmer, a Californian, who is a superior language poet. But, it's hard, it's really hard to tell. Our best poets now are the Nobel Prize–winning Mexican, Octavio Paz, whom I began publishing in 1963; Robert Creeley; and Denise Levertov, who has just brought out her fourteenth book with New Directions. Her autobiographical prose book *Tesserae* is a gem.

DLB: There were two figures associated with the early years of New Directions I'd like to know more about. One is Delmore Schwartz. People who knew him have told me that he was the most brilliant literary talker they ever heard.

Laughlin: Delmore was a pungent but humorous talker. There's a fine biography of him by James Atlas. Have you read the Schwartz-Laughlin letters volume in the Norton Series? It gives a sympathetic portrait of Delmore. Poor Delmore, what a sad story. At the end of his life he turned against me. He turned against all of his friends except Dwight

McDonald. I'd sent him for a few months to a shrink, but it didn't do any good. It didn't help him; he was gone. Schizophrenic, I guess.

DLB: The other is a kind of shadowy figure, Robert Lowry.

Laughlin: Bob was a clear-cut case of manic depressive illness. He worked for me here in the little office down the road for a while, and he was a sweet person. We published his first little novel, *Casualty*, which was about the war, a good book. He lived in Cincinnati, and his family put him in the hospital. Then he'd start again. He was living in cheap hotels and trying to reestablish his talent. A sad case.

DLB: The literary game is littered with sad cases.

Laughlin: That's true. He'd send me pages of balderdash, by the great writer James Robert Lowry, just balderdash. I think the point to make if you're writing about the more recent years of New Directions is that it has become a group effort of consenting souls. Griselda Ohannessian, Peggy Fox, Laurie Callaghan, and Dan Altman are all professionals who love what they're doing. Barbara Epler and Declan Spring are very promising. It's a democratic cooperative now. Quite different from my juvenile ego trip. They share my tastes. They speak for me. I was getting more and more interested in my own writing. And we needed more people. So we have six people at 80 Eighth Avenue in New York, and then all of the distribution is done by Norton.

DLB: What is the most recent book you claim credit for picking, for spotting?

Laughlin: This was a battle. Guy Davenport's *A Table of Green Fields*. Some of the staff were troubled by an element of pederasty, but all agreed that the stories are the work of a genius. We're now doing a volume of Davenport's drawings and paintings. Few know that his gift extends to original art. I'm equally impressed by Anne Carson. She's on quite a different track from Davenport, but just as original. Her style is disjunctive. Her work is dissociational. She disobeys the rules of syntax, but she is full of feeling.

I wish my memory were better. I never kept a diary except through the texts of my poems. So much detail has been lost but not all. The poet Gregory Corso called me up one morning, about 2:30 A.M.. He wakes me up; the operator says, "Hello, he has to talk to you." I said, "What is it, Gregory?" He says, "James, I was just thinking, would you bequeath me your teeth?" And then he hung up.

DLB: No response seems possible. Thank you, sir.

Reading Series in New York City

Kelli Rae Patton
Unterberg Poetry Center

We go to readings to hear an author interpret his or her work in the way that only he or she can. We go to readings to catch a glimpse of the person who crafted a story or poem we particularly like. We go to readings to revisit a work of fiction or a collection of poetry that once engaged us or to discover a new writer we might be compelled to read. It makes sense that performance space for public readings is at a premium in New York City, the home of many poets and writers, and the hub of much publishing activity. I have no doubt that one could devote every night of the week to attending literary events in upper and lower Manhattan. This is cheering because it testifies to a widespread interest in literature.

But taken in perspective, surely New York City represents an anomaly in this respect in the United States: where in Middle America, after all, will you find readings nightly, much less in such a diversity of venues? The sheer number of reading series would seem to support New York's claim to being the cultural capital of the country. What follows represents an eclectic and (perhaps) haphazard survey of sites at which readings were held in 1996 in Manhattan.

How does one find out about literary readings in a city as massive and rich in cultural opportunities as New York? Though many series rely heavily on word-of-mouth, publications such as the monthly *Poetry Calendar* and the weeklies *New York, The New Yorker, Time Out,* and the *Village Voice* include comprehensive listings of literary activities throughout the city, ensuring ample audience turnout for most readings. Well-established literary organizations such as the Poetry Project at St. Mark's Church (downtown) and the Unterberg Poetry Center at the Ninety-second Street Y (uptown) produce their own newsletters, *Poetry Project Newsletter* and *Po-et-iks,* respectively, and hold creative writing classes in addition to their popular reading series. And, despite their names, both series include readings by poets and prose writers.

The Poetry Project at St. Mark's Church, which showcases the talent of writers downtown, recently celebrated its thirtieth anniversary. Though many readings are held throughout the year at the organization's East Tenth Street location, the Poetry Project joined forces with the behemoth Union Square Barnes & Noble in the summer to sponsor picnic-style readings in Union Square Park. The young and tanned Paul Auster read from his forthcoming memoir, and Susan Sontag read from her acclaimed novel *The Volcano Lover* (1992). The reading attracted the usual literary audience and all manner of dog walkers and dawdling pedestrians. Under the cosponsorship of the Poetry Project and Barnes & Noble, other such outdoor events took place later in the summer.

Funding cutbacks for arts organizations have made such collaborations increasingly necessary. More and more, organizations such as the Poetry Project and the Poetry Center depend on contributions from individuals, corporations, and foundations to combat the loss of funding from sources such as the federal government. Midyear, the Academy of American Poets, whose labyrinthine offices are in SoHo, agreed to cosponsor the poetry readings held at the Poetry Center, whose labyrinthine offices are on the Upper East Side. Before moving downtown and before joining with the Poetry Center, the Academy of American Poets held readings in the auditorium at the Guggenheim Museum uptown. The Main Reading Series at the Poetry Center runs from October to May and typically consists of readings by playwrights, prose writers, and poets. The organization also presents a distinguished lecture series highlighting literary biography, the Sunday morning gathering known as Biographers & Brunch.

The Poetry Center is something of an institution in New York—and in the literary world in general. Established in 1939, it has played host to some of the most famous and celebrated writers of the twentieth century. The readings are held in the nine-hundred-seat Kaufmann Concert Hall and the smaller Buttenwieser Hall. Neil Simon opened the 1996–1997 season in late September. In October Harold Pinter read to near-capacity crowds. Pinter has stage presence and possesses a gravelly, deep voice that suggests his training as an actor. He read from his most recent play, *Ashes to Ashes* (1996), in his fifth appearance at the Poetry Center. In an interview afterward that more closely resembled an Elizabethan bearbaiting, Professor Austin

Cover for a program from the best-known literary reading series in New York City

Quigley of Columbia University questioned Pinter about his career as a writer. Pinter was by turns candid and near-belligerent. When questions from the public were taken, one of the first came from a young woman who asked Pinter his opinion of sadomasochism.

Several factors contribute to the proliferation of reading series in the city, not least of which is the number of major trade publishing houses located in Manhattan. New York City is a main stop on most authors' promotional book tours. Directors of reading series, in conjunction with book publicists, often schedule readings to coincide with book publication dates; public appearances by authors sometimes result in an increase in book sales, which proves beneficial to all involved. In autumn the august publishing firm of Farrar, Straus & Giroux took the literary reading to a new height, that of the literary event. Highbrow and star-studded readings marked the fiftieth anniversary of FSG: who else can count so many Nobel laureates among its ranks? On two successive evenings in mid September, Town Hall on West Forth-third Street packed in enthusiastic crowds for a poetry reading and a fiction reading by Farrar, Straus & Giroux writers. Robert A. Giroux and Jonathan Galassi (who does double duty as both an FSG editor and as president of the Academy of American Poets) introduced the poetry evening in which John Ashbery, Frank Bidart, James Fenton, Thom Gunn, Seamus Heaney, Paul Muldoon, Robert Pinsky, Derek Walcott, C. K. Williams, Charles Wright, and Adam Zagajewski took part (and, in spirit, the late Nobel winner Joseph Brodsky). The fiction reading, introduced by Roger Straus, featured Rosellen Brown, Ian Frazier, David Grossman, Jamaica Kincaid, Madeleine L'Engle, John McPhee, Grace Paley, Susan Sontag, Calvin Trillin, Scott Turow, and Tom Wolfe.

Some readings have the aura of a rock concert about them, with rabid fans lined up hours beforehand to secure a seat. Readings of this magnitude are infrequent—even in New York City, where everything operates on a larger scale—just as few authors command large advances for their published work. Literary celebrities draw especially eccentric crowds, sometimes toting shopping bags chock-full of books to be signed by the author. But there are enthusiastic audiences for relative newcomers to the literary scene too: the actor-cum-author Ethan Hawke, whose debut novel *The Hottest State* was published in the fall, commanded an enormous turnout at his October reading at Rizzoli Bookstore on West Broadway in SoHo. Of course, a prepubescent crowd of females constituted a big portion of the audience, despite the unfavorable reviews of the book. If Ethan Hawke garnered the attention of the paparazzi, Dominican American writer Junot Díaz attracted the attention of the literati. Díaz was surely the literary starchild of the year; 1996 saw the publication of his first collection of short stories, *Drown,* which generally received strong reviews. Díaz, like Hawke, also participated in Rizzoli's reading series, as did George Garrett.

Bookstores large and small continued to sponsor readings to celebrate the publication of new books and to generate revenue in 1996. Nearly all links of the Barnes & Noble chain throughout the city host reading series, and many of the independent bookshops do as well. The Barnes & Noble on Union Square occupies three floors and has plenty of room for readings. This store, which celebrated its first anniversary in 1996, held a variety of readings, including one in honor of the controversial *Granta* "Best of Young American Novelists" issue. The selection process *Granta* employed generated dubiety in New York and elsewhere, as literary spectators questioned the objectivity of the decisions. About a dozen of these "best" showed up to read from their work. Jonathan Franzen, a writer associated with *The New Yorker,* read an affecting story about a young man and his overbearing family.

Books & Co., one of the more homey and intelligent bookstores in New York, is wedged into a small two-story space on Madison Avenue and hosts a reading series of some note. An independent store that managed to survive 1996 despite the swelling population of superbookstores, Books & Co. will close in mid 1997. Poets Charles Simic and John Yau provided a memorable reading there in the fall—memorable as much for its literary impact as the sweltering condition of the fully packed house. The young poet Matthew Rohrer, whose first book, *A Hummock in the Malookas* (1996), won a National Poetry Series Award, introduced the readers. Fans crammed upstairs in between the philosophy and poetry sections to hear selections from Simic's new book, *Walking the Black Cat,* and some lengthy poems by Yau. Books & Co. consistently offered very literary readings, versus the more popular readings occasionally held at the larger chain stores. The closing of the store will leave a gap in the cultural landscape of Manhattan's Upper East Side; no other bookstores currently offer such interesting readings in that neighborhood. Downtown, A Different Light Bookstore showcased the work of bisexual, gay, and lesbian writers in 1996 as usual. The bookstore typically offered an overwhelming array of events from which to choose each month. In February, J. S. Marcus, one of the writers on the *Granta* list, read from his first novel, *The Captain's Fire* (1996), at the Chelsea bookstore. Boston's Stephen McCauley gave a reading in March from his latest novel, *The Man of the House* (1996). Posman Books, with locations both uptown and downtown near Columbia University and New York University, hosted readings on selected Sundays during the year.

Universities and schools attract a range of writers, from the experimental to the traditional. At either end of Manhattan, schools with well-respected creative writing programs—Columbia and NYU—played host to writers such as Miroslav Holub and Charles Simic, as well as many others. The New School on West Twelfth Street launched a master of fine arts program in 1996 under the direction of the writer Robert Polito. In the early fall the New School auditorium was the site of a panel discussion titled "True Confessions: The Age of the Literary Memoir," featuring James Atlas, Susan Cheever, Joyce Carol Oates, Phyllis Rose, and Luc Sante. This program was a joint venture of the Unterberg Poetry Center and the New School.

A handful of other venues throughout Manhattan sponsor literary events akin to readings. The Manhattan Theatre Club (celebrating its twenty-fifth season in 1996–1997) and Symphony Space are two of the larger and more well known venues. Readings, enactments, and panel discussions filled out the Theatre Club's season at City Center in the West Fifties. In the past year Symphony Space, which is located on the Upper West Side, continued its acclaimed Selected Shorts series, which consists of actors reading short stories. Raucous crowds still attend the poetry slams, which started nearly a decade ago in the East Village, held at places such as the Nuyorican Poets Cafe.

One of this reviewer's favored spots for literary happenings is the headquarters of the Poetry Society of America at the National Arts Club on tony Gramercy Park. The National Arts Club has the feel of a well-appointed Edith Wharton drawing room. Each year PSA sponsors a Writer's Exchange program that enables writers from other states to travel to New York City to present a reading there. Fiction writer Sue Monk Kidd and poet Quitman Marshall of South Carolina were the recipients of this honor in 1996.

Teachers & Writers Collaborative, which has long sponsored creative writing programs in the public schools across the five boroughs, has been the site of the Poetry City series for the past year in its Union Square headquarters. This series features writers who are up-and-coming as well as more-established ones. On a night in November poets Kevin Young, a 1995 winner of the National Poetry Series Award, and Janice Lowe read from their work. Though the Poetry City series is relatively new, it manages to attract a crowd to the airy and recently refurbished Teachers & Writers Collaborative.

Many of the directors of reading series around the city are writers themselves, those who have a vested interest in seeing the proliferation and dissemination of good writing. Poet Karl Kirchwey, artistic director of the Unterberg Poetry Center, has been involved in the literary programming at the Ninety-second Street YM-YWHA on the Upper East Side for well over ten years. Kirchwey and other administrators of reading series see the artistic and financial support of writers as primary functions of their programs. According to the Poetry Center's "Mission Statement," the organization is committed to

> featuring a wide-ranging and diverse roster of authors..., including established and new writers, American and international writers, and also pursuing aesthetic diversity in its programs, which consist of readings as well as panel discussions, one-on-one discussions and cross-genre programming, and diversity in its forms, presenting poets, fiction writers, playwrights and essayists.

While reading series can easily be categorized as large or small, for-profit or not-for-profit, regularly occurring or sporadic, traditional or avant-garde, all seem to have a similar mission in common: to celebrate the written word as an aural art as well. What unites the many reading series held around the city is a love of literature and a delight in the peculiar embodying of a piece of writing that takes place in public performance.

Conversations with Rare Book Dealers (Publishers) III: An Interview with Otto Penzler

DLB: You are the General Motors of mystery literature, the only person in the history of the field who has had a bookshop, a publishing company, and a journal, and also you've been a major figure in the collecting field. Did you have a master plan, or did these things just sort of happen one after another?

Penzler: They just happened. I suppose after the snowball was already rolling down the hill, I recognized that I was in a unique position and decided that this was a good thing and tried to solidify that position. But almost everything happened accidentally. The easiest way to describe my good fortune or fortuitousness is when I needed a bigger office for the Mysterious Press, my publishing company, I had to move out of my apartment in the Bronx because there was no room for secretarial or any other kind of help. So I tried to rent a larger apartment in Manhattan—couldn't afford the rent. Wound up buying this building instead because it was during a time when real estate values were very, very low. Once I had the building I said, "Well, gee, wouldn't it be fun with all this space to open a bookshop?" So there was no plan to open a bookshop; it just presented itself.

DLB: You're unusual among publishers in that you're really a bookman. You started as a bookman, and then you became a publisher. Let's talk about how you became a bookman. What turned you on to collecting? What were the seminal early experiences?

Penzler: Well, first of all, I think it's genetic. I think we're born collectors. Collectors collect things when they're four years old and six years old, whether it's marbles or baseball cards or whatever. That was true of me, and once I was able to read and was able to actually own a book or two I would never give those books up. Those books became part of me. Plus I was collecting books when I was a very young child. I didn't think of it as being a collector. I just knew that I wanted to be surrounded by books, because I loved them.

DLB: Did you start with comic books?

Penzler: I didn't. I did collect them, but without the enthusiasm that a lot of other children have. I started with books.

DLB: How could you collect books in the Bronx?

Penzler: Well, my mother and stepfather owned a little delicatessen at Saint Anne's Avenue and 138th Street, right next to which was a moving and storage company that had shelves of books for ten cents apiece. That's where I was introduced to Tarzan of the Apes and Dr. Fu Manchu and Sherlock Holmes. And I loved those books. Probably my allowance per week was ten cents, and that was enough to buy one of those books, which I read over and over. But when I became seriously interested in collecting, there were two events—two people—who had a profound effect. One was a bookseller who ran a wonderful little bookshop in a townhouse on 41st Street—James F. Drake. I don't remember his name; he was an elderly gentleman. James Drake was his father, who had founded the company. I was a copyboy at the *Daily News* in those days. I was earning forty-two bucks a week.

DLB: Let's put a date on this, please.

Penzler: 1962, '63, '64. It was thirty-seven dollars take-home. I had about five dollars a week to spend on books, and I would go in and look around and find a book for five dollars. It was still possible in those days.

DLB: Can you remember a five-dollar book?

Penzler: I can't remember specifically. I remember buying some Rudyard Kiplings. Rudyard Kipling was out of favor; I don't even know if he's back in favor or not, but it was possible to buy inexpensive Kipling in those days. And I was collecting English and American literature. At that point I had not yet focused on detective fiction. And every week I'd go in and buy a five-dollar book. I'd give him a five-dollar bill and he'd give me a book.

Otto Penzler

Then one week there were two books. One was five dollars and one was ten dollars; it was the same title. I started to take the five-dollar book and he said, "You know, you would do much better to buy this book." And I said, "Well why? It's the same book and it's twice as much money." And then he explained to me about condition, and why it was a better deal and a better buy. And this went on for years. He would put a book away for me, and I would come in and he would say, "Well, I have something for you." And it was twenty dollars. I was earning a little more money, so I could maybe afford that, or I could have a layaway plan. And a couple of times he did something—and I only learned about it years later—he'd put a book away for me and he'd say, "This is absolutely for you. It's twenty dollars." And I later learned it was something like a $500 book. I had no idea at the time, but I knew it was a wonderful book that I really wanted, and I could never have afforded a $500 book in those days.

DLB: You said that there were two people. Drake was one. Who was the other?

Penzler: Frazer Clark. My old friend Frazer Clark. Well, I wrote a couple of pieces for a little magazine called something like *The Booklover's Companion.* I'm not sure that was the name, since it was thirty-five years ago. I think I wrote an article about Stephen Crane. And he wrote back. He wrote to me, saying that he really liked the article very much and he was very impressed with it, and did I collect Stephen Crane, and that sort of thing. And we began a correspondence. We met in New York, and we talked about books. It was midsixties. I can't put a year on it beyond that. I know I started working at ABC Sports in 1969, but I was still at the *Daily News* then. I had been there a few years and remained there a few years after that, so I know it was sort of midsixties. We talked about books for hours, and he really instilled in me the appreciation of great books in the same way that Mr. Drake had and told me that nothing was impossible—that when he started collecting Nathaniel Hawthorne everyone told him that it was impossible, that all the great Hawthorne books had been bought. And, of course, he put together one of the great collections of all time. He encouraged me, and I went to visit him at his house in a Detroit suburb, Bloomfield Hills. And I stayed overnight. His wife cooked a wonderful dinner, and as I left he said, "Here, I want you to take a couple of books away with you." And you know what he gave me? Two first editions, both fine copies, both in fine dust jackets. *Red Harvest,* the first Dashiell Hammett book, and *The Big Sleep,* the first Raymond Chandler book. To this day, I have the *Red Harvest* that he gave me, and it's still the finest copy I have ever seen in my life. He gave them to me. That was Fraze. And that did something for me that actually has turned out to be very important in the bookshop for me over the years.

Every now and then people come in with their kids to the bookshop, and the kids love the books. Sometimes the parents love the books with them, and they say, "Well, what would you like? Find a book that you'd like and I'll get it for you." And other parents—I know this will come as a stupefying notion for you—but other parents will say, "Come on, hurry up, we have to go. You spend too much time in bookshops; let's get outta here." And the kid will want to buy a book, or will want the parents to buy a book, and they'll say, "No, you have books at home." And I give books to the kids. And every time I do, I feel very noble, but I also always remember Frazer Clark.

DLB: Not many people have acknowledged Frazer Clark's many benefactions. When he was in a position to do so, he was extraordinarily generous to other book collectors.

Penzler: Generous not only with giving things, but giving warm advice and encouragement. Those were the two great influences of my book-collecting life. I remember them fondly. Hardly a week goes by without my thinking of them.

DLB: We have the kid in the delicatessen, and next door there were the shelves of junk books, and you're reading *Tarzan,* and you're reading God knows what else. What hooked you on mystery fiction? Can you remember the first mystery novel you read?

Penzler: Probably *The Hound of the Baskervilles.* I can't say it was an epiphany. I remember the first short story that I read was "The Red-Headed League," and Sherlock Holmes had great resonance for me.

DLB: Why?

Penzler: Oh, who can explain these chemical reactions?

DLB: Well, you're a boy in the Bronx. How does Queen Victoria's London grab you?

Penzler: Who knew? I can't even say, because you know, I lived a blessed life. I mean, it was a very very poor neighborhood, but nobody knew we were poor, because everybody else was. We didn't see television shows where people tried to sell you everything in the world. And if we saw movies with rich people it was a fantasy world—didn't mean anything. Didn't connect to us. So I never felt deprived, never felt that we didn't have as much as we needed. What did you need in those days? Fifteen cents for a Spalding. And a dime for a book. You didn't need any more. So we were perfectly happy.

I don't know why that era spoke to me the way it did. I think I loved the eccentricity of Holmes. I loved the fact that he knew everything. I loved his quiet arrogance. I loved the romanticism of gaslight, of the fog, and fireplaces. That was very far away to me—that could have been Uganda to my world. You know, we didn't have any of those things. And I loved it. I loved immersing myself in that world. And I loved other books from that era. I read so many things in the Victorian era: Wilkie Collins. . . .

DLB: You went through a Raffles period, didn't you?

Penzler: Well, I always liked all the adventure stuff.

DLB: I remember Fraze saying that he met this guy who was collecting Raffles, and he wanted to find something for him.

Penzler: I also collected The Saint at the same time. I liked the gentleman adventurers and the rogues. I loved Raffles more than almost anybody, except for Sherlock Holmes. But I loved Nayland Smith and the Fu Manchu books; I loved John Buchan. All of those sort of careless, carefree adventurers who laughed in the face of danger, and would just go off and do anything, and save the world yet again. I loved those things. Still do.

DLB: Now here you are working as a copyboy at the *New York Daily News* taking home thirty-seven bucks a week. What gave you the idea you could become a publisher?

Penzler: Well, I wasn't a copyboy anymore. By that time I had got to where I was writing the Reasoner report for Harry Reasoner at ABC. The best and the worst job I ever had. I couldn't get away from it, because I had to come up with all the ideas for the shows, five days a week. And it was just agony for me to come up with something new, a new slant on a news commentary every day of the week. I quit after six months, and they said, "Well, fine, but would you stay till we can hire a replacement?" Took something like eight more months for them to find somebody, and I just couldn't wait to get away. I began publishing and freelance writing at the same time. I had always harbored the hope that I could write something good or something important. By that time I had already written books like *The Ency-*

129 West 56th Street, New York City

clopedia of Mystery and Detection, for which I won an Edgar, and a couple of other books about mystery fiction.

DLB: What was the first book you published?

Penzler: The first book that I published was a book called *The Adventures of Herlock Sholmes,* by Peter Todd. It was a series of Sherlockian parodies that ran in the *Boys' Weekly* magazine in 1915 and 1916 that had never been in book form. And I collected them in book form. Did a thousand copies.

DLB: Did you sell them?

Penzler: Every one. And then I did a book called *Kek Huuygens, Smuggler* with Robert L. Fish. I knew Bob Fish because he was in the Baker Street Irregulars, and I said, "Could I do a book of yours?" He gave me these stories, and I published that. And he introduced me to his friend Robert Bloch, who wrote *Psycho,* and he said, "Otto, this is the guy you should be publishing; forget about me." I said, "Well, I'd give anything to do that." And Bob said, "Well, what would you give me?" I said, "I'll give you a thousand dollars." He said, "Fine, you got a book. I'll send you a collection of stories." Which he did. And then I did *Asimov's Sherlockian Limericks.* Isaac Asimov was told by his doctor that he was overworking–he had to go on a cruise. We were just talking, and he said, "God, I can't believe I'm gonna be on this stinking boat for eight days. I have nothing to do." And I said, "I have an idea. Write a limerick for each of the sixty Sherlock Holmes stories." He said, "Oh, you saved my life." And I sent him a copy of *The Complete Sherlock Holmes,* and he took it. That was the next book. You know, by then we were starting to have success. We went up to three thousand copies with Asimov.

DLB: How were you selling these? Word-of-mouth? Advertisements? Or were you walking around from bookstore to bookstore hustling?

Penzler: Not quite, although I did write letters endlessly in the age before computers. You know, each one had to be written over and over again. I typed letters for hours into the night–to every mystery store that I knew and every collector that I knew–and developed a mailing list. I typed the envelopes and stuffed them with little flyers and filled orders from my home. Sat there and typed invoices on my typewriter and wrapped the books and brought 'em to the post office–just did everything. And it was fine when we were only selling 20 or 30 books a week, but all of a sudden I was selling 100 books a week, or 150, and I couldn't keep up. That's when I started looking for other quarters. The first book came out in '76, and it was about two years or two and a half years later that I bought this building at 129 West 56th Street, in July of '78.

DLB: Who was your first discovery as a publisher?

Penzler: Probably first and greatest was James Ellroy. I didn't entirely discover him in the sense that he'd already had a paperback original out when I signed him for Mysterious Press.

DLB: How'd you find him?

Penzler: He walked into the bookshop one day and said, "You're Otto Penzler. I'm James Ellroy, and I'm the next great one." And I said, "You'll forgive me if I reserve judgment." He said, "I have a paperback original out." I said, "If it were any good it would have been a hardcover." He said, "I am good. They just didn't appreciate me. I have another one on the way, and I have a manuscript." And I said, "Well, let me look at it." At that time I also had a literary agency in partnership with Nat Sobel, called the Mysterious Literary Agency. I said, "I'll take you on as a client, and I'll try to sell the book. But first, you gotta do some rewrites." Well, he did rewrites for a year.

DLB: Was that *Blood on the Moon?*

Penzler: Blood on the Moon. It was called "L.A. Death Trip" when it first came to me. We removed about two-thirds of the violence, and it's still the most violent book I've ever read. You should have seen it in its original form—it was just over the top. But he was willing to do the work, and it came out. We did not sell a thousand copies in hardcover; so, naturally I signed him up for two more books. The three of them combined did not sell five thousand copies. I knew he was going to be important. I just saw this raw talent. He was learning to write novels by writing them. He had no discipline; he didn't know what to do. He had never had any training, never any formal education. He just had this innate, visceral talent. And in the next book, he said, "I'm not going to write a book in the series. I want to write the book that I'm now ready to write." And it was called *The Black Dahlia,* which became a *New York Times* best-seller.

DLB: Apart from the fiction you published, you wrote at least one reference book that has become a standard in the reference field. Will you tell us how that came into being?

Penzler: The Encyclopedia of Mystery and Detection. I had been brought in to work on a book by my friend Chris Steinbrunner—a book called *Detectionary,* which was a sort of a slight encyclopedia or dictionary of famous mystery-fiction characters and cases. Little one-paragraph plot synopses of famous books and famous movies and two or three paragraphs about famous detectives, in which we treated them as real-life characters. That was commissioned by the Hammermill Paper Company, which they had been doing for some years to show off one of their papers. They had done, oh, a cookbook to show this paper that you could wash, because it was always getting dirty in the kitchen, and they did this to show off their dictionary paper, which they then gave away to publishing companies. Instead of showing them a piece of paper, they showed them this book so that they could really get some good promotional value out of it. We were paid a small flat fee to write this book. It was just great fun. There were five of us who worked on it, and when it was delivered they said, "Gee, the writing here is very erratic. Some of it's really boring. I'd like the person who wrote these several articles to rewrite the whole book." It was me, and I rewrote the whole book so that it had a sort of similar style. One of the people who got that book as a giveaway was at McGraw-Hill, and he called Chris and said, "Could you write a real book like this, you know, with real scholarship and real dates and solid bibliographies and all of that?" And he said yes, and he called me and said, "Do you want to do this? I'll write the films, the television, and radio stuff; you write the books and the authors." And I said, "Deal." It took us about four years. Three years to write it, and then a year with Toby Worth, the copy editor at McGraw-Hill. I met with her every Wednesday afternoon for a year. She would go over it line by line, word by word, letter by letter, saying "How do you know this film was released in 1937? I checked the Library of Congress Archives and it says 1936." And she did that with everything in the book. That's why to this day, with twenty years more of scholarship and erudition, that book remains arguably the most accurate book published in the mystery field. Very, very few flaws in that book.

DLB: Is it still in print?

Penzler: No. It's really outdated now; I mean, it was published in 1976. Twenty years ago. It doesn't have people like Elmore Leonard or Robert Parker or Sue Grafton—the major names of today—because they came along afterward.

DLB: Time for a new edition.

Penzler: Well, publishers have been after me for years and have waved lots of money in front of

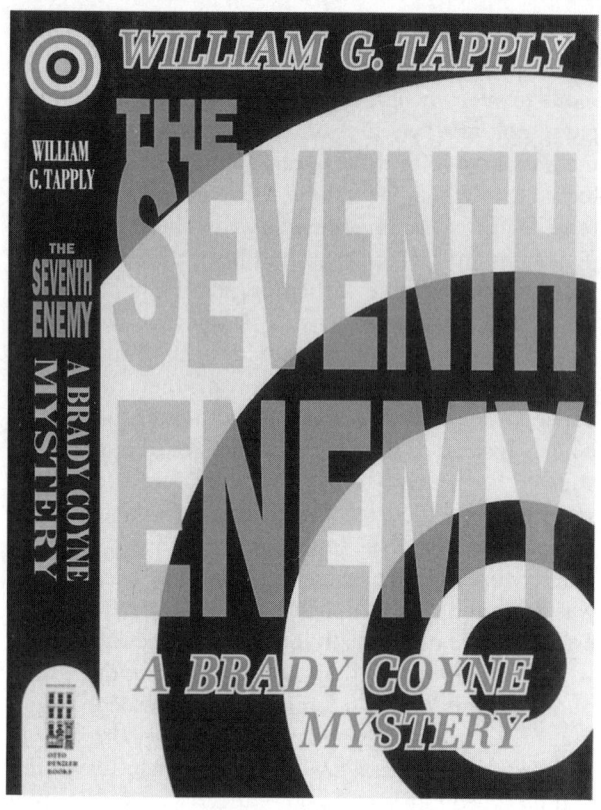

Dust jacket for a recent volume published by Otto Penzler Books

me, but not enough, not enough. I can't afford to rewrite that book. Too much time.

DLB: Why did you dispose of the Mysterious Press?

Penzler: I sold the Mysterious Press for two reasons. The first, perhaps obviously, is money. My wife wanted a house in the country, and this seemed a pretty good way to pay for it. The second reason, and equally important, is Warner Books made me an offer I couldn't refuse. Our publishing deal had been for a five-year contract, and as it was nearing expiration the president of Warner said they wanted to continue with me and the Mysterious Press, but they wanted to renegotiate the terms of the deal. Our arrangement had been that I got a share of the profits and some fees to cover expenses and overhead, and it was their feeling that I was making too much money. Naturally, I didn't share that feeling, so I refused their offer. Then they said, "Okay, then we'll buy the company." I won't go into all the negotiations, as they lasted through many months of frustration, anger, fear, saber-rattling, and all the rest, but we finally ended it on a note of some acrimony. Macmillan had wanted to buy the company for a great deal more than Warner, and I wanted to sell it to them but couldn't for various legal difficulties. Warner wanted me to continue to run the Mysterious Press; I didn't want to if I didn't own it. They suggested a ten-year contract. I suggested they just let me go. We compromised—I ran it for one year, then was a consultant for two more years, though I wasn't consulted at all during the second year. Mostly, it was a noncompete agreement, so I couldn't work for any other publisher. The day that contract expired I made a deal with Macmillan to start a new publishing firm, Otto Penzler Books.

DLB: What happened? It's not around now, is it?

Penzler: No. When Simon & Schuster bought Macmillan, they closed down quite a few imprints, including mine. I have to admit they behaved well, however, and published just about all the books that I had under contract, so the authors weren't simply cut loose. They weren't very well-published, if the truth be told, but at least their books came out and they were paid.

DLB: What's the best single mystery book you own? If the shop were on fire, which book would you grab?

Penzler: Far from the most valuable, but the one that I love most is—you may know that E. W. Hornung, who wrote the Raffles stories, was the brother-in-law of Arthur Conan Doyle—I have a copy of one of the Raffles first editions that Hornung inscribed to Conan Doyle. It said, "The sincerest form of flattery." That book is nearest and dearest to my heart. I love my inscribed *Big Sleep*. I love my dedication copy of Wilkie Collins's *After Dark,* inscribed to W. E. Herrick.

DLB: What's the book that's missing? What's the book you've been looking for for forty years? What's the book that every time you walk into a bookshop you hope against hope you're going to find?

Penzler: The one book that I would most love to own is my favorite novel of all time, which is *The Woman in White*. I would like to own the English three-decker. I own the American edition, which is the true first, but the English three-decker in original cloth in a fine copy is the one book that I think about all the time. And the agony of it is that I had one, and I had to sell it. I couldn't afford to keep it. I also had an inscribed copy that was rebound, which

I also had to sell. I couldn't afford to keep it at the time.

DLB: How are you fixed for manuscripts?

Penzler: I have a lot of them. I have an Agatha Christie manuscript. I have a Father Brown manuscript. I have Elmore Leonard. I have James Ellroy, James M. Cain, Ross Macdonald.

DLB: What is the scope of your library?

Penzler: The collection is out of hand. While I started with some special areas, such as nineteenth-century fiction, short-story collections, and American hard-boiled fiction of the 1920s, '30s, and '40s, it started to expand its parameters. As I became friends with many authors, I collected all of their books, and then a few other things seemed interesting, and then a few nice opportunities presented themselves, and now I'm out of control. I decided about a decade ago to try to collect it all. As with any serious collection, I acquire only first editions and the best copies available. And I'm fortunate enough to be in a place where I can get a lot of books inscribed, so I do. The office here is pretty big, so it holds about sixteen to eighteen thousand volumes. The house that my then-wife and I built is really a nice cozy cottage appended to a huge library which, if it ever gets completely built, will hold about sixty-five thousand volumes. I no longer know exactly what I have, as there are hundreds of cartons in the basement at the house, but I guess it's a good bit more than forty thousand volumes. Easily the biggest (and probably the best) collection of mystery fiction ever assembled. I'm quite willing to concede that rationality has long since vanished and obsession took over.

DLB: You've done so many splendid things in the world of books, publishing, bibliography, bibliophily, and obviously you're the kind of man who has six pots boiling on the stove at one time. What can we look forward to from you?

Penzler: Well, the things that I'm sort of working on at the moment are, first of all, putting another bookshop in London, on Marylebone High Street. Rare books, used books, new books. That's been taking quite a lot of my time.

DLB: Is this going to be an American going over there and showing the Brits how to run a mystery bookstore?

Penzler: I kind of hate to say that. It sounds arrogant as hell, but we both know it's true. So, that's been taking quite a lot of energy; I've been editing some anthologies, which is much more time-consuming than one would have imagined—mostly original stories on a theme. One came out last Valentine's Day called *Murder for Love,* every contributing author having been a best-seller. And the next one is *Murder for Revenge,* and then the third one will be *Obsession.*

DLB: These are original stories that you've commissioned?

Penzler: All original, yeah. Then for Dove I'm doing a book called *The Hundred Greatest Mystery Stories of All Time*. Big, fat volume. And I started with Houghton Mifflin a series called *Best American Mystery Stories.*

DLB: When will the first one come out?

Penzler: It'll come out in the fall of '97; that's for the calendar year of '96. I think that's going to be a good thing for the mystery field.

DLB: Why hasn't this happened before?

Penzler: Well, my agent and I were having lunch one day, and he said, "I have an idea," and he suggested this, and I said, "Boy, I'd love to do that. That would be great. I think this is good for mysteries." He called Houghton Mifflin, and they agreed ten minutes later. They thought it was a good idea too. The guest editor of the first book will be Robert B. Parker. Each volume will have a guest editor, just as the *Best American Short Stories* does.

DLB: And the guest editor will do the actual picking?

Penzler: I will bring it down to fifty stories, and the guest editor will cut them to twenty. Then the other thirty will be honorable mention. I'm very excited about it. I really like it. Another new thing that I'm doing, a new form almost, is for an English publishing house called Orion, run by Anthony Cheetham, who used to own Century Hutchinson Arrow. I'm editing a series of original novellas. Each book will be about fifteen thousand words—little paperback original novellas—which sell for the equivalent of a buck apiece in England.

DLB: And you're commissioning each one?

Penzler: Yeah, and editing them. There'll be twenty-five books in that first series. Probably they'll all be printed at roughly the same time for economies of scale, to keep them at sixty pence apiece. And there are wonderful writers in the series already. Ed McBain has written a story, and Peter Lovesy, H. R. F. Keating, Eric Lustbader.

DLB: Extraordinary. Am I right in detecting a certain fastidious avoidance of the spy novel on your part?

Penzler: No, not at all. The first book I signed for Otto Penzler Books was a spy novel. It's called *Maestro*, by John Gardner. Wonderful book. I like espionage fiction. The difficulty these days is twofold. One is there are very few good ones being written with the great enemy, the Soviet Union, no longer there. It's very difficult for an author to have America as the great conscience of the world when it doesn't have a suitable enemy. You know, it seems like America is a bully if it takes on Iraq or some lesser power. And there is no great power in the world anymore, except for the United States. To try to get Americans or other readers interested in Latin America or the Mideast is difficult. It's not the same as the great ogre of the Soviet Union. So they don't sell very well. John le Carré is an exception, of course. But there was a time when Len Deighton would immediately make the best-seller list, and there were great sales for John Gardner. A lot of other writers—Eric Ambler, even—they don't sell anymore. And so as a publisher, it would have to be something really exceptional for me to publish a spy novel now.

DLB: The name of this publication, after all, is the *Dictionary of Literary Biography Yearbook*. Can we talk about biographies of mystery writers? Do you prefer the term *mystery writer* or *crime writer*?

Penzler: Well, the Brits use the word *crime writer*. Americans use the term *mystery writer*. I think *crime writer* is probably more accurate.

DLB: Well, the Brits also use *thriller,* but *thriller* seems to be a moving target, because they also treat spy novels as thrillers.

Penzler: I just define mystery as any book in which a crime or the threat of a crime is central to the theme.

DLB: Mystery-writer biography.

Penzler: Well, with Otto Penzler Books I published *The Life of John Dickson Carr*. I published the biography of Jim Thompson. I published the biography of Cornell Woolrich at the Mysterious Press; an eight-hundred-page book, which won the Edgar. Certainly worthy of a major biography is E. W. Hornung, and it's now being written by Richard Lancelyn Green. He wrote the best bibliography that I've ever seen in my life, the Oxford University Press *Bibliography of Arthur Conan Doyle*. The most extraordinary, perfect bibliography. Actually identified the number of copies printed of almost all of the books. Hard to do.

DLB: You have no plans for writing a biography?

Penzler: I don't. I think one of the problems in my life is that I have to earn money. There was a time a few years ago, before my divorce, when I could really think about getting back into some scholarly pursuits which really interest me. That is not possible for me—for the next six years, anyway. After that I'll see. I don't know. Right now I have to be concerned with earning a living. And you can't do it writing biographies, as you know.

DLB: Have you spotted someone you regard as the coming Chandler or the coming Hammett? We all have someone, from time to time, whose first novel makes us say, "This one is going to be one of the great ones." Have you got one?

Penzler: Michael Connolly. *Black Echo.*

DLB: How many books has he done?

Penzler: Five.

DLB: What makes Otto run?

Penzler: Well, what you could say is "Why mystery fiction? What's the great attraction in mystery fiction? You've devoted twenty-five years of your life to this genre—why?" I don't know if I have a good answer, but I have an answer. And I don't think I knew it at the time. You asked me how I got into this and then we got off on a tangent. This is a digression from the real answer, but the collecting of it really became the motivation for all the rest of it, because at that time there was so little competition for mystery fiction. It was easy to collect good books for very little money. In the 1960s and 1970s nobody collected mystery fiction. Of course, now it's the hottest collected field in the world. But at that time it wasn't, and I was able to buy books very, very inexpensively.

DLB: What shops did you . . . ?

Penzler: Well, the Fourth Avenue Booksellers, of course. You know, that was the great treasure trove. You'd go to the Strand, Schulte's, Biblo and Tannen, Pageant—and you could buy good books for fifty cents or a dollar. And even things like the college sales—where you could go on a Saturday afternoon, two days after it opened, and go to the mystery section and still find terrific books because the mystery section wasn't very popular. This is how I discovered a first edition of Anna Katharine Green's *The Leavenworth Case.*

DLB: Was that the biggest sleeper you ever found?

Penzler: That was the one, yeah, sure. It's now a couple-thousand-dollar book. And it was literally a quarter; it was sitting there. I think with agony of returning books that I already had, like *The Thin Man.* I must have seen five copies of *The Thin Man* in a two-year period for fifty cents or a dollar. And I said, "No, no, I already have it. I don't need another one." Who knew? So I started to collect this field because I realized in short order that I couldn't collect all of American and English literature, which is what I had set out to do. But then when I saw the prices of the English Romantic poets, for example, and the other stuff, I realized it was hopeless. So I started to focus a little bit on Sherlock Holmes first, and then other mystery-fiction things, which I started to read as well and really liked. I was an English major in college and read all that serious stuff, and loved reading mystery fiction for fun. And then, I began thinking mystery fiction really needed something better than it was getting. It needed better books, better-quality materials, someone to treat it with the reverence that it deserved—with writers like Ross Macdonald working and not really being given the respect that I thought they deserved. And then, as time went by I really started to think about what the great attraction to mystery fiction was for me. Why did I think it deserved all of this credit? And why did it appeal to me so much, when say, science fiction didn't, or Westerns didn't, or whatever didn't? I came to the conclusion, without realizing it at the time, that it had to do with a philosophical premise—which was political conservatism—that the mystery form embodied. And by that I mean that the mystery novel is a very conservative form. In a mystery novel there is a world, a society, which is ripped apart by an antisocial act—murder. And then the rest of the book is spent trying to restore that social fabric that existed at the beginning of the book, by which I mean it sets out to capture the criminal and pluck him away from that society and punish him for what he did to it. It's a very conservative notion. I had no idea that this appealed to me while I was reading these books and enjoying them. I still never read a book with that thought in mind, but I know that underlying my own philosophy, my own politics, all of that pertains. And the mystery fiction fulfills that.

Falsifying Hemingway

The wrong way to publish authors' correspondence is to correct and cosmetize the letters for the sake of readability. The editors who perpetrate such atrocities argue that they have the right – even the duty – to gussy up the letters in order to protect dead authors from embarrassment. But as Ernest Hemingway liked to say, "The worms won't mind."

The proper way to publish authors' correspondence is to print the letters exactly as written, retaining all the errors or putative errors of the original document. Because an author's letters are normally not intended for publication – although some authors clearly conduct their correspondence in the expectation of eventual publication – the printed letter text should convey the writer's state of mind and replicate his natural expression. Even so, the intention of providing an exact transcription of a letter is not fully achievable – except by means of facsimile. In transcribing holograph letters the editor is compelled to make many interpretations and judgments: Is an undotted *i* an *i* or an *e*? Is it a hyphen or a dash? Is it a regular dash or a long dash? Are dash lengths meaningful? Is it a capital letter or a large lowercase letter? Are the variable spaces between words intentional and meaningful? Many of these problems also arise in transcribing typed letters, as well as new problems: What to do about strikeovers? What about parts of words that did not get inked – e.g., "lov r."

Mindful that to correct is to falsify, the competent editor endeavors to protect an author's published letters from the ravages of fancied correction. It is not easy to be an honest editor because the enemies of accurate transcription commit their violations in the sure and certain knowledge that anything they do to a published letter is permissible or even laudatory. It is difficult to anticipate the actions of people who don't know how to do their jobs.

In 1996 Scribner published *The Only Thing That Counts,* my edition of the Ernest Hemingway/Maxwell Perkins correspondence. Over my protests Scribner permitted *The New Yorker* to butcher letters for prebook publication ("Three Words" [24 June and 1 July 1996], 73–77). I was informed by my Scribner editor that someone at that magazine with the title of "the grammarian" demanded the power to alter Hemingway's punctuation – or else no *New Yorker* deal. "Tell 'em to go away," I instructed her. "But you don't understand the power of *The New Yorker,*" she told me.

The magazine's headnote stipulated: "The following are excerpts from letters recently released from private collections; the punctuation has been modified." Two lies, one partial lie, and one deception. The letters are not in private collections; they are not recently released; much more than punctuation modification has transpired; and the deletions are concealed.

Ernest Hemingway resisted editorial tampering and fought for his words. Accordingly, when *The New Yorker* printed Hemingway's letters about his endeavor to have *A Farewell to Arms* published as written, that magazine arrogated the right to falsify the documents. Scribner, the relic of Charles Scribner's Sons, capitulated.

– M.J.B.

Hemingway (16 February 1929)

    I know, on the other hand, that you will not want to print in a magazine certain words and, you say, certain passages. In that event what I ask is that when omission are made a blank or some sign of omission be made that isnt to be confused with the dots that writers employ when they wish to avoid biting on the nail and writing a hard part of a book to do.
    Still the dots may be that sign--I'm not Unreasonable--I know we both have to be careful because we have the same interest i.e. (literature or whatever you call it) and I know that you yourself are shooting for the same thing I am. And I tell you that emasculation is a small operation and we dont want to perform it without realizing it.
    Anyway enough of talking--I am not satisfied with the last page and will change it--but the change will in no way affect the serializability--(what a word.)

The New Yorker

    I know, on the other hand, that you will not want to print in a magazine certain words and, you say, certain passages.

Hemingway (16 February 1929)

--He seemed strong as a tarpon ½ hour later Charles hooked a Mako shark out in the Gulf and fought him 3 hrs and 20 minutes--The shark jumped like a tarpon--Straight up high clear of the water--He lost him--after finally bringing him to gaff (and having the heaviest tarpon rod break) when the hook straightened--all the other boats came around to see the fight--We were way out where the big tankers were passing--Mike gaffed him 3 or 4 times but he couldn't hold him. I never saw a fish jump more beautifully--We hoped of course he was a Marlin--he was bigger than the one at Casa Marina--but he was a shark of a sort nobody had ever seen around here--So we called him a Mako--the kind they have in New Zealand that jumps so wonderfully--

The New Yorker

He seemed strong as a tarpon. ½ hour later Charles hooked a Mako shark out in the Gulf and fought him 3 hours and 20 minutes. The shark jumped like a tarpon--straight up high clear of the water.

Portion of Hemingway's 3 October 1929 letter to Perkins

Hemingway (3 October 1929)

　　It looks as though the jacket designer had been so wrapped up in the beautiful artistic effort on the front that she had tried to eliminate if possible the title and author's name as they wouldn't intrude on the conception of that nude figure with those so horrible legs and those belly muscles like Wladek Zbyszko's who is labeled Cleon (a character in the book I presume or the spirit of no sex appeal) and the big shouldered lad with the prominent nipples who is holding the broken axle (signifying no doubt the defeat of The Horse Drawn Vehicle)

The New Yorker

　　It looks as though the jacket designer had been so wrapped up in the beautiful artistic effort on the front that she had tried to eliminate if possible the title and author's name so they wouldn't intrude on the conception of that nude figure with those so horrible legs and those belly muscles and the big shouldered lad with the prominent nipples who is holding the broken axle (signifying no doubt the defeat of The Horse Drawn Vehicle).

Hemingway (26 July 1929)

　　I'm sick of all of it. Of course I have nothing to complain of. You have been swell (what a lousy word to mean so much) consistently. But I am sick of writing; of the disaster of a family debacle; of the shit-i-ness of critics--(Harry Hansen IE Naughty Ernest in the World which Wister just sent me) of damned near everything but Pauline to get back to Key West and Wyoming--Paris has been--nasty enough.

The New Yorker

　　I'm sick of all of it. Of course I have nothing to complain of. You have been swell (what a lousy word to mean so much) consistently. But I am sick of writing; of the disaster of a family debacle; of the shit-i-ness of critics--of damn near everything but Pauline and to get back to Key West and Wyoming--Paris has been--nasty enough.

Die Fürstliche Bibliothek Corvey

Nancy Emery and Charles Egleston
University of Colorado at Boulder

Die Fürstliche Bibliothek Corvey, popularly known as Bibliothek Corvey, is located in Schloss Corvey, two kilometers from Höxter, North Rhine–Westphalia, on the banks of the Weser River. It is a collection of seventy-two thousand works, many of them rare or unique English, French, and German publications from the late eighteenth and early nineteenth centuries.

Schloss Corvey has its origins in a monastery, although the Bibliothek Corvey holds no books or manuscripts from either of the two monastic libraries that have been on the site. In the early ninth century Saxony had come under the dominion of the Emperor Charlemagne, who, in order to ensure the prevalence of Christianity, commissioned many ecclesiastical institutions. Among these was Corbeia Nova (Corvey), an imperial Benedictine monastery. In 822, during the reign of Emperor Louis (Louis the Pious), land was given to Saint Adalhard (the first abbot). One of Adalhard's first tasks was to design a building, and he chose as his model the Abbey of Corbie, located on the Somme River in Picardy in the north of France. It was from here that the founding monks had come.

Louis gave the Abbey of Corvey the relics of Saint Stephen (hence, Saint Stephen's Church, which is on the site). In 836 Abbot Hilduin of Saint-Denis transferred the relics of Saint Vitus from Saint-Denis in France to the abbey, and it was from Corvey that the cult of Saint Vitus spread over Saxony and north Germany (Saint Vitus is patron saint of actors and dancers, and the protector of epileptics and against all injuries that beasts can do to mankind).

Among the significant persons associated with Corvey in the ninth and tenth centuries are Wala, Adalhard's brother and abbot after his death; Saint Ansgar, the apostle to the Nordic countries; Saint Bruno of Cologne (the son of Henry I of Germany and Saint Matilda), who was the abbot of Corvey, later archbishop of Cologne, and ultimately chancellor of the Holy Roman Empire under his brother Kaiser Otto I; and Widukind of Corvey, who wrote one of the first histories of the Saxons.

The Abbey of Corvey and the Abbey of Bursfeld, which also is located on the Weser, participated together in the German monastic reformation of 1505 known as the Bursfeld Congregation. Unlike Bursfeld, which for a time had coabbots, one Catholic and one Lutheran, and which finally became Lutheran, the Abbey of Corvey remained Catholic during the Protestant Reformation.

The earlier library of the monastery and many of the buildings were destroyed in 1635 during the Thirty Years' War. A second monastic library was founded in 1714; its collections were dispersed in 1803. F. W. Hall's *A Companion to Classical Texts* (1913) reports that some manuscripts formerly held at Corvey are at Wolfenbüttel, Paderborn, and Marburg. Lists of Corvey manuscripts are in volumes 4 and 8 of *Archiv der Gesellschaft für ältere deutsche Geschichtskunde* (1820–1872) and in Paul Joachim Georg Lehmann's "Corveyer Studien" (*Abhandlungen der bayrischen Akademie der Wissenschaften; Philosophisch-Philologische und Historische Klasse*, 1919).

Notable manuscripts with a Corvey provenance include the *Annales Corbeienses* and the *Translatio S. Viti*. The best-known manuscript associated with Corvey is the only surviving copy of the first six books of the *Annals* of Tacitus, the so-called *Codex Mediceus I,* which was written in the ninth century. In 1509 the manuscript was brought to Rome, where Pope Leo X (Giovanni de' Medici) had the first complete edition of the *Annals* printed in 1515. During the Renaissance the *Annals* was a major influence on, among others, Niccolò Machiavelli, who made use of it in *The Prince* (1532). *Codex Mediceus I* is now in the Laurentian Library in Florence.

Rebuilding of the abbey buildings began in 1699 and continued into the middle of the eighteenth century. By the end of the century interest in monastic vocations among the German nobility had declined; as an imperial abbey, Corvey was only supposed to accept noblemen as novices. In 1794 the abbey dissolved, and the property was converted into a princely bishopric and remained so until 1803. In 1820 the complex became the property

Schloss Corvey

of Victor Amadeus, Count of Hessen-Rotenburg, as a compensation asset of the Peace of Lunéville (1801). At his castle in Rotenburg on the Fulda River, he and his wife, Elise von Hohenlohe-Langenburg, had gathered a collection of approximately 1,400 volumes in 1790, almost exclusively in French. By 1796 this number had grown to 6,728; it is now considered to be the nucleus of the Corvey collection. The couple became avid collectors of German, French, and English belles lettres and popular fiction, buying from different booksellers in Germany and during their travels to England and France. Although they remained at Rotenburg, their ultimate collection of approximately 36,000 volumes was moved to Corvey in the years 1825–1833.

After the death of Victor Amadeus, Schloss Corvey was bequeathed to Prince Victor of Hohenlohe-Waldenburg-Schillingsfürst and his brother Chlodwig, the sons of Elise's sister, Constanze, Princess of Hohenlohe-Schillingsfürst. Victor (the first duke of Ratibor and prince of Corvey) had the collection at Corvey organized according to the accepted practice of the time, but this arrangement was considered arcane by 1840. In that year Victor charged his administrator and chamberlain, Dedié, with reorganization of the library. Dedié used a classification scheme then in place at the public library in Hamburg and finished his cataloguing in 1850.

In February 1860 Prince Victor met the German nationalist August Heinrich Hoffman (Hoffman von Fallersleben) in Berlin. Hoffman, a poet, philologist, and historian of literature, is best known for his authorship of the poem that became the German national anthem in 1922, "Deutschland, Deutschland, über alles." Prince Victor had learned of Hoffman's talents through Princess Marie of Sayn-Wittgenstein, who was married to Victor's brother, Prince Konstantin von Hohenlohe-Schillingsfürst. Prince Victor appointed Hoffman librarian in May 1860.

Hoffman von Fallersleben remained librarian until his death and is buried in Saint Stephen's churchyard. He accomplished two major projects for the library during his tenure. He expanded the collection to around seventy-two thousand volumes, including among the acquisitions scholarly materials (for example, in history and the sciences), and he made refinements to the catalogue left by Dedié.

Bibliothek Corvey contains many items that are richly bound so as to fit the tastes of its royal proprietors and thus is an excellent repository for the study of binding in the late eighteenth and nineteenth centuries (see *Prachteinbände des Historismus aus der Fürstlichen Bibliothek zu Corvey*, 1995), but the primary importance for scholarship of the library is that it provides an accurate cross section of popular reading in Europe during that period. The subjects and genres of the collection vary, yet there are particular strengths in fiction and drama. In belles lettres and popular fiction, nearly 50 percent of the holdings in French and German are found nowhere else. The library has more than 600 German novels published from 1815 to 1830 and 154 German drama collections published from 1805 to 1832. Ten percent of the latter are held only in the Bibliothek Corvey.

The Bibliothek Corvey contains 58 percent of the total of the English belles lettres and popular fiction published from 1780 to 1829 and 93 percent of English publications of this type published from 1818 to 1829. Corvey's English-language fiction holdings during the period 1780 to 1830 far exceed comparable collections in either the United Kingdom or the United States, both in size and uniqueness. According to John Graham's *Novels in English* (1983), "the Corvey collection contains 2,107 novels in English, almost exclusively first editions." He notes that very few of these novels were published before 1786.

A large percentage of the fiction found in the Bibliothek Corvey is considered by most literary scholars to be trivial, popular work. However, the library's holdings provide much insight into the context of classics of the Romantic period. Social historians and sociologists have found the collection to be an excellent indicator of the literary tastes of the time. The collection opens large new areas of research, including studies of the process of the publication of novels and other aspects of the publishing industry, such as the interdependence of "higher" literature (read by a relatively small population) and popular literature (read by the masses).

In 1985 the state of North Rhine–Westphalia, through the Universität-Gesamthochschule Paderborn and Franz-Albrecht Metternich-Sándor, Prince of Ratibor and Corvey, agreed to cooperate to improve access to the Fürstliche Bibliothek Corvey for the academic community. According to Hartmut Steinecke, a professor at the university, the goal for access has been twofold: complete cataloguing and reproduction of the library's holdings.

A contract was made between Belser Verlag and the Universität-Gesamthochschule Paderborn to reproduce selected titles from the fiction and belles lettres collections of Corvey; subsequently, the microfilm division of Belser has been separated from the parent company and is now known as Belser Wissenschaftlicher Dienst. Books from Bibliothek Corvey were transported to the library at Paderborn, and a microfilming studio was set up there in addition to terminals for data entry that were connected to the Hochschul-Bibliothekszentrum (HBZ), of which Paderborn is a member.

Cataloguing of the collection is being done on HBZ to facilitate access to Bibliothek Corvey materials, as many North Rhine–Westphalian libraries participate in shared cataloguing through the system. Funding for cataloguing has been provided by the Thyssen-Stiftung, the Deutsche Forschungsgemeinschaft, the Ministerium für Wissenschaft und Forschung des Landes Nordrhein-Westfalen, the Rektorate und Forschungskommissionen der Universität-Gesamthochschule Paderborn, and the Förderverein Fürstliche Bibliothek Corvey. At this time the cataloguing is 94 percent complete (a printed catalogue is planned). The Bibliothek Corvey holdings of imprints in English before the end of 1801 (304 titles) are catalogued in the English Short Title Catalog (ESTC). These particular catalogue records are available through the Research Libraries Network (RLIN).

The Universität-Gesamthochschule Paderborn connection with the Bibliothek Corvey continues to be maintained through the Projekt Fürstliche Bibliothek Corvey, which has sponsored two symposia to date (1990 and 1993; the proceedings of the first were published in *Die Fürstliche Bibliothek Corvey*, 1992). The project sponsors a printed journal (*Corvey-Journal*) and a facsimile series (*Edition Corvey*), which is available from Belser Wissenschaftlicher Dienst. The current director of the Projekt Fürstliche Bibliothek Corvey is Rainer Schöwerling, a professor of English at the University of Paderborn.

Titles published to date in the Belser series are Willibald Alexis's *Schloss Avalon*, E. T. A. Hoffmann's *Die Vision auf dem Schlachtfelde bei Dresden*, Albert Montémont's *Voyage aux Alpes et en Italie*, Auguste Ricard's *La grisette*, Carl Töpfer's *Muck-Kobold und Peter Meffert*, Hippolyte Vallée's *Le souterrain de la forêt des Ardennes*, R. Plumer Ward's *De Vere*, Voltaire's *Memnon*, Daniel Lessmann's *Luise von Halling*, August Klingemann's *Faust*, and Theodor Ernst's *Verrath und Rache*.

Belser Wissenschaftlicher Dienst's microfilming of much of the best of the fiction collection of the Bibliothek Corvey was completed in 1991. Their *Edition Corvey* set (ISBN 3-628-9099-9, as distinguished from a facsimile series, *Edition Corvey*) comprises 9,572 titles from nearly 30,000 volumes written in English, German, and French (28,202 microfiche). Holdings are searchable on the Internet at their website: (http://germ2.uik.ac.at/edition-corvey/e/). The microfiche is full text. For the convenience of scholars with special interests, the set also is available in parts as language editions (English language, 3,261 titles; German language, 2,653 titles; and French language, 3,658 titles); as genres within those languages (novel, short story, short prose, drama, poetry, essay); as compilations by literary period; and as women's literature. For libraries, the publisher can provide catalogue records in MARC format. Belser Wissenschaftlicher Dienst has published theme-specific bibliographies in conjunction with their microfiche set; for example, *German Language Women's Literature of the 18th & 19th Century: Bibliography from the Edition Corvey* (1994).

One current user of the Belser Wissenschaftlicher Dienst microfiche is the Sheffield Hallam Corvey Project, which is based at Sheffield Hallam University. Its home page (15 November 1996) says, "the initial stage of this project . . . focuses upon writings by women and on issues of gender and genre. The intention is to enable us to create a fuller picture of the diversity of women's writing in this period, the ways in which such writings draw on or reformulate existing models, the ways in which they engage with contemporary intellectual and cultural debates, [and] the changing ideologies of femininity as realised by women's writings of the period." The project's director is Judy Simmons, a professor at Sheffield Hallam.

Georg Olms Verlag has produced two full-text microfiche sets of nonfiction from Corvey, *Microedition der Sachliteratur* (ten thousand volumes mainly focused on history and geography) and *Anti-Napoleonische Pamphlete* (pamphlets in opposition to Napoleon and the Napoleonic Wars).

Schloss Corvey remains in the possession of the descendants of Prince Victor. The buildings (including the library) are open for tours by the public, and there is a museum that features exhibits on the history of the buildings and the library. Scholars desiring access to the Bibliothek Corvey collection should contact Projekt Fürstliche Bibliothek Corvey (Universität-Gesamthochschule Paderborn, 33098 Paderborn, Warburger Strasse 100, Germany).

The Mercantile Library of New York

Harold Augenbraum
Director, Mercantile Library

The Mercantile Library of New York was founded in 1820 by William Wood, a native of Boston who devoted a great deal of time and money to the establishment of institutions for the benefit of clerks, apprentices, sailors, and prisoners. He is often credited with first developing and putting to practical use the plans devised by Benjamin Franklin for the proliferation of lending libraries.

Wood had been a dealer in glass and earthenware in Boston at the beginning of the nineteenth century and afterward did business in London. He founded a library for clerks in Liverpool and, a few months after he returned to the United States, founded the Mercantile Library of New York. He also later helped found similar institutions in Brooklyn, Philadelphia, Albany, New Orleans, and other places. In New York he convinced several young businessmen that they could improve themselves and their situation by investing a few hundred dollars that would keep their employees away from taverns and billiard rooms. Meetings were held at the Tontine Coffee House on Wall Street; funds were raised from public-spirited merchants; books were purchased and contributed; and the Library was opened on 2 February 1821 at 49 Fulton Street. In later years the Library would reside in Cliff Street (off Fulton), the corner of Nassau and Beekman Streets (Clinton Hall), Astor Place (on the site of the old Italian Opera House), and, finally, beginning in 1932, its present location at 17 East Forty-seventh Street, in the heart of midtown Manhattan, in an eight-story building designed by the firm of Henry Otis Chapman. Funding for Clinton Hall was raised by the Clinton Hall Association, a newly formed group of older merchants with a desire to help their younger colleagues, who sold shares to raise construction funds; subsequent buildings were financed by the sale of the previous one.

The mid nineteenth century was the heyday of membership libraries. Hundreds of these libraries had been founded in the second quarter of the century, from Boston to Saint Louis, and they continued to flourish. By 1870 the Mercantile Library was the fourth largest library in the United States, with a collection of more than 140,000 volumes and a membership of almost eleven thousand; it was visited by more than one thousand people a day and circulated hundreds of thousands of books, newspapers, and periodicals a year—including twelve thousand by delivery wagon and an elaborate system of small depositories similar to postal boxes. Novels were the most requested; and multiples of the most popular works were purchased and put into the collection (Benjamin Disraeli's *Lothair* was represented by 700 copies; Mark Twain's *Innocents Abroad* by 115, and Louisa May Alcott's *Little Women* by 250). By the 1860s evening classes in modern languages, elocution, music, and phonography (a shorthand system based on sound) were held at the Library, and a small museum was established.

In the early 1870s the Library's Board of Trustees initiated a series of lectures at the Library and off-site that established it as one of the foremost sources of public cultural programs in New York, especially in literary topics. Bret Harte and Henry Ward Beecher were popular speakers; Twain also lectured several times during the decade. Despite such successes, there was great dissent over the cost of the talks, which frequently lost money, and they were discontinued, not to be revived to any great extent until the 1990s.

By the 1890s, however, free (tax-supported) public libraries began to appear. In New York City the Astor, Tilden, and Lenox libraries merged and became The New York Public Library, in the process also becoming the foremost public research library in the city. With the availability of Andrew Carnegie's foresighted capital grants, the city's branch library system grew. The impact on the Mercantile Library of New York and its sister membership libraries was enormous. Most membership libraries began to drift into oblivion: some were absorbed into stronger institutions; others simply closed, and their collections were dispersed. The Mercantile Library survived as a general circulating collection through the foresight of its founders and the Clinton Hall Association's fund-raising. In effect, the current building is the Library's chief asset

"The Merc," 17 East Forty-seventh Street, New York

and has been used as a source of support through rents, a practice maintained by several other membership libraries around the country.

The early twentieth century was a time of collection building, particularly in fiction and the new subgenre of mystery and suspense. The portion of the Library's collection acquired during its first eighty years today numbers about fifteen thousand titles, two-thirds fiction and the remainder nonfiction. Duplicates have been deaccessioned. The fiction collection, in particular, which numbers about eighty-five hundred titles, is a rich representation of "mid-list" fiction from the century, including many hard-to-find Victorian yellowbacks, such as an 1880 reprint of Jane W. Loudon's *The Mummy*; various editions of classic works by authors such as Charles Dickens, William Makepeace Thackeray, Nathaniel Hawthorne, and Mrs. Humphry Ward; and novels by authors well known in the last century but obscure today, such as Edgar Wallace, Charlotte Mary Yonge, and Anna E. Dickinson. That collection is extraordinary in that about 95 percent of its volumes remain in their original binding. That they have been little used in the past century has increased their longevity, though many are brittle and need cleaning. According to a recent survey, about 25 percent of these titles do not appear on the national computer databases of RLIN (Research Libraries Information Network) or OCLC (Ohio Computer Library Center).

The Library also holds a young-reader's collection of 1,750 titles, which include excellent examples of The Boy Traveller series. This segment of the nineteenth-century collection remains in excellent condition, with gilt bindings and engravings that increase its potential for research. The Library has also retained a collection of more than 5,000 nonfiction titles published during the nineteenth century, including literary biographies, belles lettres, poetry, and drama.

As far back as the middle of the nineteenth century, a focus on contemporary fiction has been one of the hallmarks of The Mercantile Library of New York. The Library's current goal is to build one of the finest English-language fiction collections in the United States. The collections policy for fiction is to acquire from both small and large presses, with continuing emphasis on mainstream fiction, works by lesser-known authors, and the literature of displaced minority groups. Since 1991, when the Library presented a symposium on U.S. Latino literature, the collection of fiction by and criticism about U.S. Latinos has expanded greatly. African American fiction and Asian American fiction (a reflection of a recent series of monthly seminars for New York City high-school teachers, held at the Library once a month) are also a focus.

As part of the Library's building of the fiction collection, it has also concentrated on acquiring mystery and suspense fiction. Assisting in this effort is the Mystery Writers of America, whose national headquarters is located at the Library. The collection of mystery and detective fiction at The Mercantile Library of New York includes about eight thousand titles published between 1901 and 1996, expanding each year by about five hundred titles acquired by purchase and gift; the Library's collection is particularly strong in works published before 1960. The late Ellen Nehr made extensive use of the collection in compiling the *Doubleday Crime Club Companion 1928–1991* (Offspring Press, 1992). Members continue to mine the collection for works by once-popular writers, represented by such hard-to-find authors as Clayton Rawson (series character, "The Great Merlini") and Samuel Hopkins Adams ("Average Jones"); and early, unreprinted editions of works by Craig Rice, Alice Tilton ("Leonidas Witheral"; also known as Phoebe Atwood Taylor),

Frank Presnell, Hammond Innes, P. Wilder, Richard Sale, Edgar Wallace (the Just Men series), Gladys Mitchell, Hilda Lawrence, C. and G. Little, Harold Q. Masur, Mabel Seeley, Frank Packard, Francis Lynde, Harrison Stevens, Lenore Glen Offord, M.V. Heberden, Paul Pine, Alan Green, Carter Dickson (also known as John Dickson Carr), Cleve F. Adams, David Dodge, Robert Finnegan, Anthony Wynne, Fergus Hume, Harry Stephen Keeler, E. W. Hornung, Freeman Wills Crofts, R. Austin Freeman, Eden Phillpotts, and Arthur B. Reeves.

During the late 1980s the dominance of the public library system and forty years of The Mercantile Library's overspending its budget resulted in financial difficulties that almost sent it into bankruptcy. By 1987 the situation had become so acute that the Board of Directors voted to close from June to September, with the objective of refocusing and reopening with a revised mission. Nevertheless, in the following two years the situation deteriorated. Increased membership dues—from $45 to $150—established to try to balance the Library's budget, resulted in steep drops in membership, and the Library was forced to close again. Whether this situation was to be permanent or simply another attempt to right the Library's ship was uncertain. When a *New York Times* article about the Library's closing appeared on 10 August 1989, it seemed to ring the Library's death knell, after almost 170 years. However, the Library's chairman, J. Richard Edmondson, was unwilling to surrender. He and other members of the Board negotiated an eighteen-month commitment from the Clinton Hall Association, which had continued to maintain its support of the Library and had often provided deficit relief, to pledge enough financial support to maintain the Library's general operations, while a new focus was developed.

At that point, because fiction represented 70 percent of the Library's circulation, the Board decided to re-create the Library as a literary center-*cum*-literature library, with other literary organizations recruited to become partners, use the Library's building for literary programming, and give the Library broader appeal among the general public. The current director, Harold Augenbraum, was hired to carry out this plan.

Beginning in 1990, the last year the Clinton Hall Association was asked to provide deficit relief, the Library refashioned itself into a literary center based on a humanities approach to the study and enjoyment of fiction as an art form. The Library itself was divided into three divisions: the Library, the Literary Center, and the Center for World Literature.

The circulating library consisted of the collection and The Reading Room. The collection was weeded of thousands of volumes that were no longer appropriate for a circulating collection, especially outdated technical books. A new periodical-subscriptions policy was initiated to reflect the new, literary mission. The collection began to build and attract new members whose principal interest was fiction, from the most popular to the most esoteric. By acquiring a broad, comprehensive array of fiction, the collection was being prepared for later use as both a circulating and research collection, based on the idea that fiction for both entertainment and research is meant to be read at leisure.

To create The Literary Center, the Library's Board of Trustees and Director recruited small literary organizations, some of which wished to use the Library's Reading Room for their public programs and be included in the Library's publicity and marketing. The Trollope Society of America, dedicated to the publishing and study of the works of Anthony Trollope, took up residence in 1990, followed by the Mystery Writers of America, the Augustus Saint-Gaudens Memorial, Playwrights Preview Productions, The Touchstone Center, and Arethusa, which is dedicated to presenting public programs centered on Italian and Italian American culture. The Writers Studio, a room reserved for writers in which carrels, lockers, and electric outlets for laptop computers are provided, boasts such alumni as double O. Henry Award winner Janice Eidus ("Elvis, Axl and Me"), nominee for the PEN-Hemingway Award A. J. Verdelle (*The Good Negress*), mystery writer William Caunitz (*Cleopatra Gold*), biographer John Loughery (*John Sloan*), and poet-novelist Sapphire (*Push*).

The Center for World Literature has been the most visible aspect of the Library's current work. Beginning in 1990 with a series of readings and talks by and about mystery writers, the Library's literary programming for the public and teachers has expanded each year and now encompasses lectures, panel discussions, symposia, book discussions, and publications. The first major effort came in 1991. Under a grant from the National Endowment for the Humanities, one of four NEH grants awarded to the Library in the past five years, the Center organized and presented "Bendíceme, América," a symposium, five panel discussions, and two books on Latino literature of the United States. Programs included talks by Rudolfo A. Anaya (*Bless Me, Ultima*), Rolando Hinojosa Smith (the Klail City Death Trip series), Gary Soto (*Baseball in April*), Nicholasa Mohr (*El Bronx Remembered*), and Ilan Stavans (*The Hispanic Condition*). The bibliography *Latinos in Eng-*

lish, edited by Library Director Harold Augenbraum and distributed to fifty-two hundred libraries, reached out across the country.

The following year the Library presented a series of public programs on French and Francophone literature, with the *nouveau roman* novelist Alain Robbe-Grillet the most prominent of the speakers. In 1993, for the sesquicentennial of Henry James, under a grant from the New York Council for the Humanities, the Library not only presented a series of lectures, including talks by prominent Jamesians R. W. B. Lewis (*The Jameses*) and Adeline Tintner (*The Cosmopolitan World of Henry James*), but also produced a script-in-hand reading by professional actors of James's first full-length original play, *Guy Domville,* a rarely seen work that was booed and hissed at its London premiere in 1895.

From the fall of 1995 to the spring of 1996 the Library commemorated the centenary of the birth of American critic Edmund Wilson by presenting "The Last of the Public Intellectuals: Edmund Wilson and the American Century," an extraordinary series of panel discussions on Wilson's life and career. Organized by biographer Lewis Dabney, the series included talks by Denis Donoghue, Morris Dickstein, Louis Menand, Arthur M. Schlesinger Jr., Daniel Aaron, James Sanders, Elizabeth Hardwick, and Wilson's publisher Roger Straus. Additional talks later presented at Princeton University were combined with those that took place at the Library and edited by Dabney for *Edmund Wilson: Centennial Reflections,* to be published in one volume in 1998 by the Library in association with Princeton University Press.

In 1996 the Library focused on the life and work of two extraordinary fiction writers. To commemorate the centenary of the birth of F. Scott Fitzgerald, in conjunction with the F. Scott Fitzgerald Society, the Library presented "The Romantic American: F. Scott Fitzgerald and His World," a series of four lectures by, among others, Matthew J. Bruccoli and John Callihan.

This event was followed by a series of public readings, talks, and panels titled "The Enchanter: The Work and Life of Vladimir Nabokov." Underwritten by the New York Council for the Humanities, the series included talks on Nabokov's émigré years, a reading of two of his plays, talks by his son Dmitri and biographer Brian Boyd, and a rare reading of the novel *The Enchanter* by Dmitri Nabokov and actor Michael Tolan, which took place in a Barnes and Noble superstore and was attended by almost two hundred people.

This was also the first year of the Florence Gould Distinguished Lecture in French Literature. On 28 February 1996 "Another Baudelaire (American?)" was delivered by poet and translator Richard Howard, who speculated on what manner of poetry Charles-Pierre Baudelaire would have created had he been influenced by Walt Whitman rather than by Edgar Allan Poe. The second Gould Lecture, scheduled for March 1997, is to be delivered by Canadian short-story writer and longtime Paris resident Mavis Gallant, who will speak on The Dreyfus Affair in literature, with a focus on the works of Marcel Proust and Anatole France. The talks will be collected and published every three years, with the first volume scheduled for 1998.

In 1995, under a grant from the National Endowment for the Humanities, the Library presented "The Latino Cultural Thread: Five Centuries of Latino Literature," its first series of monthly seminars for New York City high-school teachers of English. This led to a grant from Lucy and Phil Suarez to publish *Teaching U.S. Latino Literature*, a collection of essays by prominent American scholars and guides for U.S. high-school teachers on how to integrate U.S. Latino literature into the public-school curriculum. This series was followed by "Intersections: Asian-American Literature in Context," which began in the fall of 1996 and will continue to the middle of 1997.

The past year also brought a change in the Library's leadership, as J. Richard Edmondson retired and the mantle of governance was assumed by Mrs. James Benenson Jr., the first female chair in the Library's history. Though Edmondson's tenure had left the Library in healthy financial condition, Benenson set out to create the Library's first permanent endowment. Using the Library's 175th anniversary as a springboard, she and the other board members are developing ways to increase the public's awareness of the Library's rich collection, commission an institutional history, and develop strategic plans for the coming century. Current plans include the restoration of the Library's first floor, which deteriorated badly during the years of financial difficulty, to its original 1932 appearance. The Library has contracted with Beyer Blinder Belle, New York's foremost architectural firm specializing in historic restoration, to design and oversee the restoration.

As the Library and its divisions develop, its Board has stated that its goal will be to foster the development of an ongoing public discourse on fiction, with the active participation of academic and other scholars whose significant work can be brought into the public eye.

The American Trust for the British Library

Andrew Digby and Roy Sully
British Library

The American Trust for the British Library (ATBL) is a charitable body set up to develop the collections of American books and newspapers in the British Library. The trust was established in 1979 by Lord Eccles, chairman of the board of the British Library from 1973 to 1978; bibliophile Arthur Houghton; and Douglas Bryant, librarian emeritus at Harvard University. Bryant became executive director of the trust. The same year he oversaw the formation of an advisory council composed of American academics and librarians.

The collections that the British Museum Department of Printed Books (which became the British Library in 1973) has founded are rich in American materials. In the 1830s the department adopted an omnium-gatherum acquisitions policy and attempted to collect "all the books worth reading" no matter where they were published. In 1846 Anthony Panizzi, keeper of printed books, commissioned an American book agent, Henry Stevens, to "sweep America for books." By the time Panizzi retired in 1866, the museum had, as a result of their efforts, the foremost collection of American books outside of the United States. But after Panizzi's departure, the museum could not secure the funds to continue acquisition on the same scale, nor was it able to until the end of World War II. The war brought a further setback for the museum: in the course of an air raid in May 1941 bombs destroyed a large area of the book stacks. More than 250,000 books were lost, including many volumes of important American material.

By the late 1970s the gaps in the British Library's collections had become a matter of serious concern. It was to this issue, therefore, that the trust turned its attention. To determine the extent of the problem, the trust commissioned a series of subject bibliographies. Each bibliography, compiled by an expert in the field, listed important books, periodicals, and newspapers published in the United States between 1880 and 1950. The trust raised money for the bibliographies to be checked in London against the library's collections. What emerged was that the collections of material published by academic and trade publishers were good, but that there were significant gaps in the collections of works published by small presses and publishers from the Midwest and West Coast.

Once the library had a picture of the strengths and weaknesses of the collections, the trust was ready to embark on the second stage of its acquisitions program. Filling the gaps in the collections by buying from the rare book market would have proved too costly and time-consuming. Instead, the trust arranged for American libraries to prepare microfilm copies from works in their own collections of the books the British Library needed. The trust also raised the money for this work to be undertaken. It was an arrangement from which the British Library and its American partners benefited. The British Library acquired texts of rare books that would have been otherwise impossible to obtain; its partners were able to prepare microfilm surrogate copies of material that was in a poor state of repair, thus extending their conservation programs. Once a master negative microfilm had been produced for the trust, further positive copies could be taken from it for use by other libraries and scholars. The program thereby made many important and rare American books widely available for the first time. All material filmed for the British Library was reported to the United States Register of Microfilm Masters.

The British Library has to date received 13,200 titles on microfilm and will have received a further 2,500 by the time the acquisition project is completed. The trust has also provided the money for the library to replace more than 5,000 of the books lost in the war. With the trust's support the library has also acquired other rare and important material, including several works of American Judaica for the Oriental and India Office Collections and several files of newspapers for the Newspaper Library. It has also been instrumental in arranging for donations to be made to the library, most notably an important collection of Rudyard Kipling's American first editions. In the seventeen years since its foundation, the trust has raised more than

The cherished main reading room of the British Library in the British Museum on Great Russell Street, Bloomsbury

$5 million for the British Library from individuals and from charitable institutions, including the Mellon Foundation, the Pew Memorial Trust, and the Hewlett Foundation.

While the initial program to restore the North American collections is largely complete, the trust continues to raise funds for and publicize the work of the library in the United States. In particular it is about to begin a program of activity to coincide with a British Library touring exhibition, "Let There Be Light," which traces the life and achievement of William Tyndale, the first man to translate the Bible into English from original sources. The exhibition will be seen at the Huntington Library, the New York Public Library, and the Library of Congress.

The trust publishes a newsletter three times a year. For information about the benefits of membership contact The British Library Development Office, 96 Euston Road, London NW1 2DB, England, telephone +44 171 412 7023, or fax +44 171 412 7268, or E-mail roy.sully@bl.uk.

The McKenzie Trust

Ian Gadd
Pembroke College, Oxford

and

Martin Moonie
Somerville College, Oxford

For the past decade D. F. McKenzie has been professor of bibliography and textual criticism at the University of Oxford and a fellow of Pembroke College. Spanning almost forty years, his list of more than fifty books, lectures, and articles covers virtually every aspect of bibliography and gives a particularly comprehensive treatment of the English book trade in the seventeenth century. Professor McKenzie's writing has decisively steered bibliography into the wider arenas of cultural history; the title of his 1985 Panizzi Lectures at the British Library, "The Sociology of Texts," has been adopted as a key phrase for an increasingly dominant field of humanities research. Moreover, his energy, generosity, and enthusiasm as a teacher have fired a generation of researchers at the University of Oxford and at Victoria University of Wellington, New Zealand, where he formerly taught for some thirty years. He retired in June 1996 and is currently Emeritus Professor of Bibliography and Textual Criticism and a supernumerary fellow of the University of Oxford.

THE MCKENZIE TRUST INAUGURATION

On 5 June 1996 Professor McKenzie's sixty-fifth birthday, the McKenzie Trust was unveiled at a surprise lecture and party held at the University of Oxford's English Faculty. Barry Bloomfield, chairman of the trust, explained the aims of its organizers to an audience of more than 250 scholars and students from across the world, including many of Professor McKenzie's friends and colleagues. The trust was set up "to establish an annual public lecture, delivered at Oxford University by a distinguished scholar on the history of the book, scholarly editing, bibliography and the sociology of texts, and to establish the D. F. McKenzie Prize for excellence in teaching as an annual award in recognition of distinction in teaching at this University"—both fields

D. F. McKenzie

in which Professor McKenzie excels. Bloomfield spoke for many participants when he closed his remarks by saying that Professor McKenzie's work and teaching have "imported humanity back into bibliography–to our mutual benefit."

He then introduced the inaugural McKenzie Lecturer, Dr. David McKitterick, fellow and librarian of Trinity College, Cambridge, who delivered a paper on the Cambridge University Press in the eighteenth and nineteenth centuries, drawing, in subject and approach, on much of Professor McKenzie's work on the press. The lecture was followed by a drinks party hosted by the English Faculty and a dinner held at New College.

D. F. McKenzie: A Brief Biography

A native of New Zealand, Professor McKenzie received his early education at Victoria University College in Wellington, where he received his M.A. in 1957, during which time he was also a public servant at the New Zealand Post Office. On the completion of his Ph.D. at Corpus Christi College, Cambridge, in 1961, he returned to New Zealand to teach. He was made professor of English language and literature at the Victoria University of Wellington in 1969, a position he held until 1987. He was then appointed Reader in Textual Criticism at the University of Oxford and was made professor in 1989. At Oxford, Professor McKenzie has been instrumental in establishing the English Faculty's master's course in research methods: an understanding of the sociology of texts and the physical processes of textual production are key components of the course. He was also co-organizer of a companion course teaching students the hands-on techniques involved in hand-press printing–an interest that he also developed in New Zealand, where from 1961 to 1986 he was the founder-manager of the Wai-te-ata Press in Wellington.

During his career Professor McKenzie has held several of highly prestigious lectureships and readerships. He was the Sandars Reader in Bibliography at Cambridge in 1975–1976, the British Library's Panizzi Lecturer in 1985, and the Lyell Reader in Bibliography at Oxford in 1987–1988. Additionally, he was president of the Bibliography Society for 1982–1983 and received their Gold medal for bibliography in 1990. In 1986 he was made a fellow of the British Academy. In 1997 Professor McKenzie will deliver the Clark Lectures at Cambridge.

His study of the Cambridge University Press between 1696 and 1712, begun when he was a doctoral student, led him to challenge many of the assumptions made by bibliographers regarding early modern printing houses. The resulting article, "Printers of the Mind," remains one of the most important bibliographical essays of the later twentieth century.

His exhaustive compilation of the apprenticeship records of the London Stationers' Company has produced three invaluable volumes of indexes for some seventeen thousand apprentices enrolled by the Stationers Company between 1605 and 1800. He has also made several studies and surveys of the London book trade, including his influential Sandars Lectures, "The London Book Trade in the Later Seventeenth Century," delivered in Cambridge in 1976. He has published widely on William Congreve and on printing in New Zealand.

Presently, Professor McKenzie is working on an edition of Congreve, along with an edition of Liber A, one of the key records of the Stationers' Company. He is also one of the volume editors for the *History of the Book in Britain* project. A collection of his writings, edited by Peter McDonald and Michael Suarez, including the previously unpublished Sandars and Lyell Lectures, is set to be published at the end of 1997.

The Future of the McKenzie Trust

As of November 1996 the McKenzie Trust had already raised more than two-thirds of its target of £25,000, and it continues to look for further contributions. This year's D. F. McKenzie Lecture will be delivered by Roger Chartier, directeur d'Études at the École des Hautes Études en Science Sociales in Paris; he will also chair a seminar at Oxford's English Faculty in early June 1997. The first recipient of the McKenzie Prize for Excellence in Teaching will be announced at the same time.

Contributions are handled by the secretary of the trust, Michael Suarez, at Campion Hall, Oxford, OX1 1QS. For further information about the trust or about Professor McKenzie's career and publications, along with words and images from the events of 5 June 1996, a dedicated Web site can be accessed on http://users.ox.ac.uk/~pemb0049/dfmhome.html.

The St. John's College Robert Graves Trust

Patrick Quinn
Nene College, Northampton

The St. John's College Robert Graves Trust was founded in 1994 in order to advance the awareness of both the general public and the academic community about the work of the late Robert Graves. The trust is registered as a United Kingdom charity, and all the trustees are members of St. John's College, Oxford, where Graves studied and took his B.Litt. degree. It is also the college where Graves was later a fellow, when he was elected to the chair of poetry at Oxford. The current trustees are Professor William Hayes, president of St. John's College; Dr. Anthony Boyce, bursar; and Professor John Kelly, senior English fellow.

The trustees have delegated the day-to-day running of the trust to the Management Committee, which includes a representative of the Graves family. The constituent Management Committee was formed by William Graves, son and executor of Robert Graves; Professor John Kelly; and Patrick Quinn, professor of English literature at Nene College, Northampton, and general editor of the Carcanet Press twenty-four-volume Robert Graves Programme. The Management Committee has now expanded to include Dr. Robert Bertholf, curator of the Poetry/Rare Books Collection at SUNY, Buffalo, and Professor Robin Alston, professor of library studies, University College, London.

The trust aims to act as a clearinghouse for projects, studies, and theses on Robert Graves's life and works. To this end it hopes to promote the integrated cataloguing of letters, manuscripts, recordings, photographs, and other extant Gravesiana in institutions in Britain, the United States, and elsewhere so that interested persons can determine what material is available.

The St. John's College Robert Graves Trust hopes to produce a catalogue of Robert Graves holdings worldwide in order to simplify the work of scholars, writers, biographers, librarians, and rare-book archivists, all looking at particular aspects of his work. Although there are similar projects, the idea of a single author database is relatively new. Generally databases concern holdings in particular collections. This database would concern the total Graves opus and would detail all known manuscripts and letters and their locations. This location register of Graves manuscripts and letters would be complemented with a location register of photographs and paintings, sound recordings, and radio and television broadcasts. The location register would be backed by a writings register with transcriptions of his works and available letters and a microfiche library of all his manuscripts and letters.

The ten centers with major holdings of Graves manuscripts, letters, and documents are:

University Libraries, SUNY at Buffalo
Lilly Library, Indiana University
Morris Library, University of Southern Illinois
Harry Ransom Humanities Research Center, University of Texas at Austin
Berg Collection, New York Public Library
University of Victoria Library, Victoria, British Columbia
St. John's College, Oxford
Eton College, England
Canellun, Deia, Mallorca (RG's Library)
MacFarlin Library, University of Tulsa.

There are also many other minor library collections. The manuscripts are mostly in safekeeping, but there are still important sets of Graves's letters in the hands of the recipients.

Many photographers, including Douglas Glass, Tom Weedon, Tom Blau, Toby Molenar, and Daniel Farson, have taken pictures of Graves. Representative copies of these are already on file in the National Portrait Gallery Archives. There are also several drawings and paintings. However, there are many amateur snapshots that should also be collected and copied. Recordings and radio and television material are relatively abundant both in Britain and America, and some material may also be available in Spain and other countries.

The trust further aims to locate manuscripts and letters still in private hands and ensure their safekeeping and preservation. The trust is willing to advise holders wishing to sell their material about reputable establishments that are willing to buy Gravesiana. The trust is pleased to accept donations to augment the already important collection of Rob-

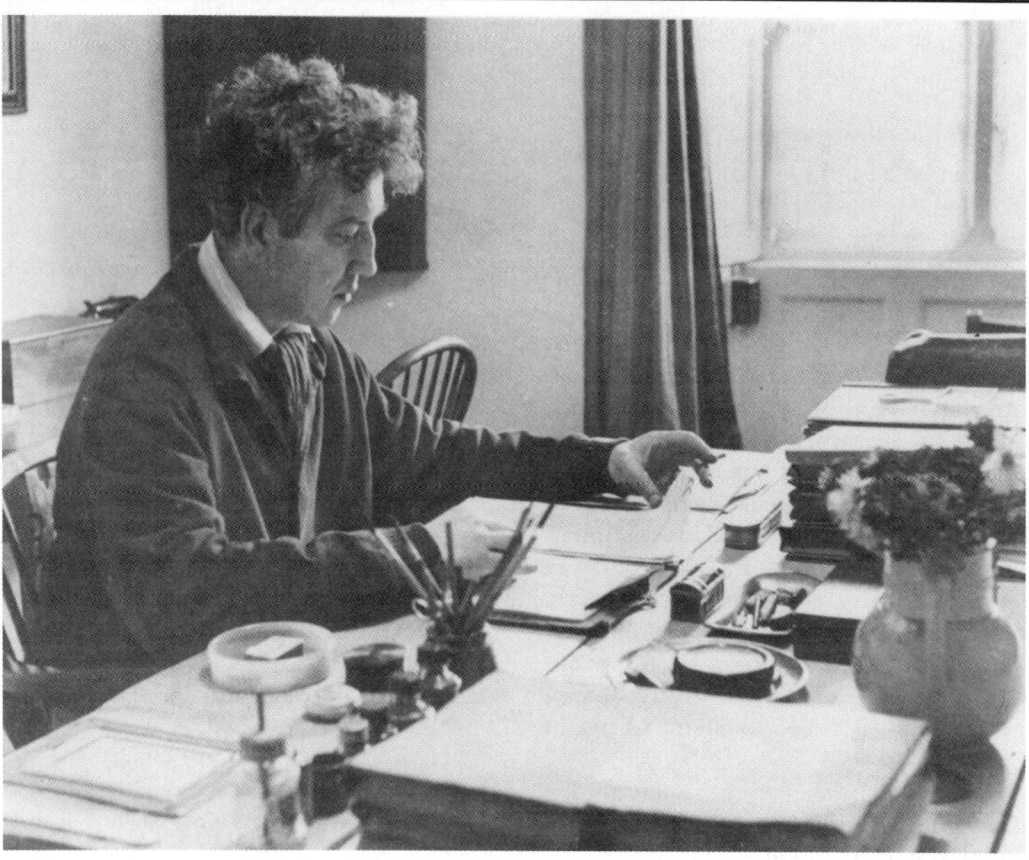

Robert Graves

ert Graves correspondence that is housed in the St. John's Library, where it is looked after by professional archivists. It is the aim of the trust to hold information as a text, copy, or original for every extant letter Graves wrote. By keeping a part-time secretary at St. John's, the trust hopes to assist scholars wishing general information about our holdings as well as personally assisting scholars coming to Oxford by making our records and manuscripts accessible for research.

Finally, the trust lends its support to the Robert Graves Society and its endeavors such as *Gravesiana* as well as assisting with the organization of conferences dedicated to Robert Graves. The trust also works in close conjunction with the general editor and publisher (Carcanet Press in Manchester) of the twenty-four-volume Robert Graves Programme, volumes of which will appear yearly from 1995 to 2004.

The trust address is The St. John's College Robert Graves Trust, St. John's College, Oxford OX1 3JP England. The telephone number for inquiries is 44-1865-277384.

THE ROBERT GRAVES SOCIETY

With the success of the two Robert Graves centenary conferences in Oxford and Deia in 1995, various delegates to the conferences felt there should be a society formed to promote the writings of Graves actively. The society would in some ways support the workings of the St. John's College Robert Graves Trust; however, it would be independent of the trust and would concern itself mostly with offering a center where scholars and readers throughout the world could discover a shared interest in Robert Graves and his circle. The society recognized that interest in Graves and his work embraced a broad range of enthusiasms, for Graves is seen by various people under various categories: anthropologist, mythographer, folklorist, literary critic, novelist, poet, and even military historian.

The desire to create a Robert Graves Society had perhaps always been an impulse of the academic network that had defended Graves's reputation since his death in 1985. Nevertheless, it required the energy and publicity of the centenary conferences and events to galvanize people from the various interest groups around this new undertaking. By the end of the Oxford conference in August 1995, it had become clear that the dynamism generated by the years' events could only be properly and effectively channeled through the creation of an active literary society. At an open meeting on 12 August 1995 the society was formed. The first decision was to make Graves's widow, Beryl Graves, the honorary president. Patrick Quinn was elected president of the society, and it was decided that two branches were needed in order to disseminate information to members more effectively. The North American branch was to include members in the Americas and Asia, while the European branch was to include members in Europe, Australia, and Africa. Dr. Robert Bertholf at the University of Buffalo was elected as the North American vice president, and Dr. Robert Davis at St. Andrews College, Glasgow, was elected European vice president.

The Robert Graves Society exists to promote understanding and appreciation of the achievements of Graves and to enable its members to deepen their explorations of his writings and his career. It pursues these goals in the belief that knowledge of Graves's work expands the boundaries of our understanding of twentieth-century literature and some of the central concepts and experiences that have informed it. The mission of the society, in part, is to place the works of Graves in a broader historical context that takes full account of the literary culture that influenced him and was influenced by him during and after his long, productive lifetime.

The main purpose of the society will be the encouragement of the study and enjoyment of the writings of Graves, and this purpose will be furthered by the establishment through the society of a forum for the sharing and exchange of scholarly and artistic responses to the many aspects of Graves's work as a writer and man of ideas. The society will facilitate discussion, analysis, study, and celebration of the life and work of Graves through support of research programs, conferences, and regional events such as seminars and lectures. The society will exist to serve its members in relation to their shared interest in Graves and will seek opportunities to bring them together at local, national, and international gatherings. It will also act as a clearinghouse for information about Graves-related publications, disseminate information for and from book-seekers, maintain a site on the World Wide Web with up-to-the-minute announcements (http://www.nene.ac.uk/graves/graves.html), and maintain E-mail lists for discussion groups on an endless variety of Gravesian topics.

One of the key objectives of the society will be the encouragement of publications of unpublished primary material, such as Graves's diaries from 1935 to 1939, which offer a study of his life in Deia before the outbreak of the Spanish Civil War and his return to England after seven years of exile. Further, the society hopes to encourage the publication of critical material on Graves through its literary journal, *Gravesiana: The Journal of the Robert Graves Society,* as well as networking with other literary journals. The society works in close cooperation with the St. John's College Robert Graves Trust and is, in turn, fully endorsed by the trust.

Membership in the Robert Graves Society will be open to any individual with an interest in Graves. The annual fee for membership is fifteen pounds sterling or twenty-five U.S. dollars, with one-third of the fees returned to the St. John's College Robert Graves Trust, and the remainder used to support the activities of the society, including publication and distribution of the journal, the ongoing work of the database project, and the promotion of the society throughout the world. Membership forms and information can be obtained from Dr. Robert Bertholf, Poetry/Rare Books Collection, SUNY Buffalo, 420 Capen Hall, Box 602200, Buffalo, New York, 14620-2200 USA, or Dr. Robert Davis, St. Andrew's College, Glasgow University School of Education, Duntocher Road, Bearsden, Glasgow G61 4QA UK.

GRAVESIANA

Gravesiana is the offspring of *Focus on Robert Graves and His Contemporaries,* which appeared at irregular intervals between 1971 and 1995. The seventeen issues of *Focus* were of great importance in keeping Graves's critical reputation alive during the final period of his life, but the journal had begun to concentrate too much on the literature of the Great War and the emergence of modernism to represent the large and varied works of scholarship being done on Graves and his circle. As a result of the centenary conferences, the executive committee of the Robert Graves Society felt it incumbent to establish a literary vehicle to publicize its activities and exclusively promote scholarship on Graves.

Patrick Quinn was asked by the Graves Society to edit the new journal, and Ian Firla was asked to become his deputy editor. In December 1995 work began on getting the journal ready for publication. The mission statement is as follows:

> *Gravesiana* will appear twice a year, in June and December. The journal is dedicated to publishing scholarly articles concerning Robert Graves and the members of his circle. The editors hope also to publish timely reviews of the latest studies about Graves or Graves re-

articles concerning Robert Graves and the members of his circle. The editors hope also to publish timely reviews of the latest studies about Graves or Graves related material. The journal features a biographical section in each issue which will highlight Graves' personal relationships with various literary personages as well as offering personal glimpses of the man behind the poetry, fiction, mythography, and criticism.

Further, as the journal is the literary production of the Robert Graves Society, it will contain information concerning the two branches of the society and publish news from the society's Branch Vice-Presidents, Robert Bertholf in America and Robert Davis in Scotland. *Gravesiana* will also contain material about what individual members are doing, publish information about forthcoming conferences and lectures of interest to those engaged in Graves studies, and act as a clearing house for questions and topical discussion concerning matters Gravesian.

The journal encourages submissions from everyone who reads its pages: the intention is to create a lively and intellectually challenging journal which in some degree reflects the protean spirit of the man to whom it is dedicated.

The first two numbers of the journal appeared in 1996 and were distributed to seven hundred subscribers. The journal costs ten pounds (fifteen U.S. dollars) an issue or annual subscriptions can be obtained for fifteen pounds (twenty-five U.S. dollars); an annual subscription includes a yearlong membership to the Robert Graves Society. Subscription information is available from the society's vice presidents, Dr. Robert Bertholf, Poetry/Rare Books Collection, SUNY Buffalo, 420 Capen Hall, Box 602200, Buffalo, New York, 14620-2200 USA, or Dr. Robert Davis, St. Andrew's College, Glasgow University School of Education, Duntocher Road, Bearsden, Glasgow G61 4QA UK.

Information about submitting articles or projects can be obtained from Patrick Quinn or Ian Firla, Department of English, Nene College, Park Campus, Northampton, NN2 7AL England, or questions can be E-mailed to patrick.quinn@nene.ac.uk.

The Canadian Publishers' Records Database

Carole Gerson
Simon Fraser University

The Canadian Publishers' Records Database currently contains some twelve hundred detailed entries describing archival collections relating to the history of secular English-language book publication in what is now Canada from the beginnings until 1980. Access is through the Simon Fraser University Library at http://www.lib.sfu.ca. Choose "Databases," then click on "SFU Library and Community Resources."

Canada's publishing history extends back nearly 250 years. The first printing presses arrived in Nova Scotia in the early 1750s and in central Canada in 1764, subsequent to the conclusion of the Seven Years' War. Later in the century, Loyalists fleeing the American Revolution brought literary culture and printing equipment to the remaining portions of British North America. As the colonies expanded, so did their local print culture; newspapers quickly became one of the major modes of expression and communication, as frequently noted by visitors to Upper and Lower Canada, who remarked on the proliferation and vituperation of the local press.

The development of book publishing in Canada parallels that in other countries in that distinctions between publishing and other aspects of textual production such as printing, distributing, and bookselling did not emerge until well into the nineteenth century. Unlike their counterparts in other English-speaking countries, Canada's book publishers have always operated under the shadow of major imperial powers: first as a political colony of the British crown and subsequently as an economic colony of the ever-expanding American publishing industry. Their inability to secure international copyrights and consequently to compete internationally has left Canadian publishers perennially insecure. Symptomatic of this situation is the fact that before the 1960s no Canadian bookseller was able to operate by selling only books produced in Canada.

On the one hand, the role of Canadian publishers in fostering Canadian culture, and concomitantly Canadian national identity, became increasingly significant during the middle years of the twentieth century. This trend peaked with the nationalism and prosperity of the 1960s and 1970s, which saw the founding of dozens of small literary presses, many of whose authors have since achieved international reputations. On the other hand, since the middle of the nineteenth century Canadian authors of even modest ambition have known that publishing in Britain and the United States is essential to the achievement of any degree of recognition and financial solvency. Consequently, Canadian presses have seldom been able to benefit fully from the success of the best-known Canadian authors, even those they may have launched. Rather, their survival has almost always depended on the bread-and-butter production of locally necessary items such as Canadian-focused textbooks, city directories, and regional reference works, or on their work as agents or distributors for foreign (mostly American and British) publishers.

In its struggle for survival the Canadian publishing industry has proven remarkably tenacious, thereby enacting the endurance theme that some cultural critics regard as an intrinsic component of Canadian identity. While many of the country's professional writers have not been able to afford to publish at home, thousands of others, whose only goal is to see their work in print, have relied on local presses to issue their poems, stories, sermons, and other writings, often at the author's own expense. In addition, local publishers are necessarily the primary producers of materials of primarily local interest, including guidebooks, community histories, and works on Canada. Consequently the arguments raised in recent years on behalf of studying publishers as "architects of culture," in Wayne Templeton's phrase, hold as strongly in Canada as elsewhere. Scholars in many disciplines who research the construction of culture are increasingly aware of the role of publishers as gatekeepers and of the need to fill what John Sutherland has described as the "Hole at the Centre" of traditional literary studies.

Evidence of growing attention to these concerns with regard to facilitating the study of American and British publishing appears in the survey of

```
CPRD id:         0220
Main Entry:      Contact Press
Title:           Contact Press [fonds]
Title Note:      Title has been assigned after interview with archivist.
Repository:
                 University of Toronto.
                 Thomas Fisher Rare Book Library
Telephone:       (416) 978-5285
Fax:             (416) 978-7653
Address:
                 University of Toronto
                 120 St. George Street
                 Toronto, Ontario
                 M5S 1A5
Geog. Zone:      Metropolitan Toronto
Dates for Creators:
                 Birth or Creation Date:       1952
                 Incorporation Date:           No information available.
                 Death or Termination Date:    1967
                 Inclusive Dates of Creator:   1952-1967
Dates for Records:
                 Start Date of Records:        1959
                 End Date of Records:          1966
                 Creation Dates of Records:    1959-1966
Fonds Id:        Ms. Coll. 69
Extent:          1 box
Lin. Measure:    1 m approx.
Media:           Includes textual records.
Admn History:    Contact Press was founded as a poets' co-operative in
                 Montreal in 1952 by Raymond Souster, Irving Layton, and
                 Louis Dudek. They began to publish the experimental poetry
                 magazine, CONTACT, in the same year. The Press's first
                 publication, CERBERUS, was a book jointly written by
                 Souster, Layton, and Dudek. Layton left the Press in 1956
                 and Peter Miller joined in 1959. Contact Press published
                 primarily the work of young Canadian poets, including D.G.
                 Jones, Eli Mandel, Al Purdy, Margaret Atwood, Gwendolyn
                 MacEwen, Octavio Paz, and Anne Hebert.
Records:         The records of Contact Press consist of editorial
                 correspondence (1959-1966) which deals with editorial policy
                 and discusses submitted manuscripts. Accompanying some of
                 the letters are distribution lists of bookstores, lists of
                 Contact Press publications, three orders from bookstores,
                 and three accounts of the Press.
Keywords:
                 small press publishing
                 literature--poetry
```

Printout for Contact Press from the Canadian Publishers' Records Database

American publishers' records recently launched by Professor Beth Luey of Arizona State University, in the Location Register of archival sources for the study of the British book trade compiled by Professor Alexis Weedon of Luton University, and in Nan Albinski's "Guide to the Archives of Publishers, Journals, and Literary Agents in North American Libraries" (*DLB Yearbook: 1993*), which, despite its title, contains no references to Canadian publishing or repositories. In Canada, such experts as Jean-Pierre Wallot, former national archivist of Canada, and I. S. MacLaren, affiliated with the Research Institute for Comparative Literature at the University of Alberta, have both called attention to the need for greater attention to publishing history in English-Canadian cultural studies. To quote further from Templeton: "A study of publishing in Canada is not so much a survey of Canadian publishing as a statement on Canadian culture, of which publishing, writing, and national self-awareness become points of a vital triangle of mediation with publishing at the apex. A nation's self-awareness, in the cultural context, is created by writers . . . a nation's literature in turn is created by publishers."

A thorough study of Canadian literature addresses not just books and authors, but also the particular ways that books and authors have been created in this country. Who gets published, why, and how? What are the procedures and difficulties of getting into print, and what kinds of editorial and market factors intervene in these processes? How do Canadian publishers function in often marginal conditions? What have been the effects of changes in copyright legislation, censorship, and government policy? In French Canada, such questions have been actively addressed through the Groupe de recherche sur l'édition littéraire, founded in 1981 at L'Université de Sherbrooke, and L'Association québécoise pour l'étude de l'imprimé, founded in 1987. In English-speaking Canada, the first stage has been the Publishers' Papers Project, located in the Canadian Centre for Studies in Publishing at Simon Fraser University, which has produced the Canadian Publishers' Records Database. Under the direction of Dr. Carole Gerson of the SFU English Department, and Ann Cowan, codirector of the Canadian Centre for Studies in Publishing at SFU, the research, development, and implementation of the database have been supervised by archivist Dr. Laura Millar.

With the ongoing support of SFU and several substantial grants from the Social Sciences and Humanities Research Council of Canada (SSHRC), the project has developed a database that describes more than twelve hundred archival collections relating to secular English-language book publishing in what is now Canada. In order to accommodate effectively both a broad time spectrum (since 1751) and Canada's vast geographical expanse, the collection and processing of the data were divided into several phases. The database's current cutoff date of 1980 allows for the inclusion of publishers, individuals, and associations founded or active before that date, many of whom remained engaged long afterwards. The next phase, now under way, will add records to cover the period 1980-1995.

Included in the Canadian Publishers' Records Database are the archival and in-house records of publishing companies; the personal papers of individuals involved in publishing, such as authors, illustrators, agents, and editors; the papers of scholarly researchers who have worked on Canadian publishing; relevant government records; and the papers of related groups, such as publishers' and writers' associations. Each collection is described in its own computer record of up to thirty-eight fields. The field designations follow the Canadian archival descriptive standard, Rules for Archival Description, to identify the record's title, dates, scope and contents, extent, and location, as well as the types of publishing represented, the availability of finding aids, bibliographical references, and a detailed administrative history or biographical sketch, as appropriate.

All material recorded in the database is currently located in Canada – in public institutions, publishers' offices, or private hands. Indeed, one of the unique features of this project is its inclusion of material stored on-site on publishers' premises. While these records are usually not as accessible to researchers as are those in public libraries and archives, knowledge of their whereabouts can lead to important research discoveries. Project archivist Dr. Laura Millar points out that the description of records still held by creating agencies is a recognition of the archival realities of the late twentieth century. Fewer and fewer public repositories can afford to acquire and preserve large bodies of corporate records, while at the same time more businesses are choosing to retain their own archival materials rather than transfer ownership away. With the advent of computer networks this "post-custodial" archival environment can be accommodated; it matters less where the records are housed if information about their existence can be made available to researchers. The Canadian Publishers' Records Database has served as a pioneer in recognizing and acknowledging the reality that, as the twenty-first century approaches, the traditional world of centralized preservation of archival records is being replaced by the development of networks of public and private archival repositories.

Furthermore, a significant result of this project has been to alert publishers to the importance of preserving their records. Each publisher surveyed in the collection of project data has received a copy of *Archival Gold,* by Laura Coles (now Millar), a book that specifically addresses the cultural and historical value of publishers' papers and advises publishers on the preservation of their records. Because publishers necessarily focus more prominently on the current business aspects of their enterprise than on its historical value, Canada's early publishing records have largely disappeared because of fire, flood, storage difficulties, business failures, and the past tendency of historians and archivists to privilege the collection and study of political records over attention to personal and business papers, most of the material in the database postdates World War I. In effect, the development of the Canadian Publishers' Records Database participates in the recent reconstruction of notions of historical and cultural value that in the past two decades has shifted attention to both the larger structures that dominate social organization and the participation of ordinary individuals in these structures.

This project was generated by an inquiry from Carole Gerson, a literary researcher seeking Canadian women writers' correspondence with their publishers, to Ann Cowan, one of the founders of the Canadian Centre for Studies in Publishing. Recognizing the absence of an important research resource, Cowan's response was: "Let's do it ourselves." In 1989 the project began with a President's Research Grant from Simon Fraser University that enabled an initial mail survey that established the existence of a sufficient body of publishing-related records to justify further investigation.

The subsequent two-year grant (1990-1992) issued under the now-discontinued Canadian Research Tools Strategic Grants Program of the SSHRC supported the development and testing of the project's methodology in two geographical regions: the province of British Columbia and metropolitan Toronto (currently the center of the Canadian publishing industry), as well as several selected major repositories, including the National Archives of Canada in Ottawa. During this phase the program developed research-gathering instruments (questionnaires and interviews), selected and defined database software (InMagic Plus) and structure, developed a project manual, and trained student personnel.

An important outcome of this initiation phase was the decision to have researchers visit major repositories in person, as publishing has often not been recognized as a distinct subject, and consequently the identification of publishing-related collections sometimes requires considerable sleuthing. As well, the project established two databases in addition to the Main Database, which houses the records now available to dial-up users. They are an Administrative Database recording the name, address, and situation of every publisher, archives, and institution identified, whether or not they were ultimately deemed relevant to the project (some 1,800 in all), and a Deadends Database listing 575 publishers or institutions for which no current address or contact information could be found. With these two databases, as well as the finding aids and other materials gathered during the course of the research, the project office at the Canadian Centre for Studies in Publishing at SFU now houses a research center accessible to visitors and to enquiries from afar (ccsp-cprd@sfu.ca). The project manual is also available for consultation by those contemplating similar projects.

The second phase, supported by additional funds from the SSHRC, received through SFU as well as through the SSHRC's regular Research Grants program, produced a thorough cross-Canada survey of archival repositories, institutions, agencies, organizations, and publishers' offices that was conducted from 1992 through 1994. This work proved to be labor-intensive, over its course employing some fifteen students and graduates who visited publishers' offices and major public repositories, wrote up and entered the data, and proofread and edited the entries. All entries went through several editing stages and were then confirmed with their sources before inclusion in the final database.

There then followed a third, consolidation phase as the project's databases were given final edits and the Main Database was made publicly accessible in October 1995 through the assistance of library and computing services staff at SFU. With a renewed grant to update the project to include the publishing-related records of companies, individuals, and organizations that begin in the period 1980–1995, the project's fourth phase is under way. Because the project's terms of reference are now well known, it can forgo on-site visits to repositories and publishers; consequently this phase can be conducted at considerably less expense through conventional and electronic communication channels, including regular mail, electronic mail, telephone, and fax. A future phase may include adding material relating to Canadian book publishing that is housed in institutions outside of Canada, as well as the possibility of integrating this database with similar databases as such projects proliferate.

As with many large undertakings, this project's methodology evolved somewhat as the research progressed. In this case, the major adjustment reflects the rapid worldwide transition over the past five years from paper to electronic formats for information storage and retrieval. When originally conceived, the goal of the project was to generate published regional reports describing publishing-related archival materials. The database fields were consequently designed to permit the production of printed indexes. But after several years it became obvious that printed guides would be obsolete; moreover, a single database with universal dial-in access is not only less cumbersome and less expensive to produce, but also permits frequent corrections and updating. However, because this format requires rigorous editorial standards to ensure that keyword searches will be accurate and inclusive, it is still necessary to maintain careful proofreading procedures. A less momentous adjustment concerns the handling of authors' papers and other collections that lack itemized finding aids, whose inclusion of material relating directly to book publishing can be difficult to ascertain without an item-by-item perusal of the entire collection. In such cases, the "scope and contents" field includes the qualification that the amount of material relating directly to book publishing has not been determined.

With the Canadian Publishers' Records Database, researchers can now identify records relating to specific publishers, the publishing-related records of individual authors, records concerning specific types of publishing (for example, children's books, travel writing, and education publishing), as well as records relating to a geographical region or those held in a specific repository. Searches may also be limited by date (for example, thirty-one collections include material from the eighteenth century). As it is possible to ascertain the status of particular records in publishers' offices as well as institutions, a researcher may track the records of a particular company through various combinations and mergers or the various companies with which a key individual has been associated. Although only a small proportion of authors can be specifically named in the descriptions of large publishers such as Macmillan or McClelland and Stewart, the researcher working on a particular writer will know the firms with which she or he published and from the author's personal papers may also know about dealings with other publishers.

The database informs the researcher whether these publishers' records have survived, identifies their location if they are indeed extant, and indicates how they can be accessed. One note of caution needs to be sounded regarding the use of this database for researching the larger patterns of Canadian publishing. Because of its accessibility, it can be tempting to use data from the CPRD to analyze the Canadian publishing industry itself, by calculating the numbers of publishers in different regions, for example, or the kinds of publishing performed by Canadian firms. However, the finding that nearly half the records (561) concern the production of literature, whereas there are 126 related to educational publishing, 101 related to the production of scientific books, and 30 concerning the publication of books on business matters, does not describe the output of Canadian publishers, but rather the records deemed worthy of preservation. Clearly, the papers of authors and literary presses have received priority in the acquisition practices of libraries and archives, and because university archives preserve the papers of professors, scholarly publishing is well represented, with 192 records. Canada has also produced its share of cookbooks and sports writing, but the records of their authors and producers tend to disappear.

The Canadian Publishers' Records Database is a comprehensive research tool, prepared according to rigorous editorial and archival standards, which addresses a research audience of many disciplines. Publishers, for their part, are learning about the acquisition practices of archives and are encouraged to maintain and preserve their records. By facilitating investigation in an increasingly important area of scholarly activity, this database streamlines the often time-consuming and haphazard processes of conducting archival research in a country whose geographical expanse often complicates the plans and projects of its scholars.

References:

Richard Giguère and Jacques Michon, "An Approach to the History of Publishing: Twentieth-Century Quebec," translated by Larry Shouldice, *Book Research Quarterly,* 6 (Spring 1990): 55–64;

I. S. MacLaren, "Introduction: Entering the Z Zone," in *Questions of Funding, Publishing and Distribution/Questions d'édition et de diffusion,* edited by MacLaren and C. Potvin (Edmonton: Research Institute for Comparative Literature, 1989), pp. xiii–xxi;

Laura Coles Millar, *Archival Gold: Managing & Preserving Publishers' Records* (Vancouver: Canadian Centre for Studies in Publishing, 1989);

George Parker, *The Beginnings of the Book Trade in Canada* (Toronto: University of Toronto Press, 1985);

John Sutherland, "Publishing History: The Hole at the Centre of Literary Sociology," in *Literature and Social Practice,* edited by Philippe Desan, Priscilla Parkhurst Ferguson, and Wendy Griswold (Chicago: University of Chicago Press, 1989), pp. 267–282;

Jean-Pierre Wallot, foreword to *Archival Gold,* by Millar (Vancouver: Canadian Centre for Studies in Publishing, 1989), pp. vii–viii;

Alexis Weedon, *British Book Trade Archives 1830–1939: A Location Register* (Bristol: Simon Eliot & Michael Turner, 1996).

The Book Arts Press at the University of Virginia

Terry Belanger
Alderman Library, University of Virginia

The Book Arts Press (BAP) was founded by Terry Belanger at Columbia University in 1972 as an institute supporting various programs concerned with the history of books and printing, descriptive bibliography, the antiquarian book trade, and rare-book and special collections. When Belanger became University Professor and Honorary Curator of Special Collections at the University of Virginia (UVa) in 1992, the BAP and its collections moved with him to Charlottesville.

At UVa, the BAP supports courses concerning the history of the book and related subjects. It carries on publication and exhibition programs; it runs an annual summer institute, Rare Book School; and it sponsors public lectures—notably the annual Sol. M. Malkin Lecture in Bibliography.

The BAP's collection of printing presses and equipment includes a full-scale reproduction of a wooden common press (of the sort Benjamin Franklin might have used), a nineteenth-century Washington iron handpress (such presses could be broken down and loaded onto a Conestoga wagon), and a twentieth-century flatbed cylinder proof press (a Vandercook SP-15, favorite of modern private-press letterpress printers). Other equipment includes two hundred cases of printing type (among them the forty-eight-case Annenberg collection of wood type), an assortment of chases and furniture, a large standing press, a Brand etching press, and various pieces of hand bookbinding equipment.

The BAP owns about fifteen thousand books and five thousand prints dating from the fifteenth century to the present. Many of the books—including a large collection assembled to illustrate the history of cloth bookbindings—are on display in glass-fronted bookcases in the Dome Room of the Rotunda (the original library of the university), located a short distance from Alderman Library on the central grounds of the university. Other collections are kept in the BAP's classroom and studio rooms, which, together with the pressroom, make up the BAP's suite in Alderman Library.

Rare Book School (RBS) classes make heavy use of the BAP's collections. This monthlong summer institute annually attracts about 250 students, who come for one or more five-day courses taught by an international roster of specialists in the history of the manuscript book, typography, papermaking, printing, book illustration, leather and cloth bookbinding, publishing, descriptive bibliography, rare-book and special collections curatorship, book collecting, and related subjects. While they are at UVa, many RBS students and faculty members live on the lawn in rooms designed by Thomas Jefferson, the founder of the university.

The physical arrangement of the BAP's book and print collection supports both classroom and independent study. The books are generally shelved by date (rather than by author or subject) to show the chronological development of vellum, leather, cloth, and paper bindings. Most of the prints are filed by technique (rather than by artist or engraver) to facilitate the identification of illustration processes. Other BAP collection arrangements assist the study of various formats, genres, materials, and physical features, such as sewing structures, endpapers, and dust jackets. An unusual feature of some of these collections is the presence of multiple copies (sometimes as many as a dozen or more) of the same (or almost the same) book—a duplication valuable not only for facilitating group viewing in the classroom but also for demonstrating the bibliographical principle that "almost exactly the same" can be another way of saying "quite different."

The BAP also maintains a library of about a thousand recently published books on various aspects of the history of the book: papermaking, typefounding, typography, printing, illustration, binding, publishing, bookselling, collecting, the antiquarian book trade, and related subjects. This noncirculating reference collection ensures that the most useful books for the BAP's purposes are always close at hand. Supplementing this library are much larger holdings on the same subjects in the Al-

Students examining folio sheets in a University of Virginia summer Rare Book School class

derman Library stacks and in various UVa special collections.

Since 1992, the BAP has been responsible for an ongoing series of exhibitions on display in the Dome Room of the Rotunda. The books and other materials shown in these exhibitions are drawn both from the BAP's own collections and from the superb special collections of the UVa library, supplemented by occasional loans from other institutions and individuals.

The subject of recent Dome Room exhibitions has ranged from the evolution of nineteenth-century bookbinding styles to novels and children's books with Charlottesville settings. Books Go to War: Armed Services Editions in World War II, an exhibition that opened in the spring of 1996, was the first Rotunda exhibition to have a student curator (Daniel J. Miller, class of 1996).

The Book Arts Press is supported by a 600-member friends group, the Friends of the Book Arts Press. Since 1976 individual friends have contributed nearly $500,000 to the BAP, as well as many gifts in kind. In addition, more than two hundred libraries have donated unwanted, damaged, and defective books (or parts of books), both old and new, to the BAP's collections. The BAP's relationship to these gifts tends to resemble that of the Bedouins to their camels: little goes to waste. For more information about the BAP and its activities, write to: The Book Arts Press, 114 Alderman Library, University of Virginia, Charlottesville, Va. 22903; telephone: (804) 924-8851; fax: (804) 924-8824; E-mail: books@virginia.edu.

The Glass Key and Other Dashiell Hammett Mysteries

Mark Sutcliffe

In spring 1962 *The Book Collector* published an article by Roger Stoddard discussing the literary career of Dashiell Hammett under the series heading "Some Uncollected Authors." Few collectors of modern first editions, let alone collectors of crime fiction, will fail to see the irony; for today, the first printings of the key works of Dashiell Hammett rank alongside those of Ernest Hemingway and F. Scott Fitzgerald as some of the most highly prized books of the twentieth century—with four- and five-figure price tags reflecting the demand. Yet despite this dramatic rise in interest and the many biographical and bibliographical works that have accompanied it, there are several important questions that have remained unanswered, the most intriguing of which surround Hammett's fourth novel, *The Glass Key*.

Considered by Julian Symons to be "the peak of the crime writer's art in the twentieth century" (*Mortal Consequences,* 1973), *The Glass Key,* like *Red Harvest, The Dain Curse,* and *The Maltese Falcon* before it, was first published in serial form (March–June 1930) by the pulp magazine *Black Mask*. Hammett himself thought it "not so bad" and that "the clews" were "nicely placed." But whether or not the reader is sharp enough to identify these clues (a hat and a stick, in particular), the real mystery is not who killed Taylor Henry in "dark China Street," nor is it the true significance of the key itself (which is not, or course, a physical object, but a metaphysical one, appearing only in a dream). The question whose answer has eluded collectors, scholars, and dealers alike is simply this: why did *The Glass Key* receive its first book publication in England?

In 1930, after writing sixty-five pages of a new book (*The Thin Man*), Hammett, together with his publisher, Alfred A. Knopf, for some reason decided to postpone the publication of *The Glass Key*, scheduled for fall, until the following spring. During the course of that same year, Knopf also decided to discontinue operations at his London of-

Dust jacket for the first English edition of The Glass Key *(London: Knopf, 1931); photography by Virginia Berry*

fice and, following fierce but friendly negotiations, agreed to transfer the British publishing rights for Hammett, Willa Cather, Carl Van Vechten, and others to the London publisher Cassell, early in 1931. Putting these two facts together—the postponement of American publication and the planned closure of the London office—one could infer that Alfred A. Knopf believed it would be in his best interests to publish *The Glass Key* in England before he transferred the rights to Cassell, thus ensuring it carried the Knopf imprint.

The Glass Key was produced at the end of 1930 by the firm that had printed the English edition of *The Maltese Falcon* earlier that year, Richard Clay and Sons of Bungay, Suffolk. Following the recent discovery of a contemporary records book in one of Clay's storage rooms, some interesting facts concerning the printing history of *The Glass Key* have come to light. The size of the first printing, for example, hitherto unknown, reveals itself in the Sales Day Book (entry for 19 December 1930) to be three thousand copies. This same entry records that there were ten proofs and that the total cost to Knopf was £83 3s. 6d., which included £45 18s. for "composing in small Pica type," £1 7s. 6d. "extra for small type," £3 16s. 9d. for "corrections & alterations & supplying . . . proofs," £29 11s. 9d. for "printing from type," and £2 9s. 6d. "extra for title in green and black." On 12 January 1931 Clay's sent Knopf a second invoice, this time for "composing and printing 2,700 copies" of the "jacket wrapper" in "3 colours and black." Eight days later, on 20 January 1931, *The Glass Key* was published in London at 7s. 6d.

The book was bound in blue cloth, lettered on the spine in red (THE / GLASS / KEY / DASHIELL / HAMMETT / KNOPF), with a white key on the front board and on the spine and a red Knopf Borzoi seal on the back board. There is also a variant binding that has the Cassell imprint at the foot of the spine, the "key" motifs in reddish-orange instead of white, and the back board blank. Until now, precedence between these two bindings has not been established, but with the discovery of the Sales Day Book it can be stated conclusively that the Knopf bindings precede the Cassell bindings.

What, then, is the status of the Cassell-bound copies? Are they first-printing binding variants? Or are they second, or even later, printings? It is possible that unbound first-printing sheets were bought and bound by Cassell after they acquired the rights early in 1931; supporting this theory is the intriguing fact from Clay's Sales Day Book that only twenty-seven hundred jackets were printed—when there were three thousand books. It is equally possible that Cassell-bound copies are second or later printings: the Sales Day Book entry for 30 April 1931 records invoicing Cassell for "making moulds of the standing type" (the purpose of which would be to preserve the original setting, thereby allowing the printing plates to be melted down), and the copy in the Bodleian Library, which is a Cassell-bound copy, is stamped "Jun 6 1931." I am inclined to favor the first theory, believing the Cassell binding to be a binding variant of the first printing.

The British Black Mask, *featuring the first installment of* The Glass Key. *These issues were most likely imported American sheets with the addition of British advertisements and an overprinted sterling cover price.*

While the information gleaned from the Sales Day Book is as significant as it is interesting, it does nothing to shed light on another point concerning the English first edition of *The Glass Key:* the Knopf dust jacket. The Pittsburgh Hammett bibliography says simply, "Not seen," and since this collector first embarked on his quest to acquire a copy he has spoken to not a single book dealer, on either side of the Atlantic, who can recall even having heard of one, let alone seen one. There is, however, at least one extant example of this great rarity, and it owes its survival to a piece of legislation passed shortly before World War I.

Under the terms of the British Copyright Act of 1911, publishers have been obliged by law to deposit with selected libraries a copy of any book they offer for sale in the United Kingdom; these "copyright libraries" include the British Library, the Bodleian Library, the University of Dublin's Trinity College Library, the National Library of Scotland, the National Library of Wales, and Cambridge Uni-

The dust jacket for the first English edition of The Thin Man *(London: Barker, 1934)*

versity Library. It has been common policy to discard dust jackets before adding the books to the collections, but both the National Library of Wales and Cambridge University Library have retained the jackets on works of fiction, believing them to be a part of the bibliographical entity. While the National Library of Wales does not have a copy of the Knopf first English edition of *The Glass Key,* Cambridge University Library does; for the past sixty-six years it has been sitting on a shelf in a darkened room in what library staff refer to as "the tower," complete with its original jacket.

Those familiar with the *Black Mask* serialization of the book will immediately recognize the predominantly white dust jacket's front-panel illustration by J. W. Schlaikjer. Taken straight from the cover of the March 1930 issue, the painting depicts the story's protagonist, Ned Beaumont, above the words "A 'different' detective thriller by the author of *THE MALTESE FALCON,* etc." The spine of the jacket is lettered in black with a red Knopf Borzoi seal, and the back panel lists "Other Borzoi Novels of Crime and Suspense," including Hammett's first three books, accompanied by extracts of reviews from the *Glasgow Herald,* the *Sunday Express,* and *Truth.*

Cambridge University Library also holds copies of two other Hammett titles whose dust jackets are rarely seen—the first English editions of *The Dain Curse* (Knopf, 1930) and *The Thin Man* (Arthur Barker, 1934). The front and spine panels of the former are identical to those of the American first edition, showing F. H. Horvarth's illustration against an orange background of a strange-looking bird and a sinister figure with a hangman's noose; the inside flap of the English edition, however, prints the plot outline from the back panel of the American jacket and adds a review of *Red Harvest* from the *Birmingham* (England) *Post* above the price, 7s. 6d. The back panel lists five "New Borzoi Novels," and the back flap is blank. The creamy yellow dust jacket for the Barker edition of *The Thin Man,* which is printed in black, dark blue, and green, features a sitting-room scene with Nick and Nora Charles on the front panel, a shadowy thin man on the spine panel, and a blurb followed by the price, 7/6 NET, on the front flap. The back panel is blank save for the publisher's seal, and the back flap lists "THREE MORE THRILLERS."

The Dain Curse was the first of Hammett's novels to be printed specifically for U.K. distribution, and it has been widely thought that when *Red Harvest* was published in Britain the books were simply imported copies of the first American printing. That is only partly true, as the jacketed copy in the National Library of Wales proves. The book itself is, indeed, a copy of the first American printing, but the dust jacket was printed for the British market. The front panel of the cream jacket announces "Something new in crime fiction!" above the title and author's name, which are printed over a mountain of red skull-and-crossbones motifs. The spine is plain, with title, author, and publisher in black and a skull and crossbones and the Borzoi seal in red; the front flap has a plot outline, and the back flap is blank. On the back panel are reviews of three "exciting tales."

With the discovery of this dust jacket in the National Library of Wales it can be stated that the first American printing was distributed in at least three different jackets: the British jacket, the jacket described in Richard Layman's *Dashiell Hammett: A Descriptive Bibliography* (1979), and a jacket that is the same as the latter in all respects save that on the back panel, in place of the seventeen lines of reviews from *The Bookman, The Outlook,* and *The Chicago Post,* there is an eighteen-line plot outline. Later printings carried reviews of *The Dain Curse*.

Of the two American jackets for *Red Harvest*, the example with the eighteen-line plot outline is likely to be the earliest, and there are three points to support this claim: *Red Harvest* was published on 1 February 1929, and the issue of *The Outlook* featuring the review that appears on the Pittsburgh bibliography jacket was published on 13 February; in the body of the complete review from *The Outlook*, *Red Harvest* is described as being "A thriller that lives up to the blurb on the jacket"; and the issue of *The Bookman* from which Herbert Asbury's review was quoted appeared in March.

The completist might set his sights on collecting all three states of the *Red Harvest* jacket—thereby, to boot, acquiring the book as it first appeared in Britain. But if that is the goal, it will not be an easy one; for if collecting the first American printings of Hammett's five great novels in their original dust jackets is a difficult task, collecting the British editions is a Herculean one. My own attempts have, to date, proved fruitless.

Cream-colored dust jacket printed in black and red for the British publication of Red Harvest. *Knopf imported copies of the first American printing for sale in the U.K.*

Kingsley Amis
(16 April 1922 – 22 October 1995)

Merritt Moseley
University of North Carolina at Asheville

See also the Amis entries in *DLB 15: British Novelists, 1930–1959; DLB 27: Poets of Great Britain and Ireland, 1945–1960; DLB 100: Modern British Essayists, Second Series;* and *DLB 139: British Short-Fiction Writers, 1945–1980.*

BOOKS: *Bright November* (London: Fortune, 1947);

Lucky Jim (London: Gollancz, 1954; Garden City, N.Y.: Doubleday, 1954);

That Uncertain Feeling (London: Gollancz, 1955; New York: Harcourt, Brace, 1956);

A Case of Samples: Poems 1946–1956 (London: Gollancz, 1956; New York: Harcourt, Brace, 1957);

Socialism and the Intellectuals (London: Fabian Society, 1957);

I Like It Here (London: Gollancz, 1958; New York: Harcourt, Brace, 1958);

New Maps of Hell: A Survey of Science Fiction (New York: Harcourt, Brace, 1960; London: Gollancz, 1961);

Take a Girl Like You (London: Gollancz, 1960; New York: Harcourt, Brace, 1961);

My Enemy's Enemy (London: Gollancz, 1962; New York: Harcourt, Brace, 1962);

One Fat Englishman (London: Gollancz, 1963; New York: Harcourt, Brace, 1964);

The James Bond Dossier (London: Cape, 1965; New York: New American Library, 1965);

The Egyptologists, by Amis and Robert Conquest (London: Cape, 1965; New York: Random House, 1966);

The Anti-Death League (London: Gollancz, 1966; New York: Harcourt, Brace, 1966);

A Look Round the Estate: Poems 1957–1967 (London: Cape, 1967; New York: Harcourt, Brace, 1968);

Colonel Sun, as Robert Markham (London: Cape, 1968; New York: Harper & Row, 1968);

Lucky Jim's Politics (London: Conservative Political Centre, 1968);

I Want It Now (London: Cape, 1968; New York: Harcourt, Brace, 1969);

Sir Kingsley Amis (photograph by Johnny Boylan)

The Green Man (London: Cape, 1969; New York: Harcourt, Brace, 1970);

What Became of Jane Austen? and Other Questions (London: Cape, 1970; New York: Harcourt, Brace, 1971);

Girl, 20 (London: Cape, 1971; New York: Harcourt, Brace, 1972);

On Drink (London: Cape, 1972; New York: Harcourt, Brace, 1973);

The Riverside Villas Murder (London: Cape, 1973; New York: Harcourt, Brace, 1973);

Ending Up (London: Cape, 1974; New York: Harcourt, Brace, 1974);

Rudyard Kipling and His World (London: Thames & Hudson, 1975; New York: Scribners, 1976);

The Alteration (London: Cape, 1976; New York: Viking, 1977);

Jake's Thing (London: Hutchinson, 1978; New York: Viking, 1978);

Collected Poems 1944-1979 (London: Hutchinson, 1979; New York: Viking, 1980);

Russian Hide-and-Seek (London: Hutchinson, 1980);

Collected Short Stories (London: Hutchinson, 1980);

Every Day Drinking (London: Hutchinson, 1983);

How's Your Glass? (London: Weidenfeld & Nicolson, 1984);

Stanley and the Women (London: Hutchinson, 1984; New York: Summit, 1985);

The Old Devils (London: Hutchinson, 1986; New York: Summit, 1987);

The Crime of the Century (London: Dent, 1987; New York: Mysterious Press, 1989);

The Amis Anthology (London: Hutchinson, 1988);

Difficulties with Girls (London: Hutchinson, 1988; New York: Summit, 1989);

The Folks That Live on the Hill (London: Hutchinson, 1990; New York: Summit, 1990);

The Pleasure of Poetry, edited by Amis (London: Cassell, 1990);

The Amis Collection (London: Hutchinson, 1990);

Memoirs (London: Hutchinson, 1991; New York: Summit, 1991);

The Russian Girl (London: Hutchinson, 1992; New York: Viking/Penguin, 1994);

You Can't Do Both (London: Hutchinson, 1994);

The Biographer's Moustache (London: Flamingo, 1995).

OTHER: *Spectrum: A Science Fiction Anthology,* edited by Amis and Robert Conquest (London: Gollancz, 1961-1965; New York: Harcourt, Brace, 1962-1967);

Harold's Years: Impressions from the New Statesman and Spectator, edited by Amis (London: Quartet, 1977);

The New Oxford Book of Light Verse, edited by Amis (London & New York: Oxford University Press, 1978);

The Faber Popular Reciter, edited by Amis (London & Boston: Faber, 1978);

The Golden Age of Science Fiction, edited by Amis (London: Hutchinson, 1981);

The Great British Songbook, edited by Amis and James Cochrane (London: Pavilion/M. Joseph, 1986).

When he died on 22 October 1995 at the age of seventy-three, Sir Kingsley Amis was simultaneously one of England's most admired novelists and one of its most controversial. Though he had been in declining health for some time before he died as a result of a stroke following a serious fall, he was mentally vigorous and artistically productive. He wrote for several hours every day. His last novel, *The Biographer's Moustache,* appeared in autumn 1995.

It was his twenty-fourth novel, counting one co-authored with Robert Conquest and a James Bond book published under the name Robert Markham. In addition he had published three books of his own short fiction; four books of literary criticism; three books about drinking; two miscellaneous collections of his essays; anthologies of short stories, popular verse, songs, and science fiction; several volumes of poetry; and his *Memoirs* (1991). Amis was, as Paul Fussell explains, "a man of letters in the old sense, a writer conspicuous for complex literary knowledge and subtle taste as well as for vigorous views on politics and society"; if he was not fully accepted in this role it is because he lacked pomposity and because his writing was often funny. He received during his long career the tribute of popularity (most of his books were still in print at the time of his death) and honors both literary and national. He received the Booker Prize in 1986 for his novel *The Old Devils*—twice before he had been one of the finalists—and he was made a CBE (commander of the Order of the British Empire) in 1981 and a knight in 1990.

Kingsley William Amis was born on 16 April 1922 in Norbury, South London. He was the only son of William Robert Amis, a clerk for Colman's Mustard, and Rosa Annie Lucas Amis, usually called Peggy. He has written about his childhood both in his memoirs and in semi-autobiographical novels such as *The Riverside Villas Murder* (1973) and *You Can't Do Both* (1994); though certainly not deprived or mistreated, he felt oppressed by his parents (perhaps because he was their only child). His father's impact on him, as recorded in his *Memoirs,* was narrow: William Amis worried that his son would masturbate; he disagreed violently with Kingsley about music; and he bored the boy. His mother spoiled him. Kingsley was schooled at St. Hilda's, which, though a girls' school, took boys in the lower forms, then at Norbury College, then at the City of London School, a good institution where he soon won a scholarship to save his father the costs of his fees. When World War II broke out his whole school was evacuated to Wiltshire and combined with Marlborough School, an exclusive public school whose students remained aloof from Amis and his urban companions.

In 1941 Amis matriculated at Oxford as a member of St. John's College. He read English, developing a distaste for many of the classic books, acquired a taste for jazz, experimented with sex, and joined the Communist Party. After four terms he joined the army; after training at Catterick Camp in Yorkshire and assignments elsewhere in England, he reached France in June 1944 and remained there until being demobilized in October 1945, when he returned to Oxford. There he met Hilary Bardwell; they were married on 21 January 1948; their first child, Philip (named after Philip Larkin), was born on 16 August 1948. Two more children, Martin and Sally, would be born in 1949 and 1954.

Amis had received his B.A. in 1947 and proceeded to a B.Litt., which he intended to prepare him for a career as an academic. Unfortunately he failed this degree. Failure seems not to have harmed his career prospects. In 1949 he became a lecturer in English at the University College of Swansea in Wales, where he would remain until 1961.

His writing career had already begun. Though he had tried fiction before attending Oxford, it was as a poet that Amis first saw print. He published verse in little magazines; in 1949 he coedited an anthology called *Oxford Poetry*. His own first book of poems was *Bright November,* published in 1947 by the Fortune Press, a somewhat disreputable firm run by a man named Caton, whom Amis would travesty as "L. S. Caton" in novel after novel and ultimately kill off in *The Anti-Death League* (1966). Amis published another four books of poems, then a collection of his work up to 1967, and finally his *Collected Poems 1944–1979* in 1979. His output of verse declined as his career as a novelist gathered momentum.

His first novel, begun in 1947, was "The Legacy"; as described by Eric Jacobs, it is derivative of a public-school novel called *The Senior Commoner* and describes the dilemma of a man called Kingsley Amis, forced to choose between the woman he loves and worldly comfort with another woman. It was submitted to publishers but never accepted. In 1946, though, Amis had visited his friend Philip Larkin at Leicester University; they went together to the Senior Common Room, and Amis describes his reaction in his *Memoirs:*

> I looked round a couple of times and said to myself, "Christ, somebody ought to do something with this." Not that it was awful—well, only a bit; it was strange and sort of *developed,* a whole mode of existence no one had got on to from outside, like the SS in 1940, say. I would do something with it.

The result, finally published in 1954, was his first novel, *Lucky Jim,* a comic masterpiece, one of the most important novels of the middle twentieth century, and the book with which, despite all the ones that followed, Amis continues to be most identified. This "campus novel" tells the story of Jim Dixon, a lecturer in history at an unnamed Midlands university, and his twin predicaments: his extremely shaky job prospects—he is not good at his job, does not respect it or try very hard at it, and hates his superiors but has nothing else to do—and his involuntary entanglement in a relationship with a neurotic colleague, Margaret Peel. The novel is rich in satire on Jim's milieu and on his own collaboration with it. It is a comedy and has a happy, almost Cinderella-like, ending. *Lucky Jim* was a critical success and led reviewers and cultural critics to label Amis as an "Angry Young Man"—placing him in a "school" that was also held to include John Osborne, Colin Wilson, John Wain, John Braine, and even Iris Murdoch. Though the inappropriateness of the grouping has become more apparent as the years go by, Amis was at this time something of a radical, discontented with the establishment, with privilege, and with Conservative government.

During the remainder of his time at Swansea, he published three more novels: *That Uncertain Feeling* (1955), *I Like It Here* (1958), and *Take a Girl Like You* (1960). The first two of these were in his familiar comic vein, with *I Like It Here* somewhat more insubstantial; both of them, though, center on the familiar Amis situation of a protagonist facing a moral choice, often presented comically but with an underlying seriousness. In *Take a Girl Like You,* an analysis of selfishness concerning an innocent young teacher from the North, his protagonist, who is raped by a more sophisticated southerner whom she loves despite his shortcomings, Amis's work achieved a greater depth and scope. In 1958–1959 he had been a visiting fellow at Princeton University; he considered settling in the United States, which he admired at that time, but decided against it. He was asked to deliver a course of lectures on science fiction, which became *New Maps of Hell: A Survey of Science Fiction* (1960).

In 1961 Amis left Swansea, though he was fond of the town and spent some time there almost every year, up to the last year of his life—it was in Swansea that he suffered the fall that put him in the hospital for his final stay. Amis had been appointed to a fellowship at Peterhouse College, Cambridge; he stayed there only two years, disappointed with the place—primarily because of his colleagues, he explained, not his students. When he left Cambridge he gave up his teaching career, though he did spend a short time (also unsatisfactorily) at Vanderbilt University in 1967–1968. He continued to practice literary criticism, though in a nonacademic vein, and review books; but after leaving Cambridge he was a full-time writer for the rest of his life.

In the 1960s Amis published an interesting variety of books. His novels included *One Fat Englishman* (1963), a searing treatment of an obnoxious and anti-American English publisher visiting some of the scenes of Amis's time in Princeton; *The Egyptologists* (1965), a fantasy about an antifemale conspiracy, co-authored with his friend Robert Conquest; *The Anti-Death League,* a thoughtful novel about men in the peacetime army faced with a Communist Chinese threat; *Colonel Sun* (1968), a James Bond thriller written under the pseudonym Robert Markham; and *I Want It Now* (1968), a forceful satire on 1960s hedonism, media celebrity, and southern American racism. During the decade he also published *A Look Round the Estate: Poems 1957–1967* (1967); *My Enemy's Enemy,* a book of short stories (1962); and *The James Bond Dossier* (1965), a critical overview of Ian Fleming's work that also seems to have provided the background to Amis's efforts in *Colonel Sun.*

During this period Amis began to become more conservative. Though his communist period had ended during the war, he had been determined to vote Labour and to remain a socialist; events during the 1950s, particularly the Soviet invasion of Hungary, as well as a growing opposition to the kind of people he thought dominated English socialism and their antipatriotism moved Amis to the Right. In the 1960s he was preoccupied with the international communist threat; both *Colonel Sun* and *The Anti-Death League* pivot on a terrible nuclear threat from the Communist Chinese. In 1964 he cast his last vote for Labour; in 1967 he began voting Conservative. Amis came to believe that leftist beliefs were largely the result of envy, and he grew to oppose many of the causes of English socialists and leftist intellectuals. He was a consistent opponent of changes in education in the name of democratization, including wider access to the universities and an end to selective grammar schools; he also consistently opposed public funding of the arts.

Perhaps his most trenchant fictional commentary on the 1960s came in his 1971 novel, *Girl, 20.* A study of an aging leftist whose politics come to seem largely a matter of abandoning personal responsibility and emulating the young, this novel is a splendid portrayal of a certain type of celebrity. Sir John Vandervane's socialism coexists perfectly with private wealth and privilege; his "liberation" amounts to self-indulgence and a benign neglect of his own family; Amis cruelly catches his attempts to be youthful and proletarian by using language carelessly.

In 1969 Amis had published *The Green Man,* a dark fantasy about an alcoholic innkeeper who does battle with a sinister autochthonous being, a seventeenth-century magus, and his own bodily weaknesses and has a conversation with God. *Ending Up* (1974) is a short and penetrating ensemble novel about a group of old people living together; Amis has said that it was a projection into the future of the members of his own extended family. Also concerned with the effects of growing old and the fear of death, like *The Green Man* and *Ending Up,* is *Jake's Thing* (1978), a tragicomic study of an aging Oxford don who has suddenly become impotent.

Amis published two "genre" novels during this decade. *The Riverside Villas Murder* is set in the 1930s and recaptures both the flavor of that period's detective novels and something of the author's childhood, which is presented in some of the details of his protagonist, the teenaged Peter Furneaux. *The Alteration* (1976) is a "time romance," or alternative history, which is set in England in the present day but assumes a dramatic change in the past—in this case, that there was no Protestant Reformation—to trace out the different way life would have worked out. The title refers both to this alteration in history and to the proposed castration of a boy soprano, in the service of the Church.

Amis has always written about love and about the relations between men and women. In *Jake's Thing* he provided ammunition for the first time to critics who accused him of misogyny. The charge is complicated; the evidence supporting it is contradictory. Eric Jacobs traces this strain in *Jake's Thing,* and the later *Stanley and the Women* (1984) to the conditions of Amis's second marriage. Amis had met the novelist Elizabeth Jane Howard in 1962 and had left his first wife, Hilary, for her in the summer of 1963. They were married in 1965 and lived together until 1980 but were divorced in 1983. Amis came to believe that the marriage had become hopeless by 1970. He strongly disliked his second wife, and this feeling carried over into his novels, producing characters who seem to feel a hatred for women as a group, an attitude the novels do not seem to disavow.

During the 1970s Amis published his first collection of nonfiction, *What Became of Jane Austen? and Other Questions* (1970). This volume includes spirited criticism, particularly of overpraised authors and trends that seemed phony to Amis (Dylan Thomas, modernism in most of the arts, fashionable leftism), as well as such sacred cows as Jane Austen and John Keats. There was also some autobiography and some personal discussion of Amis's father and his religious background. Amis's novels had begun to be more seriously concerned with religious questions: *The Anti-Death League, The Green Man,* and *The*

Amis with his children—Sally, Martin, and Philip—in the early 1960s

Alteration all have religious themes at their cores. He continued to contribute works in a variety of genres, including, in the 1970s, books about drinking; radio and television plays; two anthologies of poetry; occasional polemical papers about education; and a book on Rudyard Kipling, which forcefully defends the Victorian writer, as he had earlier defended Ian Fleming, against charges of racism, imperialism, and sadism. Amis also brought out his *Collected Poems 1944–1979*.

In the 1980s Amis continued the anthologizing and consolidating project, publishing his *Collected Short Stories* (1980) and *The Amis Anthology* (1988); he wrote another detective story, *The Crime of the Century* (1987), which was originally published in installments in a newspaper, with readers invited to supply a conclusion, and another time romance of sorts, *Russian Hide-and-Seek* (1980). This novel stipulates that the Russians conquered Britain in the 1970s and is set another fifty years in the future, depicting a bleak and servile population controlled and despised by a ruthless Russian military.

Amis's "straight" novels of the 1980s were *Stanley and the Women,* a study of the impact on a man's life of his son's madness and the unreasonability of most of the women he knows, and *The Old Devils* (1986), a return to the Welsh setting of *That Uncertain Feeling,* which won the Booker Prize for the best British novel of the year; and *Difficulties with Girls* (1988), which follows Jenny Bunn and Patrick Standish, the central couple of *Take a Girl Like You,* into the 1960s. The later work does not seem to shed much light either on the Standishes or on the 1960s. *The Old Devils,* though, is a more profound book, both sad and funny, again on the subject of age. It follows a group of friends in Wales who drink and think about lost opportunities, as they warily welcome back a retiring writer who combines professional (and thus bogus) "Welshness" with bland cruelty and blind selfishness, in about equal parts.

The 1990s saw no diminishment of Amis's output. The keynote may have been an increasing turn toward his own life as the material for his books. In 1990 he published *The Amis Collection,* another gathering of essays, reviews, and poems; his *Memoirs* appeared in 1991, attracting considerable attention—much of it hostile because of the author's sharp and unforgiving portraits of people he did not like and even of those he counted as friends.

In his fiction, too, Amis was more forthcoming. *The Folks That Live on the Hill* (1990) is a portrait of a group of people, mostly dysfunctional, revolving around a philanthrope; but the setting is much like Amis's location in Primrose Hill, where he was living in unusual circumstances: since 1981 he had been sharing a house—actually a series of different houses—with his former wife Hilary and her third husband, Lord Kilmarnock. This arrangement lasted until Amis's death.

His softened attitude toward Hilary is evidenced by the dedication and poem in the *Memoirs,*

"Instead of an Epilogue," and by the wistfulness of his depiction of the marriage in his 1994 novel, *You Can't Do Both,* by far Amis's most straightforward use of his own life in fiction. *The Russian Girl* (1992) is also about love and infidelity, this time involving an English academic and an émigré poet whose charms inspire the protagonist to pretend that her poetry is good rather than rubbish. In *You Can't Do Both*–the title making explicit the series of dilemmas that shape the protagonist's progress–Amis deals with the pinched atmosphere of his childhood and the casual unfaithfulness that (as Eric Jacobs's biography, authorized and assisted by Amis, makes explicit) had destroyed the author's first marriage.

Kingsley Amis: A Biography appeared in 1995. Its author, a journalist, had spent several years with Amis and, clearly with Amis's approval, presented a portrait with warts and all. It opens with a vignette of Amis's average day, showing his domestic helplessness, his excessive drinking, his scabrous jokes, his erotic fantasies about Margaret Thatcher, his rudeness to bores, and his fondness for televised soap operas and cop shows in preference to almost any reading. Jacobs says, "Amis is as concerned with his reputation as any writer, and he would certainly like his own novels to be included in the popular English canon." Clearly he expected his reputation as a writer to survive the authorized revelations about his imperfections as a man.

Amis's last completed novel, *The Biographer's Moustache,* is about the relationship between a biographer and his famous subject; it seems unlike the Amis/Jacobs relation, as, in the novel, the famous and wealthy Jimmie Fane sets out to use the obscure and penniless Gordon Scott-Thompson, with paradoxical results. Like most of Amis's works, it has love and sex at the center of it; like most of them, too, it punctures pretension and affectation. Jimmie Fane is a humbug reminiscent of Roy Vandervane in *Girl, 20,* and this novel is a worthy book to finish the Amis oeuvre.

In the aftermath of Amis's death, there was an expected outpouring of tributes. Malcolm Bradbury, referring in particular to *Lucky Jim,* commented: "He was writing in a new and fresh way–it's a great comedy. He absolutely captured the tone of the times in the way that Martin Amis has done for this generation." Melvyn Bragg called him "one of the masters of comic fiction," and Keith Waterhouse called him "a great storyteller, although he was much more than that." Auberon Waugh said that "He did not give a damn what other people thought about him, and he said what he thought." Eric Pace, looking beyond the comic (as Amis always hoped that people would do), called him "a moral satirist who has been compared with Pope and Swift."

More unexpected was a dispute between Eric Jacobs and Amis's family, led by his son, novelist Martin Amis. Jacobs had been cooperating with Amis on a collection of the author's letters for the purposes of publication, a project in which Jacobs has been replaced; the reason is his own decision to seek publication, within seven days of the novelist's death, of a diary he kept during Amis's last illness. It eventually appeared in the *Sunday Times* and made melancholy reading, being much concerned with Amis's depression, his inability to work, and his drunkenness.

It was perhaps an unworthy picture to leave in the public eye–even if it could be argued that Amis did not give a damn what other people thought about him. When the circumstances of his death and the temper of his political opinions have been forgotten, Kingsley Amis may well be recalled as the author of *Lucky Jim*–and that is an enviable reputation. More justly, he will be remembered as a significant poet, a serious critic of literature and society, and one of the foremost novelists of his generation.

Interviews:

Peter Firchow, ed., *The Writer's Place: Interviews on the Literary Situation in Contemporary Britain* (Minneapolis: University of Minnesota Press, 1974), pp. 15–38;

Clive James, "Profile 4: Kingsley Amis," *New Review,* 1 (July 1974): 21–28;

Michael Barber, "The Art of Fiction LIX: Kingsley Amis," *Paris Review,* 64 (1975): 39–72;

Dale Salwak, "Kingsley Amis: Mimic and Moralist," in his *Interviews with Britain's Angry Young Men* (San Bernardino, Cal.: Borgo, 1984).

Bibliographies:

J. B. Gohn, *Kingsley Amis: A Checklist* (Kent, Ohio: Kent State University Press, 1976);

D. F. Salwak, *Kingsley Amis: A Reference Guide* (Boston: G. K. Hall, 1978).

References:

Walter Allen, *Tradition and Dream: The English and American Novel from the Twenties to Our Time* (London: Phoenix House, 1964);

Malcolm Bradbury, " 'No, Not Bloomsbury': The Comic Fiction of Kingsley Amis," in his *No, Not Bloomsbury* (New York: Columbia University Press, 1988);

Richard Bradford, *Kingsley Amis* (London: Edward Arnold, 1989);

Caroline Davies, "Cantankerous, Irascible and Rude—We'll Miss the Old Devil, Say Sir Kingsley's Friends," *Electronic Telegraph,* 23 October 1995, p. 1;

Paul Fussell, *The Anti-Egotist: Kingsley Amis, Man of Letters* (New York & Oxford: Oxford University Press, 1994);

Philip Gardner, *Kingsley Amis* (Boston: Twayne, 1981);

Robert Hewison, *In Anger: British Culture in the Cold War 1945–60* (New York: Oxford University Press, 1981);

D. A. N. Jones, "Kingsley Amis," *Grand Street,* 4 (Spring 1985): 206–214;

David Lodge, "The Modern, the Contemporary, and the Importance of Being Amis," in his *Language of Fiction* (New York: Columbia University Press, 1966);

John McDermott, *Kingsley Amis: An English Moralist* (New York: St. Martin's Press, 1989);

William D. Montalbano, "Kingsley Amis; British Novelist and Poet," *Los Angeles Times,* 23 October 1995, p. A18;

Merritt Moseley, *Understanding Kingsley Amis* (Columbia: University of South Carolina Press, 1993);

Rubin Rabinovitz, *The Reaction Against Experiment in the English Novel, 1950–1960* (New York: Columbia University Press, 1967);

Dale Salwak, *Kingsley Amis, Modern Novelist* (Lanham, Md.: Barnes & Noble, 1992);

Salwak, ed., *Kingsley Amis in Life and Letters* (London: Macmillan, 1990);

Terry Teachout, "A Touch of Class," *New Criterion,* 7 (November 1988): 8–17;

Keith Wilson, "Jim, Jake and the Years Between: The Will to Stasis in the Contemporary British Novel," *Ariel,* 13 (January 1982): 55–69;

James Wolcott, "Kingsley's Ransom," *New Yorker,* 71 (30 October 1995): 52–58.

Joseph Mitchell
(27 July 1908 - 24 May 1996)

Raymond J. Rundus
University of North Carolina at Pembroke

BOOKS: *My Ears Are Bent* (New York: Sheridan House, 1938);

McSorley's Wonderful Saloon (New York: Duell, Sloan & Pearce, 1943);

Old Mr. Flood (New York: Duell, Sloan & Pearce, 1948);

The Bottom of the Harbor (Boston: Little, Brown, 1960; New York: Modern Library/Random House, 1994);

Joe Gould's Secret (New York: Viking, 1965; New York: Modern Library/Random House, 1996);

Up in the Old Hotel (New York: Pantheon, 1992).

OTHER: "The Mohawks in High Steel," the introductory essay in Edmund Wilson's *Apologies to the Iroquois* (New York: Farrar, Straus & Cudahy, 1959).

SELECTED PERIODICAL PUBLICATIONS–
UNCOLLECTED: "They Got Married in Elkton," *The New Yorker* (11 November 1933): 36-43;

"Home Girl," *The New Yorker* (3 March 1934): 25-29;

"Mrs. Bright and Shining Star Chibby," *The New Yorker* (5 November 1938): 19-21.

At his death in New York City two months short of his eighty-eighth birthday, Joseph Mitchell, a native son of rural southeastern North Carolina, left behind an impressive record. Not only was he considered by peers to be the paragon among writer-reporters of his generation, he was also an influential presence ("Read the master," new staff writers at *The New Yorker* were sometimes told). Except for a short stint as a mariner, he had been a reporter for the *World,* the *Herald Tribune,* and the *World-Telegram* from 1929 to 1938. Active as a staff writer on *The New Yorker* from 1938 until 1964 (when the last piece under his byline appeared), Joseph Mitchell was promised in 1992 by new editor Tina Brown that he could maintain his *New Yorker* office for as long as he wished.

The eldest of the six children born to Averette Nance ("A. N.") and Elizabeth Amanda (Parker) Mitchell, Joseph Mitchell was born in rural Robeson county and raised along with his three sisters and two brothers in the family home on Church Street in Fairmont, a tobacco, cotton, and textile community in the coastal plains area of southeastern North Carolina. The population of Fairmont remains near the roughly twenty-seven hundred residents it had at the time Joseph Mitchell graduated from the local high school in the spring of 1925.

The Mitchells were and still are prominent residents of southern Robeson County; the patriarch of the family, Nazareth Mitchell (first name spelled variously in legal documents of the time) had come to the area from Virginia shortly after the Revolutionary War ended and through marriage and other means had acquired a considerable amount of acreage in the county. A. N. Mitchell maintained a similar sense of proprietorship; he sold only one piece of land during his business life and lived to regret having done so. Joseph Mitchell inherited significant property from both his father and mother and likely could have lived fairly comfortably on the revenues from these estates without relying on the vagaries of the writing profession.

A. N. Mitchell was known for his integrity and fair dealing and served for a time as mayor of Fairmont. A. N. Mitchell and Sons, with offices on Thompson Street in Fairmont, continues to be a thriving enterprise in the warehouse and agriculture businesses of a town that is a major tobacco market in the Border Belt. For some time A. N. Mitchell had hoped that Joseph would continue in the family business. In particular, he thought that Joseph might be well suited to become a cotton trader. Young Joseph, however, was aware that he could never make the quick and accurate calculations needed in that occupation. As he once said about his difficult but life-forming relationship with his father, "I was always learning that everything I did was wrong." As a consequence of these concerns about his son's future, A. N. realized that Joseph had in his farm work exhibited another talent. After watching how well Joseph worked with a black foreman named Lonzo McNair, in butchering a wide variety of livestock, A. N. told him, "Son, you really know how to butcher. You ought to be a surgeon." His premed curriculum at Chapel Hill, however, quickly proved to be unsuited to his interests and talents; Dean Addison Hibbard told him, "You are not cut out to be a doctor."

The young scholar's interests then shifted to English, philosophy, theater, and journalism. He learned much from an English professor about the *Canterbury Tales,* and from Paul Green he learned about philosophy, the Russian writers, and theater, soon trying his hand at playwriting and at being a drama critic for the campus newspaper, the *Daily Tar Heel.* He left Chapel Hill to return to Fairmont in the spring of 1929 without a degree (he could never satisfy either the chemistry or the mathematics requirements), but with a record of reporting for a local newspaper about campus activities and professors, of having been published in the *Carolina Magazine,* and of having a story accepted for an early issue of *American Caravan.* Most importantly, perhaps, he was working on a story about the Fairmont tobacco market that would soon be accepted for publication in the Sunday magazine of the *Herald Tribune.*

His father having advised him that he would not have to return to Chapel Hill to complete his baccalaureate requirements, a relieved, exultant Joseph Mitchell returned to the family home in Fairmont in plenty of time to celebrate his twenty-first birthday. That summer hiatus was interrupted by an emergency appendectomy and an extended convalescence, during which time he read James Bryce's *The American Commonwealth* (1888). Inspired by the British historian and diplomat's theories and his example, Joseph Mitchell determined that he would seek a career as a political journalist.

Before leaving for New York, he heard ringing in his ears a mixed sort of blessing from his father, who had inquired, "What do you want to do, going around and sticking your nose in other people's business?" But he had also been supplied a modest financial stake. Arriving on 25 October 1929 (four days before the stock market collapse that heralded the beginning of the Great Depression), Joseph Mitchell went immediately to his interview with City Editor Stanley Walker at the *Herald Tribune,* which had recently published his story on the Fairmont tobacco market. The Texan and the North Carolinian hit it off immediately, and while Walker could not place him on his own newspaper, he did get him a cub reporter's job on the *New York World* and would later prove to be a reliable and friendly contact at *The New Yorker,* where Walker spent some time working for Harold Ross.

The aspiring reporter took to heart Walker's advice—to immerse himself in the life of the great city. At the end of Mitchell's life, he was widely regarded as the great historian of Manhattan's human landscape and its infrastructure. He was one of the founders of the South Street Seaport Museum and member of its Restoration Committee and had been appointed by Mayor Koch as a commissioner of the New York City Landmarks Preservation committee. For a time he was a vestryman in Grace Church, vice president and secretary of the American Academy and Institute of Arts and Letters, and member as well of such diverse groups as the Society of Architectural Historians, the Society of Industrial Archeology, the Friends of Cast-Iron Architecture, the James Joyce Society, the Gypsy Lore Society (spending several years on its board), and the Century Association.

When Mitchell did join the *Tribune* in 1931, his career was sharply terminated as the consequence of his bitter resentment of an editorial decision made by publisher Ogden Reid. Reid had decided to reassign a story Mitchell had been covering to a more experienced reporter; Mitchell's bitterness culminated in his throwing an inkwell at a wall in Reid's office. Fired from the newspaper that nurtured so many aspiring writers, Joseph Mitchell began a short-lived career as a deck boy on a "worn-out Hog Island freighter, the *City of Fairbury,*" an experience recounted in his 1938 collection of newspaper pieces, *My Ears Are Bent.*

Returning to New York, he soon married photographer Therese Dagny Engelstad Jacobsen (who died in 1980) and began a seven-year stint with the *World-Telegram.* Where his earlier beat had been mostly street crime, he now became a skillful interviewer, using his legendary listening skills in sessions with such celebrities as Jimmy Durante, George Bernard Shaw, Nicholas Murray Butler, Tallulah Bankhead, Joe Louis, Billy Sunday, Noel Coward, and Eleanor Roosevelt, as well as exotic dancers and other "marginal" characters of the day. He also did some reporting of the Lindbergh kidnapping trial and the 1936 New York World's Fair. What he learned from these interviews in part, as he said in his lengthy introduction to *My Ears Are Bent,* was that the people he found least interesting were society women, industrial leaders, distinguished authors, ministers, explorers, movie actors (except W. C. Fields and Stepin Fetchit), and "any actress under the age of thirty-five." He preferred instead to talk to "anthropologists, farmers, prostitutes, psychiatrists, and an occasional bartender."

Joseph Mitchell's great arena, however, was as an active staff writer for twenty-six years on *The New Yorker*. When he first reported for work on 26 September 1938, he had been actively shopping unsolicited fictional and factual pieces to the magazine for several years. He was also a particularly close friend of A. J. Liebling, who had already been on staff for three years. They had come to like and respect each other from their time together on the *World-Telegram*. In a "friendship based upon literary argument," as Mitchell later termed it, they had discovered a mutual interest in and respect for such earlier literary journalists as Daniel Defoe, Henry Fielding, William Cobbett, Pierce Egan, and George Borrow; they particularly aimed to recapture the low-life subject matter and to some extent the styles of the three nineteenth-century writers. Mitchell would present a memorable eulogy at the final services for his agnostic friend, a tribute that says perhaps more about how Mitchell valued the literary life than what it reveals about his closest colleague and friend.

The archives of *The New Yorker* provide insights into the backstage operations of the magazine: especially during the tenure of the magazine's first and most memorable editor, Harold Ross. For at least five years before he caught on with the magazine permanently, Joseph Mitchell had been submitting both short fiction and (sometimes assigned) factual pieces, the earliest of which was published in 1933. Editor and writer St. Clair McKelway, a fellow North Carolinian, had become an important link in the aspiring writer's profession; but still the road to acceptance of his work by *The New Yorker* was a bumpy one. One of the early problems Mitchell experienced in getting his "casuals" published was that both Ross and McKelway believed he was having trouble making his contributions either distinctly fictional or distinctly factual. About a year before Mitchell joined the staff, in an internal memo to McKelway, Harold Ross suggested that "this man ought to be encouraged to write fiction for us, but . . . the little coincidences of a newspaper reporter aren't stories. . . . They are well done and the two pieces taken from this man are *damn* good."

In the four years from 1938 to 1942 Joseph Mitchell published (up to "Professor Sea Gull" on 12 December 1942) thirty signed pieces, a heroic pace. These mostly factual pieces, preceding his 1943 collection, *McSorley's Wonderful Saloon,* constitute the bulk of his work and most of what appeared in the first half of *Up in the Old Hotel* (1992). That his productivity dropped off sharply after the publication of his first profile of Joe Gould and then ended completely (so far as contributions to *The New Yorker* were concerned) after his second series of profiles about Gould in 1964 suggests much about the most perplexing and frustrating experience in Joseph Mitchell's adult writing life.

With the publication of *McSorley's Wonderful Saloon* in the middle of World War II, Mitchell's work was beginning to receive its first important recognition from members of the literary establishment, most notably by Malcolm Cowley in the *New Republic*. *McSorley's Wonderful Saloon* was also welcomed abroad, having been reviewed in London newspapers by John Betjeman, Peter Quennell, and A. A. Milne.

The articles in the 1943 collection integrated time, character, and place in an intimate, complex, and yet seamless fashion, reminding the well-read reader of the *Tatler* and *Spectator* essays of Joseph Addison and Richard Steele. By bringing into literary life such human beings as Mazie P. Gordon, Commodore Dutch, museum keeper and shameless self-promoter Charles Eugene Cassell ("Captain Charlie"), Jane Barnell (the bearded "Lady Olga"), and Joseph Ferdinand Gould (author of an "Oral History of the World"), Mitchell had now created his own illuminated tapestry of urban pilgrims, each seeking his or her private shrine or corner of happiness. Mitchell eventually came to observe, "It turns out, when I look at these things, just about everybody is me. I didn't know it at the time, but I interviewed people just like me." Moreover, not only here but in his later pieces can be found universal traits, archetypes such as the "Bountiful Mother" (Mazie), "Martyrs" (Lady Olga and Joe Gould), "Magicians" (Commodore Dutch and Captain Charlie), and "Kings" or "Warriors" (John McSorley, Gypsy leader Johnny Nikanov, and Captain Archie Clock).

Joseph Mitchell's next book did not sell as well, nor was it as widely reviewed as his 1943 collection, but it is the book that comes most profoundly out of the psychic life of the family; it echoes and mirrors the market culture of his native Fairmont; and it expresses a prescient personal philosophy. Dedicated to his father, *Old Mr. Flood* (1948) presents (aside from the monolithic presence of Joe Gould) his most fully realized character, Fulton Fish Market habitué Hugh Griffin Flood, to whom the author gives his own birthday (July 27) and who believes that living long is the best revenge: he aspires to become 115 years old. He will be able to credit such longevity largely to his "seafoodetarian" diet, but also to a simple but fulfilled life.

Dust jacket for the 1994 reprint of Joseph Mitchell's fourth collection

The three parts that make up *Old Mr. Flood* had been published in *The New Yorker* (all of Mitchell's books since *My Ears Are Bent* included only pieces originally published in that magazine) in 1944 and 1945 (the third part, though titled "Mr. Flood's Party," had no connection, Mr. Mitchell has said, to Edwin Arlington Robinson's poem with that title). Mr. Flood, the author notes, is "not one man; combined in him are aspects of several old men who work or hang out in Fulton Fish Market, or who did in the past." And the author quite candidly acknowledges that he has become something more than just a reporter: "I wanted these stories to be truthful rather than factual, but they are solidly based on facts."

Mr. Flood shares with his creator many common traits: love of good food (especially seafood), distaste for the hybridization and mechanization of contemporary culture, respect for fine architectural work, love of good company and lively conversation, and detestation of racial and ethnic prejudices. Like his creator, Hugh Flood is also depressed by newness; he and Joseph Mitchell take comfort in the milieu of the Fulton Fish Market for that reason; it is richly redolent in history, even in its odors.

While the subject matter of his writing—the low-life environment of Manhattan, and particularly the denizens of the Lower East Side—remained a constant in Mitchell's prose, he had at age forty with *Old Mr. Flood* reached a mature elegiac phase. As Noel Perrin observed in a 1983 *Sewanee Review* article, the past had now fully entered Mitchell's work and was "there to explain the present." The stories about Mr. Flood, ninety-five and more at the end of the book, are also, as Christopher Carduff pointed out in 1992, "about more than the ancient lore of fishmongers and fish-eaters and the crotchets of an old man. They are about, above all else, how to live the right life."

After the publication of *Old Mr. Flood,* Joseph Mitchell wrote only eight more *New Yorker* articles in the next sixteen years. But during these next years he published in book form the works that form his most lasting contributions to literature: the essay "The Mohawks in High Steel" (which first appeared in *The New Yorker* in 1949 and then became the lead essay in Edmund Wilson's *Apologies to the Iroquois* ten years later); *The Bottom of the Harbor* in 1960 (which collected six nonfiction pieces); and *Joe Gould's Secret* in 1965 (which collected again "Professor Sea Gull" from *McSorley's Wonderful Saloon* along with one article in two parts titled "Joe Gould's Secret" that had appeared in consecutive issues in September 1964).

Noel Perrin remarked that "the Dominant note of the book is a kind of sad gallantry, as in Robinson's poem." This quality can be observed again in Mitchell's last three printed works, with the one exception of "The Rats on the Waterfront," which appears as the first story in *The Bottom of the Harbor* and is the only "Reporter at Large" piece in the collection. Here again, as with Mr. Flood and his aging friends and/or fellow residents of the Hartford House, Mitchell presents in these stories "wildly imaginative mythomaniacs," men profoundly apprehensive of what the future may bring. More often, though, there is a quiet dignity and a stoic heroism among Mitchell's entropic heroes, whom the present has passed by but who continue to defy death. And a figure of ultimately tragic stature is finally brought forward in Joseph Ferdinand Gould, Joseph Mitchell's alter ego, nemesis, and finally his deconstructor, "who came to the city in 1916 and ducked and dodged and held on as hard as he could for over thirty-five years."

"The Mohawks in High Steel" (17 September 1949) was Joseph Mitchell's last published "Reporter at Large" piece; it offers the reader an awareness and an understanding of the culture and the daily lives of a group of nomadic riveters and high steel workers, Caughnawaga Indians from near Montreal who have established a colony in the North Gowanus neighborhood of Brooklyn. A trait (among many others) Mitchell shared with Joe Gould was his deep empathy for Native American culture. Gould informed him that "Indians are the only true aristocrats I've ever known. They ought to run the country, and we ought to be put on reservations."

Joseph Mitchell's admiration for Native Americans derived from his familiarity with the Lumbees–the largest non–federally recognized "tribe" in the United States–who comprise the greatest proportion of Robeson County's triracial makeup. His father had instilled in him a deep respect for these now nearly fully amalgamated peoples; he would sometimes take Joseph to the library of what is now the University of North Carolina at Pembroke, a totally Indian institution of higher learning until 1953, when it voluntarily "desegregated." Mitchell became an admirer and then a friend of its current Lumbee chancellor and a great respecter of Lumbee historian Adolph Dial. His study of the Caughnawagas took him to their Canadian reservation, and in "The Mohawks in High Steel" he offers a richly intimate appreciation of their religion, their foods, and, as workers in a dangerous environment, their physical and mental dexterity and close bonding.

The six stories in *The Bottom of the Harbor* provide the best examples of Joseph Mitchell's "graveyard humor," a cast of his mind that, he says in the "Author's Note" to *Up in the Old Hotel,* he did not comprehend in his temperament until he was re-reading the stories in preparing the anthology. While Mitchell may have been congenitally morose or gloomy (a well-known anecdote has Harold Ross confessing to him, "I'm no goddam little ray of sunshine myself"), such a temperament was enhanced by his longtime fascination with the Mexican street artist José Guadalupe Posada (introduced to him by Frida Kahlo in 1933). Posada (1852–1913) was, like Mitchell's most memorable personages, also a street character, one who eked out his living in part by selling broadsides; the majority of the engravings, as Norman Sims has Mitchell describing them in an essay on Mitchell, are of "animated skeletons mimicking living human beings engaged in many kinds of human activities, mimicking them and mocking them. . . ."

In all of the stories in *The Bottom of the Harbor,* then, the reader will expect to find, in either subtle or overt form, evidence of a macabre kind of humor: the stealthiness of anxiety-plagued rats slinking down the streets of Manhattan in search of their next meal; Leroy Poole's nightmare about the graveyard of lost ships and the lost (or murdered) men on the bottom of New York Harbor; the serio-comic allegory of Mitchell's and Louie Morino's visit to the long-abandoned upper floors of the old hotel; the ghastly human remains brought on board ship by dragger captain Ellery Thompson's ship; the cemetery in Edgewater, New Jersey, now completely surrounded by a factory; and finally the badly overgrown but tranquil Sandy Ground cemetery on Staten Island where George Hunter will eventually be laid to rest.

Five years after the publication of *The Bottom of the Harbor, New Yorker* colleague and friend Stanley Edgar Hyman wrote an article in the *New Leader* that did much to compel a larger understanding of the sweep and scope of Joseph Mitchell's work. Speaking of Mitchell as the "paragon of reporters," Hyman pointed out that he could no longer just be reckoned as a master reporter. In "The Art of Joseph Mitchell," Hyman asserted that "his importance as a writer, a unique figure in our literature, [has] been generally overlooked. . . . Mitchell is a formidable prose stylist and a master rhetorician; he is a reporter only in the sense that Defoe is a reporter, a humorist only in the sense that Faulkner is a humorist."

Picking up twenty-three years later on the epithet used by Hyman, Perrin wrote in "A Kind of Writing for Which No Name Exists" in 1988 that, while discounting "The Rats on the Waterfront" as the one piece of the six that "some lesser person might have written," the remaining five pieces in *The Bottom of the Harbor* constitute

> a kind of writing for which there is no name. Each tells a story, and is dramatic; each is both wildly funny and so sad you can hardly bear it; each tells its story so much in the words of its characters that it feels like a kind of apotheosis of oral history. . . . Mitchell has the gift of making roses bloom in the darkest and most unexpected places.

Commenting, too, on the enhanced perspective of the stories in *The Bottom of the Harbor,* William Zinsser said in *On Writing Well* (1994) that this collection "still strikes me as one of the best of all American nonfiction books."

While the connection between fiction and nonfiction has become increasingly blurred and while those who come to fiction from journalism face augmented challenges, the greatest challenge to Joseph Mitchell as a creative genius derived from his at-

Dust jacket for the collection of Mitchell's writings published in 1992

tempts to shelter his own psyche from the encroachments of the street character who became his most conflicted and perplexing literary recreation. As Joe Gould said to him in early 1943, "You wanted to write a story ["Professor Sea Gull"] about me, and you did, and you'll have to take the consequences."

Those consequences for Joseph Mitchell were ultimately profound. He told David Streitfeld in the summer of 1992 as *Up in the Old Hotel* was being read to wide acclaim that "talking to Joe Gould all those years he became me in a way, if you see what I mean." The world for Joseph Mitchell by this time had changed from a kind of cosmic joke (though the "straight men" in the joke could be treated elegiacally) to a perception of the human condition as ultimately tragic: as one observer put it, the myths of art and poverty for him had been deconstructed. His bonding with Gould from 1942 until Professor Sea Gull's death in a state hospital in 1957 had led Mitchell to a deep and disturbing awareness of both the wellspring of creativity and the consequences of its drying up. Joe Gould had become a figure of archetypal scope:

In my eyes, he was an ancient, enigmatic, spectral figure, a banished man. I never saw him without thinking of the Ancient Mariner or of the Wandering Jew or of the Flying Dutchman, or of a silent old man called Swamp Jackson who lived alone in a shack on the edge of a swamp near the small farming town in the South that I come from and wandered widely on foot on the back roads of the countryside at night, or of one of those men I used to puzzle over when I read the Bible as a child, who, for transgressions that seemed mysterious to me, had been "cast out."

The most remarkable piece of writing in the corpus of Joseph Mitchell's considerable achievements is *Joe Gould's Secret,* and the most remarkable passage within that book is an eleven-hundred-word stream-of-consciousness paragraph, in which the author reflects upon his long-abandoned Joycean novel as he comes to a sympathetic understanding as to why Joe Gould for all of these years had perpetrated the fraud of his "Oral History of the World," a masterpiece supposedly over a million words and growing. If a writer-cum-bum like Gould could have been befriended by, been patronized by, influenced, and/or been published by Marianne Moore, Horace

Gregory, E. E. Cummings, Ezra Pound, William Saroyan, Malcolm Cowley, and Dylan Thomas, what would Joseph Mitchell, realizing how badly he had been duped, feel like, remembering that his daughter Elizabeth was fond of teasing him about his varied and many successes: "Daddy, you're still fooling them, aren't you?"

In that remarkable paragraph Joseph Mitchell comes to recognize that listening to the fire-and-brimstone evangelists in the Fairmont of his youth had given him "a lasting liking for the cryptic and the ambiguous and the incantatory and the disconnected and the extravagant and the oracular and the apocalyptic." In the story of Joe Gould's life he may well have found all of these elements. Dawn Powell, in reviewing *Joe Gould's Secret* in 1965, offered a perceptive conclusion:

> If that strange little goblin creature named Joe Gould had never existed, Joseph Mitchell would have been compelled to invent him. The wonderful life story here reads as if Mitchell had been haunted for years by such a figure, then encountering him in real life must spend 22 years piling up evidence to prove to himself that the man is real.

During the period of twenty-seven years between the publication of *Joe Gould's Secret* and the appearance of the Pantheon edition of *Up in the Old Hotel*, Mitchell continued to keep regular hours in his office at *The New Yorker* and devoted much of his time and energy to various civic projects. For many years he had been rumored at times to be working on a biographical study of Ann Honeycutt (at one time James Thurber's lover and a prominent figure in Manhattan literary and artistic circles) and at other times to be working on a book about the Fulton Fish Market.

Some of his best pieces are increasingly being anthologized in literary and journalism texts. And for *Up in the Old Hotel* to be included in the "Canonical Prophecy" of Harold Bloom's *The Western Canon: The Books and the School of the Ages* marked how far his work had come in the esteem of the literary establishment.

Interview:

Personal interview with Joseph Mitchell in his *New Yorker* office and at the Century Association on 19 June 1995 (amplified by telephone conversations).

References:

Christopher Carduff, "Fish-eating, whiskey, death & rebirth," *The New Criterion,* 2 (November 1992): 12–22;

Stanley Edgar Hyman, "The Art of Joseph Mitchell," in *The Critic's Credentials* (New York: Atheneum, 1978);

Benjamin Ivry, "Joseph Mitchell's Secret," *New York,* 20 (9 February 1987): 20;

Noel Perrin, "Paragon of Reporters: Joseph Mitchell," *Sewanee Review,* 91 (Spring 1983): 167–184;

"Postscript: Joseph Mitchell (Three generations of New Yorker writers remember the city's incomparable chronicler)," *New Yorker,* 72 (10 June 1996): 78–83;

Dawn Powell, "Baa, baa black sheep," *New York Herald Tribune Book Week,* 19 September 1965, p. 8;

Norman Sims, "Joseph Mitchell and *The New Yorker* Nonfiction Writers," in *Literary Journalism in the Twentieth Century,* edited by Sims (New York: Oxford University Press, 1990);

Sanford Smoller, "Rosebushes and Bones: Joseph Mitchell's Enduring Values" *Pembroke Magazine* 26 (1994): 10–31;

David Streitfeld, "The Subjective Observer," *New York Newsday,* 27 August 1992, II: 60–61;

William Zinsser, *On Writing Well,* fifth edition, revised and updated (New York: Harper Perennial, 1994).

Orville Prescott
(1906 – 1996)

Peter S. Prescott

Thirty years after its publication, a women's group in New York City staged a symposium to celebrate Doris Lessing's *The Golden Notebook* (1962). Though none of the speakers was old enough to have noticed the reception of Lessing's novel at its birthing, the chatter began with reflexive denunciations of Orville Prescott's critical review of the book, thirty years earlier, in *The New York Times*.

A curious business. Who but an injured author, nursing her wrath to keep it warm, remembers a newspaper review from so long ago, or for that matter the name of a newspaper critic who had by then been silent for nearly thirty years? George Bernard Shaw's theater and music reviews are still read by some in the critical fraternity; perhaps Max Beerbohm's theater reviews; H. L. Mencken on books; James Agee and Pauline Kael on film; possibility Virgil Thompson on music. Then one must pause for breath. Edmund Wilson's book reviews do not enter the lists because he wrote at any length he pleased and only about subjects that interested him, disdaining to notice the week's offering of books demanding a critic's consideration. That obligation—to deal day after day with a substantial book that is published that day—becomes in time fatiguing.

No journalistic critic whose influence outlasts his season ever shows a trace of fatigue. The fatigue inevitably arrives, and when it does, a good critic plucks the ribbon from his typewriter—after first checking that his last review is as vigorous as all his others. Energy and a distinctively personal tone of voice: these are necessary commodities if a critic is to be remembered for some decades. He must also be willing to beat against the tide of fashionable opinion. Shaw denounced Johannes Brahms in all his work; Prescott presumed to criticize the more undisciplined fiction of William Faulkner and John O'Hara. He wrote that Faulkner was "probably the most gifted of contemporary American novelists, [but he] has never learned to master the rudiments of his craft." He called O'Hara's *A Rage to Live* "an abomination, a travesty of false values and decadent sentimentality." Such a verdict concentrates an author's mind wonderfully: O'Hara told Random House that henceforth his annual volume was to be

Orville Prescott (photograph by Nickolas Muray)

published on Thanksgiving Day, when Prescott didn't review, but the number-two book critic, an O'Hara partisan, did.

My father didn't care for novels about sex, or for Marxist-feminist fulminations by Lessing or anyone else. Perhaps, had he been cautious, he would have ducked and weaved, but in fact the critic never gave a thought to his critics. In life, he was a prudent man—"courtly" is the word that acquaintances offered on the occasion of his death—careful never to give offense, or even to raise his voice, but in his work he forsook caution. He had firm ideas of what literature is, or should be, and he wrote plainly, vigorously. A really good novel, he wrote, "must be written with sound craftsmanship, must create interesting characters and involve them in a significant situation, must reflect the special personality and point of view of its author." A great novel needed something more: "a feeling of passionate participation in life. . . . [a] belief in the essential dignity of man."

You might agree with him, and most of his readers did, or you might wince at his judgments, but you were never in doubt about what he thought. (The jacket of his first book, *In My Opinion* [1952], shows the shadow of a judge's gavel falling on an open book.) And there was a cumulative effect: many people who disagreed with my father's perspective read him because they knew his reviews would tell them what *they* would think of the book at hand.

The consensus in the literary precincts (even if some would exclaim "Alas!") remains what it always was: that Orville Prescott was the most influential journalistic book critic of his time. LeBaron Barker, once Doubleday's executive editor, called him "the top critic in the country" and added: "In almost forty years of publishing I can think of no other critic who has been able, as he has done, to affect the course of reading in this country." He was, Alfred A. Knopf wrote in *The Borzoi Quarterly,* the only critic "who today can command a hearing for a book which really excites him—who has proved his ability to communicate to thousands of readers a desire to share that excitement. For the rest . . . every publisher can give you endless evidence to prove how little effect reviews have on the sale of books."

Hiram Haydn, once editor in chief at Random House and later cofounder of Atheneum publishers, put it succinctly in his memoir, *Words and Faces* (1975), published nine years after Prescott's retirement from the *Times:* "No other reviewer exercised so much influence on the sales of books as he did. For his reviews were electric: they crackled with the energy of his convictions. Many readers thought and said that 'prejudices' would be the more accurate word. Never mind: he said what he thought; he said it pungently and directly. As a result, he had no apathetic readers. And often he did as much to accelerate interest in a book he loathed as in one he fiercely admired—so provocative were his strictures. Only in the case of his reviews could a publisher watch the 'ticker tape' and measure the impact on orders of an exuberant or bristling Prescott review. The phone would ring and the jobbers would reorder: fifty copies, one hundred, even five hundred. He was the only book reviewer with a power comparable to that of the *Times* theater critic."

No critic of integrity cares to sell books; he is not an adjunct to a publisher's publicity department. A critic aspires not to power, which if it comes to him is usually an embarrassment, but to influence. My father may have been the only critic ever named in a Broadway comedy to get an early laugh from the audience: Jean Kerr's *Mary, Mary* opens with a woman saying into a telephone: "We all wish Orville Prescott would write a novel."

He was aware that in his era, when publication dates were strictly adhered to, authors would go to Times Square on the eve of their book's publication to snatch an early copy of the paper. Would there be a review by Prescott? Well, there would at least be a Prescott review. For a quarter century, from 1942 to 1966, he was the principal book critic at the *Times,* and for most of that time, he wrote four book columns a week, a feat unimaginable today. (Admittedly, my father never had to endure the time-devouring, spirit-depleting confrontations with junior editors determined to geld his sentences that now consume much of a critic's working day.) Four distinctively personal reviews a week, year after year, launched from journalism's bulliest pulpit: influence necessarily follows.

Equally necessarily, he had his detractors, particularly among the coterie known as the New York Intellectuals, and among them particularly the aging socialists and disappointed Marxists. A critic who does not draw detractors should look for another line of work. My father was especially a target for the intellectuals because in his criticism he often swam against the modernist tide that had secured the beach long before he put his feet in the surf. He did not actually reject modernism in the arts; rather, he refused to change his critical perspective to accommodate it. He and my mother collected contemporary paintings: rather elegant figurative paintings—the sort that our century had worked strenuously to dismiss. Once he told me that Pablo Picasso was a fraud. I never asked, but I suspect that he never came to terms with Paul Cézanne or Henri Matisse. He cared for no kind of music at all.

It was in the nature of his job that his range must be extraordinary: he reviewed everyone from F. A. Hayek to Robert A. Heinlein. By refusing to follow fashion, Orville Prescott introduced readers of *The New York Times* to hundreds of authors of excellent books that were at first not much talked of elsewhere. He argued the case for the historical novel as literary art, citing the work of Marguerite Yourcenar, L. H. Myers, and Hope Muntz, when others, forgetting that Leo Tolstoy's *War and Peace* (1863-1869) is a historical novel, denounced the entire species as commercial trash. He recommended the early novels of C. P. Snow and Joyce Cary. He praised the realistic novels of J. P. Marquand, James Gould Cozzens, A. B. Guthrie, Louis Auchincloss, Conrad Richter, John Hersey, Rumer Godden, Anne Tyler, and Evelyn Waugh. He applauded Phyllis McGinley's witty, sinuous poems of suburban matrimony with affectionate doggerel:

> When winter moans and my blood runs thinly
>   And solemn books drift drear and deep
> I take my ease with Phyllis McGinley
>   And send her part of my heart to keep.

He counted many authors among his friends, but they knew they couldn't rely on him for a favorable review. If he didn't like a friend's book, he wouldn't review it, and if the book was so important that he couldn't avoid reviewing it, his review would contain no judgment at all. The silence of his disapproval rang like a cannonball ricocheting down Publishers' Row.

Wearily brushing aside a bleating protest from an interviewer that Prescott couldn't be taken seriously, Alfred Knopf replied that "for the *Times* audience he was absolutely perfect. I think Prescott was underestimated by the Intellectuals; I think he was much more intelligent than he was given credit for being. I think he was much more widely read. I don't think he was particularly interested in fiction. You could tell from what he wrote after he retired that he was interested in history and biography."

So he was. My father retired to write a book, *Princes of the Renaissance* (1969), an account of Italy in turmoil in the fifteenth century, of how the Medici, the Sforzas, the d'Estes, the Gonzagas, Pope Julius II, and others sorted matters out, usually with blood. It was followed by *Lords of Italy* (1972), a more remarkable book in that it examines the even bloodier history of medieval Italy, about which little had been written. These are popular histories so soundly researched that academic historians of the period asked my father to help them in their own efforts. He produced two anthologies of historical writing: *The Undying Past* (1961) and *History as Literature* (1971). He edited The Crossroads of World History series for Doubleday and served as a founding adviser for *The Dictionary of Literary Biography*.

Born in 1906 in Cleveland, Ohio, he published an autobiography fifty years later. In *The Five-Dollar Gold Piece* (1956) he relates that when he was six his grandmother offered him that small gold coin if he would learn to read. That shameless bribe, he wrote, "marked the most important point of my life." Toward the end of his days, a five-dollar gold piece (not the original) encased in lucite stood on a table in his home in New Canaan, Connecticut. For his sins, his son followed him into his trade, much as the farrier's son would have taken up shoeing horses two centuries ago.

RECOLLECTION BY SAM VAUGHAN

As a young editor in the 1960s, I had the luck to sit in with a group of editors at Doubleday who were to plan a new series of books in history. There, excellent editors much my senior—such as LeBaron R. Barker, Kenneth D. McCormick, and Walter I. Bradbury—would meet regularly to think up ideas for books to be commissioned and authors to be approached. It was, if I recall, a series grandly titled The Crossroads of World History. The advisory (or "outside") editor engaged for the series was Orville Prescott.

I didn't know him but had read him, of course; he was a mainstay of daily book reviewing in America. I was quite certain that we wouldn't get on. Not that it mattered, because I was very much the junior editor of the group, knew too little about history, thought that I would have to hold, or bite, my tongue, and in general prepared myself to be the Odd Man Out. Mr. Prescott's views on books and, occasionally, book publishers were on display several times each week in the august pages of *The New York Times,* and clearly he was an old fuddy-duddy with a taste for "clean" books and a marked dislike for certain kinds that were, even then, beginning to crowd the lists. The Age of the Hooligan had not yet arrived, but it wasn't far off.

Some years earlier, as Doubleday's advertising manager, I had worked up a pro-and-con ad to promote the popular novel by Robert Ruark about the Mau Mau or native uprisings in Kenya, *Something of Value* (1955). Such ads were rare to the point of nonexistence. We proposed to list the good things reviewers had to say about the novel in one column on the left and the negative criticisms on the right, in almost the same space—an act of boldness, or publishing bravado, and, not incidentally, because some of the things the antireviewers were saying about the book could cause even more people to buy the book than the admiring critics. Chief among the naysayers was Prescott.

The New York Times, reviewing the unusual ad in advance for acceptability, refused to print it. "Why?" we asked, in high dudgeon, concealing our low motives. Some of the things written against the novel, they answered, were not fit to print. We pointed out that it was the novel, not the reviews, that some detractors thought not fit to print, but anyhow, it was their reviewer, Mr. Prescott, who had led the pack, and they had already printed that much. They gave in, as I recall, but the fuss added fuel to an already ignited best-seller.

Well, I puffed up like a pouter pigeon, pleased with myself at having bested *The New York Times* and so, later, got ready to meet their cranky critic. Prescott turned out to be a plainspoken middle-aged man, unfailingly polite, even to a kid editor who didn't know the War of the Roses from Four Roses. He proved to be a person of rare qualities. His great reserve was coupled with great enthusiasm. Erect, immaculately dressed, with a pencil-thin mustache and a touch of gray over the ears, he was respectful of other people's opinions. The other advisory editors on similar committees were apt to be too quick, promiscuous with ideas and nominations, ready to play favorites. "Bill" Prescott, as they called him, was not to be stampeded, insisted on being thoughtful, and suffered from an affliction of character called high standards.

What impressed me beyond that was the genuine love he had, and they all had, for good history and good writing. There wasn't a snob in the bunch. They welcomed serious journalists; they welcomed novelists who wanted to write history; and they welcomed autodidacts or academics who had somehow slipped through the net and could write clear prose. They admired the good story well told and backed by excellent—but not overly apparent—research. (My mentors taught me an expression I still sometimes use with an author, the kind of superconscientious, don't waste a drop school: "Your research is showing.") They also shared a respect for the beauties of the straightforward English sentence. Needless to say, I became a fan of and then, to my astonishment, a friend of the man they called Bill Prescott.

In approaching the output, or outpourings, of authors and publishers, he seemed to have some strange notion that writing, and beyond that disseminating what was written, involved issues of moral responsibility. This underlying assumption is not something that takes up much time in current gatherings of authors and publishers. His revulsion against the truly vulgar would today seem an outrageously archaic notion at a time when Neanderthals ride the best-seller lists, earn the staggering paychecks in sports, business, and showbiz. Not the good vulgar, in the broad sense, which animates some of popular culture, but the meretricious, pervasive pap which threatens not only forests and landfills but at times seems to crowd out the good, better, and best. The novel, for instance, which is the hack author's prequel to the action-thriller movie in which everyone and everything is blown up until you could, in Dorothy Parker's words, "thwow up." Profanity as part of art, to use another example, has been replaced by the ubiquitous, wholesale profaning of language, symbols, and persons; schlock and would-be shock is used as a substitute for real writing, imagination, selectivity, sensitivity, and genuine drama.

The other night, walking in New York with my wife, there was a fellow staggering down the street, spewing the four-letter F-word at machine-gun pace. "Must be a screenwriter," I said to her, and she laughed in agreement.

But now I am writing pure fuddy-duddy but lacking Mr. Prescott's eloquence even in anger.

These days, most days, we miss Bill Prescott. Even more, we need him.

STATEMENT BY LOUIS S. AUCHINCLOSS

Prescott's intense interest in current American writing enabled him to describe with considerable clarity even work that he disliked. His great value as a regular critic was that you could get a vivid idea of what a book was like from his digest of it. You could tell whether or not it would be your affair even if it was evidently not Prescott's. That seems to me the first job of a book reviewer.

Letter from London

Julian Evans

Time is the great editor, as writers know—though by the mysterious nature of the proposition, they tend not to live to see precisely how time toils. It can be a frustrating truth for all concerned, since publishers, as well as writers, can never know precisely where they are. Both have their specific, if inadequate, remedies: the writer to have another crack, the publisher to judge and sell copies with the means he or she disposes of, patient in their faith in both this book and next crack. That was, of course, the Hogarth Press's way with Virginia Woolf before *To the Lighthouse* launched her to popular fame, Heinemann's with Graham Greene before *Stamboul Train,* Editions de Minuit's with the first printing of *Waiting for Godot*—which in its first five years sold 150 copies a year. That was their faith, imperfect but willing; times have changed, and we now have the additional element of the market in all its possibilities, including that of a promo budget for a book that may be higher than either production costs or author's advance, even both. Like our politicians and our desktop hardware, however, the market is both good servant and atrocious master. Before discoursing on literary life—a funny phrase, incidentally, one of the few in the language capable of being both tautology and oxymoron—in Britain in the year past, I want to pass on a story on another axis that has made me wonder, about times, about the market. A longtime French friend translated into more than a dozen languages including English (I happen to have been his British translator), sent this spring a collection of stories to an agent in New York who shall, as they say, remain nameless, though she had been personally recommended to my friend. A reply eventually came back after two months: the writer did not linger on the stories' contents but merely pointed out sorrowfully that the market's demands lay elsewhere. An interesting development: has the market—not time or editorial judgment—now become the great editor? In that case, *quis custodiet mercatum?*

In London, as in New York, we go on believing in the novel the way we believe in God: when the pilot's voice floats down from the flight deck to report a fault in his instruments we hope to high heaven He exists. The market too believes, naturally enough, because it cannot ignore that there are notable novelists who have notable successes. In London, I'm glad to report, the novel *business* is healthy. The proof is that everyone writes them. MPs, journalists, TV scriptwriters, pop stars, comedians, supermodels—possibly we can expect in the future to add chefs, footballers, and other neocelebrities to the list. Defiantly, the novel continues to offer the lure of a little well-marketed immortality. Interesting cross-fertilisations have sprouted: the market's indifference to the short story has led to even more novelists flocking to comfortably indentured newspaper columns; well-paid journalists, for reasons possibly connected with the transience of their prose, seem to ache to be novelists. Strangely, food is a dominant motif. Thus, this year has seen the curious triple spectacle of enfant terrible novelist Will Self dining weekly at the expense of the *Observer;* the *Times* restaurant critic A. A. Gill publishing a (terribly infantile) novel; and John Lanchester, ex-deputy editor of the *London Review of Books,* producing a novel about a gourmet and murderer. Of the novels, Lanchester's *The Debt to Pleasure* is infinitely the racier and more intelligent, and of the columns Self's infinitely the more innovative. It was a relief that it was Gill who claimed amid howls of critical derision that "The truth is there is no difference between novels and journalism." Coming from him, the remark may turn out to have done the cause of the novel incalculable good.

Certainly the business is healthy. There is no shortage of variety, and we have to count readers' enthusiasm high on the list of qualifiers. The fiction tables at Waterstones and Dillon's testify like supermarket shelves; though "Novels of the year" is a slightly cramping appellation—I would rather register my personal enthusiasm for those to be kept on my own shelf, the ones worth a more defining rereading. Soon, I'm certain, I will reread *High Latitudes,* James Buchan's ironic morality tale of money and female passion, an extraordinary successor to his Cold War spyfest, *Heart's Journey in Winter.* Renewable pleasure is the only test: I have read the earlier book three times and possibly, the handicap of Buchan's cleverness notwithstanding in a culture in which good manners demand concealment, we don't have a better writer among us. I would single out Graham Swift's Booker Prize–winning *Last Orders,* his demotic melodrama of four drinking pals on a burial party, and a resonant first novel by Seamus Deane, *Reading in the Dark;* both tapestries, the one a colloquial hanging of English past and present needfully more peaceable than the other, since any tapestry of Irishness cannot bury, or escape, the island's haunted and haunting past. But would I be alone in brooding that as a literary form our novel is somehow restless and subdued? We readers do try: we listen; we do our duty by each Thursday's and Sunday's coverage in the prints. We read what is, but cannot reread what wasn't. Perhaps the brood *is* taking over: would one want more than three books a year? In thirty years a hundred recent novels worth taking off the shelves again would be a respectable count, not to mention close to the nursing home's baggage limit. There is one other I am not decided on: Lawrence Norfolk's luxurious, overspilling fable *The Pope's Rhinoceros*—undecided only because the interior lives of ants, rats, and fish may be a species too many.

We have known for this short century that novels must vie for an audience and kindness and periodically fail with grace. First there were movies, then TV and media acceleration. (Milan Kundera's contention that slowness equals humanity, humour, love, and sex sadly won't retard the spin of the Darwinian wheel.) And now we've got millennial angst—we appear to need something more reliable than fiction. There has been, still is, a heyday of science writing: it is Richard Dawkins and Stephen Jay Gould who have filled book festival tents and literary talk programmes this year, although my own assessment of Dawkins's gleeful assertion that "DNA just is, and we dance to its music" is that it is about as much use to us as a spelling checker to Pushkin in the composition of *Eugene Onegin*. Happily, a consoling row erupted in the summer at Oxford between Dawkins and the theologian Keith Ward, author of *God, Chance and Necessity,* who took on the reductionists on their own territory and sort of won: won, let's say, the right to be heard—which is after all the market's greatest bouquet.

In fact it has been a turbulent year, what with Martin Amis resigning in fury from the *Sunday Times* after it serialised his father's biographer's diary of Kingsley Amis's last days and Anthony Julius's "J'accuse" against T. S. Eliot for his anti-Semitism. Both disputes generated more heat than light, though that is of course the prerogative of a media that has time for mediation but for which debate is unconsciously long, and in any case the subject was perfect for the former, unnecessary for the latter. They tried to do the same to Philip Larkin a few years back on account of alleged racism: there is a borough in London called Islington where most of these self-appointed guardians of our frail susceptibilities live. A serendipitous discovery of long-lost Eliot poems was offered us by way of another excellent consolation: *The Inventions of the March Hare,* containing the poet's juvenilia and bawdy songs. (So quick was one newspaper to cover the find as frontpage news here that its hapless literary editor printed as proof of the Nobel laureate's hitherto unknown saltiness an apparently original piece of Eliot bawdy, only to find it was a traditional shanty to which Eliot had merely altered a few words.)

Yes, Eliot was one of the year's greatest rediscoveries—I only wish that one year soon Wyndham Lewis, an authentic English genius and Eliot's companion in modernist greatness, will be refound with equal generosity. (A small part of Eliot's eminence, I'm sure, came from his understanding of the social mechanics of gaining a reputation. Compare the rewards of truth with those of diplomacy: compare Lewis's 1948, writing art reviews for the *Listener* from his derelict flat in Notting Hill Gate and going blind, with Eliot's, honoured the same year with an Order of Merit, a Légion d'Honneur, and the Nobel Prize.)

It has been a good year for that other great branch of truth, the literary biography. The flood has slowed and falling author's advances—floating readers having perhaps drifted to science—have frightened off that seemingly inexhaustible supply of gourmet opportunists dining out on a greater artist's life. I was once more puritanical about biographies than now. The domestic details *are* salutary; there is the continuing joy of a preserved anecdote. James Knowlson's distinguished life of Samuel Beckett, *Damned to Fame,* contains both the Irishman's generosity (the jacket he gave to a tramp, without emptying the pockets) and desert wit: when a woman asked him if he minded that she had named her dog after him, he replied, "Don't worry about me. What about the dog?" We come, then, and inevitably, to Virginia Woolf; personally, I cannot abide the whole of the Bloomsbury Group—let us say it once and for all: there is not a single joke in Bloomsbury, *anywhere,* and the intellectual armlock their memory still exerts on us pains me to undignified fury—but Hermione Lee's meticulous *Virginia Woolf* contains her achievement with tact and admitted inspiration: Woolf cannot be faulted for her remarkable industry. Let words go to two sad geniuses too: *Ford Madox Ford* by Max Saunders, and *Rosebud* by David Thomson, both the best kind of guides to lead you back to the work of both Ford and Orson Welles.

From a quiet street in South Kensington I have watched this year roll past. Everything I have needed has been very close to hand: the busyness of the global village has passed me by, along with its dubious implications, and this parish newsletter, albeit partisan, has restricted itself to parochial reflections. Sufficient unto the day, and the place, are the books thereof, technically speaking: my reading of Richard Ford and E. Annie Proulx, my rereading of F. Scott Fitzgerald, André Malraux and André Gide have no place (though *Gatsby* is still as grand a view, as firm a bend in the road as ever was). I have brooded about the market, about books past and present, out of print, or that no one will take note of. In the East there are native censors; in the West, sometimes the market's trick is indifference. God may be dead, replaced by the vast interconnecting hubbub of the commercial media, but we still need messengers with a message to deliver. I nominate two from afar: *I, Fellini,* Federico Fellini's magnificent autobiography (and Charlotte Chandler's who wrote down his words), and *Dear Writer, Dear Actress,* the love letters of Anton Chekhov and Olga Knipper. Different messages, different countries, different centuries; same rapture.

The Booker Prize

Merritt Moseley
University of North Carolina at Asheville

The Booker Prize, Britain's foremost award for literary fiction, was created in 1968 and first awarded in 1969. It annually aims to recognize the best full-length novel written in English, published first in the United Kingdom, and written by a citizen of the United Kingdom, the British Commonwealth, the Republic of South Africa, or the Republic of Ireland. In practice almost any novel written in English by anyone other than an American is eligible for consideration. The winner receives £20,000 (more than $30,000) in prize money and can usually expect to earn much more by the increased sales resulting from the award. When the Booker Prize was established, the model was France's Prix Goncourt, which, in contrast, is worth a paltry Fr 50 ($10). The Booker was lucrative from the beginning, at £5,000; it has been raised three times, most recently to £20,000 in 1989. Since its creation there have been many new prizes created for British authors: the Betty Trask Award for a romantic novel (£25,000), the Whitbread Book of the Year (£17,500), the *Sunday Express* Award for an explicitly "readable"—that is, usually more traditional—novel (£20,000), the Commonwealth Writers Prize Best Book Award (£10,000), the *Irish Times*-Aer Lingus International Fiction Prize (IR£25,000), the Cohen British Literature Prize (£30,000), and, most recently established, the Orange Prize, which is limited to women authors and carries an award of £30,000. It is unarguable, though, that the Booker Prize remains the most prestigious award for a British novelist, using the term *British* in its large, inclusive sense. This may be because of its priority, its publicity, or its history.

In the years since the first award in 1969, the Booker Prize has been given to some of the best-known of postwar novelists—Kingsley Amis, Salman Rushdie, William Golding—and to some much more obscure, Keri Hulme and Stanley Middleton being good examples. The list of outstanding novelists who have *not* won it is equally impressive: among the outstanding writers now in their forties, Martin Amis, Julian Barnes, and Ian McEwan. Though in principle the award exists to recognize the best novel of the year that meets the qualifications—arguably an impossible aim to begin with—the Booker has a complicated mixed agenda. Over its twenty-nine-year run it has dramatically increased the sales of some books and authors nominated; it has successfully advertised its sponsor company, Booker-McConnell PLC, a grocery conglomerate; it has produced a lively argument about political correctness and diversity among nominees and judges; and, perhaps most consistently and successfully of all, it has generated plenty of controversy. This controversy is at least a discussion of books, authors, and literary values, and in this way the Booker has foregrounded contemporary fiction writing. The awards ceremony is televised live in Britain; in the months after the shortlist (five or six finalists) is announced, bookmakers take thousands or millions of pounds in bets on the outcome.

Richard Todd, whose *Consuming Fictions: The Booker Prize and Fiction in Britain Today* (London: Bloomsbury, 1996) is the first considered study of the award, goes so far as to declare that

> controversy has in many respects actually been the making of the Booker Prize.... It is surely evident that it is precisely by "getting it wrong" that the Booker survives. Each year's Booker shortlist is to a greater or lesser extent by definition a disappointment.... Every decision, in other words, must be more or less contentious.

The controversy is most often about the books chosen for the shortlist; of course, the one novel chosen as winner must be more contentious. According to Todd's estimate, there were between 4,500 and 7,000 eligible books published yearly during the 1980s. Only 100 to 120 of them were considered each year by the panel of judges, and only 5 or 6 of these were named to the shortlist. The potential for disagreement is obvious.

There is also controversy about the judges and their procedures. Celebrity judges with no connection to academia or the literary world (Mrs. Harold Wilson, the actress Joanna Lumley) have participated, to the bemusement of observers. Somehow

Graham Swift (photograph by Mark Douet)

the judges always reveal how the voting went, and it sometimes goes quite oddly. With five judges there should be no ties. Once there were three votes for one novel and two votes for a second, so they gave the prize to yet a third novel, which was nobody's favorite. As Todd points out, the decision to announce the shortlist early in the autumn, some two months before the winner is chosen (even though the judges ballot on the night of the announcement and there is no good reason to wait two months before they do it), has been a clever way of increasing tension, betting, controversy, and probably book sales. Since, after all this time, the panel must decide rather quickly on the night of the announcement with a live television audience and guests at a glittering banquet at London's Guildhall waiting for word, it is not surprising that they sometimes act oddly. The judges do not always seem to follow their own rules. The requirements for eligibility specify that a book must have been published first in the United Kingdom; in 1994 one of the shortlisted books was *Knowledge of Angels,* by Jill Paton Walsh, which had appeared first in the United States before being self-published in England. My copy of Rohinton Mistry's *A Fine Balance* (Toronto: McClelland & Stewart, 1995; New York: Knopf, London: Faber & Faber, 1996), one of the 1996 finalists, suggests it was first published in 1995 in Canada.

Todd shows that the Booker Prize, though founded in 1968, did not seize the public imagination until about 1980. In that year there was a perceived "two-horse race" between Anthony Burgess's *Earthly Powers* and William Golding's *Rites of Passage,* which eventually won. There have been other such two-book shoot-outs since then, sometimes settled (rather unsatisfactorily) by giving the award to two novels, as happened most recently in 1992, when Michael Ondaatje's *The English Patient* and Barry Unsworth's *Sacred Hunger* shared the prize. Most recently the excitement has been about omissions from the shortlist; about perceived bias among the judges; about the behavior of the winning authors; about the behavior of the judges; and about the tantalizing question of who will be the first author to win it for a second time—something that has never happened, though many have been nominated again and again (Salman Rushdie, for instance, has been nominated three more times since he won the prize in 1981 with *Midnight's Children*).

The award in 1996 was surprisingly uncontentious. Of course it was easy to think of overlooked books—in any year when only six books are short-listed and seven thousand are eligible it will be surprising if worthy candidates are not left out—and there was some of the usual grumbling about political correctness and possible bias against novelists from England. The six authors on the shortlist included three men and three women—two English writers, one Scottish, one Ulster Irishman, one Canadian, and one born in India and now resident in Canada. The big names were undoubtedly Margaret Atwood, who, having been short-listed twice before and having twenty-five books to her credit seems likely to be the first Canadian to win the Booker; Beryl Bainbridge, short-listed three times previously; and Graham Swift, whose novel *Waterland* was short-listed in 1981. The oddsmakers declared his 1996 book, *Last Orders* (New York: Knopf, 1996; London: Picador, 1996), the favorite—and it won. The other three nominees were Seamus Deane, an Irishman who lives in the United States, for his first novel, *Reading in the Dark* (London: Cape, 1996; New York: Knopf, 1997); Shena Mackay, for *The Orchard on Fire* (Wakefield, R. I.: Moyer Bell, 1996; London: Heinemann, 1996); and Rohinton Mistry, for *A Fine Balance.* Mistry was short-listed once before, for his only other novel, *Such A Long Journey.*

The list of six novels drew some querulous commentary on the retrospective stance they all shared. This is a perennial complaint too. Commentators often fret about the high proportion of historical fiction represented among the Booker finalists. Unfortunately, nominated novels that deal with the reality of today are often depressing and dismal: the 1994 winner, Scottish novelist James Kelman's *How Late It Was, How Late,* is a murky and tedious account of the squalid adventures of a drunken Glaswegian vagrant, told in a Scots dialect that is about one-third obscenities. Kelman has collected some comparisons to Joyce that are wildly inaccurate. Recent finalists *Serenity House,* by Christopher Hope, and *In Every Face I Meet,* by Justin Cartwright, also deal with contemporary life: *Serenity House* is a study of murder posing as care of the elderly, while Cartwright shows a good man's life brutally destroyed by the social and legal system.

None of the 1996 nominees is merry, but none is as unpleasant as the Kelman book. They are all, in important ways, backward-looking (not necessarily a bad thing, after all). Atwood's *Alias Grace* (New York: Doubleday, 1996; Toronto: McClelland & Stewart, 1996; London: Bloomsbury, 1996) recounts a real-life nineteenth-century murder case, while Bainbridge's *Every Man for Himself* (London: Duckworth, 1996; New York: Carroll & Graf, 1996) tells the story of the sinking of the *Titanic;* Deane's *Reading in the Dark* is about sectarian troubles in Derry in the 1940s, with a brief coda showing the protagonist in the 1970s; Swift's *Last Orders* has a present-day setting but is retrospective, returning frequently to World War II; *The Orchard on Fire* begins with its adult narrator in the present day but (like *Reading in the Dark*) is fundamentally a study of childhood and adolescence; and *A Fine Balance* is a historical novel about recent history—the crisis provoked by Indira Gandhi in 1975. Like Swift and Deane, Mistry uses the ostensible "present" of his novel as a starting point for rich tunnelings back through the lives of his characters and their families.

The announcement that Swift had won the Booker was well received, partly because it was widely expected and partly (perhaps) because this list contained no great book, so almost any could have been accepted as the winner. One of the judges, A. N. Wilson, who is a literary editor and a prolific novelist, did cause some astonishment when, a few days after the award, he denounced it in his column in London's *Evening Standard.* The voting was rigged, he declared; the chair, Carmen Callil, had written her speech (which made a point of the prize being won by an English author) before the judges had voted; and, for good measure, he went on to describe the sponsors of the prize, the Booker company, as "a somewhat sleazy food chain." The source of Wilson's pique was the decision of the other judges to remove Bainbridge's *Every Man for Himself,* Wilson's favorite, from consideration as they narrowed the list to three. Along with it went *A Fine Balance* and *The Orchard on Fire.* The first division contained *Last Orders; Reading in the Dark,* which finished a close second; and *Alias Grace,* which took the imaginary bronze.

Turning first to the also-rans, and considering quality rather than politics or other extraliterary qualities, it is impossible to share Wilson's outrage; on the other hand, there is not a great difference in quality among the books. Bainbridge, not primarily known as a historical novelist, does a fine job creating a convincing atmosphere for the doomed *Titanic* voyage. She has created as her narrator an imaginary American named Morgan, nephew to J. P. Morgan, and involved him in various intrigues with mysterious beauties, Jewish arrivistes, and plutocrats traveling on the voyage. The task she has set for herself in keeping her novel interesting is not a light one. Surely every reader knows that the *Titanic* will hit an iceberg and sink, and the first-person narration makes it predictable that Morgan will be among the survivors; the suspense, then, is sustained by minor things. We wonder if Morgan will be among those men who put on women's clothing to get into the lifeboats. He behaves impeccably in the end, and, despite the implications of her title, Bainbridge depicts many people behaving nobly and unselfishly in the crisis.

Every Man for Himself is a short book, despite its world-historical significance. *The Orchard on Fire* is another short book and is almost purely personal. Despite some effective scene setting that evokes the 1950s in southern England, this is the story of one year in the life of a young girl. Her family having moved from London to a village in Kent to run a tearoom, April Harlency soon finds a best friend, the child of a local publican's family. April and Ruby share secrets, including a secret retreat in the orchard, where they find an abandoned railway car and make a fort. Their friendship is imperiled by adults: by Ruby's coarse and abusive parents, who resent the friendship; by a suspected molester named Rodney Pegg; and by the biggest problem for April, which is her involvement with an elderly man of a higher class named Mr. Greenidge. Mr. Greenidge is a dirty old man; he persuades himself—and to a certain extent April—that he is in love with her, and his attentions, resentments, and importunings make her life miserable. *The Orchard on Fire* is a sort of coming-of-age novel, and a fine one. It is unmelodramatic and often funny but is a convincing picture of April's season of maturing and loss.

The best novel on the 1995 Booker shortlist was Salman Rushdie's *The Moor's Last Sigh,* which did not win (Pat Barker's *The Ghost Road* did). In 1996 the best novel was Rohinton Mistry's *A Fine Balance.* This is striking, because both are long, ambitious, panoramic books, concerned with Indian history and its recent tragic politics, and both novelists are expatriates from Bombay. Rushdie is a sort of fantasist or magic realist, while Mistry is a traditionalist writer akin to the great nineteenth-century realists (his book carries an epigraph from Balzac).

A Fine Balance begins with three men traveling together on a train. Two of them are tailors going to a new job; the third is a college student (studying refrigeration and air-conditioning) going to his new lodgings. They get off at the same place and discover that the landlady of Maneck Kohlah, the student, is also the employer of Ishvar and Omprakash Darji, the tailors. Her name is Dina Dalal. These four become the main characters in a dense, engrossing, and heartbreaking story. Mistry has grand ambitions, which, without grandiosity or pomposity, produce not only the scope of the book (it is about the same length as *Alias Grace*) but also its social grasp and its moral dimension.

The Darjis are from a village family of untouchables, some of whom have been barbarously treated and even murdered for daring to aspire above their caste assignment of disposing of dead animals and working in leather. Having become tailors by apprenticing with a Moslem, the two of them become embroiled in the sectarian violence at the time of Partition. Ishvar and Omprakash are then thrown into the maelstrom of poverty, utter corruption, and casual murder that, according to this novel, is the atmosphere of Mrs. Gandhi's "emergency" suspension of the rule of law. That the Beggarmaster, a man who arranges for children to undergo amputations and disfigurements so they can be successful mendicants, is one of the more benign presences in this book indicates the level of public moral anarchy. Private people—particularly the central four but also others, including their families and one of the chief beggars—are good. It is just life that is bad. *A Fine Balance,* more than any other book on this year's Booker list, is an unforgettable novel. Maggie Gee has declared that Margaret Atwood is our modern-day George Eliot or Charles Dickens; that slot is more firmly filled by Mistry.

Turning to the novels favored by the judges: Margaret Atwood's *Alias Grace* is intelligent but strangely lifeless, a disappointment from this author. Perhaps beginning from a real story hampered her. Grace Marks, an uneducated immigrant Irish servant, was convicted in 1843 of murdering her employer, Thomas Kinnear, and his housekeeper, Nancy Montgomery, in collaboration with a fellow servant named James McDermott. The two had fled the farm together immediately after the murder and were apprehended in New York. McDermott was executed, and Grace's sentence was commuted; after spending thirty years in confinement, sometimes in a lunatic asylum, sometimes as a sort of "trusty" servant to the prison governor, she was released and returned to New York.

Aside from much uninformed but vehement commentary on the case at the time, this is about all that is known of Grace Marks. Atwood has invented a psychology for her, created a friend and confidante and assigned to this young woman the false name Grace used for herself, and, most crucial, has created a young American doctor who wants to try out his newfangled psychological ideas by interviewing Grace. This gives her great scope for bringing in mesmerism, sexual politics (through Dr. Jordan's emotions regarding Grace and an entanglement he slides into with his landlady), and the U.S. Civil War. This is not to say that she simplifies matters. Nineteenth-century observers were dramatically divided over Grace Marks's character: was she an innocent falsely accused of participating in a murder, an unwilling participant coerced by McDermott, or the mastermind who (as McDermott claimed) invented the plan to kill two people and ran off with him afterward as his mistress? *Alias Grace* never provides unequivocal answers, and that is to the good. It does answer the question of what happened to Grace Marks after she got out of prison in a way that will strike some readers as anticlimactic and sentimental.

Seamus Deane grew up in Derry, in Northern Ireland, against a background of violence and republican agitation. *Reading in the Dark* is based closely on his own history; the characters' names are unchanged from his own family, and, while working on the novel for ten years, he realized that it could not be published until his mother had died. It is an evocative, moving tale of a boy coming of age by learning terrible family secrets, including an understanding of what has driven his mother mad.

The novel is rich. It includes a mythical background—for instance, the boy visits the Field of the Disappeared, where the souls of the dead who did not receive a proper Christian burial reappear as birds on Samhain, Christmas, and Saint Brigid's Day. There is a legendary background—"the old fort of Grianan," where legendary warriors supposedly sleep below and real-life terrors persist. There is a miraculous dimension, in which ghosts appear. And there is the ever-present historical background—feuds, murders, troubles with the police, and family betrayals. There is much in *Reading in the Dark* that reminds one of Joyce—particularly the Joyce of *A Portrait* but also of Stephen Dedalus's remark in *Ulysses* that "history is a nightmare from which I am trying to awake." The events take place in the 1940s, though the narrator is an adult looking back on his childhood experience. The life is hard: unemployment, poverty, brutality, relatives dying of tuberculosis, children killed in the street. "Everybody hated the police," the narrator calmly remarks. It ends in 1971 with the Troubles, renewed fighting between the republicans and the British state.

Deane, who teaches at Notre Dame University, is a poet (a schoolmate of Nobel Prize-winner Seamus Heaney), and the prose of this novel is poetic in the best sense. Not flowery, his writing is direct and strong. The narrator tells us how he liked to read novels when he was a boy: "I'd switch off the light, get back in bed, and lie there, the book still open, re-imagining all I had read, the various ways the plot might unravel, the novel opening into endless possibilities in the dark."

Booker winner *Last Orders* begins in a pub in London and ends fifty miles or so away on a jetty in Margate. In between is a complex journey. Four men have met as requested by the deceased Jack Dodds—three friends and his son—to carry out his last wishes: he asked that his ashes be scattered into the sea. His widow, Amy, has refused to accompany them. Along the way, particularly as they take their time reaching Margate—stopping in several pubs, visiting a war memorial in Chatham, bickering and worrying about each other—it becomes a sort of trip into memory. The narration is divided among the four men—Ray, Lenny, Vince, and Vic—with some sparse contribution by Amy. Delving into the past becomes a process of revealing secrets, as in *Reading in the Dark*. The secrets they keep from each other constitute barriers; by the end there is a sort of rough reconciliation. Only the readers learn all the history that explains how these four men are related to each other and to Jack and Amy Dodds.

Swift is from South London—though of middle-class origins—and has made some attempt to reproduce the sounds of Cockney voices in the short chapters of narration:

> Canterbury Cathedral. I ask you. I should've kept my big mouth shut.

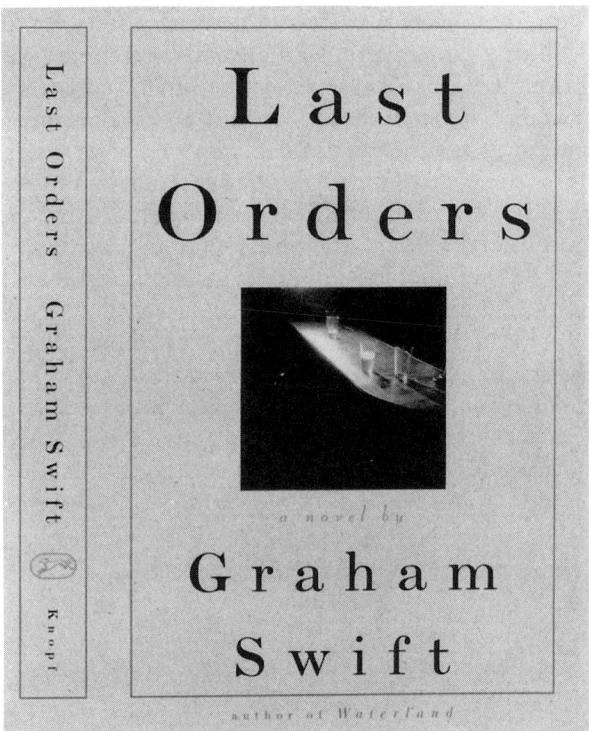

Dust jacket for the American edition of the 1996 Booker Prize-winning novel

> Still, dose of holiness'll do us good, I suppose, the way things were going.
>
> So glory be. Lift up your hearts for Lenny.

There is some dissonance when the characters also say things such as "Lenny edges out of his seat, all abashed and obedient" or "But the outcast and the outlawed have to die too, the shunned and forgotten, and somewhere there's a reluctant relative who has to step uneasily forward." Perhaps these men could have these thoughts, even if they would never have them in these words; but then why make a point of using their kind of words? Better to eschew the attempt to use dialect, as Seamus Deane and Shena Mackay do; on the other hand, Deane's adult narrator is university educated, and Mackay's is an English teacher. Swift filters his voice through those of a vegetable stallholder, a shady car dealer, an undertaker, and a man who makes his living betting on horses. It is an earnest attempt, and harder than choosing the voice of an English teacher, but the results are thin.

Carmen Callil a forceful figure in British publishing (she founded Virago Books) and the chair of the Booker committee, reviewed *Last Orders* when it came out—a controversial practice for Booker judges—and gave it high praise. When it won, there were further satisfied reflections that it would "give pleasure" to thousands and thousands of readers, that it was reflective of contemporary Britain, and even that it was written by an English author. The award looks a bit like a midcourse correction, an answer to all those who claim the Booker is reserved for the difficult, unreadable, gloomy, and/or exotic book, the foreign author. *Last Orders* can be thought of as an old-fashioned novel that is also right up-to-date.

V. S. Naipaul, a former Booker winner (for *In a Free State,* 1971), has denounced the prize: "The Booker is murder. . . . It is useless. I have no regard for it at all." Naipaul is surely above envy. A. N. Wilson declares that "It would be a good idea if we could announce that there would be no more Booker Prizes from now on." His reason is that too many mediocre books have won it, too many good ones have not, and the process of compromise assures that the best book will not win. Both he and Naipaul probably underestimate the value of the prize to the winner, and even to the short-listed authors, in sales, in esteem, and in visibility. And when Naipaul goes on to argue that "Absolutely nothing would be lost if it withered away and died," he ignores one simple, if slightly crass, fact: the prize makes people talk about it for two months every year, makes people discuss authors and books, gets ordinary people to place bets on novels and stay up late to watch a novelist receive an award (and five others lose); it thus, even in an ordinary year like 1996, contributes to the health of fiction.

Editorial

There are five principal terms of bibliographical description: *edition, printing* (or *impression*), *issue,* and *state*. These terms are used so carelessly by booksellers and by alleged bibliographers that students and tyro collectors are frequently confused and discouraged. Although there is a certain amount of intentional chicanery in the promiscuous application of "first issue" or "first state" to every book in a dealer's catalogue, most misdescription of books—especially in the field of modern firsts—results from incompetence or laziness. There is no alibi for a professional bookman to abuse terms that have precise meanings as formulated by Fredson Bowers in *Principles of Bibliographical Description* (1949).

An *edition* consists of all the copies of a book printed from one setting of type or from printing plates made from the typesetting. All the printings from a particular typesetting are subsumed within the edition. Thus, copies from the tenth printing or twentieth printing from the initial typesetting belong to the first edition. What amateurs assume to be the "first edition" is really the first printing of the first edition.

A *printing* or *impression*—the terms are interchangeable—consists of all the copies printed at one time, i.e., without removing the type or plates from the press. The first printing is usually the collector's desideratum of the first edition.

Thus, the first edition of *The Great Gatsby* was set in type and plated in 1925. The first printing was published by Scribners on 10 April 1925. It can be identified by six readings that were emended in the second printing:

60.16 chatter *emended to* echolalia

119.22 northern [southern

165.16 it's [its

165.29 away [away.

205.9-10 sick in tired [sickantired

211.7-8 Union Street station [Union Station

The Scribners second printing of the first edition was printed in August 1925.

The third printing of *Gatsby* was printed by Chatto and Windus in 1926. These copies constitute the first English printing, but they are the third printing from the Scribners first-edition plates.

The Modern Library used the Scribners plates in 1934 to produce the fourth printing of the first edition. In 1942 Scribners used their plates to manufacture the fifth printing. New Directions produced the sixth printing in 1946, and Grosset and Dunlap produced the seventh printing in 1949. The publisher's imprint on the title page has no bearing on the precise use of *edition*. One may refer carelessly to "the Grossett and Dunlap edition of *Gatsby*," but it is the sixth printing of the first edition.

The most misunderstood and most abused bibliographical terms are *issue* and *state*. In bad hands "first issue" and "first state" are used interchangeably to designate something early and therefore expensive. Correctly applied, *issue* and *state* occur only within a single printing. There are no states or issues in any printings of the first edition of *Gatsby*.

States result when the printed pages of some copies of a single printing are altered either during the course of printing or after the printing is completed. Stop-press correction of one or more words creates *states:* the first state with the original reading and the second state with the emended reading. The correction may be performed by cancellation: removing pages and inserting emended replacement pages, which are called "tip-ins." There are two states of the first printing of Fitzgerald's *Taps at Reveille* resulting from the cancellation of pp. 349-352. There can be no second state unless there is a first state. There can be no first state unless there is a second state.

Issues are created by an alteration of the pages—affecting the conditions of publication or sale—of some copies of a printing. Usually *issues* result from title-page alterations. Fitzgerald's *The Beautiful and Damned* was published with the Scribners title page and with the title page of Canadian publisher Copp Clark. The presumption is that the prelims, or preliminary material (the first gathering of leaves), were printed with title pages for Scribners and Copp Clark; therefore, two issues resulted: an American issue and a Canadian issue—which may have been issued simultaneously. There can be no second issue without a first issue. There can be no first issue without a second issue. Binding variants—different cloths or different cloth colors or changes in the stamping—have no bearing on *edition, printing, state,* or *issue*. Binding variants are binding variants. It may be possible to determine the priority of a particular binding variant used for

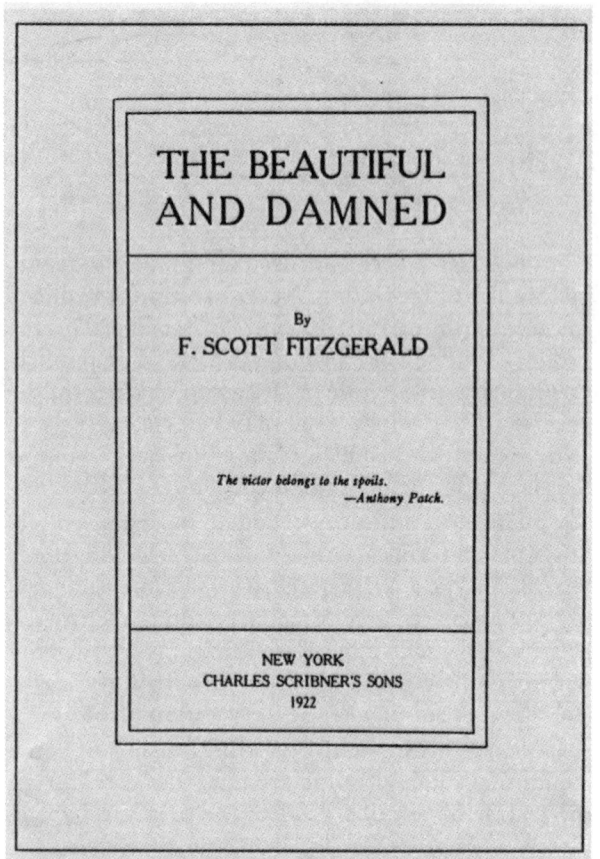 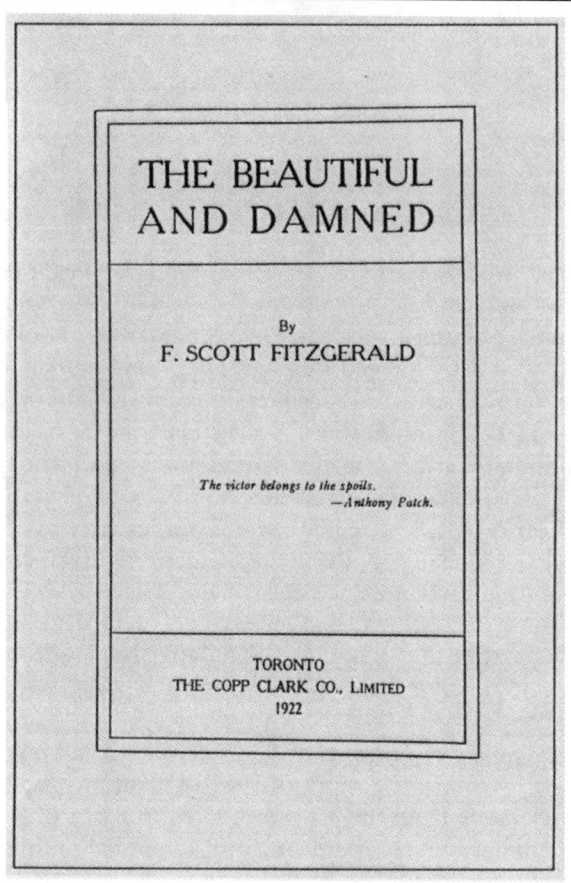

The American and Canadian issues of the first printing, presumably issued simultaneously

part of printing, but bindings have no connection with text. *Binding issues* are possible—for example, parts of a printing may be bound in paper and cloth to create *binding issues*. But this term is potentially treacherous and should be applied with reluctance.

A dust jacket—which may be more valuable than the book it accompanies—has no bearing on the edition, printing, state, or issue of the book. But there may be editions, printings, states, or issues of jackets themselves. The first printing of the *Gatsby* jacket was printed with a lower-case "j" in "Jay Gatsby." Some of these jackets were hand corrected, thereby creating two states of the first printing of the jacket. The error was corrected when the jacket was reprinted. The English dust jacket exists with and without a circular label lowering the price; these are best described as issues because the label changes the conditions of sale. There is no way to determine that the dust jacket now on a volume was always on that volume. Jackets are frequently swapped. The description of a book and its dust jacket are independent of each other.

—M.J.B.

Statement of Correction to "The Jack Kerouac Revival"

New York University Programs Director Helen Kelly moves to block Jan Kerouac and Jacques Kirouac from approaching the podium at the opening of the Jack Kerouac Conference, 5 June 1995. Anne Waldman looks on from the stage. Photo © 1995 by Gerald Nicosia (used with permission)

There are very serious errors of fact and interpretation that appear in Matt Theado's "The Jack Kerouac Revival" in *DLB Yearbook 1995*. Referring to myself and Jan Kerouac together, Theado states: "Nicosia and Jan Kerouac had been making a strong statement on the West Coast on behalf of *their* legal case against the Sampas family [italics mine]" (p. 277). I have never in my life sued the Sampas family or anyone else. The lawsuit was Jan Kerouac's. I supported Jan Kerouac as a friend during those last trying years of her legal battles and kidney failure, and it was my friendship with Jan Kerouac that put me in opposition to Mr. Sampas.

The second serious error is Mr. Theado's statement that Allen Ginsberg asked the crowd at NYU if anyone "thought Nicosia should be allowed to speak . . ." (p. 278). This could not have happened because I never asked to speak that day, 5 June 1995, at NYU.

On Friday, 2 June 1995, Jan Kerouac and I gave a press conference in midtown Manhattan, several miles from NYU. In our statement to the press, we suggested that it would be proper for NYU to welcome us and to allow us to speak. By Monday, 5 June, it was perfectly clear that the university was not going to extend that invitation to either Jan or myself. I never had any intention of trying to defy the university. But Jan felt that as Jack Kerouac's daughter, she had a moral right to speak there.

I came to the opening of the Jack Kerouac Conference on 5 June to support Jan Kerouac. Jan Kerouac and her Quebec cousin, Jacques Kirouac, approached the stage to ask Allen Ginsberg for a few minutes at the podium. I sat watching. When

287

they were being escorted out by police, I stood up and yelled to Allen to allow Jan to speak. That is when they had me removed as well. There would have been no reason for Allen to ask the audience if Gerald Nicosia should be allowed to speak since I was not the person asking for that right. Certainly I would not have been hauled out by police for a statement I made at a press conference three days earlier!

Moreover, the "setting up camp" in Washington Square and "fielding reporters' questions" Mr. Theado refers to (p. 277) took place *after* we were removed by police, not before, as he implies.

Mr. Theado claims I put on a benefit "nominally to raise money" for Jan Kerouac's dialysis, whereas it was secretly also to raise money for her lawsuit against the Kerouac Estate (p. 277). Quite the contrary, the fundraising events were billed from the very beginning, in all publicity, as a means "to pay the health costs of her [Jan Kerouac's] kidney failure *and* the legal costs of her fight to preserve her father's literary archive [italics mine]."

As for the errors of interpretation, Theado states that in the 1970s and 1980s "no one accords Kerouac the one title he pursued from his youngest years: writer" and that "in the early 1970s . . . practically no one was reading him" (p. 273). Many people I knew in graduate school at the University of Illinois in Chicago in 1972 were reading Kerouac, including myself, and we certainly took Kerouac seriously as a writer. The late U. of I. Professor Paul Carroll was a great champion of the literary merits of Kerouac's work. More absurdly, after referring to the first six Kerouac biographies, including my *Memory Babe* (1983), Theado claims that these works also concentrated on his life, while Kerouac's works "were customarily considered uneven, unliterary, or just unreadable" (p. 274). In fact, *Memory Babe* attempted to rank Kerouac as the equal of Herman Melville and made a case for Kerouac's being the American equivalent of James Joyce and Marcel Proust. Theado also states that "no major literature anthology yet includes Kerouac" (p. 273), when in fact Kerouac's work has been anthologized in dozens of major anthologies, from Donald Allen's *The New American Poetry* (Grove, 1960) to Paul Hoover's *Postmodern American Poetry* (Norton, 1994).

Theado finds Kerouac Estate Representative John Sampas responsible "for promoting Kerouac's reputation" and claims that ever since Sampas's appointment in 1991 "the Kerouac world has been growing" (pp. 274-275); yet Theado gives no credit at all for this revival to the major Kerouac conferences that took place in Boulder, Colorado, in 1982; in Plymouth, England, in June 1987; in Quebec City in October 1987; and in Lowell, Massachusetts, in June 1988. In my opinion, the work of the hundreds of participants in those conferences was far more important to Kerouac's establishment as a major writer than the marketing of his collectible items or the sale of a few unpublished manuscripts by Mr. Sampas.

– Gerald Nicosia

MATT THEADO'S RESPONSE

I mistakenly state in "The Jack Kerouac Revival" that Gerald Nicosia and Jan Kerouac pursued a legal case against the Sampas family. I regret the error.

Literary Awards and Honors Announced in 1996

ACADEMY OF AMERICAN POETS LITERARY AWARDS

ACADEMY FELLOWSHIP
Jay Wright.

HAROLD MORTON LANDON TRANSLATION AWARD
Guy Davenport, *7 Greeks* (New Directions).

JAMES LAUGHLIN AWARD
David Rivard, *Wise Poison* (Graywolf Press).

LENORE MARSHALL – *NATION* MAGAZINE PRIZE FOR POETRY
Charles Wright, *Chickamauga* (Farrar, Straus, and Giroux).

RAIZISS/DE PALCHI TRANSLATION AWARD
This Strange Joy: Selected Poems of Sandro Penna, translated by W. S. Di Piero (Ohio State University Press).

RAIZISS/DE PALCHI TRANSLATION FELLOWSHIP
Anthony Molino, tr. (in progress), *Esercizi di tipologia*, by Valerio Magrelli.

TANNING PRIZE
Adrienne Rich.

WALT WHITMAN AWARD
Joshua Clover, *Madonna Anno Domini* (Louisiana State University Press).

AMERICAN ACADEMY AND INSTITUTE OF ARTS AND LETTERS AWARDS

ACADEMY MEMBERS
Bennett L. Carter, Vija Celmins, Louise Glück, Daniel Urban Kiley, Kenneth Koch, John Russell, Oliver Sacks, Sidney Simon, Edmund White, Elie Wiesel.

FOREIGN HONORARY MEMBERS
Madgalena Abakanowicz, Bei Dao, Brian Friel, Arvo Paert, Aldo Rossi.

ACADEMY AWARDS IN LITERATURE
Whitney Balliett, Carol Brightman, Robert Fagles, Robert Hughes, August Kleinzahler, Larry Kramer, Paul Muldoon, David Quammen.

WITTER BYNNER PRIZE FOR POETRY
Lucie Brock-Broido.

E. M. FORSTER AWARD IN LITERATURE
Jim Crace.

SUE KAUFMAN PRIZE FOR FIRST FICTION
Peter Landesman, *The Raven* (Baskerville).

ROME FELLOWSHIP IN LITERATURE
Randall Kenan.

RICHARD AND HINDA ROSENTHAL FOUNDATION AWARD IN LITERATURE
David Long, *Blue Spruce* (Scribners).

HAROLD D. VURSELL MEMORIAL AWARD IN LITERATURE
A. J. Verdelle.

MORTON DAUWEN ZABEL AWARD IN LITERATURE
J. D. Landis, *Lying in Bed* (Algonquin).

AMERICAN BOOK AWARDS

Sherman Alexie, *Reservation Blues* (Atlantic Monthly Press).

Stephanie Cowell, *The Physician of London* (Norton).

Chitra Banerjee Divakaruni, *Arranged Marriage* (Anchor).

Maria Espinosa, *Longing* (Arte Publico).

William Gass, *The Tunnel* (Knopf).

Kimiko Hahn, *The Unbearable Heart* (Kaya Productions).

E. J. Miller Laino, *Girl Hurt* (Alice James Books).

Chang-rae Lee, *Native Speaker* (Riverhead Books).

James W. Loewen, *Lies My Teacher Told Me: Everything Your American History Textbook Got Wrong* (New Press).

Glenn C. Loury, *One By One from the Inside Out* (Free Press).

Agate Nesaule, *Woman in Amber: Healing the Trauma of War and Exile* (Soho Press).

Joe Sacco, *Palestine* (Fantagraphics).

Ron Sakolsky and Fred Wei-han Ho, eds., *Sounding Off! Music as Subversion/Resistance/Revolution* (Autonomedia).

Arthur Sze, *Archipelago* (Copper Canyon Press).

Robert Viscusi, *Astoria* (Guernica Editions).

CHILDREN'S BOOK AWARD
Paul Owen Lewis, *Storm Boy* (Beyond Words Publishing).

EDITOR/PUBLISHER AWARD
Alexander Taylor and Judith Doyle, Curbstone Press.

LIFETIME ACHIEVEMENT AWARD
Janice Mirikitani.

AMERICAN LIBRARY ASSOCIATION AWARDS

JOHN NEWBERY MEDAL
Karen Cushman, *The Midwife's Apprentice* (Clarion).

RANDOLPH CALDECOTT MEDAL
Peggy Rathmann, *Officer Buckle and Gloria* (Putnam).

MARGARET R. EDWARDS AWARD FOR OUTSTANDING LITERATURE FOR YOUNG ADULTS
Judy Blume.

CORETTA SCOTT KING AUTHOR AWARD
Virginia Hamilton, *Her Stories* (Scholastic).

MILDRED L. BATCHELDER AWARD
Uri Orlev, *The Lady With the Hat,* translated from Hebrew by Hillel Halkin (Houghton Mifflin).

HANS CHRISTIAN ANDERSEN PRIZE

AUTHOR AWARD
Uri Orlev.

ILLUSTRATOR AWARD
Klaus Ensikat.

ANTHONY (BOUCHERCON)

BEST NOVEL
Mary Willis Walker, *Under the Beetle's Cellar* (Doubleday).

BEST FIRST NOVEL
Virginia Lanier, *Death in Bloodhound Red* (Pineapple).

BEST PAPERBACK ORIGINAL
Harlan Coben, *Deal Breaker* (Dell).

AUDUBON MEDAL

The Very Reverend James Parks Morton.

IRMA S. AND JAMES H. BLACK AWARD FOR EXCELLENCE IN CHILDREN'S LITERATURE

Connie N. Wooldridge, *Wicked Jack,* illustrated by Will Hillenbrand (Holiday).

JAMES TAIT BLACK MEMORIAL PRIZES

BIOGRAPHY
Gitta Sereny, *Albert Speer: His Battle With Truth* (Macmillan).

FICTION
Christopher Priest, *The Prestige* (Touchstone).

REBEKAH JOHNSON BOBBITT NATIONAL PRIZE FOR POETRY

Kenneth Koch, *One Train* (Knopf).

THE BOOKER PRIZES

THE BOOKER PRIZE FOR FICTION
Graham Swift, *Last Orders* (Knopf; Picador).

BOOKER RUSSIAN NOVEL PRIZE
Andrei Sergeev, *Stamp Album* (*Druzhba Naradov*, 7–8).

THE LITTLE BOOKER
Sergei Gandlevsky, *Trepanning* (*Znamya*, no. 1, 1995).

BOSTON GLOBE–HORN BOOK AWARDS

FICTION
Avi, *Poppy,* illustrated by Brian Floca (Orchard/Jackson).

NONFICTION
Andrea Warren, *Orphan Train Rider* (Houghton Mifflin).

PICTURE BOOK
Amy Hest, *In the Rain with Baby Duck,* illustrated by Jill Barton (Candlewick).

JOHN BURROUGHS MEDAL

Bill Green, *Water, Ice and Stone* (Harmony/Crown).

THE CHRISTOPHER AWARDS

BOOKS

Ruth-Alice von Bismarck and Ulrich Kabitz, eds., *Love Letters from Cell 92: The Correspondence Between Dietrich Bonhoeffer and Maria von Wedemeyer 1943–45,* translated by John Brownjohn (Abingdon).

David Herbert Donald, *Lincoln* (Simon and Schuster).

Lawrence Martin Jenco, O.S.M., *Bound to Forgive: The Pilgrimage to Reconciliation of a Beirut Hostage* (Ave Maria Press).

Jonathan Kozol, *Amazing Grace: The Lives of Children and the Conscience of a Nation* (Crown).

Jeff Leeland, *One Small Sparrow* (Multnomah Books).

Eric Lomax, *The Railway Man* (Norton).

Kathryn Watterson, *Not By the Sword* (Simon and Schuster).

BOOKS FOR YOUNG PEOPLE

Eleanor Ayer, with Helen Waterford and Alfons Heck, *Parallel Journeys* (Atheneum).

Lee Bennett Hopkins, *Been to Yesterdays,* illustrated by Charlene Rendeiro (Wordsong/Boyds Mills Press).

Betsy Harvey Kraft, *Mother Jones* (Clarion Books).

Susan Wojciechowski, *The Christmas Miracle of Jonathan Toomey,* illustrated by P. J. Lynch (Candlewick).

ARTHUR C. CLARKE AWARD

Paul J. McAuley, *Fairyland* (Gollancz).

COMMONWEALTH WRITER'S PRIZES

BEST BOOK

Rohinton Mistry, *A Fine Balance* (McClelland and Stewart; Faber and Faber).

BEST FIRST PUBLISHED BOOK

Vikram Chandra, *Red Earth and Pouring Rain* (Penguin India; Faber and Faber).

CRIME WRITERS ASSOCIATION AWARDS

CARTIER DIAMOND DAGGER

Harry Keating.

GOLDEN DAGGER

Ben Elton, *Popcorn* (Simon and Schuster).

GOLDEN DAGGER FOR NONFICTION

Antonia Fraser, *Faith and Treason: The Story of the Gunpowder Plot* (Doubleday).

SILVER DAGGER

Peter Lovesey, *Bloodhounds* (Warner).

LAST LAUGH

Janet Evanovich, *Two for the Dough* (Scribners).

JOHN DOS PASSOS PRIZE FOR LITERATURE

Helena María Viramontes.

GEORGE M. FREEDLEY MEMORIAL AWARD

Lyle Leverich, *Tom: The Unknown Tennessee Williams* (Crown).

LIONEL GELBER PRIZE

V. Zubok and C. Pleshakov, *Inside the Kremlin's Cold War* (Harvard University Press).

GOLDEN KITE AWARDS

FICTION

Christopher Paul Curtis, *The Watsons Go To Birmingham – 1963* (Delacorte).

NONFICTION

Natalie S. Bober, *Abigail Adams* (Atheneum).

PICTURE ILLUSTRATION

Dennis Nolan and Lauren Mills, *Fairy Wings* (Little, Brown).

GOVERNOR GENERAL'S LITERARY AWARDS

CHILDREN'S LITERATURE – ILLUSTRATION

Eric Beddows, *The Rooster's Gift,* text by Pam Conrad (Groundwood Books).

CHILDREN'S LITERATURE – TEXT

Paul Yee, *Ghost Train* (Groundwood Books).

Gilles Tibo, *Noémie – Le Secret de Madame Lumbago* (Éditions Québec/Amérique).

DRAMA

Colleen Wagner, *The Monument* (Playwrights Canada Press).

Normand Chaurette, *Le Passage de l'Indiana* (Leméac Éditeur/Actes Sud-Papiers).

FICTION
   Guy Vanderhaeghe, *The Englishman's Boy* (McClelland and Stewart).
   Marie-Claire Blais, *Soifs* (Éditions du Boréal).
NONFICTION
   John Ralston Saul, *The Unconscious Civilization* (House of Anansi Press).
   Michel Freitag, *Le Naufrage de l'université – Et autres essais d'épistémologie politique* (Nuit blanche éditeur/Éditions La Découverte).
POETRY
   E. D. Blodgett, *Apostrophes: Woman at a Piano* (Buschek Books).
   Serge Patrice Thibodeau, *Le Quatuor de l'errance*, followed by *La Traversée du désert* (Éditions de l'Hexagone).
TRANSLATION
   Linda Gaboriau, *Stone and Ashes* (Coach House Press); English version of Daniel Danis's *Cendres de cailloux* (Leméac Éditeur).
   Christiane Teasdale, *Systèmes de survie – Dialogue sur les fondements moraux du commerce et de la politique* (Éditions du Boréal); French version of Jane Jacobs's *Systems of Survival: A Dialogue on the Moral Foundations of Commerce and Politics* (Random House).

GRAND PRIX DES BIENNALES INTERNATIONALES DE POESIE

John Ashbery.

DRUE HEINZ LITERATURE PRIZE

Edith Pearlman, *Vaquita and Other Stories* (University of Pittsburgh Press).

DAVID HIGHAM PRIZE FOR FICTION

Vikram Chandra, *Red Earth and Pouring Rain* (Penguin India; Faber and Faber).

HUGO AWARDS

BEST NOVEL
   Neal Stephenson, *The Diamond Age* (Bantam/Spectra).
BEST NOVELLA
   Allen Steele, "The Death of Captain Future" (*Isaac Asimov's Science Fiction Magazine*).
BEST NOVELETTE
   James Patrick Kelly, "Think Like a Dinosaur" (*Isaac Asimov's Science Fiction Magazine*).
BEST SHORT STORY
   Maureen F. McHugh, "The Lincoln Train" (*Magazine of Fantasy & Science Fiction*).

IMPAC DUBLIN LITERARY AWARD

David Malouf, *Remembering Babylon* (Chatto and Windus/Pantheon).

INGERSOLL PRIZES

T. S. ELIOT AWARD
   Richard Wilbur.
RICHARD WEAVER AWARD
   David Hackett Fischer.

KIRIYAMA PACIFIC RIM BOOK PRIZE

Alan Brown, *Audrey Hepburn's Neck* (Pocket).

LANNAN LITERARY AWARDS

LIFETIME ACHIEVEMENT AWARD FOR POETRY
R. S. Thomas.
POETRY
   Anne Carson, Lucille Clifton, Donald Justice.
FICTION
   Howard Norman, Tim Pears, William Trevor, David Foster Wallace.
NONFICTION
   David Abram, Charles Bowden.

THE LIBRARY ASSOCIATION LITERARY AWARDS

BESTERMAN MEDAL
   Peter J. Seddon, *A Football Compendium: A Comprehensive Guide to the Literature of Association Football* (British Library).
CARNEGIE MEDAL
   Philip Pullman, *His Dark Materials: Book 1 Northern Lights* (Scholastic).
KATE GREENAWAY MEDAL
   P. J. Lynch, *The Christmas Miracle of Jonathan Toomey* (Walker Books).
McCOLVIN MEDAL
   Roger Kain and Richard Oliver, *The Tithe Maps of England and Wales* (Cambridge University Press).
WHEATLEY MEDAL
   Ruth Richardson and Robert Thorne, *The Builder Illustrations Index* (Hutton and Rostron).

RUTH LILLY PRIZE FOR POETRY

Gerald Stern.

THE JOHN D. AND CATHERINE T. MACARTHUR FOUNDATION FELLOWS

James Roger Prior Angel, Joaquin G. Avila, Allan Bérubé, Barbara Block, Joan Breton Connelly, Thomas L. Daniel, Martin Daniel Eakes, Rebecca Goldstein, Robert Greenstein, Richard Howard, John Jesurun, Richard E. Lenski, Louis Massiah, Vonnie C. McLoyd, Thylias Moss, Eiko and Koma Otake, Nathan Seiberg, Anna Deavere Smith, Dorothy Stoneman, William Strickland.

MACAVITY (MYSTERY READERS' INTERNATIONAL)

BEST NOVEL
Mary Willis Walker, *Under the Beetle's Cellar* (Doubleday).

BEST FIRST NOVEL
Dianne Day, *The Strange Files of Fremont Jones* (Doubleday).

MEDAL OF HONOR FOR LITERATURE

E. L. Doctorow.

MODERN LANGUAGE ASSOCIATION PRIZES

PRIZE FOR A FIRST BOOK
Elaine Hadley, *Melodramatic Tactics: Theatricalized Dissent in the English Marketplace, 1800–1885* (Stanford University Press).

PRIZE FOR INDEPENDENT SCHOLARS
Nora Sayre, *Previous Convictions: A Journey through the 1950s* (Rutgers University Press).

JAMES RUSSELL LOWELL PRIZE
Rey Chow, *Primitive Passions: Visuality, Sexuality, Ethnography, and Contemporary Chinese Cinema* (Columbia University Press).

ALDO AND JEAN SCAGLIONE PRIZE FOR A TRANSLATION OF A LITERARY WORK
David Ball, *Darkness Moves: An Henri Michaux Anthology, 1927–1984* (University of California Press); and Rosa Chacel, *Memoirs of Leticia Valle* (University of Nebraska Press).

MINA P. SHAUGHNESSY PRIZE
Ross Talarico, *Spreading the Word: Poetry and the Survival of Community in America* (Duke University Press).

HARRIET MONROE PRIZE

James Dickey.

NATIONAL BOOK AWARDS

FICTION
Andrea Barrett, *Ship Fever and Other Stories* (Norton).

NONFICTION
James Carroll, *An American Requiem: God, My Father, and the War that Came Between Us* (Houghton Mifflin).

POETRY
Hayden Carruth, *Scrambled Eggs & Whiskey, Poems 1991–1995* (Copper Canyon Press).

YOUNG PEOPLE'S LITERATURE
Victor Martinez, *Parrot in the Oven: Mi Vida* (HarperCollins/Joanna Cotler Books).

MEDAL FOR DISTINGUISHED CONTRIBUTION TO AMERICAN LETTERS
Toni Morrison.

NATIONAL BOOK CRITICS CIRCLE AWARDS

FICTION
Stanley Elkin, *Mrs. Ted Bliss* (Hyperion).

NONFICTION
Jonathan Harr, *A Civil Action* (Random House).

BIOGRAPHY
Robert Polito, *Savage Art: A Biography of Jim Thompson* (Knopf).

POETRY
William Matthews, *Time & Money* (Houghton Mifflin).

CRITICISM
Robert Darnton, *The Forbidden Best-Sellers of Pre-Revolutionary France* (Norton).

NONA BALAKIAN CITATION FOR EXCELLENCE IN REVIEWING
Laurie Stone.

IVAN SANDROF AWARD FOR LIFETIME ACHIEVEMENT IN PUBLISHING
Alfred Kazin, Elizabeth Hardwick.

NATIONAL JEWISH BOOK AWARDS

SANDRA BRAND AND ARIK WEINTRAUB AWARD FOR AUTOBIOGRAPHY AND MEMOIR
Binjamin Wilkomirski, *Fragments: Memories of a Wartime Childhood* (Schocken).

NCR BOOK AWARD

Eric Lomax, *The Railway Man* (Cape; Norton).

NEUSTADT INTERNATIONAL PRIZE FOR LITERATURE

Assia Djebar.

NOBEL PRIZE FOR LITERATURE

Wisława Szymborska.

SCOTT O'DELL AWARD FOR HISTORICAL FICTION

Theodore Taylor, *The Bomb* (Harcourt Brace).

PEN AMERICAN CENTER LITERARY AWARDS

PEN/MARTHA ALBRAND AWARD FOR FIRST NONFICTION

Mary Karr, *The Liars' Club: A Memoir* (Viking).

PEN/BOOK-OF-THE-MONTH CLUB TRANSLATION PRIZE

Wisława Szymborska, *View With a Grain of Sand,* translated from the Polish by Stanisław Barańczak and Clare Cavanagh (Harcourt Brace).

PEN/FAULKNER AWARD FOR FICTION

Richard Ford, *Independence Day* (Knopf).

ERNEST HEMINGWAY/PEN AWARD FOR FIRST FICTION

Chang-rae Lee, *Native Speaker* (Riverhead).

PEN AWARD FOR POETRY IN TRANSLATION

Guy Davenport, ed. and trans., *7 Greeks* (New Directions).

PEN/SPIELVOGEL-DIAMONSTEIN AWARD FOR THE ART OF THE ESSAY

Thomas Nagel, *Other Minds* (Oxford).

GREGORY KOLOVAKOS AWARDS

Jean Franco, Suzanne Jill Levine.

PEN/VOELCKER AWARD FOR POETRY

Franz Wright.

RENATO POGGIOLI TRANSLATION AWARD FOR A WORK IN PROGRESS

Louise Rozier, tr., *Little Jesus of Sicily,* by Fortunato Pasqualino.

PEN CENTER USA WEST LITERARY AWARDS

BODY OF WORK IN CRITICISM
Pauline Kael.

CHILDREN'S LITERATURE
Cynthia Rylant, *The Van Gogh Cafe* (Harcourt Brace).

DRAMA
Philip Kan Gotanda, *Ballad of Yachiyo* (Berkeley Repertory).

Simon Levy, *Tender is the Night* (The Fountain Theatre).

FICTION
Pete Dexter, *The Paperboy* (Random House).

FREEDOM TO WRITE AWARDS
Dapo Olorunyomi, exiled Nigerian journalist and editor.

Zeta, weekly newspaper in Tijuana.

JOURNALISM
George Cothran, "Shut Up, Little Man" (*SF Weekly*).

LIFETIME ACHIEVEMENT AWARD IN NONFICTION
Betty Friedan.

NONFICTION
Al Young, *Drowning in the Sea of Love* (Ecco Press).

POETRY
Carl Rakosi, *Poems 1923–1941* (Sun and Moon Press).

SCREENPLAY
William Broyles Jr. and Al Reinert, *Apollo 13* (Universal).

TELEPLAY
Charles Fuller, *Zooman* (Showtime).

John Hopkins, *Hiroshima* (Showtime).

TRANSLATION
Red Pine, *Guide to Capturing a Plum Blossom* (Mercury House).

ANTOINETTE PERRY AWARDS

PLAY
Terrence McNally, *Master Class.*

MUSICAL
Johnathan Larson, *Rent.*

PHI BETA KAPPA BOOK AWARDS

CHRISTIAN GAUSS AWARD
Paul Alpers, *What is Pastoral?* (University of Chicago Press).

RALPH WALDO EMERSON AWARD
Eloise Quiñones-Keber, *Codex Telleriano-Remensis: Ritual, Divination, and History in a Pictorial Aztec Manuscript* (University of Texas Press).

EDGAR ALLAN POE AWARDS

GRAND MASTER AWARD
Dick Francis.

BEST MYSTERY NOVEL
Dick Francis, *Come To Grief* (Putnam).

BEST FIRST MYSTERY NOVEL BY AN AMERICAN AUTHOR
David Housewright, *Penance* (Foul Play/Countryman).

BEST ORIGINAL PAPERBACK MYSTERY NOVEL
William Heffernan, *Tarnished Blue* (Onyx).

BEST FACT CRIME BOOK
Pete Earley, *Circumstantial Evidence* (Bantam).

BEST CRITICAL/BIOGRAPHICAL WORK
Robert Polito, *Savage Art: A Biography of Jim Thompson* (Knopf).

BEST YOUNG ADULT MYSTERY
Rob MacGregor, *Prophecy Rock* (Simon and Schuster).

BEST JUVENILE MYSTERY
Nancy Springer, *Looking for Jamie Bridger* (Dial).

BEST MYSTERY SHORT STORY
Jean B. Cooper, "The Judge's Boy" (*Ellery Queen*).

BEST MYSTERY MOTION PICTURE
The Usual Suspects, screenplay by Christopher McQuarrie (Gramercy Pictures, PolyGram, Bad Hat Harry, Blue Parrot).

ROBERT L. FISH MEMORIAL AWARD
James Sarafin, "The Word for Breaking August Sky" (*Alfred Hitchcock Mystery Magazine*).

ELLERY QUEEN AWARD
Jacques Barzun.

RAVEN AWARD
The Library of America for their publication of the collected writings of Raymond Chandler.

PULITZER PRIZES

BIOGRAPHY
Jack Miles, *God: A Biography* (Knopf).

CRITICISM
Robert Campbell.

DRAMA
Jonathan Larson, *Rent.*

FICTION
Richard Ford, *Independence Day* (Knopf).

GENERAL NONFICTION
Tina Rosenberg, *The Haunted Land: Facing Europe's Ghosts After Communism* (Random House).

HISTORY
Alan Taylor, *William Cooper's Town: Power and Persuasion on the Frontier of the Early American Republic* (Knopf).

POETRY
Jorie Graham, *The Dream of the Unified Field: Selected Poems, 1974–1994* (Ecco Press).

QPB/NEW VISIONS AWARD

Bia Lowe, *Wild Ride* (HarperCollins).

QPB/JOE SAVAGO NEW VOICES AWARD

Bruce Olds, *Raising Holy Hell* (Holt).

REA AWARD FOR THE SHORT STORY

Andre Dubus.

ROYAL SOCIETY OF LITERATURE AWARDS

W. H. HEINEMANN AWARD FOR LITERATURE
Theo Richmond, *Konin* (Pantheon).

Tony Harrison, for *The Shadow of Hiroshima and Other Film/Poems* (Faber and Faber) and *Permanently Bard* (Bloodaxe).

WINIFRED HOLTBY AWARD
Paul Watkins, *Archangel* (Random House).

IAN SAINT JAMES LITERARY AWARD

Joshua Davidson.

SHAMUS (PRIVATE-EYE WRITERS)

BEST NOVEL
S. J. Rozen, *Concourse* (St. Martin's Press).

BEST PAPERBACK ORIGINAL
  William Jaspersohn, *Native Angels* (Bantam).
BEST FIRST P.I. NOVEL
  Richard Barre, *The Innocents* (Walker).

W. H. SMITH LITERARY AWARD

Simon Schama, *Landscape and Memory* (HarperCollins).

SOCIETY OF AMERICAN HISTORIANS AWARDS

FRANCIS PARKMAN PRIZE
  Robert D. Richardson Jr., *Emerson: The Mind on Fire* (University of California Press).

SOCIETY OF AUTHORS AWARDS

THE CHOLMONDELEY AWARDS
  Elizabeth Bartlett, Dorothy Nimmo, Peter Scupham, Iain Crichton Smith.
THE ERIC GREGORY AWARDS
  Sue Butler, Cathy Cullis, Jane Griffiths, Jane Holland, Chris Jones, Sinead Morrissey, Kate Thomas.
THE MARGARET RHONDDA AWARD
  Laura Spinney.
THE SOMERSET MAUGHAM AWARDS
  Katherine Pierpont, *Truffle Beds* (Faber and Faber); and Alan Warner, *Morven Caller* (Vintage).
THE TRAVELLING SCHOLARSHIPS
  Stewart Conn, Annette Kobak, Theo Richmond.
THE McKITTERICK PRIZE
  Stephen Blanchard, *Gagarin and I* (Vintage).
THE SAGITTARIUS PRIZE
  Samuel Lock, *As Luck Would Have It* (Cape).
THE BETTY TRASK PRIZE AND AWARDS
  John Lanchester, *The Debt to Pleasure* (Macmillan) (main prize); Meera Syal, *Anita and Me* (Flamingo); Rhidian Brook, *The Testimony of Taliesin Jones* (Flamingo); Louis Caron Buss, *The Luxury of Exile* (Cape).

BRAM STOKER AWARDS

BEST NOVELETTE
  Stephen King, "Lunch at the Gotham Café," from *Dark Love* (Roc).
BEST NOVEL
  Joyce Carol Oates, *Zombie* (Dutton).

TEXAS INSTITUTE OF LETTERS LITERARY AWARDS

BOOK PUBLISHERS OF TEXAS AWARD FOR BEST BOOK FOR YOUNG PEOPLE
  Diane Stevens, *Liza's Blue Moon* (Greenwillow Books).
BRAZOS BOOKSTORE AWARD FOR BEST SHORT STORY
  Paul Christensen, "Water" (*Madison Review*).
CARR P. COLLINS AWARD FOR NONFICTION
  Mary Karr, *The Liars' Club: A Memoir* (Viking).
SOEURETTE DIEHL FRASER AWARD FOR TRANSLATION
  Dick Gerdes, *The Fourth World* (University of Nebraska Press); English translation of Diamela Eltit's *El Quarto Mundo*.
FRIENDS OF THE DALLAS PUBLIC LIBRARY AWARD FOR THE BOOK MAKING THE MOST SIGNIFICANT CONTRIBUTION TO KNOWLEDGE
  Ben Huseman, *Wild River, Timeless Canyon* (Amon Carter Museum).
JESSE JONES AWARD FOR FICTION
  Paul Scott Malone, *In an Arid Land: Thirteen Stories of Texas* (Texas Christian University Press).
NATALIE ORNISH AWARD FOR POETRY
  Betty Adcock, *The Difficult Years* (Louisiana State University Press).
LON TINKLE AWARD FOR CAREER ACHIEVEMENT
  William Humphrey.
STEVEN TURNER AWARD FOR FIRST BOOK OF FICTION
  Jewel Mogan, *Beyond Telling* (Ontario Review Press).

WHITBREAD PRIZES

BIOGRAPHY
  Roy Jenkins, *Gladstone* (Macmillan).
BOOK OF THE YEAR
  Kate Atkinson, *Behind the Scenes at the Museum* (Doubleday).
FIRST NOVEL
  Kate Atkinson, *Behind the Scenes at the Museum* (Doubleday).
NOVEL
  Salman Rushdie, *The Moor's Last Sigh* (Cape).

POETRY

Bernard O'Donoghue, *Gunpowder* (Chatto and Windus).

WHITING WRITERS' AWARDS

Anderson Ferrell, Cristina Garcia, Molly Gloss, Brigit Pegeen Kelly, Brian Kiteley, Chris Offutt, Elizabeth Spires, Patricia Storace, Judy Troy, A. J. Verdelle.

YALE SERIES OF YOUNGER POETS

Winner for 1995 – Ellen Hinsey, *Cities of Memory* (Yale University Press).

Winner for 1996 – Talvikki Ansel, *My Shining Archipelago* (to be published in 1997).

Necrology

Adams, Robert M. – 16 December 1996
Albert, Marvin H. – 24 March 1996
Atwater, James David – 1 March 1996
Bazin, Hervé – 17 February 1996
Beich, Albert – 30 March 1996
Bell, Quentin – 16 December 1996
Blackwood, Lady Caroline – 14 February 1996
Blau, Raphael – 31 March 1996
Blumenberg, Hans – 28 March 1996
Bombeck, Erma – 22 April 1996
Bowen, Roger – 16 February 1996
Breslin, James E. B. – 6 January 1996
Broccoli, Albert R. – 27 June 1996
Brodkey, Harold – 26 January 1996
Brodsky, Joseph – 28 January 1996
Bufalino, Gesualdo – 16 June 1996
Burden, Carter – 23 January 1996
Caesar, Irving – 17 December 1996
Caunitz, William J. – 20 July 1996
Cobb, Richard – 15 January 1996
Comstock, James – 22 May 1996
Condon, Richard – 9 April 1996
Conrad, Pam – 22 January 1996
Cool, Pat – 5 January 1996
Cousins, Margaret – 30 July 1996
Daley, Brian – 18 February 1996
Davenport, Marcia – 16 January 1996
Davis, Arthur P. – 21 April 1996
Dearmer, Geoffrey – 18 August 1996
Donoso, José – 7 December 1996
Duras, Marguerite – 3 March 1996
Eberhart, Mignon – 8 October 1996
Eigner, Larry – 3 February 1996
Elward, James – 30 August 1996
Endo, Shusako – 29 September 1996
Fischer, Konrad – 24 November 1996
Flower, Desmond – 7 January 1997
Ford, Jesse Hill – 1 June 1996
Frenaye, Frances – 12 April 1996
Gilfillan, Merrill – 12 July 1996
Goodman, Charles – 23 February 1996
Green, Hannah – 16 October 1996
Harper, Rev. Ralph – 24 May 1996
Hawkes, Jacquetta – 18 March 1996
Heißenbüttel, Helmut – 19 September 1996
Hercules, Frank E. M. – 6 May 1996
Hirsch, Samuel – 15 February 1996
Hogan, William – 21 February 1996

Howard, Frank – 26 January 1996
Howes, Barbara – 24 February 1996
Huncke, Herbert – 8 August 1996
Izzi, Eugene "Guy" – 7 December 1996
Jackson, Brinck – 29 August 1996
Johnson, Walter J. – 15 December 1996
Keane, Molly – 22 April 1996
Kennebeck, Edwin – 2 May 1996
Kenyon, John – 6 January 1996
Kerouac, Jan – 5 June 1996
Kerr, Joan Paterson – 21 November 1996
Kerr, Walter – 9 October 1996
Kubly, Herbert – 7 August 1996
Lavin, Mary – 25 March 1996
Leary, Timothy – 31 May 1996
Le Sueur, Meridel – 14 November 1996
Lifson, David S. – 5 November 1996
Livingston, Myra Cohn – 23 August 1996
Lyon, Peter – 14 October 1996
MacCaig, Norman – 23 January 1996
Maclean, Sorley – 24 November 1996
MacRae, Jane Miller – 9 January 1996
McHugh, Arona – 22 May 1996
Milne, Christopher Robin – 20 April 1996
Mitchell, Joseph – 24 May 1996
Mitford, Jessica – 23 July 1996
Morris, Richard – 27 April 1996
Muir, Kenneth – 30 September 1996
Myer, Anton – 19 January 1996
Norman, Charles – 10 September 1996
O'Neal, Charles – 1 September 1996
Overholser, Wayne D. – August 1996
Poulin, Al – 5 June 1996
Prescott, Orville – 28 April 1996
Price, Eugenia – 28 May 1996
Ragan, Samuel T. – 11 May 1996
Rey, Margret E. – 21 December 1996
Robbins, Jhan – 27 September 1996
Rosenberg, Samuel – 5 January 1996
Rosenthal, M. L. – 21 July 1996
Sagan, Carl – 20 December 1996
Samuels, Ernest – 12 February 1996
Santos, Bienvenido – 7 January 1996
Schoenbaum, S – 27 March 1996
Shapiro, Lois – 17 January 1996
Starbuck, George – 15 August 1996
Stein, Gordon – 27 August 1996
Steiner, Paul – 7 March 1996

Sutherland, Sir James – 24 February 1996
Taylor, Martin – 16 June 1996
Tesich, Steve – 1 July 1996
Thompson, Paul – 9 February 1996
Travers, P. L. (Pamela Lyndon) – 23 April 1996
Trilling, Diana – 23 October 1996
Uno, Chiyo – 10 June 1996
Wilcox, Collin – 12 July 1996
Williams, Garth – 8 May 1996
Wolf, Dan – 11 April 1996

Checklist: Contributions to Literary History and Biography

This list is a selection of new books on various aspects of literary and cultural history, including biographies, memoirs, and correspondence of literary people and their associates.

Ahearn, Barry, ed. *Pound/Cummings: The Correspondence of Ezra Pound and E. E. Cummings.* Ann Arbor: University of Michigan Press, 1996.

Baker, Carlos. *Emerson among the Eccentrics.* New York: Viking, 1996.

Barker, Andrew. *Telegrams from the Soul: Peter Altenberg and the Culture of fin-de-siècle Vienna.* Columbia, S.C.: Camden House, 1996.

Bell, Quentin. *Bloomsbury Recalled.* New York: Columbia University Press, 1995.

Burns, Edward, and Ulla E. Dydo, with William Rice, eds. *The Letters of Gertrude Stein and Thornton Wilder.* New Haven: Yale University Press, 1996.

Caramello, Charles. *Henry James, Gertrude Stein, and the Biographical Act.* Chapel Hill: University of North Carolina Press, 1996.

Davenport-Hines, Richard. *Auden.* New York: Pantheon, 1996.

Davison, Peter Hobley. *George Orwell: A Literary Life.* New York: St. Martin's Press, 1996.

Everett, Patricia R., ed. *A History of Having a Great Many Times Not Continued to Be Friends: The Correspondence Between Mabel Dodge and Gertrude Stein, 1911–1934.* Albuquerque: University of New Mexico Press, 1996.

Fisher, Clive. *Cyril Connolly: The Life and Times of England's Most Controversial Literary Critic.* New York: St. Martin's Press, 1996.

Freedman, Ralph. *Life of a Poet: Rainer Maria Rilke.* New York: Farrar, Straus & Giroux, 1996.

Gibson, James. *Thomas Hardy: A Literary Life.* New York: St. Martin's Press, 1996.

Gordon, Lois. *The World of Samuel Beckett 1906–1946.* New Haven: Yale University Press, 1996.

Graham, Kenneth. *Henry James: A Literary Life.* New York: St. Martin's Press, 1995.

Grant, Judith Skelton. *Robertson Davies: Man of Myth.* New York: Viking, 1996.

Hobbs, Catherine, ed. *Nineteenth-Century Women Learn to Write.* Charlottesville: University Press of Virginia, 1995.

Lang, Cecil Y., ed. *The Letters of Matthew Arnold.* Charlottesville: University Press of Virginia, 1996.

Marotti, Maria-Ornella, ed. *Italian Women Writers from the Renaissance to the Present: Revising the Canon.* University Park: Pennsylvania State University Press, 1996.

McElrath, Joseph R. Jr., and Robert C. Leitz, eds. *"To Be an Author": Letters of Charles W. Chesnutt, 1889–1905.* Princeton: Princeton University Press, 1997.

McIntyre, Ian. *Dirt & Deity: A Life of Robert Burns.* London: HarperCollins, 1996.

Millgate, Michael, ed. *Letters of Emma and Florence Hardy.* Oxford: Clarendon Press / New York: Oxford University Press, 1996.

Mishkin, Tracy, ed. *Literary Influence and African-American Writers.* New York: Garland, 1996.

Morris, Roy. *Ambrose Bierce: Alone in Bad Company.* New York: Crown, 1996.

Murray, Nicholas. *A Life of Matthew Arnold.* London: Hodder & Stoughton, 1996.

Myers, Jeffrey. *Robert Frost: A Biography.* Boston: Houghton Mifflin, 1996.

Peters, Sally. *Bernard Shaw: The Ascent of the Superman.* New Haven: Yale University Press, 1996.

Polizzotti, Mark. *Revolution of the Mind: The Life of André Breton.* New York: Farrar, Straus & Giroux, 1995.

Powell, Hugh. *Fervor and Fiction: Therese von Bacheracht and Her Works.* Columbia, S.C.: Camden House, 1996.

Quinlan, Kieran. *Walker Percy: The Last Catholic Novelist.* Baton Rouge: Louisiana State University Press, 1996.

Reynolds, Barbara, ed. *The Letters of Dorothy L. Sayers, 1899–1936: The Making of a Detective Novelist.* New York: St. Martin's Press, 1996.

Robertson-Lorant, Laurie. *Melville: A Biography.* New York: Clarkson Potter, 1996.

Sarton, May. *At Eighty-Two: A Journal.* New York: Norton, 1996.

Sloan, John. *John Davidson, First of the Moderns: A Literary Biography.* Oxford: Oxford University Press, 1996.

Smith, Grahame. *Charles Dickens: A Literary Life.* New York: St. Martin's Press, 1996.

Springer, Haskell, ed. *America and the Sea: A Literary History.* Athens: University of Georgia Press, 1995.

Steiner, George. *No Passion Spent: Essays, 1978–1995.* New Haven: Yale University Press, 1996.

Twain, Mark. *Huckleberry Finn: A Comprehensive Edition.* New York: Random House, 1996.

Venclova, Tomas. *Aleksander Wat: Life and Art of an Iconoclast.* New Haven: Yale University Press, 1996.

Weintraub, Stanley, ed. *Shaw's People: Victoria to Churchill.* University Park: Pennsylvania State University Press, 1996.

Williams, John. *William Wordsworth: A Literary Life.* New York: St. Martin's Press, 1996.

Wilson, John Howard. *Evelyn Waugh: A Literary Biography, 1903–1924.* Rutherford, N.J.: Fairleigh Dickinson University Press, 1996.

Witemeyer, Hugh, ed. *Pound/Williams: Selected Letters of Ezra Pound and William Carlos Williams.* New York: New Directions, 1996.

Contributors

Harold Augenbraum	*Mercantile Library*
Stanisław Barańczak	*Harvard University*
Judith S. Baughman	*University of South Carolina*
Terry Belanger	*Alderman Library, University of Virginia*
Matthew J. Bruccoli	*University of South Carolina*
Andrew Digby	*British Library*
Charles Egleston	*University of Colorado at Boulder*
Nancy Emery	*University of Colorado at Boulder*
Julian Evans	*London, England*
William Foltz	*University of Hawaii*
Philip Furia	*University of North Carolina at Wilmington*
Ian Gadd	*Pembroke College, Oxford*
George Garrett	*University of Virginia*
Carole Gerson	*Simon Fraser University*
Caroline C. Hunt	*College of Charleston*
Howard Kissel	*New York Daily News*
Richard Layman	*Columbia, South Carolina*
Martin Moonie	*Somerville College, Oxford*
Merritt Moseley	*University of North Carolina at Asheville*
Kelli Rae Patton	*Unterberg Poetry Center*
Peter S. Prescott	*New Canaan, Connecticut*
Patrick Quinn	*Nene College, Northampton*
Raymond J. Rundus	*University of North Carolina at Pembroke*
David R. Slavitt	*University of Pennsylvania*
Ernest Suarez	*Catholic University of America*
Roy Sully	*British Library*
Mark Sutcliffe	*Ilkley, West Yorkshire*

Cumulative Index

Dictionary of Literary Biography, Volumes 1-177
Dictionary of Literary Biography Yearbook, 1980-1996
Dictionary of Literary Biography Documentary Series, Volumes 1-14

Cumulative Index

DLB before number: *Dictionary of Literary Biography,* Volumes 1-177
Y before number: *Dictionary of Literary Biography Yearbook,* 1980-1996
DS before number: *Dictionary of Literary Biography Documentary Series,* Volumes 1-14

A

Abbey Press DLB-49

The Abbey Theatre and Irish Drama, 1900-1945 .... DLB-10

Abbot, Willis J. 1863-1934 .... DLB-29

Abbott, Jacob 1803-1879 .... DLB-1

Abbott, Lee K. 1947- DLB-130

Abbott, Lyman 1835-1922 .... DLB-79

Abbott, Robert S. 1868-1940 .... DLB-29, 91

Abelard, Peter circa 1079-1142 ..... DLB-115

Abelard-Schuman DLB-46

Abell, Arunah S. 1806-1888 .... DLB-43

Abercrombie, Lascelles 1881-1938 ... DLB-19

Aberdeen University Press Limited DLB-106

Abish, Walter 1931- DLB-130

Ablesimov, Aleksandr Onisimovich 1742-1783 .... DLB-150

Abraham à Sancta Clara 1644-1709 .... DLB-168

Abrahams, Peter 1919- DLB-117

Abrams, M. H. 1912- DLB-67

Abrogans circa 790-800 .... DLB-148

Abschatz, Hans Aßmann von 1646-1699 .... DLB-168

Abse, Dannie 1923- DLB-27

Academy Chicago Publishers DLB-46

Accrocca, Elio Filippo 1923- DLB-128

Ace Books DLB-46

Achebe, Chinua 1930- DLB-117

Achtenberg, Herbert 1938- DLB-124

Ackerman, Diane 1948- DLB-120

Ackroyd, Peter 1949- DLB-155

Acorn, Milton 1923-1986 .... DLB-53

Acosta, Oscar Zeta 1935?- DLB-82

Actors Theatre of Louisville DLB-7

Adair, James 1709?-1783? DLB-30

Adam, Graeme Mercer 1839-1912 ... DLB-99

Adame, Leonard 1947- DLB-82

Adamic, Louis 1898-1951 .... DLB-9

Adams, Alice 1926- Y-86

Adams, Brooks 1848-1927 .... DLB-47

Adams, Charles Francis, Jr. 1835-1915 .... DLB-47

Adams, Douglas 1952- Y-83

Adams, Franklin P. 1881-1960 ..... DLB-29

Adams, Henry 1838-1918 .... DLB-12, 47

Adams, Herbert Baxter 1850-1901 ... DLB-47

Adams, J. S. and C. [publishing house] DLB-49

Adams, James Truslow 1878-1949 ... DLB-17

Adams, John 1735-1826 .... DLB-31

Adams, John Quincy 1767-1848 ..... DLB-37

Adams, Léonie 1899-1988 .... DLB-48

Adams, Levi 1802-1832 .... DLB-99

Adams, Samuel 1722-1803 .... DLB-31, 43

Adams, Thomas 1582 or 1583-1652 .... DLB-151

Adams, William Taylor 1822-1897 ... DLB-42

Adamson, Sir John 1867-1950 ...... DLB-98

Adcock, Arthur St. John 1864-1930 .... DLB-135

Adcock, Betty 1938- DLB-105

Adcock, Betty, *Certain Gifts* DLB-105

Adcock, Fleur 1934- DLB-40

Addison, Joseph 1672-1719 .... DLB-101

Ade, George 1866-1944 .... DLB-11, 25

Adeler, Max (see Clark, Charles Heber)

Adonias Filho 1915-1990 .... DLB-145

Advance Publishing Company DLB-49

AE 1867-1935 .... DLB-19

Ælfric circa 955-circa 1010 .... DLB-146

Aeschines circa 390 B.C.-circa 320 B.C. DLB-176

Aeschylus 525-524 B.C.-456-455 B.C. DLB-176

Aesthetic Poetry (1873), by Walter Pater DLB-35

After Dinner Opera Company Y-92

Afro-American Literary Critics: An Introduction DLB-33

Agassiz, Jean Louis Rodolphe 1807-1873 .... DLB-1

Agee, James 1909-1955 ...... DLB-2, 26, 152

The Agee Legacy: A Conference at the University of Tennessee at Knoxville Y-89

Aguilera Malta, Demetrio 1909-1981 .... DLB-145

Ai 1947- DLB-120

Aichinger, Ilse 1921- DLB-85

Aidoo, Ama Ata 1942- DLB-117

Aiken, Conrad 1889-1973 .... DLB-9, 45, 102

Aiken, Joan 1924- DLB-161

Aikin, Lucy 1781-1864 .... DLB-144, 163

Ainsworth, William Harrison 1805-1882 .... DLB-21

Aitken, George A. 1860-1917 ...... DLB-149

Aitken, Robert [publishing house] ... DLB-49

Akenside, Mark 1721-1770 .... DLB-109

Akins, Zoë 1886-1958 .... DLB-26

Alabaster, William 1568-1640 ...... DLB-132

Alain-Fournier 1886-1914 .... DLB-65

Alarcón, Francisco X. 1954- DLB-122

Alba, Nanina 1915-1968 .... DLB-41

Albee, Edward 1928- DLB-7

Albert the Great circa 1200-1280 ... DLB-115

Alberti, Rafael 1902- DLB-108

Albertinus, Aegidius circa 1560-1620 .... DLB-164

Alcaeus born circa 620 B.C. DLB-176

Alcott, Amos Bronson 1799-1888 ..... DLB-1

Cumulative Index

Alcott, Louisa May
    1832-1888 .... DLB-1, 42, 79; DS-14

Alcott, William Andrus 1798-1859 .... DLB-1

Alcuin circa 732-804 .... DLB-148

Alden, Henry Mills 1836-1919 ...... DLB-79

Alden, Isabella 1841-1930 .... DLB-42

Alden, John B. [publishing house] ... DLB-49

Alden, Beardsley and Company DLB-49

Aldington, Richard
    1892-1962 .... DLB-20, 36, 100, 149

Aldis, Dorothy 1896-1966 .... DLB-22

Aldiss, Brian W. 1925- DLB-14

Aldrich, Thomas Bailey
    1836-1907 .... DLB-42, 71, 74, 79

Alegría, Ciro 1909-1967 .... DLB-113

Alegría, Claribel 1924- DLB-145

Aleixandre, Vicente 1898-1984 ..... DLB-108

Aleramo, Sibilla 1876-1960 .... DLB-114

Alexander, Charles 1868-1923 ...... DLB-91

Alexander, Charles Wesley
    [publishing house] DLB-49

Alexander, James 1691-1756 .... DLB-24

Alexander, Lloyd 1924- DLB-52

Alexander, Sir William, Earl of Stirling
    1577?-1640 .... DLB-121

Alexie, Sherman 1966- DLB-175

Alexis, Willibald 1798-1871 .... DLB-133

Alfred, King 849-899 .... DLB-146

Alger, Horatio, Jr. 1832-1899 .... DLB-42

Algonquin Books of Chapel Hill DLB-46

Algren, Nelson 1909-1981 ... DLB-9; Y-81, 82

Allan, Andrew 1907-1974 .... DLB-88

Allan, Ted 1916- DLB-68

Allbeury, Ted 1917- DLB-87

Alldritt, Keith 1935- DLB-14

Allen, Ethan 1738-1789 .... DLB-31

Allen, Frederick Lewis 1890-1954 ... DLB-137

Allen, Gay Wilson
    1903-1995 .... DLB-103; Y-95

Allen, George 1808-1876 .... DLB-59

Allen, George [publishing house] ... DLB-106

Allen, George, and Unwin Limited ... DLB-112

Allen, Grant 1848-1899 .... DLB-70, 92

Allen, Henry W. 1912- Y-85

Allen, Hervey 1889-1949 .... DLB-9, 45

Allen, James 1739-1808 .... DLB-31

Allen, James Lane 1849-1925 .... DLB-71

Allen, Jay Presson 1922- DLB-26

Allen, John, and Company DLB-49

Allen, Paula Gunn 1939- DLB-175

Allen, Samuel W. 1917- DLB-41

Allen, Woody 1935- DLB-44

Allende, Isabel 1942- DLB-145

Alline, Henry 1748-1784 .... DLB-99

Allingham, Margery 1904-1966 ..... DLB-77

Allingham, William 1824-1889 ..... DLB-35

Allison, W. L. [publishing house] DLB-49

The *Alliterative Morte Arthure* and
    the *Stanzaic Morte Arthur*
    circa 1350-1400 .... DLB-146

Allott, Kenneth 1912-1973 .... DLB-20

Allston, Washington 1779-1843 ...... DLB-1

Almon, John [publishing house] DLB-154

Alonzo, Dámaso 1898-1990 .... DLB-108

Alsop, George 1636-post 1673 ...... DLB-24

Alsop, Richard 1761-1815 .... DLB-37

Altemus, Henry, and Company DLB-49

Altenberg, Peter 1885-1919 .... DLB-81

Altolaguirre, Manuel 1905-1959 ... DLB-108

Aluko, T. M. 1918- DLB-117

Alurista 1947- DLB-82

Alvarez, A. 1929- DLB-14, 40

Amadi, Elechi 1934- DLB-117

Amado, Jorge 1912- DLB-113

Ambler, Eric 1909- DLB-77

*America: or, a Poem on the Settlement of the
    British Colonies* (1780?), by Timothy
    Dwight DLB-37

American Conservatory Theatre DLB-7

American Fiction and the 1930s DLB-9

American Humor: A Historical Survey
    East and Northeast
    South and Southwest
    Midwest
    West DLB-11

The American Library in Paris Y-93

American News Company DLB-49

The American Poets' Corner: The First
    Three Years (1983-1986) Y-86

American Proletarian Culture:
    The 1930s DS-11

American Publishing Company DLB-49

American Stationers' Company DLB-49

American Sunday-School Union DLB-49

American Temperance Union DLB-49

American Tract Society DLB-49

The American Trust for the British Library
    .... Y-96

The American Writers Congress
    (9-12 October 1981) Y-81

The American Writers Congress: A Report
    on Continuing Business Y-81

Ames, Fisher 1758-1808 .... DLB-37

Ames, Mary Clemmer 1831-1884 .... DLB-23

Amini, Johari M. 1935- DLB-41

Amis, Kingsley 1922-1995
    .... DLB-15, 27, 100, 139, Y-96

Amis, Martin 1949- DLB-14

Ammons, A. R. 1926- DLB-5, 165

Amory, Thomas 1691?-1788 .... DLB-39

Anaya, Rudolfo A. 1937- DLB-82

Ancrene Riwle circa 1200-1225 ...... DLB-146

Andersch, Alfred 1914-1980 .... DLB-69

Anderson, Margaret 1886-1973 ... DLB-4, 91

Anderson, Maxwell 1888-1959 .... DLB-7

Anderson, Patrick 1915-1979 .... DLB-68

Anderson, Paul Y. 1893-1938 .... DLB-29

Anderson, Poul 1926- DLB-8

Anderson, Robert 1750-1830 .... DLB-142

Anderson, Robert 1917- DLB-7

Anderson, Sherwood
    1876-1941 .... DLB-4, 9, 86; DS-1

Andreae, Johann Valentin
    1586-1654 .... DLB-164

Andreas-Salomé, Lou 1861-1937 .... DLB-66

Andres, Stefan 1906-1970 .... DLB-69

Andreu, Blanca 1959- DLB-134

Andrewes, Lancelot
    1555-1626 .... DLB-151, 172

Andrews, Charles M. 1863-1943 ..... DLB-17

Andrews, Miles Peter ?-1814 .... DLB-89

Andrian, Leopold von 1875-1951 .... DLB-81

Andrić, Ivo 1892-1975 .... DLB-147

Andrieux, Louis (see Aragon, Louis)

Andrus, Silas, and Son DLB-49

Angell, James Burrill 1829-1916 ..... DLB-64

Angell, Roger 1920- DLB-171

Angelou, Maya 1928- DLB-38

Anger, Jane flourished 1589 .... DLB-136

Angers, Félicité (see Conan, Laure)

Anglo-Norman Literature in the Development of Middle English Literature.... DLB-146

The Anglo-Saxon Chronicle circa 890-1154................ DLB-146

The "Angry Young Men"......... DLB-15

Angus and Robertson (UK) Limited.................... DLB-112

Anhalt, Edward 1914-........... DLB-26

Anners, Henry F. [publishing house]... DLB-49

Annolied between 1077 and 1081.... DLB-148

Anselm of Canterbury 1033-1109... DLB-115

Anstey, F. 1856-1934............ DLB-141

Anthony, Michael 1932-......... DLB-125

Anthony, Piers 1934-........... DLB-8

Anthony Burgess's *99 Novels:* An Opinion Poll.................. Y-84

Antin, David 1932-............. DLB-169

Antin, Mary 1881-1949............. Y-84

Anton Ulrich, Duke of Brunswick-Lüneburg 1633-1714................... DLB-168

Antschel, Paul (see Celan, Paul)

Anyidoho, Kofi 1947-........... DLB-157

Anzaldúa, Gloria 1942-......... DLB-122

Anzengruber, Ludwig 1839-1889... DLB-129

Apess, William 1798-1839........ DLB-175

Apodaca, Rudy S. 1939-......... DLB-82

Apollonius Rhodius third century B.C. DLB-176

Apple, Max 1941-............... DLB-130

Appleton, D., and Company....... DLB-49

Appleton-Century-Crofts.......... DLB-46

Applewhite, James 1935-........ DLB-105

Apple-wood Books................ DLB-46

Aquin, Hubert 1929-1977........ DLB-53

Aquinas, Thomas 1224 or 1225-1274.................. DLB-115

Aragon, Louis 1897-1982......... DLB-72

Aratus of Soli circa 315 B.C.-circa 239 B.C. DLB-176

Arbor House Publishing Company.................... DLB-46

Arbuthnot, John 1667-1735....... DLB-101

Arcadia House.................... DLB-46

Arce, Julio G. (see Ulica, Jorge)

Archer, William 1856-1924........ DLB-10

Archilochhus mid seventh century B.C.E. DLB-176

The Archpoet circa 1130?-?....... DLB-148

Archpriest Avvakum (Petrovich) 1620?-1682.................. DLB-150

Arden, John 1930-.............. DLB-13

Arden of Faversham................ DLB-62

Ardis Publishers.................... Y-89

Ardizzone, Edward 1900-1979..... DLB-160

Arellano, Juan Estevan 1947-..... DLB-122

The Arena Publishing Company.... DLB-49

Arena Stage..................... DLB-7

Arenas, Reinaldo 1943-1990....... DLB-145

Arensberg, Ann 1937-.............. Y-82

Arguedas, José María 1911-1969.... DLB-113

Argueta, Manilio 1936-.......... DLB-145

Arias, Ron 1941-................ DLB-82

Aristophanes circa 446 B.C.-circa 446 B.C.-circa 386 B.C................ DLB-176

Aristotle 384 B.C.-322 B.C........ DLB-176

Arland, Marcel 1899-1986......... DLB-72

Arlen, Michael 1895-1956... DLB-36, 77, 162

Armah, Ayi Kwei 1939-.......... DLB-117

Der arme Hartmann ?-after 1150.................. DLB-148

Armed Services Editions.......... DLB-46

Armstrong, Richard 1903-....... DLB-160

Arndt, Ernst Moritz 1769-1860...... DLB-90

Arnim, Achim von 1781-1831....... DLB-90

Arnim, Bettina von 1785-1859...... DLB-90

Arno Press....................... DLB-46

Arnold, Edwin 1832-1904........... DLB-35

Arnold, Matthew 1822-1888..... DLB-32, 57

Arnold, Thomas 1795-1842......... DLB-55

Arnold, Edward [publishing house]............ DLB-112

Arnow, Harriette Simpson 1908-1986.................. DLB-6

Arp, Bill (see Smith, Charles Henry)

Arpino, Giovanni 1927-1987....... DLB-177

Arreola, Juan José 1918-......... DLB-113

Arrian circa 89-circa 155.......... DLB-176

Arrowsmith, J. W. [publishing house]............ DLB-106

Arthur, Timothy Shay 1809-1885........ DLB-3, 42, 79; DS-13

The Arthurian Tradition and Its European Context..................... DLB-138

Artmann, H. C. 1921-........... DLB-85

Arvin, Newton 1900-1963......... DLB-103

As I See It, by Carolyn Cassady..... DLB-16

Asch, Nathan 1902-1964........ DLB-4, 28

Ash, John 1948-................ DLB-40

Ashbery, John 1927-..... DLB-5, 165; Y-81

Ashendene Press.................. DLB-112

Asher, Sandy 1942-............... Y-83

Ashton, Winifred (see Dane, Clemence)

Asimov, Isaac 1920-1992........ DLB-8; Y-92

Askew, Anne circa 1521-1546...... DLB-136

Asselin, Olivar 1874-1937.......... DLB-92

Asturias, Miguel Angel 1899-1974................... DLB-113

Atheneum Publishers.............. DLB-46

Atherton, Gertrude 1857-1948.... DLB-9, 78

Athlone Press..................... DLB-112

Atkins, Josiah circa 1755-1781...... DLB-31

Atkins, Russell 1926-............. DLB-41

The Atlantic Monthly Press........ DLB-46

Attaway, William 1911-1986........ DLB-76

Atwood, Margaret 1939-.......... DLB-53

Aubert, Alvin 1930-.............. DLB-41

Aubert de Gaspé, Phillipe-Ignace-François 1814-1841.................... DLB-99

Aubert de Gaspé, Phillipe-Joseph 1786-1871.................... DLB-99

Aubin, Napoléon 1812-1890........ DLB-99

Aubin, Penelope 1685-circa 1731.... DLB-39

Aubrey-Fletcher, Henry Lancelot (see Wade, Henry)

Auchincloss, Louis 1917-...... DLB-2; Y-80

Auden, W. H. 1907-1973........ DLB-10, 20

Audio Art in America: A Personal Memoir....................... Y-85

Auerbach, Berthold 1812-1882..... DLB-133

Auernheimer, Raoul 1876-1948..... DLB-81

Augustine 354-430............... DLB-115

Austen, Jane 1775-1817........... DLB-116

Austin, Alfred 1835-1913.......... DLB-35

Austin, Mary 1868-1934......... DLB-9, 78

Austin, William 1778-1841......... DLB-74

Author-Printers, 1476–1599....... DLB-167

The Author's Apology for His Book (1684), by John Bunyan......... DLB-39

An Author's Response, by Ronald Sukenick................ Y-82

Authors and Newspapers Association................... DLB-46

Cumulative Index

Authors' Publishing Company DLB-49

Avalon Books DLB-46

Avancini, Nicolaus 1611-1686 . . . . . . DLB-164

Avendaño, Fausto 1941- DLB-82

Averroës 1126-1198 . . . . DLB-115

Avery, Gillian 1926- DLB-161

Avicenna 980-1037 . . . . DLB-115

Avison, Margaret 1918- DLB-53

Avon Books DLB-46

Awdry, Wilbert Vere 1911- DLB-160

Awoonor, Kofi 1935- DLB-117

Ayckbourn, Alan 1939- DLB-13

Aymé, Marcel 1902-1967 . . . . DLB-72

Aytoun, Sir Robert 1570-1638 . . . . . DLB-121

Aytoun, William Edmondstoune
    1813-1865 . . . . DLB-32, 159

B

B. V. (see Thomson, James)

Babbitt, Irving 1865-1933 . . . . DLB-63

Babbitt, Natalie 1932- DLB-52

Babcock, John [publishing house] DLB-49

Babrius circa 150-200 . . . . DLB-176

Baca, Jimmy Santiago 1952- DLB-122

Bache, Benjamin Franklin
    1769-1798 . . . . DLB-43

Bachmann, Ingeborg 1926-1973 . . . . DLB-85

Bacon, Delia 1811-1859 . . . . DLB-1

Bacon, Francis 1561-1626 . . . . DLB-151

Bacon, Roger circa
    1214/1220-1292 . . . . DLB-115

Bacon, Sir Nicholas
    circa 1510-1579 . . . . DLB-132

Bacon, Thomas circa 1700-1768 . . . . DLB-31

Badger, Richard G.,
    and Company DLB-49

Bage, Robert 1728-1801 . . . . DLB-39

Bagehot, Walter 1826-1877 . . . . DLB-55

Bagley, Desmond 1923-1983 . . . . DLB-87

Bagnold, Enid 1889-1981 . . . . . . DLB-13, 160

Bagryana, Elisaveta 1893-1991 . . . . . DLB-147

Bahr, Hermann 1863-1934 . . . . . DLB-81, 118

Bailey, Alfred Goldsworthy
    1905- DLB-68

Bailey, Francis [publishing house] . . . DLB-49

Bailey, H. C. 1878-1961 . . . . DLB-77

Bailey, Jacob 1731-1808 . . . . DLB-99

Bailey, Paul 1937- DLB-14

Bailey, Philip James 1816-1902 . . . . . DLB-32

Baillargeon, Pierre 1916-1967 . . . . . . DLB-88

Baillie, Hugh 1890-1966 . . . . DLB-29

Baillie, Joanna 1762-1851 . . . . DLB-93

Bailyn, Bernard 1922- DLB-17

Bainbridge, Beryl 1933- DLB-14

Baird, Irene 1901-1981 . . . . DLB-68

Baker, Augustine 1575-1641 . . . . DLB-151

Baker, Carlos 1909-1987 . . . . DLB-103

Baker, David 1954- DLB-120

Baker, Herschel C. 1914-1990 . . . . . . DLB-111

Baker, Houston A., Jr. 1943- DLB-67

Baker, Samuel White 1821-1893 . . . . DLB-166

Baker, Walter H., Company
    ("Baker's Plays") DLB-49

The Baker and Taylor Company DLB-49

Balaban, John 1943- DLB-120

Bald, Wambly 1902- DLB-4

Balde, Jacob 1604-1668 . . . . DLB-164

Balderston, John 1889-1954 . . . . DLB-26

Baldwin, James
    1924-1987 . . . . DLB-2, 7, 33; Y-87

Baldwin, Joseph Glover
    1815-1864 . . . . DLB-3, 11

Baldwin, Richard and Anne
    [publishing house] DLB-170

Baldwin, William
    circa 1515-1563 . . . . DLB-132

Bale, John 1495-1563 . . . . DLB-132

Balestrini, Nanni 1935- DLB-128

Ballantine Books DLB-46

Ballantyne, R. M. 1825-1894 . . . . DLB-163

Ballard, J. G. 1930- DLB-14

Ballerini, Luigi 1940- DLB-128

Ballou, Maturin Murray
    1820-1895 . . . . DLB-79

Ballou, Robert O.
    [publishing house] DLB-46

Balzac, Honoré de 1799-1855 . . . . . . DLB-119

Bambara, Toni Cade 1939- DLB-38

Bancroft, A. L., and
    Company DLB-49

Bancroft, George
    1800-1891 . . . . DLB-1, 30, 59

Bancroft, Hubert Howe
    1832-1918 . . . . DLB-47, 140

Bangs, John Kendrick
    1862-1922 . . . . DLB-11, 79

Banim, John 1798-1842 . . . DLB-116, 158, 159

Banim, Michael 1796-1874 . . . . DLB-158, 159

Banks, John circa 1653-1706 . . . . DLB-80

Banks, Russell 1940- DLB-130

Bannerman, Helen 1862-1946 . . . . . . DLB-141

Bantam Books DLB-46

Banti, Anna 1895-1985 . . . . DLB-177

Banville, John 1945- DLB-14

Baraka, Amiri
    1934- DLB-5, 7, 16, 38; DS-8

Barbauld, Anna Laetitia
    1743-1825 . . . . DLB-107, 109, 142, 158

Barbeau, Marius 1883-1969 . . . . DLB-92

Barber, John Warner 1798-1885 . . . . DLB-30

Bàrberi Squarotti, Giorgio
    1929- DLB-128

Barbey d'Aurevilly, Jules-Amédée
    1808-1889 . . . . DLB-119

Barbour, John circa 1316-1395 . . . . . DLB-146

Barbour, Ralph Henry
    1870-1944 . . . . DLB-22

Barbusse, Henri 1873-1935 . . . . DLB-65

Barclay, Alexander
    circa 1475-1552 . . . . DLB-132

Barclay, E. E., and Company DLB-49

Bardeen, C. W.
    [publishing house] DLB-49

Barham, Richard Harris
    1788-1845 . . . . DLB-159

Baring, Maurice 1874-1945 . . . . DLB-34

Baring-Gould, Sabine 1834-1924 . . . . DLB-156

Barker, A. L. 1918- DLB-14, 139

Barker, George 1913-1991 . . . . DLB-20

Barker, Harley Granville
    1877-1946 . . . . DLB-10

Barker, Howard 1946- DLB-13

Barker, James Nelson 1784-1858 . . . . DLB-37

Barker, Jane 1652-1727 . . . . DLB-39, 131

Barker, Lady Mary Anne
    1831-1911 . . . . DLB-166

Barker, William
    circa 1520-after 1576 . . . . DLB-132

Barker, Arthur, Limited DLB-112

Barkov, Ivan Semenovich
    1732-1768 . . . . DLB-150

Barks, Coleman 1937- DLB-5

Barlach, Ernst 1870-1938 ......DLB-56, 118

Barlow, Joel 1754-1812 ....DLB-37

Barnard, John 1681-1770..........DLB-24

Barne, Kitty (Mary Catherine Barne)
    1883-1957 ....DLB-160

Barnes, Barnabe 1571-1609........DLB-132

Barnes, Djuna 1892-1982 ......DLB-4, 9, 45

Barnes, Jim 1933- DLB-175

Barnes, Julian 1946- Y-93

Barnes, Margaret Ayer 1886-1967 ....DLB-9

Barnes, Peter 1931- DLB-13

Barnes, William 1801-1886 ....DLB-32

Barnes, A. S., and Company DLB-49

Barnes and Noble Books DLB-46

Barnet, Miguel 1940- DLB-145

Barney, Natalie 1876-1972..........DLB-4

Barnfield, Richard 1574-1627 ......DLB-172

Baron, Richard W.,
    Publishing Company DLB-46

Barr, Robert 1850-1912 ....DLB-70, 92

Barral, Carlos 1928-1989..........DLB-134

Barrax, Gerald William
    1933- DLB-41, 120

Barrès, Maurice 1862-1923 ....DLB-123

Barrett, Eaton Stannard
    1786-1820 ....DLB-116

Barrie, J. M. 1860-1937....DLB-10, 141, 156

Barrie and Jenkins DLB-112

Barrio, Raymond 1921- DLB-82

Barrios, Gregg 1945- DLB-122

Barry, Philip 1896-1949 ....DLB-7

Barry, Robertine (see Françoise)

Barse and Hopkins DLB-46

Barstow, Stan 1928- DLB-14, 139

Barth, John 1930- DLB-2

Barthelme, Donald
    1931-1989............DLB-2; Y-80, 89

Barthelme, Frederick 1943- Y-85

Bartholomew, Frank 1898-1985 ....DLB-127

Bartlett, John 1820-1905..........DLB-1

Bartol, Cyrus Augustus 1813-1900....DLB-1

Barton, Bernard 1784-1849 ....DLB-96

Barton, Thomas Pennant
    1803-1869 ....DLB-140

Bartram, John 1699-1777..........DLB-31

Bartram, William 1739-1823........DLB-37

Basic Books......................DLB-46

Basille, Theodore (see Becon, Thomas)

Bass, T. J. 1932- Y-81

Bassani, Giorgio 1916- DLB-128, 177

Basse, William circa 1583-1653.....DLB-121

Bassett, John Spencer 1867-1928.....DLB-17

Bassler, Thomas Joseph (see Bass, T. J.)

Bate, Walter Jackson 1918- DLB-67, 103

Bateman, Christopher
    [publishing house] DLB-170

Bateman, Stephen
    circa 1510-1584..............DLB-136

Bates, H. E. 1905-1974...........DLB-162

Bates, Katharine Lee 1859-1929 .....DLB-71

Batsford, B. T.
    [publishing house] DLB-106

Battiscombe, Georgina 1905- DLB-155

The Battle of Maldon circa 1000......DLB-146

Bauer, Bruno 1809-1882 ....DLB-133

Bauer, Wolfgang 1941- DLB-124

Baum, L. Frank 1856-1919 ....DLB-22

Baum, Vicki 1888-1960 ....DLB-85

Baumbach, Jonathan 1933- Y-80

Bausch, Richard 1945- DLB-130

Bawden, Nina 1925- DLB-14, 161

Bax, Clifford 1886-1962 ......DLB-10, 100

Baxter, Charles 1947- DLB-130

Bayer, Eleanor (see Perry, Eleanor)

Bayer, Konrad 1932-1964 ....DLB-85

Baynes, Pauline 1922- DLB-160

Bazin, Hervé 1911- DLB-83

Beach, Sylvia 1887-1962 ....DLB-4

Beacon Press......................DLB-49

Beadle and Adams DLB-49

Beagle, Peter S. 1939- Y-80

Beal, M. F. 1937- Y-81

Beale, Howard K. 1899-1959........DLB-17

Beard, Charles A. 1874-1948........DLB-17

A Beat Chronology: The First Twenty-five
    Years, 1944-1969 ....DLB-16

Beattie, Ann 1947- Y-82

Beattie, James 1735-1803 ....DLB-109

Beauchemin, Nérée 1850-1931 ......DLB-92

Beauchemin, Yves 1941- DLB-60

Beaugrand, Honoré 1848-1906 ......DLB-99

Beaulieu, Victor-Lévy 1945- DLB-53

Beaumont, Francis circa 1584-1616
    and Fletcher, John 1579-1625....DLB-58

Beaumont, Sir John 1583?-1627 ....DLB-121

Beaumont, Joseph 1616–1699 ......DLB-126

Beauvoir, Simone de
    1908-1986 ....DLB-72; Y-86

Becher, Ulrich 1910- DLB-69

Becker, Carl 1873-1945 ....DLB-17

Becker, Jurek 1937- DLB-75

Becker, Jurgen 1932- DLB-75

Beckett, Samuel
    1906-1989............DLB-13, 15; Y-90

Beckford, William 1760-1844 ....DLB-39

Beckham, Barry 1944- DLB-33

Becon, Thomas circa 1512-1567....DLB-136

Beddoes, Thomas 1760-1808.......DLB-158

Beddoes, Thomas Lovell
    1803-1849 ....DLB-96

Bede circa 673-735 ....DLB-146

Beecher, Catharine Esther
    1800-1878 ....DLB-1

Beecher, Henry Ward
    1813-1887 ....DLB-3, 43

Beer, George L. 1872-1920 ....DLB-47

Beer, Johann 1655-1700 ....DLB-168

Beer, Patricia 1919- DLB-40

Beerbohm, Max 1872-1956 .....DLB-34, 100

Beer-Hofmann, Richard
    1866-1945 ....DLB-81

Beers, Henry A. 1847-1926 ....DLB-71

Beeton, S. O. [publishing house]....DLB-106

Bégon, Elisabeth 1696-1755........DLB-99

Behan, Brendan 1923-1964 ....DLB-13

Behn, Aphra 1640?-1689....DLB-39, 80, 131

Behn, Harry 1898-1973 ....DLB-61

Behrman, S. N. 1893-1973........DLB-7, 44

Belaney, Archibald Stansfeld (see Grey Owl)

Belasco, David 1853-1931 ....DLB-7

Belford, Clarke and Company DLB-49

Belitt, Ben 1911- DLB-5

Belknap, Jeremy 1744-1798......DLB-30, 37

Bell, Clive 1881-1964 ....DS-10

Bell, Gertrude Margaret Lowthian
    1868-1926 ....DLB-174

Bell, James Madison 1826-1902......DLB-50

Bell, Marvin 1937- DLB-5

Bell, Millicent 1919- DLB-111

Bell, Quentin 1910- DLB-155

Bell, Vanessa 1879-1961 ....DS-10

Bell, George, and Sons...........DLB-106

Bell, Robert [publishing house]......DLB-49

Bellamy, Edward 1850-1898 ....DLB-12

Bellamy, John [publishing house]...DLB-170

Bellamy, Joseph 1719-1790 ....DLB-31

Bellezza, Dario 1944- DLB-128

La Belle Assemblée 1806-1837........DLB-110

Belloc, Hilaire
    1870-1953........DLB-19, 100, 141, 174

Bellow, Saul
    1915- DLB-2, 28; Y-82; DS-3

Belmont Productions DLB-46

Bemelmans, Ludwig 1898-1962......DLB-22

Bemis, Samuel Flagg 1891-1973 .....DLB-17

Bemrose, William
    [publishing house] DLB-106

Benchley, Robert 1889-1945 ....DLB-11

Benedetti, Mario 1920- DLB-113

Benedictus, David 1938- DLB-14

Benedikt, Michael 1935- DLB-5

Benét, Stephen Vincent
    1898-1943.............DLB-4, 48, 102

Benét, William Rose 1886-1950 .....DLB-45

Benford, Gregory 1941- Y-82

Benjamin, Park 1809-1864 ....DLB-3, 59, 73

Benlowes, Edward 1602-1676 ......DLB-126

Benn, Gottfried 1886-1956..........DLB-56

Benn Brothers Limited............DLB-106

Bennett, Arnold
    1867-1931..........DLB-10, 34, 98, 135

Bennett, Charles 1899- DLB-44

Bennett, Gwendolyn 1902- DLB-51

Bennett, Hal 1930- DLB-33

Bennett, James Gordon 1795-1872 ...DLB-43

Bennett, James Gordon, Jr.
    1841-1918 ....DLB-23

Bennett, John 1865-1956 ....DLB-42

Bennett, Louise 1919- DLB-117

Benoit, Jacques 1941- DLB-60

Benson, A. C. 1862-1925 ....DLB-98

Benson, E. F. 1867-1940 .....DLB-135, 153

Benson, Jackson J. 1930- DLB-111

Benson, Robert Hugh 1871-1914 ...DLB-153

Benson, Stella 1892-1933........DLB-36, 162

Bent, James Theodore 1852-1897...DLB-174

Bent, Mabel Virginia Anna ?-? DLB-174

Bentham, Jeremy 1748-1832 ...DLB-107, 158

Bentley, E. C. 1875-1956 ....DLB-70

Bentley, Richard
    [publishing house] DLB-106

Benton, Robert 1932- and Newman,
    David 1937- DLB-44

Benziger Brothers DLB-49

Beowulf circa 900-1000
    or 790-825...................DLB-146

Beresford, Anne 1929- DLB-40

Beresford, John Davys
    1873-1947 ....DLB-162

Beresford-Howe, Constance
    1922- DLB-88

Berford, R. G., Company DLB-49

Berg, Stephen 1934- DLB-5

Bergengruen, Werner 1892-1964 ....DLB-56

Berger, John 1926- DLB-14

Berger, Meyer 1898-1959...........DLB-29

Berger, Thomas 1924- DLB-2; Y-80

Berkeley, Anthony 1893-1971.......DLB-77

Berkeley, George 1685-1753 ....DLB-31, 101

The Berkley Publishing
    Corporation DLB-46

Berlin, Lucia 1936- DLB-130

Bernal, Vicente J. 1888-1915........DLB-82

Bernanos, Georges 1888-1948.......DLB-72

Bernard, Harry 1898-1979..........DLB-92

Bernard, John 1756-1828 ....DLB-37

Bernard of Chartres
    circa 1060-1124?..............DLB-115

Bernari, Carlo 1909-1992..........DLB-177

Bernhard, Thomas
    1931-1989................DLB-85, 124

Bernstein, Charles 1950- DLB-169

Berriault, Gina 1926- DLB-130

Berrigan, Daniel 1921- DLB-5

Berrigan, Ted 1934-1983........DLB-5, 169

Berry, Wendell 1934- DLB-5, 6

Berryman, John 1914-1972 ....DLB-48

Bersianik, Louky 1930- DLB-60

Berthelet, Thomas
    [publishing house] DLB-170

Berto, Giuseppe 1914-1978 ....DLB-177

Bertolucci, Attilio 1911- DLB-128

Berton, Pierre 1920- DLB-68

Besant, Sir Walter 1836-1901 ......DLB-135

Bessette, Gerard 1920- DLB-53

Bessie, Alvah 1904-1985 ....DLB-26

Bester, Alfred 1913-1987 ....DLB-8

The Bestseller Lists: An Assessment Y-84

Betham-Edwards, Matilda Barbara (see Edwards, Matilda Barbara Betham-)

Betjeman, John 1906-1984.....DLB-20; Y-84

Betocchi, Carlo 1899-1986.........DLB-128

Bettarini, Mariella 1942- DLB-128

Betts, Doris 1932- Y-82

Beveridge, Albert J. 1862-1927 .....DLB-17

Beverley, Robert
    circa 1673-1722 ....DLB-24, 30

Beyle, Marie-Henri (see Stendhal)

Bianco, Margery Williams
    1881-1944 ....DLB-160

Bibaud, Adèle 1854-1941...........DLB-92

Bibaud, Michel 1782-1857..........DLB-99

Bibliographical and Textual Scholarship
    Since World War II...............Y-89

The Bicentennial of James Fenimore
    Cooper: An International
    Celebration.....................Y-89

Bichsel, Peter 1935- DLB-75

Bickerstaff, Isaac John
    1733-circa 1808..................DLB-89

Biddle, Drexel [publishing house]....DLB-49

Bidermann, Jacob
    1577 or 1578-1639 ....DLB-164

Bidwell, Walter Hilliard
    1798-1881 ....DLB-79

Bienek, Horst 1930- DLB-75

Bierbaum, Otto Julius 1865-1910 ....DLB-66

Bierce, Ambrose
    1842-1914?.......DLB-11, 12, 23, 71, 74

Bigelow, William F. 1879-1966......DLB-91

Biggle, Lloyd, Jr. 1923- DLB-8

Bigiaretti, Libero 1905-1993 ....DLB-177

Biglow, Hosea (see Lowell, James Russell)

Bigongiari, Piero 1914- DLB-128

Billinger, Richard 1890-1965.......DLB-124

Billings, John Shaw 1898-1975 .....DLB-137

Billings, Josh (see Shaw, Henry Wheeler)

Binding, Rudolf G. 1867-1938.......DLB-66

Bingham, Caleb 1757-1817 .... DLB-42

Bingham, George Barry 1906-1988 .... DLB-127

Bingley, William [publishing house] DLB-154

Binyon, Laurence 1869-1943 .... DLB-19

Biographia Brittanica DLB-142

Biographical Documents I Y-84

Biographical Documents II Y-85

Bioren, John [publishing house] DLB-49

Bioy Casares, Adolfo 1914- DLB-113

Bird, Isabella Lucy 1831-1904 ...... DLB-166

Bird, William 1888-1963 .... DLB-4

Birken, Sigmund von 1626-1681 .... DLB-164

Birney, Earle 1904- DLB-88

Birrell, Augustine 1850-1933 .... DLB-98

Bisher, Furman 1918- DLB-171

Bishop, Elizabeth 1911-1979 ..... DLB-5, 169

Bishop, John Peale 1892-1944 ... DLB-4, 9, 45

Bismarck, Otto von 1815-1898 ..... DLB-129

Bisset, Robert 1759-1805 .... DLB-142

Bissett, Bill 1939- DLB-53

Bitzius, Albert (see Gotthelf, Jeremias)

Black, David (D. M.) 1941- DLB-40

Black, Winifred 1863-1936 .... DLB-25

Black, Walter J. [publishing house] DLB-46

The Black Aesthetic: Background DS-8

The Black Arts Movement, by Larry Neal DLB-38

Black Theaters and Theater Organizations in America, 1961-1982: A Research List DLB-38

Black Theatre: A Forum [excerpts] DLB-38

Blackamore, Arthur 1679-? DLB-24, 39

Blackburn, Alexander L. 1929- Y-85

Blackburn, Paul 1926-1971 .... DLB-16; Y-81

Blackburn, Thomas 1916-1977 ...... DLB-27

Blackmore, R. D. 1825-1900 .... DLB-18

Blackmore, Sir Richard 1654-1729 .... DLB-131

Blackmur, R. P. 1904-1965 .... DLB-63

Blackwell, Basil, Publisher DLB-106

Blackwood, Algernon Henry 1869-1951 .... DLB-153, 156

Blackwood, Caroline 1931- DLB-14

Blackwood, William, and Sons, Ltd. DLB-154

Blackwood's Edinburgh Magazine 1817-1980 .... DLB-110

Blair, Eric Arthur (see Orwell, George)

Blair, Francis Preston 1791-1876 .... DLB-43

Blair, James circa 1655-1743 .... DLB-24

Blair, John Durburrow 1759-1823 ... DLB-37

Blais, Marie-Claire 1939- DLB-53

Blaise, Clark 1940- DLB-53

Blake, Nicholas 1904-1972 .... DLB-77 (see Day Lewis, C.)

Blake, William 1757-1827 .... DLB-93, 154, 163

The Blakiston Company DLB-49

Blanchot, Maurice 1907- DLB-72

Blanckenburg, Christian Friedrich von 1744-1796 .... DLB-94

Blaser, Robin 1925- DLB-165

Bledsoe, Albert Taylor 1809-1877 .... DLB-3, 79

Blelock and Company DLB-49

Blennerhassett, Margaret Agnew 1773-1842 .... DLB-99

Bles, Geoffrey [publishing house] DLB-112

Blessington, Marguerite, Countess of 1789-1849 .... DLB-166

The Blickling Homilies circa 971 .... DLB-146

Blish, James 1921-1975 .... DLB-8

Bliss, E., and E. White [publishing house] DLB-49

Bliven, Bruce 1889-1977 .... DLB-137

Bloch, Robert 1917-1994 .... DLB-44

Block, Rudolph (see Lessing, Bruno)

Blondal, Patricia 1926-1959 .... DLB-88

Bloom, Harold 1930- DLB-67

Bloomer, Amelia 1818-1894 .... DLB-79

Bloomfield, Robert 1766-1823 .... DLB-93

Bloomsbury Group DS-10

Blotner, Joseph 1923- DLB-111

Bloy, Léon 1846-1917 .... DLB-123

Blume, Judy 1938- DLB-52

Blunck, Hans Friedrich 1888-1961 ... DLB-66

Blunden, Edmund 1896-1974 .... DLB-20, 100, 155

Blunt, Lady Anne Isabella Noel 1837-1917 .... DLB-174

Blunt, Wilfrid Scawen 1840-1922 .... DLB-19, 174

Bly, Nellie (see Cochrane, Elizabeth)

Bly, Robert 1926- DLB-5

Blyton, Enid 1897-1968 .... DLB-160

Boaden, James 1762-1839 .... DLB-89

Boas, Frederick S. 1862-1957 ...... DLB-149

The Bobbs-Merrill Archive at the Lilly Library, Indiana University ... Y-90

The Bobbs-Merrill Company DLB-46

Bobrov, Semen Sergeevich 1763?-1810 .... DLB-150

Bobrowski, Johannes 1917-1965 ..... DLB-75

Bodenheim, Maxwell 1892-1954 ... DLB-9, 45

Bodenstedt, Friedrich von 1819-1892 .... DLB-129

Bodini, Vittorio 1914-1970 .... DLB-128

Bodkin, M. McDonnell 1850-1933 .... DLB-70

Bodley Head DLB-112

Bodmer, Johann Jakob 1698-1783 .... DLB-97

Bodmershof, Imma von 1895-1982 ... DLB-85

Bodsworth, Fred 1918- DLB-68

Boehm, Sydney 1908- DLB-44

Boer, Charles 1939- DLB-5

Boethius circa 480-circa 524 .... DLB-115

Boethius of Dacia circa 1240-? DLB-115

Bogan, Louise 1897-1970 ...... DLB-45, 169

Bogarde, Dirk 1921- DLB-14

Bogdanovich, Ippolit Fedorovich circa 1743-1803 .... DLB-150

Bogue, David [publishing house] ... DLB-106

Böhme, Jakob 1575-1624 .... DLB-164

Bohn, H. G. [publishing house] DLB-106

Bohse, August 1661-1742 .... DLB-168

Boie, Heinrich Christian 1744-1806 .... DLB-94

Bok, Edward W. 1863-1930 .... DLB-91

Boland, Eavan 1944- DLB-40

Bolingbroke, Henry St. John, Viscount 1678-1751 .... DLB-101

Böll, Heinrich 1917-1985 ...... Y-85, DLB-69

Bolling, Robert 1738-1775 .... DLB-31

Bolotov, Andrei Timofeevich 1738-1833 .... DLB-150

Bolt, Carol 1941- DLB-60

Bolt, Robert 1924- DLB-13

Bolton, Herbert E. 1870-1953 .... DLB-17

Bonaventura DLB-90

Bonaventure circa 1217-1274 ...... DLB-115

Bonaviri, Giuseppe 1924- DLB-177

Bond, Edward 1934- DLB-13

Bond, Michael 1926- DLB-161

Bonnin, Gertrude Simmons (see Zitkala-Ša)

Boni, Albert and Charles
[publishing house] DLB-46

Boni and Liveright DLB-46

Robert Bonner's Sons.............. DLB-49

Bonsanti, Alessandro 1904-1984.... DLB-177

Bontemps, Arna 1902-1973....... DLB-48, 51

The Book Arts Press at the University
of Virginia Y-96

The Book League of America DLB-46

Book Reviewing in America: I Y-87

Book Reviewing in America: II Y-88

Book Reviewing in America: III........ Y-89

Book Reviewing in America: IV........ Y-90

Book Reviewing in America: V Y-91

Book Reviewing in America: VI........ Y-92

Book Reviewing in America: VII........ Y-93

Book Reviewing in America: VIII....... Y-94

Book Reviewing in America and the
Literary Scene Y-95

Book Reviewing and the
Literary Scene Y-96

Book Supply Company DLB-49

The Book Trade History Group Y-93

The Booker Prize Y-96

The Booker Prize
Address by Anthony Thwaite,
Chairman of the Booker Prize Judges
Comments from Former Booker
Prize Winners................... Y-86

Boorde, Andrew circa 1490-1549 ... DLB-136

Boorstin, Daniel J. 1914- DLB-17

Booth, Mary L. 1831-1889......... DLB-79

Booth, Philip 1925- Y-82

Booth, Wayne C. 1921- DLB-67

Borchardt, Rudolf 1877-1945 ...... DLB-66

Borchert, Wolfgang
1921-1947 .... DLB-69, 124

Borel, Pétrus 1809-1859 .... DLB-119

Borges, Jorge Luis
1899-1986 .... DLB-113; Y-86

Börne, Ludwig 1786-1837 .... DLB-90

Borrow, George
1803-1881 .... DLB-21, 55, 166

Bosch, Juan 1909- DLB-145

Bosco, Henri 1888-1976........... DLB-72

Bosco, Monique 1927- DLB-53

Boston, Lucy M. 1892-1990 .... DLB-161

Boswell, James 1740-1795 ..... DLB-104, 142

Botev, Khristo 1847-1876 .... DLB-147

Botta, Anne C. Lynch 1815-1891 ..... DLB-3

Bottomley, Gordon 1874-1948 ...... DLB-10

Bottoms, David 1949- DLB-120; Y-83

Bottrall, Ronald 1906- DLB-20

Boucher, Anthony 1911-1968 .... DLB-8

Boucher, Jonathan 1738-1804 .... DLB-31

Boucher de Boucherville, George
1814-1894 .... DLB-99

Boudreau, Daniel (see Coste, Donat)

Bourassa, Napoléon 1827-1916...... DLB-99

Bourget, Paul 1852-1935 .... DLB-123

Bourinot, John George 1837-1902 ... DLB-99

Bourjaily, Vance 1922- DLB-2, 143

Bourne, Edward Gaylord
1860-1908 .... DLB-47

Bourne, Randolph 1886-1918 .... DLB-63

Bousoño, Carlos 1923- DLB-108

Bousquet, Joë 1897-1950 .... DLB-72

Bova, Ben 1932- Y-81

Bovard, Oliver K. 1872-1945 .... DLB-25

Bove, Emmanuel 1898-1945 .... DLB-72

Bowen, Elizabeth 1899-1973 .... DLB-15, 162

Bowen, Francis 1811-1890........ DLB-1, 59

Bowen, John 1924- DLB-13

Bowen, Marjorie 1886-1952........ DLB-153

Bowen-Merrill Company DLB-49

Bowering, George 1935- DLB-53

Bowers, Claude G. 1878-1958....... DLB-17

Bowers, Edgar 1924- DLB-5

Bowers, Fredson Thayer
1905-1991 .... DLB-140; Y-91

Bowles, Paul 1910- DLB-5, 6

Bowles, Samuel III 1826-1878 .... DLB-43

Bowles, William Lisles 1762-1850 ... DLB-93

Bowman, Louise Morey
1882-1944 .... DLB-68

Boyd, James 1888-1944 .... DLB-9

Boyd, John 1919- DLB-8

Boyd, Thomas 1898-1935 .... DLB-9

Boyesen, Hjalmar Hjorth
1848-1895.......... DLB-12, 71; DS-13

Boyle, Kay
1902-1992........ DLB-4, 9, 48, 86; Y-93

Boyle, Roger, Earl of Orrery
1621-1679 .... DLB-80

Boyle, T. Coraghessan 1948- Y-86

Brackenbury, Alison 1953- DLB-40

Brackenridge, Hugh Henry
1748-1816 .... DLB-11, 37

Brackett, Charles 1892-1969 .... DLB-26

Brackett, Leigh 1915-1978........ DLB-8, 26

Bradburn, John
[publishing house] DLB-49

Bradbury, Malcolm 1932- DLB-14

Bradbury, Ray 1920- DLB-2, 8

Bradbury and Evans............... DLB-106

Braddon, Mary Elizabeth
1835-1915............. DLB-18, 70, 156

Bradford, Andrew 1686-1742 .... DLB-43, 73

Bradford, Gamaliel 1863-1932 ...... DLB-17

Bradford, John 1749-1830 .... DLB-43

Bradford, Roark 1896-1948......... DLB-86

Bradford, William 1590-1657 .... DLB-24, 30

Bradford, William III
1719-1791 .... DLB-43, 73

Bradlaugh, Charles 1833-1891 ...... DLB-57

Bradley, David 1950- DLB-33

Bradley, Marion Zimmer 1930- DLB-8

Bradley, William Aspenwall
1878-1939 .... DLB-4

Bradley, Ira, and Company......... DLB-49

Bradley, J. W., and Company....... DLB-49

Bradstreet, Anne
1612 or 1613-1672 .... DLB-24

Bradwardine, Thomas circa
1295-1349 .... DLB-115

Brady, Frank 1924-1986.......... DLB-111

Brady, Frederic A.
[publishing house] DLB-49

Bragg, Melvyn 1939- DLB-14

Brainard, Charles H.
[publishing house] DLB-49

Braine, John 1922-1986 .... DLB-15; Y-86

Braithwait, Richard 1588-1673 ..... DLB-151

Braithwaite, William Stanley
1878-1962 .... DLB-50, 54

Braker, Ulrich 1735-1798.......... DLB-94

Bramah, Ernest 1868-1942 ....DLB-70

Branagan, Thomas 1774-1843 ....DLB-37

Branch, William Blackwell
 1927- DLB-76

Branden Press DLB-46

Brassey, Lady Annie (Allnutt)
 1839-1887 ....DLB-166

Brathwaite, Edward Kamau
 1930- DLB-125

Brault, Jacques 1933- DLB-53

Braun, Volker 1939- DLB-75

Brautigan, Richard
 1935-1984 ....DLB-2, 5; Y-80, 84

Braxton, Joanne M. 1950- DLB-41

Bray, Anne Eliza 1790-1883 ....DLB-116

Bray, Thomas 1656-1730 ....DLB-24

Braziller, George
 [publishing house] DLB-46

The Bread Loaf Writers'
 Conference 1983 ....Y-84

The Break-Up of the Novel (1922),
 by John Middleton Murry DLB-36

Breasted, James Henry 1865-1935 ...DLB-47

Brecht, Bertolt 1898-1956 ......DLB-56, 124

Bredel, Willi 1901-1964 ....DLB-56

Breitinger, Johann Jakob
 1701-1776 ....DLB-97

Bremser, Bonnie 1939- DLB-16

Bremser, Ray 1934- DLB-16

Brentano, Bernard von
 1901-1964 ....DLB-56

Brentano, Clemens 1778-1842 ....DLB-90

Brentano's DLB-49

Brenton, Howard 1942- DLB-13

Breton, André 1896-1966 ....DLB-65

Breton, Nicholas
 circa 1555-circa 1626 ....DLB-136

The Breton Lays
 1300-early fifteenth century DLB-146

Brewer, Warren and Putnam DLB-46

Brewster, Elizabeth 1922- DLB-60

Bridgers, Sue Ellen 1942- DLB-52

Bridges, Robert 1844-1930 .....DLB-19, 98

Bridie, James 1888-1951 ....DLB-10

Briggs, Charles Frederick
 1804-1877 ....DLB-3

Brighouse, Harold 1882-1958 ....DLB-10

Bright, Mary Chavelita Dunne
 (see Egerton, George)

Brimmer, B. J., Company DLB-46

Brines, Francisco 1932- DLB-134

Brinley, George, Jr. 1817-1875 .....DLB-140

Brinnin, John Malcolm 1916- DLB-48

Brisbane, Albert 1809-1890 ....DLB-3

Brisbane, Arthur 1864-1936 ....DLB-25

British Academy DLB-112

The British Library and the Regular
 Readers' Group Y-91

The British Critic 1793-1843 ....DLB-110

The British Review and London
 Critical Journal 1811-1825 ....DLB-110

Brito, Aristeo 1942- DLB-122

Broadway Publishing Company DLB-46

Broch, Hermann 1886-1951 ....DLB-85, 124

Brochu, André 1942- DLB-53

Brock, Edwin 1927- DLB-40

Brockes, Barthold Heinrich
 1680-1747 ....DLB-168

Brod, Max 1884-1968 ....DLB-81

Brodber, Erna 1940- DLB-157

Brodhead, John R. 1814-1873 ....DLB-30

Brodkey, Harold 1930- DLB-130

Broeg, Bob 1918- DLB-171

Brome, Richard circa 1590-1652 .....DLB-58

Brome, Vincent 1910- DLB-155

Bromfield, Louis 1896-1956 ....DLB-4, 9, 86

Broner, E. M. 1930- DLB-28

Bronk, William 1918- DLB-165

Bronnen, Arnolt 1895-1959 ....DLB-124

Brontë, Anne 1820-1849 ....DLB-21

Brontë, Charlotte 1816-1855 ....DLB-21, 159

Brontë, Emily 1818-1848 ....DLB-21, 32

Brooke, Frances 1724-1789 ......DLB-39, 99

Brooke, Henry 1703?-1783 ....DLB-39

Brooke, L. Leslie 1862-1940 ....DLB-141

Brooke, Margaret, Ranee of Sarawak
 1849-1936 ....DLB-174

Brooke, Rupert 1887-1915 ....DLB-19

Brooker, Bertram 1888-1955 ....DLB-88

Brooke-Rose, Christine 1926- DLB-14

Brookner, Anita 1928- Y-87

Brooks, Charles Timothy
 1813-1883 ....DLB-1

Brooks, Cleanth 1906-1994 ....DLB-63; Y-94

Brooks, Gwendolyn
 1917- DLB-5, 76, 165

Brooks, Jeremy 1926- DLB-14

Brooks, Mel 1926- DLB-26

Brooks, Noah 1830-1903 .....DLB-42; DS-13

Brooks, Richard 1912-1992 ....DLB-44

Brooks, Van Wyck
 1886-1963 ....DLB-45, 63, 103

Brophy, Brigid 1929- DLB-14

Brossard, Chandler 1922-1993 ......DLB-16

Brossard, Nicole 1943- DLB-53

Broster, Dorothy Kathleen
 1877-1950 ....DLB-160

Brother Antoninus (see Everson, William)

Brougham and Vaux, Henry Peter
 Brougham, Baron
 1778-1868 ....DLB-110, 158

Brougham, John 1810-1880 ....DLB-11

Broughton, James 1913- DLB-5

Broughton, Rhoda 1840-1920 ....DLB-18

Broun, Heywood 1888-1939 ....DLB-29, 171

Brown, Alice 1856-1948 ....DLB-78

Brown, Bob 1886-1959 ....DLB-4, 45

Brown, Cecil 1943- DLB-33

Brown, Charles Brockden
 1771-1810 ....DLB-37, 59, 73

Brown, Christy 1932-1981 ....DLB-14

Brown, Dee 1908- Y-80

Brown, Frank London 1927-1962 ....DLB-76

Brown, Fredric 1906-1972 ....DLB-8

Brown, George Mackay
 1921- DLB-14, 27, 139

Brown, Harry 1917-1986 ....DLB-26

Brown, Marcia 1918- DLB-61

Brown, Margaret Wise
 1910-1952 ....DLB-22

Brown, Morna Doris (see Ferrars, Elizabeth)

Brown, Oliver Madox
 1855-1874 ....DLB-21

Brown, Sterling
 1901-1989 ....DLB-48, 51, 63

Brown, T. E. 1830-1897 ....DLB-35

Brown, William Hill 1765-1793 .....DLB-37

Brown, William Wells
 1814-1884 ....DLB-3, 50

Browne, Charles Farrar
 1834-1867 ....DLB-11

Browne, Francis Fisher
 1843-1913 ....DLB-79

Browne, Michael Dennis
1940- DLB-40

Browne, Sir Thomas 1605-1682 .... DLB-151

Browne, William, of Tavistock
1590-1645 .... DLB-121

Browne, Wynyard 1911-1964 .... DLB-13

Browne and Nolan DLB-106

Brownell, W. C. 1851-1928 .... DLB-71

Browning, Elizabeth Barrett
1806-1861 .... DLB-32

Browning, Robert
1812-1889 .... DLB-32, 163

Brownjohn, Allan 1931- DLB-40

Brownson, Orestes Augustus
1803-1876 .... DLB-1, 59, 73

Bruccoli, Matthew J. 1931- DLB-103

Bruce, Charles 1906-1971 .... DLB-68

Bruce, Leo 1903-1979 .... DLB-77

Bruce, Philip Alexander
1856-1933 .... DLB-47

Bruce Humphries
[publishing house] DLB-46

Bruce-Novoa, Juan 1944- DLB-82

Bruckman, Clyde 1894-1955 .... DLB-26

Bruckner, Ferdinand 1891-1958 .... DLB-118

Brundage, John Herbert (see Herbert, John)

Brutus, Dennis 1924- DLB-117

Bryant, Arthur 1899-1985 .... DLB-149

Bryant, William Cullen
1794-1878 .... DLB-3, 43, 59

Bryce Echenique, Alfredo
1939- DLB-145

Bryce, James 1838-1922 .... DLB-166

Brydges, Sir Samuel Egerton
1762-1837 .... DLB-107

Bryskett, Lodowick 1546?-1612 .... DLB-167

Buchan, John 1875-1940 .... DLB-34, 70, 156

Buchanan, George 1506-1582 ...... DLB-132

Buchanan, Robert 1841-1901 .... DLB-18, 35

Buchman, Sidney 1902-1975 .... DLB-26

Buchner, Augustus 1591-1661 ...... DLB-164

Büchner, Georg 1813-1837 .... DLB-133

Bucholtz, Andreas Heinrich
1607-1671 .... DLB-168

Buck, Pearl S. 1892-1973 .... DLB-9, 102

Bucke, Charles 1781-1846 .... DLB-110

Bucke, Richard Maurice
1837-1902 .... DLB-99

Buckingham, Joseph Tinker 1779-1861 and
Buckingham, Edwin
1810-1833 .... DLB-73

Buckler, Ernest 1908-1984 .... DLB-68

Buckley, William F., Jr.
1925- DLB-137; Y-80

Buckminster, Joseph Stevens
1784-1812 .... DLB-37

Buckner, Robert 1906- DLB-26

Budd, Thomas ?-1698 .... DLB-24

Budrys, A. J. 1931- DLB-8

Buechner, Frederick 1926- Y-80

Buell, John 1927- DLB-53

Buffum, Job [publishing house] DLB-49

Bugnet, Georges 1879-1981 .... DLB-92

Buies, Arthur 1840-1901 .... DLB-99

Building the New British Library
at St Pancras Y-94

Bukowski, Charles
1920-1994 .... DLB-5, 130, 169

Bulger, Bozeman 1877-1932 .... DLB-171

Bullein, William
between 1520 and 1530-1576 ... DLB-167

Bullins, Ed 1935- DLB-7, 38

Bulwer-Lytton, Edward (also Edward Bulwer)
1803-1873 .... DLB-21

Bumpus, Jerry 1937- Y-81

Bunce and Brother DLB-49

Bunner, H. C. 1855-1896 .... DLB-78, 79

Bunting, Basil 1900-1985 .... DLB-20

Bunyan, John 1628-1688 .... DLB-39

Burch, Robert 1925- DLB-52

Burciaga, José Antonio 1940- DLB-82

Bürger, Gottfried August
1747-1794 .... DLB-94

Burgess, Anthony 1917-1993 .... DLB-14

Burgess, Gelett 1866-1951 .... DLB-11

Burgess, John W. 1844-1931 .... DLB-47

Burgess, Thornton W.
1874-1965 .... DLB-22

Burgess, Stringer and Company DLB-49

Burick, Si 1909-1986 .... DLB-171

Burk, John Daly circa 1772-1808 ... DLB-37

Burke, Edmund 1729?-1797 .... DLB-104

Burke, Kenneth 1897-1993 ..... DLB-45, 63

Burlingame, Edward Livermore
1848-1922 .... DLB-79

Burnet, Gilbert 1643-1715 .... DLB-101

Burnett, Frances Hodgson
1849-1924 ...... DLB-42, 141; DS-13, 14

Burnett, W. R. 1899-1982 .... DLB-9

Burnett, Whit 1899-1973 and
Martha Foley 1897-1977 .... DLB-137

Burney, Fanny 1752-1840 .... DLB-39

Burns, Alan 1929- DLB-14

Burns, John Horne 1916-1953 .... Y-85

Burns, Robert 1759-1796 .... DLB-109

Burns and Oates DLB-106

Burnshaw, Stanley 1906- DLB-48

Burr, C. Chauncey 1815?-1883 ..... DLB-79

Burroughs, Edgar Rice 1875-1950 .... DLB-8

Burroughs, John 1837-1921 .... DLB-64

Burroughs, Margaret T. G.
1917- DLB-41

Burroughs, William S., Jr.
1947-1981 .... DLB-16

Burroughs, William Seward
1914- DLB-2, 8, 16, 152; Y-81

Burroway, Janet 1936- DLB-6

Burt, Maxwell S. 1882-1954 .... DLB-86

Burt, A. L., and Company DLB-49

Burton, Hester 1913- DLB-161

Burton, Isabel Arundell
1831-1896 .... DLB-166

Burton, Miles (see Rhode, John)

Burton, Richard Francis
1821-1890 .... DLB-55, 166

Burton, Robert 1577-1640 .... DLB-151

Burton, Virginia Lee 1909-1968 .... DLB-22

Burton, William Evans
1804-1860 .... DLB-73

Burwell, Adam Hood 1790-1849 .... DLB-99

Bury, Lady Charlotte
1775-1861 .... DLB-116

Busch, Frederick 1941- DLB-6

Busch, Niven 1903-1991 .... DLB-44

Bushnell, Horace 1802-1876 .... DS-13

Bussieres, Arthur de 1877-1913 ... DLB-92

Butler, Juan 1942-1981 .... DLB-53

Butler, Octavia E. 1947- DLB-33

Butler, Robert Olen 1945- DLB-173

Butler, Samuel 1613-1680 ..... DLB-101, 126

Butler, Samuel 1835-1902 ... DLB-18, 57, 174

Butler, William Francis
1838-1910 .... DLB-166

Butler, E. H., and Company DLB-49

Butor, Michel 1926- DLB-83

Butter, Nathaniel
  [publishing house] DLB-170

Butterworth, Hezekiah 1839-1905 ... DLB-42

Buttitta, Ignazio 1899- DLB-114

Buzzati, Dino 1906-1972 .... DLB-177

Byars, Betsy 1928- DLB-52

Byatt, A. S. 1936- DLB-14

Byles, Mather 1707-1788 .... DLB-24

Bynneman, Henry
  [publishing house] DLB-170

Bynner, Witter 1881-1968 .... DLB-54

Byrd, William circa 1543-1623 .... DLB-172

Byrd, William II 1674-1744 .... DLB-24, 140

Byrne, John Keyes (see Leonard, Hugh)

Byron, George Gordon, Lord
  1788-1824 .... DLB-96, 110

C

Caballero Bonald, José Manuel
  1926- DLB-108

Cabañero, Eladio 1930- DLB-134

Cabell, James Branch
  1879-1958 .... DLB-9, 78

Cabeza de Baca, Manuel
  1853-1915 .... DLB-122

Cabeza de Baca Gilbert, Fabiola
  1898- DLB-122

Cable, George Washington
  1844-1925 .... DLB-12, 74; DS-13

Cabrera, Lydia 1900-1991 .... DLB-145

Cabrera Infante, Guillermo
  1929- DLB-113

Cadell [publishing house] DLB-154

Cady, Edwin H. 1917- DLB-103

Caedmon flourished 658-680 ...... DLB-146

Caedmon School circa 660-899 ..... DLB-146

Cahan, Abraham
  1860-1951 .... DLB-9, 25, 28

Cain, George 1943- DLB-33

Caldecott, Randolph 1846-1886 .... DLB-163

Calder, John
  (Publishers), Limited DLB-112

Caldwell, Ben 1937- DLB-38

Caldwell, Erskine 1903-1987 ..... DLB-9, 86

Caldwell, H. M., Company DLB-49

Calhoun, John C. 1782-1850 .... DLB-3

Calisher, Hortense 1911- DLB-2

A Call to Letters and an Invitation
  to the Electric Chair,
  by Siegfried Mandel DLB-75

Callaghan, Morley 1903-1990 .... DLB-68

Callahan, S. Alice 1868-1894 ...... DLB-175

Callaloo Y-87

Callimachus circa 305 B.C.-240 B.C.
  .... DLB-176

Calmer, Edgar 1907- DLB-4

Calverley, C. S. 1831-1884 .... DLB-35

Calvert, George Henry
  1803-1889 .... DLB-1, 64

Cambridge Press DLB-49

Cambridge Songs (Carmina Cantabrigensia)
  circa 1050 .... DLB-148

Cambridge University Press DLB-170

Camden, William 1551-1623 .... DLB-172

Camden House: An Interview with
  James Hardin Y-92

Cameron, Eleanor 1912- DLB-52

Cameron, George Frederick
  1854-1885 .... DLB-99

Cameron, Lucy Lyttelton
  1781-1858 .... DLB-163

Cameron, William Bleasdell
  1862-1951 .... DLB-99

Camm, John 1718-1778 .... DLB-31

Campana, Dino 1885-1932 .... DLB-114

Campbell, Gabrielle Margaret Vere
  (see Shearing, Joseph, and Bowen, Marjorie)

Campbell, James Dykes
  1838-1895 .... DLB-144

Campbell, James Edwin
  1867-1896 .... DLB-50

Campbell, John 1653-1728 .... DLB-43

Campbell, John W., Jr.
  1910-1971 .... DLB-8

Campbell, Roy 1901-1957 .... DLB-20

Campbell, Thomas
  1777-1844 .... DLB-93, 144

Campbell, William Wilfred
  1858-1918 .... DLB-92

Campion, Edmund 1539-1581 ...... DLB-167

Campion, Thomas
  1567-1620 .... DLB-58, 172

Camus, Albert 1913-1960 .... DLB-72

The Canadian Publishers' Records
  Database Y-96

Canby, Henry Seidel 1878-1961 ..... DLB-91

Candelaria, Cordelia 1943- DLB-82

Candelaria, Nash 1928- DLB-82

Candour in English Fiction (1890),
  by Thomas Hardy DLB-18

Canetti, Elias 1905-1994 .... DLB-85, 124

Canham, Erwin Dain
  1904-1982 .... DLB-127

Canitz, Friedrich Rudolph Ludwig von
  1654-1699 .... DLB-168

Cankar, Ivan 1876-1918 .... DLB-147

Cannan, Gilbert 1884-1955 .... DLB-10

Cannell, Kathleen 1891-1974 .... DLB-4

Cannell, Skipwith 1887-1957 .... DLB-45

Canning, George 1770-1827 .... DLB-158

Cannon, Jimmy 1910-1973 .... DLB-171

Cantwell, Robert 1908-1978 .... DLB-9

Cape, Jonathan, and Harrison Smith
  [publishing house] DLB-46

Cape, Jonathan, Limited DLB-112

Capen, Joseph 1658-1725 .... DLB-24

Capes, Bernard 1854-1918 .... DLB-156

Capote, Truman
  1924-1984 .... DLB-2; Y-80, 84

Caproni, Giorgio 1912-1990 .... DLB-128

Cardarelli, Vincenzo 1887-1959 .... DLB-114

Cárdenas, Reyes 1948- DLB-122

Cardinal, Marie 1929- DLB-83

Carew, Jan 1920- DLB-157

Carew, Thomas
  1594 or 1595-1640 .... DLB-126

Carey, Henry
  circa 1687-1689-1743 .... DLB-84

Carey, Mathew 1760-1839 ...... DLB-37, 73

Carey and Hart DLB-49

Carey, M., and Company DLB-49

Carlell, Lodowick 1602-1675 .... DLB-58

Carleton, William 1794-1869 ...... DLB-159

Carleton, G. W.
  [publishing house] DLB-49

Carlile, Richard 1790-1843 .... DLB-110, 158

Carlyle, Jane Welsh 1801-1866 ...... DLB-55

Carlyle, Thomas 1795-1881 .... DLB-55, 144

Carman, Bliss 1861-1929 .... DLB-92

Carmina Burana circa 1230 .... DLB-138

Carnero, Guillermo 1947- DLB-108

Carossa, Hans 1878-1956 .... DLB-66

Carpenter, Humphrey 1946- DLB-155

Carpenter, Stephen Cullen
  ?-1820? DLB-73

315

Carpentier, Alejo 1904-1980 . . . . DLB-113

Carrier, Roch 1937- DLB-53

Carrillo, Adolfo 1855-1926 . . . . DLB-122

Carroll, Gladys Hasty 1904- DLB-9

Carroll, John 1735-1815 . . . . DLB-37

Carroll, John 1809-1884 . . . . DLB-99

Carroll, Lewis 1832-1898 . . . . . . DLB-18, 163

Carroll, Paul 1927- DLB-16

Carroll, Paul Vincent 1900-1968 . . . . DLB-10

Carroll and Graf Publishers DLB-46

Carruth, Hayden 1921- DLB-5, 165

Carryl, Charles E. 1841-1920 . . . . DLB-42

Carswell, Catherine 1879-1946 . . . . . . DLB-36

Carter, Angela 1940-1992 . . . . DLB-14

Carter, Elizabeth 1717-1806 . . . . DLB-109

Carter, Henry (see Leslie, Frank)

Carter, Hodding, Jr. 1907-1972 . . . . DLB-127

Carter, Landon 1710-1778 . . . . DLB-31

Carter, Lin 1930- Y-81

Carter, Martin 1927- DLB-117

Carter and Hendee DLB-49

Carter, Robert, and Brothers DLB-49

Cartwright, John 1740-1824 . . . . DLB-158

Cartwright, William circa
  1611-1643 . . . . DLB-126

Caruthers, William Alexander
  1802-1846 . . . . DLB-3

Carver, Jonathan 1710-1780 . . . . DLB-31

Carver, Raymond
  1938-1988 . . . . DLB-130; Y-84, 88

Cary, Joyce 1888-1957 . . . . DLB-15, 100

Cary, Patrick 1623?-1657 . . . . DLB-131

Casey, Juanita 1925- DLB-14

Casey, Michael 1947- DLB-5

Cassady, Carolyn 1923- DLB-16

Cassady, Neal 1926-1968 . . . . DLB-16

Cassell and Company DLB-106

Cassell Publishing Company DLB-49

Cassill, R. V. 1919- DLB-6

Cassity, Turner 1929- DLB-105

Cassius Dio circa 155/164-post 229
  . . . . DLB-176

Cassola, Carlo 1917-1987 . . . . DLB-177

The Castle of Perseverance
  circa 1400-1425 . . . . DLB-146

Castellano, Olivia 1944- DLB-122

Castellanos, Rosario 1925-1974 . . . . DLB-113

Castillo, Ana 1953- DLB-122

Castlemon, Harry (see Fosdick, Charles Austin)

Caswall, Edward 1814-1878 . . . . DLB-32

Catacalos, Rosemary 1944- DLB-122

Cather, Willa
  1873-1947 . . . . DLB-9, 54, 78; DS-1

Catherine II (Ekaterina Alekseevna), "The
  Great," Empress of Russia
  1729-1796 . . . . DLB-150

Catherwood, Mary Hartwell
  1847-1902 . . . . DLB-78

Catledge, Turner 1901-1983 . . . . DLB-127

Cattafi, Bartolo 1922-1979 . . . . DLB-128

Catton, Bruce 1899-1978 . . . . DLB-17

Causley, Charles 1917- DLB-27

Caute, David 1936- DLB-14

Cavendish, Duchess of Newcastle,
  Margaret Lucas 1623-1673 . . . . . DLB-131

Cawein, Madison 1865-1914 . . . . DLB-54

The Caxton Printers, Limited DLB-46

Caxton, William
  [publishing house] DLB-170

Cayrol, Jean 1911- DLB-83

Cecil, Lord David 1902-1986 . . . . . . DLB-155

Celan, Paul 1920-1970 . . . . DLB-69

Celaya, Gabriel 1911-1991 . . . . DLB-108

Céline, Louis-Ferdinand
  1894-1961 . . . . DLB-72

The Celtic Background to Medieval English
  Literature DLB-146

Center for Bibliographical Studies and
  Research at the University of
  California, Riverside Y-91

The Center for the Book in the Library
  of Congress Y-93

Center for the Book Research Y-84

Centlivre, Susanna 1669?-1723 . . . . . DLB-84

The Century Company DLB-49

Cernuda, Luis 1902-1963 . . . . DLB-134

Cervantes, Lorna Dee 1954- DLB-82

Chacel, Rosa 1898- DLB-134

Chacón, Eusebio 1869-1948 . . . . DLB-82

Chacón, Felipe Maximiliano
  1873-? DLB-82

Chadwyck-Healey's Full-Text Literary Databases: Editing Commercial Databases of
  Primary Literary Texts Y-95

Challans, Eileen Mary (see Renault, Mary)

Chalmers, George 1742-1825 . . . . DLB-30

Chaloner, Sir Thomas
  1520-1565 . . . . DLB-167

Chamberlain, Samuel S.
  1851-1916 . . . . DLB-25

Chamberland, Paul 1939- DLB-60

Chamberlin, William Henry
  1897-1969 . . . . DLB-29

Chambers, Charles Haddon
  1860-1921 . . . . DLB-10

Chambers, W. and R.
  [publishing house] DLB-106

Chamisso, Albert von
  1781-1838 . . . . DLB-90

Champfleury 1821-1889 . . . . DLB-119

Chandler, Harry 1864-1944 . . . . DLB-29

Chandler, Norman 1899-1973 . . . . . . DLB-127

Chandler, Otis 1927- DLB-127

Chandler, Raymond 1888-1959 . . . . DS-6

Channing, Edward 1856-1931 . . . . . . DLB-17

Channing, Edward Tyrrell
  1790-1856 . . . . DLB-1, 59

Channing, William Ellery
  1780-1842 . . . . DLB-1, 59

Channing, William Ellery, II
  1817-1901 . . . . DLB-1

Channing, William Henry
  1810-1884 . . . . DLB-1, 59

Chaplin, Charlie 1889-1977 . . . . DLB-44

Chapman, George
  1559 or 1560 - 1634 . . . . . . DLB-62, 121

Chapman, John DLB-106

Chapman, William 1850-1917 . . . . . . DLB-99

Chapman and Hall DLB-106

Chappell, Fred 1936- DLB-6, 105

Chappell, Fred, A Detail
  in a Poem DLB-105

Charbonneau, Jean 1875-1960 . . . . . . DLB-92

Charbonneau, Robert 1911-1967 . . . . DLB-68

Charles, Gerda 1914- DLB-14

Charles, William
  [publishing house] DLB-49

The Charles Wood Affair:
  A Playwright Revived Y-83

Charlotte Forten: Pages from
  her Diary DLB-50

Charteris, Leslie 1907-1993 . . . . DLB-77

Charyn, Jerome 1937- Y-83

Chase, Borden 1900-1971 . . . . DLB-26

Chase, Edna Woolman 1877-1957 ...... DLB-91

Chase-Riboud, Barbara 1936- DLB-33

Chateaubriand, François-René de 1768-1848 ...... DLB-119

Chatterton, Thomas 1752-1770 .... DLB-109

Chatto and Windus DLB-106

Chaucer, Geoffrey 1340?-1400 ..... DLB-146

Chauncy, Charles 1705-1787 ...... DLB-24

Chauveau, Pierre-Joseph-Olivier 1820-1890 ...... DLB-99

Chávez, Denise 1948- DLB-122

Chávez, Fray Angélico 1910- DLB-82

Chayefsky, Paddy 1923-1981 ...... DLB-7, 44; Y-81

Cheever, Ezekiel 1615-1708 ...... DLB-24

Cheever, George Barrell 1807-1890 ...... DLB-59

Cheever, John 1912-1982 ...... DLB-2, 102; Y-80, 82

Cheever, Susan 1943- Y-82

Cheke, Sir John 1514-1557 ...... DLB-132

Chelsea House DLB-46

Cheney, Ednah Dow (Littlehale) 1824-1904 ...... DLB-1

Cheney, Harriet Vaughn 1796-1889 ...... DLB-99

Cherry, Kelly 1940 ...... Y-83

Cherryh, C. J. 1942- Y-80

Chesnutt, Charles Waddell 1858-1932 ...... DLB-12, 50, 78

Chester, Alfred 1928-1971 ...... DLB-130

Chester, George Randolph 1869-1924 ...... DLB-78

The Chester Plays circa 1505-1532; revisions until 1575 ...... DLB-146

Chesterfield, Philip Dormer Stanhope, Fourth Earl of 1694-1773 ...... DLB-104

Chesterton, G. K. 1874-1936 ... DLB-10, 19, 34, 70, 98, 149

Chettle, Henry circa 1560-circa 1607 ...... DLB-136

Chew, Ada Nield 1870-1945 ...... DLB-135

Cheyney, Edward P. 1861-1947 ..... DLB-47

Chiara, Piero 1913-1986 ...... DLB-177

Chicano History DLB-82

Chicano Language DLB-82

Child, Francis James 1825-1896 ...... DLB-1, 64

Child, Lydia Maria 1802-1880 ...... DLB-1, 74

Child, Philip 1898-1978 ...... DLB-68

Childers, Erskine 1870-1922 ...... DLB-70

Children's Book Awards and Prizes DLB-61

Children's Illustrators, 1800-1880 ...... DLB-163

Childress, Alice 1920-1994 ...... DLB-7, 38

Childs, George W. 1829-1894 ...... DLB-23

Chilton Book Company DLB-46

Chinweizu 1943- DLB-157

Chitham, Edward 1932- DLB-155

Chittenden, Hiram Martin 1858-1917 ...... DLB-47

Chivers, Thomas Holley 1809-1858 ...... DLB-3

Chopin, Kate 1850-1904 ...... DLB-12, 78

Chopin, Rene 1885-1953 ...... DLB-92

Choquette, Adrienne 1915-1973 ..... DLB-68

Choquette, Robert 1905- DLB-68

The Christian Publishing Company DLB-49

Christie, Agatha 1890-1976 ...... DLB-13, 77

Christus und die Samariterin circa 950 ...... DLB-148

Chulkov, Mikhail Dmitrievich 1743?-1792 ...... DLB-150

Church, Benjamin 1734-1778 ...... DLB-31

Church, Francis Pharcellus 1839-1906 ...... DLB-79

Church, William Conant 1836-1917 ...... DLB-79

Churchill, Caryl 1938- DLB-13

Churchill, Charles 1731-1764 ...... DLB-109

Churchill, Sir Winston 1874-1965 ...... DLB-100

Churchyard, Thomas 1520?-1604 ...... DLB-132

Churton, E., and Company DLB-106

Chute, Marchette 1909-1994 ...... DLB-103

Ciardi, John 1916-1986 ...... DLB-5; Y-86

Cibber, Colley 1671-1757 ...... DLB-84

Cima, Annalisa 1941- DLB-128

Cirese, Eugenio 1884-1955 ...... DLB-114

Cisneros, Sandra 1954- DLB-122, 152

City Lights Books DLB-46

Cixous, Hélène 1937- DLB-83

Clampitt, Amy 1920-1994 ...... DLB-105

Clapper, Raymond 1892-1944 ...... DLB-29

Clare, John 1793-1864 ...... DLB-55, 96

Clarendon, Edward Hyde, Earl of 1609-1674 ...... DLB-101

Clark, Alfred Alexander Gordon (see Hare, Cyril)

Clark, Ann Nolan 1896- DLB-52

Clark, Catherine Anthony 1892-1977 ...... DLB-68

Clark, Charles Heber 1841-1915 ...... DLB-11

Clark, Davis Wasgatt 1812-1871 .... DLB-79

Clark, Eleanor 1913- DLB-6

Clark, J. P. 1935- DLB-117

Clark, Lewis Gaylord 1808-1873 ...... DLB-3, 64, 73

Clark, Walter Van Tilburg 1909-1971 ...... DLB-9

Clark, C. M., Publishing Company DLB-46

Clarke, Austin 1896-1974 ...... DLB-10, 20

Clarke, Austin C. 1934- DLB-53, 125

Clarke, Gillian 1937- DLB-40

Clarke, James Freeman 1810-1888 ...... DLB-1, 59

Clarke, Pauline 1921- DLB-161

Clarke, Rebecca Sophia 1833-1906 ...... DLB-42

Clarke, Robert, and Company DLB-49

Clarkson, Thomas 1760-1846 ...... DLB-158

Claudius, Matthias 1740-1815 ...... DLB-97

Clausen, Andy 1943- DLB-16

Claxton, Remsen and Haffelfinger DLB-49

Clay, Cassius Marcellus 1810-1903 ...... DLB-43

Cleary, Beverly 1916- DLB-52

Cleaver, Vera 1919- and Cleaver, Bill 1920-1981 ...... DLB-52

Cleland, John 1710-1789 ...... DLB-39

Clemens, Samuel Langhorne 1835-1910 ...... DLB-11, 12, 23, 64, 74

Clement, Hal 1922- DLB-8

Clemo, Jack 1916- DLB-27

Cleveland, John 1613-1658 ...... DLB-126

Cliff, Michelle 1946- DLB-157

Clifford, Lady Anne 1590-1676 .... DLB-151

Clifford, James L. 1901-1978 ...... DLB-103

Clifford, Lucy 1853?-1929 ... DLB-135, 141

Clifton, Lucille 1936- DLB-5, 41	Collier, Mary 1690-1762 .... DLB-95	Connolly, James B. 1868-1957 ...... DLB-78
Clode, Edward J. [publishing house] DLB-46	Collier, Robert J. 1876-1918 .... DLB-91	Connor, Ralph 1860-1937 .... DLB-92
Clough, Arthur Hugh 1819-1861 .... DLB-32	Collier, P. F. [publishing house] DLB-49	Connor, Tony 1930- DLB-40
Cloutier, Cécile 1930- DLB-60	Collin and Small DLB-49	Conquest, Robert 1917- DLB-27
Clutton-Brock, Arthur 1868-1924 .... DLB-98	Collingwood, W. G. 1854-1932 .... DLB-149	Conrad, Joseph 1857-1924 .... DLB-10, 34, 98, 156
Coates, Robert M. 1897-1973 .... DLB-4, 9, 102	Collins, An floruit circa 1653 ...... DLB-131	Conrad, John, and Company DLB-49
Coatsworth, Elizabeth 1893- DLB-22	Collins, Merle 1950- DLB-157	Conroy, Jack 1899-1990 .... Y-81
Cobb, Charles E., Jr. 1943- DLB-41	Collins, Mortimer 1827-1876 .... DLB-21, 35	Conroy, Pat 1945- DLB-6
Cobb, Frank I. 1869-1923 .... DLB-25	Collins, Wilkie 1824-1889... DLB-18, 70, 159	The Consolidation of Opinion: Critical Responses to the Modernists DLB-36
Cobb, Irvin S. 1876-1944 .... DLB-11, 25, 86	Collins, William 1721-1759 .... DLB-109	Constable, Henry 1562-1613 .... DLB-136
Cobbett, William 1763-1835 .... DLB-43, 107	Collins, William, Sons and Company DLB-154	Constable and Company Limited DLB-112
Cobbledick, Gordon 1898-1969 .... DLB-171	Collins, Isaac [publishing house] DLB-49	Constable, Archibald, and Company DLB-154
Cochran, Thomas C. 1902- DLB-17	Collyer, Mary 1716?-1763? DLB-39	Constant, Benjamin 1767-1830 ..... DLB-119
Cochrane, Elizabeth 1867-1922 ..... DLB-25	Colman, Benjamin 1673-1747 ...... DLB-24	Constant de Rebecque, Henri-Benjamin de (see Constant, Benjamin)
Cockerill, John A. 1845-1896 .... DLB-23	Colman, George, the Elder 1732-1794 .... DLB-89	
Cocteau, Jean 1889-1963 .... DLB-65	Colman, George, the Younger 1762-1836 .... DLB-89	Constantine, David 1944- DLB-40
Coderre, Emile (see Jean Narrache)	Colman, S. [publishing house] DLB-49	Constantin-Weyer, Maurice 1881-1964 .... DLB-92
Coffee, Lenore J. 1900?-1984 .... DLB-44	Colombo, John Robert 1936- DLB-53	Contempo Caravan: Kites in a Windstorm Y-85
Coffin, Robert P. Tristram 1892-1955 .... DLB-45	Colquhoun, Patrick 1745-1820 ..... DLB-158	A Contemporary Flourescence of Chicano Literature Y-84
Cogswell, Fred 1917- DLB-60	Colter, Cyrus 1910- DLB-33	
Cogswell, Mason Fitch 1761-1830 .... DLB-37	Colum, Padraic 1881-1972 .... DLB-19	The Continental Publishing Company DLB-49
Cohen, Arthur A. 1928-1986 .... DLB-28	Colvin, Sir Sidney 1845-1927 ..... DLB-149	A Conversation with Chaim Potok Y-84
Cohen, Leonard 1934- DLB-53	Colwin, Laurie 1944-1992 .... Y-80	Conversations with Editors Y-95
Cohen, Matt 1942- DLB-53	Comden, Betty 1919- and Green, Adolph 1918- DLB-44	Conversations with Publishers I: An Interview with Patrick O'Connor Y-84
Colden, Cadwallader 1688-1776 .... DLB-24, 30	Comi, Girolamo 1890-1968 .... DLB-114	Conversations with Publishers II: An Interview with Charles Scribner III Y-94
Cole, Barry 1936- DLB-14	The Comic Tradition Continued [in the British Novel] DLB-15	Conversations with Publishers III: An Interview with Donald Lamm Y-95
Cole, George Watson 1850-1939 .... DLB-140	Commager, Henry Steele 1902- DLB-17	Conversations with Publishers IV: An Interview with James Laughlin Y-96
Colegate, Isabel 1931- DLB-14	The Commercialization of the Image of Revolt, by Kenneth Rexroth..... DLB-16	
Coleman, Emily Holmes 1899-1974 .... DLB-4	Community and Commentators: Black Theatre and Its Critics DLB-38	Conversations with Rare Book Dealers I: An Interview with Glenn Horowitz..... Y-90
Coleman, Wanda 1946- DLB-130	Compton-Burnett, Ivy 1884?-1969 .... DLB-36	Conversations with Rare Book Dealers II: An Interview with Ralph Sipper........ Y-94
Coleridge, Hartley 1796-1849 ...... DLB-96	Conan, Laure 1845-1924 .... DLB-99	Conversations with Rare Book Dealers (Publishers) III: An Interview with Otto Penzler Y-96
Coleridge, Mary 1861-1907..... DLB-19, 98	Conde, Carmen 1901- DLB-108	
Coleridge, Samuel Taylor 1772-1834 .... DLB-93, 107	Conference on Modern Biography...... Y-85	
Colet, John 1467-1519 .... DLB-132	Congreve, William 1670-1729 .... DLB-39, 84	The Conversion of an Unpolitical Man, by W. H. Bruford.............. DLB-66
Colette 1873-1954 .... DLB-65	Conkey, W. B., Company DLB-49	Conway, Moncure Daniel 1832-1907 .... DLB-1
Colette, Sidonie Gabrielle (see Colette)	Connell, Evan S., Jr. 1924- DLB-2; Y-81	Cook, Ebenezer circa 1667-circa 1732 .... DLB-24
Colinas, Antonio 1946- DLB-134	Connelly, Marc 1890-1980 ..... DLB-7; Y-80	
Collier, John 1901-1980 .... DLB-77	Connolly, Cyril 1903-1974 .... DLB-98	Cook, Edward Tyas 1857-1919 .... DLB-149

Cook, Michael 1933- DLB-53

Cook, David C., Publishing Company DLB-49

Cooke, George Willis 1848-1923 .... DLB-71

Cooke, Increase, and Company DLB-49

Cooke, John Esten 1830-1886 .... DLB-3

Cooke, Philip Pendleton 1816-1850 .... DLB-3, 59

Cooke, Rose Terry 1827-1892 .... DLB-12, 74

Cook-Lynn, Elizabeth 1930- DLB-175

Coolbrith, Ina 1841-1928 .... DLB-54

Cooley, Peter 1940- DLB-105

Cooley, Peter, Into the Mirror DLB-105

Coolidge, Susan (see Woolsey, Sarah Chauncy)

Coolidge, George [publishing house] DLB-49

Cooper, Giles 1918-1966 .... DLB-13

Cooper, James Fenimore 1789-1851 ... DLB-3

Cooper, Kent 1880-1965 .... DLB-29

Cooper, Susan 1935- DLB-161

Cooper, William [publishing house] DLB-170

Coote, J. [publishing house] DLB-154

Coover, Robert 1932- DLB-2; Y-81

Copeland and Day DLB-49

Copland, Robert 1470?-1548 .... DLB-136

Coppard, A. E. 1878-1957 .... DLB-162

Coppel, Alfred 1921- Y-83

Coppola, Francis Ford 1939- DLB-44

Copway, George (Kah-ge-ga-gah-bowh) 1818-1869 .... DLB-175

Corazzini, Sergio 1886-1907 .... DLB-114

Corbett, Richard 1582-1635 .... DLB-121

Corcoran, Barbara 1911- DLB-52

Corelli, Marie 1855-1924 .... DLB-34, 156

Corle, Edwin 1906-1956 .... Y-85

Corman, Cid 1924- DLB-5

Cormier, Robert 1925- DLB-52

Corn, Alfred 1943- DLB-120; Y-80

Cornish, Sam 1935- DLB-41

Cornish, William circa 1465-circa 1524 .... DLB-132

Cornwall, Barry (see Procter, Bryan Waller)

Cornwallis, Sir William, the Younger circa 1579-1614 .... DLB-151

Cornwell, David John Moore (see le Carré, John)

Corpi, Lucha 1945- DLB-82

Corrington, John William 1932- DLB-6

Corrothers, James D. 1869-1917 ..... DLB-50

Corso, Gregory 1930- DLB-5, 16

Cortázar, Julio 1914-1984 .... DLB-113

Cortez, Jayne 1936- DLB-41

Corvinus, Gottlieb Siegmund 1677-1746 .... DLB-168

Corvo, Baron (see Rolfe, Frederick William)

Cory, Annie Sophie (see Cross, Victoria)

Cory, William Johnson 1823-1892 .... DLB-35

Coryate, Thomas 1577?-1617 .... DLB-151, 172

Cosin, John 1595-1672 .... DLB-151

Cosmopolitan Book Corporation DLB-46

Costain, Thomas B. 1885-1965 .... DLB-9

Coste, Donat 1912-1957 .... DLB-88

Costello, Louisa Stuart 1799-1870 .. DLB-166

Cota-Cárdenas, Margarita 1941- DLB-122

Cotter, Joseph Seamon, Sr. 1861-1949 .... DLB-50

Cotter, Joseph Seamon, Jr. 1895-1919 .... DLB-50

Cottle, Joseph [publishing house] ... DLB-154

Cotton, Charles 1630-1687 .... DLB-131

Cotton, John 1584-1652 .... DLB-24

Coulter, John 1888-1980 .... DLB-68

Cournos, John 1881-1966 .... DLB-54

Cousins, Margaret 1905- DLB-137

Cousins, Norman 1915-1990 .... DLB-137

Coventry, Francis 1725-1754 ...... DLB-39

Coverdale, Miles 1487 or 1488-1569 .... DLB-167

Coverly, N. [publishing house] DLB-49

Covici-Friede DLB-46

Coward, Noel 1899-1973 .... DLB-10

Coward, McCann and Geoghegan DLB-46

Cowles, Gardner 1861-1946 .... DLB-29

Cowles, Gardner ("Mike"), Jr. 1903-1985 .... DLB-127, 137

Cowley, Abraham 1618-1667 .... DLB-131, 151

Cowley, Hannah 1743-1809 .... DLB-89

Cowley, Malcolm 1898-1989 .... DLB-4, 48; Y-81, 89

Cowper, William 1731-1800 .... DLB-104, 109

Cox, A. B. (see Berkeley, Anthony)

Cox, James McMahon 1903-1974 .... DLB-127

Cox, James Middleton 1870-1957 .... DLB-127

Cox, Palmer 1840-1924 .... DLB-42

Coxe, Louis 1918-1993 .... DLB-5

Coxe, Tench 1755-1824 .... DLB-37

Cozzens, James Gould 1903-1978 .... DLB-9; Y-84; DS-2

Crabbe, George 1754-1832 .... DLB-93

Crackanthorpe, Hubert 1870-1896 .... DLB-135

Craddock, Charles Egbert (see Murfree, Mary N.)

Cradock, Thomas 1718-1770 .... DLB-31

Craig, Daniel H. 1811-1895 .... DLB-43

Craik, Dinah Maria 1826-1887 .... DLB-35, 136

Cranch, Christopher Pearse 1813-1892 .... DLB-1, 42

Crane, Hart 1899-1932 .... DLB-4, 48

Crane, R. S. 1886-1967 .... DLB-63

Crane, Stephen 1871-1900 ... DLB-12, 54, 78

Crane, Walter 1845-1915 .... DLB-163

Cranmer, Thomas 1489-1556 ...... DLB-132

Crapsey, Adelaide 1878-1914 .... DLB-54

Crashaw, Richard 1612 or 1613-1649 .... DLB-126

Craven, Avery 1885-1980 .... DLB-17

Crawford, Charles 1752-circa 1815 .... DLB-31

Crawford, F. Marion 1854-1909 ..... DLB-71

Crawford, Isabel Valancy 1850-1887 .... DLB-92

Crawley, Alan 1887-1975 .... DLB-68

Crayon, Geoffrey (see Irving, Washington)

Creamer, Robert W. 1922- DLB-171

Creasey, John 1908-1973 .... DLB-77

Creative Age Press DLB-46

Creech, William [publishing house] DLB-154

Creede, Thomas [publishing house] DLB-170

Creel, George 1876-1953 .... DLB-25

Creeley, Robert 1926- DLB-5, 16, 169

Creelman, James 1859-1915 ....DLB-23

Cregan, David 1931- DLB-13

Creighton, Donald Grant
   1902-1979 ....DLB-88

Cremazie, Octave 1827-1879........DLB-99

Crémer, Victoriano 1909?- DLB-108

Crescas, Hasdai
   circa 1340-1412?.............DLB-115

Crespo, Angel 1926- DLB-134

Cresset Press DLB-112

Cresswell, Helen 1934- DLB-161

Crèvecoeur, Michel Guillaume Jean de
   1735-1813 ....DLB-37

Crews, Harry 1935- DLB-6, 143

Crichton, Michael 1942- Y-81

A Crisis of Culture: The Changing Role
   of Religion in the New Republic
   ....DLB-37

Crispin, Edmund 1921-1978 ....DLB-87

Cristofer, Michael 1946- DLB-7

"The Critic as Artist" (1891), by
   Oscar Wilde DLB-57

"Criticism In Relation To Novels" (1863),
   by G. H. Lewes DLB-21

Crnjanski, Miloš 1893-1977........DLB-147

Crockett, David (Davy)
   1786-1836 ....DLB-3, 11

Croft-Cooke, Rupert (see Bruce, Leo)

Crofts, Freeman Wills
   1879-1957 ....DLB-77

Croker, John Wilson
   1780-1857 ....DLB-110

Croly, George 1780-1860.........DLB-159

Croly, Herbert 1869-1930 ....DLB-91

Croly, Jane Cunningham
   1829-1901 ....DLB-23

Crompton, Richmal 1890-1969.....DLB-160

Crosby, Caresse 1892-1970.........DLB-48

Crosby, Caresse 1892-1970 and Crosby,
   Harry 1898-1929 ....DLB-4

Crosby, Harry 1898-1929 ....DLB-48

Cross, Gillian 1945- DLB-161

Cross, Victoria 1868-1952........DLB-135

Crossley-Holland, Kevin
   1941- DLB-40, 161

Crothers, Rachel 1878-1958 ....DLB-7

Crowell, Thomas Y., Company DLB-49

Crowley, John 1942- Y-82

Crowley, Mart 1935- DLB-7

Crown Publishers DLB-46

Crowne, John 1641-1712 ....DLB-80

Crowninshield, Edward Augustus
   1817-1859 ....DLB-140

Crowninshield, Frank 1872-1947 ....DLB-91

Croy, Homer 1883-1965 ....DLB-4

Crumley, James 1939- Y-84

Cruz, Victor Hernández 1949- DLB-41

Csokor, Franz Theodor
   1885-1969 ....DLB-81

Cuala Press DLB-112

Cullen, Countee 1903-1946 ...DLB-4, 48, 51

Culler, Jonathan D. 1944- DLB-67

The Cult of Biography
   Excerpts from the Second Folio Debate:
   "Biographies are generally a disease of
   English Literature" – Germaine Greer,
   Victoria Glendinning, Auberon Waugh,
   and Richard Holmes............Y-86

Cumberland, Richard 1732-1811 ....DLB-89

Cummings, Constance Gordon
   1837-1924 ....DLB-174

Cummings, E. E. 1894-1962 .....DLB-4, 48

Cummings, Ray 1887-1957 ....DLB-8

Cummings and Hilliard DLB-49

Cummins, Maria Susanna
   1827-1866 ....DLB-42

Cundall, Joseph
   [publishing house] DLB-106

Cuney, Waring 1906-1976.........DLB-51

Cuney-Hare, Maude 1874-1936 .....DLB-52

Cunningham, Allan
   1784-1842 ....DLB-116, 144

Cunningham, J. V. 1911- DLB-5

Cunningham, Peter F.
   [publishing house] DLB-49

Cunquiero, Alvaro 1911-1981......DLB-134

Cuomo, George 1929- Y-80

Cupples and Leon.................DLB-46

Cupples, Upham and Company DLB-49

Cuppy, Will 1884-1949 ....DLB-11

Curll, Edmund
   [publishing house] DLB-154

Currie, James 1756-1805 ....DLB-142

Currie, Mary Montgomerie Lamb Singleton,
   Lady Currie (see Fane, Violet)

Cursor Mundi circa 1300 ....DLB-146

Curti, Merle E. 1897- DLB-17

Curtis, Anthony 1926- DLB-155

Curtis, Cyrus H. K. 1850-1933......DLB-91

Curtis, George William
   1824-1892 ....DLB-1, 43

Curzon, Robert 1810-1873 ....DLB-166

Curzon, Sarah Anne 1833-1898 .....DLB-99

Cynewulf circa 770-840 ....DLB-146

Czepko, Daniel 1605-1660........DLB-164

D

D. M. Thomas: The Plagiarism
   Controversy....................Y-82

Dabit, Eugène 1898-1936..........DLB-65

Daborne, Robert circa 1580-1628....DLB-58

Dacey, Philip 1939- DLB-105

Dacey, Philip, Eyes Across Centuries:
   Contemporary Poetry and "That
   Vision Thing" DLB-105

Dach, Simon 1605-1659..........DLB-164

Daggett, Rollin M. 1831-1901 ....DLB-79

D'Aguiar, Fred 1960- DLB-157

Dahl, Roald 1916-1990 ....DLB-139

Dahlberg, Edward 1900-1977 ....DLB-48

Dahn, Felix 1834-1912 ....DLB-129

Dale, Peter 1938- DLB-40

Daley, Arthur 1904-1974..........DLB-171

Dall, Caroline Wells (Healey)
   1822-1912 ....DLB-1

Dallas, E. S. 1828-1879 ....DLB-55

The Dallas Theater Center DLB-7

D'Alton, Louis 1900-1951 ....DLB-10

Daly, T. A. 1871-1948 ....DLB-11

Damon, S. Foster 1893-1971 ....DLB-45

Damrell, William S.
   [publishing house] DLB-49

Dana, Charles A. 1819-1897 ......DLB-3, 23

Dana, Richard Henry, Jr
   1815-1882....................DLB-1

Dandridge, Ray Garfield DLB-51

Dane, Clemence 1887-1965........DLB-10

Danforth, John 1660-1730 ....DLB-24

Danforth, Samuel, I 1626-1674 ......DLB-24

Danforth, Samuel, II 1666-1727 .....DLB-24

Dangerous Years: London Theater,
   1939-1945 ....DLB-10

Daniel, John M. 1825-1865 ....DLB-43

Daniel, Samuel
  1562 or 1563-1619 .... DLB-62

Daniel Press DLB-106

Daniells, Roy 1902-1979 .... DLB-68

Daniels, Jim 1956- DLB-120

Daniels, Jonathan 1902-1981 ...... DLB-127

Daniels, Josephus 1862-1948 .... DLB-29

Dannay, Frederic 1905-1982 and
  Manfred B. Lee 1905-1971 ..... DLB-137

Danner, Margaret Esse 1915- DLB-41

Danter, John [publishing house] DLB-170

Dantin, Louis 1865-1945 .... DLB-92

Danzig, Allison 1898-1987 .... DLB-171

D'Arcy, Ella circa 1857-1937 ...... DLB-135

Darley, George 1795-1846 .... DLB-96

Darwin, Charles 1809-1882 ..... DLB-57, 166

Darwin, Erasmus 1731-1802 .... DLB-93

Daryush, Elizabeth 1887-1977 ...... DLB-20

Dashkova, Ekaterina Romanovna
  (née Vorontsova) 1743-1810 .... DLB-150

Dashwood, Edmée Elizabeth Monica
  de la Pasture (see Delafield, E. M.)

Daudet, Alphonse 1840-1897 ...... DLB-123

d'Aulaire, Edgar Parin 1898- and
  d'Aulaire, Ingri 1904- DLB-22

Davenant, Sir William
  1606-1668 .... DLB-58, 126

Davenport, Guy 1927- DLB-130

Davenport, Robert ?-? DLB-58

Daves, Delmer 1904-1977 .... DLB-26

Davey, Frank 1940- DLB-53

Davidson, Avram 1923-1993 .... DLB-8

Davidson, Donald 1893-1968 .... DLB-45

Davidson, John 1857-1909 .... DLB-19

Davidson, Lionel 1922- DLB-14

Davie, Donald 1922- DLB-27

Davie, Elspeth 1919- DLB-139

Davies, Sir John 1569-1626 .... DLB-172

Davies, John, of Hereford
  1565?-1618 .... DLB-121

Davies, Rhys 1901-1978 .... DLB-139

Davies, Robertson 1913- DLB-68

Davies, Samuel 1723-1761 .... DLB-31

Davies, Thomas 1712?-1785 ... DLB-142, 154

Davies, W. H. 1871-1940 ..... DLB-19, 174

Davies, Peter, Limited DLB-112

Daviot, Gordon 1896?-1952 .... DLB-10
  (see also Tey, Josephine)

Davis, Charles A. 1795-1867 .... DLB-11

Davis, Clyde Brion 1894-1962 ...... DLB-9

Davis, Dick 1945- DLB-40

Davis, Frank Marshall 1905-? DLB-51

Davis, H. L. 1894-1960 .... DLB-9

Davis, John 1774-1854 .... DLB-37

Davis, Lydia 1947- DLB-130

Davis, Margaret Thomson 1926- ... DLB-14

Davis, Ossie 1917- DLB-7, 38

Davis, Paxton 1925-1994 .... Y-94

Davis, Rebecca Harding
  1831-1910 .... DLB-74

Davis, Richard Harding
  1864-1916 .... DLB-12, 23, 78, 79; DS-13

Davis, Samuel Cole 1764-1809 ...... DLB-37

Davison, Peter 1928- DLB-5

Davys, Mary 1674-1732 .... DLB-39

DAW Books DLB-46

Dawson, Ernest 1882-1947 .... DLB-140

Dawson, Fielding 1930- DLB-130

Dawson, William 1704-1752 .... DLB-31

Day, Angel flourished 1586 .... DLB-167

Day, Benjamin Henry 1810-1889 .... DLB-43

Day, Clarence 1874-1935 .... DLB-11

Day, Dorothy 1897-1980 .... DLB-29

Day, Frank Parker 1881-1950 .... DLB-92

Day, John circa 1574-circa 1640 ..... DLB-62

Day, John [publishing house] DLB-170

Day Lewis, C. 1904-1972 .... DLB-15, 20
  (see also Blake, Nicholas)

Day, Thomas 1748-1789 .... DLB-39

Day, The John, Company DLB-46

Day, Mahlon [publishing house] DLB-49

Deacon, William Arthur
  1890-1977 .... DLB-68

Deal, Borden 1922-1985 .... DLB-6

de Angeli, Marguerite 1889-1987 .... DLB-22

De Angelis, Milo 1951- DLB-128

De Bow, James Dunwoody Brownson
  1820-1867 .... DLB-3, 79

de Bruyn, Günter 1926- DLB-75

de Camp, L. Sprague 1907- DLB-8

The Decay of Lying (1889),
  by Oscar Wilde [excerpt] DLB-18

Dedication, *Ferdinand Count Fathom* (1753),
  by Tobias Smollett DLB-39

Dedication, *The History of Pompey the Little*
  (1751), by Francis Coventry DLB-39

Dedication, *Lasselia* (1723), by Eliza
  Haywood [excerpt] DLB-39

Dedication, *The Wanderer* (1814),
  by Fanny Burney DLB-39

Dee, John 1527-1609 .... DLB-136

Deeping, George Warwick
  1877-1950 .... DLB 153

Defense of *Amelia* (1752), by
  Henry Fielding DLB-39

Defoe, Daniel 1660-1731 .... DLB-39, 95, 101

de Fontaine, Felix Gregory
  1834-1896 .... DLB-43

De Forest, John William
  1826-1906 .... DLB-12

DeFrees, Madeline 1919- DLB-105

DeFrees, Madeline, The Poet's Kaleidoscope:
  The Element of Surprise in the Making
  of the Poem DLB-105

de Graff, Robert 1895-1981 .... Y-81

de Graft, Joe 1924-1978 .... DLB-117

De Heinrico circa 980? DLB-148

Deighton, Len 1929- DLB-87

DeJong, Meindert 1906-1991 .... DLB-52

Dekker, Thomas
  circa 1572-1632 .... DLB-62, 172

Delacorte, Jr., George T.
  1894-1991 .... DLB-91

Delafield, E. M. 1890-1943 .... DLB-34

Delahaye, Guy 1888-1969 .... DLB-92

de la Mare, Walter
  1873-1956 .... DLB-19, 153, 162

Deland, Margaret 1857-1945 .... DLB-78

Delaney, Shelagh 1939- DLB-13

Delany, Martin Robinson
  1812-1885 .... DLB-50

Delany, Samuel R. 1942- DLB-8, 33

de la Roche, Mazo 1879-1961 ...... DLB-68

Delbanco, Nicholas 1942- DLB-6

De León, Nephtal 1945- DLB-82

Delgado, Abelardo Barrientos
  1931- DLB-82

De Libero, Libero 1906-1981 ...... DLB-114

DeLillo, Don 1936- DLB-6, 173

de Lisser H. G. 1878-1944 .... DLB-117

Dell, Floyd 1887-1969 .... DLB-9

Dell Publishing Company DLB-46

delle Grazie, Marie Eugene
    1864-1931 . . . . DLB-81

Deloney, Thomas died 1600 . . . . DLB-167

Deloria, Ella C. 1889-1971 . . . . DLB-175

Deloria, Vine, Jr. 1933- DLB-175

del Rey, Lester 1915-1993 . . . . DLB-8

Del Vecchio, John M. 1947- DS-9

de Man, Paul 1919-1983 . . . . DLB-67

Demby, William 1922- DLB-33

Deming, Philander 1829-1915 . . . . . . DLB-74

Demorest, William Jennings
    1822-1895 . . . . DLB-79

De Morgan, William 1839-1917 . . . . DLB-153

Demosthenes 384 B.C.-322 B.C. DLB-176

Denham, Henry
    [publishing house] DLB-170

Denham, Sir John
    1615-1669 . . . . DLB-58, 126

Denison, Merrill 1893-1975 . . . . DLB-92

Denison, T. S., and Company DLB-49

Dennie, Joseph
    1768-1812 . . . . DLB-37, 43, 59, 73

Dennis, John 1658-1734 . . . . DLB-101

Dennis, Nigel 1912-1989 . . . . DLB-13, 15

Dent, Tom 1932- DLB-38

Dent, J. M., and Sons DLB-112

Denton, Daniel circa 1626-1703 . . . . . DLB-24

DePaola, Tomie 1934- DLB-61

De Quincey, Thomas
    1785-1859 . . . . DLB-110, 144

Derby, George Horatio
    1823-1861 . . . . DLB-11

Derby, J. C., and Company DLB-49

Derby and Miller DLB-49

Derleth, August 1909-1971 . . . . DLB-9

The Derrydale Press DLB-46

Derzhavin, Gavriil Romanovich
    1743-1816 . . . . DLB-150

Desaulniers, Gonsalve
    1863-1934 . . . . DLB-92

Desbiens, Jean-Paul 1927- DLB-53

des Forêts, Louis-Rene 1918- DLB-83

DesRochers, Alfred 1901-1978 . . . . . . DLB-68

Desrosiers, Léo-Paul 1896-1967 . . . . . DLB-68

Dessì, Giuseppe 1909-1977 . . . . DLB-177

Destouches, Louis-Ferdinand
    (see Céline, Louis-Ferdinand)

De Tabley, Lord 1835-1895 . . . . DLB-35

Deutsch, Babette 1895-1982 . . . . DLB-45

Deutsch, André, Limited DLB-112

Deveaux, Alexis 1948- DLB-38

The Development of the Author's Copyright
    in Britain DLB-154

The Development of Lighting in the Staging
    of Drama, 1900-1945 . . . . DLB-10

de Vere, Aubrey 1814-1902 . . . . DLB-35

Devereux, second Earl of Essex, Robert
    1565-1601 . . . . DLB-136

The Devin-Adair Company DLB-46

De Voto, Bernard 1897-1955 . . . . DLB-9

De Vries, Peter 1910-1993 . . . . . . DLB-6; Y-82

Dewdney, Christopher 1951- DLB-60

Dewdney, Selwyn 1909-1979 . . . . DLB-68

DeWitt, Robert M., Publisher DLB-49

DeWolfe, Fiske and Company DLB-49

Dexter, Colin 1930- DLB-87

de Young, M. H. 1849-1925 . . . . DLB-25

Dhlomo, H. I. E. 1903-1956 . . . . . . DLB-157

Dhuoda circa 803-after 843 . . . . DLB-148

The Dial Press DLB-46

Diamond, I. A. L. 1920-1988 . . . . . . DLB-26

Di Cicco, Pier Giorgio 1949- DLB-60

Dick, Philip K. 1928-1982 . . . . DLB-8

Dick and Fitzgerald DLB-49

Dickens, Charles
    1812-1870 . . . . . DLB-21, 55, 70, 159, 166

Dickinson, Peter 1927- DLB-161

Dickey, James
    1923-1997 . . . . DLB-5; Y-82, 93; DS-7

James Dickey, American Poet Y-96

Dickey, William 1928-1994 . . . . DLB-5

Dickinson, Emily 1830-1886 . . . . DLB-1

Dickinson, John 1732-1808 . . . . DLB-31

Dickinson, Jonathan 1688-1747 . . . . . DLB-24

Dickinson, Patric 1914- DLB-27

Dickinson, Peter 1927- DLB-87

Dicks, John [publishing house] DLB-106

Dickson, Gordon R. 1923- DLB-8

*Dictionary of Literary Biography
    Yearbook* Awards Y-92, 93

The Dictionary of National Biography
    . . . . DLB-144

Didion, Joan 1934- . . . DLB-2, 173; Y-81, 86

Di Donato, Pietro 1911- DLB-9

Die Fürstliche Bibliothek Corvey Y-96

Diego, Gerardo 1896-1987 . . . . DLB-134

Digges, Thomas circa 1546-1595 . . . DLB-136

Dillard, Annie 1945- Y-80

Dillard, R. H. W. 1937- DLB-5

Dillingham, Charles T.,
    Company DLB-49

The Dillingham, G. W.,
    Company DLB-49

Dilly, Edward and Charles
    [publishing house] DLB-154

Dilthey, Wilhelm 1833-1911 . . . . DLB-129

Dingelstedt, Franz von
    1814-1881 . . . . DLB-133

Dintenfass, Mark 1941- Y-84

Diogenes, Jr. (see Brougham, John)

Diogenes Laertius circa 200 . . . . DLB-176

DiPrima, Diane 1934- DLB-5, 16

Disch, Thomas M. 1940- DLB-8

Disney, Walt 1901-1966 . . . . DLB-22

Disraeli, Benjamin 1804-1881 . . . . DLB-21, 55

D'Israeli, Isaac 1766-1848 . . . . DLB-107

Ditzen, Rudolf (see Fallada, Hans)

Dix, Dorothea Lynde 1802-1887 . . . . . DLB-1

Dix, Dorothy (see Gilmer,
    Elizabeth Meriwether)

Dix, Edwards and Company DLB-49

Dixie, Florence Douglas
    1857-1905 . . . . DLB-174

Dixon, Paige (see Corcoran, Barbara)

Dixon, Richard Watson
    1833-1900 . . . . DLB-19

Dixon, Stephen 1936- DLB-130

Dmitriev, Ivan Ivanovich
    1760-1837 . . . . DLB-150

Dobell, Sydney 1824-1874 . . . . DLB-32

Döblin, Alfred 1878-1957 . . . . DLB-66

Dobson, Austin
    1840-1921 . . . . DLB-35, 144

Doctorow, E. L.
    1931- DLB-2, 28, 173; Y-80

Documents on Sixteenth-Century
    Literature DLB-167, 172

Dodd, William E. 1869-1940 . . . . DLB-17

Dodd, Anne [publishing house] DLB-154

Dodd, Mead and Company DLB-49

Doderer, Heimito von 1896-1968 . . . . DLB-85

Dodge, Mary Mapes
    1831?-1905 . . . . DLB-42, 79; DS-13

Dodge, B. W., and Company DLB-46

Dodge Publishing Company DLB-49

Dodgson, Charles Lutwidge
  (see Carroll, Lewis)

Dodsley, Robert 1703-1764 .... DLB-95

Dodsley, R. [publishing house] DLB-154

Dodson, Owen 1914-1983 .... DLB-76

Doesticks, Q. K. Philander, P. B.
  (see Thomson, Mortimer)

Doheny, Carrie Estelle
  1875-1958 .... DLB-140

Domínguez, Sylvia Maida
  1935- DLB-122

Donahoe, Patrick
  [publishing house] DLB-49

Donald, David H. 1920- DLB-17

Donaldson, Scott 1928- DLB-111

Doni, Rodolfo 1919- DLB-177

Donleavy, J. P. 1926- DLB-6, 173

Donnadieu, Marguerite (see Duras, Marguerite)

Donne, John 1572-1631 .... DLB-121, 151

Donnelley, R. R., and Sons
  Company DLB-49

Donnelly, Ignatius 1831-1901 .... DLB-12

Donohue and Henneberry DLB-49

Donoso, José 1924- DLB-113

Doolady, M. [publishing house] DLB-49

Dooley, Ebon (see Ebon)

Doolittle, Hilda 1886-1961 .... DLB-4, 45

Doplicher, Fabio 1938- DLB-128

Dor, Milo 1923- DLB-85

Doran, George H., Company DLB-46

Dorgelès, Roland 1886-1973 .... DLB-65

Dorn, Edward 1929- DLB-5

Dorr, Rheta Childe 1866-1948 ..... DLB-25

Dorris, Michael 1945- DLB-175

Dorset and Middlesex, Charles Sackville,
  Lord Buckhurst,
  Earl of 1643-1706 .... DLB-131

Dorst, Tankred 1925- DLB-75, 124

Dos Passos, John
  1896-1970 .... DLB-4, 9; DS-1

John Dos Passos: A Centennial
  Commemoration Y-96

Doubleday and Company DLB-49

Dougall, Lily 1858-1923 .... DLB-92

Doughty, Charles M.
  1843-1926 .... DLB-19, 57, 174

Douglas, Gavin 1476-1522 .... DLB-132

Douglas, Keith 1920-1944 .... DLB-27

Douglas, Norman 1868-1952 .... DLB-34

Douglass, Frederick
  1817?-1895 .... DLB-1, 43, 50, 79

Douglass, William circa
  1691-1752 .... DLB-24

Dourado, Autran 1926- DLB-145

Dove, Rita 1952- DLB-120

Dover Publications DLB-46

Doves Press DLB-112

Dowden, Edward 1843-1913 .... DLB-35, 149

Dowell, Coleman 1925-1985 ...... DLB-130

Dowland, John 1563-1626 .... DLB-172

Downes, Gwladys 1915- DLB-88

Downing, J., Major (see Davis, Charles A.)

Downing, Major Jack (see Smith, Seba)

Dowriche, Anne
  before 1560-after 1613 .... DLB-172

Dowson, Ernest 1867-1900 ..... DLB-19, 135

Doxey, William
  [publishing house] DLB-49

Doyle, Sir Arthur Conan
  1859-1930 .... DLB-18, 70, 156

Doyle, Kirby 1932- DLB-16

Drabble, Margaret 1939- DLB-14, 155

Drach, Albert 1902- DLB-85

The Dramatic Publishing
  Company DLB-49

Dramatists Play Service DLB-46

Drant, Thomas
  early 1540s?-1578 .... DLB-167

Draper, John W. 1811-1882 .... DLB-30

Draper, Lyman C. 1815-1891 .... DLB-30

Drayton, Michael 1563-1631 .... DLB-121

Dreiser, Theodore
  1871-1945 .... DLB-9, 12, 102, 137; DS-1

Drewitz, Ingeborg 1923-1986 .... DLB-75

Drieu La Rochelle, Pierre
  1893-1945 .... DLB-72

Drinkwater, John 1882-1937
  .... DLB-10, 19, 149

Droste-Hülshoff, Annette von
  1797-1848 .... DLB-133

The Drue Heinz Literature Prize
  Excerpt from "Excerpts from a Report
  of the Commission," in David
  Bosworth's *The Death of Descartes*
  An Interview with David
  Bosworth Y-82

Drummond, William Henry
  1854-1907 .... DLB-92

Drummond, William, of Hawthornden
  1585-1649 .... DLB-121

Dryden, Charles 1860?-1931 .... DLB-171

Dryden, John 1631-1700 ... DLB-80, 101, 131

Drží, Marin circa 1508-1567 .... DLB-147

Duane, William 1760-1835 .... DLB-43

Dubé, Marcel 1930- DLB-53

Dubé, Rodolphe (see Hertel, François)

Dubie, Norman 1945- DLB-120

Du Bois, W. E. B.
  1868-1963 .... DLB-47, 50, 91

Du Bois, William Pène 1916- DLB-61

Dubus, Andre 1936- DLB-130

Ducharme, Réjean 1941- DLB-60

Dučić, Jovan 1871-1943 .... DLB-147

Duck, Stephen 1705?-1756 .... DLB-95

Duckworth, Gerald, and
  Company Limited DLB-112

Dudek, Louis 1918- DLB-88

Duell, Sloan and Pearce DLB-46

Duff Gordon, Lucie 1821-1869 ..... DLB-166

Duffield and Green DLB-46

Duffy, Maureen 1933- DLB-14

Dugan, Alan 1923- DLB-5

Dugard, William
  [publishing house] DLB-170

Dugas, Marcel 1883-1947 .... DLB-92

Dugdale, William
  [publishing house] DLB-106

Duhamel, Georges 1884-1966 .... DLB-65

Dujardin, Edouard 1861-1949 ...... DLB-123

Dukes, Ashley 1885-1959 .... DLB-10

Du Maurier, George 1834-1896 .... DLB-153

Dumas, Alexandre, père
  1802-1870 .... DLB-119

Dumas, Henry 1934-1968 .... DLB-41

Dunbar, Paul Laurence
  1872-1906 .... DLB-50, 54, 78

Dunbar, William
  circa 1460-circa 1522 ...... DLB-132, 146

Duncan, Norman 1871-1916 .... DLB-92

Duncan, Quince 1940- DLB-145

Duncan, Robert 1919-1988 .... DLB-5, 16

Duncan, Ronald 1914-1982 .... DLB-13

Duncan, Sara Jeannette
  1861-1922 .... DLB-92

Dunigan, Edward, and Brother DLB-49

Dunlap, John 1747-1812 .... DLB-43

Dunlap, William 1766-1839 .... DLB-30, 37, 59

Dunn, Douglas 1942- DLB-40

Dunn, Stephen 1939- DLB-105

Dunn, Stephen, The Good, The Not So Good............ DLB-105

Dunne, Finley Peter 1867-1936 .... DLB-11, 23

Dunne, John Gregory 1932- Y-80

Dunne, Philip 1908-1992 .... DLB-26

Dunning, Ralph Cheever 1878-1930 .... DLB-4

Dunning, William A. 1857-1922..... DLB-17

Duns Scotus, John circa 1266-1308............... DLB-115

Dunsany, Lord (Edward John Moreton Drax Plunkett, Baron Dunsany) 1878-1957......... DLB-10, 77, 153, 156

Dunton, John [publishing house] ... DLB-170

Dupin, Amantine-Aurore-Lucile (see Sand, George)

Durand, Lucile (see Bersianik, Louky)

Duranty, Walter 1884-1957......... DLB-29

Duras, Marguerite 1914- DLB-83

Durfey, Thomas 1653-1723......... DLB-80

Durrell, Lawrence 1912-1990........... DLB-15, 27; Y-90

Durrell, William [publishing house] DLB-49

Dürrenmatt, Friedrich 1921-1990 .... DLB-69, 124

Dutton, E. P., and Company........ DLB-49

Duvoisin, Roger 1904-1980......... DLB-61

Duyckinck, Evert Augustus 1816-1878 .... DLB-3, 64

Duyckinck, George L. 1823-1863 ..... DLB-3

Duyckinck and Company DLB-49

Dwight, John Sullivan 1813-1893 ..... DLB-1

Dwight, Timothy 1752-1817 .... DLB-37

Dybek, Stuart 1942- DLB-130

Dyer, Charles 1928- DLB-13

Dyer, George 1755-1841 .... DLB-93

Dyer, John 1699-1757............... DLB-95

Dyer, Sir Edward 1543-1607....... DLB-136

Dylan, Bob 1941- DLB-16

E

Eager, Edward 1911-1964 .... DLB-22

Eames, Wilberforce 1855-1937..... DLB-140

Earle, James H., and Company...... DLB-49

Earle, John 1600 or 1601-1665..... DLB-151

Early American Book Illustration, by Sinclair Hamilton DLB-49

Eastlake, William 1917- DLB-6

Eastman, Carol ?- DLB-44

Eastman, Charles A. (Ohiyesa) 1858-1939 .... DLB-175

Eastman, Max 1883-1969........... DLB-91

Eaton, Daniel Isaac 1753-1814 ..... DLB-158

Eberhart, Richard 1904- DLB-48

Ebner, Jeannie 1918- DLB-85

Ebner-Eschenbach, Marie von 1830-1916 .... DLB-81

Ebon 1942- DLB-41

Ecbasis Captivi circa 1045 .... DLB-148

Ecco Press DLB-46

Eckhart, Meister circa 1260-circa 1328 .... DLB-115

The Eclectic Review 1805-1868....... DLB-110

Edel, Leon 1907- DLB-103

Edes, Benjamin 1732-1803.......... DLB-43

Edgar, David 1948- DLB-13

Edgeworth, Maria 1768-1849.......... DLB-116, 159, 163

The Edinburgh Review 1802-1929 .... DLB-110

Edinburgh University Press DLB-112

The Editor Publishing Company DLB-49

Editorial Statements DLB-137

Edmonds, Randolph 1900- DLB-51

Edmonds, Walter D. 1903- DLB-9

Edschmid, Kasimir 1890-1966....... DLB-56

Edwards, Amelia Anne Blandford 1831-1892 .... DLB-174

Edwards, Jonathan 1703-1758....... DLB-24

Edwards, Jonathan, Jr. 1745-1801.... DLB-37

Edwards, Junius 1929- DLB-33

Edwards, Matilda Barbara Betham- 1836-1919 .... DLB-174

Edwards, Richard 1524-1566........ DLB-62

Edwards, James [publishing house] DLB-154

Effinger, George Alec 1947- DLB-8

Egerton, George 1859-1945........ DLB-135

Eggleston, Edward 1837-1902........ DLB-12

Eggleston, Wilfred 1901-1986....... DLB-92

Ehrenstein, Albert 1886-1950 .... DLB-81

Ehrhart, W. D. 1948- DS-9

Eich, Günter 1907-1972........ DLB-69, 124

Eichendorff, Joseph Freiherr von 1788-1857 .... DLB-90

1873 Publishers' Catalogues DLB-49

Eighteenth-Century Aesthetic Theories DLB-31

Eighteenth-Century Philosophical Background.................... DLB-31

Eigner, Larry 1927- DLB-5

Eikon Basilike 1649................. DLB-151

Eilhart von Oberge circa 1140-circa 1195 .... DLB-148

Einhard circa 770-840 .... DLB-148

Eisenreich, Herbert 1925-1986 ...... DLB-85

Eisner, Kurt 1867-1919 .... DLB-66

Eklund, Gordon 1945- Y-83

Ekwensi, Cyprian 1921- DLB-117

Eld, George [publishing house] DLB-170

Elder, Lonne III 1931- DLB-7, 38, 44

Elder, Paul, and Company DLB-49

Elements of Rhetoric (1828; revised, 1846), by Richard Whately [excerpt].... DLB-57

Elie, Robert 1915-1973............. DLB-88

Elin Pelin 1877-1949............... DLB-147

Eliot, George 1819-1880 ..... DLB-21, 35, 55

Eliot, John 1604-1690............. DLB-24

Eliot, T. S. 1888-1965 ..... DLB-7, 10, 45, 63

Eliot's Court Press DLB-170

Elizabeth I 1533-1603.............. DLB-136

Elizondo, Salvador 1932- DLB-145

Elizondo, Sergio 1930- DLB-82

Elkin, Stanley 1930- DLB-2, 28; Y-80

Elles, Dora Amy (see Wentworth, Patricia)

Ellet, Elizabeth F. 1818?-1877....... DLB-30

Elliot, Ebenezer 1781-1849 .... DLB-96

Elliot, Frances Minto (Dickinson) 1820-1898 .... DLB-166

Elliott, George 1923- DLB-68

Elliott, Janice 1931- DLB-14

Elliott, William 1788-1863........... DLB-3

Elliott, Thomes and Talbot......... DLB-49

Ellis, Edward S. 1840-1916 .... DLB-42

Ellis, Frederick Staridge [publishing house] DLB-106

The George H. Ellis Company DLB-49

Ellison, Harlan 1934- DLB-8

Ellison, Ralph Waldo
  1914-1994 .... DLB-2, 76; Y-94

Ellmann, Richard
  1918-1987 .... DLB-103; Y-87

The Elmer Holmes Bobst Awards in Arts
  and Letters Y-87

Elyot, Thomas 1490?-1546 .... DLB-136

Emanuel, James Andrew 1921- DLB-41

Emecheta, Buchi 1944- DLB-117

The Emergence of Black Women
  Writers DS-8

Emerson, Ralph Waldo
  1803-1882 .... DLB-1, 59, 73

Emerson, William 1769-1811 .... DLB-37

Emin, Fedor Aleksandrovich
  circa 1735-1770 .... DLB-150

Empedocles fifth century B.C. DLB-176

Empson, William 1906-1984 .... DLB-20

The End of English Stage Censorship,
  1945-1968 .... DLB-13

Ende, Michael 1929- DLB-75

Engel, Marian 1933-1985 .... DLB-53

Engels, Friedrich 1820-1895 .... DLB-129

Engle, Paul 1908- DLB-48

English Composition and Rhetoric (1866),
  by Alexander Bain [excerpt] DLB-57

The English Language:
  410 to 1500 .... DLB-146

The English Renaissance of Art (1908),
  by Oscar Wilde DLB-35

Enright, D. J. 1920- DLB-27

Enright, Elizabeth 1909-1968 .... DLB-22

L'Envoi (1882), by Oscar Wilde..... DLB-35

Epictetus circa 55-circa 125-130 .... DLB-176

Epicurus 342/341 B.C.-271/270 B.C.
  .... DLB-176

Epps, Bernard 1936- DLB-53

Epstein, Julius 1909- and
  Epstein, Philip 1909-1952 .... DLB-26

Equiano, Olaudah
  circa 1745-1797 .... DLB-37, 50

Eragny Press DLB-112

Erasmus, Desiderius 1467-1536 .... DLB-136

Erba, Luciano 1922- DLB-128

Erdrich, Louise 1954- DLB-152, 178

Erichsen-Brown, Gwethalyn Graham
  (see Graham, Gwethalyn)

Eriugena, John Scottus
  circa 810-877 .... DLB-115

Ernest Hemingway's Toronto Journalism
  Revisited: With Three Previously
  Unrecorded Stories Y-92

Ernst, Paul 1866-1933 .... DLB-66, 118

Erskine, Albert 1911-1993 .... Y-93

Erskine, John 1879-1951 .... DLB-9, 102

Ervine, St. John Greer 1883-1971 .... DLB-10

Eschenburg, Johann Joachim
  1743-1820 .... DLB-97

Escoto, Julio 1944- DLB-145

Eshleman, Clayton 1935- DLB-5

Espriu, Salvador 1913-1985 .... DLB-134

Ess Ess Publishing Company DLB-49

Essay on Chatterton (1842), by
  Robert Browning DLB-32

Essex House Press DLB-112

Estes, Eleanor 1906-1988 .... DLB-22

Estes and Lauriat DLB-49

Etherege, George 1636-circa 1692 ... DLB-80

Ethridge, Mark, Sr. 1896-1981 ..... DLB-127

Ets, Marie Hall 1893- DLB-22

Etter, David 1928- DLB-105

Ettner, Johann Christoph
  1654-1724 .... DLB-168

Eudora Welty: Eye of the Storyteller ... Y-87

Eugene O'Neill Memorial Theater
  Center DLB-7

Eugene O'Neill's Letters: A Review Y-88

Eupolemius
  flourished circa 1095 .... DLB-148

Euripides circa 484 B.C.-407/406 B.C.
  .... DLB-176

Evans, Caradoc 1878-1945 .... DLB-162

Evans, Donald 1884-1921 .... DLB-54

Evans, George Henry 1805-1856 .... DLB-43

Evans, Hubert 1892-1986 .... DLB-92

Evans, Mari 1923- DLB-41

Evans, Mary Ann (see Eliot, George)

Evans, Nathaniel 1742-1767 .... DLB-31

Evans, Sebastian 1830-1909 .... DLB-35

Evans, M., and Company DLB-46

Everett, Alexander Hill
  790-1847 .... DLB-59

Everett, Edward 1794-1865 .... DLB-1, 59

Everson, R. G. 1903- DLB-88

Everson, William 1912-1994 ..... DLB-5, 16

Every Man His Own Poet; or, The
  Inspired Singer's Recipe Book (1877),
  by W. H. Mallock DLB-35

Ewart, Gavin 1916- DLB-40

Ewing, Juliana Horatia
  1841-1885 .... DLB-21, 163

The Examiner 1808-1881 .... DLB-110

Exley, Frederick
  1929-1992 .... DLB-143; Y-81

Experiment in the Novel (1929),
  by John D. Beresford........... DLB-36

Eyre and Spottiswoode DLB-106

Ezzo ?-after 1065................. DLB-148

F

"F. Scott Fitzgerald: St. Paul's Native Son
  and Distinguished American Writer":
  University of Minnesota Conference,
  29-31 October 1982............... Y-82

Faber, Frederick William
  1814-1863 .... DLB-32

Faber and Faber Limited DLB-112

Faccio, Rena (see Aleramo, Sibilla)

Fagundo, Ana María 1938- DLB-134

Fair, Ronald L. 1932- DLB-33

Fairfax, Beatrice (see Manning, Marie)

Fairlie, Gerard 1899-1983 .... DLB-77

Fallada, Hans 1893-1947 .... DLB-56

Falsifying Hemingway................ Y-96

Fancher, Betsy 1928- Y-83

Fane, Violet 1843-1905 .... DLB-35

Fanfrolico Press.................. DLB-112

Fanning, Katherine 1927 .... DLB-127

Fanshawe, Sir Richard
  1608-1666 .... DLB-126

Fantasy Press Publishers DLB-46

Fante, John 1909-1983 .... DLB-130; Y-83

Al-Farabi circa 870-950 .... DLB-115

Farah, Nuruddin 1945- DLB-125

Farber, Norma 1909-1984 .... DLB-61

Farigoule, Louis (see Romains, Jules)

Farjeon, Eleanor 1881-1965........ DLB-160

Farley, Walter 1920-1989 .... DLB-22

Farmer, Penelope 1939- DLB-161

Farmer, Philip José 1918- DLB-8

Farquhar, George circa 1677-1707 ... DLB-84

Farquharson, Martha (see Finley, Martha)

325

Farrar, Frederic William 1831-1903 .... DLB-163

Farrar and Rinehart DLB-46

Farrar, Straus and Giroux DLB-46

Farrell, James T. 1904-1979 .... DLB-4, 9, 86; DS-2

Farrell, J. G. 1935-1979 .... DLB-14

Fast, Howard 1914- DLB-9

Faulkner, William 1897-1962 .... DLB-9, 11, 44, 102; DS-2; Y-86

Faulkner, George [publishing house] DLB-154

Fauset, Jessie Redmon 1882-1961 .... DLB-51

Faust, Irvin 1924- DLB-2, 28; Y-80

Fawcett Books DLB-46

Fearing, Kenneth 1902-1961 .... DLB-9

Federal Writers' Project........... DLB-46

Federman, Raymond 1928- Y-80

Feiffer, Jules 1929- DLB-7, 44

Feinberg, Charles E. 1899-1988 .... Y-88

Feind, Barthold 1678-1721......... DLB-168

Feinstein, Elaine 1930- DLB-14, 40

Feldman, Irving 1928- DLB-169

Felipe, Léon 1884-1968 .... DLB-108

Fell, Frederick, Publishers DLB-46

Felltham, Owen 1602?-1668 ... DLB-126, 151

Fels, Ludwig 1946- DLB-75

Felton, Cornelius Conway 1807-1862 .... DLB-1

Fennario, David 1947- DLB-60

Fenno, John 1751-1798............. DLB-43

Fenno, R. F., and Company DLB-49

Fenoglio, Beppe 1922-1963 .... DLB-177

Fenton, Geoffrey 1539?-1608 ...... DLB-136

Fenton, James 1949- DLB-40

Ferber, Edna 1885-1968 ...... DLB-9, 28, 86

Ferdinand, Vallery III (see Salaam, Kalamu ya)

Ferguson, Sir Samuel 1810-1886..... DLB-32

Ferguson, William Scott 1875-1954 .... DLB-47

Fergusson, Robert 1750-1774 ..... DLB-109

Ferland, Albert 1872-1943 .... DLB-92

Ferlinghetti, Lawrence 1919- DLB-5, 16

Fern, Fanny (see Parton, Sara Payson Willis)

Ferrars, Elizabeth 1907- DLB-87

Ferré, Rosario 1942- DLB-145

Ferret, E., and Company DLB-49

Ferrier, Susan 1782-1854 .... DLB-116

Ferrini, Vincent 1913- DLB-48

Ferron, Jacques 1921-1985......... DLB-60

Ferron, Madeleine 1922- DLB-53

Fetridge and Company............ DLB-49

Feuchtersleben, Ernst Freiherr von 1806-1849 .... DLB-133

Feuchtwanger, Lion 1884-1958 ...... DLB-66

Feuerbach, Ludwig 1804-1872 ..... DLB-133

Fichte, Johann Gottlieb 1762-1814 .... DLB-90

Ficke, Arthur Davison 1883-1945.... DLB-54

Fiction Best-Sellers, 1910-1945 .... DLB-9

Fiction into Film, 1928-1975: A List of Movies Based on the Works of Authors in *British Novelists,* 1930-1959 .... DLB-15

Fiedler, Leslie A. 1917- DLB-28, 67

Field, Edward 1924- DLB-105

Field, Edward, The Poetry File..... DLB-105

Field, Eugene 1850-1895 ...... DLB-23, 42, 140; DS-13

Field, John 1545?-1588............ DLB-167

Field, Marshall, III 1893-1956...... DLB-127

Field, Marshall, IV 1916-1965...... DLB-127

Field, Marshall, V 1941- DLB-127

Field, Nathan 1587-1619 or 1620 .... DLB-58

Field, Rachel 1894-1942.......... DLB-9, 22

A Field Guide to Recent Schools of American Poetry Y-86

Fielding, Henry 1707-1754............. DLB-39, 84, 101

Fielding, Sarah 1710-1768 .... DLB-39

Fields, James Thomas 1817-1881 ..... DLB-1

Fields, Julia 1938- DLB-41

Fields, W. C. 1880-1946 .... DLB-44

Fields, Osgood and Company DLB-49

Fifty Penguin Years Y-85

Figes, Eva 1932- DLB-14

Figuera, Angela 1902-1984 .... DLB-108

Filmer, Sir Robert 1586-1653 ..... DLB-151

Filson, John circa 1753-1788 .... DLB-37

Finch, Anne, Countess of Winchilsea 1661-1720 .... DLB-95

Finch, Robert 1900- DLB-88

Findley, Timothy 1930- DLB-53

Finlay, Ian Hamilton 1925- DLB-40

Finley, Martha 1828-1909 .... DLB-42

Finn, Elizabeth Anne (McCaul) 1825-1921 .... DLB-166

Finney, Jack 1911- DLB-8

Finney, Walter Braden (see Finney, Jack)

Firbank, Ronald 1886-1926 .... DLB-36

Firmin, Giles 1615-1697............ DLB-24

First Edition Library/Collectors' Reprints, Inc..................... Y-91

First International F. Scott Fitzgerald Conference..................... Y-92

First Strauss "Livings" Awarded to Cynthia Ozick and Raymond Carver An Interview with Cynthia Ozick An Interview with Raymond Carver........................ Y-83

Fischer, Karoline Auguste Fernandine 1764-1842 .... DLB-94

Fish, Stanley 1938- DLB-67

Fishacre, Richard 1205-1248 .... DLB-115

Fisher, Clay (see Allen, Henry W.)

Fisher, Dorothy Canfield 1879-1958 .... DLB-9, 102

Fisher, Leonard Everett 1924- DLB-61

Fisher, Roy 1930- DLB-40

Fisher, Rudolph 1897-1934 ..... DLB-51, 102

Fisher, Sydney George 1856-1927 ... DLB-47

Fisher, Vardis 1895-1968............ DLB-9

Fiske, John 1608-1677 .... DLB-24

Fiske, John 1842-1901 .... DLB-47, 64

Fitch, Thomas circa 1700-1774 ...... DLB-31

Fitch, William Clyde 1865-1909 ...... DLB-7

FitzGerald, Edward 1809-1883 ...... DLB-32

Fitzgerald, F. Scott 1896-1940 ..... DLB-4, 9, 86; Y-81; DS-1

F. Scott Fitzgerald Centenary Celebrations Y-96

Fitzgerald, Penelope 1916- DLB-14

Fitzgerald, Robert 1910-1985 .... Y-80

Fitzgerald, Thomas 1819-1891 ...... DLB-23

Fitzgerald, Zelda Sayre 1900-1948 ...... Y-84

Fitzhugh, Louise 1928-1974......... DLB-52

Fitzhugh, William circa 1651-1701................. DLB-24

Flanagan, Thomas 1923- Y-80

Flanner, Hildegarde 1899-1987 ...... DLB-48

Flanner, Janet 1892-1978 .... DLB-4

Flaubert, Gustave 1821-1880....... DLB-119

Flavin, Martin 1883-1967.......... DLB-9

Fleck, Konrad (flourished circa 1220) DLB-138

Flecker, James Elroy 1884-1915 .. DLB-10, 19

Fleeson, Doris 1901-1970 .... DLB-29

Fleißer, Marieluise 1901-1974 ... DLB-56, 124

Fleming, Ian 1908-1964 .... DLB-87

Fleming, Paul 1609-1640 .... DLB-164

The Fleshly School of Poetry and Other Phenomena of the Day (1872), by Robert Buchanan DLB-35

The Fleshly School of Poetry: Mr. D. G. Rossetti (1871), by Thomas Maitland (Robert Buchanan) DLB-35

Fletcher, Giles, the Elder 1546-1611 .... DLB-136

Fletcher, Giles, the Younger 1585 or 1586-1623 .... DLB-121

Fletcher, J. S. 1863-1935 .... DLB-70

Fletcher, John (see Beaumont, Francis)

Fletcher, John Gould 1886-1950 ... DLB-4, 45

Fletcher, Phineas 1582-1650 .... DLB-121

Flieg, Helmut (see Heym, Stefan)

Flint, F. S. 1885-1960 .... DLB-19

Flint, Timothy 1780-1840 .... DLB-734

Florio, John 1553?-1625 .... DLB-172

Foix, J. V. 1893-1987 .... DLB-134

Foley, Martha (see Burnett, Whit, and Martha Foley)

Folger, Henry Clay 1857-1930 ..... DLB-140

Folio Society DLB-112

Follen, Eliza Lee (Cabot) 1787-1860 ... DLB-1

Follett, Ken 1949- Y-81, DLB-87

Follett Publishing Company DLB-46

Folsom, John West [publishing house] DLB-49

Fontane, Theodor 1819-1898 ...... DLB-129

Fonvisin, Denis Ivanovich 1744 or 1745-1792 .... DLB-150

Foote, Horton 1916- DLB-26

Foote, Samuel 1721-1777 .... DLB-89

Foote, Shelby 1916- DLB-2, 17

Forbes, Calvin 1945- DLB-41

Forbes, Ester 1891-1967 .... DLB-22

Forbes and Company DLB-49

Force, Peter 1790-1868 .... DLB-30

Forché, Carolyn 1950- DLB-5

Ford, Charles Henri 1913- DLB-4, 48

Ford, Corey 1902-1969 .... DLB-11

Ford, Ford Madox 1873-1939 .... DLB-34, 98, 162

Ford, Jesse Hill 1928- DLB-6

Ford, John 1586-? DLB-58

Ford, R. A. D. 1915- DLB-88

Ford, Worthington C. 1858-1941 .... DLB-47

Ford, J. B., and Company DLB-49

Fords, Howard, and Hulbert DLB-49

Foreman, Carl 1914-1984 .... DLB-26

Forester, Frank (see Herbert, Henry William)

Fornés, María Irene 1930- DLB-7

Forrest, Leon 1937- DLB-33

Forster, E. M. 1879-1970 ...... DLB-34, 98, 162; DS-10

Forster, Georg 1754-1794 .... DLB-94

Forster, John 1812-1876 .... DLB-144

Forster, Margaret 1938- DLB-155

Forsyth, Frederick 1938- DLB-87

Forten, Charlotte L. 1837-1914 ..... DLB-50

Fortini, Franco 1917- DLB-128

Fortune, T. Thomas 1856-1928 ..... DLB-23

Fosdick, Charles Austin 1842-1915 .... DLB-42

Foster, Genevieve 1893-1979 .... DLB-61

Foster, Hannah Webster 1758-1840 .... DLB-37

Foster, John 1648-1681 .... DLB-24

Foster, Michael 1904-1956 .... DLB-9

Foulis, Robert and Andrew / R. and A. [publishing house] DLB-154

Fouqué, Caroline de la Motte 1774-1831 .... DLB-90

Fouqué, Friedrich de la Motte 1777-1843 .... DLB-90

Four Essays on the Beat Generation, by John Clellon Holmes DLB-16

Four Seas Company DLB-46

Four Winds Press DLB-46

Fournier, Henri Alban (see Alain-Fournier)

Fowler and Wells Company DLB-49

Fowles, John 1926- DLB-14, 139

Fox, John, Jr. 1862 or 1863-1919 .... DLB-9; DS-13

Fox, Paula 1923- DLB-52

Fox, Richard Kyle 1846-1922 .... DLB-79

Fox, William Price 1926- DLB-2; Y-81

Fox, Richard K. [publishing house] DLB-49

Foxe, John 1517-1587 .... DLB-132

Fraenkel, Michael 1896-1957 .... DLB-4

France, Anatole 1844-1924 .... DLB-123

France, Richard 1938- DLB-7

Francis, Convers 1795-1863 .... DLB-1

Francis, Dick 1920- DLB-87

Francis, Jeffrey, Lord 1773-1850 .... DLB-107

Francis, C. S. [publishing house] DLB-49

François 1863-1910 .... DLB-92

François, Louise von 1817-1893 .... DLB-129

Francke, Kuno 1855-1930 .... DLB-71

Frank, Bruno 1887-1945 .... DLB-118

Frank, Leonhard 1882-1961 .... DLB-56, 118

Frank, Melvin (see Panama, Norman)

Frank, Waldo 1889-1967 .... DLB-9, 63

Franken, Rose 1895?-1988 .... Y-84

Franklin, Benjamin 1706-1790 .... DLB-24, 43, 73

Franklin, James 1697-1735 .... DLB-43

Franklin Library DLB-46

Frantz, Ralph Jules 1902-1979 .... DLB-4

Franzos, Karl Emil 1848-1904 ...... DLB-129

Fraser, G. S. 1915-1980 .... DLB-27

Fraser, Kathleen 1935- DLB-169

Frattini, Alberto 1922- DLB-128

Frau Ava ?-1127 .... DLB-148

Frayn, Michael 1933- DLB-13, 14

Frederic, Harold 1856-1898 .... DLB-12, 23; DS-13

Freeling, Nicolas 1927- DLB-87

Freeman, Douglas Southall 1886-1953 .... DLB-17

Freeman, Legh Richmond 1842-1915 .... DLB-23

Freeman, Mary E. Wilkins 1852-1930 .... DLB-12, 78

Freeman, R. Austin 1862-1943 ...... DLB-70

Freidank circa 1170-circa 1233 ..... DLB-138

Freiligrath, Ferdinand 1810-1876 ... DLB-133

French, Alice 1850-1934 ..... DLB-74; DS-13

French, David 1939- DLB-53

French, James [publishing house] DLB-49

French, Samuel [publishing house] ... DLB-49

Samuel French, Limited DLB-106

Freneau, Philip 1752-1832 ...... DLB-37, 43

Freni, Melo 1934- DLB-128

Freshfield, Douglas W. 1845-1934 ....DLB-174

Freytag, Gustav 1816-1895 ....DLB-129

Fried, Erich 1921-1988.............DLB-85

Friedman, Bruce Jay 1930- DLB-2, 28

Friedrich von Hausen circa 1171-1190...............DLB-138

Friel, Brian 1929- DLB-13

Friend, Krebs 1895?-1967? DLB-4

Fries, Fritz Rudolf 1935- DLB-75

Fringe and Alternative Theater in Great Britain................DLB-13

Frisch, Max 1911-1991 ....DLB-69, 124

Frischmuth, Barbara 1941- DLB-85

Fritz, Jean 1915- DLB-52

Fromentin, Eugene 1820-1876......DLB-123

From The Gay Science, by E. S. Dallas....................DLB-21

Frost, A. B. 1851-1928...............DS-13

Frost, Robert 1874-1963 ......DLB-54; DS-7

Frothingham, Octavius Brooks 1822-1895 ....DLB-1

Froude, James Anthony 1818-1894.............DLB-18, 57, 144

Fry, Christopher 1907- DLB-13

Fry, Roger 1866-1934 ....DS-10

Frye, Northrop 1912-1991.......DLB-67, 68

Fuchs, Daniel 1909-1993..........DLB-9, 26, 28; Y-93

Fuentes, Carlos 1928- DLB-113

Fuertes, Gloria 1918- DLB-108

The Fugitives and the Agrarians: The First Exhibition Y-85

Fulbecke, William 1560-1603? DLB-172

Fuller, Charles H., Jr. 1939- DLB-38

Fuller, Henry Blake 1857-1929 .....DLB-12

Fuller, John 1937- DLB-40

Fuller, Roy 1912-1991 ....DLB-15, 20

Fuller, Samuel 1912- DLB-26

Fuller, Sarah Margaret, Marchesa D'Ossoli 1810-1850........DLB-1, 59, 73

Fuller, Thomas 1608-1661.........DLB-151

Fullerton, Hugh 1873-1945 ....DLB-171

Fulton, Len 1934- Y-86

Fulton, Robin 1937- DLB-40

Furbank, P. N. 1920- DLB-155

Furman, Laura 1945- Y-86

Furness, Horace Howard 1833-1912 ....DLB-64

Furness, William Henry 1802-1896 ...DLB-1

Furthman, Jules 1888-1966 ....DLB-26

The Future of the Novel (1899), by Henry James DLB-18

Fyleman, Rose 1877-1957 ....DLB-160

G

The G. Ross Roy Scottish Poetry Collection at the University of South Carolina...................Y-89

Gadda, Carlo Emilio 1893-1973 ....DLB-177

Gaddis, William 1922- DLB-2

Gág, Wanda 1893-1946 ....DLB-22

Gagnon, Madeleine 1938- DLB-60

Gaine, Hugh 1726-1807............DLB-43

Gaine, Hugh [publishing house] DLB-49

Gaines, Ernest J. 1933- DLB-2, 33, 152; Y-80

Gaiser, Gerd 1908-1976 ....DLB-69

Galarza, Ernesto 1905-1984........DLB-122

Galaxy Science Fiction Novels DLB-46

Gale, Zona 1874-1938 ....DLB-9, 78

Galen of Pergamon 129-after 210...DLB-176

Gall, Louise von 1815-1855........DLB-133

Gallagher, Tess 1943- DLB-120

Gallagher, Wes 1911- DLB-127

Gallagher, William Davis 1808-1894 ....DLB-73

Gallant, Mavis 1922- DLB-53

Gallico, Paul 1897-1976..........DLB-9, 171

Galsworthy, John 1867-1933..........DLB-10, 34, 98, 162

Galt, John 1779-1839 ....DLB-99, 116

Galton, Sir Francis 1822-1911......DLB-166

Galvin, Brendan 1938- DLB-5

Gambit.........................DLB-46

Gamboa, Reymundo 1948- DLB-122

Gammer Gurton's Needle DLB-62

Gannett, Frank E. 1876-1957 ....DLB-29

Gaos, Vicente 1919-1980..........DLB-134

García, Lionel G. 1935- DLB-82

García Lorca, Federico 1898-1936 ....DLB-108

García Márquez, Gabriel 1928- DLB-113

Gardam, Jane 1928- DLB-14, 161

Garden, Alexander circa 1685-1756................DLB-31

Gardiner, Margaret Power Farmer (see Blessington, Marguerite, Countess of)

Gardner, John 1933-1982 ......DLB-2; Y-82

Garfield, Leon 1921- DLB-161

Garis, Howard R. 1873-1962........DLB-22

Garland, Hamlin 1860-1940..............DLB-12, 71, 78

Garneau, Francis-Xavier 1809-1866 ....DLB-99

Garneau, Hector de Saint-Denys 1912-1943 ....DLB-88

Garneau, Michel 1939- DLB-53

Garner, Alan 1934- DLB-161

Garner, Hugh 1913-1979...........DLB-68

Garnett, David 1892-1981 ....DLB-34

Garnett, Eve 1900-1991 ....DLB-160

Garraty, John A. 1920- DLB-17

Garrett, George 1929- DLB-2, 5, 130, 152; Y-83

Garrick, David 1717-1779 ....DLB-84

Garrison, William Lloyd 1805-1879 ....DLB-1, 43

Garro, Elena 1920- DLB-145

Garth, Samuel 1661-1719..........DLB-95

Garve, Andrew 1908- DLB-87

Gary, Romain 1914-1980..........DLB-83

Gascoigne, George 1539?-1577.....DLB-136

Gascoyne, David 1916- DLB-20

Gaskell, Elizabeth Cleghorn 1810-1865............DLB-21, 144, 159

Gaspey, Thomas 1788-1871 ....DLB-116

Gass, William Howard 1924- DLB-2

Gates, Doris 1901- DLB-22

Gates, Henry Louis, Jr. 1950- DLB-67

Gates, Lewis E. 1860-1924..........DLB-71

Gatto, Alfonso 1909-1976 ....DLB-114

Gaunt, Mary 1861-1942..........DLB-174

Gautier, Théophile 1811-1872......DLB-119

Gauvreau, Claude 1925-1971 ....DLB-88

The *Gawain*-Poet flourished circa 1350-1400 .....DLB-146

Gay, Ebenezer 1696-1787 ....DLB-24

Gay, John 1685-1732 ....DLB-84, 95

The Gay Science (1866), by E. S. Dallas [excerpt] DLB-21

328

Gayarré, Charles E. A. 1805-1895 ... DLB-30

Gaylord, Edward King
1873-1974 .... DLB-127

Gaylord, Edward Lewis 1919- DLB-127

Gaylord, Charles
[publishing house] DLB-49

Geddes, Gary 1940- DLB-60

Geddes, Virgil 1897- DLB-4

Gedeon (Georgii Andreevich Krinovsky)
circa 1730-1763................ DLB-150

Geibel, Emanuel 1815-1884........ DLB-129

Geiogamah, Hanay 1945- DLB-175

Geis, Bernard, Associates.......... DLB-46

Geisel, Theodor Seuss
1904-1991 .... DLB-61; Y-91

Gelb, Arthur 1924- DLB-103

Gelb, Barbara 1926- DLB-103

Gelber, Jack 1932- DLB-7

Gelinas, Gratien 1909- DLB-88

Gellert, Christian Fuerchtegott
1715-1769 .... DLB-97

Gellhorn, Martha 1908- Y-82

Gems, Pam 1925- DLB-13

A General Idea of the College of Mirania (1753),
by William Smith [excerpts] DLB-31

Genet, Jean 1910-1986 .... DLB-72; Y-86

Genevoix, Maurice 1890-1980....... DLB-65

Genovese, Eugene D. 1930- DLB-17

Gent, Peter 1942- Y-82

Geoffrey of Monmouth
circa 1100-1155................ DLB-146

George, Henry 1839-1897 .... DLB-23

George, Jean Craighead 1919- DLB-52

Georgslied 896?................... DLB-148

Gerhardie, William 1895-1977 ..... DLB-36

Gerhardt, Paul 1607-1676 .... DLB-164

Gérin, Winifred 1901-1981 .... DLB-155

Gérin-Lajoie, Antoine 1824-1882 .... DLB-99

German Drama 800-1280........ DLB-138

German Drama from Naturalism
to Fascism: 1889-1933 .... DLB-118

German Literature and Culture from
Charlemagne to the Early Courtly
Period DLB-148

German Radio Play, The.......... DLB-124

German Transformation from the Baroque
to the Enlightenment, The DLB-97

The Germanic Epic and Old English Heroic
Poetry: *Widseth, Waldere,* and *The
Fight at Finnsburg*............. DLB-146

Germanophilism, by Hans Kohn DLB-66

Gernsback, Hugo 1884-1967..... DLB-8, 137

Gerould, Katharine Fullerton
1879-1944 .... DLB-78

Gerrish, Samuel [publishing house] .. DLB-49

Gerrold, David 1944- DLB-8

The Ira Gershwin Centenary Y-96

Gersonides 1288-1344 .... DLB-115

Gerstäcker, Friedrich 1816-1872.... DLB-129

Gerstenberg, Heinrich Wilhelm von
1737-1823 .... DLB-97

Gervinus, Georg Gottfried
1805-1871 .... DLB-133

Geßner, Salomon 1730-1788 .... DLB-97

Geston, Mark S. 1946- DLB-8

Al-Ghazali 1058-1111 .... DLB-115

Gibbon, Edward 1737-1794........ DLB-104

Gibbon, John Murray 1875-1952 .... DLB-92

Gibbon, Lewis Grassic (see Mitchell,
James Leslie)

Gibbons, Floyd 1887-1939......... DLB-25

Gibbons, Reginald 1947- DLB-120

Gibbons, William ?-? DLB-73

Gibson, Charles Dana 1867-1944 ..... DS-13

Gibson, Charles Dana 1867-1944 ..... DS-13

Gibson, Graeme 1934- DLB-53

Gibson, Margaret 1944- DLB-120

Gibson, Margaret Dunlop
1843-1920 .... DLB-174

Gibson, Wilfrid 1878-1962 .... DLB-19

Gibson, William 1914- DLB-7

Gide, André 1869-1951 .... DLB-65

Giguère, Diane 1937- DLB-53

Giguère, Roland 1929- DLB-60

Gil de Biedma, Jaime 1929-1990.... DLB-108

Gil-Albert, Juan 1906- DLB-134

Gilbert, Anthony 1899-1973 .... DLB-77

Gilbert, Michael 1912- DLB-87

Gilbert, Sandra M. 1936- DLB-120

Gilbert, Sir Humphrey
1537-1583 .... DLB-136

Gilchrist, Alexander
1828-1861 .... DLB-144

Gilchrist, Ellen 1935- DLB-130

Gilder, Jeannette L. 1849-1916 ...... DLB-79

Gilder, Richard Watson
1844-1909 .... DLB-64, 79

Gildersleeve, Basil 1831-1924 .... DLB-71

Giles, Henry 1809-1882............ DLB-64

Giles of Rome circa 1243-1316 ..... DLB-115

Gilfillan, George 1813-1878........ DLB-144

Gill, Eric 1882-1940 .... DLB-98

Gill, William F., Company DLB-49

Gillespie, A. Lincoln, Jr.
1895-1950 .... DLB-4

Gilliam, Florence ?-?................ DLB-4

Gilliatt, Penelope 1932-1993 .... DLB-14

Gillott, Jacky 1939-1980........... DLB-14

Gilman, Caroline H. 1794-1888 ... DLB-3, 73

Gilman, W. and J.
[publishing house] DLB-49

Gilmer, Elizabeth Meriwether
1861-1951 .... DLB-29

Gilmer, Francis Walker
1790-1826 .... DLB-37

Gilroy, Frank D. 1925- DLB-7

Gimferrer, Pere (Pedro) 1945- DLB-134

Gingrich, Arnold 1903-1976 .... DLB-137

Ginsberg, Allen 1926- DLB-5, 16, 169

Ginzburg, Natalia 1916-1991...... DLB-177

Ginzkey, Franz Karl 1871-1963...... DLB-81

Gioia, Dana 1950- DLB-120

Giono, Jean 1895-1970........... DLB-72

Giotti, Virgilio 1885-1957 .... DLB-114

Giovanni, Nikki 1943- DLB-5, 41

Gipson, Lawrence Henry
1880-1971 .... DLB-17

Girard, Rodolphe 1879-1956........ DLB-92

Giraudoux, Jean 1882-1944......... DLB-65

Gissing, George 1857-1903 ..... DLB-18, 135

Giudici, Giovanni 1924- DLB-128

Giuliani, Alfredo 1924- DLB-128

Gladstone, William Ewart
1809-1898 .... DLB-57

Glaeser, Ernst 1902-1963.......... DLB-69

Glancy, Diane 1941- DLB-175

Glanville, Brian 1931- DLB-15, 139

Glapthorne, Henry 1610-1643?...... DLB-58

Glasgow, Ellen 1873-1945 .... DLB-9, 12

Glaspell, Susan 1876-1948...... DLB-7, 9, 78

Glass, Montague 1877-1934........ DLB-11

The Glass Key and Other Dashiell Hammett Mysteries Y-96

Glassco, John 1909-1981 .... DLB-68

Glauser, Friedrich 1896-1938 .... DLB-56

F. Gleason's Publishing Hall DLB-49

Gleim, Johann Wilhelm Ludwig 1719-1803 .... DLB-97

Glendinning, Victoria 1937- DLB-155

Glover, Richard 1712-1785 .... DLB-95

Glück, Louise 1943- DLB-5

Glyn, Elinor 1864-1943 .... DLB-153

Gobineau, Joseph-Arthur de 1816-1882 .... DLB-123

Godbout, Jacques 1933- DLB-53

Goddard, Morrill 1865-1937 .... DLB-25

Goddard, William 1740-1817 .... DLB-43

Godden, Rumer 1907- DLB-161

Godey, Louis A. 1804-1878......... DLB-73

Godey and McMichael............ DLB-49

Godfrey, Dave 1938- DLB-60

Godfrey, Thomas 1736-1763........ DLB-31

Godine, David R., Publisher DLB-46

Godkin, E. L. 1831-1902 .... DLB-79

Godolphin, Sidney 1610-1643...... DLB-126

Godwin, Gail 1937- DLB-6

Godwin, Mary Jane Clairmont 1766-1841 .... DLB-163

Godwin, Parke 1816-1904 .... DLB-3, 64

Godwin, William 1756-1836 ... DLB-39, 104, 142, 158, 163

Godwin, M. J., and Company...... DLB-154

Goering, Reinhard 1887-1936...... DLB-118

Goes, Albrecht 1908- DLB-69

Goethe, Johann Wolfgang von 1749-1832 .... DLB-94

Goetz, Curt 1888-1960............ DLB-124

Goffe, Thomas circa 1592-1629 ..... DLB-58

Goffstein, M. B. 1940- DLB-61

Gogarty, Oliver St. John 1878-1957 .... DLB-15, 19

Goines, Donald 1937-1974 .... DLB-33

Gold, Herbert 1924- DLB-2; Y-81

Gold, Michael 1893-1967......... DLB-9, 28

Goldbarth, Albert 1948- DLB-120

Goldberg, Dick 1947- DLB-7

Golden Cockerel Press........... DLB-112

Golding, Arthur 1536-1606 .... DLB-136

Golding, William 1911-1993.... DLB-15, 100

Goldman, William 1931- DLB-44

Goldsmith, Oliver 1730?-1774 ... DLB-39, 89, 104, 109, 142

Goldsmith, Oliver 1794-1861 ...... DLB-99

Goldsmith Publishing Company..... DLB-46

Gollancz, Victor, Limited DLB-112

Gómez-Quiñones, Juan 1942- DLB-122

Gomme, Laurence James [publishing house] DLB-46

Goncourt, Edmond de 1822-1896... DLB-123

Goncourt, Jules de 1830-1870 ...... DLB-123

Gonzales, Rodolfo "Corky" 1928- DLB-122

González, Angel 1925- DLB-108

Gonzalez, Genaro 1949- DLB-122

Gonzalez, Ray 1952- DLB-122

González de Mireles, Jovita 1899-1983 .... DLB-122

González-T., César A. 1931- DLB-82

Goodbye, Gutenberg? A Lecture at the New York Public Library, 18 April 1995 .... Y-95

Goodison, Lorna 1947- DLB-157

Goodman, Paul 1911-1972 .... DLB-130

The Goodman Theatre DLB-7

Goodrich, Frances 1891-1984 and Hackett, Albert 1900- DLB-26

Goodrich, Samuel Griswold 1793-1860................ DLB-1, 42, 73

Goodrich, S. G. [publishing house]... DLB-49

Goodspeed, C. E., and Company DLB-49

Goodwin, Stephen 1943- Y-82

Googe, Barnabe 1540-1594 .... DLB-132

Gookin, Daniel 1612-1687.......... DLB-24

Gordon, Caroline 1895-1981......... DLB-4, 9, 102; Y-81

Gordon, Giles 1940- DLB-14, 139

Gordon, Lyndall 1941- DLB-155

Gordon, Mary 1949- DLB-6; Y-81

Gordone, Charles 1925- DLB-7

Gore, Catherine 1800-1861 .... DLB-116

Gorey, Edward 1925- DLB-61

Gorgias of Leontini circa 485 B.C.-376 B.C. DLB-176

Görres, Joseph 1776-1848 .... DLB-90

Gosse, Edmund 1849-1928 ..... DLB-57, 144

Gosson, Stephen 1554-1624........ DLB-172

Gotlieb, Phyllis 1926- DLB-88

Gottfried von Straßburg died before 1230.............. DLB-138

Gotthelf, Jeremias 1797-1854 ...... DLB-133

Gottschalk circa 804/808-869 ...... DLB-148

Gottsched, Johann Christoph 1700-1766 .... DLB-97

Götz, Johann Nikolaus 1721-1781 .... DLB-97

Gould, Wallace 1882-1940.......... DLB-54

Govoni, Corrado 1884-1965 .... DLB-114

Gower, John circa 1330-1408 ...... DLB-146

Goyen, William 1915-1983 ..... DLB-2; Y-83

Goytisolo, José Augustín 1928- DLB-134

Gozzano, Guido 1883-1916 .... DLB-114

Grabbe, Christian Dietrich 1801-1836 .... DLB-133

Gracq, Julien 1910- DLB-83

Grady, Henry W. 1850-1889........ DLB-23

Graf, Oskar Maria 1894-1967 ...... DLB-56

Graf Rudolf between circa 1170 and circa 1185 .... DLB-148

Grafton, Richard [publishing house] DLB-170

Graham, George Rex 1813-1894..... DLB-73

Graham, Gwethalyn 1913-1965 ..... DLB-88

Graham, Jorie 1951- DLB-120

Graham, Katharine 1917- DLB-127

Graham, Lorenz 1902-1989......... DLB-76

Graham, Philip 1915-1963......... DLB-127

Graham, R. B. Cunninghame 1852-1936............ DLB-98, 135, 174

Graham, Shirley 1896-1977......... DLB-76

Graham, W. S. 1918- DLB-20

Graham, William H. [publishing house] DLB-49

Graham, Winston 1910- DLB-77

Grahame, Kenneth 1859-1932 .... DLB-34, 141

Grainger, Martin Allerdale 1874-1941 .... DLB-92

Gramatky, Hardie 1907-1979 .... DLB-22

Grand, Sarah 1854-1943 .... DLB-135

Grandbois, Alain 1900-1975 .... DLB-92

Grange, John circa 1556-? DLB-136

Granich, Irwin (see Gold, Michael)

Grant, Duncan 1885-1978............ DS-10

Grant, George 1918-1988........... DLB-88

Grant, George Monro 1835-1902 .... DLB-99

Grant, Harry J. 1881-1963 .... DLB-29

Grant, James Edward 1905-1966..... DLB-26

Grass, Günter 1927- DLB-75, 124

Grasty, Charles H. 1863-1924....... DLB-25

Grau, Shirley Ann 1929- DLB-2

Graves, John 1920- Y-83

Graves, Richard 1715-1804 .... DLB-39

Graves, Robert
  1895-1985 .... DLB-20, 100; Y-85

Gray, Asa 1810-1888 .... DLB-1

Gray, David 1838-1861 .... DLB-32

Gray, Simon 1936- DLB-13

Gray, Thomas 1716-1771 .... DLB-109

Grayson, William J. 1788-1863.... DLB-3, 64

The Great Bibliographers Series........ Y-93

The Great War and the Theater, 1914-1918
  [Great Britain]................. DLB-10

Greeley, Horace 1811-1872....... DLB-3, 43

Green, Adolph (see Comden, Betty)

Green, Duff 1791-1875............. DLB-43

Green, Gerald 1922- DLB-28

Green, Henry 1905-1973 .... DLB-15

Green, Jonas 1712-1767 .... DLB-31

Green, Joseph 1706-1780 .... DLB-31

Green, Julien 1900- DLB-4, 72

Green, Paul 1894-1981 ...... DLB-7, 9; Y-81

Green, T. and S.
  [publishing house] DLB-49

Green, Timothy
  [publishing house] DLB-49

Greenaway, Kate 1846-1901 .... DLB-141

Greenberg: Publisher DLB-46

Green Tiger Press................ DLB-46

Greene, Asa 1789-1838 .... DLB-11

Greene, Benjamin H.
  [publishing house] DLB-49

Greene, Graham 1904-1991
  .... DLB-13, 15, 77, 100, 162; Y-85, Y-91

Greene, Robert 1558-1592..... DLB-62, 167

Greenhow, Robert 1800-1854 .... DLB-30

Greenough, Horatio 1805-1852 ..... DLB-1

Greenwell, Dora 1821-1882 .... DLB-35

Greenwillow Books................ DLB-46

Greenwood, Grace (see Lippincott, Sara Jane Clarke)

Greenwood, Walter 1903-1974 ...... DLB-10

Greer, Ben 1948- DLB-6

Greflinger, Georg 1620?-1677...... DLB-164

Greg, W. R. 1809-1881 .... DLB-55

Gregg Press DLB-46

Gregory, Isabella Augusta
  Persse, Lady 1852-1932......... DLB-10

Gregory, Horace 1898-1982 .... DLB-48

Gregory of Rimini
  circa 1300-1358................ DLB-115

Gregynog Press................... DLB-112

Greiffenberg, Catharina Regina von
  1633-1694 .... DLB-168

Grenfell, Wilfred Thomason
  1865-1940 .... DLB-92

Greve, Felix Paul (see Grove, Frederick Philip)

Greville, Fulke, First Lord Brooke
  1554-1628 .... DLB-62, 172

Grey, Lady Jane 1537-1554 .... DLB-132

Grey Owl 1888-1938 .... DLB-92

Grey, Zane 1872-1939 .... DLB-9

Grey Walls Press DLB-112

Grier, Eldon 1917- DLB-88

Grieve, C. M. (see MacDiarmid, Hugh)

Griffin, Bartholomew
  flourished 1596............... DLB-172

Griffin, Gerald 1803-1840 .... DLB-159

Griffith, Elizabeth 1727?-1793 ... DLB-39, 89

Griffiths, Trevor 1935- DLB-13

Griffiths, Ralph
  [publishing house] DLB-154

Griggs, S. C., and Company DLB-49

Griggs, Sutton Elbert 1872-1930..... DLB-50

Grignon, Claude-Henri 1894-1976 ... DLB-68

Grigson, Geoffrey 1905- DLB-27

Grillparzer, Franz 1791-1872....... DLB-133

Grimald, Nicholas
  circa 1519-circa 1562 .... DLB-136

Grimké, Angelina Weld
  1880-1958 .... DLB-50, 54

Grimm, Hans 1875-1959 .... DLB-66

Grimm, Jacob 1785-1863 .... DLB-90

Grimm, Wilhelm 1786-1859 .... DLB-90

Grimmelshausen, Johann Jacob Christoffel von
  1621 or 1622-1676 .... DLB-168

Grimshaw, Beatrice Ethel
  1871-1953 .... DLB-174

Grindal, Edmund
  1519 or 1520-1583 .... DLB-132

Griswold, Rufus Wilmot
  1815-1857 .... DLB-3, 59

Gross, Milt 1895-1953 .... DLB-11

Grosset and Dunlap DLB-49

Grossman Publishers DLB-46

Grosseteste, Robert
  circa 1160-1253................ DLB-115

Grosvenor, Gilbert H. 1875-1966.... DLB-91

Groth, Klaus 1819-1899........... DLB-129

Groulx, Lionel 1878-1967 .... DLB-68

Grove, Frederick Philip 1879-1949... DLB-92

Grove Press..................... DLB-46

Grubb, Davis 1919-1980 .... DLB-6

Gruelle, Johnny 1880-1938 .... DLB-22

Grymeston, Elizabeth
  before 1563-before 1604 .... DLB-136

Gryphius, Andreas 1616-1664...... DLB-164

Gryphius, Christian 1649-1706..... DLB-168

Guare, John 1938- DLB-7

Guerra, Tonino 1920- DLB-128

Guest, Barbara 1920- DLB-5

Guèvremont, Germaine
  1893-1968 .... DLB-68

Guidacci, Margherita 1921-1992.... DLB-128

Guide to the Archives of Publishers, Journals,
  and Literary Agents in North American Libraries Y-93

Guillén, Jorge 1893-1984 .... DLB-108

Guilloux, Louis 1899-1980 .... DLB-72

Guilpin, Everard
  circa 1572-after 1608? DLB-136

Guiney, Louise Imogen 1861-1920... DLB-54

Guiterman, Arthur 1871-1943....... DLB-11

Günderrode, Caroline von
  1780-1806 .... DLB-90

Gundulić, Ivan 1589-1638 .... DLB-147

Gunn, Bill 1934-1989 .... DLB-38

Gunn, James E. 1923- DLB-8

Gunn, Neil M. 1891-1973 .... DLB-15

Gunn, Thom 1929- DLB-27

Gunnars, Kristjana 1948- DLB-60

Günther, Johann Christian
  1695-1723 .... DLB-168

Gurik, Robert 1932- DLB-60

Gustafson, Ralph 1909- DLB-88

Gütersloh, Albert Paris 1887-1973 ... DLB-81

Guthrie, A. B., Jr. 1901- DLB-6

Guthrie, Ramon 1896-1973 .... DLB-4

The Guthrie Theater DLB-7

Gutzkow, Karl 1811-1878 .... DLB-133

Guy, Ray 1939- DLB-60

Guy, Rosa 1925- DLB-33

Guyot, Arnold 1807-1884........... DS-13

Gwynne, Erskine 1898-1948 .... DLB-4

Gyles, John 1680-1755 .... DLB-99

Gysin, Brion 1916- DLB-16

H

H. D. (see Doolittle, Hilda)

Habington, William 1605-1654..... DLB-126

Hacker, Marilyn 1942- DLB-120

Hackett, Albert (see Goodrich, Frances)

Hacks, Peter 1928- DLB-124

Hadas, Rachel 1948- DLB-120

Hadden, Briton 1898-1929......... DLB-91

Hagedorn, Friedrich von
1708-1754 .... DLB-168

Hagelstange, Rudolf 1912-1984...... DLB-69

Haggard, H. Rider
1856-1925............ DLB-70, 156, 174

Haggard, William 1907-1993 .... Y-93

Hahn-Hahn, Ida Gräfin von
1805-1880 .... DLB-133

Haig-Brown, Roderick 1908-1976.... DLB-88

Haight, Gordon S. 1901-1985 ...... DLB-103

Hailey, Arthur 1920- DLB-88; Y-82

Haines, John 1924- DLB-5

Hake, Edward
flourished 1566-1604 .... DLB-136

Hake, Thomas Gordon 1809-1895 ... DLB-32

Hakluyt, Richard 1552?-1616 ..... DLB-136

Halbe, Max 1865-1944 .... DLB-118

Haldane, J. B. S. 1892-1964....... DLB-160

Haldeman, Joe 1943- DLB-8

Haldeman-Julius Company DLB-46

Hale, E. J., and Son DLB-49

Hale, Edward Everett
1822-1909.............. DLB-1, 42, 74

Hale, Janet Campbell 1946- DLB-175

Hale, Kathleen 1898- DLB-160

Hale, Leo Thomas (see Ebon)

Hale, Lucretia Peabody
1820-1900 .... DLB-42

Hale, Nancy 1908-1988.... DLB-86; Y-80, 88

Hale, Sarah Josepha (Buell)
1788-1879............... DLB-1, 42, 73

Hales, John 1584-1656 .... DLB-151

Haley, Alex 1921-1992............ DLB-38

Haliburton, Thomas Chandler
1796-1865 .... DLB-11, 99

Hall, Anna Maria 1800-1881....... DLB-159

Hall, Donald 1928- DLB-5

Hall, Edward 1497-1547 .... DLB-132

Hall, James 1793-1868 .... DLB-73, 74

Hall, Joseph 1574-1656 ...... DLB-121, 151

Hall, Samuel [publishing house] DLB-49

Hallam, Arthur Henry 1811-1833.... DLB-32

Halleck, Fitz-Greene 1790-1867 ...... DLB-3

Haller, Albrecht von 1708-1777 .... DLB-168

Hallmann, Johann Christian
1640-1704 or 1716? DLB-168

Hallmark Editions................ DLB-46

Halper, Albert 1904-1984........... DLB-9

Halperin, John William 1941- DLB-111

Halstead, Murat 1829-1908 .... DLB-23

Hamann, Johann Georg 1730-1788... DLB-97

Hamburger, Michael 1924- DLB-27

Hamilton, Alexander 1712-1756..... DLB-31

Hamilton, Alexander 1755?-1804.... DLB-37

Hamilton, Cicely 1872-1952 .... DLB-10

Hamilton, Edmond 1904-1977 .... DLB-8

Hamilton, Elizabeth 1758-1816 ... DLB-116, 158

Hamilton, Gail (see Corcoran, Barbara)

Hamilton, Ian 1938- DLB-40, 155

Hamilton, Patrick 1904-1962........ DLB-10

Hamilton, Virginia 1936- DLB-33, 52

Hamilton, Hamish, Limited........ DLB-112

Hammett, Dashiell 1894-1961 .... DS-6

Dashiell Hammett:
An Appeal in TAC.............. Y-91

Hammon, Jupiter 1711-died between
1790 and 1806 .... DLB-31, 50

Hammond, John ?-1663 .... DLB-24

Hamner, Earl 1923- DLB-6

Hampton, Christopher 1946- DLB-13

Handel-Mazzetti, Enrica von
1871-1955 .... DLB-81

Handke, Peter 1942- DLB-85, 124

Handlin, Oscar 1915- DLB-17

Hankin, St. John 1869-1909........ DLB-10

Hanley, Clifford 1922- DLB-14

Hannah, Barry 1942- DLB-6

Hannay, James 1827-1873 .... DLB-21

Hansberry, Lorraine 1930-1965 ... DLB-7, 38

Hapgood, Norman 1868-1937....... DLB-91

Happel, Eberhard Werner
1647-1690 .... DLB-168

Harcourt Brace Jovanovich DLB-46

Hardenberg, Friedrich von (see Novalis)

Harding, Walter 1917- DLB-111

Hardwick, Elizabeth 1916- DLB-6

Hardy, Thomas 1840-1928 ... DLB-18, 19, 135

Hare, Cyril 1900-1958 .... DLB-77

Hare, David 1947- DLB-13

Hargrove, Marion 1919- DLB-11

Häring, Georg Wilhelm Heinrich (see Alexis, Willibald)

Harington, Donald 1935- DLB-152

Harington, Sir John 1560-1612 ..... DLB-136

Harjo, Joy 1951- DLB-120, 175

Harlow, Robert 1923- DLB-60

Harman, Thomas
flourished 1566-1573 .... DLB-136

Harness, Charles L. 1915- DLB-8

Harnett, Cynthia 1893-1981 .... DLB-161

Harper, Fletcher 1806-1877........ DLB-79

Harper, Frances Ellen Watkins
1825-1911 .... DLB-50

Harper, Michael S. 1938- DLB-41

Harper and Brothers DLB-49

Harraden, Beatrice 1864-1943...... DLB-153

Harrap, George G., and Company
Limited DLB-112

Harriot, Thomas 1560-1621 .... DLB-136

Harris, Benjamin ?-circa 1720.... DLB-42, 43

Harris, Christie 1907- DLB-88

Harris, Frank 1856-1931 .... DLB-156

Harris, George Washington
1814-1869 .... DLB-3, 11

Harris, Joel Chandler
1848-1908 .... DLB-11, 23, 42, 78, 91

Harris, Mark 1922- DLB-2; Y-80

Harris, Wilson 1921- DLB-117

Harrison, Charles Yale
1898-1954 .... DLB-68

Harrison, Frederic 1831-1923 ...... DLB-57

Harrison, Harry 1925- DLB-8
Harrison, Jim 1937- Y-82
Harrison, Mary St. Leger Kingsley (see Malet, Lucas)
Harrison, Paul Carter 1936- DLB-38
Harrison, Susan Frances 1859-1935 ....DLB-99
Harrison, Tony 1937- DLB-40
Harrison, William 1535-1593 ......DLB-136
Harrison, James P., Company........DLB-49
Harrisse, Henry 1829-1910 ....DLB-47
Harsdörffer, Georg Philipp 1607-1658 ....DLB-164
Harsent, David 1942- DLB-40
Hart, Albert Bushnell 1854-1943 ....DLB-17
Hart, Julia Catherine 1796-1867 .....DLB-99
The Lorenz Hart Centenary Y-95
Hart, Moss 1904-1961 ....DLB-7
Hart, Oliver 1723-1795 ....DLB-31
Hart-Davis, Rupert, Limited DLB-112
Harte, Bret 1836-1902....DLB-12, 64, 74, 79
Harte, Edward Holmead 1922- DLB-127
Harte, Houston Harriman 1927- DLB-127
Hartlaub, Felix 1913-1945 ....DLB-56
Hartlebon, Otto Erich 1864-1905 ....DLB-118
Hartley, L. P. 1895-1972 ....DLB-15, 139
Hartley, Marsden 1877-1943 ....DLB-54
Hartling, Peter 1933- DLB-75
Hartman, Geoffrey H. 1929- DLB-67
Hartmann, Sadakichi 1867-1944 ....DLB-54
Hartmann von Aue circa 1160-circa 1205 ....DLB-138
Harvey, Gabriel 1550?-1631 ....DLB-167
Harvey, Jean-Charles 1891-1967.....DLB-88
Harvill Press Limited DLB-112
Harwood, Lee 1939- DLB-40
Harwood, Ronald 1934- DLB-13
Haskins, Charles Homer 1870-1937 ....DLB-47
Hass, Robert 1941- DLB-105
The Hatch-Billops Collection DLB-76
Hathaway, William 1944- DLB-120
Hauff, Wilhelm 1802-1827 ....DLB-90
A Haughty and Proud Generation (1922), by Ford Madox Hueffer DLB-36

Haugwitz, August Adolph von 1647-1706 ....DLB-168
Hauptmann, Carl 1858-1921 ....DLB-66, 118
Hauptmann, Gerhart 1862-1946 ....DLB-66, 118
Hauser, Marianne 1910- Y-83
Hawes, Stephen 1475?-before 1529 ....DLB-132
Hawker, Robert Stephen 1803-1875 ....DLB-32
Hawkes, John 1925- DLB-2, 7; Y-80
Hawkesworth, John 1720-1773 .....DLB-142
Hawkins, Sir Anthony Hope (see Hope, Anthony)
Hawkins, Sir John 1719-1789 ....DLB-104, 142
Hawkins, Walter Everette 1883-?....DLB-50
Hawthorne, Nathaniel 1804-1864 ....DLB-1, 74
Hay, John 1838-1905 ....DLB-12, 47
Hayden, Robert 1913-1980 ....DLB-5, 76
Haydon, Benjamin Robert 1786-1846 ....DLB-110
Hayes, John Michael 1919- DLB-26
Hayley, William 1745-1820.....DLB-93, 142
Haym, Rudolf 1821-1901..........DLB-129
Hayman, Robert 1575-1629........DLB-99
Hayman, Ronald 1932- DLB-155
Hayne, Paul Hamilton 1830-1886...............DLB-3, 64, 79
Hays, Mary 1760-1843 ....DLB-142, 158
Haywood, Eliza 1693?-1756 ....DLB-39
Hazard, Willis P. [publishing house]...DLB-49
Hazlitt, William 1778-1830 ....DLB-110, 158
Hazzard, Shirley 1931- Y-82
Head, Bessie 1937-1986 ....DLB-117
Headley, Joel T. 1813-1897 ..DLB-30; DS-13
Heaney, Seamus 1939- DLB-40
Heard, Nathan C. 1936- DLB-33
Hearn, Lafcadio 1850-1904.....DLB-12, 78
Hearne, John 1926- DLB-117
Hearne, Samuel 1745-1792 ....DLB-99
Hearst, William Randolph 1863-1951 ....DLB-25
Hearst, William Randolph, Jr 1908-1993...................DLB-127
Heath, Catherine 1924- DLB-14
Heath, Roy A. K. 1926- DLB-117

Heath-Stubbs, John 1918- DLB-27
Heavysege, Charles 1816-1876......DLB-99
Hebbel, Friedrich 1813-1863.......DLB-129
Hebel, Johann Peter 1760-1826......DLB-90
Hébert, Anne 1916- DLB-68
Hébert, Jacques 1923- DLB-53
Hecht, Anthony 1923- DLB-5, 169
Hecht, Ben 1894-1964 ....DLB-7, 9, 25, 26, 28, 86
Hecker, Isaac Thomas 1819-1888.....DLB-1
Hedge, Frederic Henry 1805-1890 ....DLB-1, 59
Hefner, Hugh M. 1926- DLB-137
Hegel, Georg Wilhelm Friedrich 1770-1831 ....DLB-90
Heidish, Marcy 1947- Y-82
Heißenbüttel 1921- DLB-75
Hein, Christoph 1944- DLB-124
Heine, Heinrich 1797-1856 ....DLB-90
Heinemann, Larry 1944- DS-9
Heinemann, William, Limited......DLB-112
Heinlein, Robert A. 1907-1988 ....DLB-8
Heinrich Julius of Brunswick 1564-1613 ....DLB-164
Heinrich von dem Türlîn flourished circa 1230 ....DLB-138
Heinrich von Melk flourished after 1160 ....DLB-148
Heinrich von Veldeke circa 1145-circa 1190 ....DLB-138
Heinrich, Willi 1920- DLB-75
Heiskell, John 1872-1972..........DLB-127
Heinse, Wilhelm 1746-1803 ....DLB-94
Heinz, W. C. 1915- DLB-171
Hejinian, Lyn 1941- DLB-165
Heliand circa 850 ....DLB-148
Heller, Joseph 1923- DLB-2, 28; Y-80
Heller, Michael 1937- DLB-165
Hellman, Lillian 1906-1984.....DLB-7; Y-84
Hellwig, Johann 1609-1674 ....DLB-164
Helprin, Mark 1947- Y-85
Helwig, David 1938- DLB-60
Hemans, Felicia 1793-1835 ....DLB-96
Hemingway, Ernest 1899-1961 ....DLB-4, 9, 102; Y-81, 87; DS-1
Hemingway: Twenty-Five Years Later Y-85

Hémon, Louis 1880-1913 . . . . DLB-92

Hemphill, Paul 1936- Y-87

Hénault, Gilles 1920- DLB-88

Henchman, Daniel 1689-1761 . . . . DLB-24

Henderson, Alice Corbin
    1881-1949 . . . . DLB-54

Henderson, Archibald
    1877-1963 . . . . DLB-103

Henderson, David 1942- DLB-41

Henderson, George Wylie
    1904- DLB-51

Henderson, Zenna 1917-1983 . . . . DLB-8

Henisch, Peter 1943- DLB-85

Henley, Beth 1952- Y-86

Henley, William Ernest
    1849-1903 . . . . DLB-19

Henniker, Florence 1855-1923 . . . . . DLB-135

Henry, Alexander 1739-1824 . . . . DLB-99

Henry, Buck 1930- DLB-26

Henry VIII of England
    1491-1547 . . . . DLB-132

Henry, Marguerite 1902- DLB-22

Henry, O. (see Porter, William Sydney)

Henry of Ghent
    circa 1217-1229 - 1293 . . . . DLB-115

Henry, Robert Selph 1889-1970 . . . . . DLB-17

Henry, Will (see Allen, Henry W.)

Henryson, Robert
    1420s or 1430s-circa 1505 . . . . . . DLB-146

Henschke, Alfred (see Klabund)

Hensley, Sophie Almon 1866-1946 . . . DLB-99

Henson, Lance 1944- DLB-175

Henty, G. A. 1832?-1902 . . . . DLB-18, 141

Hentz, Caroline Lee 1800-1856 . . . . DLB-3

Heraclitus flourished circa 500 B.C.
    . . . . DLB-176

Herbert, Agnes circa 1880-1960 . . . . DLB-174

Herbert, Alan Patrick 1890-1971 . . . . DLB-10

Herbert, Edward, Lord, of Cherbury
    1582-1648 . . . . DLB-121, 151

Herbert, Frank 1920-1986 . . . . DLB-8

Herbert, George 1593-1633 . . . . DLB-126

Herbert, Henry William
    1807-1858 . . . . DLB-3, 73

Herbert, John 1926- DLB-53

Herbert, Mary Sidney, Countess of Pembroke
    (see Sidney, Mary)

Herbst, Josephine 1892-1969 . . . . DLB-9

Herburger, Gunter 1932- DLB-75, 124

Hercules, Frank E. M. 1917- DLB-33

Herder, Johann Gottfried
    1744-1803 . . . . DLB-97

Herder, B., Book Company DLB-49

Herford, Charles Harold
    1853-1931 . . . . DLB-149

Hergesheimer, Joseph
    1880-1954 . . . . DLB-9, 102

Heritage Press DLB-46

Hermann the Lame 1013-1054 . . . . . DLB-148

Hermes, Johann Timotheus
    1738-1821 . . . . DLB-97

Hermlin, Stephan 1915- DLB-69

Hernández, Alfonso C. 1938- DLB-122

Hernández, Inés 1947- DLB-122

Hernández, Miguel 1910-1942 . . . . . DLB-134

Hernton, Calvin C. 1932- DLB-38

"The Hero as Man of Letters: Johnson,
    Rousseau, Burns" (1841), by Thomas
    Carlyle [excerpt] DLB-57

The Hero as Poet. Dante; Shakspeare (1841),
    by Thomas Carlyle DLB-32

Herodotus circa 484 B.C.-circa 420 B.C.
    . . . . DLB-176

Heron, Robert 1764-1807 . . . . DLB-142

Herrera, Juan Felipe 1948- DLB-122

Herrick, Robert 1591-1674 . . . . DLB-126

Herrick, Robert 1868-1938 . . . . DLB-9, 12, 78

Herrick, William 1915- Y-83

Herrick, E. R., and Company DLB-49

Herrmann, John 1900-1959 . . . . DLB-4

Hersey, John 1914-1993 . . . . DLB-6

Hertel, François 1905-1985 . . . . DLB-68

Hervé-Bazin, Jean Pierre Marie (see Bazin,
    Hervé)

Hervey, John, Lord 1696-1743 . . . . . DLB-101

Herwig, Georg 1817-1875 . . . . DLB-133

Herzog, Emile Salomon Wilhelm (see Maurois,
    André)

Hesiod eighth century B.C. DLB-176

Hesse, Hermann 1877-1962 . . . . DLB-66

Hewat, Alexander
    circa 1743-circa 1824 . . . . DLB-30

Hewitt, John 1907- DLB-27

Hewlett, Maurice 1861-1923 . . . . DLB-34, 156

Heyen, William 1940- DLB-5

Heyer, Georgette 1902-1974 . . . . DLB-77

Heym, Stefan 1913- DLB-69

Heyse, Paul 1830-1914 . . . . DLB-129

Heytesbury, William
    circa 1310-1372 or 1373 . . . . DLB-115

Heyward, Dorothy 1890-1961 . . . . DLB-7

Heyward, DuBose
    1885-1940 . . . . DLB-7, 9, 45

Heywood, John 1497?-1580? DLB-136

Heywood, Thomas
    1573 or 1574-1641 . . . . DLB-62

Hibbs, Ben 1901-1975 . . . . DLB-137

Hichens, Robert S. 1864-1950 . . . . . . DLB-153

Hickman, William Albert
    1877-1957 . . . . DLB-92

Hidalgo, José Luis 1919-1947 . . . . . . DLB-108

Hiebert, Paul 1892-1987 . . . . DLB-68

Hierro, José 1922- DLB-108

Higgins, Aidan 1927- DLB-14

Higgins, Colin 1941-1988 . . . . DLB-26

Higgins, George V. 1939- DLB-2; Y-81

Higginson, Thomas Wentworth
    1823-1911 . . . . DLB-1, 64

Highwater, Jamake 1942?- . . . DLB-52; Y-85

Hijuelos, Oscar 1951- DLB-145

Hildegard von Bingen
    1098-1179 . . . . DLB-148

Das Hildesbrandslied circa 820 . . . . DLB-148

Hildesheimer, Wolfgang
    1916-1991 . . . . DLB-69, 124

Hildreth, Richard
    1807-1865 . . . . DLB-1, 30, 59

Hill, Aaron 1685-1750 . . . . DLB-84

Hill, Geoffrey 1932- DLB-40

Hill, "Sir" John 1714?-1775 . . . . DLB-39

Hill, Leslie 1880-1960 . . . . DLB-51

Hill, Susan 1942- DLB-14, 139

Hill, Walter 1942- DLB-44

Hill and Wang DLB-46

Hill, George M., Company DLB-49

Hill, Lawrence, and Company,
    Publishers DLB-46

Hillberry, Conrad 1928- DLB-120

Hilliard, Gray and Company DLB-49

Hills, Lee 1906- DLB-127

Hillyer, Robert 1895-1961 . . . . DLB-54

Hilton, James 1900-1954 . . . . DLB-34, 77

Hilton, Walter died 1396 . . . . DLB-146

Hilton and Company DLB-49

Himes, Chester
 1909-1984 .... DLB-2, 76, 143

Hindmarsh, Joseph
 [publishing house] DLB-170

Hine, Daryl 1936- DLB-60

Hingley, Ronald 1920- DLB-155

Hinojosa-Smith, Rolando
 1929- DLB-82

Hippel, Theodor Gottlieb von
 1741-1796 .... DLB-97

Hippocrates of Cos flourished circa 425 B.C.
 .... DLB-176

Hirsch, E. D., Jr. 1928- DLB-67

Hirsch, Edward 1950- DLB-120

The History of the Adventures of Joseph Andrews
 (1742), by Henry Fielding
 [excerpt] DLB-39

Hoagland, Edward 1932- DLB-6

Hoagland, Everett H., III 1942- DLB-41

Hoban, Russell 1925- DLB-52

Hobbes, Thomas 1588-1679 .... DLB-151

Hobby, Oveta 1905- DLB-127

Hobby, William 1878-1964 .... DLB-127

Hobsbaum, Philip 1932- DLB-40

Hobson, Laura Z. 1900- DLB-28

Hoby, Thomas 1530-1566 .... DLB-132

Hoccleve, Thomas
 circa 1368-circa 1437 .... DLB-146

Hochhuth, Rolf 1931- DLB-124

Hochman, Sandra 1936- DLB-5

Hodder and Stoughton, Limited DLB-106

Hodgins, Jack 1938- DLB-60

Hodgman, Helen 1945- DLB-14

Hodgskin, Thomas 1787-1869 ..... DLB-158

Hodgson, Ralph 1871-1962 .... DLB-19

Hodgson, William Hope
 1877-1918 .... DLB-70, 153, 156

Hoffenstein, Samuel 1890-1947 ..... DLB-11

Hoffman, Charles Fenno
 1806-1884 .... DLB-3

Hoffman, Daniel 1923- DLB-5

Hoffmann, E. T. A. 1776-1822 ..... DLB-90

Hoffmanswaldau, Christian Hoffman von
 1616-1679 .... DLB-168

Hofmann, Michael 1957- DLB-40

Hofmannsthal, Hugo von
 1874-1929 .... DLB-81, 118

Hofstadter, Richard 1916-1970 ..... DLB-17

Hogan, Desmond 1950- DLB-14

Hogan, Linda 1947- DLB-175

Hogan and Thompson DLB-49

Hogarth Press DLB-112

Hogg, James 1770-1835.... DLB-93, 116, 159

Hohberg, Wolfgang Helmhard Freiherr von
 1612-1688 .... DLB-168

Hohl, Ludwig 1904-1980 .... DLB-56

Holbrook, David 1923- DLB-14, 40

Holcroft, Thomas
 1745-1809 .... DLB-39, 89, 158

Holden, Jonathan 1941- DLB-105

Holden, Jonathan, Contemporary
 Verse Story-telling DLB-105

Holden, Molly 1927-1981 .... DLB-40

Hölderlin, Friedrich 1770-1843 ...... DLB-90

Holiday House DLB-46

Holinshed, Raphael died 1580 ..... DLB-167

Holland, J. G. 1819-1881 .... DS-13

Holland, Norman N. 1927- DLB-67

Hollander, John 1929- DLB-5

Holley, Marietta 1836-1926 .... DLB-11

Hollingsworth, Margaret 1940- DLB-60

Hollo, Anselm 1934- DLB-40

Holloway, Emory 1885-1977 .... DLB-103

Holloway, John 1920- DLB-27

Holloway House Publishing
 Company DLB-46

Holme, Constance 1880-1955 .... DLB-34

Holmes, Abraham S. 1821?-1908 .... DLB-99

Holmes, John Clellon 1926-1988 .... DLB-16

Holmes, Oliver Wendell
 1809-1894 .... DLB-1

Holmes, Richard 1945- DLB-155

Holroyd, Michael 1935- DLB-155

Holst, Hermann E. von
 1841-1904 .... DLB-47

Holt, John 1721-1784 .... DLB-43

Holt, Henry, and Company DLB-49

Holt, Rinehart and Winston DLB-46

Holthusen, Hans Egon 1913- DLB-69

Hölty, Ludwig Christoph Heinrich
 1748-1776 .... DLB-94

Holz, Arno 1863-1929 .... DLB-118

Home, Henry, Lord Kames (see Kames, Henry
 Home, Lord)

Home, John 1722-1808 .... DLB-84

Home, William Douglas 1912- DLB-13

Home Publishing Company DLB-49

Homer circa eighth-seventh centuries B.C.
 .... DLB-176

Homes, Geoffrey (see Mainwaring, Daniel)

Honan, Park 1928- DLB-111

Hone, William 1780-1842 ..... DLB-110, 158

Hongo, Garrett Kaoru 1951- DLB-120

Honig, Edwin 1919- DLB-5

Hood, Hugh 1928- DLB-53

Hood, Thomas 1799-1845 .... DLB-96

Hook, Theodore 1788-1841 .... DLB-116

Hooker, Jeremy 1941- DLB-40

Hooker, Richard 1554-1600 .... DLB-132

Hooker, Thomas 1586-1647 .... DLB-24

Hooper, Johnson Jones
 1815-1862 .... DLB-3, 11

Hope, Anthony 1863-1933 .... DLB-153, 156

Hopkins, Gerard Manley
 1844-1889 .... DLB-35, 57

Hopkins, John (see Sternhold, Thomas)

Hopkins, Lemuel 1750-1801 .... DLB-37

Hopkins, Pauline Elizabeth
 1859-1930 .... DLB-50

Hopkins, Samuel 1721-1803 .... DLB-31

Hopkins, John H., and Son DLB-46

Hopkinson, Francis 1737-1791 ...... DLB-31

Horgan, Paul 1903- DLB-102; Y-85

Horizon Press DLB-46

Horne, Frank 1899-1974 .... DLB-51

Horne, Richard Henry (Hengist)
 1802 or 1803-1884 .... DLB-32

Hornung, E. W. 1866-1921 .... DLB-70

Horovitz, Israel 1939- DLB-7

Horton, George Moses
 1797?-1883? DLB-50

Horváth, Ödön von
 1901-1938 .... DLB-85, 124

Horwood, Harold 1923- DLB-60

Hosford, E. and E.
 [publishing house] DLB-49

Hoskyns, John 1566-1638 .... DLB-121

Hotchkiss and Company DLB-49

Hough, Emerson 1857-1923 .... DLB-9

Houghton Mifflin Company DLB-49

Houghton, Stanley 1881-1913 ...... DLB-10

Household, Geoffrey 1900-1988 .... DLB-87

335

Housman, A. E. 1859-1936 .... DLB-19

Housman, Laurence 1865-1959 ...... DLB-10

Houwald, Ernst von 1778-1845 ..... DLB-90

Hovey, Richard 1864-1900 .... DLB-54

Howard, Donald R. 1927-1987 ..... DLB-111

Howard, Maureen 1930- Y-83

Howard, Richard 1929- DLB-5

Howard, Roy W. 1883-1964 .... DLB-29

Howard, Sidney 1891-1939 .... DLB-7, 26

Howe, E. W. 1853-1937 .... DLB-12, 25

Howe, Henry 1816-1893 .... DLB-30

Howe, Irving 1920-1993 .... DLB-67

Howe, Joseph 1804-1873 .... DLB-99

Howe, Julia Ward 1819-1910 .... DLB-1

Howe, Percival Presland
 1886-1944 .... DLB-149

Howe, Susan 1937- DLB-120

Howell, Clark, Sr. 1863-1936 .... DLB-25

Howell, Evan P. 1839-1905 .... DLB-23

Howell, James 1594?-1666 .... DLB-151

Howell, Warren Richardson
 1912-1984 .... DLB-140

Howell, Soskin and Company DLB-46

Howells, William Dean
 1837-1920 .... DLB-12, 64, 74, 79

Howitt, William 1792-1879 and
 Howitt, Mary 1799-1888 .... DLB-110

Hoyem, Andrew 1935- DLB-5

Hoyers, Anna Ovena 1584-1655 .... DLB-164

Hoyos, Angela de 1940- DLB-82

Hoyt, Palmer 1897-1979 .... DLB-127

Hoyt, Henry [publishing house] DLB-49

Hrabanus Maurus 776?-856 .... DLB-148

Hrotsvit of Gandersheim
 circa 935-circa 1000 .... DLB-148

Hubbard, Elbert 1856-1915 .... DLB-91

Hubbard, Kin 1868-1930 .... DLB-11

Hubbard, William circa 1621-1704 ... DLB-24

Huber, Therese 1764-1829 .... DLB-90

Huch, Friedrich 1873-1913 .... DLB-66

Huch, Ricarda 1864-1947 .... DLB-66

Huck at 100: How Old Is
 Huckleberry Finn? Y-85

Huddle, David 1942- DLB-130

Hudgins, Andrew 1951- DLB-120

Hudson, Henry Norman
 1814-1886 .... DLB-64

Hudson, W. H.
 1841-1922 .... DLB-98, 153, 174

Hudson and Goodwin DLB-49

Huebsch, B. W.
 [publishing house] DLB-46

Hughes, David 1930- DLB-14

Hughes, John 1677-1720 .... DLB-84

Hughes, Langston
 1902-1967 .... DLB-4, 7, 48, 51, 86

Hughes, Richard 1900-1976 .... DLB-15, 161

Hughes, Ted 1930- DLB-40, 161

Hughes, Thomas 1822-1896 .... DLB-18, 163

Hugo, Richard 1923-1982 .... DLB-5

Hugo, Victor 1802-1885 .... DLB-119

Hugo Awards and Nebula Awards DLB-8

Hull, Richard 1896-1973 .... DLB-77

Hulme, T. E. 1883-1917 .... DLB-19

Humboldt, Alexander von
 1769-1859 .... DLB-90

Humboldt, Wilhelm von
 1767-1835 .... DLB-90

Hume, David 1711-1776 .... DLB-104

Hume, Fergus 1859-1932 .... DLB-70

Hummer, T. R. 1950- DLB-120

Humorous Book Illustration DLB-11

Humphrey, William 1924- DLB-6

Humphreys, David 1752-1818 ...... DLB-37

Humphreys, Emyr 1919- DLB-15

Huncke, Herbert 1915- DLB-16

Huneker, James Gibbons
 1857-1921 .... DLB-71

Hunold, Christian Friedrich
 1681-1721 .... DLB-168

Hunt, Irene 1907- DLB-52

Hunt, Leigh 1784-1859 .... DLB-96, 110, 144

Hunt, Violet 1862-1942 .... DLB-162

Hunt, William Gibbes 1791-1833 .... DLB-73

Hunter, Evan 1926- Y-82

Hunter, Jim 1939- DLB-14

Hunter, Kristin 1931- DLB-33

Hunter, Mollie 1922- DLB-161

Hunter, N. C. 1908-1971 .... DLB-10

Hunter-Duvar, John 1821-1899 ..... DLB-99

Huntington, Henry E.
 1850-1927 .... DLB-140

Hurd and Houghton DLB-49

Hurst, Fannie 1889-1968 .... DLB-86

Hurst and Blackett DLB-106

Hurst and Company DLB-49

Hurston, Zora Neale
 1901?-1960 .... DLB-51, 86

Husson, Jules-François-Félix (see Champfleury)

Huston, John 1906-1987 .... DLB-26

Hutcheson, Francis 1694-1746 ...... DLB-31

Hutchinson, Thomas
 1711-1780 .... DLB-30, 31

Hutchinson and Company
 (Publishers) Limited DLB-112

Hutton, Richard Holt 1826-1897 .... DLB-57

Huxley, Aldous
 1894-1963 .... DLB-36, 100, 162

Huxley, Elspeth Josceline 1907- DLB-77

Huxley, T. H. 1825-1895 .... DLB-57

Huyghue, Douglas Smith
 1816-1891 .... DLB-99

Huysmans, Joris-Karl 1848-1907 .... DLB-123

Hyman, Trina Schart 1939- DLB-61

I

Iavorsky, Stefan 1658-1722 .... DLB-150

Ibn Bajja circa 1077-1138 .... DLB-115

Ibn Gabirol, Solomon
 circa 1021-circa 1058 .... DLB-115

The Iconography of Science-Fiction
 Art DLB-8

Iffland, August Wilhelm
 1759-1814 .... DLB-94

Ignatow, David 1914- DLB-5

Ike, Chukwuemeka 1931- DLB-157

Iles, Francis (see Berkeley, Anthony)

The Illustration of Early German
 Literary Manuscripts,
 circa 1150-circa 1300 .... DLB-148

Imbs, Bravig 1904-1946 .... DLB-4

Imbuga, Francis D. 1947- DLB-157

Immermann, Karl 1796-1840 ...... DLB-133

Inchbald, Elizabeth 1753-1821 ... DLB-39, 89

Inge, William 1913-1973 .... DLB-7

Ingelow, Jean 1820-1897 .... DLB-35, 163

Ingersoll, Ralph 1900-1985 .... DLB-127

The Ingersoll Prizes Y-84

Ingoldsby, Thomas (see Barham, Richard
 Harris)

Ingraham, Joseph Holt 1809-1860 ..... DLB-3

Inman, John 1805-1850 . . . . DLB-73

Innerhofer, Franz 1944- DLB-85

Innis, Harold Adams 1894-1952 . . . . . DLB-88

Innis, Mary Quayle 1899-1972 . . . . . . DLB-88

International Publishers Company . . . DLB-46

An Interview with David Rabe Y-91

An Interview with George Greenfield, Literary Agent Y-91

An Interview with James Ellroy Y-91

An Interview with Peter S. Prescott Y-86

An Interview with Russell Hoban Y-90

An Interview with Tom Jenks Y-86

Introduction to Paul Laurence Dunbar, Lyrics of Lowly Life (1896), by William Dean Howells DLB-50

Introductory Essay: *Letters of Percy Bysshe Shelley* (1852), by Robert Browning DLB-32

Introductory Letters from the Second Edition of *Pamela* (1741), by Samuel Richardson DLB-39

Irving, John 1942- DLB-6; Y-82

Irving, Washington 1783-1859 . . . . . DLB-3, 11, 30, 59, 73, 74

Irwin, Grace 1907- DLB-68

Irwin, Will 1873-1948 . . . . DLB-25

Isherwood, Christopher 1904-1986 . . . . DLB-15; Y-86

The Island Trees Case: A Symposium on School Library Censorship
An Interview with Judith Krug
An Interview with Phyllis Schlafly
An Interview with Edward B. Jenkinson
An Interview with Lamarr Mooneyham
An Interview with Harriet Bernstein Y-82

Islas, Arturo 1938-1991 . . . . DLB-122

Ivers, M. J., and Company DLB-49

Iyayi, Festus 1947- DLB-157

J

Jackmon, Marvin E. (see Marvin X)

Jacks, L. P. 1860-1955 . . . . DLB-135

Jackson, Angela 1951- DLB-41

Jackson, Helen Hunt 1830-1885 . . . . DLB-42, 47

Jackson, Holbrook 1874-1948 . . . . DLB-98

Jackson, Laura Riding 1901-1991 . . . . DLB-48

Jackson, Shirley 1919-1965 . . . . DLB-6

Jacob, Piers Anthony Dillingham (see Anthony, Piers)

Jacobi, Friedrich Heinrich 1743-1819 . . . . DLB-94

Jacobi, Johann Georg 1740-1841 . . . . . DLB-97

Jacobs, Joseph 1854-1916 . . . . DLB-141

Jacobs, W. W. 1863-1943 . . . . DLB-135

Jacobs, George W., and Company . . . DLB-49

Jacobson, Dan 1929- DLB-14

Jaggard, William [publishing house] DLB-170

Jahier, Piero 1884-1966 . . . . DLB-114

Jahnn, Hans Henny 1894-1959 . . . . DLB-56, 124

Jakes, John 1932- Y-83

James, C. L. R. 1901-1989 . . . . DLB-125

James, George P. R. 1801-1860 . . . . . DLB-116

James, Henry 1843-1916 . . . . . . DLB-12, 71, 74; DS-13

James, John circa 1633-1729 . . . . DLB-24

The James Jones Society Y-92

James, M. R. 1862-1936 . . . . DLB-156

James, P. D. 1920- DLB-87

James Joyce Centenary: Dublin, 1982 . . . Y-82

James Joyce Conference Y-85

James VI of Scotland, I of England 1566-1625 . . . . DLB-151, 172

James, U. P. [publishing house] DLB-49

Jameson, Anna 1794-1860 . . . . . . DLB-99, 166

Jameson, Fredric 1934- DLB-67

Jameson, J. Franklin 1859-1937 . . . . . . DLB-17

Jameson, Storm 1891-1986 . . . . DLB-36

Janés, Clara 1940- DLB-134

Jaramillo, Cleofas M. 1878-1956 . . . . DLB-122

Jarman, Mark 1952- DLB-120

Jarrell, Randall 1914-1965 . . . . DLB-48, 52

Jarrold and Sons DLB-106

Jasmin, Claude 1930- DLB-60

Jay, John 1745-1829 . . . . DLB-31

Jefferies, Richard 1848-1887 . . . . DLB-98, 141

Jeffers, Lance 1919-1985 . . . . DLB-41

Jeffers, Robinson 1887-1962 . . . . DLB-45

Jefferson, Thomas 1743-1826 . . . . DLB-31

Jelinek, Elfriede 1946- DLB-85

Jellicoe, Ann 1927- DLB-13

Jenkins, Elizabeth 1905- DLB-155

Jenkins, Robin 1912- DLB-14

Jenkins, William Fitzgerald (see Leinster, Murray)

Jenkins, Herbert, Limited DLB-112

Jennings, Elizabeth 1926- DLB-27

Jens, Walter 1923- DLB-69

Jensen, Merrill 1905-1980 . . . . DLB-17

Jephson, Robert 1736-1803 . . . . DLB-89

Jerome, Jerome K. 1859-1927 . . . . DLB-10, 34, 135

Jerome, Judson 1927-1991 . . . . DLB-105

Jerome, Judson, Reflections: After a Tornado DLB-105

Jerrold, Douglas 1803-1857 . . . . DLB-158, 159

Jesse, F. Tennyson 1888-1958 . . . . DLB-77

Jewett, Sarah Orne 1849-1909 . . . . DLB-12, 74

Jewett, John P., and Company DLB-49

The Jewish Publication Society DLB-49

Jewitt, John Rodgers 1783-1821 . . . . . DLB-99

Jewsbury, Geraldine 1812-1880 . . . . . DLB-21

Jhabvala, Ruth Prawer 1927- DLB-139

Jiménez, Juan Ramón 1881-1958 . . . . DLB-134

Joans, Ted 1928- DLB-16, 41

John, Eugenie (see Marlitt, E.)

John of Dumbleton circa 1310-circa 1349 . . . . DLB-115

John Edward Bruce: Three Documents DLB-50

John O'Hara's Pottsville Journalism Y-88

John Steinbeck Research Center Y-85

John Webster: The Melbourne Manuscript Y-86

Johns, Captain W. E. 1893-1968 . . . . DLB-160

Johnson, B. S. 1933-1973 . . . . DLB-14, 40

Johnson, Charles 1679-1748 . . . . DLB-84

Johnson, Charles R. 1948- DLB-33

Johnson, Charles S. 1893-1956 . . . DLB-51, 91

Johnson, Denis 1949- DLB-120

Johnson, Diane 1934- Y-80

Johnson, Edgar 1901- DLB-103

Johnson, Edward 1598-1672 . . . . DLB-24

Johnson E. Pauline (Tekahionwake) 1861-1913 . . . . DLB-175

Johnson, Fenton 1888-1958 . . . . . . DLB-45, 50

Johnson, Georgia Douglas 1886-1966 . . . . DLB-51

Johnson, Gerald W. 1890-1980 . . . . . . DLB-29

Johnson, Helene 1907- DLB-51

337

Johnson, James Weldon 1871-1938 ... DLB-51

Johnson, John H. 1918- ... DLB-137

Johnson, Linton Kwesi 1952- ... DLB-157

Johnson, Lionel 1867-1902 ... DLB-19

Johnson, Nunnally 1897-1977 ... DLB-26

Johnson, Owen 1878-1952 ... Y-87

Johnson, Pamela Hansford 1912- ... DLB-15

Johnson, Pauline 1861-1913 ... DLB-92

Johnson, Ronald 1935- ... DLB-169

Johnson, Samuel 1696-1772 ... DLB-24

Johnson, Samuel 1709-1784 ... DLB-39, 95, 104, 142

Johnson, Samuel 1822-1882 ... DLB-1

Johnson, Uwe 1934-1984 ... DLB-75

Johnson, Benjamin [publishing house] ... DLB-49

Johnson, Benjamin, Jacob, and Robert [publishing house] ... DLB-49

Johnson, Jacob, and Company ... DLB-49

Johnson, Joseph [publishing house] ... DLB-154

Johnston, Annie Fellows 1863-1931 ... DLB-42

Johnston, Basil H. 1929- ... DLB-60

Johnston, Denis 1901-1984 ... DLB-10

Johnston, George 1913- ... DLB-88

Johnston, Sir Harry 1858-1927 ... DLB-174

Johnston, Jennifer 1930- ... DLB-14

Johnston, Mary 1870-1936 ... DLB-9

Johnston, Richard Malcolm 1822-1898 ... DLB-74

Johnstone, Charles 1719?-1800? ... DLB-39

Johst, Hanns 1890-1978 ... DLB-124

Jolas, Eugene 1894-1952 ... DLB-4, 45

Jones, Alice C. 1853-1933 ... DLB-92

Jones, Charles C., Jr. 1831-1893 ... DLB-30

Jones, D. G. 1929- ... DLB-53

Jones, David 1895-1974 ... DLB-20, 100

Jones, Diana Wynne 1934- ... DLB-161

Jones, Ebenezer 1820-1860 ... DLB-32

Jones, Ernest 1819-1868 ... DLB-32

Jones, Gayl 1949- ... DLB-33

Jones, Glyn 1905- ... DLB-15

Jones, Gwyn 1907- ... DLB-15, 139

Jones, Henry Arthur 1851-1929 ... DLB-10

Jones, Hugh circa 1692-1760 ... DLB-24

Jones, James 1921-1977 ... DLB-2, 143

Jones, Jenkin Lloyd 1911- ... DLB-127

Jones, LeRoi (see Baraka, Amiri)

Jones, Lewis 1897-1939 ... DLB-15

Jones, Madison 1925- ... DLB-152

Jones, Major Joseph (see Thompson, William Tappan)

Jones, Preston 1936-1979 ... DLB-7

Jones, Rodney 1950- ... DLB-120

Jones, Sir William 1746-1794 ... DLB-109

Jones, William Alfred 1817-1900 ... DLB-59

Jones's Publishing House ... DLB-49

Jong, Erica 1942- ... DLB-2, 5, 28, 152

Jonke, Gert F. 1946- ... DLB-85

Jonson, Ben 1572?-1637 ... DLB-62, 121

Jordan, June 1936- ... DLB-38

Joseph, Jenny 1932- ... DLB-40

Joseph, Michael, Limited ... DLB-112

Josephson, Matthew 1899-1978 ... DLB-4

Josephus, Flavius 37-100 ... DLB-176

Josiah Allen's Wife (see Holley, Marietta)

Josipovici, Gabriel 1940- ... DLB-14

Josselyn, John ?-1675 ... DLB-24

Joudry, Patricia 1921- ... DLB-88

Jovine, Giuseppe 1922- ... DLB-128

Joyaux, Philippe (see Sollers, Philippe)

Joyce, Adrien (see Eastman, Carol)

Joyce, James 1882-1941 ... DLB-10, 19, 36, 162

Judd, Sylvester 1813-1853 ... DLB-1

Judd, Orange, Publishing Company ... DLB-49

Judith circa 930 ... DLB-146

Julian of Norwich 1342-circa 1420 ... DLB-1146

Julian Symons at Eighty ... Y-92

June, Jennie (see Croly, Jane Cunningham)

Jung, Franz 1888-1963 ... DLB-118

Jünger, Ernst 1895- ... DLB-56

Der jüngere Titurel circa 1275 ... DLB-138

Jung-Stilling, Johann Heinrich 1740-1817 ... DLB-94

Justice, Donald 1925- ... Y-83

The Juvenile Library (see Godwin, M. J., and Company)

K

Kacew, Romain (see Gary, Romain)

Kafka, Franz 1883-1924 ... DLB-81

Kahn, Roger 1927- ... DLB-171

Kaiser, Georg 1878-1945 ... DLB-124

Kaiserchronik circca 1147 ... DLB-148

Kalechofsky, Roberta 1931- ... DLB-28

Kaler, James Otis 1848-1912 ... DLB-12

Kames, Henry Home, Lord 1696-1782 ... DLB-31, 104

Kandel, Lenore 1932- ... DLB-16

Kanin, Garson 1912- ... DLB-7

Kant, Hermann 1926- ... DLB-75

Kant, Immanuel 1724-1804 ... DLB-94

Kantemir, Antiokh Dmitrievich 1708-1744 ... DLB-150

Kantor, Mackinlay 1904-1977 ... DLB-9, 102

Kaplan, Fred 1937- ... DLB-111

Kaplan, Johanna 1942- ... DLB-28

Kaplan, Justin 1925- ... DLB-111

Kapnist, Vasilii Vasilevich 1758?-1823 ... DLB-150

Karadií, Vuk Stefanovií 1787-1864 ... DLB-147

Karamzin, Nikolai Mikhailovich 1766-1826 ... DLB-150

Karsch, Anna Louisa 1722-1791 ... DLB-97

Kasack, Hermann 1896-1966 ... DLB-69

Kaschnitz, Marie Luise 1901-1974 ... DLB-69

Kaštelan, Jure 1919-1990 ... DLB-147

Kästner, Erich 1899-1974 ... DLB-56

Kattan, Naim 1928- ... DLB-53

Katz, Steve 1935- ... Y-83

Kauffman, Janet 1945- ... Y 86

Kauffmann, Samuel 1898-1971 ... DLB-127

Kaufman, Bob 1925- ... DLB-16, 41

Kaufman, George S. 1889-1961 ... DLB-7

Kavanagh, P. J. 1931- ... DLB-40

Kavanagh, Patrick 1904-1967 ... DLB-15, 20

Kaye-Smith, Sheila 1887-1956 ... DLB-36

Kazin, Alfred 1915- ... DLB-67

Keane, John B. 1928- ... DLB-13

Keary, Annie 1825-1879 ... DLB-163

Keating, H. R. F. 1926- ... DLB-87

Keats, Ezra Jack 1916-1983 ... DLB-61

Keats, John 1795-1821 ... DLB-96, 110

DLB Yearbook 1996 — Cumulative Index

Keble, John 1792-1866 . . . . DLB-32, 55

Keeble, John 1944- Y-83

Keeffe, Barrie 1945- DLB-13

Keeley, James 1867-1934 . . . . DLB-25

W. B. Keen, Cooke and Company DLB-49

Keillor, Garrison 1942- Y-87

Keith, Marian 1874?-1961 . . . . DLB-92

Keller, Gary D. 1943- DLB-82

Keller, Gottfried 1819-1890 . . . . DLB-129

Kelley, Edith Summers 1884-1956 . . . . DLB-9

Kelley, William Melvin 1937- DLB-33

Kellogg, Ansel Nash 1832-1886 . . . . . DLB-23

Kellogg, Steven 1941- DLB-61

Kelly, George 1887-1974 . . . . DLB-7

Kelly, Hugh 1739-1777 . . . . DLB-89

Kelly, Robert 1935- DLB-5, 130, 165

Kelly, Piet and Company DLB-49

Kelmscott Press DLB-112

Kemble, Fanny 1809-1893 . . . . DLB-32

Kemelman, Harry 1908- DLB-28

Kempe, Margery circa 1373-1438 . . . . DLB-146

Kempner, Friederike 1836-1904 . . . . DLB-129

Kempowski, Walter 1929- DLB-75

Kendall, Claude [publishing company] DLB-46

Kendell, George 1809-1867 . . . . DLB-43

Kenedy, P. J., and Sons DLB-49

Kennedy, Adrienne 1931- DLB-38

Kennedy, John Pendleton 1795-1870 . . . DLB-3

Kennedy, Leo 1907- DLB-88

Kennedy, Margaret 1896-1967 . . . . . . DLB-36

Kennedy, Patrick 1801-1873 . . . . DLB-159

Kennedy, Richard S. 1920- DLB-111

Kennedy, William 1928- DLB-143; Y-85

Kennedy, X. J. 1929- DLB-5

Kennelly, Brendan 1936- DLB-40

Kenner, Hugh 1923- DLB-67

Kennerley, Mitchell [publishing house] DLB-46

Kenny, Maurice 1929- DLB-175

Kent, Frank R. 1877-1958 . . . . DLB-29

Kenyon, Jane 1947- DLB-120

Keough, Hugh Edmund 1864-1912 . . DLB-171

Keppler and Schwartzmann DLB-49

Kerner, Justinus 1776-1862 . . . . DLB-90

Kerouac, Jack 1922-1969 . . . . DLB-2, 16; DS-3

The Jack Kerouac Revival Y-95

Kerouac, Jan 1952- DLB-16

Kerr, Orpheus C. (see Newell, Robert Henry)

Kerr, Charles H., and Company DLB-49

Kesey, Ken 1935- DLB-2, 16

Kessel, Joseph 1898-1979 . . . . DLB-72

Kessel, Martin 1901- DLB-56

Kesten, Hermann 1900- DLB-56

Keun, Irmgard 1905-1982 . . . . DLB-69

Key and Biddle DLB-49

Keynes, John Maynard 1883-1946 . . . . DS-10

Keyserling, Eduard von 1855-1918 . . DLB-66

Khan, Ismith 1925- DLB-125

Khemnitser, Ivan Ivanovich 1745-1784 . . . . DLB-150

Kheraskov, Mikhail Matveevich 1733-1807 . . . . DLB-150

Khvostov, Dmitrii Ivanovich 1757-1835 . . . . DLB-150

Kidd, Adam 1802?-1831 . . . . DLB-99

Kidd, William [publishing house] DLB-106

Kiely, Benedict 1919- DLB-15

Kieran, John 1892-1981 . . . . DLB-171

Kiggins and Kellogg DLB-49

Kiley, Jed 1889-1962 . . . . DLB-4

Kilgore, Bernard 1908-1967 . . . . DLB-127

Killens, John Oliver 1916- DLB-33

Killigrew, Anne 1660-1685 . . . . DLB-131

Killigrew, Thomas 1612-1683 . . . . . . DLB-58

Kilmer, Joyce 1886-1918 . . . . DLB-45

Kilwardby, Robert circa 1215-1279 . . . . DLB-115

Kincaid, Jamaica 1949- DLB-157

King, Clarence 1842-1901 . . . . DLB-12

King, Florence 1936 . . . . Y-85

King, Francis 1923- DLB-15, 139

King, Grace 1852-1932 . . . . DLB-12, 78

King, Henry 1592-1669 . . . . DLB-126

King, Stephen 1947- DLB-143; Y-80

King, Thomas 1943- DLB-175

King, Woodie, Jr. 1937- DLB-38

King, Solomon [publishing house] . . . DLB-49

Kinglake, Alexander William 1809-1891 . . . . DLB-55, 166

Kingsley, Charles 1819-1875 . . . . DLB-21, 32, 163

Kingsley, Mary Henrietta 1862-1900 . . . . DLB-174

Kingsley, Henry 1830-1876 . . . . DLB-21

Kingsley, Sidney 1906- DLB-7

Kingsmill, Hugh 1889-1949 . . . . DLB-149

Kingston, Maxine Hong 1940- DLB-173; Y-80

Kingston, William Henry Giles 1814-1880 . . . . DLB-163

Kinnell, Galway 1927- DLB-5; Y-87

Kinsella, Thomas 1928- DLB-27

Kipling, Rudyard 1865-1936 . . . . DLB-19, 34, 141, 156

Kipphardt, Heinar 1922-1982 . . . . . . DLB-124

Kirby, William 1817-1906 . . . . DLB-99

Kircher, Athanasius 1602-1680 . . . . . DLB-164

Kirk, John Foster 1824-1904 . . . . DLB-79

Kirkconnell, Watson 1895-1977 . . . . . DLB-68

Kirkland, Caroline M. 1801-1864 . . . . DLB-3, 73, 74; DS-13

Kirkland, Joseph 1830-1893 . . . . DLB-12

Kirkman, Francis [publishing house] DLB-170

Kirkpatrick, Clayton 1915- DLB-127

Kirkup, James 1918- DLB-27

Kirouac, Conrad (see Marie-Victorin, Frère)

Kirsch, Sarah 1935- DLB-75

Kirst, Hans Hellmut 1914-1989 . . . . . DLB-69

Kitcat, Mabel Greenhow 1859-1922 . . . . DLB-135

Kitchin, C. H. B. 1895-1967 . . . . DLB-77

Kizer, Carolyn 1925- DLB-5, 169

Klabund 1890-1928 . . . . DLB-66

Klaj, Johann 1616-1656 . . . . DLB-164

Klappert, Peter 1942- DLB-5

Klass, Philip (see Tenn, William)

Klein, A. M. 1909-1972 . . . . DLB-68

Kleist, Ewald von 1715-1759 . . . . DLB-97

Kleist, Heinrich von 1777-1811 . . . . . DLB-90

Klinger, Friedrich Maximilian 1752-1831 . . . . DLB-94

Klopstock, Friedrich Gottlieb 1724-1803 . . . . DLB-97

Klopstock, Meta 1728-1758 . . . . DLB-97

Kluge, Alexander 1932- DLB-75

Knapp, Joseph Palmer 1864-1951 .... DLB-91

Knapp, Samuel Lorenzo
    1783-1838 .... DLB-59

Knapton, J. J. and P.
    [publishing house] DLB-154

Kniazhnin, Iakov Borisovich
    1740-1791 .... DLB-150

Knickerbocker, Diedrich (see Irving, Washington)

Knigge, Adolph Franz Friedrich Ludwig,
    Freiherr von 1752-1796 .... DLB-94

Knight, Damon 1922- DLB-8

Knight, Etheridge 1931-1992 .... DLB-41

Knight, John S. 1894-1981 .... DLB-29

Knight, Sarah Kemble 1666-1727 .... DLB-24

Knight, Charles, and Company DLB-106

Knight-Bruce, G. W. H.
    1852-1896 .... DLB-174

Knister, Raymond 1899-1932 .... DLB-68

Knoblock, Edward 1874-1945 .... DLB-10

Knopf, Alfred A. 1892-1984 .... Y-84

Knopf, Alfred A.
    [publishing house] DLB-46

Knorr von Rosenroth, Christian
    1636-1689 .... DLB-168

Knowles, John 1926- DLB-6

Knox, Frank 1874-1944 .... DLB-29

Knox, John circa 1514-1572 .... DLB-132

Knox, John Armoy 1850-1906 .... DLB-23

Knox, Ronald Arbuthnott
    1888-1957 .... DLB-77

Kober, Arthur 1900-1975 .... DLB-11

Kocbek, Edvard 1904-1981 .... DLB-147

Koch, Howard 1902- DLB-26

Koch, Kenneth 1925- DLB-5

Koenigsberg, Moses 1879-1945 ..... DLB-25

Koeppen, Wolfgang 1906- DLB-69

Koertge, Ronald 1940- DLB-105

Koestler, Arthur 1905-1983 .... Y-83

Kokoschka, Oskar 1886-1980 ..... DLB-124

Kolb, Annette 1870-1967 .... DLB-66

Kolbenheyer, Erwin Guido
    1878-1962 .... DLB-66, 124

Kolleritsch, Alfred 1931- DLB-85

Kolodny, Annette 1941- DLB-67

Komarov, Matvei
    circa 1730-1812 .... DLB-150

Komroff, Manuel 1890-1974 .... DLB-4

Komunyakaa, Yusef 1947- DLB-120

Konigsburg, E. L. 1930- DLB-52

Konrad von Würzburg
    circa 1230-1287 .... DLB-138

Konstantinov, Aleko 1863-1897 .... DLB-147

Kooser, Ted 1939- DLB-105

Kopit, Arthur 1937- DLB-7

Kops, Bernard 1926?- DLB-13

Kornbluth, C. M. 1923-1958 .... DLB-8

Körner, Theodor 1791-1813 .... DLB-90

Kornfeld, Paul 1889-1942 .... DLB-118

Kosinski, Jerzy 1933-1991 ...... DLB-2; Y-82

Kosovel, Sreúko 1904-1926 .... DLB-147

Kostrov, Ermil Ivanovich
    1755-1796 .... DLB-150

Kotzebue, August von 1761-1819 ... DLB-94

Kotzwinkle, William 1938- DLB-173

Kovaúií, Ante 1854-1889 .... DLB-147

Kraf, Elaine 1946- Y-81

Kranjúevií, Silvije Strahimir
    1865-1908 .... DLB-147

Krasna, Norman 1909-1984 .... DLB-26

Kraus, Karl 1874-1936 .... DLB-118

Krauss, Ruth 1911-1993 .... DLB-52

Kreisel, Henry 1922- DLB-88

Kreuder, Ernst 1903-1972 .... DLB-69

Kreymborg, Alfred 1883-1966 .... DLB-4, 54

Krieger, Murray 1923- DLB-67

Krim, Seymour 1922-1989 .... DLB-16

Krlea, Miroslav 1893-1981 .... DLB-147

Krock, Arthur 1886-1974 .... DLB-29

Kroetsch, Robert 1927- DLB-53

Krutch, Joseph Wood 1893-1970 .... DLB-63

Krylov, Ivan Andreevich
    1769-1844 .... DLB-150

Kubin, Alfred 1877-1959 .... DLB-81

Kubrick, Stanley 1928- DLB-26

Kudrun circa 1230-1240 .... DLB-138

Kuffstein, Hans Ludwig von
    1582-1656 .... DLB-164

Kuhlmann, Quirinus 1651-1689 .... DLB-168

Kuhnau, Johann 1660-1722 .... DLB-168

Kumin, Maxine 1925- DLB-5

Kunene, Mazisi 1930- DLB-117

Kunitz, Stanley 1905- DLB-48

Kunjufu, Johari M. (see Amini, Johari M.)

Kunnert, Gunter 1929- DLB-75

Kunze, Reiner 1933- DLB-75

Kupferberg, Tuli 1923- DLB-16

Kürnberger, Ferdinand
    1821-1879 .... DLB-129

Kurz, Isolde 1853-1944 .... DLB-66

Kusenberg, Kurt 1904-1983 .... DLB-69

Kuttner, Henry 1915-1958 .... DLB-8

Kyd, Thomas 1558-1594 .... DLB-62

Kyffin, Maurice
    circa 1560?-1598 .... DLB-136

Kyger, Joanne 1934- DLB-16

Kyne, Peter B. 1880-1957 .... DLB-78

L

L. E. L. (see Landon, Letitia Elizabeth)

Laberge, Albert 1871-1960 .... DLB-68

Laberge, Marie 1950- DLB-60

Lacombe, Patrice (see Trullier-Lacombe, Joseph Patrice)

Lacretelle, Jacques de 1888-1985 .... DLB-65

Lacy, Sam 1903- DLB-171

Ladd, Joseph Brown 1764-1786 ..... DLB-37

La Farge, Oliver 1901-1963 .... DLB-9

Lafferty, R. A. 1914- DLB-8

La Flesche, Francis 1857-1932 ..... DLB-175

La Guma, Alex 1925-1985 .... DLB-117

Lahaise, Guillaume (see Delahaye, Guy)

Lahontan, Louis-Armand de Lom d'Arce,
    Baron de 1666-1715? DLB-99

Laing, Kojo 1946- DLB-157

Laird, Carobeth 1895- Y-82

Laird and Lee DLB-49

Lalonde, Michèle 1937- DLB-60

Lamantia, Philip 1927- DLB-16

Lamb, Charles
    1775-1834 .... DLB-93, 107, 163

Lamb, Lady Caroline 1785-1828 ... DLB-116

Lamb, Mary 1764-1847 .... DLB-163

Lambert, Betty 1933-1983 .... DLB-60

Lamming, George 1927- DLB-125

L'Amour, Louis 1908?- Y-80

Lampman, Archibald 1861-1899 .... DLB-92

Lamson, Wolffe and Company DLB-49

Lancer Books DLB-46

Landesman, Jay 1919- and
  Landesman, Fran 1927- DLB-16

Landolfi, Tommaso 1908-1979 ..... DLB-177

Landon, Letitia Elizabeth 1802-1838 . DLB-96

Landor, Walter Savage
  1775-1864 .... DLB-93, 107

Landry, Napoléon-P. 1884-1956 ..... DLB-92

Lane, Charles 1800-1870 .... DLB-1

Lane, Laurence W. 1890-1967 ...... DLB-91

Lane, M. Travis 1934- DLB-60

Lane, Patrick 1939- DLB-53

Lane, Pinkie Gordon 1923- DLB-41

Lane, John, Company DLB-49

Laney, Al 1896-1988 .... DLB-4, 171

Lang, Andrew 1844-1912 ...... DLB-98, 141

Langevin, André 1927- DLB-60

Langgässer, Elisabeth 1899-1950 .... DLB-69

Langhorne, John 1735-1779 .... DLB-109

Langland, William
  circa 1330-circa 1400 .... DLB-146

Langton, Anna 1804-1893 .... DLB-99

Lanham, Edwin 1904-1979 .... DLB-4

Lanier, Sidney 1842-1881 .... DLB-64; DS-13

Lanyer, Aemilia 1569-1645 .... DLB-121

Lapointe, Gatien 1931-1983 .... DLB-88

Lapointe, Paul-Marie 1929- DLB-88

Lardner, John 1912-1960 .... DLB-171

Lardner, Ring
  1885-1933 .... DLB-11, 25, 86, 171

Lardner, Ring, Jr. 1915- DLB-26

Lardner 100: Ring Lardner
  Centennial Symposium Y-85

Larkin, Philip 1922-1985 .... DLB-27

La Roche, Sophie von 1730-1807 .... DLB-94

La Rocque, Gilbert 1943-1984 ...... DLB-60

Laroque de Roquebrune, Robert (see Roquebrune, Robert de)

Larrick, Nancy 1910- DLB-61

Larsen, Nella 1893-1964 .... DLB-51

Lasker-Schüler, Else
  1869-1945 .... DLB-66, 124

Lasnier, Rina 1915- DLB-88

Lassalle, Ferdinand 1825-1864 ..... DLB-129

Lathrop, Dorothy P. 1891-1980 ..... DLB-22

Lathrop, George Parsons
  1851-1898 .... DLB-71

Lathrop, John, Jr. 1772-1820 ...... DLB-37

Latimer, Hugh 1492?-1555 .... DLB-136

Latimore, Jewel Christine McLawler
  (see Amini, Johari M.)

Latymer, William 1498-1583 .... DLB-132

Laube, Heinrich 1806-1884 .... DLB-133

Laughlin, James 1914- DLB-48

Laumer, Keith 1925- DLB-8

Lauremberg, Johann 1590-1658 .... DLB-164

Laurence, Margaret 1926-1987 ...... DLB-53

Laurentius von Schnüffis
  1633-1702 .... DLB-168

Laurents, Arthur 1918- DLB-26

Laurie, Annie (see Black, Winifred)

Laut, Agnes Christiana 1871-1936 ... DLB-92

Lavater, Johann Kaspar 1741-1801 ... DLB-97

Lavin, Mary 1912- DLB-15

Lawes, Henry 1596-1662 .... DLB-126

Lawless, Anthony (see MacDonald, Philip)

Lawrence, D. H.
  1885-1930 ...... DLB-10, 19, 36, 98, 162

Lawrence, David 1888-1973 .... DLB-29

Lawrence, Seymour 1926-1994 .... Y-94

Lawson, John ?-1711 .... DLB-24

Lawson, Robert 1892-1957 .... DLB-22

Lawson, Victor F. 1850-1925 .... DLB-25

Layard, Sir Austen Henry
  1817-1894 .... DLB-166

Layton, Irving 1912- DLB-88

LaZamon flourished circa 1200 .... DLB-146

Lazareví, Laza K. 1851-1890 ...... DLB-147

Lea, Henry Charles 1825-1909 ..... DLB-47

Lea, Sydney 1942- DLB-120

Lea, Tom 1907- DLB-6

Leacock, John 1729-1802 .... DLB-31

Leacock, Stephen 1869-1944 .... DLB-92

Lead, Jane Ward 1623-1704 .... DLB-131

Leadenhall Press DLB-106

Leapor, Mary 1722-1746 .... DLB-109

Lear, Edward 1812-1888 ... DLB-32, 163, 166

Leary, Timothy 1920-1996 .... DLB-16

Leary, W. A., and Company DLB-49

Léautaud, Paul 1872-1956 .... DLB-65

Leavitt, David 1961- DLB-130

Leavitt and Allen DLB-49

Le Blond, Mrs. Aubrey
  1861-1934 .... DLB-174

le Carré, John 1931- DLB-87

Lécavelé, Roland (see Dorgeles, Roland)

Lechlitner, Ruth 1901- DLB-48

Leclerc, Félix 1914- DLB-60

Le Clézio, J. M. G. 1940- DLB-83

Lectures on Rhetoric and Belles Lettres (1783),
  by Hugh Blair [excerpts] DLB-31

Leder, Rudolf (see Hermlin, Stephan)

Lederer, Charles 1910-1976 .... DLB-26

Ledwidge, Francis 1887-1917 ...... DLB-20

Lee, Dennis 1939- DLB-53

Lee, Don L. (see Madhubuti, Haki R.)

Lee, George W. 1894-1976 .... DLB-51

Lee, Harper 1926- DLB-6

Lee, Harriet (1757-1851) and
  Lee, Sophia (1750-1824) DLB-39

Lee, Laurie 1914- DLB-27

Lee, Li-Young 1957- DLB-165

Lee, Manfred B. (see Dannay, Frederic, and
  Manfred B. Lee)

Lee, Nathaniel circa 1645 - 1692 ..... DLB-80

Lee, Sir Sidney 1859-1926 .... DLB-149

Lee, Sir Sidney, "Principles of Biography," in
  Elizabethan and Other Essays DLB-149

Lee, Vernon
  1856-1935 .... DLB-57, 153, 156, 174

Lee and Shepard DLB-49

Le Fanu, Joseph Sheridan
  1814-1873 .... DLB-21, 70, 159

Leffland, Ella 1931- Y-84

le Fort, Gertrud von 1876-1971 ..... DLB-66

Le Gallienne, Richard 1866-1947 ..... DLB-4

Legaré, Hugh Swinton
  1797-1843 .... DLB-3, 59, 73

Legaré, James M. 1823-1859 .... DLB-3

The Legends of the Saints and a Medieval
  Christian Worldview DLB-148

Léger, Antoine-J. 1880-1950 .... DLB-88

Le Guin, Ursula K. 1929- DLB-8, 52

Lehman, Ernest 1920- DLB-44

Lehmann, John 1907- DLB-27, 100

Lehmann, Rosamond 1901-1990 ..... DLB-15

Lehmann, Wilhelm 1882-1968 ...... DLB-56

Lehmann, John, Limited DLB-112

Leiber, Fritz 1910-1992 .... DLB-8

Leibniz, Gottfried Wilhelm
  1646-1716 .... DLB-168

Leicester University Press DLB-112

Leinster, Murray 1896-1975 .... DLB-8

Leisewitz, Johann Anton 1752-1806 .... DLB-94

Leitch, Maurice 1933- DLB-14

Leithauser, Brad 1943- DLB-120

Leland, Charles G. 1824-1903 .... DLB-11

Leland, John 1503?-1552 .... DLB-136

Lemay, Pamphile 1837-1918 .... DLB-99

Lemelin, Roger 1919- DLB-88

Lemon, Mark 1809-1870 .... DLB-163

Le Moine, James MacPherson 1825-1912 .... DLB-99

Le Moyne, Jean 1913- DLB-88

L'Engle, Madeleine 1918- DLB-52

Lennart, Isobel 1915-1971 .... DLB-44

Lennox, Charlotte 1729 or 1730-1804 .... DLB-39

Lenox, James 1800-1880 .... DLB-140

Lenski, Lois 1893-1974 .... DLB-22

Lenz, Hermann 1913- DLB-69

Lenz, J. M. R. 1751-1792 .... DLB-94

Lenz, Siegfried 1926- DLB-75

Leonard, Elmore 1925- DLB-173

Leonard, Hugh 1926- DLB-13

Leonard, William Ellery 1876-1944 .... DLB-54

Leonowens, Anna 1834-1914 ... DLB-99, 166

LePan, Douglas 1914- DLB-88

Leprohon, Rosanna Eleanor 1829-1879 .... DLB-99

Le Queux, William 1864-1927 ...... DLB-70

Lerner, Max 1902-1992 .... DLB-29

Lernet-Holenia, Alexander 1897-1976 .... DLB-85

Le Rossignol, James 1866-1969 ...... DLB-92

Lescarbot, Marc circa 1570-1642 .... DLB-99

LeSeur, William Dawson 1840-1917 .... DLB-92

LeSieg, Theo. (see Geisel, Theodor Seuss)

Leslie, Frank 1821-1880 .... DLB-43, 79

Leslie, Frank, Publishing House DLB-49

Lesperance, John 1835?-1891 .... DLB-99

Lessing, Bruno 1870-1940 .... DLB-28

Lessing, Doris 1919- DLB-15, 139; Y-85

Lessing, Gotthold Ephraim 1729-1781 .... DLB-97

Lettau, Reinhard 1929- DLB-75

Letter from Japan Y-94

Letter from London Y-96

Letter to [Samuel] Richardson on *Clarissa* (1748), by Henry Fielding DLB-39

Lever, Charles 1806-1872 .... DLB-21

Leverson, Ada 1862-1933 .... DLB-153

Levertov, Denise 1923- DLB-5, 165

Levi, Peter 1931- DLB-40

Levi, Primo 1919-1987 .... DLB-177

Levien, Sonya 1888-1960 .... DLB-44

Levin, Meyer 1905-1981 .... DLB-9, 28; Y-81

Levine, Norman 1923- DLB-88

Levine, Philip 1928- DLB-5

Levis, Larry 1946- DLB-120

Levy, Amy 1861-1889 .... DLB-156

Levy, Benn Wolfe 1900-1973 .... DLB-13; Y-81

Lewald, Fanny 1811-1889 .... DLB-129

Lewes, George Henry 1817-1878 .... DLB-55, 144

Lewis, Agnes Smith 1843-1926 ..... DLB-174

Lewis, Alfred H. 1857-1914 .... DLB-25

Lewis, Alun 1915-1944 .... DLB-20, 162

Lewis, C. Day (see Day Lewis, C.)

Lewis, C. S. 1898-1963 .... DLB-15, 100, 160

Lewis, Charles B. 1842-1924 .... DLB-11

Lewis, Henry Clay 1825-1850 .... DLB-3

Lewis, Janet 1899- Y-87

Lewis, Matthew Gregory 1775-1818 .... DLB-39, 158

Lewis, R. W. B. 1917- DLB-111

Lewis, Richard circa 1700-1734 ..... DLB-24

Lewis, Sinclair 1885-1951 .... DLB-9, 102; DS-1

Lewis, Wilmarth Sheldon 1895-1979 .... DLB-140

Lewis, Wyndham 1882-1957 .... DLB-15

Lewisohn, Ludwig 1882-1955 .... DLB-4, 9, 28, 102

Lezama Lima, José 1910-1976 ...... DLB-113

The Library of America DLB-46

The Licensing Act of 1737 .... DLB-84

Lichfield, Leonard I [publishing house] DLB-170

Lichtenberg, Georg Christoph 1742-1799 .... DLB-94

Lieb, Fred 1888-1980 .... DLB-171

Liebling, A. J. 1904-1963 .... DLB-4, 171

Lieutenant Murray (see Ballou, Maturin Murray)

Lighthall, William Douw 1857-1954 .... DLB-92

Lilar, Françoise (see Mallet-Joris, Françoise)

Lillo, George 1691-1739 .... DLB-84

Lilly, J. K., Jr. 1893-1966 .... DLB-140

Lilly, Wait and Company DLB-49

Lily, William circa 1468-1522 ...... DLB-132

Limited Editions Club DLB-46

Lincoln and Edmands DLB-49

Lindsay, Jack 1900- Y-84

Lindsay, Sir David circa 1485-1555 .... DLB-132

Lindsay, Vachel 1879-1931 .... DLB-54

Linebarger, Paul Myron Anthony (see Smith, Cordwainer)

Link, Arthur S. 1920- DLB-17

Linn, John Blair 1777-1804 .... DLB-37

Lins, Osman 1924-1978 .... DLB-145

Linton, Eliza Lynn 1822-1898 .... DLB-18

Linton, William James 1812-1897 .... DLB-32

Lintot, Barnaby Bernard [publishing house] DLB-170

Lion Books DLB-46

Lionni, Leo 1910- DLB-61

Lippincott, Sara Jane Clarke 1823-1904 .... DLB-43

Lippincott, J. B., Company DLB-49

Lippmann, Walter 1889-1974 .... DLB-29

Lipton, Lawrence 1898-1975 .... DLB-16

Liscow, Christian Ludwig 1701-1760 .... DLB-97

Lish, Gordon 1934- DLB-130

Lispector, Clarice 1925-1977 .... DLB-113

The Literary Chronicle and Weekly Review 1819-1828 .... DLB-110

Literary Documents: William Faulkner and the People-to-People Program Y-86

Literary Documents II: *Library Journal* Statements and Questionnaires from First Novelists Y-87

Literary Effects of World War II [British novel] DLB-15

Literary Prizes [British] DLB-15

Literary Research Archives: The Humanities Research Center, University of Texas.................Y-82

Literary Research Archives II: Berg Collection of English and American Literature of the New York Public Library....................Y-83

Literary Research Archives III: The Lilly Library.................Y-84

Literary Research Archives IV: The John Carter Brown Library....Y-85

Literary Research Archives V: Kent State Special Collections......Y-86

Literary Research Archives VI: The Modern Literary Manuscripts Collection in the Special Collections of the Washington University Libraries..............Y-87

Literary Research Archives VII: The University of Virginia Libraries.....................Y-91

Literary Research Archives VIII: The Henry E. Huntington Library.....................Y-92

"Literary Style" (1857), by William Forsyth [excerpt].............DLB-57

Literatura Chicanesca: The View From Without DLB-82

Literature at Nurse, or Circulating Morals (1885), by George Moore..............DLB-18

Littell, Eliakim 1797-1870..........DLB-79

Littell, Robert S. 1831-1896........DLB-79

Little, Brown and Company........DLB-49

Littlewood, Joan 1914-............DLB-13

Lively, Penelope 1933-......DLB-14, 161

Liverpool University Press........DLB-112

The Lives of the Poets.................DLB-142

Livesay, Dorothy 1909-............DLB-68

Livesay, Florence Randal 1874-1953...................DLB-92

Livings, Henry 1929-.............DLB-13

Livingston, Anne Howe 1763-1841....................DLB-37

Livingston, Myra Cohn 1926-......DLB-61

Livingston, William 1723-1790......DLB-31

Livingstone, David 1813-1873......DLB-166

Liyong, Taban lo (see Taban lo Liyong)

Lizárraga, Sylvia S. 1925-..........DLB-82

Llewellyn, Richard 1906-1983.......DLB-15

Lloyd, Edward [publishing house]............DLB-106

Lobel, Arnold 1933-...............DLB-61

Lochridge, Betsy Hopkins (see Fancher, Betsy)

Locke, David Ross 1833-1888....DLB-11, 23

Locke, John 1632-1704........DLB-31, 101

Locke, Richard Adams 1800-1871...DLB-43

Locker-Lampson, Frederick 1821-1895....................DLB-35

Lockhart, John Gibson 1794-1854..........DLB-110, 116 144

Lockridge, Ross, Jr. 1914-1948..............DLB-143; Y-80

Locrine and *Selimus*.................DLB-62

Lodge, David 1935-..............DLB-14

Lodge, George Cabot 1873-1909....DLB-54

Lodge, Henry Cabot 1850-1924.....DLB-47

Lodge, Thomas 1558-1625........DLB-172

Loeb, Harold 1891-1974...........DLB-4

Loeb, William 1905-1981..........DLB-127

Lofting, Hugh 1886-1947..........DLB-160

Logan, James 1674-1751.......DLB-24, 140

Logan, John 1923-................DLB-5

Logan, William 1950-............DLB-120

Logau, Friedrich von 1605-1655....DLB-164

Logue, Christopher 1926-.........DLB-27

Lohenstein, Daniel Casper von 1635-1683....................DLB-168

Lomonosov, Mikhail Vasil'evich 1711-1765....................DLB-150

London, Jack 1876-1916......DLB-8, 12, 78

The London Magazine 1820-1829.....DLB-110

Long, Haniel 1888-1956............DLB-45

Long, Ray 1878-1935.............DLB-137

Long, H., and Brother.............DLB-49

Longfellow, Henry Wadsworth 1807-1882..................DLB-1, 59

Longfellow, Samuel 1819-1892.......DLB-1

Longford, Elizabeth 1906-........DLB-155

Longinus circa first century........DLB-176

Longley, Michael 1939-...........DLB-40

Longman, T. [publishing house]....DLB-154

Longmans, Green and Company....DLB-49

Longmore, George 1793?-1867......DLB-99

Longstreet, Augustus Baldwin 1790-1870...............DLB-3, 11, 74

Longworth, D. [publishing house]...DLB-49

Lonsdale, Frederick 1881-1954......DLB-10

A Look at the Contemporary Black Theatre Movement...................DLB-38

Loos, Anita 1893-1981.....DLB-11, 26; Y-81

Lopate, Phillip 1943-...............Y-80

López, Diana (see Isabella, Ríos)

Loranger, Jean-Aubert 1896-1942....DLB-92

Lorca, Federico García 1898-1936..DLB-108

Lord, John Keast 1818-1872........DLB-99

The Lord Chamberlain's Office and Stage Censorship in England.........DLB-10

Lorde, Audre 1934-1992...........DLB-41

Lorimer, George Horace 1867-1939....................DLB-91

Loring, A. K. [publishing house].....DLB-49

Loring and Mussey................DLB-46

Lossing, Benson J. 1813-1891.......DLB-30

Lothar, Ernst 1890-1974...........DLB-81

Lothrop, Harriet M. 1844-1924......DLB-42

Lothrop, D., and Company........DLB-49

Loti, Pierre 1850-1923.............DLB-123

Lott, Emeline ?-?.................DLB-166

The Lounger, no. 20 (1785), by Henry Mackenzie....................DLB-39

Lounsbury, Thomas R. 1838-1915...DLB-71

Louÿs, Pierre 1870-1925..........DLB-123

Lovelace, Earl 1935-.............DLB-125

Lovelace, Richard 1618-1657......DLB-131

Lovell, Coryell and Company.......DLB-49

Lovell, John W., Company.........DLB-49

Lover, Samuel 1797-1868.........DLB-159

Lovesey, Peter 1936-..............DLB-87

Lovingood, Sut (see Harris, George Washington)

Low, Samuel 1765-?..............DLB-37

Lowell, Amy 1874-1925........DLB-54, 140

Lowell, James Russell 1819-1891............DLB-1, 11, 64, 79

Lowell, Robert 1917-1977.......DLB-5, 169

Lowenfels, Walter 1897-1976........DLB-4

Lowndes, Marie Belloc 1868-1947...DLB-70

Lownes, Humphrey [publishing house]............DLB-170

Lowry, Lois 1937-................DLB-52

Lowry, Malcolm 1909-1957.........DLB-15

Lowther, Pat 1935-1975............DLB-53

Loy, Mina 1882-1966...........DLB-4, 54

Lozeau, Albert 1878-1924..........DLB-92

Lubbock, Percy 1879-1965.........DLB-149

Lucas, E. V. 1868-1938....DLB-98, 149, 153

Lucas, Fielding, Jr.
[publishing house] DLB-49

Luce, Henry R. 1898-1967..........DLB-91

Luce, John W., and Company.......DLB-46

Lucian circa 120-180 ....DLB-176

Lucie-Smith, Edward 1933- DLB-40

Lucini, Gian Pietro 1867-1914 .....DLB-114

Ludlum, Robert 1927- Y-82

Ludus de Antichristo circa 1160 ......DLB-148

Ludvigson, Susan 1942- DLB-120

Ludwig, Jack 1922- DLB-60

Ludwig, Otto 1813-1865 ....DLB-129

Ludwigslied 881 or 882 ....DLB-148

Luera, Yolanda 1953- DLB-122

Luft, Lya 1938- DLB-145

Luke, Peter 1919- DLB-13

Lupton, F. M., Company...........DLB-49

Lupus of Ferrières
circa 805-circa 862 ....DLB-148

Lurie, Alison 1926- DLB-2

Luzi, Mario 1914- DLB-128

L'vov, Nikolai Aleksandrovich
1751-1803 ....DLB-150

Lyall, Gavin 1932- DLB-87

Lydgate, John circa 1370-1450 .....DLB-146

Lyly, John circa 1554-1606 .....DLB-62, 167

Lynch, Patricia 1898-1972 ....DLB-160

Lynch, Richard
flourished 1596-1601 ....DLB-172

Lynd, Robert 1879-1949 ....DLB-98

Lyon, Matthew 1749-1822..........DLB-43

Lysias circa 459 B.C.-circa 380 B.C.
..........................DLB-176

Lytle, Andrew 1902-1995 ......DLB-6; Y-95

Lytton, Edward (see Bulwer-Lytton, Edward)

Lytton, Edward Robert Bulwer
1831-1891 ....DLB-32

Maass, Joachim 1901-1972..........DLB-69

Mabie, Hamilton Wright
1845-1916 ....DLB-71

Mac A'Ghobhainn, Iain (see Smith, Iain Crichton)

MacArthur, Charles
1895-1956...............DLB-7, 25, 44

Macaulay, Catherine 1731-1791 ....DLB-104

Macaulay, David 1945- DLB-61

Macaulay, Rose 1881-1958 ....DLB-36

Macaulay, Thomas Babington
1800-1859 ....DLB-32, 55

Macaulay Company DLB-46

MacBeth, George 1932- DLB-40

Macbeth, Madge 1880-1965.........DLB-92

MacCaig, Norman 1910- DLB-27

MacDiarmid, Hugh 1892-1978 ......DLB-20

MacDonald, Cynthia 1928- DLB-105

MacDonald, George
1824-1905 ....DLB-18, 163

MacDonald, John D.
1916-1986 ....DLB-8; Y-86

MacDonald, Philip 1899?-1980......DLB-77

Macdonald, Ross (see Millar, Kenneth)

MacDonald, Wilson 1880-1967......DLB-92

Macdonald and Company
(Publishers)..................DLB-112

MacEwen, Gwendolyn 1941- DLB-53

Macfadden, Bernarr
1868-1955 ....DLB-25, 91

MacGregor, John 1825-1892 ......DLB-166

MacGregor, Mary Esther (see Keith, Marian)

Machado, Antonio 1875-1939......DLB-108

Machado, Manuel 1874-1947 ......DLB-108

Machar, Agnes Maule 1837-1927 ....DLB-92

Machen, Arthur Llewelyn Jones
1863-1947 ....DLB-36, 156

MacInnes, Colin 1914-1976.........DLB-14

MacInnes, Helen 1907-1985 ....DLB-87

Mack, Maynard 1909- DLB-111

Mackall, Leonard L. 1879-1937 ....DLB-140

MacKaye, Percy 1875-1956 ....DLB-54

Macken, Walter 1915-1967 ....DLB-13

Mackenzie, Alexander 1763-1820....DLB-99

Mackenzie, Compton
1883-1972.................DLB-34, 100

Mackenzie, Henry 1745-1831 ....DLB-39

Mackey, Nathaniel 1947- DLB-169

Mackey, William Wellington
1937- DLB-38

Mackintosh, Elizabeth (see Tey, Josephine)

Mackintosh, Sir James
1765-1832 ....DLB-158

Maclaren, Ian (see Watson, John)

Macklin, Charles 1699-1797 ....DLB-89

MacLean, Katherine Anne 1925- DLB-8

MacLeish, Archibald
1892-1982..........DLB-4, 7, 45; Y-82

MacLennan, Hugh 1907-1990.......DLB-68

Macleod, Fiona (see Sharp, William)

MacLeod, Alistair 1936- DLB-60

Macleod, Norman 1906-1985 ....DLB-4

Macmillan and Company DLB-106

The Macmillan Company DLB-49

Macmillan's English Men of Letters,
First Series (1878-1892)........DLB-144

MacNamara, Brinsley 1890-1963 ....DLB-10

MacNeice, Louis 1907-1963 .....DLB-10, 20

MacPhail, Andrew 1864-1938.......DLB-92

Macpherson, James 1736-1796 .....DLB-109

Macpherson, Jay 1931- DLB-53

Macpherson, Jeanie 1884-1946 ......DLB-44

Macrae Smith Company DLB-46

Macrone, John
[publishing house] DLB-106

MacShane, Frank 1927- DLB-111

Macy-Masius DLB-46

Madden, David 1933- DLB-6

Maddow, Ben 1909-1992..........DLB-44

Maddux, Rachel 1912-1983 ....Y-93

Madgett, Naomi Long 1923- DLB-76

Madhubuti, Haki R.
1942- DLB-5, 41; DS-8

Madison, James 1751-1836 ....DLB-37

Maginn, William 1794-1842 ...DLB-110, 159

Mahan, Alfred Thayer 1840-1914....DLB-47

Maheux-Forcier, Louise 1929- DLB-60

Mahin, John Lee 1902-1984.........DLB-44

Mahon, Derek 1941- DLB-40

Maikov, Vasilii Ivanovich
1728-1778 ....DLB-150

Mailer, Norman
1923- DLB-2, 16, 28; Y-80, 83; DS-3

Maillet, Adrienne 1885-1963........DLB-68

Maimonides, Moses 1138-1204 ....DLB-115

Maillet, Antonine 1929- DLB-60

Maillu, David G. 1939- DLB-157

Main Selections of the Book-of-the-Month
Club, 1926-1945 ....DLB-9

Main Trends in Twentieth-Century Book Clubs
DLB-46

Mainwaring, Daniel 1902-1977......DLB-44

Mair, Charles 1838-1927 .... DLB-99

Mais, Roger 1905-1955 .... DLB-125

Major, Andre 1942- DLB-60

Major, Clarence 1936- DLB-33

Major, Kevin 1949- DLB-60

Major Books DLB-46

Makemie, Francis circa 1658-1708 ... DLB-24

The Making of a People, by
J. M. Ritchie DLB-66

Maksimović, Desanka 1898-1993 ... DLB-147

Malamud, Bernard
1914-1986 ..... DLB-2, 28, 152; Y-80, 86

Malet, Lucas 1852-1931 .... DLB-153

Malleson, Lucy Beatrice (see Gilbert, Anthony)

Mallet-Joris, Françoise 1930- DLB-83

Mallock, W. H. 1849-1923 ...... DLB-18, 57

Malone, Dumas 1892-1986 .... DLB-17

Malone, Edmond 1741-1812 .... DLB-142

Malory, Sir Thomas
circa 1400-1410 - 1471 .... DLB-146

Malraux, André 1901-1976 .... DLB-72

Malthus, Thomas Robert
1766-1834 .... DLB-107, 158

Maltz, Albert 1908-1985 .... DLB-102

Malzberg, Barry N. 1939- DLB-8

Mamet, David 1947- DLB-7

Manaka, Matsemela 1956- DLB-157

Manchester University Press DLB-112

Mandel, Eli 1922- DLB-53

Mandeville, Bernard 1670-1733 .... DLB-101

Mandeville, Sir John
mid fourteenth century DLB-146

Mandiargues, André Pieyre de
1909- DLB-83

Manfred, Frederick 1912-1994 .... DLB-6

Mangan, Sherry 1904-1961 .... DLB-4

Mankiewicz, Herman 1897-1953 ..... DLB-26

Mankiewicz, Joseph L. 1909-1993 ... DLB-44

Mankowitz, Wolf 1924- DLB-15

Manley, Delarivière
1672?-1724 .... DLB-39, 80

Mann, Abby 1927- DLB-44

Mann, Heinrich 1871-1950 ..... DLB-66, 118

Mann, Horace 1796-1859 .... DLB-1

Mann, Klaus 1906-1949 .... DLB-56

Mann, Thomas 1875-1955 .... DLB-66

Mann, William D'Alton
1839-1920 .... DLB-137

Manning, Marie 1873?-1945 .... DLB-29

Manning and Loring DLB-49

Mannyng, Robert
flourished 1303-1338 .... DLB-146

Mano, D. Keith 1942- DLB-6

Manor Books DLB-46

Mansfield, Katherine 1888-1923 .... DLB-162

Manzini, Gianna 1896-1974 .... DLB-177

Mapanje, Jack 1944- DLB-157

March, William 1893-1954 .... DLB-9, 86

Marchand, Leslie A. 1900- DLB-103

Marchant, Bessie 1862-1941 ...... DLB-160

Marchessault, Jovette 1938- DLB-60

Marcus, Frank 1928- DLB-13

Marden, Orison Swett
1850-1924 .... DLB-137

Marechera, Dambudzo
1952-1987 .... DLB-157

Marek, Richard, Books DLB-46

Mares, E. A. 1938- DLB-122

Mariani, Paul 1940- DLB-111

Marie-Victorin, Frère 1885-1944 ..... DLB-92

Marin, Biagio 1891-1985 .... DLB-128

Marinković, Ranko 1913- DLB-147

Marinetti, Filippo Tommaso
1876-1944 .... DLB-114

Marion, Frances 1886-1973 .... DLB-44

Marius, Richard C. 1933- Y-85

The Mark Taper Forum............. DLB-7

Mark Twain on Perpetual Copyright ... Y-92

Markfield, Wallace 1926- DLB-2, 28

Markham, Edwin 1852-1940 .... DLB-54

Markle, Fletcher 1921-1991.... DLB-68; Y-91

Marlatt, Daphne 1942- DLB-60

Marlitt, E. 1825-1887 .... DLB-129

Marlowe, Christopher 1564-1593 .... DLB-62

Marlyn, John 1912- DLB-88

Marmion, Shakerley 1603-1639 ..... DLB-58

Der Marner
before 1230-circa 1287......... DLB-138

The *Marprelate Tracts* 1588-1589 .... DLB-132

Marquand, John P. 1893-1960 ... DLB-9, 102

Marqués, René 1919-1979 .... DLB-113

Marquis, Don 1878-1937 .... DLB-11, 25

Marriott, Anne 1913- DLB-68

Marryat, Frederick 1792-1848 .. DLB-21, 163

Marsh, George Perkins
1801-1882 .... DLB-1, 64

Marsh, James 1794-1842 .... DLB-1, 59

Marsh, Capen, Lyon and Webb DLB-49

Marsh, Ngaio 1899-1982 .... DLB-77

Marshall, Edison 1894-1967 .... DLB-102

Marshall, Edward 1932- DLB-16

Marshall, Emma 1828-1899......... DLB-163

Marshall, James 1942-1992 .... DLB-61

Marshall, Joyce 1913- DLB-88

Marshall, Paule 1929- DLB-33, 157

Marshall, Tom 1938- DLB-60

Marsilius of Padua
circa 1275-circa 1342 .... DLB-115

Marson, Una 1905-1965 .... DLB-157

Marston, John 1576-1634 ...... DLB-58, 172

Marston, Philip Bourke 1850-1887 ... DLB-35

Martens, Kurt 1870-1945 .... DLB-66

Martien, William S.
[publishing house] DLB-49

Martin, Abe (see Hubbard, Kin)

Martin, Charles 1942- DLB-120

Martin, Claire 1914- DLB-60

Martin, Jay 1935- DLB-111

Martin, Johann (see Laurentius von Schnüffis)

Martin, Violet Florence (see Ross, Martin)

Martin du Gard, Roger 1881-1958 ... DLB-65

Martineau, Harriet
1802-1876 .... DLB-21, 55, 159, 163, 166

Martínez, Eliud 1935- DLB-122

Martínez, Max 1943- DLB-82

Martyn, Edward 1859-1923......... DLB-10

Marvell, Andrew 1621-1678 .... DLB-131

Marvin X 1944- DLB-38

Marx, Karl 1818-1883 .... DLB-129

Marzials, Theo 1850-1920 .... DLB-35

Masefield, John
1878-1967......... DLB-10, 19, 153, 160

Mason, A. E. W. 1865-1948 .... DLB-70

Mason, Bobbie Ann
1940- DLB-173; Y-87

Mason, William 1725-1797 .... DLB-142

Mason Brothers................... DLB-49

Massey, Gerald 1828-1907......... DLB-32

Massinger, Philip 1583-1640 . . . . DLB-58

Masson, David 1822-1907 . . . . DLB-144

Masters, Edgar Lee 1868-1950 . . . . . . DLB-54

Mastronardi, Lucio 1930-1979 . . . . . DLB-177

Mather, Cotton
1663-1728 . . . . DLB-24, 30, 140

Mather, Increase 1639-1723 . . . . DLB-24

Mather, Richard 1596-1669 . . . . DLB-24

Matheson, Richard 1926- DLB-8, 44

Matheus, John F. 1887- DLB-51

Mathews, Cornelius
1817?-1889 . . . . DLB-3, 64

Mathews, John Joseph
1894-1979 . . . . DLB-175

Mathews, Elkin
[publishing house] DLB-112

Mathias, Roland 1915- DLB-27

Mathis, June 1892-1927 . . . . DLB-44

Mathis, Sharon Bell 1937- DLB-33

Matoš, Antun Gustav 1873-1914 . . . DLB-147

The Matter of England
1240-1400 . . . . DLB-146

The Matter of Rome
early twelfth to late fifteenth
century DLB-146

Matthews, Brander
1852-1929 . . . . DLB-71, 78; DS-13

Matthews, Jack 1925- DLB-6

Matthews, William 1942- DLB-5

Matthiessen, F. O. 1902-1950 . . . . DLB-63

Matthiessen, Peter 1927- DLB-6, 173

Maugham, W. Somerset
1874-1965 . . . . . DLB-10, 36, 77, 100, 162

Maupassant, Guy de 1850-1893 . . . . DLB-123

Mauriac, Claude 1914- DLB-83

Mauriac, François 1885-1970 . . . . . . DLB-65

Maurice, Frederick Denison
1805-1872 . . . . DLB-55

Maurois, André 1885-1967 . . . . DLB-65

Maury, James 1718-1769 . . . . DLB-31

Mavor, Elizabeth 1927- DLB-14

Mavor, Osborne Henry (see Bridie, James)

Maxwell, William 1908- Y-80

Maxwell, H. [publishing house] DLB-49

Maxwell, John [publishing house] . . . DLB-106

May, Elaine 1932- DLB-44

May, Karl 1842-1912 . . . . DLB-129

May, Thomas 1595 or 1596-1650 . . . . DLB-58

Mayer, Bernadette 1945- DLB-165

Mayer, Mercer 1943- DLB-61

Mayer, O. B. 1818-1891 . . . . DLB-3

Mayes, Herbert R. 1900-1987 . . . . . . DLB-137

Mayes, Wendell 1919-1992 . . . . DLB-26

Mayfield, Julian 1928-1984 . . . . DLB-33; Y-84

Mayhew, Henry 1812-1887 . . . . . . DLB-18, 55

Mayhew, Jonathan 1720-1766 . . . . . . DLB-31

Mayne, Jasper 1604-1672 . . . . DLB-126

Mayne, Seymour 1944- DLB-60

Mayor, Flora Macdonald
1872-1932 . . . . DLB-36

Mayrocker, Friederike 1924- DLB-85

Mazrui, Ali A. 1933- DLB-125

Mauranií, Ivan 1814-1890 . . . . DLB-147

Mazursky, Paul 1930- DLB-44

McAlmon, Robert 1896-1956 . . . . . DLB-4, 45

McArthur, Peter 1866-1924 . . . . DLB-92

McBride, Robert M., and
Company DLB-46

McCaffrey, Anne 1926- DLB-8

McCarthy, Cormac 1933- DLB-6, 143

McCarthy, Mary 1912-1989 . . . . DLB-2; Y-81

McCay, Winsor 1871-1934 . . . . DLB-22

McClane, Albert Jules 1922-1991 . . . DLB-171

McClatchy, C. K. 1858-1936 . . . . DLB-25

McClellan, George Marion
1860-1934 . . . . DLB-50

McCloskey, Robert 1914- DLB-22

McClung, Nellie Letitia 1873-1951 . . . DLB-92

McClure, Joanna 1930- DLB-16

McClure, Michael 1932- DLB-16

McClure, Phillips and Company DLB-46

McClure, S. S. 1857-1949 . . . . DLB-91

McClurg, A. C., and Company DLB-49

McCluskey, John A., Jr. 1944- DLB-33

McCollum, Michael A. 1946- Y-87

McConnell, William C. 1917- DLB-88

McCord, David 1897- DLB-61

McCorkle, Jill 1958- Y-87

McCorkle, Samuel Eusebius
1746-1811 . . . . DLB-37

McCormick, Anne O'Hare
1880-1954 . . . . DLB-29

McCormick, Robert R. 1880-1955 . . . DLB-29

McCourt, Edward 1907-1972 . . . . . . DLB-88

McCoy, Horace 1897-1955 . . . . DLB-9

McCrae, John 1872-1918 . . . . DLB-92

McCullagh, Joseph B. 1842-1896 . . . . DLB-23

McCullers, Carson
1917-1967 . . . . DLB-2, 7, 173

McCulloch, Thomas 1776-1843 . . . . . DLB-99

McDonald, Forrest 1927- DLB-17

McDonald, Walter
1934- DLB-105, DS-9

McDonald, Walter, Getting Started:
Accepting the Regions You Own—
or Which Own You DLB-105

McDougall, Colin 1917-1984 . . . . DLB-68

McDowell, Obolensky DLB-46

McEwan, Ian 1948- DLB-14

McFadden, David 1940- DLB-60

McFall, Frances Elizabeth Clarke
(see Grand, Sarah)

McFarlane, Leslie 1902-1977 . . . . DLB-88

McFee, William 1881-1966 . . . . DLB-153

McGahern, John 1934- DLB-14

McGee, Thomas D'Arcy
1825-1868 . . . . DLB-99

McGeehan, W. O. 1879-1933 . . . DLB-25, 171

McGill, Ralph 1898-1969 . . . . DLB-29

McGinley, Phyllis 1905-1978 . . . . DLB-11, 48

McGirt, James E. 1874-1930 . . . . DLB-50

McGlashan and Gill DLB-106

McGough, Roger 1937- DLB-40

McGraw-Hill DLB-46

McGuane, Thomas 1939- DLB-2; Y-80

McGuckian, Medbh 1950- DLB-40

McGuffey, William Holmes
1800-1873 . . . . DLB-42

McIlvanney, William 1936- DLB-14

McIlwraith, Jean Newton
1859-1938 . . . . DLB-92

McIntyre, James 1827-1906 . . . . DLB-99

McIntyre, O. O. 1884-1938 . . . . DLB-25

McKay, Claude
1889-1948 . . . . DLB-4, 45, 51, 117

The David McKay Company DLB-49

McKean, William V. 1820-1903 . . . . . DLB-23

The McKenzie Trust Y-96

McKinley, Robin 1952- DLB-52

McLachlan, Alexander 1818-1896 . . . DLB-99

McLaren, Floris Clark 1904-1978 . . . . DLB-68

McLaverty, Michael 1907- DLB-15

McLean, John R. 1848-1916 .... DLB-23

McLean, William L. 1852-1931 ...... DLB-25

McLennan, William 1856-1904 ..... DLB-92

McLoughlin Brothers DLB-49

McLuhan, Marshall 1911-1980 ..... DLB-88

McMaster, John Bach 1852-1932 .... DLB-47

McMurtry, Larry
1936- DLB-2, 143; Y-80, 87

McNally, Terrence 1939- DLB-7

McNeil, Florence 1937- DLB-60

McNeile, Herman Cyril
1888-1937 .... DLB-77

McNickle, D'Arcy 1904-1977 ...... DLB-175

McPherson, James Alan 1943- DLB-38

McPherson, Sandra 1943- Y-86

McWhirter, George 1939- DLB-60

McWilliams, Carey 1905-1980 ..... DLB-137

Mead, L. T. 1844-1914 .... DLB-141

Mead, Matthew 1924- DLB-40

Mead, Taylor ?- DLB-16

Meany, Tom 1903-1964 .... DLB-171

Mechthild von Magdeburg
circa 1207-circa 1282 .... DLB-138

Medill, Joseph 1823-1899 .... DLB-43

Medoff, Mark 1940- DLB-7

Meek, Alexander Beaufort
1814-1865 .... DLB-3

Meeke, Mary ?-1816? DLB-116

Meinke, Peter 1932- DLB-5

Mejia Vallejo, Manuel 1923- DLB-113

Melançon, Robert 1947- DLB-60

Mell, Max 1882-1971 .... DLB-81, 124

Mellow, James R. 1926- DLB-111

Meltzer, David 1937- DLB-16

Meltzer, Milton 1915- DLB-61

Melville, Elizabeth, Lady Culross
circa 1585-1640 .... DLB-172

Melville, Herman 1819-1891 ...... DLB-3, 74

Memoirs of Life and Literature (1920),
by W. H. Mallock [excerpt] DLB-57

Menander 342-341 B.C.-circa 292-291 B.C.
.... DLB-176

Menantes (see Hunold, Christian Friedrich)

Mencke, Johann Burckhard
1674-1732 .... DLB-168

Mencken, H. L.
1880-1956 .... DLB-11, 29, 63, 137

Mencken and Nietzsche: An Unpublished Excerpt from H. L. Mencken's *My Life as Author and Editor* Y-93

Mendelssohn, Moses 1729-1786 ..... DLB-97

Méndez M., Miguel 1930- DLB-82

The Mercantile Library of
New York Y-96

Mercer, Cecil William (see Yates, Dornford)

Mercer, David 1928-1980 .... DLB-13

Mercer, John 1704-1768 .... DLB-31

Meredith, George
1828-1909 .... DLB-18, 35, 57, 159

Meredith, Louisa Anne
1812-1895 .... DLB-166

Meredith, Owen (see Lytton, Edward Robert Bulwer)

Meredith, William 1919- DLB-5

Mergerle, Johann Ulrich
(see Abraham ä Sancta Clara)

Mérimée, Prosper 1803-1870 .... DLB-119

Merivale, John Herman
1779-1844 .... DLB-96

Meriwether, Louise 1923- DLB-33

Merlin Press DLB-112

Merriam, Eve 1916-1992 .... DLB-61

The Merriam Company DLB-49

Merrill, James
1926-1995 .... DLB-5, 165; Y-85

Merrill and Baker DLB-49

The Mershon Company DLB-49

Merton, Thomas 1915-1968 ... DLB-48; Y-81

Merwin, W. S. 1927- DLB-5, 169

Messner, Julian [publishing house] ... DLB-46

Metcalf, J. [publishing house] DLB-49

Metcalf, John 1938- DLB-60

The Methodist Book Concern DLB-49

Methuen and Company DLB-112

Mew, Charlotte 1869-1928 ..... DLB-19, 135

Mewshaw, Michael 1943- Y-80

Meyer, Conrad Ferdinand
1825-1898 .... DLB-129

Meyer, E. Y. 1946- DLB-75

Meyer, Eugene 1875-1959 .... DLB-29

Meyer, Michael 1921- DLB-155

Meyers, Jeffrey 1939- DLB-111

Meynell, Alice
1847-1922 .... DLB-19, 98

Meynell, Viola 1885-1956 .... DLB-153

Meyrink, Gustav 1868-1932 .... DLB-81

Michaels, Leonard 1933- DLB-130

Micheaux, Oscar 1884-1951 .... DLB-50

Michel of Northgate, Dan
circa 1265-circa 1340 .... DLB-146

Micheline, Jack 1929- DLB-16

Michener, James A. 1907?- DLB-6

Micklejohn, George
circa 1717-1818 .... DLB-31

Middle English Literature:
An Introduction DLB-146

The Middle English Lyric DLB-146

Middle Hill Press DLB-106

Middleton, Christopher 1926- DLB-40

Middleton, Richard 1882-1911 .... DLB-156

Middleton, Stanley 1919- DLB-14

Middleton, Thomas 1580-1627 ..... DLB-58

Miegel, Agnes 1879-1964 .... DLB-56

Miles, Josephine 1911-1985 .... DLB-48

Milius, John 1944- DLB-44

Mill, James 1773-1836 .... DLB-107, 158

Mill, John Stuart 1806-1873 .... DLB-55

Millar, Kenneth
1915-1983 .... DLB-2; Y-83; DS-6

Millar, Andrew
[publishing house] DLB-154

Millay, Edna St. Vincent
1892-1950 .... DLB-45

Miller, Arthur 1915- DLB-7

Miller, Caroline 1903-1992 .... DLB-9

Miller, Eugene Ethelbert 1950- DLB-41

Miller, Heather Ross 1939- DLB-120

Miller, Henry 1891-1980 ..... DLB-4, 9; Y-80

Miller, J. Hillis 1928- DLB-67

Miller, James [publishing house] DLB-49

Miller, Jason 1939- DLB-7

Miller, May 1899- DLB-41

Miller, Paul 1906-1991 .... DLB-127

Miller, Perry 1905-1963 .... DLB-17, 63

Miller, Sue 1943- DLB-143

Miller, Vassar 1924- DLB-105

Miller, Walter M., Jr. 1923- DLB-8

Miller, Webb 1892-1940 .... DLB-29

Millhauser, Steven 1943- DLB-2

Millican, Arthenia J. Bates
1920- DLB-38

Mills and Boon DLB-112

Milman, Henry Hart 1796-1868 ..... DLB-96

Milne, A. A.
    1882-1956......... DLB-10, 77, 100, 160

Milner, Ron 1938- DLB-38

Milner, William
    [publishing house] DLB-106

Milnes, Richard Monckton (Lord Houghton)
    1809-1885 .... DLB-32

Milton, John 1608-1674....... DLB-131, 151

The Minerva Press............... DLB-154

Minnesang circa 1150-1280 .... DLB-138

Minns, Susan 1839-1938 .... DLB-140

Minor Illustrators, 1880-1914 ...... DLB-141

Minor Poets of the Earlier Seventeenth
    Century..................... DLB-121

Minton, Balch and Company DLB-46

Mirbeau, Octave 1848-1917 .... DLB-123

Mirk, John died after 1414?........ DLB-146

Miron, Gaston 1928- DLB-60

A Mirror for Magistrates DLB-167

Mitchel, Jonathan 1624-1668........ DLB-24

Mitchell, Adrian 1932- DLB-40

Mitchell, Donald Grant
    1822-1908............... DLB-1; DS-13

Mitchell, Gladys 1901-1983......... DLB-77

Mitchell, James Leslie 1901-1935 .... DLB-15

Mitchell, John (see Slater, Patrick)

Mitchell, John Ames 1845-1918 ..... DLB-79

Mitchell, Joseph 1908-1996........... Y-96

Mitchell, Julian 1935- DLB-14

Mitchell, Ken 1940- DLB-60

Mitchell, Langdon 1862-1935 .... DLB-7

Mitchell, Loften 1919- DLB-38

Mitchell, Margaret 1900-1949 .... DLB-9

Mitchell, W. O. 1914- DLB-88

Mitchison, Naomi Margaret (Haldane)
    1897- DLB-160

Mitford, Mary Russell
    1787-1855 .... DLB-110, 116

Mittelholzer, Edgar 1909-1965 ..... DLB-117

Mitterer, Erika 1906- DLB-85

Mitterer, Felix 1948- DLB-124

Mitternacht, Johann Sebastian
    1613-1679 .... DLB-168

Mizener, Arthur 1907-1988 .... DLB-103

Modern Age Books............... DLB-46

"Modern English Prose" (1876),
    by George Saintsbury DLB-57

The Modern Language Association of America
    Celebrates Its Centennial Y-84

The Modern Library DLB-46

"Modern Novelists – Great and Small" (1855),
    by Margaret Oliphant DLB-21

"Modern Style" (1857), by Cockburn
    Thomson [excerpt]............. DLB-57

The Modernists (1932), by Joseph Warren
    Beach DLB-36

Modiano, Patrick 1945- DLB-83

Moffat, Yard and Company DLB-46

Moffet, Thomas 1553-1604 .... DLB-136

Mohr, Nicholasa 1938- DLB-145

Moix, Ana María 1947- DLB-134

Molesworth, Louisa 1839-1921..... DLB-135

Möllhausen, Balduin 1825-1905 .... DLB-129

Momaday, N. Scott 1934- DLB-143, 175

Monkhouse, Allan 1858-1936 .... DLB-10

Monro, Harold 1879-1932......... DLB-19

Monroe, Harriet 1860-1936...... DLB-54, 91

Monsarrat, Nicholas 1910-1979 ..... DLB-15

Montagu, Lady Mary Wortley
    1689-1762 .... DLB-95, 101

Montague, John 1929- DLB-40

Montale, Eugenio 1896-1981....... DLB-114

Monterroso, Augusto 1921- DLB-145

Montgomerie, Alexander
    circa 1550?-1598.............. DLB-167

Montgomery, James
    1771-1854 .... DLB-93, 158

Montgomery, John 1919- DLB-16

Montgomery, Lucy Maud
    1874-1942.............. DLB-92; DS-14

Montgomery, Marion 1925- DLB-6

Montgomery, Robert Bruce (see Crispin, Edmund)

Montherlant, Henry de 1896-1972 ... DLB-72

The Monthly Review 1749-1844 ...... DLB-110

Montigny, Louvigny de 1876-1955... DLB-92

Montoya, José 1932- DLB-122

Moodie, John Wedderburn Dunbar
    1797-1869 .... DLB-99

Moodie, Susanna 1803-1885 .... DLB-99

Moody, Joshua circa 1633-1697 ..... DLB-24

Moody, William Vaughn
    1869-1910 .... DLB-7, 54

Moorcock, Michael 1939- DLB-14

Moore, Catherine L. 1911- DLB-8

Moore, Clement Clarke 1779-1863 .. DLB-42

Moore, Dora Mavor 1888-1979 ..... DLB-92

Moore, George
    1852-1933......... DLB-10, 18, 57, 135

Moore, Marianne
    1887-1972............... DLB-45; DS-7

Moore, Mavor 1919- DLB-88

Moore, Richard 1927- DLB-105

Moore, Richard, The No Self, the Little Self,
    and the Poets DLB-105

Moore, T. Sturge 1870-1944 .... DLB-19

Moore, Thomas 1779-1852..... DLB-96, 144

Moore, Ward 1903-1978 .... DLB-8

Moore, Wilstach, Keys and
    Company.................... DLB-49

The Moorland-Spingarn Research
    Center DLB-76

Moorman, Mary C. 1905-1994 ..... DLB-155

Moraga, Cherríe 1952- DLB-82

Morales, Alejandro 1944- DLB-82

Morales, Mario Roberto 1947- DLB-145

Morales, Rafael 1919- DLB-108

Morality Plays: *Mankind* circa 1450-1500 and
    Everyman circa 1500 .... DLB-146

Morante, Elsa 1912-1985.......... DLB-177

Moravia, Alberto 1907-1990 .... DLB-177

Mordaunt, Elinor 1872-1942....... DLB-174

More, Hannah
    1745-1833....... DLB-107, 109, 116, 158

More, Henry 1614-1687.......... DLB-126

More, Sir Thomas
    1477 or 1478-1535 .... DLB-136

Moreno, Dorinda 1939- DLB-122

Morency, Pierre 1942- DLB-60

Moretti, Marino 1885-1979 .... DLB-114

Morgan, Berry 1919- DLB-6

Morgan, Charles 1894-1958 .... DLB-34, 100

Morgan, Edmund S. 1916- DLB-17

Morgan, Edwin 1920- DLB-27

Morgan, John Pierpont
    1837-1913 .... DLB-140

Morgan, John Pierpont, Jr.
    1867-1943 .... DLB-140

Morgan, Robert 1944- DLB-120

Morgan, Sydney Owenson, Lady
    1776?-1859 .... DLB-116, 158

Morgner, Irmtraud 1933- DLB-75

Morhof, Daniel Georg
    1639-1691 .... DLB-164

Morier, James Justinian
    1782 or 1783?-1849 .... DLB-116

Mörike, Eduard 1804-1875 .... DLB-133

Morin, Paul 1889-1963 .... DLB-92

Morison, Richard 1514?-1556 ..... DLB-136

Morison, Samuel Eliot 1887-1976 .... DLB-17

Moritz, Karl Philipp 1756-1793 ...... DLB-94

Moriz von Craûn
    circa 1220-1230 .... DLB-138

Morley, Christopher 1890-1957 ...... DLB-9

Morley, John 1838-1923 .... DLB-57, 144

Morris, George Pope 1802-1864 ..... DLB-73

Morris, Lewis 1833-1907 .... DLB-35

Morris, Richard B. 1904-1989 .... DLB-17

Morris, William
    1834-1896 .... DLB-18, 35, 57, 156

Morris, Willie 1934- Y-80

Morris, Wright 1910- DLB-2; Y-81

Morrison, Arthur 1863-1945 .... DLB-70, 135

Morrison, Charles Clayton
    1874-1966 .... DLB-91

Morrison, Toni
    1931- DLB-6, 33, 143; Y-81

Morrow, William, and Company DLB-46

Morse, James Herbert 1841-1923 .... DLB-71

Morse, Jedidiah 1761-1826 .... DLB-37

Morse, John T., Jr. 1840-1937 .... DLB-47

Morselli, Guido 1912-1973 .... DLB-177

Mortimer, Favell Lee 1802-1878 .... DLB-163

Mortimer, John 1923- DLB-13

Morton, Carlos 1942- DLB-122

Morton, John P., and Company DLB-49

Morton, Nathaniel 1613-1685 ...... DLB-24

Morton, Sarah Wentworth
    1759-1846 .... DLB-37

Morton, Thomas
    circa 1579-circa 1647 .... DLB-24

Moscherosch, Johann Michael
    1601-1669 .... DLB-164

Moseley, Humphrey
    [publishing house] DLB-170

Möser, Justus 1720-1794 .... DLB-97

Mosley, Nicholas 1923- DLB-14

Moss, Arthur 1889-1969 .... DLB-4

Moss, Howard 1922-1987 .... DLB-5

Moss, Thylias 1954- DLB-120

The Most Powerful Book Review in America
    [*New York Times Book Review*] Y-82

Motion, Andrew 1952- DLB-40

Motley, John Lothrop
    1814-1877 .... DLB-1, 30, 59

Motley, Willard 1909-1965 .... DLB-76, 143

Motte, Benjamin Jr.
    [publishing house] DLB-154

Motteux, Peter Anthony
    1663-1718 .... DLB-80

Mottram, R. H. 1883-1971 .... DLB-36

Mouré, Erin 1955- DLB-60

Mourning Dove (Humishuma)
    between 1882 and 1888?-1936 .... DLB-175

Movies from Books, 1920-1974 ...... DLB-9

Mowat, Farley 1921- DLB-68

Mowbray, A. R., and Company,
    Limited DLB-106

Mowrer, Edgar Ansel 1892-1977 .... DLB-29

Mowrer, Paul Scott 1887-1971 ...... DLB-29

Moxon, Edward
    [publishing house] DLB-106

Moxon, Joseph
    [publishing house] DLB-170

Mphahlele, Es'kia (Ezekiel)
    1919- DLB-125

Mtshali, Oswald Mbuyiseni
    1940- DLB-125

Mucedorus DLB-62

Mudford, William 1782-1848 ..... DLB-159

Mueller, Lisel 1924- DLB-105

Muhajir, El (see Marvin X)

Muhajir, Nazzam Al Fitnah (see Marvin X)

Mühlbach, Luise 1814-1873 .... DLB-133

Muir, Edwin 1887-1959 .... DLB-20, 100

Muir, Helen 1937- DLB-14

Mukherjee, Bharati 1940- DLB-60

Mulcaster, Richard
    1531 or 1532-1611 .... DLB-167

Muldoon, Paul 1951- DLB-40

Müller, Friedrich (see Müller, Maler)

Müller, Heiner 1929- DLB-124

Müller, Maler 1749-1825 .... DLB-94

Müller, Wilhelm 1794-1827 .... DLB-90

Mumford, Lewis 1895-1990 .... DLB-63

Munby, Arthur Joseph 1828-1910 .... DLB-35

Munday, Anthony 1560-1633 ... DLB-62, 172

Mundt, Clara (see Mühlbach, Luise)

Mundt, Theodore 1808-1861 .... DLB-133

Munford, Robert circa 1737-1783 .... DLB-31

Mungoshi, Charles 1947- DLB-157

Munonye, John 1929- DLB-117

Munro, Alice 1931- DLB-53

Munro, H. H. 1870-1916 .... DLB-34, 162

Munro, Neil 1864-1930 .... DLB-156

Munro, George
    [publishing house] DLB-49

Munro, Norman L.
    [publishing house] DLB-49

Munroe, James, and Company DLB-49

Munroe, Kirk 1850-1930 .... DLB-42

Munroe and Francis DLB-49

Munsell, Joel [publishing house] DLB-49

Munsey, Frank A. 1854-1925 .... DLB-25, 91

Munsey, Frank A., and
    Company DLB-49

Murav'ev, Mikhail Nikitich
    1757-1807 .... DLB-150

Murdoch, Iris 1919- DLB-14

Murdoch, Rupert 1931- DLB-127

Murfree, Mary N. 1850-1922 .... DLB-12, 74

Murger, Henry 1822-1861 .... DLB-119

Murger, Louis-Henri (see Murger, Henry)

Muro, Amado 1915-1971 .... DLB-82

Murphy, Arthur 1727-1805 ..... DLB-89, 142

Murphy, Beatrice M. 1908- DLB-76

Murphy, Emily 1868-1933 .... DLB-99

Murphy, John H., III 1916- DLB-127

Murphy, John, and Company DLB-49

Murphy, Richard 1927-1993 .... DLB-40

Murray, Albert L. 1916- DLB-38

Murray, Gilbert 1866-1957 .... DLB-10

Murray, Judith Sargent 1751-1820 ... DLB-37

Murray, Pauli 1910-1985 .... DLB-41

Murray, John [publishing house] ... DLB-154

Murry, John Middleton
    1889-1957 .... DLB-149

Musäus, Johann Karl August
    1735-1787 .... DLB-97

Muschg, Adolf 1934- DLB-75

The Music of *Minnesang* DLB-138

Musil, Robert 1880-1942 .... DLB-81, 124

Muspilli circa 790-circa 850 .... DLB-148

Mussey, Benjamin B., and
    Company DLB-49

Mwangi, Meja 1948- DLB-125

Myers, Gustavus 1872-1942 .... DLB-47

Myers, L. H. 1881-1944 .... DLB-15

Myers, Walter Dean 1937- DLB-33

N

Nabbes, Thomas circa 1605-1641 .... DLB-58

Nabl, Franz 1883-1974 .... DLB-81

Nabokov, Vladimir
  1899-1977 ...... DLB-2; Y-80, Y-91; DS-3

Nabokov Festival at Cornell Y-83

The Vladimir Nabokov Archive
  in the Berg Collection Y-91

Nafis and Cornish DLB-49

Naipaul, Shiva 1945-1985 .... DLB-157; Y-85

Naipaul, V. S. 1932- DLB-125; Y-85

Nancrede, Joseph
  [publishing house] DLB-49

Naranjo, Carmen 1930- DLB-145

Narrache, Jean 1893-1970 .... DLB-92

Nasby, Petroleum Vesuvius (see Locke, David Ross)

Nash, Ogden 1902-1971 .... DLB-11

Nash, Eveleigh
  [publishing house] DLB-112

Nashe, Thomas 1567-1601? DLB-167

Nast, Conde 1873-1942 .... DLB-91

Nastasijević, Momčilo 1894-1938 ... DLB-147

Nathan, George Jean 1882-1958 .... DLB-137

Nathan, Robert 1894-1985 .... DLB-9

The National Jewish Book Awards Y-85

The National Theatre and the Royal
  Shakespeare Company: The
  National Companies DLB-13

Naughton, Bill 1910- DLB-13

Naylor, Gloria 1950- DLB-173

Nazor, Vladimir 1876-1949 .... DLB-147

Ndebele, Njabulo 1948- DLB-157

Neagoe, Peter 1881-1960 .... DLB-4

Neal, John 1793-1876 .... DLB-1, 59

Neal, Joseph C. 1807-1847 .... DLB-11

Neal, Larry 1937-1981 .... DLB-38

The Neale Publishing Company DLB-49

Neely, F. Tennyson
  [publishing house] DLB-49

Negri, Ada 1870-1945 .... DLB-114

"The Negro as a Writer," by
  G. M. McClellan DLB-50

"Negro Poets and Their Poetry," by
  Wallace Thurman DLB-50

Neidhart von Reuental
  circa 1185-circa 1240 .... DLB-138

Neihardt, John G. 1881-1973 ..... DLB-9, 54

Neledinsky-Meletsky, Iurii Aleksandrovich
  1752-1828 .... DLB-150

Nelligan, Emile 1879-1941 .... DLB-92

Nelson, Alice Moore Dunbar
  1875-1935 .... DLB-50

Nelson, Thomas, and Sons [U.S.] DLB-49

Nelson, Thomas, and Sons [U.K.] .. DLB-106

Nelson, William 1908-1978 .... DLB-103

Nelson, William Rockhill
  1841-1915 .... DLB-23

Nemerov, Howard 1920-1991 ... DLB-5, 6; Y-83

Nesbit, E. 1858-1924 .... DLB-141, 153

Ness, Evaline 1911-1986 .... DLB-61

Nestroy, Johann 1801-1862 .... DLB-133

Neukirch, Benjamin 1655-1729 .... DLB-168

Neugeboren, Jay 1938- DLB-28

Neumann, Alfred 1895-1952 .... DLB-56

Neumark, Georg 1621-1681 .... DLB-164

Neumeister, Erdmann 1671-1756 ... DLB-168

Nevins, Allan 1890-1971 .... DLB-17

Nevinson, Henry Woodd
  1856-1941 .... DLB-135

The New American Library DLB-46

New Approaches to Biography: Challenges
  from Critical Theory, USC Conference
  on Literary Studies, 1990 .... Y-90

New Directions Publishing
  Corporation DLB-46

A New Edition of *Huck Finn* Y-85

New Forces at Work in the American Theatre:
  1915-1925 .... DLB-7

New Literary Periodicals:
  A Report for 1987 .... Y-87

New Literary Periodicals:
  A Report for 1988 .... Y-88

New Literary Periodicals:
  A Report for 1989 .... Y-89

New Literary Periodicals:
  A Report for 1990 .... Y-90

New Literary Periodicals:
  A Report for 1991 .... Y-91

New Literary Periodicals:
  A Report for 1992 .... Y-92

New Literary Periodicals:
  A Report for 1993 .... Y-93

The New Monthly Magazine
  1814-1884 .... DLB-110

The New *Ulysses* Y-84

The New Variorum Shakespeare Y-85

A New Voice: The Center for the Book's First
  Five Years Y-83

The New Wave [Science Fiction] DLB-8

New York City Bookshops in the 1930s and
  1940s: The Recollections of Walter
  Goldwater Y-93

Newbery, John
  [publishing house] DLB-154

Newbolt, Henry 1862-1938 .... DLB-19

Newbound, Bernard Slade (see Slade, Bernard)

Newby, P. H. 1918- DLB-15

Newby, Thomas Cautley
  [publishing house] DLB-106

Newcomb, Charles King 1820-1894 ... DLB-1

Newell, Peter 1862-1924 .... DLB-42

Newell, Robert Henry 1836-1901 .... DLB-11

Newhouse, Samuel I. 1895-1979 .... DLB-127

Newman, Cecil Earl 1903-1976 ..... DLB-127

Newman, David (see Benton, Robert)

Newman, Frances 1883-1928 .... Y-80

Newman, John Henry
  1801-1890 .... DLB-18, 32, 55

Newman, Mark [publishing house] ... DLB-49

Newnes, George, Limited DLB-112

Newsome, Effie Lee 1885-1979 ...... DLB-76

Newspaper Syndication of American
  Humor DLB-11

Newton, A. Edward 1864-1940 ..... DLB-140

Ngugi wa Thiong'o 1938- DLB-125

Niatum, Duane 1938- DLB-175

The *Nibelungenlied* and the *Klage*
  circa 1200 .... DLB-138

Nichol, B. P. 1944- DLB-53

Nicholas of Cusa 1401-1464 ...... DLB-115

Nichols, Dudley 1895-1960 .... DLB-26

Nichols, Grace 1950- DLB-157

Nichols, John 1940- Y-82

Nichols, Mary Sargeant (Neal) Gove 1810-
  11884 .... DLB-1

Nichols, Peter 1927- DLB-13

Nichols, Roy F. 1896-1973 .... DLB-17

Nichols, Ruth 1948- DLB-60

Nicholson, Norman 1914- DLB-27

Nicholson, William 1872-1949 .....DLB-141

Ní Chuilleanáin, Eiléan 1942- DLB-40

Nicol, Eric 1919- DLB-68

Nicolai, Friedrich 1733-1811 ....DLB-97

Nicolay, John G. 1832-1901 and
  Hay, John 1838-1905 ....DLB-47

Nicolson, Harold 1886-1968 ...DLB-100, 149

Nicolson, Nigel 1917- DLB-155

Niebuhr, Reinhold 1892-1971 ....DLB-17

Niedecker, Lorine 1903-1970 ....DLB-48

Nieman, Lucius W. 1857-1935 ......DLB-25

Nietzsche, Friedrich 1844-1900 .....DLB-129

Niggli, Josefina 1910- Y-80

Nightingale, Florence 1820-1910....DLB-166

Nikolev, Nikolai Petrovich
  1758-1815 ....DLB-150

Niles, Hezekiah 1777-1839 ....DLB-43

Nims, John Frederick 1913- DLB-5

Nin, Anaïs 1903-1977 ....DLB-2, 4, 152

1985: The Year of the Mystery:
  A Symposium..................Y-85

Nissenson, Hugh 1933- DLB-28

Niven, Frederick John 1878-1944 ....DLB-92

Niven, Larry 1938- DLB-8

Nizan, Paul 1905-1940 ....DLB-72

Njegoš, Petar II Petroví
  1813-1851 ....DLB-147

Nkosi, Lewis 1936- DLB-157

Nobel Peace Prize
The 1986 Nobel Peace Prize
  Nobel Lecture 1986: Hope, Despair and
  Memory
  Tributes from Abraham Bernstein,
  Norman Lamm, and
  John R. Silber..................Y-86

The Nobel Prize and Literary Politics...Y-86

Nobel Prize in Literature
The 1982 Nobel Prize in Literature
  Announcement by the Swedish Academy
  of the Nobel Prize Nobel Lecture 1982:
  The Solitude of Latin America Excerpt
  from *One Hundred Years of Solitude* The
  Magical World of Macondo A Tribute
  to Gabriel García Márquez Y-82

The 1983 Nobel Prize in Literature
  Announcement by the Swedish Academy
  Nobel Lecture 1983 The Stature of
  William Golding Y-83

The 1984 Nobel Prize in Literature
  Announcement by the Swedish Academy
  Jaroslav Seifert Through the Eyes of the
  English-Speaking Reader
  Three Poems by Jaroslav Seifert Y-84

The 1985 Nobel Prize in Literature
  Announcement by the Swedish Academy
  Nobel Lecture 1985..............Y-85

The 1986 Nobel Prize in Literature
  Nobel Lecture 1986: This Past Must Address Its Present Y-86

The 1987 Nobel Prize in Literature
  Nobel Lecture 1987..............Y-87

The 1988 Nobel Prize in Literature
  Nobel Lecture 1988..............Y-88

The 1989 Nobel Prize in Literature
  Nobel Lecture 1989..............Y-89

The 1990 Nobel Prize in Literature
  Nobel Lecture 1990..............Y-90

The 1991 Nobel Prize in Literature
  Nobel Lecture 1991..............Y-91

The 1992 Nobel Prize in Literature
  Nobel Lecture 1992..............Y-92

The 1993 Nobel Prize in Literature
  Nobel Lecture 1993..............Y-93

The 1994 Nobel Prize in Literature
  Nobel Lecture 1994..............Y-94

The 1995 Nobel Prize in Literature
  Nobel Lecture 1995..............Y-95

Nodier, Charles 1780-1844 ....DLB-119

Noel, Roden 1834-1894 ....DLB-35

Nolan, William F. 1928- DLB-8

Noland, C. F. M. 1810?-1858 ....DLB-11

Nonesuch Press..................DLB-112

Noonday Press DLB-46

Noone, John 1936- DLB-14

Nora, Eugenio de 1923- DLB-134

Nordhoff, Charles 1887-1947 ....DLB-9

Norman, Charles 1904- DLB-111

Norman, Marsha 1947- Y-84

Norris, Charles G. 1881-1945 ....DLB-9

Norris, Frank 1870-1902 ....DLB-12

Norris, Leslie 1921- DLB-27

Norse, Harold 1916- DLB-16

North, Marianne 1830-1890 ....DLB-174

North Point Press DLB-46

Nortje, Arthur 1942-1970 ....DLB-125

Norton, Alice Mary (see Norton, Andre)

Norton, Andre 1912- DLB-8, 52

Norton, Andrews 1786-1853 ....DLB-1

Norton, Caroline 1808-1877 ....DLB-21, 159

Norton, Charles Eliot 1827-1908 ..DLB-1, 64

Norton, John 1606-1663............DLB-24

Norton, Mary 1903-1992..........DLB-160

Norton, Thomas (see Sackville, Thomas)

Norton, W. W., and Company......DLB-46

Norwood, Robert 1874-1932........DLB-92

Nossack, Hans Erich 1901-1977 .....DLB-69

Notker Balbulus circa 840-912 .....DLB-148

Notker III of Saint Gall
  circa 950-1022................DLB-148

Notker von Zweifalten ?-1095......DLB-148

A Note on Technique (1926), by
  Elizabeth A. Drew [excerpts] DLB-36

Nourse, Alan E. 1928- DLB-8

Novak, Vjenceslav 1859-1905......DLB-147

Novalis 1772-1801 ....DLB-90

Novaro, Mario 1868-1944 ....DLB-114

Novás Calvo, Lino 1903-1983......DLB-145

"The Novel in [Robert Browning's] 'The Ring
  and the Book'" (1912), by
  Henry James DLB-32

The Novel of Impressionism,
  by Jethro Bithell DLB-66

Novel-Reading: *The Works of Charles Dickens,
  The Works of W. Makepeace Thackeray* (1879),
  by Anthony Trollope...........DLB-21

The Novels of Dorothy Richardson (1918), by
  May Sinclair DLB-36

Novels with a Purpose (1864), by Justin M'Carthy DLB-21

Noventa, Giacomo 1898-1960......DLB-114

Novikov, Nikolai Ivanovich
  1744-1818 ....DLB-150

Nowlan, Alden 1933-1983 ....DLB-53

Noyes, Alfred 1880-1958 ....DLB-20

Noyes, Crosby S. 1825-1908 ....DLB-23

Noyes, Nicholas 1647-1717 ....DLB-24

Noyes, Theodore W. 1858-1946.....DLB-29

N-Town Plays
  circa 1468 to early sixteenth
  century DLB-146

Nugent, Frank 1908-1965 ....DLB-44

Nugent, Richard Bruce 1906- DLB-151

Nusic, Branislav 1864-1938........DLB-147

Nutt, David [publishing house].....DLB-106

Nwapa, Flora 1931- DLB-125

Nye, Edgar Wilson (Bill)
  1850-1896 ....DLB-11, 23

Nye, Naomi Shihab 1952- DLB-120

Nye, Robert 1939- DLB-14

O

Oakes, Urian circa 1631-1681 .... DLB-24

Oates, Joyce Carol
1938- DLB-2, 5, 130; Y-81

Ober, William 1920-1993 .... Y-93

Oberholtzer, Ellis Paxson
1868-1936 .... DLB-47

Obradović, Dositej 1740?-1811 ..... DLB-147

O'Brien, Edna 1932- DLB-14

O'Brien, Fitz-James 1828-1862 ...... DLB-74

O'Brien, Kate 1897-1974 .... DLB-15

O'Brien, Tim
1946- DLB-152; Y-80; DS-9

O'Casey, Sean 1880-1964 .... DLB-10

Occom, Samson 1723-1792 .... DLB-175

Ochs, Adolph S. 1858-1935 .... DLB-25

Ochs-Oakes, George Washington
1861-1931 .... DLB-137

O'Connor, Flannery
1925-1964 ..... DLB-2, 152; Y-80; DS-12

O'Connor, Frank 1903-1966 .... DLB-162

Octopus Publishing Group DLB-112

Odell, Jonathan 1737-1818 ...... DLB-31, 99

O'Dell, Scott 1903-1989 .... DLB-52

Odets, Clifford 1906-1963 .... DLB-7, 26

Odhams Press Limited DLB-112

O'Donnell, Peter 1920- DLB-87

O'Donovan, Michael (see O'Connor, Frank)

O'Faolain, Julia 1932- DLB-14

O'Faolain, Sean 1900- DLB-15, 162

Off Broadway and Off-Off Broadway . DLB-7

Off-Loop Theatres DLB-7

Offord, Carl Ruthven 1910- DLB-76

O'Flaherty, Liam
1896-1984 .... DLB-36, 162; Y-84

Ogilvie, J. S., and Company DLB-49

Ogot, Grace 1930- DLB-125

O'Grady, Desmond 1935- DLB-40

Ogunyemi, Wale 1939- DLB-157

O'Hagan, Howard 1902-1982 ..... DLB-68

O'Hara, Frank 1926-1966 .... DLB-5, 16

O'Hara, John 1905-1970 .... DLB-9, 86; DS-2

Okara, Gabriel 1921- DLB-125

O'Keeffe, John 1747-1833 .... DLB-89

Okes, Nicholas
[publishing house] DLB-170

Okigbo, Christopher 1930-1967 .... DLB-125

Okot p'Bitek 1931-1982 .... DLB-125

Okpewho, Isidore 1941- DLB-157

Okri, Ben 1959- DLB-157

Olaudah Equiano and Unfinished Journeys:
The Slave-Narrative Tradition and
Twentieth-Century Continuities, by
Paul Edwards and Pauline T.
Wangman DLB-117

Old English Literature:
An Introduction DLB-146

Old English Riddles
eighth to tenth centuries DLB-146

Old Franklin Publishing House DLB-49

Old German Genesis and *Old German Exodus*
circa 1050-circa 1130 .... DLB-148

Old High German Charms and
Blessings DLB-148

The *Old High German Isidor*
circa 790-800 .... DLB-148

Older, Fremont 1856-1935 .... DLB-25

Oldham, John 1653-1683 .... DLB-131

Olds, Sharon 1942- DLB-120

Olearius, Adam 1599-1671 .... DLB-164

Oliphant, Laurence
1829?-1888 .... DLB-18, 166

Oliphant, Margaret 1828-1897 ...... DLB-18

Oliver, Chad 1928- DLB-8

Oliver, Mary 1935- DLB-5

Ollier, Claude 1922- DLB-83

Olsen, Tillie 1913?- DLB-28; Y-80

Olson, Charles 1910-1970 .... DLB-5, 16

Olson, Elder 1909- DLB-48, 63

Omotoso, Kole 1943- DLB-125

"On Art in Fiction "(1838),
by Edward Bulwer DLB-21

On Learning to Write Y-88

On Some of the Characteristics of Modern
Poetry and On the Lyrical Poems of
Alfred Tennyson (1831), by Arthur
Henry Hallam................. DLB-32

"On Style in English Prose" (1898), by
Frederic Harrison............. DLB-57

"On Style in Literature: Its Technical
Elements" (1885), by Robert Louis
Stevenson DLB-57

"On the Writing of Essays" (1862),
by Alexander Smith DLB-57

Ondaatje, Michael 1943- DLB-60

O'Neill, Eugene 1888-1953 .... DLB-7

Onetti, Juan Carlos 1909-1994 ..... DLB-113

Onions, George Oliver
1872-1961 .... DLB-153

Onofri, Arturo 1885-1928 .... DLB-114

Opie, Amelia 1769-1853 ...... DLB-116, 159

Opitz, Martin 1597-1639 .... DLB-164

Oppen, George 1908-1984 .... DLB-5, 165

Oppenheim, E. Phillips 1866-1946 ... DLB-70

Oppenheim, James 1882-1932 .... DLB-28

Oppenheimer, Joel 1930- DLB-5

Optic, Oliver (see Adams, William Taylor)

Orczy, Emma, Baroness
1865-1947 .... DLB-70

Origo, Iris 1902-1988 .... DLB-155

Orlovitz, Gil 1918-1973 .... DLB-2, 5

Orlovsky, Peter 1933- DLB-16

Ormond, John 1923- DLB-27

Ornitz, Samuel 1890-1957 ...... DLB-28, 44

Ortese, Anna Maria 1914- DLB-177

Ortiz, Simon J. 1941- DLB-120, 175

Ortnit and *Wolfdietrich*
circa 1225-1250............... DLB-138

Orton, Joe 1933-1967 .... DLB-13

Orwell, George 1903-1950 ..... DLB-15, 98

The Orwell Year Y-84

Ory, Carlos Edmundo de 1923- ... DLB-134

Osbey, Brenda Marie 1957- DLB-120

Osbon, B. S. 1827-1912 .... DLB-43

Osborne, John 1929-1994 .... DLB-13

Osgood, Herbert L. 1855-1918 ..... DLB-47

Osgood, James R., and
Company..................... DLB-49

Osgood, McIlvaine and
Company..................... DLB-112

O'Shaughnessy, Arthur
1844-1881 .... DLB-35

O'Shea, Patrick
[publishing house] DLB-49

Osipov, Nikolai Petrovich
1751-1799 .... DLB-150

Oskison, John Milton 1879-1947 ... DLB-175

Osofisan, Femi 1946- DLB-125

Ostenso, Martha 1900-1963 .... DLB-92

Ostriker, Alicia 1937- DLB-120

Osundare, Niyi 1947- DLB-157

Oswald, Eleazer 1755-1795 .... DLB-43

Otero, Blas de 1916-1979 .... DLB-134

Otero, Miguel Antonio
    1859-1944 .... DLB-82

Otero Silva, Miguel 1908-1985 ..... DLB-145

Otfried von Weißenburg
    circa 800-circa 875? DLB-148

Otis, James (see Kaler, James Otis)

Otis, James, Jr. 1725-1783 .... DLB-31

Otis, Broaders and Company DLB-49

Ottaway, James 1911- DLB-127

Ottendorfer, Oswald 1826-1900 ..... DLB-23

Ottieri, Ottiero 1924- DLB-177

Otto-Peters, Louise 1819-1895 ..... DLB-129

Otway, Thomas 1652-1685 .... DLB-80

Ouellette, Fernand 1930- DLB-60

Ouida 1839-1908 .... DLB-18, 156

Outing Publishing Company DLB-46

Outlaw Days, by Joyce Johnson DLB-16

Overbury, Sir Thomas
    circa 1581-1613 .... DLB-151

The Overlook Press DLB-46

Overview of U.S. Book Publishing,
    1910-1945 .... DLB-9

Owen, Guy 1925- DLB-5

Owen, John 1564-1622 .... DLB-121

Owen, John [publishing house] DLB-49

Owen, Robert 1771-1858 ...... DLB-107, 158

Owen, Wilfred 1893-1918 .... DLB-20

Owen, Peter, Limited DLB-112

The Owl and the Nightingale
    circa 1189-1199 .... DLB-146

Owsley, Frank L. 1890-1956 .... DLB-17

Oxford, Seventeenth Earl of, Edward de Vere
    1550-1604 .... DLB-172

Ozerov, Vladislav Aleksandrovich
    1769-1816 .... DLB-150

Ozick, Cynthia 1928- DLB-28, 152; Y-82

P

Pace, Richard 1482?-1536 .... DLB-167

Pacey, Desmond 1917-1975 .... DLB-88

Pack, Robert 1929- DLB-5

Packaging Papa: *The Garden of Eden* Y-86

Padell Publishing Company DLB-46

Padgett, Ron 1942- DLB-5

Padilla, Ernesto Chávez 1944- DLB-122

Page, L. C., and Company DLB-49

Page, P. K. 1916- DLB-68

Page, Thomas Nelson
    1853-1922 .... DLB-12, 78; DS-13

Page, Walter Hines 1855-1918 ... DLB-71, 91

Paget, Francis Edward
    1806-1882 .... DLB-163

Paget, Violet (see Lee, Vernon)

Pagliarani, Elio 1927- DLB-128

Pain, Barry 1864-1928 .... DLB-135

Pain, Philip ?-circa 1666 .... DLB-24

Paine, Robert Treat, Jr. 1773-1811 ... DLB-37

Paine, Thomas
    1737-1809 .... DLB-31, 43, 73, 158

Painter, George D. 1914- DLB-155

Painter, William 1540?-1594 .... DLB-136

Palazzeschi, Aldo 1885-1974 .... DLB-114

Paley, Grace 1922- DLB-28

Palfrey, John Gorham
    1796-1881 .... DLB-1, 30

Palgrave, Francis Turner
    1824-1897 .... DLB-35

Palmer, Joe H. 1904-1952 .... DLB-171

Palmer, Michael 1943- DLB-169

Paltock, Robert 1697-1767 .... DLB-39

Pan Books Limited DLB-112

Panamaa, Norman 1914- and
    Frank, Melvin 1913-1988 .... DLB-26

Pancake, Breece D'J 1952-1979 ..... DLB-130

Panero, Leopoldo 1909-1962 .... DLB-108

Pangborn, Edgar 1909-1976 .... DLB-8

"Panic Among the Philistines": A Postscript,
    An Interview with Bryan Griffin Y-81

Panneton, Philippe (see Ringuet)

Panshin, Alexei 1940- DLB-8

Pansy (see Alden, Isabella)

Pantheon Books DLB-46

Paperback Library DLB-46

Paperback Science Fiction DLB-8

Paquet, Alfons 1881-1944 .... DLB-66

Paradis, Suzanne 1936- DLB-53

Pareja Diezcanseco, Alfredo
    1908-1993 .... DLB-145

Pardoe, Julia 1804-1862 .... DLB-166

Parents' Magazine Press DLB-46

Parise, Goffredo 1929-1986 .... DLB-177

Parisian Theater, Fall 1984: Toward
    A New Baroque Y-85

Parizeau, Alice 1930- DLB-60

Parke, John 1754-1789 .... DLB-31

Parker, Dorothy
    1893-1967 .... DLB-11, 45, 86

Parker, Gilbert 1860-1932 .... DLB-99

Parker, James 1714-1770 .... DLB-43

Parker, Theodore 1810-1860 .... DLB-1

Parker, William Riley 1906-1968 ... DLB-103

Parker, J. H. [publishing house] DLB-106

Parker, John [publishing house] DLB-106

Parkman, Francis, Jr.
    1823-1893 .... DLB-1, 30

Parks, Gordon 1912- DLB-33

Parks, William 1698-1750 .... DLB-43

Parks, William [publishing house] ... DLB-49

Parley, Peter (see Goodrich, Samuel Griswold)

Parmenides late sixth-fith century B.C.
    .... DLB-176

Parnell, Thomas 1679-1718 .... DLB-95

Parr, Catherine 1513?-1548 .... DLB-136

Parrington, Vernon L.
    1871-1929 .... DLB-17, 63

Parronchi, Alessandro 1914- DLB-128

Partridge, S. W., and Company DLB-106

Parton, James 1822-1891 .... DLB-30

Parton, Sara Payson Willis
    1811-1872 .... DLB-43, 74

Pasinetti, Pier Maria 1913- DLB-177

Pasolini, Pier Paolo 1922- DLB-128, 177

Pastan, Linda 1932- DLB-5

Paston, George 1860-1936 .... DLB-149

The *Paston Letters* 1422-1509 .... DLB-146

Pastorius, Francis Daniel
    1651-circa 1720 .... DLB-24

Patchen, Kenneth 1911-1972 ..... DLB-16, 48

Pater, Walter 1839-1894 .... DLB-57, 156

Paterson, Katherine 1932- DLB-52

Patmore, Coventry 1823-1896 ... DLB-35, 98

Paton, Joseph Noel 1821-1901 .... DLB-35

Paton Walsh, Jill 1937- DLB-161

Patrick, Edwin Hill ("Ted")
    1901-1964 .... DLB-137

Patrick, John 1906- DLB-7

Pattee, Fred Lewis 1863-1950 .... DLB-71

353

Pattern and Paradigm: History as
 Design, by Judith Ryan DLB-75

Patterson, Alicia 1906-1963 .... DLB-127

Patterson, Eleanor Medill
 1881-1948 .... DLB-29

Patterson, Eugene 1923- DLB-127

Patterson, Joseph Medill
 1879-1946 .... DLB-29

Pattillo, Henry 1726-1801 .... DLB-37

Paul, Elliot 1891-1958 .... DLB-4

Paul, Jean (see Richter, Johann Paul Friedrich)

Paul, Kegan, Trench, Trubner and Company
 Limited DLB-106

Paul, Peter, Book Company DLB-49

Paul, Stanley, and Company
 Limited DLB-112

Paulding, James Kirke
 1778-1860 .... DLB-3, 59, 74

Paulin, Tom 1949- DLB-40

Pauper, Peter, Press DLB-46

Pavese, Cesare 1908-1950 ..... DLB-128, 177

Paxton, John 1911-1985 .... DLB-44

Payn, James 1830-1898............. DLB-18

Payne, John 1842-1916........... DLB-35

Payne, John Howard 1791-1852 ..... DLB-37

Payson and Clarke DLB-46

Peabody, Elizabeth Palmer
 1804-1894 .... DLB-1

Peabody, Elizabeth Palmer
 [publishing house] DLB-49

Peabody, Oliver William Bourn
 1799-1848 .... DLB-59

Peace, Roger 1899-1968........... DLB-127

Peacham, Henry 1578-1644?....... DLB-151

Peacham, Henry, the Elder
 1547-1634 .... DLB-172

Peachtree Publishers, Limited....... DLB-46

Peacock, Molly 1947- DLB-120

Peacock, Thomas Love
 1785-1866 .... DLB-96, 116

Pead, Deuel ?-1727 .... DLB-24

Peake, Mervyn 1911-1968...... DLB-15, 160

Pear Tree Press.................. DLB-112

Pearce, Philippa 1920- DLB-161

Pearson, H. B. [publishing house].... DLB-49

Pearson, Hesketh 1887-1964 .... DLB-149

Peck, George W. 1840-1916 ..... DLB-23, 42

Peck, H. C., and Theo. Bliss
 [publishing house] DLB-49

Peck, Harry Thurston
 1856-1914 .... DLB-71, 91

Peele, George 1556-1596 .... DLB-62, 167

Pegler, Westbrook 1894-1969 ...... DLB-171

Pellegrini and Cudahy DLB-46

Pelletier, Aimé (see Vac, Bertrand)

Pemberton, Sir Max 1863-1950...... DLB-70

Penguin Books [U.S.] DLB-46

Penguin Books [U.K.] DLB-112

Penn Publishing Company DLB-49

Penn, William 1644-1718........... DLB-24

Penna, Sandro 1906-1977 .... DLB-114

Penner, Jonathan 1940- Y-83

Pennington, Lee 1939- Y-82

Pepys, Samuel 1633-1703.......... DLB-101

Percy, Thomas 1729-1811 .... DLB-104

Percy, Walker 1916-1990 ... DLB-2; Y-80, 90

Percy, William 1575-1648 .... DLB-172

Perec, Georges 1936-1982 .... DLB-83

Perelman, S. J. 1904-1979 ...... DLB-11, 44

Perez, Raymundo "Tigre"
 1946- DLB-122

Peri Rossi, Cristina 1941- DLB-145

Periodicals of the Beat Generation ... DLB-16

Perkins, Eugene 1932- DLB-41

Perkoff, Stuart Z. 1930-1974 .... DLB-16

Perley, Moses Henry 1804-1862 .... DLB-99

Permabooks.................... DLB-46

Perrin, Alice 1867-1934 .... DLB-156

Perry, Bliss 1860-1954 .... DLB-71

Perry, Eleanor 1915-1981 .... DLB-44

Perry, Sampson 1747-1823 .... DLB-158

"Personal Style" (1890), by John Addington
 Symonds DLB-57

Perutz, Leo 1882-1957 .... DLB-81

Pesetsky, Bette 1932- DLB-130

Pestalozzi, Johann Heinrich
 1746-1827 .... DLB-94

Peter, Laurence J. 1919-1990........ DLB-53

Peter of Spain circa 1205-1277 ..... DLB-115

Peterkin, Julia 1880-1961 .... DLB-9

Peters, Lenrie 1932- DLB-117

Peters, Robert 1924- DLB-105

Peters, Robert, Foreword to
 Ludwig of Bavaria DLB-105

Petersham, Maud 1889-1971 and
 Petersham, Miska 1888-1960 .... DLB-22

Peterson, Charles Jacobs
 1819-1887 .... DLB-79

Peterson, Len 1917- DLB-88

Peterson, Louis 1922- DLB-76

Peterson, T. B., and Brothers DLB-49

Petitclair, Pierre 1813-1860 .... DLB-99

Petrov, Gavriil 1730-1801 .... DLB-150

Petrov, Vasilii Petrovich
 1736-1799 .... DLB-150

Petroví, Rastko 1898-1949 .... DLB-147

Petruslied circa 854?............... DLB-148

Petry, Ann 1908- DLB-76

Pettie, George circa 1548-1589 ..... DLB-136

Peyton, K. M. 1929- DLB-161

Pfaffe Konrad
 flourished circa 1172 .... DLB-148

Pfaffe Lamprecht
 flourished circa 1150 .... DLB-148

Pforzheimer, Carl H. 1879-1957.... DLB-140

Phaer, Thomas 1510?-1560........ DLB-167

Phaidon Press Limited DLB-112

Pharr, Robert Deane 1916-1992 ..... DLB-33

Phelps, Elizabeth Stuart
 1844-1911 .... DLB-74

Philander von der Linde
 (see Mencke, Johann Burckhard)

Philip, Marlene Nourbese
 1947- DLB-157

Philippe, Charles-Louis
 1874-1909 .... DLB-65

Philips, John 1676-1708 .... DLB-95

Philips, Katherine 1632-1664....... DLB-131

Phillips, Caryl 1958- DLB-157

Phillips, David Graham
 1867-1911 .... DLB-9, 12

Phillips, Jayne Anne 1952- Y-80

Phillips, Robert 1938- DLB-105

Phillips, Robert, Finding, Losing,
 Reclaiming: A Note on My
 Poems DLB-105

Phillips, Stephen 1864-1915........ DLB-10

Phillips, Ulrich B. 1877-1934........ DLB-17

Phillips, Willard 1784-1873 .... DLB-59

Phillips, William 1907- DLB-137

Phillips, Sampson and Company DLB-49

Phillpotts, Eden
 1862-1960........ DLB-10, 70, 135, 153

Philo circa 20-15 B.C.-circa A.D. 50 .... DLB-176

Philosophical Library DLB-46

"The Philosophy of Style" (1852), by Herbert Spencer DLB-57

Phinney, Elihu [publishing house] ... DLB-49

Phoenix, John (see Derby, George Horatio)

PHYLON (Fourth Quarter, 1950), The Negro in Literature: The Current Scene............ DLB-76

Physiologus circa 1070-circa 1150 .... DLB-148

Piccolo, Lucio 1903-1969........... DLB-114

Pickard, Tom 1946- DLB-40

Pickering, William [publishing house] DLB-106

Pickthall, Marjorie 1883-1922 .... DLB-92

Pictorial Printing Company......... DLB-49

Piel, Gerard 1915- DLB-137

Piercy, Marge 1936- DLB-120

Pierro, Albino 1916- DLB-128

Pignotti, Lamberto 1926- DLB-128

Pike, Albert 1809-1891............. DLB-74

Pilon, Jean-Guy 1930- DLB-60

Pinckney, Josephine 1895-1957....... DLB-6

Pindar circa 518 B.C.-circa 438 B.C. DLB-176

Pindar, Peter (see Wolcot, John)

Pinero, Arthur Wing 1855-1934 ..... DLB-10

Pinget, Robert 1919- DLB-83

Pinnacle Books DLB-46

Piñon, Nélida 1935- DLB-145

Pinsky, Robert 1940- Y-82

Pinter, Harold 1930- DLB-13

Piontek, Heinz 1925- DLB-75

Piozzi, Hester Lynch [Thrale] 1741-1821 .... DLB-104, 142

Piper, H. Beam 1904-1964........... DLB-8

Piper, Watty DLB-22

Pisar, Samuel 1929- Y-83

Pitkin, Timothy 1766-1847 .... DLB-30

The Pitt Poetry Series: Poetry Publishing Today Y-85

Pitter, Ruth 1897- DLB-20

Pix, Mary 1666-1709 .... DLB-80

Plaatje, Sol T. 1876-1932 .... DLB-125

The Place of Realism in Fiction (1895), by George Gissing............ DLB-18

Plante, David 1940- Y-83

Platen, August von 1796-1835....... DLB-90

Plath, Sylvia 1932-1963 .... DLB-5, 6, 152

Plato circa 428 B.C.-348-347 B.C. DLB-176

Platon 1737-1812 .... DLB-150

Platt and Munk Company......... DLB-46

Playboy Press DLB-46

Playford, John [publishing house] DLB-170

Plays, Playwrights, and Playgoers ... DLB-84

Playwrights and Professors, by Tom Stoppard................ DLB-13

Playwrights on the Theater DLB-80

Der Pleier flourished circa 1250 .... DLB-138

Plenzdorf, Ulrich 1934- DLB-75

Plessen, Elizabeth 1944- DLB-75

Plievier, Theodor 1892-1955........ DLB-69

Plomer, William 1903-1973..... DLB-20, 162

Plotinus 204-270 .... DLB-176

Plumly, Stanley 1939- DLB-5

Plumpp, Sterling D. 1940- DLB-41

Plunkett, James 1920- DLB-14

Plutarch circa 46-circa 120......... DLB-176

Plymell, Charles 1935- DLB-16

Pocket Books................... DLB-46

Poe, Edgar Allan 1809-1849............ DLB-3, 59, 73, 74

Poe, James 1921-1980.............. DLB-44

The Poet Laureate of the United States Statements from Former Consultants in Poetry...................... Y-86

Pohl, Frederik 1919- DLB-8

Poirier, Louis (see Gracq, Julien)

Polanyi, Michael 1891-1976 .... DLB-100

Pole, Reginald 1500-1558 .... DLB-132

Poliakoff, Stephen 1952- DLB-13

Polidori, John William 1795-1821 .... DLB-116

Polite, Carlene Hatcher 1932- DLB-33

Pollard, Edward A. 1832-1872 ...... DLB-30

Pollard, Percival 1869-1911 .... DLB-71

Pollard and Moss DLB-49

Pollock, Sharon 1936- DLB-60

Polonsky, Abraham 1910- DLB-26

Polotsky, Simeon 1629-1680 .... DLB-150

Polybius circa 200 B.C.-118 B.C. ... DLB-176

Pomilio, Mario 1921-1990 .... DLB-177

Ponce, Mary Helen 1938- DLB-122

Ponce-Montoya, Juanita 1949- DLB-122

Ponet, John 1516?-1556 .... DLB-132

Poniatowski, Elena 1933- DLB-113

Ponsonby, William [publishing house] DLB-170

Pony Stories DLB-160

Poole, Ernest 1880-1950............ DLB-9

Poole, Sophia 1804-1891 .... DLB-166

Poore, Benjamin Perley 1820-1887 .... DLB-23

Pope, Abbie Hanscom 1858-1894 .... DLB-140

Pope, Alexander 1688-1744..... DLB-95, 101

Popov, Mikhail Ivanovich 1742-circa 1790................. DLB-150

Popular Library.................. DLB-46

Porlock, Martin (see MacDonald, Philip)

Porpoise Press................... DLB-112

Porta, Antonio 1935-1989 .... DLB-128

Porter, Anna Maria 1780-1832.............. DLB-116, 159

Porter, Eleanor H. 1868-1920 .... DLB-9

Porter, Gene Stratton (see Stratton-Porter, Gene)

Porter, Henry ?-? DLB-62

Porter, Jane 1776-1850......... DLB-116, 159

Porter, Katherine Anne 1890-1980 ... DLB-4, 9, 102; Y-80; DS-12

Porter, Peter 1929- DLB-40

Porter, William Sydney 1862-1910............. DLB-12, 78, 79

Porter, William T. 1809-1858..... DLB-3, 43

Porter and Coates................ DLB-49

Portis, Charles 1933- DLB-6

Posey, Alexander 1873-1908....... DLB-175

Postans, Marianne circa 1810-1865................ DLB-166

Postl, Carl (see Sealsfield, Carl)

Poston, Ted 1906-1974 .... DLB-51

Postscript to [the Third Edition of] *Clarissa* (1751), by Samuel Richardson ... DLB-39

Potok, Chaim 1929- DLB-28, 152; Y-84

Potter, Beatrix 1866-1943 .... DLB-141

Potter, David M. 1910-1971 .... DLB-17

Potter, John E., and Company DLB-49

Pottle, Frederick A. 1897-1987 ....DLB-103; Y-87

Poulin, Jacques 1937- DLB-60

Pound, Ezra 1885-1972 ....DLB-4, 45, 63

Povich, Shirley 1905- DLB-171

Powell, Anthony 1905- DLB-15

Powers, J. F. 1917- DLB-130

Pownall, David 1938- DLB-14

Powys, John Cowper 1872-1963 .....DLB-15

Powys, Llewelyn 1884-1939 ....DLB-98

Powys, T. F. 1875-1953.........DLB-36, 162

Poynter, Nelson 1903-1978 ....DLB-127

The Practice of Biography: An Interview with Stanley Weintraub Y-82

The Practice of Biography II: An Interview with B. L. Reid..................Y-83

The Practice of Biography III: An Interview with Humphrey Carpenter.........Y-84

The Practice of Biography IV: An Interview with William Manchester Y-85

The Practice of Biography V: An Interview with Justin Kaplan................Y-86

The Practice of Biography VI: An Interview with David Herbert Donald........Y-87

The Practice of Biography VII: An Interview with John Caldwell Guilds Y-92

The Practice of Biography VIII: An Interview with Joan Mellen Y-94

The Practice of Biography IX: An Interview with Michael Reynolds............Y-95

Prados, Emilio 1899-1962 ....DLB-134

Praed, Winthrop Mackworth 1802-1839 ....DLB-96

Praeger Publishers DLB-46

Praetorius, Johannes 1630-1680 ....DLB-168

Pratolini, Vasco 1913–1991 .....DLB-177

Pratt, E. J. 1882-1964 ....DLB-92

Pratt, Samuel Jackson 1749-1814 ....DLB-39

Preface to *Alwyn* (1780), by Thomas Holcroft DLB-39

Preface to *Colonel Jack* (1722), by Daniel Defoe.................DLB-39

Preface to *Evelina* (1778), by Fanny Burney.................DLB-39

Preface to *Ferdinand Count Fathom* (1753), by Tobias Smollett................DLB-39

Preface to *Incognita* (1692), by William Congreve DLB-39

Preface to *Joseph Andrews* (1742), by Henry Fielding DLB-39

Preface to *Moll Flanders* (1722), by Daniel Defoe.................DLB-39

Preface to *Poems* (1853), by Matthew Arnold...............DLB-32

Preface to *Robinson Crusoe* (1719), by Daniel Defoe.................DLB-39

Preface to *Roderick Random* (1748), by Tobias Smollett DLB-39

Preface to *Roxana* (1724), by Daniel Defoe.................DLB-39

Preface to *St. Leon* (1799), by William Godwin...............DLB-39

Preface to Sarah Fielding's *Familiar Letters* (1747), by Henry Fielding [excerpt].....................DLB-39

Preface to Sarah Fielding's *The Adventures of David Simple* (1744), by Henry Fielding DLB-39

Preface to *The Cry* (1754), by Sarah Fielding.................DLB-39

Preface to *The Delicate Distress* (1769), by Elizabeth Griffin...............DLB-39

Preface to *The Disguis'd Prince* (1733), by Eliza Haywood [excerpt] DLB-39

Preface to *The Farther Adventures of Robinson Crusoe* (1719), by Daniel Defoe...DLB-39

Preface to the First Edition of *Pamela* (1740), by Samuel Richardson DLB-39

Preface to the First Edition of *The Castle of Otranto* (1764), by Horace Walpole DLB-39

Preface to *The History of Romances* (1715), by Pierre Daniel Huet [excerpts] DLB-39

Preface to *The Life of Charlotta du Pont* (1723), by Penelope Aubin.............DLB-39

Preface to *The Old English Baron* (1778), by Clara Reeve DLB-39

Preface to the Second Edition of *The Castle of Otranto* (1765), by Horace Walpole......................DLB-39

Preface to *The Secret History, of Queen Zarah, and the Zarazians* (1705), by Delariviere Manley......................DLB-39

Preface to the Third Edition of *Clarissa* (1751), by Samuel Richardson [excerpt]....................DLB-39

Preface to *The Works of Mrs. Davys* (1725), by Mary Davys DLB-39

Preface to Volume 1 of *Clarissa* (1747), by Samuel Richardson DLB-39

Preface to Volume 3 of *Clarissa* (1748), by Samuel Richardson DLB-39

Préfontaine, Yves 1937- DLB-53

Prelutsky, Jack 1940- DLB-61

Premisses, by Michael Hamburger ...DLB-66

Prentice, George D. 1802-1870......DLB-43

Prentice-Hall DLB-46

Prescott, Orville 1906-1996............Y-96

Prescott, William Hickling 1796-1859...............DLB-1, 30, 59

The Present State of the English Novel (1892), by George Saintsbury DLB-18

Prešeren, Francè 1800-1849........DLB-147

Preston, Thomas 1537-1598 ....DLB-62

Price, Reynolds 1933- DLB-2

Price, Richard 1723-1791..........DLB-158

Price, Richard 1949- Y-81

Priest, Christopher 1943- DLB-14

Priestley, J. B. 1894-1984 ....DLB-10, 34, 77, 100, 139; Y-84

Primary Bibliography: A Retrospective Y-95

Prime, Benjamin Young 1733-1791 ..DLB-31

Primrose, Diana floruit circa 1630 ....DLB-126

Prince, F. T. 1912- DLB-20

Prince, Thomas 1687-1758 .....DLB-24, 140

The Principles of Success in Literature (1865), by George Henry Lewes [excerpt]...DLB-57

Printz, Wolfgang Casper 1641-1717 ....DLB-168

Prior, Matthew 1664-1721..........DLB-95

Prisco, Michele 1920- DLB-177

Pritchard, William H. 1932- DLB-111

Pritchett, V. S. 1900- DLB-15, 139

Procter, Adelaide Anne 1825-1864...DLB-32

Procter, Bryan Waller 1787-1874 ....DLB-96, 144

The Profession of Authorship: Scribblers for Bread..............Y-89

The Progress of Romance (1785), by Clara Reeve [excerpt].....................DLB-39

Prokopovich, Feofan 1681?-1736 ...DLB-150

Prokosch, Frederic 1906-1989.......DLB-48

The Proletarian Novel DLB-9

Propper, Dan 1937- DLB-16

The Prospect of Peace (1778), by Joel Barlow DLB-37

Protagoras circa 490 B.C.-420 B.C. DLB-176

Proud, Robert 1728-1813.........DLB-30

Proust, Marcel 1871-1922 ....DLB-65

Prynne, J. H. 1936- DLB-40

DLB Yearbook 1996 — Cumulative Index

Przybyszewski, Stanislaw
  1868-1927 .... DLB-66

Pseudo-Dionysius the Areopagite floruit
  circa 500 .... DLB-115

The Public Lending Right in America
  Statement by Sen. Charles McC.
  Mathias, Jr. PLR and the Meaning
  of Literary Property Statements on
  PLR by American Writers Y-83

The Public Lending Right in the United Kingdom Public Lending Right: The First Year
  in the United Kingdom Y-83

The Publication of English
  Renaissance Plays.............. DLB-62

Publications and Social Movements
  [Transcendentalism]............. DLB-1

Publishers and Agents: The Columbia
  Connection..................... Y-87

A Publisher's Archives: G. P. Putnam... Y-92

Publishing Fiction at LSU Press........ Y-87

Pückler-Muskau, Hermann von
  1785-1871 .... DLB-133

Pufendorf, Samuel von
  1632-1694 .... DLB-168

Pugh, Edwin William 1874-1930 ... DLB-135

Pugin, A. Welby 1812-1852......... DLB-55

Puig, Manuel 1932-1990 .... DLB-113

Pulitzer, Joseph 1847-1911.......... DLB-23

Pulitzer, Joseph, Jr. 1885-1955 ...... DLB-29

Pulitzer Prizes for the Novel,
  1917-1945 .... DLB-9

Pulliam, Eugene 1889-1975 .... DLB-127

Purchas, Samuel 1577?-1626....... DLB-151

Purdy, Al 1918- DLB-88

Purdy, James 1923- DLB-2

Purdy, Ken W. 1913-1972 .... DLB-137

Pusey, Edward Bouverie
  1800-1882 .... DLB-55

Putnam, George Palmer
  1814-1872 .... DLB-3, 79

Putnam, Samuel 1892-1950 .... DLB-4

G. P. Putnam's Sons [U.S.] DLB-49

G. P. Putnam's Sons [U.K.] DLB-106

Puzo, Mario 1920- DLB-6

Pyle, Ernie 1900-1945 .... DLB-29

Pyle, Howard 1853-1911..... DLB-42; DS-13

Pym, Barbara 1913-1980 ..... DLB-14; Y-87

Pynchon, Thomas 1937- DLB-2, 173

Pyramid Books DLB-46

Pyrnelle, Louise-Clarke 1850-1907... DLB-42

Pythagoras circa 570 B.C.-?........ DLB-176

Q

Quad, M. (see Lewis, Charles B.)

Quarles, Francis 1592-1644........ DLB-126

The Quarterly Review
  1809-1967 .... DLB-110

Quasimodo, Salvatore 1901-1968... DLB-114

Queen, Ellery (see Dannay, Frederic, and
  Manfred B. Lee)

The Queen City Publishing House... DLB-49

Queneau, Raymond 1903-1976...... DLB-72

Quennell, Sir Peter 1905-1993...... DLB-155

Quesnel, Joseph 1746-1809 .... DLB-99

The Question of American Copyright
  in the Nineteenth Century
    Headnote
    Preface, by George Haven Putnam
  The Evolution of Copyright, by Brander
    Matthews
  Summary of Copyright Legislation in
    the United States, by R. R. Bowker
  Analysis of the Provisions of the
    Copyright Law of 1891, by
    George Haven Putnam
  The Contest for International Copyright,
    by George Haven Putnam
  Cheap Books and Good Books,
    by Brander Matthews DLB-49

Quiller-Couch, Sir Arthur Thomas
  1863-1944 .... DLB-135, 153

Quin, Ann 1936-1973.............. DLB-14

Quincy, Samuel, of Georgia ?-? DLB-31

Quincy, Samuel, of Massachusetts
  1734-1789 .... DLB-31

Quinn, Anthony 1915- DLB-122

Quintana, Leroy V. 1944- DLB-82

Quintana, Miguel de 1671-1748
  A Forerunner of Chicano
  Literature DLB-122

Quist, Harlin, Books DLB-46

Quoirez, Françoise (see Sagan, Francçise)

R

Raabe, Wilhelm 1831-1910 .... DLB-129

Rabe, David 1940- DLB-7

Raboni, Giovanni 1932- DLB-128

Rachilde 1860-1953 .... DLB-123

Racin, Koúo 1908-1943 .... DLB-147

Rackham, Arthur 1867-1939....... DLB-141

Radcliffe, Ann 1764-1823........... DLB-39

Raddall, Thomas 1903- DLB-68

Radiguet, Raymond 1903-1923...... DLB-65

Radishchev, Aleksandr Nikolaevich
  1749-1802 .... DLB-150

Radványi, Netty Reiling (see Seghers, Anna)

Rahv, Philip 1908-1973 .... DLB-137

Raimund, Ferdinand Jakob
  1790-1836 .... DLB-90

Raine, Craig 1944- DLB-40

Raine, Kathleen 1908- DLB-20

Rainolde, Richard
  circa 1530-1606................ DLB-136

Rakií, Milan 1876-1938 .... DLB-147

Ralegh, Sir Walter 1554?-1618 ..... DLB-172

Ralph, Julian 1853-1903........... DLB-23

Ralph Waldo Emerson in 1982 .... Y-82

Ramat, Silvio 1939- DLB-128

Rambler, no. 4 (1750), by Samuel Johnson
  [excerpt]..................... DLB-39

Ramée, Marie Louise de la (see Ouida)

Ramírez, Sergío 1942- DLB-145

Ramke, Bin 1947- DLB-120

Ramler, Karl Wilhelm 1725-1798.... DLB-97

Ramon Ribeyro, Julio 1929- DLB-145

Ramous, Mario 1924- DLB-128

Rampersad, Arnold 1941- DLB-111

Ramsay, Allan 1684 or 1685-1758 ... DLB-95

Ramsay, David 1749-1815.......... DLB-30

Ranck, Katherine Quintana
  1942- DLB-122

Rand, Avery and Company......... DLB-49

Rand McNally and Company DLB-49

Randall, David Anton
  1905-1975 .... DLB-140

Randall, Dudley 1914- DLB-41

Randall, Henry S. 1811-1876 .... DLB-30

Randall, James G. 1881-1953........ DLB-17

The Randall Jarrell Symposium: A Small
  Collection of Randall Jarrells
  Excerpts From Papers Delivered at
  the Randall Jarrell
  Symposium Y-86

Randolph, A. Philip 1889-1979...... DLB-91

Randolph, Anson D. F.
  [publishing house] DLB-49

Randolph, Thomas 1605-1635 .. DLB-58, 126

Random House................... DLB-46

Ranlet, Henry [publishing house] DLB-49

Ransom, John Crowe 1888-1974 .... DLB-45, 63

Ransome, Arthur 1884-1967 .... DLB-160

Raphael, Frederic 1931- DLB-14

Raphaelson, Samson 1896-1983 ..... DLB-44

Raskin, Ellen 1928-1984 .... DLB-52

Rastell, John 1475?-1536 ...... DLB-136, 170

Rattigan, Terence 1911-1977 .... DLB-13

Rawlings, Marjorie Kinnan 1896-1953 .... DLB-9, 22, 102

Raworth, Tom 1938- DLB-40

Ray, David 1932- DLB-5

Ray, Gordon Norton 1915-1986 .... DLB-103, 140

Ray, Henrietta Cordelia 1849-1916 .... DLB-50

Raymond, Henry J. 1820-1869 ... DLB-43, 79

Raymond Chandler Centenary Tributes from Michael Avallone, James Elroy, Joe Gores, and William F. Nolan Y-88

Reach, Angus 1821-1856 .... DLB-70

Read, Herbert 1893-1968 ...... DLB-20, 149

Read, Herbert, "The Practice of Biography," in *The English Sense of Humour and Other Essays* DLB-149

Read, Opie 1852-1939 .... DLB-23

Read, Piers Paul 1941- DLB-14

Reade, Charles 1814-1884 .... DLB-21

Reader's Digest Condensed Books DLB-46

Reading, Peter 1946- DLB-40

Reading Series in New York City Y-96

Reaney, James 1926- DLB-68

Rèbora, Clemente 1885-1957 ...... DLB-114

Rechy, John 1934- DLB-122; Y-82

The Recovery of Literature: Criticism in the 1990s: A Symposium Y-91

Redding, J. Saunders 1906-1988 .... DLB-63, 76

Redfield, J. S. [publishing house] DLB-49

Redgrove, Peter 1932- DLB-40

Redmon, Anne 1943- Y-86

Redmond, Eugene B. 1937- DLB-41

Redpath, James [publishing house] ... DLB-49

Reed, Henry 1808-1854 .... DLB-59

Reed, Henry 1914- DLB-27

Reed, Ishmael 1938- DLB-2, 5, 33, 169; DS-8

Reed, Sampson 1800-1880 .... DLB-1

Reed, Talbot Baines 1852-1893 ..... DLB-141

Reedy, William Marion 1862-1920 ... DLB-91

Reese, Lizette Woodworth 1856-1935 .... DLB-54

Reese, Thomas 1742-1796 .... DLB-37

Reeve, Clara 1729-1807 .... DLB-39

Reeves, James 1909-1978 .... DLB-161

Reeves, John 1926- DLB-88

Regnery, Henry, Company DLB-46

Rehberg, Hans 1901-1963 .... DLB-124

Rehfisch, Hans José 1891-1960 ..... DLB-124

Reid, Alastair 1926- DLB-27

Reid, B. L. 1918-1990 .... DLB-111

Reid, Christopher 1949- DLB-40

Reid, Forrest 1875-1947 .... DLB-153

Reid, Helen Rogers 1882-1970 ...... DLB-29

Reid, James ?-? DLB-31

Reid, Mayne 1818-1883 .... DLB-21, 163

Reid, Thomas 1710-1796 .... DLB-31

Reid, V. S. (Vic) 1913-1987 .... DLB-125

Reid, Whitelaw 1837-1912 .... DLB-23

Reilly and Lee Publishing Company DLB-46

Reimann, Brigitte 1933-1973 .... DLB-75

Reinmar der Alte circa 1165-circa 1205 .... DLB-138

Reinmar von Zweter circa 1200-circa 1250 .... DLB-138

Reisch, Walter 1903-1983 .... DLB-44

Remarque, Erich Maria 1898-1970 ... DLB-56

"Re-meeting of Old Friends": The Jack Kerouac Conference Y-82

Remington, Frederic 1861-1909 ..... DLB-12

Renaud, Jacques 1943- DLB-60

Renault, Mary 1905-1983 .... Y-83

Rendell, Ruth 1930- DLB-87

Representative Men and Women: A Historical Perspective on the British Novel, 1930-1960 .... DLB-15

(Re-)Publishing Orwell Y-86

Rettenbacher, Simon 1634-1706 .... DLB-168

Reuter, Christian 1665-after 1712 ... DLB-168

Reuter, Fritz 1810-1874 .... DLB-129

Reuter, Gabriele 1859-1941 .... DLB-66

Revell, Fleming H., Company DLB-49

Reventlow, Franziska Gräfin zu 1871-1918 .... DLB-66

Review of Reviews Office DLB-112

Review of [Samuel Richardson's] *Clarissa* (1748), by Henry Fielding DLB-39

The Revolt (1937), by Mary Colum [excerpts] DLB-36

Rexroth, Kenneth 1905-1982 .... DLB-16, 48, 165; Y-82

Rey, H. A. 1898-1977 .... DLB-22

Reynal and Hitchcock DLB-46

Reynolds, G. W. M. 1814-1879 ..... DLB-21

Reynolds, John Hamilton 1794-1852 .... DLB-96

Reynolds, Mack 1917- DLB-8

Reynolds, Sir Joshua 1723-1792 .... DLB-104

Reznikoff, Charles 1894-1976 .... DLB-28, 45

"Rhetoric" (1828; revised, 1859), by Thomas de Quincey [excerpt] ... DLB-57

Rhett, Robert Barnwell 1800-1876 ... DLB-43

Rhode, John 1884-1964 .... DLB-77

Rhodes, James Ford 1848-1927 ...... DLB-47

Rhys, Jean 1890-1979 ..... DLB-36, 117, 162

Ricardo, David 1772-1823 ..... DLB-107, 158

Ricardou, Jean 1932- DLB-83

Rice, Elmer 1892-1967 .... DLB-4, 7

Rice, Grantland 1880-1954 ..... DLB-29, 171

Rich, Adrienne 1929- DLB-5, 67

Richards, David Adams 1950- DLB-53

Richards, George circa 1760-1814 ... DLB-37

Richards, I. A. 1893-1979 .... DLB-27

Richards, Laura E. 1850-1943 .... DLB-42

Richards, William Carey 1818-1892 .... DLB-73

Richards, Grant [publishing house] DLB-112

Richardson, Charles F. 1851-1913 ... DLB-71

Richardson, Dorothy M. 1873-1957 .... DLB-36

Richardson, Jack 1935- DLB-7

Richardson, John 1796-1852 .... DLB-99

Richardson, Samuel 1689-1761 .... DLB-39, 154

Richardson, Willis 1889-1977 .... DLB-51

Riche, Barnabe 1542-1617 .... DLB-136

Richler, Mordecai 1931- DLB-53

Richter, Conrad 1890-1968 .... DLB-9

358

Richter, Hans Werner 1908- DLB-69	Rivard, Adjutor 1868-1945 .... DLB-92	Rodriguez, Richard 1944- DLB-82
Richter, Johann Paul Friedrich 1763-1825 .... DLB-94	Rive, Richard 1931-1989 .... DLB-125	Rodríguez Julia, Edgardo 1946- DLB-145
Rickerby, Joseph [publishing house] DLB-106	Rivera, Marina 1942- DLB-122	Roethke, Theodore 1908-1963 .... DLB-5
Rickword, Edgell 1898-1982 .... DLB-20	Rivera, Tomás 1935-1984 .... DLB-82	Rogers, Pattiann 1940- DLB-105
Riddell, Charlotte 1832-1906 ..... DLB-156	Rivers, Conrad Kent 1933-1968 .... DLB-41	Rogers, Samuel 1763-1855 .... DLB-93
Riddell, John (see Ford, Corey)	Riverside Press DLB-49	Rogers, Will 1879-1935 .... DLB-11
Ridge, John Rollin 1827-1867 ..... DLB-175	Rivington, James circa 1724-1802 ... DLB-43	Rohmer, Sax 1883-1959 .... DLB-70
Ridge, Lola 1873-1941 .... DLB-54	Rivington, Charles [publishing house] DLB-154	Roiphe, Anne 1935- Y-80
Ridge, William Pett 1859-1930 .... DLB-135	Rivkin, Allen 1903-1990 .... DLB-26	Rojas, Arnold R. 1896-1988 .... DLB-82
Riding, Laura (see Jackson, Laura Riding)	Roa Bastos, Augusto 1917- DLB-113	Rolfe, Frederick William 1860-1913 .... DLB-34, 156
Ridler, Anne 1912- DLB-27	Robbe-Grillet, Alain 1922- DLB-83	Rolland, Romain 1866-1944 .... DLB-65
Ridruego, Dionisio 1912-1975 ..... DLB-108	Robbins, Tom 1936- Y-80	Rolle, Richard circa 1290-1300 - 1340 .... DLB-146
Riel, Louis 1844-1885 .... DLB-99	Roberts, Charles G. D. 1860-1943 ... DLB-92	
Riemer, Johannes 1648-1714 .... DLB-168	Roberts, Dorothy 1906-1993 .... DLB-88	Rölvaag, O. E. 1876-1931 .... DLB-9
Riffaterre, Michael 1924- DLB-67	Roberts, Elizabeth Madox 1881-1941 .... DLB-9, 54, 102	Romains, Jules 1885-1972 .... DLB-65
Riggs, Lynn 1899-1954 .... DLB-175	Roberts, Kenneth 1885-1957 .... DLB-9	Roman, A., and Company DLB-49
Riis, Jacob 1849-1914 .... DLB-23	Roberts, William 1767-1849 .... DLB-142	Romano, Lalla 1906- DLB-177
Riker, John C. [publishing house] DLB-49	Roberts Brothers DLB-49	Romano, Octavio 1923- DLB-122
Riley, John 1938-1978 .... DLB-40	Roberts, James [publishing house] .. DLB-154	Romero, Leo 1950- DLB-122
Rilke, Rainer Maria 1875-1926 ..... DLB-81	Robertson, A. M., and Company DLB-49	Romero, Lin 1947- DLB-122
Rimanelli, Giose 1926- DLB-177	Robertson, William 1721-1793 ..... DLB-104	Romero, Orlando 1945- DLB-82
Rinehart and Company DLB-46	Robinson, Casey 1903-1979 .... DLB-44	Rook, Clarence 1863-1915 .... DLB-135
Ringuet 1895-1960 .... DLB-68	Robinson, Edwin Arlington 1869-1935 .... DLB-54	Roosevelt, Theodore 1858-1919 ..... DLB-47
Ringwood, Gwen Pharis 1910-1984 .... DLB-88	Robinson, Henry Crabb 1775-1867 .... DLB-107	Root, Waverley 1903-1982 .... DLB-4
Rinser, Luise 1911- DLB-69		Root, William Pitt 1941- DLB-120
Ríos, Alberto 1952- DLB-122	Robinson, James Harvey 1863-1936 .... DLB-47	Roquebrune, Robert de 1889-1978 ... DLB-68
Ríos, Isabella 1948- DLB-82	Robinson, Lennox 1886-1958 .... DLB-10	Rosa, João Guimarães 1908-1967 .... DLB-113
Ripley, Arthur 1895-1961 .... DLB-44	Robinson, Mabel Louise 1874-1962 .... DLB-22	Rosales, Luis 1910-1992 .... DLB-134
Ripley, George 1802-1880 ..... DLB-1, 64, 73	Robinson, Mary 1758-1800 .... DLB-158	Roscoe, William 1753-1831 .... DLB-163
The Rising Glory of America: Three Poems DLB-37	Robinson, Richard circa 1545-1607 .... DLB-167	Rose, Reginald 1920- DLB-26
The Rising Glory of America: Written in 1771 (1786), by Hugh Henry Brackenridge and Philip Freneau DLB-37	Robinson, Therese 1797-1870 .... DLB-59, 133	Rose, Wendy 1948- DLB-175
		Rosegger, Peter 1843-1918 .... DLB-129
		Rosei, Peter 1946- DLB-85
Riskin, Robert 1897-1955 .... DLB-26	Robison, Mary 1949- DLB-130	Rosen, Norma 1925- DLB-28
Risse, Heinz 1898- DLB-69	Roblès, Emmanuel 1914- DLB-83	Rosenbach, A. S. W. 1876-1952 .... DLB-140
Rist, Johann 1607-1667 .... DLB-164	Roccatagliata Ceccardi, Ceccardo 1871-1919 .... DLB-114	Rosenberg, Isaac 1890-1918 .... DLB-20
Ritchie, Anna Mowatt 1819-1870 ..... DLB-3		Rosenfeld, Isaac 1918-1956 .... DLB-28
Ritchie, Anne Thackeray 1837-1919 .... DLB-18	Rochester, John Wilmot, Earl of 1647-1680 .... DLB-131	Rosenthal, M. L. 1917- DLB-5
Ritchie, Thomas 1778-1854 .... DLB-43	Rock, Howard 1911-1976 .... DLB-127	Ross, Alexander 1591-1654 .... DLB-151
Rites of Passage [on William Saroyan] Y-83	Rodgers, Carolyn M. 1945- DLB-41	Ross, Harold 1892-1951 .... DLB-137
	Rodgers, W. R. 1909-1969 .... DLB-20	Ross, Leonard Q. (see Rosten, Leo)
The Ritz Paris Hemingway Award Y-85	Rodríguez, Claudio 1934- DLB-134	Ross, Martin 1862-1915 .... DLB-135

Ross, Sinclair 1908- DLB-88

Ross, W. W. E. 1894-1966 .... DLB-88

Rosselli, Amelia 1930- DLB-128

Rossen, Robert 1908-1966 .... DLB-26

Rossetti, Christina Georgina
    1830-1894 .... DLB-35, 163

Rossetti, Dante Gabriel 1828-1882 ... DLB-35

Rossner, Judith 1935- DLB-6

Rosten, Leo 1908- DLB-11

Rostenberg, Leona 1908- DLB-140

Rostovsky, Dimitrii 1651-1709 ..... DLB-150

Bertram Rota and His Bookshop Y-91

Roth, Gerhard 1942- DLB-85, 124

Roth, Henry 1906?- DLB-28

Roth, Joseph 1894-1939 .... DLB-85

Roth, Philip 1933- DLB-2, 28, 173; Y-82

Rothenberg, Jerome 1931- DLB-5

Rotimi, Ola 1938- DLB-125

Routhier, Adolphe-Basile
    1839-1920 .... DLB-99

Routier, Simone 1901-1987 .... DLB-88

Routledge, George, and Sons DLB-106

Roversi, Roberto 1923- DLB-128

Rowe, Elizabeth Singer
    1674-1737 .... DLB-39, 95

Rowe, Nicholas 1674-1718 .... DLB-84

Rowlands, Samuel
    circa 1570-1630 .... DLB-121

Rowlandson, Mary
    circa 1635-circa 1678 .... DLB-24

Rowley, William circa 1585-1626 .... DLB-58

Rowse, A. L. 1903- DLB-155

Rowson, Susanna Haswell
    circa 1762-1824 .... DLB-37

Roy, Camille 1870-1943 .... DLB-92

Roy, Gabrielle 1909-1983 .... DLB-68

Roy, Jules 1907- DLB-83

The Royal Court Theatre and the English
    Stage Company DLB-13

The Royal Court Theatre and the New Drama
DLB-10

The Royal Shakespeare Company
    at the Swan Y-88

Royall, Anne 1769-1854 .... DLB-43

The Roycroft Printing Shop DLB-49

Royster, Vermont 1914- DLB-127

Royston, Richard
    [publishing house] DLB-170

Ruark, Gibbons 1941- DLB-120

Ruban, Vasilii Grigorevich
    1742-1795 .... DLB-150

Rubens, Bernice 1928- DLB-14

Rudd and Carleton DLB-49

Rudkin, David 1936- DLB-13

Rudolf von Ems
    circa 1200-circa 1254 .... DLB-138

Ruffin, Josephine St. Pierre
    1842-1924 .... DLB-79

Ruganda, John 1941- DLB-157

Ruggles, Henry Joseph 1813-1906 ... DLB-64

Rukeyser, Muriel 1913-1980 .... DLB-48

Rule, Jane 1931- DLB-60

Rulfo, Juan 1918-1986 .... DLB-113

Rumaker, Michael 1932- DLB-16

Rumens, Carol 1944- DLB-40

Runyon, Damon 1880-1946 . DLB-11, 86, 171

Ruodlieb circa 1050-1075 .... DLB-148

Rush, Benjamin 1746-1813 .... DLB-37

Rusk, Ralph L. 1888-1962 .... DLB-103

Ruskin, John 1819-1900 .... DLB-55, 163

Russ, Joanna 1937- DLB-8

Russell, B. B., and Company DLB-49

Russell, Benjamin 1761-1845 .... DLB-43

Russell, Bertrand 1872-1970 ...... DLB-100

Russell, Charles Edward
    1860-1941 .... DLB-25

Russell, George William (see AE)

Russell, R. H., and Son DLB-49

Rutherford, Mark 1831-1913 .... DLB-18

Ryan, Michael 1946- Y-82

Ryan, Oscar 1904- DLB-68

Ryga, George 1932- DLB-60

Rymer, Thomas 1643?-1713 ...... DLB-101

Ryskind, Morrie 1895-1985 .... DLB-26

Rzhevsky, Aleksei Andreevich
    1737-1804 .... DLB-150

S

The Saalfield Publishing
    Company DLB-46

Saba, Umberto 1883-1957 .... DLB-114

Sábato, Ernesto 1911- DLB-145

Saberhagen, Fred 1930- DLB-8

Sacer, Gottfried Wilhelm
    1635-1699 .... DLB-168

Sackler, Howard 1929-1982 .... DLB-7

Sackville, Thomas 1536-1608 ...... DLB-132

Sackville, Thomas 1536-1608
    and Norton, Thomas
    1532-1584 .... DLB-62

Sackville-West, V. 1892-1962 ...... DLB-34

Sadlier, D. and J., and Company DLB-49

Sadlier, Mary Anne 1820-1903 ..... DLB-99

Sadoff, Ira 1945- DLB-120

Saenz, Jaime 1921-1986 .... DLB-145

Saffin, John circa 1626-1710 .... DLB-24

Sagan, Françoise 1935- DLB-83

Sage, Robert 1899-1962 .... DLB-4

Sagel, Jim 1947- DLB-82

Sagendorph, Robb Hansell
    1900-1970 .... DLB-137

Sahagún, Carlos 1938- DLB-108

Sahkomaapii, Piitai (see Highwater, Jamake)

Sahl, Hans 1902- DLB-69

Said, Edward W. 1935- DLB-67

Saiko, George 1892-1962 .... DLB-85

St. Dominic's Press DLB-112

Saint-Exupéry, Antoine de
    1900-1944 .... DLB-72

St. Johns, Adela Rogers 1894-1988 ... DLB-29

St. Martin's Press DLB-46

St. Omer, Garth 1931- DLB-117

Saint Pierre, Michel de 1916-1987 ... DLB-83

Saintsbury, George
    1845-1933 .... DLB-57, 149

Saki (see Munro, H. H.)

Salaam, Kalamu ya 1947- DLB-38

Salas, Floyd 1931- DLB-82

Sálaz-Marquez, Rubén 1935- DLB-122

Salemson, Harold J. 1910-1988 ..... DLB-4

Salinas, Luis Omar 1937- DLB-82

Salinas, Pedro 1891-1951 .... DLB-134

Salinger, J. D. 1919- DLB-2, 102, 173

Salkey, Andrew 1928- DLB-125

Salt, Waldo 1914- DLB-44

Salter, James 1925- DLB-130

Salter, Mary Jo 1954- DLB-120

Salustri, Carlo Alberto (see Trilussa)

Salverson, Laura Goodman
    1890-1970 .... DLB-92

Sampson, Richard Henry (see Hull, Richard)

Samuels, Ernest 1903- DLB-111

Sanborn, Franklin Benjamin
1831-1917 .... DLB-1

Sánchez, Luis Rafael 1936- DLB-145

Sánchez, Philomeno "Phil"
1917- DLB-122

Sánchez, Ricardo 1941- DLB-82

Sanchez, Sonia 1934- DLB-41; DS-8

Sand, George 1804-1876 .... DLB-119

Sandburg, Carl 1878-1967 .... DLB-17, 54

Sanders, Ed 1939- DLB-16

Sandoz, Mari 1896-1966............ DLB-9

Sandwell, B. K. 1876-1954.......... DLB-92

Sandy, Stephen 1934- DLB-165

Sandys, George 1578-1644 ..... DLB-24, 121

Sangster, Charles 1822-1893 .... DLB-99

Sanguineti, Edoardo 1930- DLB-128

Sansom, William 1912-1976 .... DLB-139

Santayana, George
1863-1952.......... DLB-54, 71; DS-13

Santiago, Danny 1911-1988........ DLB-122

Santmyer, Helen Hooven 1895-1986.... Y-84

Sapir, Edward 1884-1939........... DLB-92

Sapper (see McNeile, Herman Cyril)

Sappho circa 620 B.C.-circa 550 B.C.
.... DLB-176

Sarduy, Severo 1937- DLB-113

Sargent, Pamela 1948- DLB-8

Saro-Wiwa, Ken 1941- DLB-157

Saroyan, William
1908-1981.......... DLB-7, 9, 86; Y-81

Sarraute, Nathalie 1900- DLB-83

Sarrazin, Albertine 1937-1967 .... DLB-83

Sarris, Greg 1952- DLB-175

Sarton, May 1912- DLB-48; Y-81

Sartre, Jean-Paul 1905-1980......... DLB-72

Sassoon, Siegfried 1886-1967........ DLB-20

Saturday Review Press............. DLB-46

Saunders, James 1925- DLB-13

Saunders, John Monk 1897-1940 .... DLB-26

Saunders, Margaret Marshall
1861-1947 .... DLB-92

Saunders and Otley DLB-106

Savage, James 1784-1873 .... DLB-30

Savage, Marmion W. 1803?-1872.... DLB-21

Savage, Richard 1697?-1743 .... DLB-95

Savard, Félix-Antoine 1896-1982 .... DLB-68

Saville, (Leonard) Malcolm
1901-1982 .... DLB-160

Sawyer, Ruth 1880-1970 .... DLB-22

Sayers, Dorothy L.
1893-1957......... DLB-10, 36, 77, 100

Sayles, John Thomas 1950- DLB-44

Sbarbaro, Camillo 1888-1967 ...... DLB-114

Scannell, Vernon 1922- DLB-27

Scarry, Richard 1919-1994 .... DLB-61

Schaeffer, Albrecht 1885-1950....... DLB-66

Schaeffer, Susan Fromberg 1941- ... DLB-28

Schaff, Philip 1819-1893 .... DS-13

Schaper, Edzard 1908-1984 .... DLB-69

Scharf, J. Thomas 1843-1898........ DLB-47

Scheffel, Joseph Viktor von
1826-1886 .... DLB-129

Scheffler, Johann 1624-1677 ...... DLB-164

Schelling, Friedrich Wilhelm Joseph von
1775-1854 .... DLB-90

Scherer, Wilhelm 1841-1886 .... DLB-129

Schickele, René 1883-1940.......... DLB-66

Schiff, Dorothy 1903-1989......... DLB-127

Schiller, Friedrich 1759-1805........ DLB-94

Schirmer, David 1623-1687 .... DLB-164

Schlaf, Johannes 1862-1941 .... DLB-118

Schlegel, August Wilhelm
1767-1845 .... DLB-94

Schlegel, Dorothea 1763-1839....... DLB-90

Schlegel, Friedrich 1772-1829 ...... DLB-90

Schleiermacher, Friedrich
1768-1834 .... DLB-90

Schlesinger, Arthur M., Jr. 1917- ... DLB-17

Schlumberger, Jean 1877-1968 ...... DLB-65

Schmid, Eduard Hermann Wilhelm (see
Edschmid, Kasimir)

Schmidt, Arno 1914-1979 .... DLB-69

Schmidt, Johann Kaspar (see Stirner, Max)

Schmidt, Michael 1947- DLB-40

Schmidtbonn, Wilhelm August
1876-1952 .... DLB-118

Schmitz, James H. 1911- DLB-8

Schnabel, Johann Gottfried
1692-1760 .... DLB-168

Schnackenberg, Gjertrud 1953- ... DLB-120

Schnitzler, Arthur 1862-1931 ... DLB-81, 118

Schnurre, Wolfdietrich 1920- DLB-69

Schocken Books DLB-46

Scholartis Press.................. DLB-112

The Schomburg Center for Research
in Black Culture.............. DLB-76

Schönbeck, Virgilio (see Giotti, Virgilio)

Schönherr, Karl 1867-1943 .... DLB-118

Schoolcraft, Jane Johnston
1800-1841 .... DLB-175

School Stories, 1914-1960 .... DLB-160

Schopenhauer, Arthur 1788-1860 .... DLB-90

Schopenhauer, Johanna 1766-1838... DLB-90

Schorer, Mark 1908-1977.......... DLB-103

Schottelius, Justus Georg
1612-1676 .... DLB-164

Schouler, James 1839-1920 .... DLB-47

Schrader, Paul 1946- DLB-44

Schreiner, Olive 1855-1920..... DLB-18, 156

Schroeder, Andreas 1946- DLB-53

Schubart, Christian Friedrich Daniel
1739-1791 .... DLB-97

Schubert, Gotthilf Heinrich
1780-1860 .... DLB-90

Schücking, Levin 1814-1883 .... DLB-133

Schulberg, Budd
1914- DLB-6, 26, 28; Y-81

Schulte, F. J., and Company DLB-49

Schulze, Hans (see Praetorius, Johannes)

Schupp, Johann Balthasar
1610-1661 .... DLB-164

Schurz, Carl 1829-1906 .... DLB-23

Schuyler, George S. 1895-1977... DLB-29, 51

Schuyler, James 1923-1991 ...... DLB-5, 169

Schwartz, Delmore 1913-1966.... DLB-28, 48

Schwartz, Jonathan 1938- Y-82

Schwarz, Sibylle 1621-1638 .... DLB-164

Schwerner, Armand 1927- DLB-165

Schwob, Marcel 1867-1905 .... DLB-123

Sciascia, Leonardo 1921-1989 ...... DLB-177

Science Fantasy DLB-8

Science-Fiction Fandom and
Conventions DLB-8

Science-Fiction Fanzines: The Time
Binders....................... DLB-8

Science-Fiction Films DLB-8

Science Fiction Writers of America and the
Nebula Awards................. DLB-8

Scot, Reginald circa 1538-1599 ..... DLB-136

Scotellaro, Rocco 1923-1953 .... DLB-128

Scott, Dennis 1939-1991 .... DLB-125

Scott, Dixon 1881-1915 .... DLB-98

Scott, Duncan Campbell
1862-1947 .... DLB-92

Scott, Evelyn 1893-1963 .... DLB-9, 48

Scott, F. R. 1899-1985 .... DLB-88

Scott, Frederick George
1861-1944 .... DLB-92

Scott, Geoffrey 1884-1929 .... DLB-149

Scott, Harvey W. 1838-1910 .... DLB-23

Scott, Paul 1920-1978 .... DLB-14

Scott, Sarah 1723-1795 .... DLB-39

Scott, Tom 1918- DLB-27

Scott, Sir Walter
1771-1832 ... DLB-93, 107, 116, 144, 159

Scott, William Bell 1811-1890 ...... DLB-32

Scott, Walter, Publishing
Company Limited DLB-112

Scott, William R.
[publishing house] DLB-46

Scott-Heron, Gil 1949- DLB-41

Scribner, Charles, Jr. 1921-1995 .... Y-95

Charles Scribner's Sons DLB-49; DS-13

Scripps, E. W. 1854-1926 .... DLB-25

Scudder, Horace Elisha
1838-1902 .... DLB-42, 71

Scudder, Vida Dutton 1861-1954 .... DLB-71

Scupham, Peter 1933- DLB-40

Seabrook, William 1886-1945 .... DLB-4

Seabury, Samuel 1729-1796 .... DLB-31

Seacole, Mary Jane Grant
1805-1881 .... DLB-166

The Seafarer circa 970 .... DLB-146

Sealsfield, Charles 1793-1864 ...... DLB-133

Sears, Edward I. 1819?-1876 .... DLB-79

Sears Publishing Company DLB-46

Seaton, George 1911-1979 .... DLB-44

Seaton, William Winston
1785-1866 .... DLB-43

Secker, Martin, and Warburg
Limited DLB-112

Secker, Martin [publishing house] .. DLB-112

Second-Generation Minor Poets of the
Seventeenth Century DLB-126

Sedgwick, Arthur George
1844-1915 .... DLB-64

Sedgwick, Catharine Maria
1789-1867 .... DLB-1, 74

Sedgwick, Ellery 1872-1930 .... DLB-91

Sedley, Sir Charles 1639-1701 ...... DLB-131

Seeger, Alan 1888-1916 .... DLB-45

Seers, Eugene (see Dantin, Louis)

Segal, Erich 1937- Y-86

Seghers, Anna 1900-1983 .... DLB-69

Seid, Ruth (see Sinclair, Jo)

Seidel, Frederick Lewis 1936- Y-84

Seidel, Ina 1885-1974 .... DLB-56

Seigenthaler, John 1927- DLB-127

Seizin Press DLB-112

Séjour, Victor 1817-1874 .... DLB-50

Séjour Marcou et Ferrand, Juan Victor (see
Séjour, Victor)

Selby, Hubert, Jr. 1928- DLB-2

Selden, George 1929-1989 .... DLB-52

Selected English-Language Little Magazines
and Newspapers [France,
1920-1939] DLB-4

Selected Humorous Magazines
(1820-1950) DLB-11

Selected Science-Fiction Magazines and
Anthologies DLB-8

Self, Edwin F. 1920- DLB-137

Seligman, Edwin R. A. 1861-1939 ... DLB-47

Selous, Frederick Courteney
1851-1917 .... DLB-174

Seltzer, Chester E. (see Muro, Amado)

Seltzer, Thomas
[publishing house] DLB-46

Selvon, Sam 1923-1994 .... DLB-125

Senancour, Etienne de 1770-1846 ... DLB-119

Sendak, Maurice 1928- DLB-61

Senécal, Eva 1905- DLB-92

Sengstacke, John 1912- DLB-127

Senior, Olive 1941- DLB-157

Šenoa, August 1838-1881 .... DLB-147

"Sensation Novels" (1863), by
H. L. Manse DLB-21

Sepamla, Sipho 1932- DLB-157

Seredy, Kate 1899-1975 .... DLB-22

Sereni, Vittorio 1913-1983 .... DLB-128

Seres, William
[publishing house] DLB-170

Serling, Rod 1924-1975 .... DLB-26

Serote, Mongane Wally 1944- DLB-125

Serraillier, Ian 1912-1994 .... DLB-161

Serrano, Nina 1934- DLB-122

Service, Robert 1874-1958 .... DLB-92

Seth, Vikram 1952- DLB-120

Seton, Ernest Thompson
1860-1942 .... DLB-92; DS-13

Settle, Mary Lee 1918- DLB-6

Seume, Johann Gottfried
1763-1810 .... DLB-94

Seuss, Dr. (see Geisel, Theodor Seuss)

The Seventy-fifth Anniversary of the Armistice:
The Wilfred Owen Centenary and the
Great War Exhibit at the University of
Virginia Y-93

Sewall, Joseph 1688-1769 .... DLB-24

Sewall, Richard B. 1908- DLB-111

Sewell, Anna 1820-1878 .... DLB-163

Sewell, Samuel 1652-1730 .... DLB-24

Sex, Class, Politics, and Religion [in the
British Novel, 1930-1959] DLB-15

Sexton, Anne 1928-1974 .... DLB-5, 169

Seymour-Smith, Martin 1928- DLB-155

Shaara, Michael 1929-1988 .... Y-83

Shadwell, Thomas 1641?-1692 ..... DLB-80

Shaffer, Anthony 1926- DLB-13

Shaffer, Peter 1926- DLB-13

Shaftesbury, Anthony Ashley Cooper,
Third Earl of 1671-1713 .... DLB-101

Shairp, Mordaunt 1887-1939 .... DLB-10

Shakespeare, William
1564-1616 .... DLB-62, 172

The Shakespeare Globe Trust Y-93

Shakespeare Head Press DLB-112

Shakhovskoi, Aleksandr Aleksandrovich
1777-1846 .... DLB-150

Shange, Ntozake 1948- DLB-38

Shapiro, Karl 1913- DLB-48

Sharon Publications DLB-46

Sharp, Margery 1905-1991 .... DLB-161

Sharp, William 1855-1905 .... DLB-156

Sharpe, Tom 1928- DLB-14

Shaw, Albert 1857-1947 .... DLB-91

Shaw, Bernard 1856-1950 .... DLB-10, 57

Shaw, Henry Wheeler 1818-1885 .... DLB-11

Shaw, Joseph T. 1874-1952 .... DLB-137

Shaw, Irwin 1913-1984 .... DLB-6, 102; Y-84

Shaw, Robert 1927-1978 .... DLB-13, 14

Shaw, Robert B. 1947- DLB-120

Shawn, William 1907-1992 .... DLB-137

Shay, Frank [publishing house]...... DLB-46

Shea, John Gilmary 1824-1892 ...... DLB-30

Sheaffer, Louis 1912-1993 .... DLB-103

Shearing, Joseph 1886-1952......... DLB-70

Shebbeare, John 1709-1788 .... DLB-39

Sheckley, Robert 1928- DLB-8

Shedd, William G. T. 1820-1894 .... DLB-64

Sheed, Wilfred 1930- DLB-6

Sheed and Ward [U.S.] DLB-46

Sheed and Ward Limited [U.K.].... DLB-112

Sheldon, Alice B. (see Tiptree, James, Jr.)

Sheldon, Edward 1886-1946 .... DLB-7

Sheldon and Company............. DLB-49

Shelley, Mary Wollstonecraft
   1797-1851.......... DLB-110, 116, 159

Shelley, Percy Bysshe
   1792-1822........... DLB-96, 110, 158

Shelnutt, Eve 1941- DLB-130

Shenstone, William 1714-1763 ...... DLB-95

Shepard, Ernest Howard
   1879-1976 .... DLB-160

Shepard, Sam 1943- DLB-7

Shepard, Thomas I,
   1604 or 1605-1649 .... DLB-24

Shepard, Thomas II, 1635-1677 ..... DLB-24

Shepard, Clark and Brown DLB-49

Shepherd, Luke
   flourished 1547-1554 .... DLB-136

Sherburne, Edward 1616-1702 ..... DLB-131

Sheridan, Frances 1724-1766..... DLB-39, 84

Sheridan, Richard Brinsley
   1751-1816 .... DLB-89

Sherman, Francis 1871-1926 .... DLB-92

Sherriff, R. C. 1896-1975........... DLB-10

Sherry, Norman 1935- DLB-155

Sherwood, Mary Martha
   1775-1851 .... DLB-163

Sherwood, Robert 1896-1955 ..... DLB-7, 26

Shiel, M. P. 1865-1947 .... DLB-153

Shiels, George 1886-1949........... DLB-10

Shillaber, B.[enjamin] P.[enhallow]
   1814-1890 .... DLB-1, 11

Shine, Ted 1931- DLB-38

Ship, Reuben 1915-1975 .... DLB-88

Shirer, William L. 1904-1993 .... DLB-4

Shirinsky-Shikhmatov, Sergii Aleksandrovich
   1783-1837 .... DLB-150

Shirley, James 1596-1666........... DLB-58

Shishkov, Aleksandr Semenovich
   1753-1841 .... DLB-150

Shockley, Ann Allen 1927- DLB-33

Short, Peter
   [publishing house] DLB-170

Shorthouse, Joseph Henry
   1834-1903 .... DLB-18

Showalter, Elaine 1941- DLB-67

Shulevitz, Uri 1935- DLB-61

Shulman, Max 1919-1988 .... DLB-11

Shute, Henry A. 1856-1943 .... DLB-9

Shuttle, Penelope 1947- DLB-14, 40

Sibbes, Richard 1577-1635......... DLB-151

Sidgwick and Jackson Limited...... DLB-112

Sidney, Margaret (see Lothrop, Harriet M.)

Sidney, Mary 1561-1621 .... DLB-167

Sidney, Sir Philip 1554-1586 ...... DLB-167

Sidney's Press DLB-49

Siegfried Loraine Sassoon: A Centenary Essay
   Tributes from Vivien F. Clarke and
   Michael Thorpe.................. Y-86

Sierra, Rubén 1946- DLB-122

Sierra Club Books................. DLB-49

Siger of Brabant
   circa 1240-circa 1284 .... DLB-115

Sigourney, Lydia Howard (Huntley)
   1791-1865................ DLB-1, 42, 73

Silkin, Jon 1930- DLB-27

Silko, Leslie Marmon
   1948- DLB-143, 175

Silliman, Ron 1946- DLB-169

Silliphant, Stirling 1918- DLB-26

Sillitoe, Alan 1928- DLB-14, 139

Silman, Roberta 1934- DLB-28

Silva, Beverly 1930- DLB-122

Silverberg, Robert 1935- DLB-8

Silverman, Kenneth 1936- DLB-111

Simak, Clifford D. 1904-1988 .... DLB-8

Simcoe, Elizabeth 1762-1850........ DLB-99

Simcox, George Augustus
   1841-1905 .... DLB-35

Sime, Jessie Georgina 1868-1958..... DLB-92

Simenon, Georges
   1903-1989 .... DLB-72; Y-89

Simic, Charles 1938- DLB-105

Simic, Charles,
   Images and "Images" DLB-105

Simmel, Johannes Mario 1924- DLB-69

Simmes, Valentine
   [publishing house] DLB-170

Simmons, Ernest J. 1903-1972...... DLB-103

Simmons, Herbert Alfred 1930- DLB-33

Simmons, James 1933- DLB-40

Simms, William Gilmore
   1806-1870............ DLB-3, 30, 59, 73

Simms and M'Intyre.............. DLB-106

Simon, Claude 1913- DLB-83

Simon, Neil 1927- DLB-7

Simon and Schuster DLB-46

Simons, Katherine Drayton Mayrant
   1890-1969...................... Y-83

Simpkin and Marshall
   [publishing house] DLB-154

Simpson, Helen 1897-1940 .... DLB-77

Simpson, Louis 1923- DLB-5

Simpson, N. F. 1919- DLB-13

Sims, George 1923- DLB-87

Sims, George Robert
   1847-1922............. DLB-35, 70, 135

Sinán, Rogelio 1904- DLB-145

Sinclair, Andrew 1935- DLB-14

Sinclair, Bertrand William
   1881-1972 .... DLB-92

Sinclair, Catherine
   1800-1864 .... DLB-163

Sinclair, Jo 1913- DLB-28

Sinclair Lewis Centennial
   Conference...................... Y-85

Sinclair, Lister 1921- DLB-88

Sinclair, May 1863-1946 .... DLB-36, 135

Sinclair, Upton 1878-1968........... DLB-9

Sinclair, Upton [publishing house] ... DLB-46

Singer, Isaac Bashevis
   1904-1991.......... DLB-6, 28, 52; Y-91

Singmaster, Elsie 1879-1958 .... DLB-9

Sinisgalli, Leonardo 1908-1981..... DLB-114

Siodmak, Curt 1902- DLB-44

Sissman, L. E. 1928-1976............ DLB-5

Sisson, C. H. 1914- DLB-27

Sitwell, Edith 1887-1964 .... DLB-20

Sitwell, Osbert 1892-1969 .... DLB-100

Skármeta, Antonio 1940- DLB-145

Skeffington, William [publishing house] DLB-106

Skelton, John 1463-1529 .... DLB-136

Skelton, Robin 1925- DLB-27, 53

Skinner, Constance Lindsay 1877-1939 .... DLB-92

Skinner, John Stuart 1788-1851...... DLB-73

Skipsey, Joseph 1832-1903.......... DLB-35

Slade, Bernard 1930- DLB-53

Slater, Patrick 1880-1951 .... DLB-68

Slaveykov, Pencho 1866-1912...... DLB-147

Slavitt, David 1935- DLB-5, 6

Sleigh, Burrows Willcocks Arthur 1821-1869 .... DLB-99

A Slender Thread of Hope: The Kennedy Center Black Theatre Project DLB-38

Slesinger, Tess 1905-1945 .... DLB-102

Slick, Sam (see Haliburton, Thomas Chandler)

Sloane, William, Associates......... DLB-46

Small, Maynard and Company...... DLB-49

Small Presses in Great Britain and Ireland, 1960-1985 .... DLB-40

Small Presses I: Jargon Society......... Y-84

Small Presses II: The Spirit That Moves Us Press Y-85

Small Presses III: Pushcart Press Y-87

Smart, Christopher 1722-1771 ..... DLB-109

Smart, David A. 1892-1957 .... DLB-137

Smart, Elizabeth 1913-1986......... DLB-88

Smellie, William [publishing house] DLB-154

Smiles, Samuel 1812-1904 .... DLB-55

Smith, A. J. M. 1902-1980 .... DLB-88

Smith, Adam 1723-1790........... DLB-104

Smith, Alexander 1829-1867 ..... DLB-32, 55

Smith, Betty 1896-1972 .... Y-82

Smith, Carol Sturm 1938- Y-81

Smith, Charles Henry 1826-1903 .... DLB-11

Smith, Charlotte 1749-1806 .... DLB-39, 109

Smith, Chet 1899-1973............ DLB-171

Smith, Cordwainer 1913-1966........ DLB-8

Smith, Dave 1942- DLB-5

Smith, Dodie 1896- DLB-10

Smith, Doris Buchanan 1934- DLB-52

Smith, E. E. 1890-1965............. DLB-8

Smith, Elihu Hubbard 1771-1798 .... DLB-37

Smith, Elizabeth Oakes (Prince) 1806-1893 .... DLB-1

Smith, F. Hopkinson 1838-1915....... DS-13

Smith, George D. 1870-1920 ...... DLB-140

Smith, George O. 1911-1981 .... DLB-8

Smith, Goldwin 1823-1910 .... DLB-99

Smith, H. Allen 1907-1976 ...... DLB-11, 29

Smith, Hazel Brannon 1914- DLB-127

Smith, Henry circa 1560-circa 1591 .... DLB-136

Smith, Horatio (Horace) 1779-1849 .... DLB-116

Smith, Horatio (Horace) 1779-1849 and James Smith 1775-1839 .... DLB-96

Smith, Iain Crichton 1928- DLB-40, 139

Smith, J. Allen 1860-1924 .... DLB-47

Smith, John 1580-1631.......... DLB-24, 30

Smith, Josiah 1704-1781............ DLB-24

Smith, Ken 1938- DLB-40

Smith, Lee 1944- DLB-143; Y-83

Smith, Logan Pearsall 1865-1946 .... DLB-98

Smith, Mark 1935- Y-82

Smith, Michael 1698-circa 1771 ..... DLB-31

Smith, Red 1905-1982 .... DLB-29, 171

Smith, Roswell 1829-1892 .... DLB-79

Smith, Samuel Harrison 1772-1845 .... DLB-43

Smith, Samuel Stanhope 1751-1819 .... DLB-37

Smith, Sarah (see Stretton, Hesba)

Smith, Seba 1792-1868.......... DLB-1, 11

Smith, Sir Thomas 1513-1577 ...... DLB-132

Smith, Stevie 1902-1971............ DLB-20

Smith, Sydney 1771-1845.......... DLB-107

Smith, Sydney Goodsir 1915-1975 ... DLB-27

Smith, Wendell 1914-1972......... DLB-171

Smith, William flourished 1595-1597.......... DLB-136

Smith, William 1727-1803 .... DLB-31

Smith, William 1728-1793 .... DLB-30

Smith, William Gardner 1927-1974 .... DLB-76

Smith, William Henry 1808-1872 .... DLB-159

Smith, William Jay 1918- DLB-5

Smith, Elder and Company........ DLB-154

Smith, Harrison, and Robert Haas [publishing house] DLB-46

Smith, J. Stilman, and Company..... DLB-49

Smith, W. B., and Company........ DLB-49

Smith, W. H., and Son............ DLB-106

Smithers, Leonard [publishing house] DLB-112

Smollett, Tobias 1721-1771..... DLB-39, 104

Snellings, Rolland (see Touré, Askia Muhammad)

Snodgrass, W. D. 1926- DLB-5

Snow, C. P. 1905-1980.......... DLB-15, 77

Snyder, Gary 1930- DLB-5, 16, 165

Sobiloff, Hy 1912-1970 .... DLB-48

The Society for Textual Scholarship and TEXT Y-87

The Society for the History of Authorship, Reading and Publishing Y-92

Soffici, Ardengo 1879-1964 .... DLB-114

Sofola, 'Zulu 1938- DLB-157

Solano, Solita 1888-1975 .... DLB-4

Soldati, Mario 1906- DLB-177

Sollers, Philippe 1936- DLB-83

Solmi, Sergio 1899-1981........... DLB-114

Solomon, Carl 1928- DLB-16

Solway, David 1941- DLB-53

Solzhenitsyn and America............. Y-85

Somerville, Edith Œnone 1858-1949 .... DLB-135

Song, Cathy 1955- DLB-169

Sontag, Susan 1933- DLB-2, 67

Sophocles 497/496 B.C.-406/405 B.C. DLB-176

Sorge, Reinhard Johannes 1892-1916 .... DLB-118

Sorrentino, Gilbert 1929- DLB-5, 173; Y-80

Sotheby, William 1757-1833 .... DLB-93

Soto, Gary 1952- DLB-82

Sources for the Study of Tudor and Stuart Drama DLB-62

Souster, Raymond 1921- DLB-88

The South English Legendary circa thirteenth-fifteenth centuries DLB-146

Southerland, Ellease 1943- DLB-33

Southern Illinois University Press Y-95

Southern, Terry 1924- DLB-2

Southern Writers Between the
 Wars DLB-9

Southerne, Thomas 1659-1746 ...... DLB-80

Southey, Caroline Anne Bowles
 1786-1854 .... DLB-116

Southey, Robert
 1774-1843............ DLB-93, 107, 142

Southwell, Robert 1561?-1595...... DLB-167

Sowande, Bode 1948- DLB-157

Sowle, Tace
 [publishing house] DLB-170

Soyfer, Jura 1912-1939........... DLB-124

Soyinka, Wole 1934- DLB-125; Y-86, 87

Spacks, Barry 1931- DLB-105

Spalding, Frances 1950- DLB-155

Spark, Muriel 1918- DLB-15, 139

Sparke, Michael
 [publishing house] DLB-170

Sparks, Jared 1789-1866.......... DLB-1, 30

Sparshott, Francis 1926- DLB-60

Späth, Gerold 1939- DLB-75

Spatola, Adriano 1941-1988........ DLB-128

Spaziani, Maria Luisa 1924- DLB-128

The Spectator 1828- DLB-110

Spedding, James 1808-1881 .... DLB-144

Spee von Langenfeld, Friedrich
 1591-1635 .... DLB-164

Speght, Rachel 1597-after 1630..... DLB-126

Speke, John Hanning 1827-1864.... DLB-166

Spellman, A. B. 1935- DLB-41

Spence, Thomas 1750-1814 .... DLB-158

Spencer, Anne 1882-1975 .... DLB-51, 54

Spencer, Elizabeth 1921- DLB-6

Spencer, Herbert 1820-1903 ...... DLB-57

Spencer, Scott 1945- Y-86

Spender, J. A. 1862-1942 .... DLB-98

Spender, Stephen 1909- DLB-20

Spener, Philipp Jakob 1635-1705 ... DLB-164

Spenser, Edmund circa 1552-1599 .. DLB-167

Sperr, Martin 1944- DLB-124

Spicer, Jack 1925-1965........... DLB-5, 16

Spielberg, Peter 1929- Y-81

Spielhagen, Friedrich 1829-1911.... DLB-129

"Spielmannsepen"
 (circa 1152-circa 1500)........ DLB-148

Spier, Peter 1927- DLB-61

Spinrad, Norman 1940- DLB-8

Spires, Elizabeth 1952- DLB-120

Spitteler, Carl 1845-1924......... DLB-129

Spivak, Lawrence E. 1900- DLB-137

Spofford, Harriet Prescott
 1835-1921 .... DLB-74

Squibob (see Derby, George Horatio)

The St. John's College Robert Graves Trust
 .... Y-96

Stacpoole, H. de Vere
 1863-1951 .... DLB-153

Staël, Germaine de 1766-1817...... DLB-119

Staël-Holstein, Anne-Louise Germaine de
 (see Staël, Germaine de)

Stafford, Jean 1915-1979 .... DLB-2, 173

Stafford, William 1914- DLB-5

Stage Censorship: "The Rejected Statement"
 (1911), by Bernard Shaw
 [excerpts].................. DLB-10

Stallings, Laurence 1894-1968..... DLB-7, 44

Stallworthy, Jon 1935- DLB-40

Stampp, Kenneth M. 1912- DLB-17

Stanford, Ann 1916- DLB-5

Stankovií, Borisav ("Bora")
 1876-1927 .... DLB-147

Stanley, Henry M. 1841-1904........ DS-13

Stanley, Thomas 1625-1678........ DLB-131

Stannard, Martin 1947- DLB-155

Stansby, William
 [publishing house] DLB-170

Stanton, Elizabeth Cady 1815-1902 .. DLB-79

Stanton, Frank L. 1857-1927....... DLB-25

Stanton, Maura 1946- DLB-120

Stapledon, Olaf 1886-1950......... DLB-15

Star Spangled Banner Office DLB-49

Starkey, Thomas circa 1499-1538... DLB-132

Starkweather, David 1935- DLB-7

Statements on the Art of Poetry DLB-54

Stationers' Company of
 London, The................ DLB-170

Stead, Robert J. C. 1880-1959 ..... DLB-92

Steadman, Mark 1930- DLB-6

The Stealthy School of Criticism (1871), by
 Dante Gabriel Rossetti......... DLB-35

Stearns, Harold E. 1891-1943 ...... DLB-4

Stedman, Edmund Clarence
 1833-1908 .... DLB-64

Steegmuller, Francis 1906-1994.... DLB-111

Steel, Flora Annie
 1847-1929 .... DLB-153, 156

Steele, Max 1922- Y-80

Steele, Richard 1672-1729...... DLB-84, 101

Steele, Timothy 1948- DLB-120

Steele, Wilbur Daniel 1886-1970 .... DLB-86

Steere, Richard circa 1643-1721 ..... DLB-24

Stegner, Wallace 1909-1993 .... DLB-9; Y-93

Stehr, Hermann 1864-1940 .... DLB-66

Steig, William 1907- DLB-61

Stieler, Caspar 1632-1707 .... DLB-164

Stein, Gertrude 1874-1946 .... DLB-4, 54, 86

Stein, Leo 1872-1947 .... DLB-4

Stein and Day Publishers........... DLB-46

Steinbeck, John 1902-1968 ... DLB-7, 9; DS-2

Steiner, George 1929- DLB-67

Stendhal 1783-1842............... DLB-119

Stephen Crane: A Revaluation Virginia
 Tech Conference, 1989........... Y-89

Stephen, Leslie 1832-1904 ...... DLB-57, 144

Stephens, Alexander H. 1812-1883... DLB-47

Stephens, Ann 1810-1886 .... DLB-3, 73

Stephens, Charles Asbury
 1844?-1931 .... DLB-42

Stephens, James
 1882?-1950........... DLB-19, 153, 162

Sterling, George 1869-1926 .... DLB-54

Sterling, James 1701-1763 .... DLB-24

Sterling, John 1806-1844 .... DLB-116

Stern, Gerald 1925- DLB-105

Stern, Madeleine B. 1912- DLB-111, 140

Stern, Gerald, Living in Ruin DLB-105

Stern, Richard 1928- Y-87

Stern, Stewart 1922- DLB-26

Sterne, Laurence 1713-1768......... DLB-39

Sternheim, Carl 1878-1942 ..... DLB-56, 118

Sternhold, Thomas ?-1549 and
 John Hopkins ?-1570 .... DLB-132

Stevens, Henry 1819-1886......... DLB-140

Stevens, Wallace 1879-1955 .... DLB-54

Stevenson, Anne 1933- DLB-40

Stevenson, Lionel 1902-1973....... DLB-155

Stevenson, Robert Louis 1850-1894
 ...... DLB-18, 57, 141, 156, 174; DS-13

Stewart, Donald Ogden
 1894-1980.............. DLB-4, 11, 26

Stewart, Dugald 1753-1828 .... DLB-31

Stewart, George, Jr. 1848-1906 ..... DLB-99

365

Stewart, George R. 1895-1980 . . . . DLB-8

Stewart and Kidd Company DLB-46

Stewart, Randall 1896-1964 . . . . DLB-103

Stickney, Trumbull 1874-1904 . . . . . . DLB-54

Stifter, Adalbert 1805-1868 . . . . DLB-133

Stiles, Ezra 1727-1795 . . . . DLB-31

Still, James 1906- DLB-9

Stirner, Max 1806-1856 . . . . DLB-129

Stith, William 1707-1755 . . . . DLB-31

Stock, Elliot [publishing house] DLB-106

Stockton, Frank R. 1834-1902 . . . . DLB-42, 74; DS-13

Stoddard, Ashbel [publishing house] DLB-49

Stoddard, Richard Henry 1825-1903 . . . . DLB-3, 64; DS-13

Stoddard, Solomon 1643-1729 . . . . . . DLB-24

Stoker, Bram 1847-1912 . . . . DLB-36, 70

Stokes, Frederick A., Company DLB-49

Stokes, Thomas L. 1898-1958 . . . . DLB-29

Stokesbury, Leon 1945- DLB-120

Stolberg, Christian Graf zu 1748-1821 . . . . DLB-94

Stolberg, Friedrich Leopold Graf zu 1750-1819 . . . . DLB-94

Stone, Herbert S., and Company DLB-49

Stone, Lucy 1818-1893 . . . . DLB-79

Stone, Melville 1848-1929 . . . . DLB-25

Stone, Robert 1937- DLB-152

Stone, Ruth 1915- DLB-105

Stone, Samuel 1602-1663 . . . . DLB-24

Stone and Kimball DLB-49

Stoppard, Tom 1937- DLB-13; Y-85

Storey, Anthony 1928- DLB-14

Storey, David 1933- DLB-13, 14

Storm, Theodor 1817-1888 . . . . DLB-129

Story, Thomas circa 1670-1742 . . . . . . DLB-31

Story, William Wetmore 1819-1895 . . . DLB-1

Storytelling: A Contemporary Renaissance Y-84

Stoughton, William 1631-1701 . . . . . . DLB-24

Stow, John 1525-1605 . . . . DLB-132

Stowe, Harriet Beecher 1811-1896 . . . . DLB-1, 12, 42, 74

Stowe, Leland 1899- DLB-29

Stoyanov, Dimit"r Ivanov (see Elin Pelin)

Strabo 64 or 63 B.C.-circa A.D. 25 . . . . DLB-176

Strachey, Lytton 1880-1932 . . . . DLB-149; DS-10

Strachey, Lytton, Preface to *Eminent Victorians* DLB-149

Strahan and Company DLB-106

Strahan, William [publishing house] DLB-154

Strand, Mark 1934- DLB-5

The Strasbourg Oaths 842 . . . . DLB-148

Stratemeyer, Edward 1862-1930 . . . . . DLB-42

Strati, Saverio 1924- DLB-177

Stratton and Barnard DLB-49

Stratton-Porter, Gene 1863-1924 . . . . . DS-14

Straub, Peter 1943- Y-84

Strauß, Botho 1944- DLB-124

Strauß, David Friedrich 1808-1874 . . . . DLB-133

The Strawberry Hill Press DLB-154

Streatfeild, Noel 1895-1986 . . . . DLB-160

Street, Cecil John Charles (see Rhode, John)

Street, G. S. 1867-1936 . . . . DLB-135

Street and Smith DLB-49

Streeter, Edward 1891-1976 . . . . DLB-11

Streeter, Thomas Winthrop 1883-1965 . . . . DLB-140

Stretton, Hesba 1832-1911 . . . . DLB-163

Stribling, T. S. 1881-1965 . . . . DLB-9

Der Stricker circa 1190-circa 1250 . . DLB-138

Strickland, Samuel 1804-1867 . . . . DLB-99

Stringer and Townsend DLB-49

Stringer, Arthur 1874-1950 . . . . DLB-92

Strittmatter, Erwin 1912- DLB-69

Strode, William 1630-1645 . . . . DLB-126

Strother, David Hunter 1816-1888 . . . . DLB-3

Strouse, Jean 1945- DLB-111

Stuart, Dabney 1937- DLB-105

Stuart, Dabney, Knots into Webs: Some Autobiographical Sources DLB-105

Stuart, Jesse 1906-1984 . . . . DLB-9, 48, 102; Y-84

Stuart, Lyle [publishing house] DLB-46

Stubbs, Harry Clement (see Clement, Hal)

Stubenberg, Johann Wilhelm von 1619-1663 . . . . DLB-164

Studio DLB-112

The Study of Poetry (1880), by Matthew Arnold DLB-35

Sturgeon, Theodore 1918-1985 . . . . DLB-8; Y-85

Sturges, Preston 1898-1959 . . . . DLB-26

"Style" (1840; revised, 1859), by Thomas de Quincey [excerpt] DLB-57

"Style" (1888), by Walter Pater DLB-57

Style (1897), by Walter Raleigh [excerpt] DLB-57

"Style" (1877), by T. H. Wright [excerpt] DLB-57

"Le Style c'est l'homme" (1892), by W. H. Mallock DLB-57

Styron, William 1925- DLB-2, 143; Y-80

Suárez, Mario 1925- DLB-82

Such, Peter 1939- DLB-60

Suckling, Sir John 1609-1641? . . DLB-58, 126

Suckow, Ruth 1892-1960 . . . . DLB-9, 102

Sudermann, Hermann 1857-1928 . . . DLB-118

Sue, Eugène 1804-1857 . . . . DLB-119

Sue, Marie-Joseph (see Sue, Eugène)

Suggs, Simon (see Hooper, Johnson Jones)

Sukenick, Ronald 1932- DLB-173; Y-81

Suknaski, Andrew 1942- DLB-53

Sullivan, Alan 1868-1947 . . . . DLB-92

Sullivan, C. Gardner 1886-1965 . . . . . DLB-26

Sullivan, Frank 1892-1976 . . . . DLB-11

Sulte, Benjamin 1841-1923 . . . . DLB-99

Sulzberger, Arthur Hays 1891-1968 . . . . DLB-127

Sulzberger, Arthur Ochs 1926- DLB-127

Sulzer, Johann Georg 1720-1779 . . . . . DLB-97

Sumarokov, Aleksandr Petrovich 1717-1777 . . . . DLB-150

Summers, Hollis 1916- DLB-6

Sumner, Henry A. [publishing house] DLB-49

Surtees, Robert Smith 1803-1864 . . . . DLB-21

A Survey of Poetry Anthologies, 1879-1960 . . . . DLB-54

Surveys of the Year's Biographies

A Transit of Poets and Others: American Biography in 1982 . . . . Y-82

The Year in Literary Biography . . . Y-83–Y-96

Survey of the Year's Book Publishing

The Year in Book Publishing Y-86

Survey of the Year's Children's Books

The Year in Children's Books Y-92–Y-96

Surveys of the Year's Drama

The Year in Drama
.... Y-82–Y-85, Y-87–Y-96

The Year in London Theatre Y-92

Surveys of the Year's Fiction

The Year's Work in Fiction:
A Survey Y-82

The Year in Fiction: A Biased View Y-83

The Year in
Fiction Y-84–Y-86, Y-89, Y-94–Y-96

The Year in the
Novel Y-87, Y-88, Y-90–Y-93

The Year in Short Stories Y-87

The Year in the
Short Story Y-88, Y-90–Y-93

Survey of the Year's Literary Theory

The Year in Literary Theory Y-92–Y-93

Surveys of the Year's Poetry

The Year's Work in American
Poetry Y-82

The Year in Poetry ... Y-83–Y-92, Y-94–Y-96

Sutherland, Efua Theodora
1924- DLB-117

Sutherland, John 1919-1956 .... DLB-68

Sutro, Alfred 1863-1933 .... DLB-10

Swados, Harvey 1920-1972 .... DLB-2

Swain, Charles 1801-1874 .... DLB-32

Swallow Press DLB-46

Swan Sonnenschein Limited DLB-106

Swanberg, W. A. 1907- DLB-103

Swenson, May 1919-1989 .... DLB-5

Swerling, Jo 1897- DLB-44

Swift, Jonathan
1667-1745 .... DLB-39, 95, 101

Swinburne, A. C. 1837-1909 ..... DLB-35, 57

Swineshead, Richard floruit
circa 1350 .... DLB-115

Swinnerton, Frank 1884-1982 .... DLB-34

Swisshelm, Jane Grey 1815-1884 .... DLB-43

Swope, Herbert Bayard 1882-1958 ... DLB-25

Swords, T. and J., and Company DLB-49

Swords, Thomas 1763-1843 and
Swords, James ?-1844 .... DLB-73

Sykes, Ella C. ?-1939 .... DLB-174

Sylvester, Josuah
1562 or 1563 - 1618 .... DLB-121

Symonds, Emily Morse (see Paston, George)

Symonds, John Addington
1840-1893 .... DLB-57, 144

Symons, A. J. A. 1900-1941 .... DLB-149

Symons, Arthur
1865-1945 .... DLB-19, 57, 149

Symons, Julian
1912-1994 .... DLB-87, 155; Y-92

Symons, Scott 1933- DLB-53

A Symposium on *The Columbia History of
the Novel* Y-92

Synge, John Millington
1871-1909 .... DLB-10, 19

Synge Summer School: J. M. Synge and the
Irish Theater, Rathdrum, County Wiclow,
Ireland Y-93

Syrett, Netta 1865-1943 .... DLB-135

Szymborska, Wisława 1923- Y-96

T

Taban lo Liyong 1939?- DLB-125

Taché, Joseph-Charles 1820-1894 .... DLB-99

Tafolla, Carmen 1951- DLB-82

Taggard, Genevieve 1894-1948 ...... DLB-45

Tagger, Theodor (see Bruckner, Ferdinand)

Tait, J. Selwin, and Sons DLB-49

Tait's Edinburgh Magazine
1832-1861 .... DLB-110

The Takarazaka Revue Company Y-91

Talander (see Bohse, August)

Tallent, Elizabeth 1954- DLB-130

Talvj 1797-1870 .... DLB-59, 133

Tan, Amy 1952- DLB-173

Tapahonso, Luci 1953- DLB-175

Taradash, Daniel 1913- DLB-44

Tarbell, Ida M. 1857-1944 .... DLB-47

Tardivel, Jules-Paul 1851-1905 ...... DLB-99

Targan, Barry 1932- DLB-130

Tarkington, Booth 1869-1946 .... DLB-9, 102

Tashlin, Frank 1913-1972 .... DLB-44

Tate, Allen 1899-1979 ...... DLB-4, 45, 63

Tate, James 1943- DLB-5, 169

Tate, Nahum circa 1652-1715 ...... DLB-80

Tatian circa 830 .... DLB-148

Tavúar, Ivan 1851-1923 .... DLB-147

Taylor, Ann 1782-1866 .... DLB-163

Taylor, Bayard 1825-1878 .... DLB-3

Taylor, Bert Leston 1866-1921 ..... DLB-25

Taylor, Charles H. 1846-1921 ...... DLB-25

Taylor, Edward circa 1642-1729 ..... DLB-24

Taylor, Elizabeth 1912-1975 .... DLB-139

Taylor, Henry 1942- DLB-5

Taylor, Sir Henry 1800-1886 .... DLB-32

Taylor, Jane 1783-1824 .... DLB-163

Taylor, Jeremy circa 1613-1667 .... DLB-151

Taylor, John
1577 or 1578 - 1653 .... DLB-121

Taylor, Mildred D. ?- DLB-52

Taylor, Peter 1917-1994 .... Y-81, Y-94

Taylor, William, and Company DLB-49

Taylor-Made Shakespeare? Or Is
"Shall I Die?" the Long-Lost Text
of Bottom's Dream? Y-85

Teasdale, Sara 1884-1933 .... DLB-45

The Tea-Table (1725), by Eliza Haywood [excerpt] DLB-39

Telles, Lygia Fagundes 1924- DLB-113

Temple, Sir William 1628-1699 .... DLB-101

Tenn, William 1919- DLB-8

Tennant, Emma 1937- DLB-14

Tenney, Tabitha Gilman
1762-1837 .... DLB-37

Tennyson, Alfred 1809-1892 .... DLB-32

Tennyson, Frederick 1807-1898 ..... DLB-32

Terhune, Albert Payson 1872-1942 ... DLB-9

Terhune, Mary Virginia 1830-1922 ... DS-13

Terry, Megan 1932- DLB-7

Terson, Peter 1932- DLB-13

Tesich, Steve 1943- Y-83

Tessa, Delio 1886-1939 .... DLB-114

Testori, Giovanni 1923-1993 ... DLB-128, 177

Tey, Josephine 1896?-1952 .... DLB-77

Thacher, James 1754-1844 .... DLB-37

Thackeray, William Makepeace
1811-1863 .... DLB-21, 55, 159, 163

Thames and Hudson Limited DLB-112

Thanet, Octave (see French, Alice)

The Theater in Shakespeare's
Time DLB-62

The Theatre Guild DLB-7

Thegan and the Astronomer
flourished circa 850 .... DLB-148

Thelwall, John 1764-1834 ...... DLB-93, 158

Theocritus circa 300 B.C.-260 B.C.
.... DLB-176

367

Theodulf circa 760-circa 821 .... DLB-148

Theophrastus circa 371 B.C.-287 B.C. DLB-176

Theriault, Yves 1915-1983 .... DLB-88

Thério, Adrien 1925- DLB-53

Theroux, Paul 1941- DLB-2

Thibaudeau, Colleen 1925- DLB-88

Thielen, Benedict 1903-1965 .... DLB-102

Thiong'o Ngugi wa (see Ngugi wa Thiong'o)

Third-Generation Minor Poets of the Seventeenth Century DLB-131

Thoma, Ludwig 1867-1921 .... DLB-66

Thoma, Richard 1902- DLB-4

Thomas, Audrey 1935- DLB-60

Thomas, D. M. 1935- DLB-40

Thomas, Dylan 1914-1953 .... DLB-13, 20, 139

Thomas, Edward 1878-1917 .... DLB-19, 98, 156

Thomas, Gwyn 1913-1981 .... DLB-15

Thomas, Isaiah 1750-1831 .... DLB-43, 73

Thomas, Isaiah [publishing house] ... DLB-49

Thomas, Johann 1624-1679 .... DLB-168

Thomas, John 1900-1932 .... DLB-4

Thomas, Joyce Carol 1938- DLB-33

Thomas, Lorenzo 1944- DLB-41

Thomas, R. S. 1915- DLB-27

Thomasîn von Zerclære circa 1186-circa 1259 .... DLB-138

Thomasius, Christian 1655-1728 ... DLB-168

Thompson, David 1770-1857 .... DLB-99

Thompson, Dorothy 1893-1961 ..... DLB-29

Thompson, Francis 1859-1907 ...... DLB-19

Thompson, George Selden (see Selden, George)

Thompson, John 1938-1976 .... DLB-60

Thompson, John R. 1823-1873 .... DLB-3, 73

Thompson, Lawrance 1906-1973 ... DLB-103

Thompson, Maurice 1844-1901 .... DLB-71, 74

Thompson, Ruth Plumly 1891-1976 .... DLB-22

Thompson, Thomas Phillips 1843-1933 .... DLB-99

Thompson, William 1775-1833..... DLB-158

Thompson, William Tappan 1812-1882 .... DLB-3, 11

Thomson, Edward William 1849-1924 .... DLB-92

Thomson, James 1700-1748 .... DLB-95

Thomson, James 1834-1882 .... DLB-35

Thomson, Joseph 1858-1895 .... DLB-174

Thomson, Mortimer 1831-1875 ..... DLB-11

Thoreau, Henry David 1817-1862 .... DLB-1

Thorpe, Thomas Bangs 1815-1878 .... DLB-3, 11

Thoughts on Poetry and Its Varieties (1833), by John Stuart Mill DLB-32

Thrale, Hester Lynch (see Piozzi, Hester Lynch [Thrale])

Thucydides circa 455 B.C.-circa 395 B.C. DLB-176

Thümmel, Moritz August von 1738-1817 .... DLB-97

Thurber, James 1894-1961 .... DLB-4, 11, 22, 102

Thurman, Wallace 1902-1934 .... DLB-51

Thwaite, Anthony 1930- DLB-40

Thwaites, Reuben Gold 1853-1913 .... DLB-47

Ticknor, George 1791-1871 .... DLB-1, 59, 140

Ticknor and Fields DLB-49

Ticknor and Fields (revived) DLB-46

Tieck, Ludwig 1773-1853 .... DLB-90

Tietjens, Eunice 1884-1944 .... DLB-54

Tilney, Edmund circa 1536-1610 ... DLB-136

Tilt, Charles [publishing house] DLB-106

Tilton, J. E., and Company DLB-49

Time and Western Man (1927), by Wyndham Lewis [excerpts] DLB-36

Time-Life Books DLB-46

Times Books DLB-46

Timothy, Peter circa 1725-1782 ..... DLB-43

Timrod, Henry 1828-1867 .... DLB-3

Tinker, Chauncey Brewster 1876-1963 .... DLB-140

Tinsley Brothers DLB-106

Tiptree, James, Jr. 1915-1987 .... DLB-8

Titus, Edward William 1870-1952 .... DLB-4

Tlali, Miriam 1933- DLB-157

Todd, Barbara Euphan 1890-1976 .... DLB-160

Tofte, Robert 1561 or 1562-1619 or 1620 ..... DLB-172

Toklas, Alice B. 1877-1967 .... DLB-4

Tolkien, J. R. R. 1892-1973 .... DLB-15, 160

Toller, Ernst 1893-1939 .... DLB-124

Tollet, Elizabeth 1694-1754 .... DLB-95

Tolson, Melvin B. 1898-1966 .... DLB-48, 76

Tom Jones (1749), by Henry Fielding [excerpt] DLB-39

Tomalin, Claire 1933- DLB-155

Tomasi di Lampedusa, Giuseppe 1896-1957 .... DLB-177

Tomlinson, Charles 1927- DLB-40

Tomlinson, H. M. 1873-1958 ... DLB-36, 100

Tompkins, Abel [publishing house] .. DLB-49

Tompson, Benjamin 1642-1714 ..... DLB-24

Tonks, Rosemary 1932- DLB-14

Tonna, Charlotte Elizabeth 1790-1846 .... DLB-163

Tonson, Jacob the Elder [publishing house] DLB-170

Toole, John Kennedy 1937-1969 .... Y-81

Toomer, Jean 1894-1967 .... DLB-45, 51

Tor Books DLB-46

Torberg, Friedrich 1908-1979 .... DLB-85

Torrence, Ridgely 1874-1950 .... DLB-54

Torres-Metzger, Joseph V. 1933- DLB-122

Toth, Susan Allen 1940- Y-86

Tottell, Richard [publishing house] DLB-170

Tough-Guy Literature DLB-9

Touré, Askia Muhammad 1938- ... DLB-41

Tourgée, Albion W. 1838-1905 ...... DLB-79

Tourneur, Cyril circa 1580-1626 .... DLB-58

Tournier, Michel 1924- DLB-83

Tousey, Frank [publishing house] ... DLB-49

Tower Publications DLB-46

Towne, Benjamin circa 1740-1793 ... DLB-43

Towne, Robert 1936- DLB-44

The Townely Plays fifteenth and sixteenth centuries DLB-146

Townshend, Aurelian by 1583 - circa 1651 .... DLB-121

Tracy, Honor 1913- DLB-15

Traherne, Thomas 1637?-1674 ..... DLB-131

Traill, Catharine Parr 1802-1899 .... DLB-99

Train, Arthur 1875-1945 .... DLB-86

The Transatlantic Publishing Company DLB-49

Transcendentalists, American DS-5

Translators of the Twelfth Century:
  Literary Issues Raised and Impact
  Created DLB-115

Travel Writing, 1837-1875 .... DLB-166

Travel Writing, 1876-1909 .... DLB-174

Traven, B.
  1882? or 1890?-1969? DLB-9, 56

Travers, Ben 1886-1980 .... DLB-10

Travers, P. L. (Pamela Lyndon)
  1899- DLB-160

Trediakovsky, Vasilii Kirillovich
  1703-1769 .... DLB-150

Treece, Henry 1911-1966 .... DLB-160

Trejo, Ernesto 1950- DLB-122

Trelawny, Edward John
  1792-1881.......... DLB-110, 116, 144

Tremain, Rose 1943- DLB-14

Tremblay, Michel 1942- DLB-60

Trends in Twentieth-Century
  Mass Market Publishing DLB-46

Trent, William P. 1862-1939........ DLB-47

Trescot, William Henry
  1822-1898 .... DLB-30

Trevelyan, Sir George Otto
  1838-1928 .... DLB-144

Trevisa, John
  circa 1342-circa 1402 .... DLB-146

Trevor, William 1928- DLB-14, 139

Trierer Floyris circa 1170-1180 ...... DLB-138

Trilling, Lionel 1905-1975 .... DLB-28, 63

Trilussa 1871-1950 .... DLB-114

Trimmer, Sarah 1741-1810 .... DLB-158

Triolet, Elsa 1896-1970 .... DLB-72

Tripp, John 1927- DLB-40

Trocchi, Alexander 1925- DLB-15

Trollope, Anthony
  1815-1882............ DLB-21, 57, 159

Trollope, Frances 1779-1863.... DLB-21, 166

Troop, Elizabeth 1931- DLB-14

Trotter, Catharine 1679-1749 .... DLB-84

Trotti, Lamar 1898-1952 .... DLB-44

Trottier, Pierre 1925- DLB-60

Troupe, Quincy Thomas, Jr.
  1943- DLB-41

Trow, John F., and Company DLB-49

Truillier-Lacombe, Joseph-Patrice
  1807-1863 .... DLB-99

Trumbo, Dalton 1905-1976......... DLB-26

Trumbull, Benjamin 1735-1820 ..... DLB-30

Trumbull, John 1750-1831.......... DLB-31

Tscherning, Andreas 1611-1659 .... DLB-164

T. S. Eliot Centennial Y-88

Tucholsky, Kurt 1890-1935........ DLB-56

Tucker, Charlotte Maria
  1821-1893 .... DLB-163

Tucker, George 1775-1861 ...... DLB-3, 30

Tucker, Nathaniel Beverley
  1784-1851 .... DLB-3

Tucker, St. George 1752-1827....... DLB-37

Tuckerman, Henry Theodore
  1813-1871 .... DLB-64

Tunis, John R. 1889-1975 ...... DLB-22, 171

Tunstall, Cuthbert 1474-1559 ...... DLB-132

Tuohy, Frank 1925- DLB-14, 139

Tupper, Martin F. 1810-1889 .... DLB-32

Turbyfill, Mark 1896- DLB-45

Turco, Lewis 1934- Y-84

Turnbull, Andrew 1921-1970 ..... DLB-103

Turnbull, Gael 1928- DLB-40

Turner, Arlin 1909-1980 .... DLB-103

Turner, Charles (Tennyson)
  1808-1879 .... DLB-32

Turner, Frederick 1943- DLB-40

Turner, Frederick Jackson
  1861-1932 .... DLB-17

Turner, Joseph Addison
  1826-1868 .... DLB-79

Turpin, Waters Edward
  1910-1968 .... DLB-51

Turrini, Peter 1944- DLB-124

Tutuola, Amos 1920- DLB-125

Twain, Mark (see Clemens,
  Samuel Langhorne)

Tweedie, Ethel Brilliana
  circa 1860-1940.............. DLB-174

The 'Twenties and Berlin, by
  Alex Natan DLB-66

Tyler, Anne 1941- DLB-6, 143; Y-82

Tyler, Moses Coit 1835-1900 .... DLB-47, 64

Tyler, Royall 1757-1826 .... DLB-37

Tylor, Edward Burnett 1832-1917 ... DLB-57

Tynan, Katharine 1861-1931....... DLB-153

Tyndale, William
  circa 1494-1536.............. DLB-132

U

Udall, Nicholas 1504-1556......... DLB-62

Uhland, Ludwig 1787-1862 .... DLB-90

Uhse, Bodo 1904-1963 .... DLB-69

Ujević, Augustin ("Tin")
  1891-1955 .... DLB-147

Ulenhart, Niclas
  flourished circa 1600 .... DLB-164

Ulibarrí, Sabine R. 1919- DLB-82

Ulica, Jorge 1870-1926............ DLB-82

Ulizio, B. George 1889-1969 ...... DLB-140

Ulrich von Liechtenstein
  circa 1200-circa 1275 .... DLB-138

Ulrich von Zatzikhoven
  before 1194-after 1214......... DLB-138

Unamuno, Miguel de 1864-1936.... DLB-108

Under the Microscope (1872), by
  A. C. Swinburne............... DLB-35

Unger, Friederike Helene
  1741-1813 .... DLB-94

Ungaretti, Giuseppe 1888-1970..... DLB-114

United States Book Company....... DLB-49

Universal Publishing and Distributing
  Corporation DLB-46

The University of Iowa Writers' Workshop
  Golden Jubilee Y-86

The University of South Carolina
  Press Y-94

University of Wales Press DLB-112

"The Unknown Public" (1858), by
  Wilkie Collins [excerpt]......... DLB-57

Unruh, Fritz von 1885-1970 .... DLB-56, 118

Unspeakable Practices II: The Festival of
  Vanguard Narrative at Brown
  University.................... Y-93

Unwin, T. Fisher
  [publishing house] DLB-106

Upchurch, Boyd B. (see Boyd, John)

Updike, John
  1932- DLB-2, 5, 143; Y-80, 82; DS-3

Upton, Bertha 1849-1912.......... DLB-141

Upton, Charles 1948- DLB-16

Upton, Florence K. 1873-1922 ..... DLB-141

Upward, Allen 1863-1926 .... DLB-36

Urista, Alberto Baltazar (see Alurista)

Urzidil, Johannes 1896-1976 .... DLB-85

Urquhart, Fred 1912- DLB-139

The Uses of Facsimile Y-90

Usk, Thomas died 1388.......... DLB-146

Uslar Pietri, Arturo 1906- DLB-113

Ustinov, Peter 1921- DLB-13

Uttley, Alison 1884-1976 .... DLB-160
Uz, Johann Peter 1720-1796 .... DLB-97

V

Vac, Bertrand 1914- DLB-88
Vail, Laurence 1891-1968 .... DLB-4
Vailland, Roger 1907-1965 .... DLB-83
Vajda, Ernest 1887-1954 .... DLB-44
Valdés, Gina 1943- DLB-122
Valdez, Luis Miguel 1940- DLB-122
Valduga, Patrizia 1953- DLB-128
Valente, José Angel 1929- DLB-108
Valenzuela, Luisa 1938- DLB-113
Valeri, Diego 1887-1976 .... DLB-128
Valgardson, W. D. 1939- DLB-60
Valle, Víctor Manuel 1950- DLB-122
Valle-Inclán, Ramón del 1866-1936 .... DLB-134
Vallejo, Armando 1949- DLB-122
Vallès, Jules 1832-1885 .... DLB-123
Vallette, Marguerite Eymery (see Rachilde)
Valverde, José María 1926- DLB-108
Van Allsburg, Chris 1949- DLB-61
Van Anda, Carr 1864-1945 .... DLB-25
Van Doren, Mark 1894-1972 .... DLB-45
van Druten, John 1901-1957 ...... DLB-10
Van Duyn, Mona 1921- DLB-5
Van Dyke, Henry 1852-1933 .... DLB-71; DS-13
Van Dyke, Henry 1928- DLB-33
van Itallie, Jean-Claude 1936- DLB-7
Van Loan, Charles E. 1876-1919 ... DLB-171
Van Rensselaer, Mariana Griswold 1851-1934 .... DLB-47
Van Rensselaer, Mrs. Schuyler (see Van Rensselaer, Mariana Griswold)
Van Vechten, Carl 1880-1964...... DLB-4, 9
van Vogt, A. E. 1912- DLB-8
Vanbrugh, Sir John 1664-1726 ..... DLB-80
Vance, Jack 1916?- DLB-8
Vane, Sutton 1888-1963........... DLB-10
Vanguard Press................... DLB-46
Vann, Robert L. 1879-1940........ DLB-29
Vargas, Llosa, Mario 1936- DLB-145
Varley, John 1947- Y-81

Varnhagen von Ense, Karl August 1785-1858 .... DLB-90
Varnhagen von Ense, Rahel 1771-1833 .... DLB-90
Vásquez Montalbán, Manuel 1939- DLB-134
Vassa, Gustavus (see Equiano, Olaudah)
Vassalli, Sebastiano 1941- DLB-128
Vaughan, Henry 1621-1695 .... DLB-131
Vaughan, Thomas 1621-1666 ...... DLB-131
Vaux, Thomas, Lord 1509-1556.... DLB-132
Vazov, Ivan 1850-1921 .... DLB-147
Vega, Janine Pommy 1942- DLB-16
Veiller, Anthony 1903-1965 .... DLB-44
Velásquez-Trevino, Gloria 1949- DLB-122
Veloz Maggiolo, Marcio 1936- DLB-145
Venegas, Daniel ?-? DLB-82
Vergil, Polydore circa 1470-1555 ... DLB-132
Veríssimo, Erico 1905-1975....... DLB-145
Verne, Jules 1828-1905 .... DLB-123
Verplanck, Gulian C. 1786-1870 .... DLB-59
Very, Jones 1813-1880............. DLB-1
Vian, Boris 1920-1959 .... DLB-72
Vickers, Roy 1888?-1965.......... DLB-77
Victoria 1819-1901 .... DLB-55
Victoria Press DLB-106
Vidal, Gore 1925- DLB-6, 152
Viebig, Clara 1860-1952 .... DLB-66
Viereck, George Sylvester 1884-1962 .... DLB-54
Viereck, Peter 1916- DLB-5
Viets, Roger 1738-1811 .... DLB-99
Viewpoint: Politics and Performance, by David Edgar DLB-13
Vigil-Piñon, Evangelina 1949- DLB-122
Vigneault, Gilles 1928- DLB-60
Vigny, Alfred de 1797-1863 .... DLB-119
Vigolo, Giorgio 1894-1983 .... DLB-114
The Viking Press DLB-46
Villanueva, Alma Luz 1944- DLB-122
Villanueva, Tino 1941- DLB-82
Villard, Henry 1835-1900 .... DLB-23
Villard, Oswald Garrison 1872-1949 .... DLB-25, 91
Villarreal, José Antonio 1924- DLB-82

Villegas de Magnón, Leonor 1876-1955 .... DLB-122
Villemaire, Yolande 1949- DLB-60
Villena, Luis Antonio de 1951- DLB-134
Villiers de l'Isle-Adam, Jean-Marie Mathias Philippe-Auguste, Comte de 1838-1889 .... DLB-123
Villiers, George, Second Duke of Buckingham 1628-1687....... DLB-80
Vine Press DLB-112
Viorst, Judith ?- DLB-52
Vipont, Elfrida (Elfrida Vipont Foulds, Charles Vipont) 1902-1992..... DLB-160
Viramontes, Helena María 1954- DLB-122
Vischer, Friedrich Theodor 1807-1887 .... DLB-133
Vivanco, Luis Felipe 1907-1975 .... DLB-108
Viviani, Cesare 1947- DLB-128
Vizenor, Gerald 1934- DLB-175
Vizetelly and Company DLB-106
Voaden, Herman 1903- DLB-88
Voigt, Ellen Bryant 1943- DLB-120
Vojnovií, Ivo 1857-1929 .... DLB-147
Volkoff, Vladimir 1932- DLB-83
Volland, P. F., Company........... DLB-46
Volponi, Paolo 1924- DLB-177
von der Grün, Max 1926- DLB-75
Vonnegut, Kurt 1922- DLB-2, 8, 152; Y-80; DS-3
Voranc, Prežhov 1893-1950........ DLB-147
Voß, Johann Heinrich 1751-1826 .... DLB-90
Vroman, Mary Elizabeth circa 1924-1967................. DLB-33

W

Wace, Robert ("Maistre") circa 1100-circa 1175 .... DLB-146
Wackenroder, Wilhelm Heinrich 1773-1798 .... DLB-90
Wackernagel, Wilhelm 1806-1869 .... DLB-133
Waddington, Miriam 1917- DLB-68
Wade, Henry 1887-1969 .... DLB-77
Wagenknecht, Edward 1900- DLB-103
Wagner, Heinrich Leopold 1747-1779 .... DLB-94
Wagner, Henry R. 1862-1957..... DLB-140
Wagner, Richard 1813-1883 .... DLB-129

Wagoner, David 1926- DLB-5
Wah, Fred 1939- DLB-60
Waiblinger, Wilhelm 1804-1830 .....DLB-90
Wain, John
  1925-1994........DLB-15, 27, 139, 155
Wainwright, Jeffrey 1944- DLB-40
Waite, Peirce and Company DLB-49
Wakoski, Diane 1937- DLB-5
Walahfrid Strabo circa 808-849 ....DLB-148
Walck, Henry Z...................DLB-46
Walcott, Derek
  1930- DLB-117; Y-81, 92
Waldegrave, Robert
  [publishing house] DLB-170
Waldman, Anne 1945- DLB-16
Waldrop, Rosmarie 1935- DLB-169
Walker, Alice 1944- DLB-6, 33, 143
Walker, George F. 1947- DLB-60
Walker, Joseph A. 1935- DLB-38
Walker, Margaret 1915- DLB-76, 152
Walker, Ted 1934- DLB-40
Walker and Company DLB-49
Walker, Evans and Cogswell
  Company....................DLB-49
Walker, John Brisben 1847-1931 ....DLB-79
Wallace, Dewitt 1889-1981 and
  Lila Acheson Wallace
  1889-1984 ....DLB-137
Wallace, Edgar 1875-1932..........DLB-70
Wallace, Lila Acheson (see Wallace, Dewitt,
  and Lila Acheson Wallace)
Wallant, Edward Lewis
  1926-1962..............DLB-2, 28, 143
Waller, Edmund 1606-1687........DLB-126
Walpole, Horace 1717-1797 ....DLB-39, 104
Walpole, Hugh 1884-1941..........DLB-34
Walrond, Eric 1898-1966...........DLB-51
Walser, Martin 1927- DLB-75, 124
Walser, Robert 1878-1956..........DLB-66
Walsh, Ernest 1895-1926........DLB-4, 45
Walsh, Robert 1784-1859 ....DLB-59
Waltharius circa 825...............DLB-148
Walters, Henry 1848-1931........DLB-140
Walther von der Vogelweide
  circa 1170-circa 1230 ....DLB-138
Walton, Izaak 1593-1683..........DLB-151
Wambaugh, Joseph 1937- DLB-6; Y-83

Waniek, Marilyn Nelson 1946- ...DLB-120
Warburton, William 1698-1779 ....DLB-104
Ward, Aileen 1919- DLB-111
Ward, Artemus (see Browne, Charles Farrar)
Ward, Arthur Henry Sarsfield
  (see Rohmer, Sax)
Ward, Douglas Turner 1930- DLB-7, 38
Ward, Lynd 1905-1985 ....DLB-22
Ward, Lock and Company DLB-106
Ward, Mrs. Humphry 1851-1920....DLB-18
Ward, Nathaniel circa 1578-1652 ....DLB-24
Ward, Theodore 1902-1983 ....DLB-76
Wardle, Ralph 1909-1988 ....DLB-103
Ware, William 1797-1852 ....DLB-1
Warne, Frederick, and
  Company [U.S.] DLB-49
Warne, Frederick, and
  Company [U.K.]...............DLB-106
Warner, Charles Dudley
  1829-1900 ....DLB-64
Warner, Rex 1905- DLB-15
Warner, Susan Bogert
  1819-1885 ....DLB-3, 42
Warner, Sylvia Townsend
  1893-1978................DLB-34, 139
Warner, William 1558-1609 ....DLB-172
Warner Books....................DLB-46
Warr, Bertram 1917-1943 ....DLB-88
Warren, John Byrne Leicester (see De Tabley,
  Lord)
Warren, Lella 1899-1982..............Y-83
Warren, Mercy Otis 1728-1814 .....DLB-31
Warren, Robert Penn
  1905-1989 .....DLB-2, 48, 152; Y-80, 89
Die Wartburgkrieg
  circa 1230-circa 1280 ....DLB-138
Warton, Joseph 1722-1800 ....DLB-104, 109
Warton, Thomas 1728-1790 ...DLB-104, 109
Washington, George 1732-1799 .....DLB-31
Wassermann, Jakob 1873-1934......DLB-66
Wasson, David Atwood 1823-1887 ...DLB-1
Waterhouse, Keith 1929- DLB-13, 15
Waterman, Andrew 1940- DLB-40
Waters, Frank 1902- Y-86
Waters, Michael 1949- DLB-120
Watkins, Tobias 1780-1855 ....DLB-73
Watkins, Vernon 1906-1967 ....DLB-20

Watmough, David 1926- DLB-53
Watson, James Wreford (see Wreford, James)
Watson, John 1850-1907 ....DLB-156
Watson, Sheila 1909- DLB-60
Watson, Thomas 1545?-1592 ......DLB-132
Watson, Wilfred 1911- DLB-60
Watt, W. J., and Company DLB-46
Watterson, Henry 1840-1921 ....DLB-25
Watts, Alan 1915-1973.............DLB-16
Watts, Franklin [publishing house]...DLB-46
Watts, Isaac 1674-1748 ....DLB-95
Waugh, Auberon 1939- DLB-14
Waugh, Evelyn 1903-1966 .....DLB-15, 162
Way and Williams DLB-49
Wayman, Tom 1945- DLB-53
Weatherly, Tom 1942- DLB-41
Weaver, Gordon 1937- DLB-130
Weaver, Robert 1921- DLB-88
Webb, Frank J. ?-? DLB-50
Webb, James Watson 1802-1884 ....DLB-43
Webb, Mary 1881-1927............DLB-34
Webb, Phyllis 1927- DLB-53
Webb, Walter Prescott 1888-1963 ...DLB-17
Webbe, William ?-1591 ....DLB-132
Webster, Augusta 1837-1894........DLB-35
Webster, Charles L.,
  and Company..................DLB-49
Webster, John
  1579 or 1580-1634? DLB-58
Webster, Noah
  1758-1843 ....DLB-1, 37, 42, 43, 73
Weckherlin, Georg Rodolf
  1584-1653 ....DLB-164
Wedekind, Frank 1864-1918 ....DLB-118
Weeks, Edward Augustus, Jr.
  1898-1989 ....DLB-137
Weems, Mason Locke
  1759-1825.............DLB-30, 37, 42
Weerth, Georg 1822-1856 ....DLB-129
Weidenfeld and Nicolson DLB-112
Weidman, Jerome 1913- DLB-28
Weigl, Bruce 1949- DLB-120
Weinbaum, Stanley Grauman
  1902-1935 ....DLB-8
Weintraub, Stanley 1929- DLB-111
Weise, Christian 1642-1708........DLB-168

Weisenborn, Gunther 1902-1969 .... DLB-69, 124

Weiß, Ernst 1882-1940 .... DLB-81

Weiss, John 1818-1879 .... DLB-1

Weiss, Peter 1916-1982 .... DLB-69, 124

Weiss, Theodore 1916- DLB-5

Weisse, Christian Felix 1726-1804 ... DLB-97

Weitling, Wilhelm 1808-1871 ...... DLB-129

Welch, James 1940- DLB-175

Welch, Lew 1926-1971? DLB-16

Weldon, Fay 1931- DLB-14

Wellek, René 1903- DLB-63

Wells, Carolyn 1862-1942 .... DLB-11

Wells, Charles Jeremiah circa 1800-1879 .... DLB-32

Wells, Gabriel 1862-1946 .... DLB-140

Wells, H. G. 1866-1946 ..... DLB-34, 70, 156

Wells, Robert 1947- DLB-40

Wells-Barnett, Ida B. 1862-1931 ..... DLB-23

Welty, Eudora 1909- DLB-2, 102, 143; Y-87; DS-12

Wendell, Barrett 1855-1921 .... DLB-71

Wentworth, Patricia 1878-1961 ..... DLB-77

Werder, Diederich von dem 1584-1657 .... DLB-164

Werfel, Franz 1890-1945 .... DLB-81, 124

The Werner Company DLB-49

Werner, Zacharias 1768-1823 .... DLB-94

Wersba, Barbara 1932- DLB-52

Wescott, Glenway 1901- DLB-4, 9, 102

Wesker, Arnold 1932- DLB-13

Wesley, Charles 1707-1788 .... DLB-95

Wesley, John 1703-1791 .... DLB-104

Wesley, Richard 1945- DLB-38

Wessels, A., and Company DLB-46

Wessobrunner Gebet circa 787-815 .... DLB-148

West, Anthony 1914-1988 .... DLB-15

West, Dorothy 1907- DLB-76

West, Jessamyn 1902-1984 ..... DLB-6; Y-84

West, Mae 1892-1980 .... DLB-44

West, Nathanael 1903-1940 .... DLB-4, 9, 28

West, Paul 1930- DLB-14

West, Rebecca 1892-1983 ..... DLB-36; Y-83

West and Johnson DLB-49

Western Publishing Company DLB-46

The Westminster Review 1824-1914 ... DLB-110

Weston, Elizabeth Jane circa 1582-1612 .... DLB-172

Wetherald, Agnes Ethelwyn 1857-1940 .... DLB-99

Wetherell, Elizabeth (see Warner, Susan Bogert)

Wetzel, Friedrich Gottlob 1779-1819 .... DLB-90

Weyman, Stanley J. 1855-1928 .... DLB-141, 156

Wezel, Johann Karl 1747-1819 ...... DLB-94

Whalen, Philip 1923- DLB-16

Whalley, George 1915-1983 .... DLB-88

Wharton, Edith 1862-1937 ...... DLB-4, 9, 12, 78; DS-13

Wharton, William 1920s?- Y-80

Whately, Mary Louisa 1824-1889 .... DLB-166

What's Really Wrong With Bestseller Lists Y-84

Wheatley, Dennis Yates 1897-1977 .... DLB-77

Wheatley, Phillis circa 1754-1784 .... DLB-31, 50

Wheeler, Anna Doyle 1785-1848? DLB-158

Wheeler, Charles Stearns 1816-1843 .... DLB-1

Wheeler, Monroe 1900-1988 .... DLB-4

Wheelock, John Hall 1886-1978 ..... DLB-45

Wheelwright, John circa 1592-1679 .... DLB-24

Wheelwright, J. B. 1897-1940 .... DLB-45

Whetstone, Colonel Pete (see Noland, C. F. M.)

Whetstone, George 1550-1587 ..... DLB-136

Whicher, Stephen E. 1915-1961 .... DLB-111

Whipple, Edwin Percy 1819-1886 .... DLB-1, 64

Whitaker, Alexander 1585-1617 ..... DLB-24

Whitaker, Daniel K. 1801-1881 ..... DLB-73

Whitcher, Frances Miriam 1814-1852 .... DLB-11

White, Andrew 1579-1656 .... DLB-24

White, Andrew Dickson 1832-1918 .... DLB-47

White, E. B. 1899-1985 .... DLB-11, 22

White, Edgar B. 1947- DLB-38

White, Ethel Lina 1887-1944 .... DLB-77

White, Henry Kirke 1785-1806 ...... DLB-96

White, Horace 1834-1916 .... DLB-23

White, Phyllis Dorothy James (see James, P. D.)

White, Richard Grant 1821-1885 .... DLB-64

White, T. H. 1906-1964 .... DLB-160

White, Walter 1893-1955 .... DLB-51

White, William, and Company DLB-49

White, William Allen 1868-1944 .... DLB-9, 25

White, William Anthony Parker (see Boucher, Anthony)

White, William Hale (see Rutherford, Mark)

Whitechurch, Victor L. 1868-1933 .... DLB-70

Whitehead, Alfred North 1861-1947 .... DLB-100

Whitehead, James 1936- Y-81

Whitehead, William 1715-1785 .... DLB-84, 109

Whitfield, James Monroe 1822-1871 .... DLB-50

Whitgift, John circa 1533-1604 ..... DLB-132

Whiting, John 1917-1963 .... DLB-13

Whiting, Samuel 1597-1679 .... DLB-24

Whitlock, Brand 1869-1934 .... DLB-12

Whitman, Albert, and Company DLB-46

Whitman, Albery Allson 1851-1901 .... DLB-50

Whitman, Alden 1913-1990 .... Y-91

Whitman, Sarah Helen (Power) 1803-1878 .... DLB-1

Whitman, Walt 1819-1892 .... DLB-3, 64

Whitman Publishing Company DLB-46

Whitney, Geoffrey 1548 or 1552?-1601 .... DLB-136

Whitney, Isabella flourished 1566-1573 .... DLB-136

Whitney, John Hay 1904-1982 ..... DLB-127

Whittemore, Reed 1919- DLB-5

Whittier, John Greenleaf 1807-1892 ... DLB-1

Whittlesey House DLB-46

Who Runs American Literature? Y-94

Wideman, John Edgar 1941- DLB-33, 143

Widener, Harry Elkins 1885-1912 .... DLB-140

Wiebe, Rudy 1934- DLB-60

Wiechert, Ernst 1887-1950 .... DLB-56

Wied, Martina 1882-1957 .... DLB-85

Wiehe, Evelyn May Clowes (see Mordaunt, Elinor)

Wieland, Christoph Martin
  1733-1813 .... DLB-97

Wienbarg, Ludolf 1802-1872....... DLB-133

Wieners, John 1934- DLB-16

Wier, Ester 1910- DLB-52

Wiesel, Elie 1928- DLB-83; Y-87

Wiggin, Kate Douglas 1856-1923 .... DLB-42

Wigglesworth, Michael 1631-1705 ... DLB-24

Wilberforce, William 1759-1833.... DLB-158

Wilbrandt, Adolf 1837-1911 .... DLB-129

Wilbur, Richard 1921- DLB-5, 169

Wild, Peter 1940- DLB-5

Wilde, Oscar
  1854-1900.... DLB-10, 19, 34, 57, 141, 156

Wilde, Richard Henry
  1789-1847 .... DLB-3, 59

Wilde, W. A., Company DLB-49

Wilder, Billy 1906- DLB-26

Wilder, Laura Ingalls 1867-1957 .... DLB-22

Wilder, Thornton 1897-1975 .... DLB-4, 7, 9

Wildgans, Anton 1881-1932 .... DLB-118

Wiley, Bell Irvin 1906-1980........ DLB-17

Wiley, John, and Sons DLB-49

Wilhelm, Kate 1928- DLB-8

Wilkes, George 1817-1885.......... DLB-79

Wilkinson, Anne 1910-1961 .... DLB-88

Wilkinson, Sylvia 1940- Y-86

Wilkinson, William Cleaver
  1833-1920 .... DLB-71

Willard, Barbara 1909-1994 .... DLB-161

Willard, L. [publishing house]....... DLB-49

Willard, Nancy 1936- DLB-5, 52

Willard, Samuel 1640-1707 .... DLB-24

William of Auvergne 1190-1249 .... DLB-115

William of Conches
  circa 1090-circa 1154 .... DLB-115

William of Ockham
  circa 1285-1347................ DLB-115

William of Sherwood
  1200/1205 - 1266/1271......... DLB-115

The William Chavrat American Fiction
  Collection at the Ohio State University Libraries........................ Y-92

Williams, A., and Company DLB-49

Williams, Ben Ames 1889-1953 .... DLB-102

Williams, C. K. 1936- DLB-5

Williams, Chancellor 1905- DLB-76

Williams, Charles
  1886-1945 .... DLB-100, 153

Williams, Denis 1923- DLB-117

Williams, Emlyn 1905- DLB-10, 77

Williams, Garth 1912- DLB-22

Williams, George Washington
  1849-1891 .... DLB-47

Williams, Heathcote 1941- DLB-13

Williams, Helen Maria
  1761-1827 .... DLB-158

Williams, Hugo 1942- DLB-40

Williams, Isaac 1802-1865.......... DLB-32

Williams, Joan 1928- DLB-6

Williams, John A. 1925- DLB-2, 33

Williams, John E. 1922-1994........ DLB-6

Williams, Jonathan 1929- DLB-5

Williams, Miller 1930- DLB-105

Williams, Raymond 1921- DLB-14

Williams, Roger circa 1603-1683 .... DLB-24

Williams, Samm-Art 1946- DLB-38

Williams, Sherley Anne 1944- DLB-41

Williams, T. Harry 1909-1979 ...... DLB-17

Williams, Tennessee
  1911-1983.......... DLB-7; Y-83; DS-4

Williams, Ursula Moray 1911- DLB-160

Williams, Valentine 1883-1946...... DLB-77

Williams, William Appleman
  1921- DLB-17

Williams, William Carlos
  1883-1963........... DLB-4, 16, 54, 86

Williams, Wirt 1921- DLB-6

Williams Brothers................. DLB-49

Williamson, Jack 1908- DLB-8

Willingham, Calder Baynard, Jr.
  1922- DLB-2, 44

Williram of Ebersberg
  circa 1020-1085............... DLB-148

Willis, Nathaniel Parker
  1806-1867 ..... DLB-3, 59, 73, 74; DS-13

Willkomm, Ernst 1810-1886 .... DLB-133

Wilmer, Clive 1945- DLB-40

Wilson, A. N. 1950- DLB-14, 155

Wilson, Angus
  1913-1991............ DLB-15, 139, 155

Wilson, Arthur 1595-1652.......... DLB-58

Wilson, Augusta Jane Evans
  1835-1909 .... DLB-42

Wilson, Colin 1931- DLB-14

Wilson, Edmund 1895-1972 .... DLB-63

Wilson, Ethel 1888-1980 .... DLB-68

Wilson, Harriet E. Adams
  1828?-1863? DLB-50

Wilson, Harry Leon 1867-1939 ..... DLB-9

Wilson, John 1588-1667............ DLB-24

Wilson, John 1785-1854........... DLB-110

Wilson, Lanford 1937- DLB-7

Wilson, Margaret 1882-1973......... DLB-9

Wilson, Michael 1914-1978......... DLB-44

Wilson, Mona 1872-1954.......... DLB-149

Wilson, Thomas
  1523 or 1524-1581 .... DLB-132

Wilson, Woodrow 1856-1924 .... DLB-47

Wilson, Effingham
  [publishing house] DLB-154

Wimsatt, William K., Jr.
  1907-1975 .... DLB-63

Winchell, Walter 1897-1972 .... DLB-29

Winchester, J. [publishing house] DLB-49

Winckelmann, Johann Joachim
  1717-1768 .... DLB-97

Winckler, Paul 1630-1686 .... DLB-164

Wind, Herbert Warren 1916- DLB-171

Windet, John [publishing house].... DLB-170

Windham, Donald 1920- DLB-6

Wingate, Allan [publishing house] .. DLB-112

Winnemucca, Sarah 1844-1921 .... DLB-175

Winnifrith, Tom 1938- DLB-155

Winsloe, Christa 1888-1944 .... DLB-124

Winsor, Justin 1831-1897........... DLB-47

John C. Winston Company......... DLB-49

Winters, Yvor 1900-1968........... DLB-48

Winthrop, John 1588-1649 ...... DLB-24, 30

Winthrop, John, Jr. 1606-1676 ...... DLB-24

Wirt, William 1772-1834........... DLB-37

Wise, John 1652-1725 .... DLB-24

Wiseman, Adele 1928- DLB-88

Wishart and Company............ DLB-112

Wisner, George 1812-1849 .... DLB-43

Wister, Owen 1860-1938......... DLB-9, 78

Wither, George 1588-1667 .... DLB-121

Witherspoon, John 1723-1794...... DLB-31

Withrow, William Henry 1839-1908... DLB-99

Wittig, Monique 1935- DLB-83

Wodehouse, P. G. 1881-1975 .... DLB-34, 162

Wohmann, Gabriele 1932- DLB-75

Woiwode, Larry 1941- DLB-6

Wolcot, John 1738-1819 .... DLB-109

Wolcott, Roger 1679-1767 .... DLB-24

Wolf, Christa 1929- DLB-75

Wolf, Friedrich 1888-1953 .... DLB-124

Wolfe, Gene 1931- DLB-8

Wolfe, John [publishing house] DLB-170

Wolfe, Reyner (Reginald) [publishing house] DLB-170

Wolfe, Thomas 1900-1938 ...... DLB-9, 102; Y-85; DS-2

Wolfe, Tom 1931- DLB-152

Wolff, Helen 1906-1994 .... Y-94

Wolff, Tobias 1945- DLB-130

Wolfram von Eschenbach circa 1170-after 1220 .... DLB-138

Wolfram von Eschenbach's *Parzival*: Prologue and Book 3 .... DLB-138

Wollstonecraft, Mary 1759-1797 .... DLB-39, 104, 158

Wondratschek, Wolf 1943- DLB-75

Wood, Benjamin 1820-1900 .... DLB-23

Wood, Charles 1932- DLB-13

Wood, Mrs. Henry 1814-1887 ...... DLB-18

Wood, Joanna E. 1867-1927 .... DLB-92

Wood, Samuel [publishing house] ... DLB-49

Wood, William ?-? DLB-24

Woodberry, George Edward 1855-1930 .... DLB-71, 103

Woodbridge, Benjamin 1622-1684 ... DLB-24

Woodcock, George 1912- DLB-88

Woodhull, Victoria C. 1838-1927 .... DLB-79

Woodmason, Charles circa 1720-? ... DLB-31

Woodress, Jr., James Leslie 1916- DLB-111

Woodson, Carter G. 1875-1950 ..... DLB-17

Woodward, C. Vann 1908- DLB-17

Woodward, Stanley 1895-1965 ..... DLB-171

Wooler, Thomas 1785 or 1786-1853 .... DLB-158

Woolf, David (see Maddow, Ben)

Woolf, Leonard 1880-1969 .... DLB-100; DS-10

Woolf, Virginia 1882-1941 ..... DLB-36, 100, 162; DS-10

Woolf, Virginia, "The New Biography," *New York Herald Tribune,* 30 October 1927 .... DLB-149

Woollcott, Alexander 1887-1943 .... DLB-29

Woolman, John 1720-1772 .... DLB-31

Woolner, Thomas 1825-1892 .... DLB-35

Woolsey, Sarah Chauncy 1835-1905 .... DLB-42

Woolson, Constance Fenimore 1840-1894 .... DLB-12, 74

Worcester, Joseph Emerson 1784-1865 .... DLB-1

Worde, Wynkyn de [publishing house] DLB-170

Wordsworth, Christopher 1807-1885 .... DLB-166

Wordsworth, Dorothy 1771-1855 .... DLB-107

Wordsworth, Elizabeth 1840-1932 .... DLB-98

Wordsworth, William 1770-1850 .... DLB-93, 107

The Works of the Rev. John Witherspoon (1800-1801) [excerpts] DLB-31

A World Chronology of Important Science Fiction Works (1818-1979) DLB-8

World Publishing Company DLB-46

World War II Writers Symposium at the University of South Carolina, 12–14 April 1995 .... Y-95

Worthington, R., and Company DLB-49

Wotton, Sir Henry 1568-1639 ...... DLB-121

Wouk, Herman 1915- Y-82

Wreford, James 1915- DLB-88

Wren, Percival Christopher 1885-1941 .... DLB-153

Wrenn, John Henry 1841-1911 ..... DLB-140

Wright, C. D. 1949- DLB-120

Wright, Charles 1935- DLB-165; Y-82

Wright, Charles Stevenson 1932- DLB-33

Wright, Frances 1795-1852 .... DLB-73

Wright, Harold Bell 1872-1944 ...... DLB-9

Wright, James 1927-1980 .... DLB-5, 169

Wright, Jay 1935- DLB-41

Wright, Louis B. 1899-1984 .... DLB-17

Wright, Richard 1908-1960 .... DLB-76, 102; DS-2

Wright, Richard B. 1937- DLB-53

Wright, Sarah Elizabeth 1928- DLB-33

Writers and Politics: 1871-1918, by Ronald Gray DLB-66

Writers and their Copyright Holders: the WATCH Project............. Y-94

Writers' Forum..................... Y-85

Writing for the Theatre, by Harold Pinter DLB-13

Wroth, Lady Mary 1587-1653 ..... DLB-121

Wurlitzer, Rudolph 1937- DLB-173

Wyatt, Sir Thomas circa 1503-1542............... DLB-132

Wycherley, William 1641-1715...... DLB-80

Wyclif, John circa 1335-31 December 1384... DLB-146

Wylie, Elinor 1885-1928 .... DLB-9, 45

Wylie, Philip 1902-1971............. DLB-9

Wyllie, John Cook 1908-1968...... DLB-140

X

Xenophon circa 430 B.C.-circa 356 B.C. DLB-176

Y

Yates, Dornford 1885-1960 ..... DLB-77, 153

Yates, J. Michael 1938- DLB-60

Yates, Richard 1926-1992 ... DLB-2; Y-81, 92

Yavorov, Peyo 1878-1914 .... DLB-147

Yearsley, Ann 1753-1806 .... DLB-109

Yeats, William Butler 1865-1939 .... DLB-10, 19, 98, 156

Yep, Laurence 1948- DLB-52

Yerby, Frank 1916-1991 .... DLB-76

Yezierska, Anzia 1885-1970 .... DLB-28

Yolen, Jane 1939- DLB-52

Yonge, Charlotte Mary 1823-1901 .... DLB-18, 163

The York Cycle circa 1376-circa 1569 .... DLB-146

A Yorkshire Tragedy DLB-58

Yoseloff, Thomas [publishing house] DLB-46

Young, Al 1939- DLB-33

Young, Arthur 1741-1820 .... DLB-158

Young, Dick 1917 or 1918 - 1987 ... DLB-171

Young, Edward 1683-1765 .... DLB-95

Young, Stark 1881-1963 .... DLB-9, 102

Young, Waldeman 1880-1938 .... DLB-26

Young, William [publishing house] .. DLB-49

Young Bear, Ray A. 1950- DLB-175

Yourcenar, Marguerite 1903-1987 .... DLB-72; Y-88

"You've Never Had It So Good," Gusted by "Winds of Change": British Fiction in the 1950s, 1960s, and After DLB-14

Yovkov, Yordan 1880-1937 .... DLB-147

Z

Zachariä, Friedrich Wilhelm 1726-1777 .... DLB-97

Zamora, Bernice 1938- DLB-82

Zand, Herbert 1923-1970 .... DLB-85

Zangwill, Israel 1864-1926 ..... DLB-10, 135

Zanzotto, Andrea 1921- DLB-128

Zapata Olivella, Manuel 1920- DLB-113

Zebra Books DLB-46

Zebrowski, George 1945- DLB-8

Zech, Paul 1881-1946 .... DLB-56

Zepheria DLB-172

Zeidner, Lisa 1955- DLB-120

Zelazny, Roger 1937-1995 .... DLB-8

Zenger, John Peter 1697-1746 .... DLB-24, 43

Zesen, Philipp von 1619-1689 ...... DLB-164

Zieber, G. B., and Company DLB-49

Zieroth, Dale 1946- DLB-60

Zigler und Kliphausen, Heinrich Anshelm von 1663-1697 .... DLB-168

Zimmer, Paul 1934- DLB-5

Zingref, Julius Wilhelm 1591-1635 .... DLB-164

Zindel, Paul 1936- DLB-7, 52

Zinzendorf, Nikolaus Ludwig von 1700-1760 .... DLB-168

Zitkala-Ša 1876-1938 .... DLB-175

Zola, Emile 1840-1902 .... DLB-123

Zolotow, Charlotte 1915- DLB-52

Zschokke, Heinrich 1771-1848 ...... DLB-94

Zubly, John Joachim 1724-1781 ..... DLB-31

Zu-Bolton II, Ahmos 1936- DLB-41

Zuckmayer, Carl 1896-1977 .... DLB-56, 124

Zukofsky, Louis 1904-1978 ...... DLB-5, 165

upanúiú, Oton 1878-1949 .... DLB-147

zur Mühlen, Hermynia 1883-1951 ... DLB-56

Zweig, Arnold 1887-1968 .... DLB-66

Zweig, Stefan 1881-1942 .... DLB-81, 118

ISBN 0-8103-9972-5